Handbook of

Polyester Drug Delivery Systems

Handbook of Polyester Drug Delivery Systems

edited by

MNV Ravikumar

Published by

Pan Stanford Publishing Pte. Ltd.
Penthouse Level, Suntec Tower 3
8 Temasek Boulevard
Singapore 038988

Email: editorial@panstanford.com
Web: www.panstanford.com

British Library Cataloguing-in-Publication Data
A catalogue record for this book is available from the British Library.

Handbook of Polyester Drug Delivery Systems

ISBN 978-981-4669-65-8 (Hardcover)
ISBN 978-981-4669-66-5 (eBook)

Printed in the USA

Contents

Foreword

Polyesters are versatile building blocks for drug delivery vehicles. The wide range of functionalities available in the monomers, coupled with the ability to control the molecular weight, provide tools for controlling the properties of both particle- and materials-based system. This book is intended to provide an overview of the most exciting directions in polyester-based drug delivery systems (DDSs). The book starts with a discussion on the synthesis of polyesters by Slomkowski, with Domb presenting the use of fatty acids for generating polymers, and Ding showing how enzymes can be used to synthesize precision polyesters.

The ability to control the physical properties of polyesters enables the crafting of these polymers into a wide range of materials. Katti provides an overview of how polyesters can be used for generating micro- and nanosystems, while Cui presents the use of nanofibers for controlled release of therapeutics. Clearly, properties stem from the choice of monomers, with naturally occurring monomers being widely employed to create game-changing delivery vehicles. Sah provides an update of polylactide-*co*-glycolide (PLGA), while Salerno provides an overview of polycaprolactone-derived DDS.

The ability to tailor the properties of polyesters makes them adaptable tools for navigating the complexities found *in vivo*. Sharma presents this versatility, discussing how polyesters can be designed for compatibility in blood, while Banderas describes their use in peroral administration.

There are many ways of making therapeutics available, with most patients preferring non-injection routes. The ability to tune the properties of polyesters makes them ideal for these applications, as covered by Misra for nasal administration and Zaera for topical delivery.

Polyesters are well suited to some of the most challenging areas in DDS creation. Yallapu focuses on a specific payload, the versatile natural product curcumin. Amsden and Vertegel present the use of these polymers for protein and enzyme delivery, respectively, while Amiji looks at the challenging goal of oral gene therapy using complex polyester constructs. Blanco-Prieto discusses the use of PLGA systems for VEGF delivery, a key tool in regenerative medicine, a topic followed up by Kumbar who discusses the use of micro- and nanosystems for tissue engineering. Vyas and Almeida discuss applications of polyester particles in immunology, a use where surface tailorability is essential, as it is in Sosnik's discussion of polyester micro- and nanomaterials for fighting infectious disease.

Ravikumar has assembled a volume featuring authors from around the world, each doing cutting-edge research in the creation of polyester-based DDS. I hope you find the volume as interesting and useful as I have.

Vincent M. Rotello
Amherst, MA, USA
Summer 2016

Preface

Polymers are indispensable to mankind and have become an integral part of the current healthcare landscape, particularly the pharmaceutical industry. The advancements in combinatorial chemistry and biotechnology led to new, may be better, but complicated therapeutic strategies posing stiff challenges to the formulators. In the quest for innovative formulations attempting to meet the unmet needs in pharmaceutical space, research has taken a much more complicated path that poses a significant challenge for translation. Despite the progress made with novel materials, polyesters still remain at the helm of drug delivery technologies, with a recent example of Bydureon, a once-a-week injection of exenatide that uses polylactide-*co*-glycolide. Although there are several reviews covering the advances on polyester drug delivery, we lack a single source of reference that covers a broad spectrum of materials design, manufacturing techniques, and applications. This handbook is therefore an attempt to provide an overview of polyester drug delivery systems.

I am grateful to the authors who made this handbook possible with their comprehensive state-of-the-art reviews on the subject. The book contains 20 chapters and has more than 2000 bibliographic citations, several dozens of figures, illustrations, chemical structures, and tables.

This book is intended for a very broad and diverse audience and I hope that it will prove to be a useful tool for upper-level undergraduate and graduate students and experienced researchers engaged in materials science and medicine.

MNV Ravikumar

College Station, TX, USA
Summer 2016

Chapter 1

Overview of Synthesis of Functional Polyesters

Stanislaw Slomkowski
Center of Molecular and Macromolecular Studies, Polish Academy of Sciences, Sienkiewicza 112, 90–363 Lodz, Poland
staslomk@cbmm.lodz.pl

1.1 Introduction

Linear polyesters (most often polylactides and poly[lactide-*co*-glycolide] copolymers) found many applications in medicine [1–12]. They function usually as temporary mechanical supports, connecting elements or temporary containers, most often of nano- or micrometer size, of bioactive and in many instances very potent low molecular weight compounds. The main advantages of these polymers are related to easily adjustable polymerization conditions for obtaining polymers with controlled molecular weight, microstructure, and topology, in a way assuring required mechanical strength, convenient processability, and rate of degradation [13–17].

In addition to the above-mentioned polylactides, as well as polyglycolide and poly(lactide-*co*-glycolide) copolymers, some

Handbook of Polyester Drug Delivery Systems
Edited by MNV Ravikumar

ISBN 978-981-4669-65-8 (Hardcover), 978-981-4669-66-5 (eBook)
www.panstanford.com

other polyesters are also medically attractive, in particular poly(β-butyrolactone) and other substituted β-lactones, poly(ε-caprolactone), and copolymers containing short or long segments of these polymers. The structures of linear polyesters are shown in Fig. 1.1.

There are synthetic routes leading to polyesters with various topology, that is, composed of linear, cyclic, star-shaped, branched, and in particular dendritic macromolecules. Examples of structures of these topologically different molecules are shown in Fig. 1.2.

Since for majority of medical applications the polyester-derived devices, implants, and drug carriers are in a direct contact with tissues and body fluids (bio)degradability (i.e., ability to degrade by combination of simple chemical hydrolysis and biodegradation) is not the only important factor. For polymer–tissue interactions leading to required tissue response on molecular and cellular level, polymers with functional groups bound at particular sites of all macromolecules or, in some instances, with functional groups at surfaces of polymer objects such as nanoparticles or implants are needed. In some instances, the presence of properly chosen groups in polymer chains makes polymeric objects or individual macromolecules "invisible" (stealth) for the immune system. Some others allow for covalent binding of polyesters to cell membranes, cellular organelles, or such biomacromolecules as proteins and nucleic acids. Examples of functionalized polyester chains are shown in Fig. 1.3.

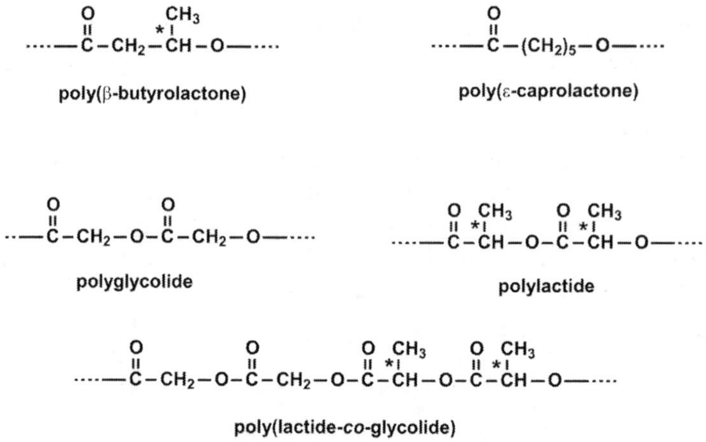

poly(β-butyrolactone)

poly(ε-caprolactone)

polyglycolide

polylactide

poly(lactide-*co*-glycolide)

Figure 1.1 Structures of linear polyesters (stars denote chiral centers).

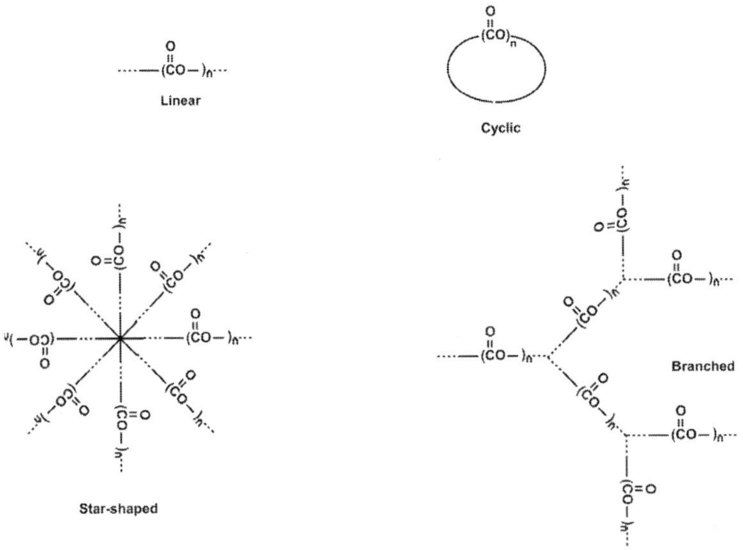

Linear

Cyclic

Star-shaped

Branched

Figure 1.2 Topology of polyesters.

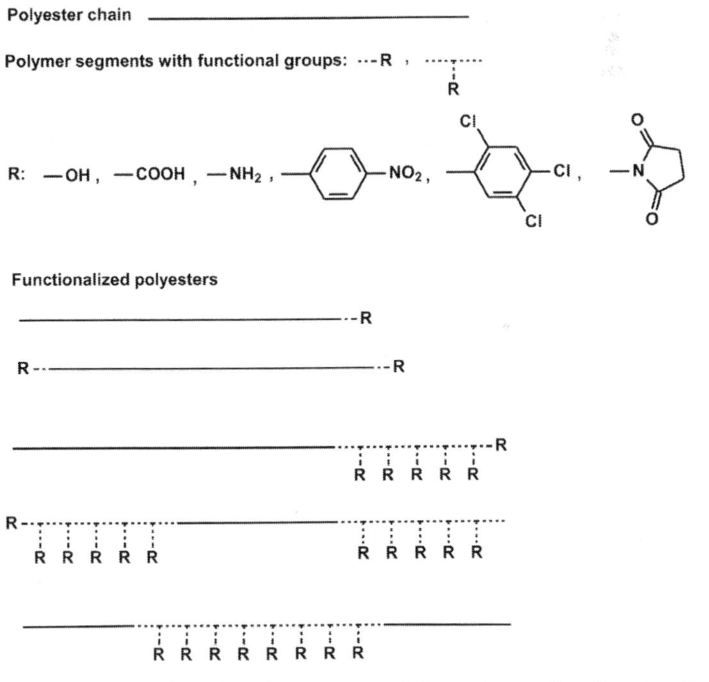

Figure 1.3 Some functional groups used for polymer functionalization and their distributions in polymer chains.

1.2 Basics of Synthesis of Polyesters

1.2.1 Synthesis by Ring-Opening Polymerization of Cyclic Monomers

Cyclic esters can be polymerized by ionic or pseudoionic (with active centers containing strongly polarized groups) mechanisms. In these polymerizations, side reactions are in principle inevitable, because active centers can react with ester groups not only in monomers but also with those along the polymer chains.

Four-membered *β*-lactones (*β*-propiolactone and its substituted derivatives) constitute the special subclass of cyclic esters. Owing to planar structure of the *β*-lactone ring and to the very large ring strain (for the extensively investigated *β*-propiolactone enthalpy of polymerization, ΔH_p equals 82.4 kJ/mol [18]), the mechanism of polymerization of these monomers differs significantly from that with larger rings.

Polymerization of *β*-propiolactone can be initiated with strong (e.g., RO^-) and weak (e.g., $RCOO^-$) nucleophiles. Extensive studies allowed for quite detailed descriptions of this process [19–23]. An attack of RO^- anions onto the carbonyl carbon atom of lactone ring in the direction of the dipole moment of the monomer molecule, leading to acyl-oxygen ring scission, is sterically hindered (see Fig. 1.4). Moreover, after the addition of RO^- anion, the carbonyl carbon atom changes its hybridization from sp^2 to sp^3. This new conformation leads to repulsion between the $\equiv CO^-$ and $\equiv COR$ groups with free electron pairs of the endocyclic oxygen atom.

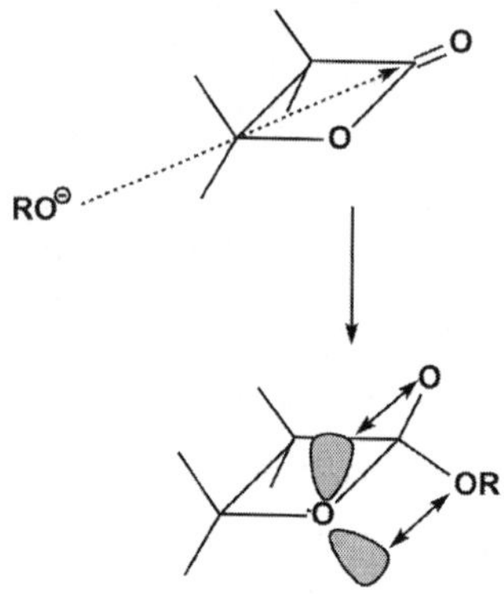

Figure 1.4 Attack of alkoxide anion on *β*-propiolactone.

The above-mentioned restrictions for the pathway leading to monomer addition with the acyl-oxygen bond scission opens the way for some contribution of the competitive route when the RO^- anion attacks the carbon atom in β position, which results in the addition of β-lactone with the carbon–oxygen bond scission and formation of carboxylate anions (see Fig. 1.5). Carboxylate anions are quite unreactive toward ester bonds in the unstrained compounds like linear esters or cyclic esters with the ring size larger than 4. Because of steric hindrance, they cannot attack carbonyl carbon atoms in β-lactones. However, having quite easy access to the alkyl carbon atom in β position of the strained β-lactones, they can add a monomer with the alkyl-oxygen bond scission, resulting in the extension of chain length and reconstruction of the carboxylic active centers.

Thus, to summarize, alkoxide anion initiators and later alkoxide propagating species add β-lactones either with alkyl-oxygen or acyl-oxygen bond scission, whereas carboxylate anions add these monomers exclusively with alkyl-oxygen bond scission. Thus, in the polymerization of β-lactones initiated with carboxylates, only carboxylate active species are present from the very beginning,

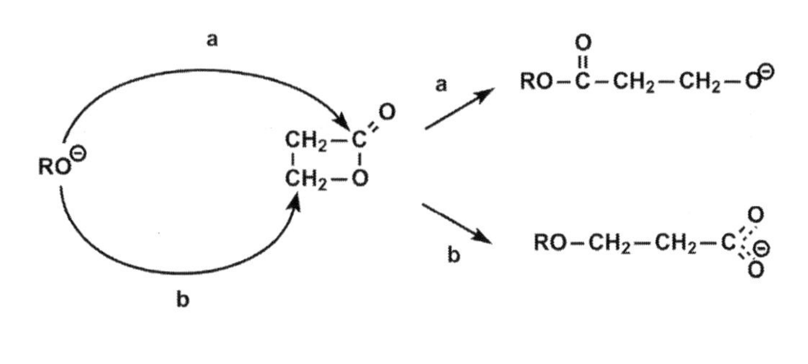

Figure 1.5 Initiation of the anionic polymerization of β-propiolactone with alkoxide and carboxylate initiators.

whereas in polymerizations initiated with alkoxides, at the beginning both kinds of active species are present (see Fig. 1.5). However, at each step, fraction of alkoxide species decreases and eventually, also in this process, only the carboxylate active centers are present in the system.

In the case of highly substituted alkoxide anions (e.g., *t*-butoxides), the direct addition to *β*-lactones is difficult due to steric hindrance and initiation proceeds with proton transfer, as shown in Fig. 1.6 [22]. Subsequent steps of polymerization consist in repeated monomer additions with alkyl-oxygen bond scission and formation of carboxylate active species during each step. The process is illustrated in Fig. 1.6.

Ring-opening polymerization of all cyclic esters with the ring size larger than 4 proceeds with the acyl-oxygen ring scission. From this group of cyclic esters, *γ*-butyrolactone does not polymerize because of thermodynamic reasons (ΔH_p^0 5.1 kJ/mol, ΔS_p^0 = −29.9 J/(mol·K) [24]. For this monomer, at typical conditions, propagation with the ring-opening is much slower than the reversible ring-closure. Polymerization could proceed only at *γ*-butyrolactone concentration exceeding the not attainable concentration equal to 3.3×10^3 mol/l. However, this monomer is suitable for copolymerization with other lactones (e.g., with *ε*-caprolactone) [25–28].

Further discussion of anionic and pseudoanionic (i.e., with covalent active species in which the metal–oxygen bond is strongly polarized) polymerization of cyclic esters will be related only to monomers with the ring size larger than 4. In these processes, propagation is accompanied with inter- and intramolecular chain transfer reactions.

Contamination of linear polymers with cyclics results in adverse changes of many properties of the synthesized polymers. The intermolecular chain transfer leads to uncontrolled broadening of molar mass distribution, which affects the mechanical properties of polymers, including their processability and degradability.

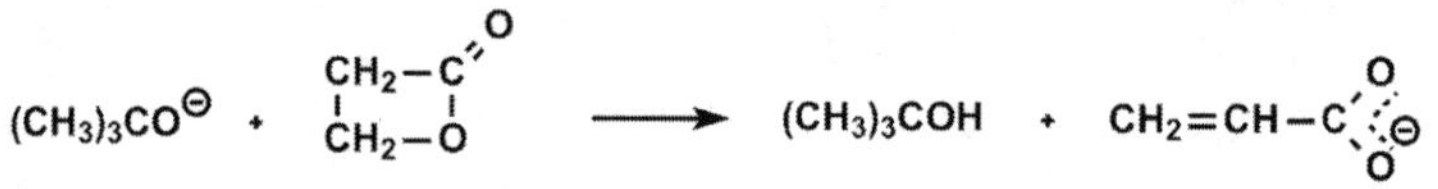

$(CH_3)_3CO^{\ominus}$ + $CH_2{-}C(=O){-}O{-}CH_2$ (ring) ⟶ $(CH_3)_3COH$ + $CH_2{=}CH{-}CO_2^{\ominus}$

Figure 1.6 Initiation of the anionic polymerization of *β*-propiolactone with *t*-butoxide anion.

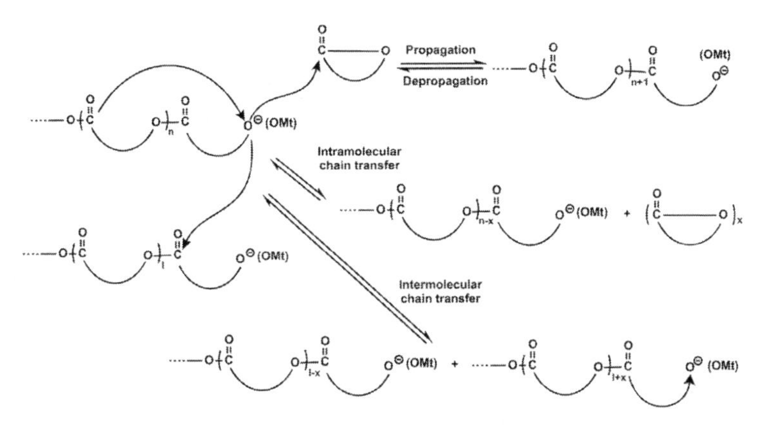

Figure 1.7 Propagation and inter- and intramolecular chain transfer reactions in polymerization of cyclic esters.

The role of unwanted transesterification reactions strongly depends on the chemical nature of the propagating species. For the polymerization of cyclic esters with selectivity, defined as a ratio of propagating and transfer rate constants (k_p/k_{tr}), the following reactivity–selectivity principle was observed: the lower the reactivity of the propagating species, the higher is the value k_p/k_{tr} [15, 25–39]. Thus, there should be possible polymerizations during which monomer would be almost completely converted into polymer before the transfer reactions play any noticeable role. In the polymerization of ε-caprolactone and lactides, the anionic monovalent metal alkoxide active species are much more reactive than the covalent alkoxides of such metals as Ca, Zn, Al, Sn, Ti, and Cr [5, 15, 16, 23, 40]. As a result, polymerizations with Na^+ and K^+ counterions lead to polymers contaminated with cyclics (due to the inter- and intramolecular transesterification reactions, respectively). Moreover, these polymers have a broad molecular weight distribution. On the contrary, the processes initiated with covalent alkoxides allow to obtain polyesters with molar masses determined simply by ratios of the initial monomer and initiator concentrations. Such polymers are free from any admixture of cyclic oligomers [29]. It is worth noting that polymerization of ε-caprolactone and L-lactide initiated with $(CH_3CH_2)_2AlOCH_2CH_3$ and $((CH_3)_2CHCH_2)_2AlOCH_3$ could be controlled molar masses up to

$M_n \approx 5 \times 10^5$ [41, 42]. In the case of poly[(L,L-lactide)] obtained in polymerization initiated with $Sn(OCH_2CH_2CH_2CH_3)_2$, the polymers with $M_n \approx 10^6$ were obtained [42].

Owing to the presence of chiral centers in polymer chains, the polylactides can be obtained in the form of poly[(L,L)-lactide], poly[(D,D)-lactide], and racemic polylactide isomers (see Fig. 1.8).

Some initiators with bulky ligands, in particular ligands with chiral structures, lead to stereospecific polymerization of lactides. For example, for up to 40% of monomer conversion during polymerization of racemic lactide initiated with ({R-(-)-2,2′-[1,l′-binaphthyl-2,2′-diylbis(nitrilomethylidyne)]diphenol}/$AlEt_3$; (R)-SALBinaphtAl-OCH_3) (structures of initiator enantiomers are shown in Fig. 1.9), the optical purity of the polymer formed is very high (higher than 80%; predominantly the (D,D)-lactide is elected). This means that the noticeable polymerization of (L,L)-lactide) occurs only when almost all (D,D)- enantiomer is polymerized [44].

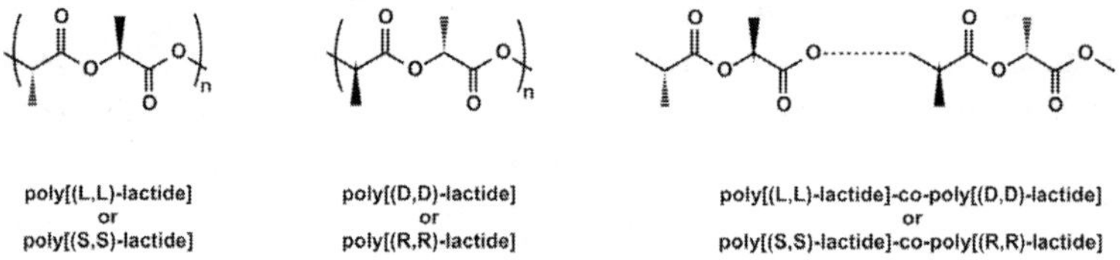

poly[(L,L)-lactide]
or
poly[(S,S)-lactide]

poly[(D,D)-lactide]
or
poly[(R,R)-lactide]

poly[(L,L)-lactide]-co-poly[(D,D)-lactide]
or
poly[(S,S)-lactide]-co-poly[(R,R)-lactide]

Figure 1.8 Microstructure of polylactides.

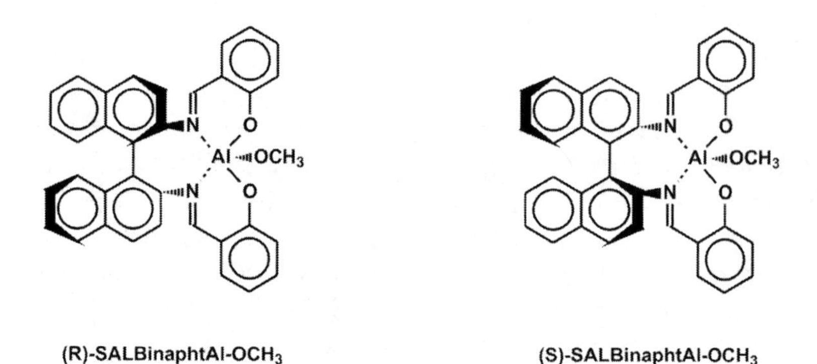

Figure 1.9 Structures of SALBinaphtAl-OCH_3 initiators of stereoelective polymerization of racemic lactide.

Stereoelective polymerization of lactides governed by chirality of the ligands did open the possibility of completing the polymerization of one enantiomer from the (D,D+L,L)-lactide racemic mixture and then, after addition of an excess of the ligand with the opposite chirality, to polymerize also the second enantiomer. By using this approach Duda *et al.* synthesized poly[(L,L)-lactide]-*grad*-poly[(D,D)-lactide] copolymer from the racemic monomer mixture [13].

During the past years, special attention was given to the synthesis of polylactides by processes initiated with compounds containing the less toxic group 2 metals. For example, $NH_2CaOCH_2CH_3$ and $NH_2CaOCH(CH_3)_2$ were used as effective initiators of the polymerization of lactide [44]. Calcium, magnesium, and zinc alkoxides containing metals complexed with the tridentate trispyrazolyl- and trisindazolylhydroborate ligands (examples are shown in Fig. 1.10), allowing reduction of aggregation of active species, were used as convenient initiators of the polymerization of lactide. The rates of polymerization for various metals increase in the following order Ca > Mg > Zn [45–47].

There is some hope that nontoxic polylactides could also be obtained by using metal-free strong bases, usually amines, and alcohol containing initiating systems. First syntheses of polylactides by using initiators belonging to the above-mentioned class were performed for 4-pyrrolidinopyridine and 4-dimethylaminopyridine [48, 49]. An extensive review of the metal-free initiators was published recently by Waymouth *et al.* [50] and Bourissou *et al.* [49]. Examples of strong bases (so-called superbases) used for initiation are shown in Fig. 1.11.

Contrary to the comprehensive knowledge amassed on the mechanism of polymerization of β-lactones initiated with alkali metal carboxylates, ε-caprolactone initiated with alkali metal and aluminum alkoxides, and polymerization of lactides with tin octoate and aluminum alkoxides, very little is known about the

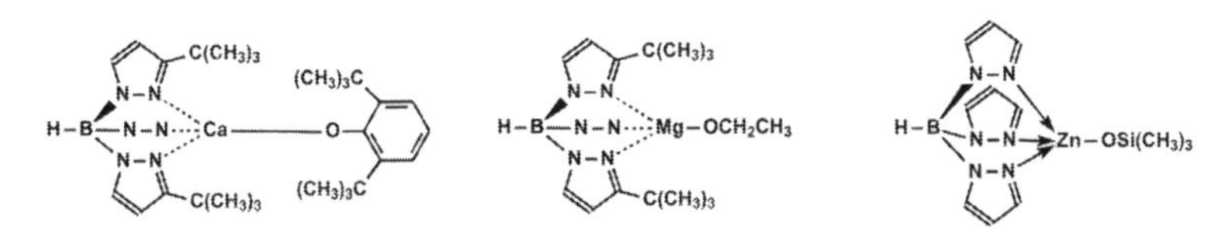

Figure 1.10 Examples of calcium, magnesium, and zinc alkoxide initiators.

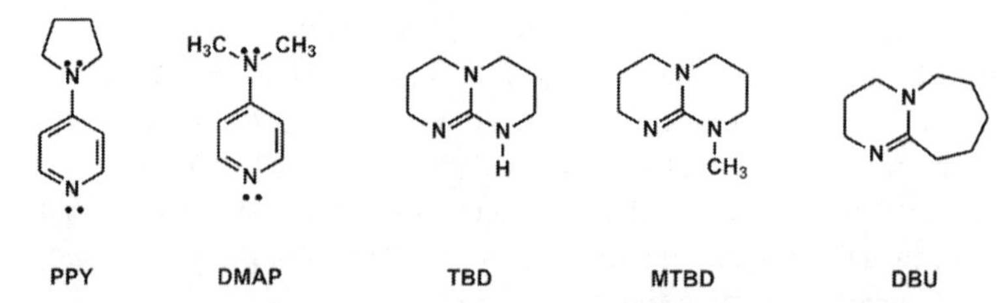

Figure 1.11 Structures of 4-pyrrolidinopyridine (PPY), 4-dimethylamino-pyridine (DMAP), 1,5,7-triazabicyclo[4.4.0] dec-5-ene (TBD), and its *N*-methylated analog (MTBD), 1,8-diazabicycloundec-7-ene (DBU).

polymerization of lactides initiated with other, mostly multivalent, metal alkoxides and with metal-free initiators. Structures of active species in these processes are usually postulated by analogy to other better-known systems.

1.2.2 Synthesis by Polycondensation

In principle, as it is shown in Fig. 1.12, the linear aliphatic polyesterscan be obtained by polycondensation of hydroxyacids.

According to Kricheldorf [51] the first report on formation of polylactide by polycondensation (a residue identified as a "water-free lactic acid") dates back to the middle of nineteenth century [52, 53]. The main problem in synthesis of aliphatic polyesters by polycondensation is related to removal of water. There were some attempts to synthesize polylactide on industrial scale by solvent based polycondensation facilitated by removal of water by azeotropic distillation. Mitsui Toatsu, a Japanese company, developed relevant technology. However, high cost and technical problems caused stopping of the production of polylactide by this method in 2008.

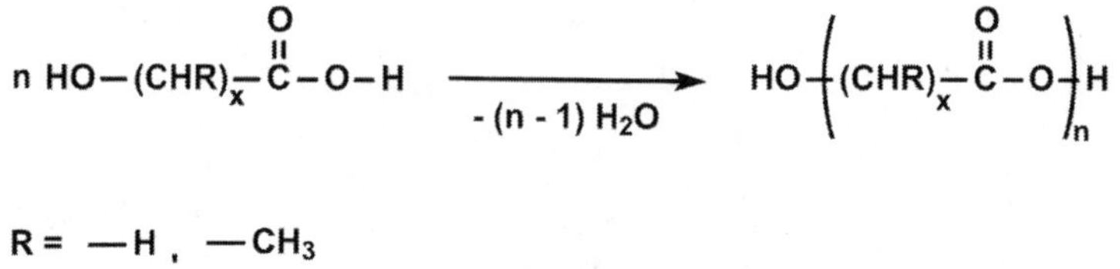

$$n\ \mathrm{HO{-}(CHR)_x{-}\overset{\overset{\displaystyle O}{\|}}{C}{-}O{-}H} \xrightarrow[-\,(n-1)\,\mathrm{H_2O}]{} \mathrm{HO}\left[\mathrm{(CHR)_x{-}\overset{\overset{\displaystyle O}{\|}}{C}{-}O}\right]_n\mathrm{H}$$

$R = -H,\ -CH_3$

Figure 1.12 Polycondensation of hydroxyacids to linear polyesters.

Nowadays, by polycondensation is used to synthesize only low molecular polylactide oligomers which are used as substrates for synthesis of lactide monomers used later for synthesis of high molecular weight polymers by ring-opening polymerization.

1.2.3 Synthesis by Enzymatic Polymerization

In nature some polyesters (e.g., polyhydroxybutyrate) are synthesized by enzymatic processes governed by a variety of esterases. Some researchers, inspired by these observations, concentrated their attention on studies of propagation by enzymatic polymerization carried out *in vitro*. According to present knowledge, the general mechanism of such process could be illustrated by Fig. 1.13.

The process consists of formation of enzyme activated monomer via enzyme acylation, with subsequent reaction of these species with –OH groups of an initiator (ROH) or a growing polymer chain. Lactones with the ring size from 4 to 17 (due to thermodynamic reasons the five-membered δ-butyrolactone cannot homopolymerize but is able to copolymerize with other lactones) were polymerized in this way [54–74]. In case of lactone monomers containing chiral active centers in the ring, stereoelective polymerization was observed. For example, in the lipase-catalyzed polymerization of 3-methyl-4-oxa-6-hexanolide the (S)-isomer-enriched polymer was produced [63]. It is worth noting that during similarly catalyzed polymerization of δ-caprolactone, the monomer with opposite (R) structure was elected [75, 76].

Lipase can catalyze not only ring-opening polymerization of lactones but also polycondensation of hydroxyacids [77, 78].

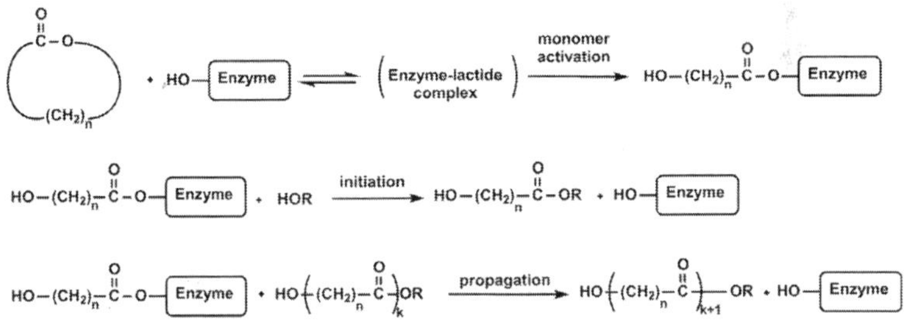

Figure 1.13 Figure of enzymatically catalyzed polymerization of cyclic esters.

Although the knowledge about enzymes and their role in polymerization of cyclic esters and polycondensation of aliphatic hydroxyacids is quite extensive, the progress in synthesis of high molecular weight polyesters by enzymatic reactions is still quite limited, M_n of synthesized polymers most often being below 25000 Da.

1.3 Polyesters with Functional Groups

1.3.1 Primary Functional Groups in Polyesters

In the case of synthesis of polyesters by polycondensation of hydroxyacids the –OH and –COOH reactive groups are inherently present in polymer chains. In the ring-opening polymerization of four-membered lactones, the reactive groups are carboxylate anions. In the polymerization of other cyclic esters (e.g., ε-caprolactone, lactides, and glycolides), the pristine functional groups are metal alkoxides in their ionized or covalent forms. However, these species are usually too reactive to remain intact in final polymers. Nevertheless, they could be used for *in situ* functionalization. Most often the propagating species are destroyed during termination by addition of small amounts of acids or water, which results in the formation of terminal –OH functions. The number and distribution of the above-described primary functional groups in synthesized macromolecules are determined by their topology.

Anionic polymerization of cyclic esters is usually accompanied with the formation of cyclic oligomers produced during intramolecular transesterification (see Fig. 1.7). Analysis of polymerizations with so-called back-biting reactions (based on attack of propagating species on reactive groups of their own chains) revealed that the fraction of cyclic oligomers strongly increases with decreasing initial monomer concentration, and below its certain limiting value the product consists almost exclusively of cyclic molecules, among which the low molecular weight oligomers constitute the main fraction. Such molecules, by virtue of their topology, do not contain any end-groups but only ester groups along their chains [34, 79, 80]. In the case of the anionic polymerization of ε-caprolactone, the limiting monomer concentration is close to 0.25 mol/l [81,82]. Below this concentration the product does not contain any significant fraction of linear polymer.

In the case of anionic and pseudoanionic polymerization of cyclic esters with propagating species having the general structure shown in Fig. 1.14, the process proceeds with kinetically reduced formation of cyclic oligomers. Therefore, termination with protic compounds yields linear polyester molecules, each of which is equipped with an –OH terminal group at one end.

Initiators yielding growing chains with propagating species at both chain ends lead to the production of polyesters with two –OH end-groups in each macromolecule. Examples of such processes include polymerizations of ε-caprolactone initiated with $(CH_3CH_2)_2AlO(CH_2CH_2O)_3Al(CH_2CH_3)_2$ [83] and polymerization of L,L-lactide initiated with bifunctional $H\text{-}(OCH(CH_2OCH(CH_3)(OCH_2CH_3)CH_2)_x\text{-}(OC(CH_3)_2CH_2CH_2C(CH_3)_2O\text{-}(CH_2CH(CH_2OCH(CH_3)(OCH_2CH_3)O)_xK$ macroinitiator [84].

Polymerizations of cyclic esters initiated with multifunctional alkoxides of $R(OMt)_n$ type or $R(OH)_n$ and $(CH_3CH_2CH_2CH_2CH(CH_2CH_3)COO)_2Sn$ (tin(II) 2-ethylhexanoate, in which R denotes an organic core and Mt metal, yield star-like macromolecules with *n* polyester arms containing terminal –OH groups.

Polylactides with three-armed chains were synthesized using trimethylolpropane/tin(II) 2-ethylhexanoate initiating system [85, 86]. Polymers with M_n up to 17,500 Da were obtained in this way.

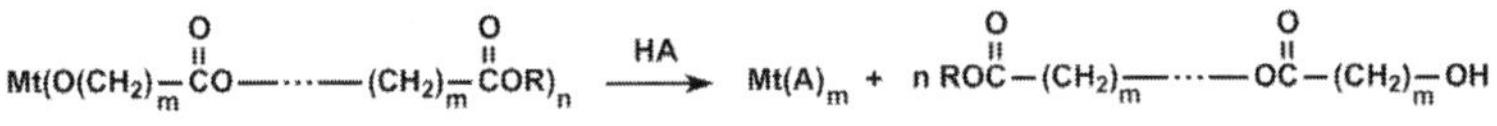

$$Mt(O(CH_2)_m\overset{O}{\overset{\|}{C}}O\text{—}\cdots\text{—}(CH_2)_m\overset{O}{\overset{\|}{C}}OR)_n \xrightarrow{HA} Mt(A)_m + n\,RO\overset{O}{\overset{\|}{C}}\text{—}(CH_2)_m\text{—}\cdots\text{—}O\overset{O}{\overset{\|}{C}}\text{—}(CH_2)_m\text{—}OH$$

Figure 1.14 Formation of linear polyester chains containing one –OH end-group per chain. Mt denotes metal atom (with or without substituents or ligands which do not participate directly in propagation) and HA, a protic terminating agent.

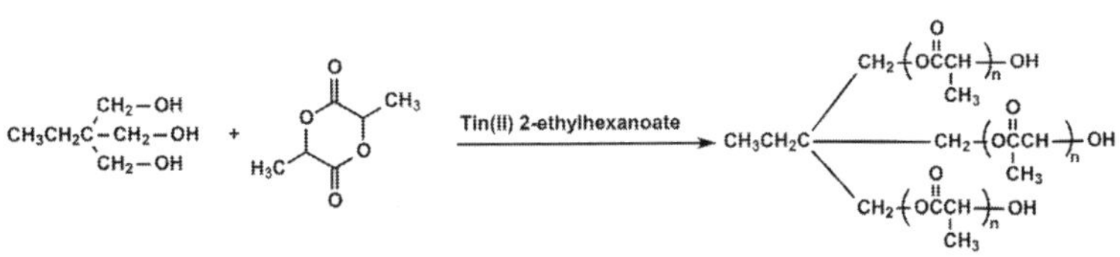

Figure 1.15 Synthesis of three-arm –OH group terminated star-like polylactide.

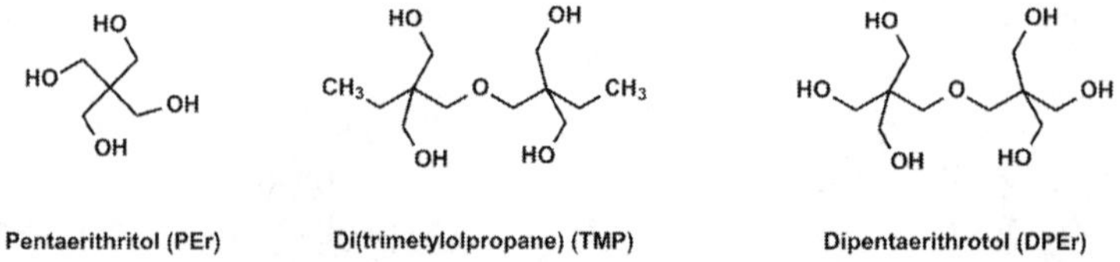

Pentaerithritol (PEr) Di(trimetylolpropane) (TMP) Dipentaerithrotol (DPEr)

Figure 1.16 Multifunctional alcohols used as initiators in the synthesis of four- and six-arm star-like polylactides.

Star-like polylactides with four or more arms were synthesized in a similar way, using alcohols, the structures of which are shown in Fig. 1.16 [74, 85, 87–89].

Dipentaeritritol and its benzylated derivatives (see Fig. 1.17) with tin(II) 2-ethylhexanoate coinitiator were used for the synthesis of star-like poly(ε-caprolactone)s with 2, 3, 4, 5, and 6 arms, each containing one terminal –OH group [90].

Hyperbranched polylactides and polylactide-containing copolymers can be synthesized similarly as dendrimers, by divergent, convergent, or a combination of these routes. Adeli *et al.* synthesized hyperbranched polylactide using hyperbranched polyglycidol as a multifunctional initiator [91].

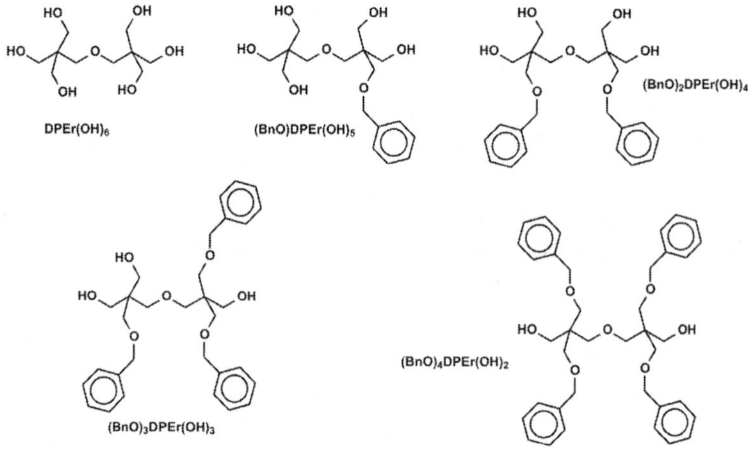

$DPEr(OH)_6$ $(BnO)DPEr(OH)_5$ $(BnO)_2DPEr(OH)_4$ $(BnO)_3DPEr(OH)_3$ $(BnO)_4DPEr(OH)_2$

Figure 1.17 Multifunctional alcohols used for synthesis of star-like multi-arm poly(ε-caprolactones).

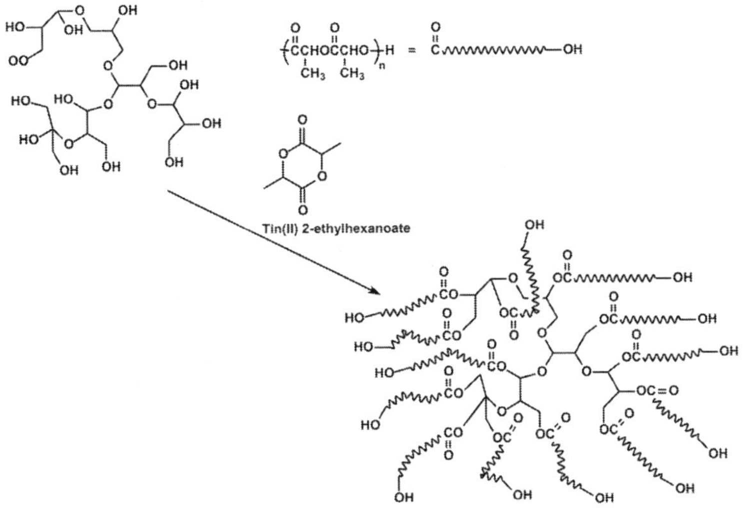

Figure 1.18 Synthesis of multi-arm polylactide from the hyperbranched polyglycidol.

Hyperbranched copolymers containing polylactide and polyglycidol segments were also synthesized by copolymerization of lactide and glycidol in a process in which the latter did function not only as a monomer (reactions involving epoxide groups) but also as an initiator (engagement of –OH functions) [92]. The process was catalyzed with tin(II) 2-ethylhexanoate. Ring-opening of epoxide ring of glycidol yielded branching points for polylactide and/or polyglycidol units. The structure of this copolymer is shown in Fig. 1.19. It is worth noting that the described branched copolymers were synthesized in bulk, at 130°C, whereas a similar process carried out in toluene, at 80°C, yielded linear polylactide with terminal epoxide groups [92].

Highly branched polylactides with terminal –OH groups could be obtained also by a combination of the ring-opening polymerization of lactide initiated with 2,2-bis(hydroxymethyl) butyric acid (TMDAP) and tin(II) 2-ethylhexanoate system. The carboxyl groups of TMDAP participated in esterification reactions, facilitated by the presence of *N, N, N′, N′*-tetramethyl-1,3-diaminopropane, a potent proton acceptor [93]. The above-mentioned synthetic route, shown in Fig. 1.20, combined both ring-opening and condensation polymerizations.

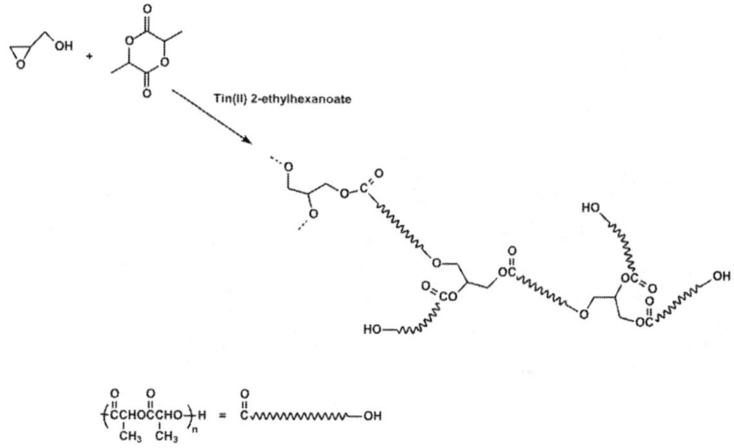

Figure 1.19 Synthetic route for branched polylactide with glycidol-derived branching points.

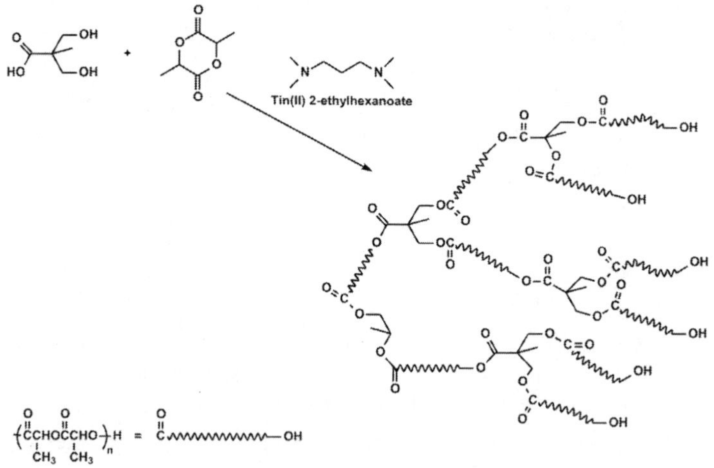

Figure 1.20 Synthesis of branched poly(L,L-lactide) with –OH groups in a process involving ring-opening polymerization of lactide and esterification reaction of carboxyl groups of TMDAP.

A special group of polylactide containing polymers with functional groups consists of copolymers built from a block of polyester and block(s) of other polymers with various functional groups. As an example, we present here copolymers with the formula polylactide-*b*-polyglycidol-*b*-polylactide [94–98]. Their synthesis is illustrated in Fig. 1. 21.

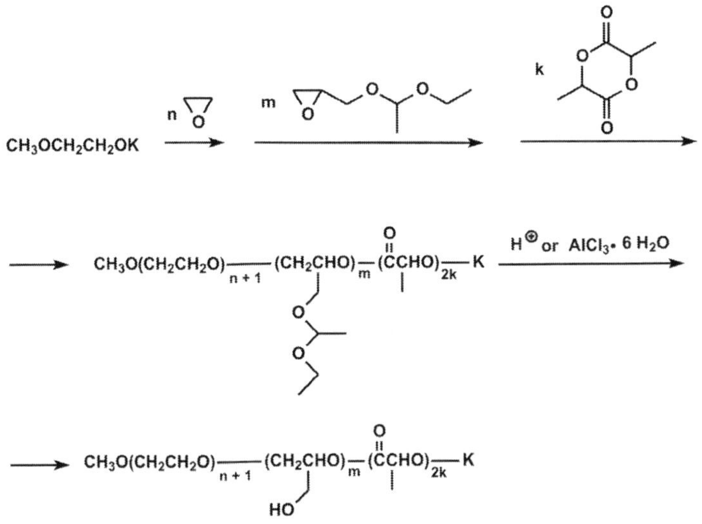

Figure 1.21 Synthesis of polylactide-*b*-polyglycidol-*b*-poly(ethylene oxide) with –OH functions in the middle block.

The synthesis proceeds in four steps. The poly(ethylene oxide) block is synthesized first. Subsequent polymerization of 1-ethoxyethylglycidyl ether (glycidol with blocked –OH group) leads to the formation of the second block containing "hidden –OH groups." In the next step the block of polylactide is synthesized. Finally, after killing active centers and deblocking hydroxyl groups, a copolymer with a terminal –OH group at the end of the polylactide block and a number of –OH groups along the middle block of polyglycidol is obtained.

1.3.2 Polyesters with Fluorescent Labels

Functional groups can be used for a variety of purposes. Fluorescent labels can help in the localization of polymeric nano- and microparticles and visualization of selected organs and tissues. Below, a few examples are given illustrating the synthesis of fluorescently labeled polyesters.

Poly(ε-caprolactone) labeled at both ends with pyrene groups was synthesized in a simple esterification reaction involving polyester terminated with –OH functions and 4-(1-pyrene)butyryl chloride (see Fig. 1.22) [99]. The reaction was carried out in the presence of triethylamine added as a scavenger of the evolving HCl.

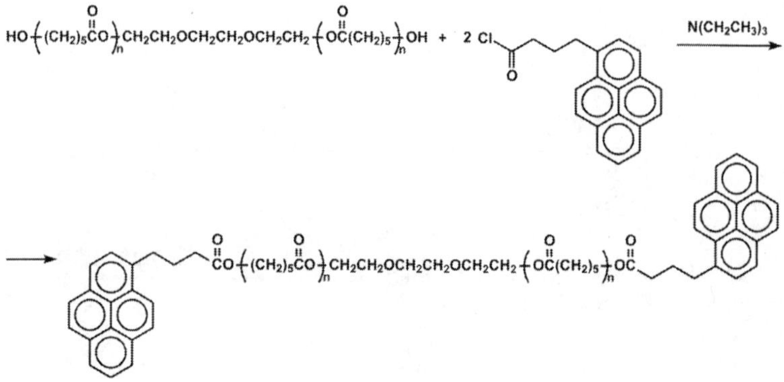

Figure 1.22 Functionalization of OH-poly(ε-caprolactone)-OH with pyrene labels.

The same procedure was used for the synthesis of poly(ε-caprolactone) star polymers with 2, 3, 4, 5, and 6 arms [90]. In Fig. 1.23 the synthesis of the 6-arm star with pyrene labels is shown as an example.

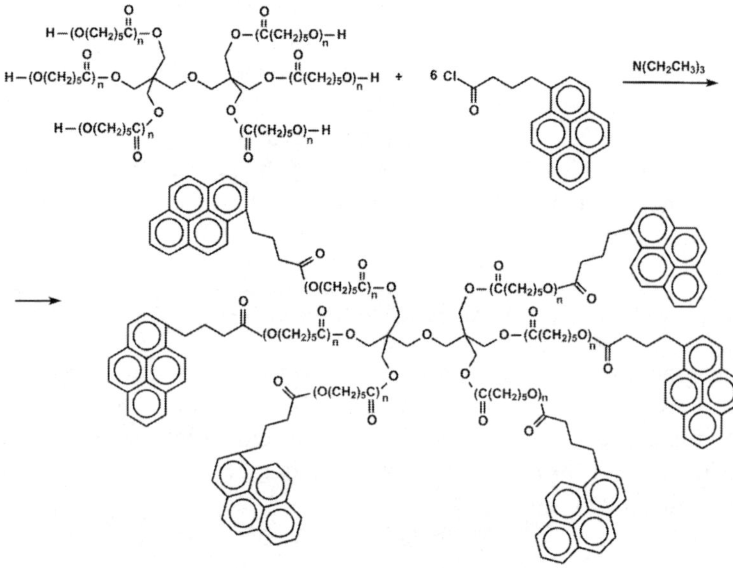

Figure 1.23 Synthesis of 6-arm star poly(ε-caprolactone) with pyrene labels.

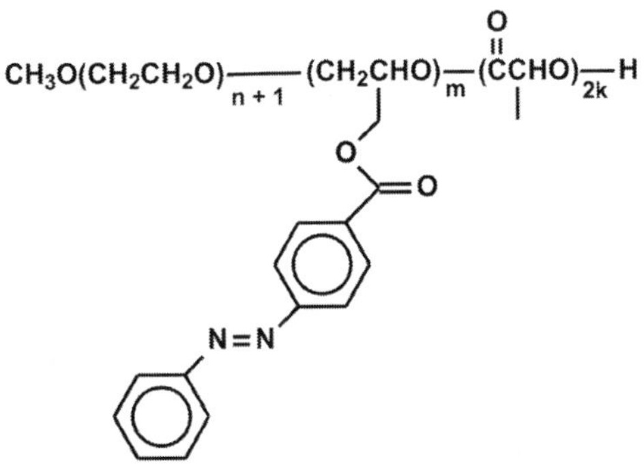

Figure 1.24 Polylactide-*b*-polyglycidol-*b*-poly(ethylene oxide) copolymers with 4-(phenyl-azo)phenyl labels.

Syntheses of polylactide-*b*-polyglycidol-*b*-poly(ethylene oxide) copolymers with 4-(phenyl-azo)phenyl moieties attached at the polyglycidol block were carried out in a similar way (see Fig. 1.24) [100].

Fluorescently labeled polyesters were obtained also by using initiators containing fluorescent labels. For example, this approach was used for the synthesis of polylactide labeled at one end with difluoroboron naphthyl-phenyl β-diketonate (Fig. 1.25) [101] and in the middle with 1,2-bis(2,4-dihydroxybenzylidene) hydrazine (Fig. 1.26) [102].

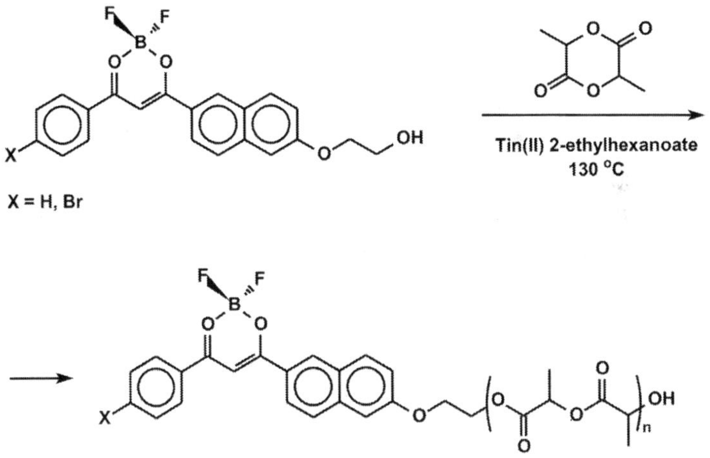

Figure 1.25 Synthesis of polylactide with difluoroboron naphthyl-phenyl β-diketonate.

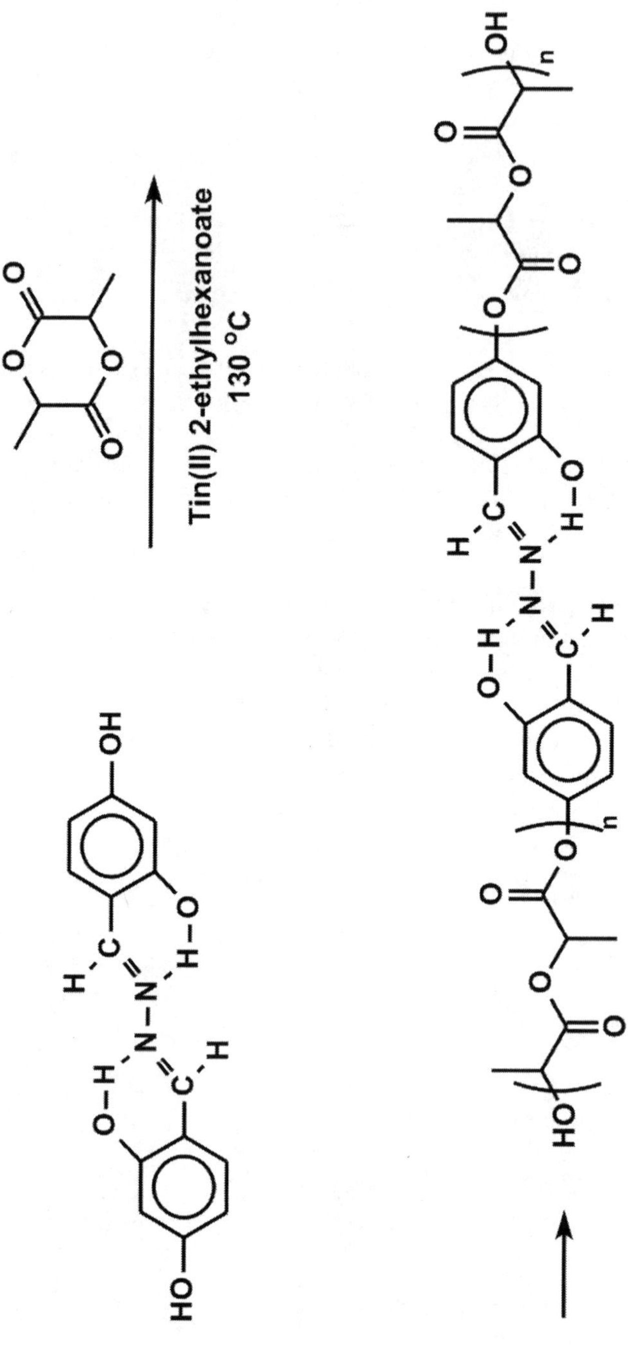

Figure 1.26 Synthesis of polylactide with the middle 1,2-bis(2,4-dihydroxybenzylidene) hydrazine label.

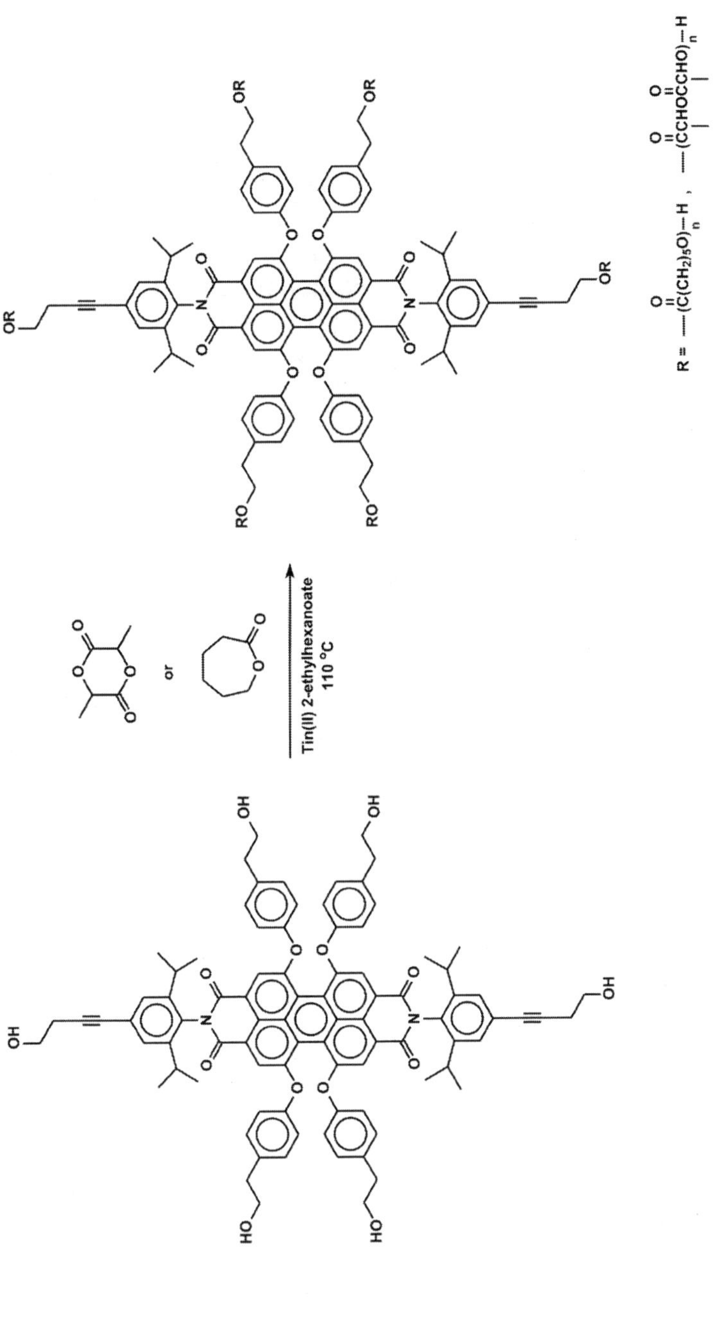

Figure 1.27 Synthesis of star-shaped polylactide and poly(ε-caprolactone) with fluorescent core.

Fluorescent probes with many –OH groups at their periphery allowed to synthesize star polylactides with fluorescent core moiety [103].

1.3.3 Polyesters with Polymerizable Groups

Polyesters containing polymerizable groups, so-called macromonomers, are used as building blocks of various complex polymer structures. These polyesters could be obtained either by using initiators with polymerizable groups which could survive conditions of synthesis of polyesters or by end-capping terminal groups of the earlier synthesized polyesters. The synthetic route based on an initiator with a polymerizable double bond was developed by Dubois *et al.* already in 1989 [104]. The process of synthesis of initiators and poly(ε-caprolactone) macromonomer is shown in Fig. 1.28.

Synthesis of poly(ε-caprolactone) macromonomer by end-capping polyester bearing the terminal –OH groups in reaction with methacrylic acid chloride, elaborated by Slomkowski, is shown in Fig. 1.29 [105].

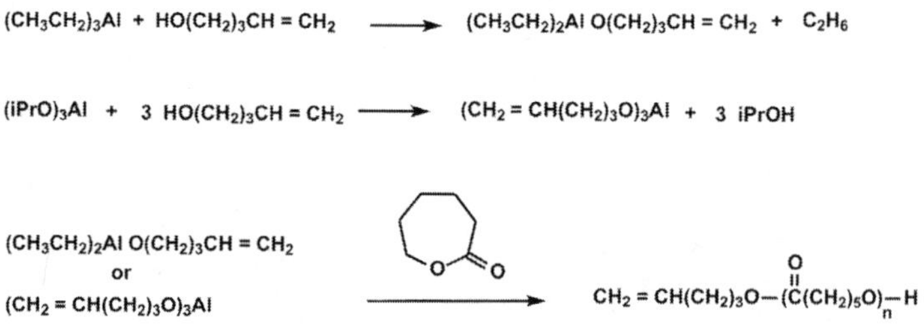

$$(CH_3CH_2)_3Al + HO(CH_2)_3CH=CH_2 \longrightarrow (CH_3CH_2)_2Al\,O(CH_2)_3CH=CH_2 + C_2H_6$$

$$(iPrO)_3Al + 3\ HO(CH_2)_3CH=CH_2 \longrightarrow (CH_2=CH(CH_2)_3O)_3Al + 3\ iPrOH$$

$$\begin{matrix}(CH_3CH_2)_2Al\,O(CH_2)_3CH=CH_2 \\ \text{or} \\ (CH_2=CH(CH_2)_3O)_3Al\end{matrix} \xrightarrow{\varepsilon\text{-caprolactone}} CH_2=CH(CH_2)_3O-(\overset{O}{\overset{\|}{C}}(CH_2)_5O)_n-H$$

Figure 1.28 Synthesis of poly(ε-caprolactone) macromonomer by using initiators with double bond moieties.

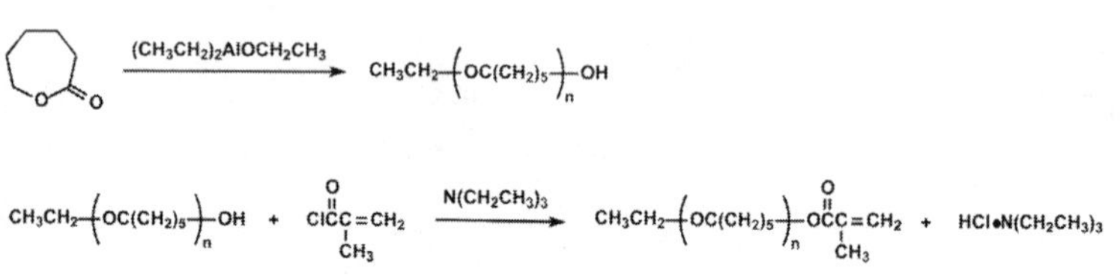

$$\varepsilon\text{-caprolactone} \xrightarrow{(CH_3CH_2)_2AlOCH_2CH_3} CH_3CH_2-(OC(CH_2)_5)_n-OH$$

$$CH_3CH_2-(OC(CH_2)_5)_n-OH + Cl\overset{O}{\overset{\|}{C}}\underset{CH_3}{\underset{|}{C}}=CH_2 \xrightarrow{N(CH_2CH_3)_3} CH_3CH_2-(OC(CH_2)_5)_n-O\overset{O}{\overset{\|}{C}}\underset{CH_3}{\underset{|}{C}}=CH_2 + HCl\bullet N(CH_2CH_3)_3$$

Figure 1.29 Synthesis of poly(ε-caprolactone)methacrylate macromonomer by end-capping method.

1.3.4 Polyesters with Terminal Chain Transfer Groups Serving as Macro-Transfer Agents in Reversible Addition–Fragmentation Chain Transfer (RAFT) Polymerization

Interesting opportunities are opened by the end-capping of polyester chains with moieties containing dithioester groups, which are suitable for participation in the reversible addition–fragmentation chain transfer polymerization (RAFT). Such macromolecules function as macromolecular chain transfer agents from the ends of which a variety of vinyl polymers could grow. Fig. 1.30 illustrates the synthesis of polylactide-*b*-poly(pentafluorophenyl methacrylate) shown as an example [106].

It is worth noting that the pentafluorophenyl ester group could be easily replaced by *N*-hydroxypropyl moiety, yielding eventually polylactide-*b*-poly((*N*-hydroxypropyl) methacrylamide) [106].

1.3.5 Polyesters with Functional Groups for Click Chemistry

Click chemistry opens convenient routes for the synthesis of functionalized polylactides. With the purpose of obtain such polymers, the alkyne substituted lactides and unsubstituted ones are used as comonomers in ring-opening copolymerization [107]. Acetylene moieties in these comonomers can be used for further functionalization by using azide containing substituents. Polylactides with aldehyde groups suitable for binding compounds with amino groups were synthesized in this way [108, 109]. Fig. 1.31 illustrates their synthesis.

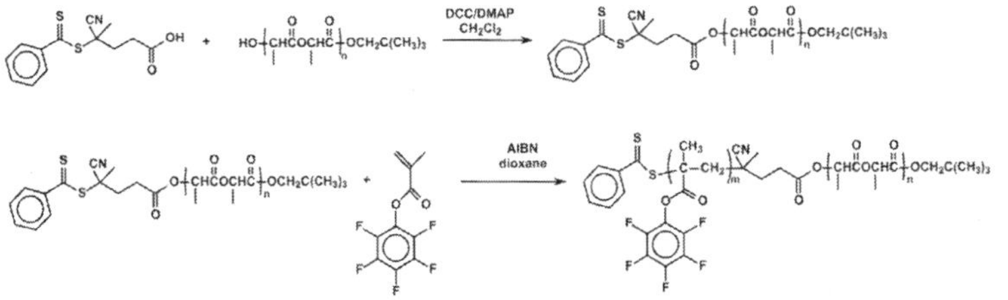

Figure 1.30 Synthesis of polylactide-*b*-poly(pentafluoromethyl mathacrylate) using a polylactide chain transfer agent with a dithioester end-group.

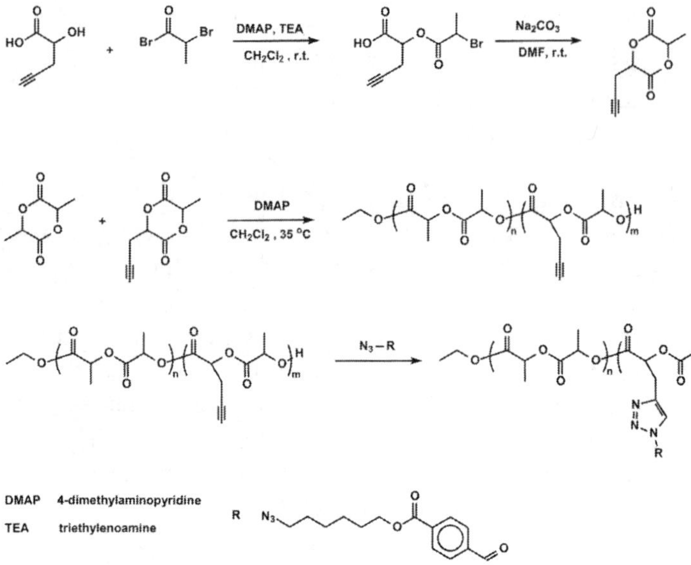

DMAP 4-dimethylaminopyridine
TEA triethylenoamine

Figure 1.31 Synthesis of alkyne containing polylactide and its modification.

1.3.6 Polyesters Modified with Succinic Anhydride and *N*-Hydroxysuccinimide

Among all kinds of functional polyesters, those with carboxylate and *N*-polyestersuccinimide end-groups deserve special attention. These groups could be used for binding compounds with hydroxyl and amine end-groups, allowing for subsequent attachment of proteins, oligo- and polysaccharides, their derivatives, as well as many others.

Polylactides and poly(ε-caprolactone) synthesized by anionic ring-opening polymerization of their parent cyclic monomers are equipped with –OH terminal hydroxyl groups which easily react with succinic anhydride, resulting in the formation of carboxyl end-groups. Synthesis of polylactide (in this particular case obtained by polycondensation) with covalently immobilized L-cysteine is presented here as an example [110]. The process is illustrated in Fig. 1.32.

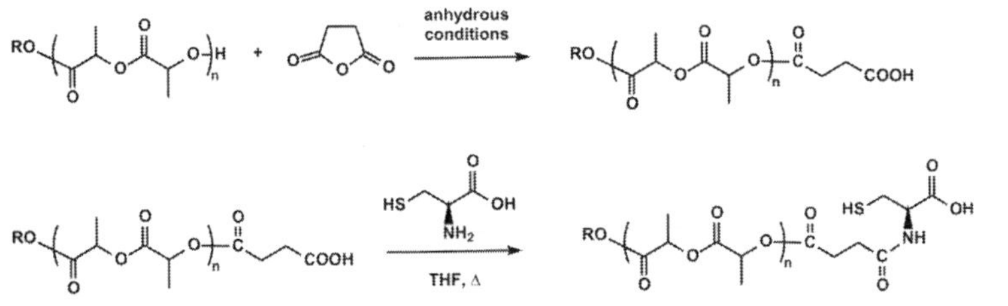

Figure 1.32 Synthesis of polylactide with L-cystein moiety.

Polyesters with *N*-succinimide end-groups are used as convenient intermediates for binding –OH group containing compounds to polyester chains. This method is conveniently used for grafting polylactide onto polysaccharides and polyaspartamide derivatives (e.g., α, β-poly(*N*-2-hydroxyethyl)-DL-aspartamide) [111, 112]. The process of grafting onto polysaccharide is illustrated in Fig. 1.33.

1.4 Conclusions

(Bio)degradable polyesters are often used as materials suitable for a large variety of medical applications. In the beginning, simple linear polymers with –OH and –COOH terminal groups were used for these purposes. Later studies resulted in the development of a large toolbox of synthetic methods allowing for the synthesis of aliphatic polyesters with controlled molar masses, topology, and tailored content and distribution of a variety of functional groups.

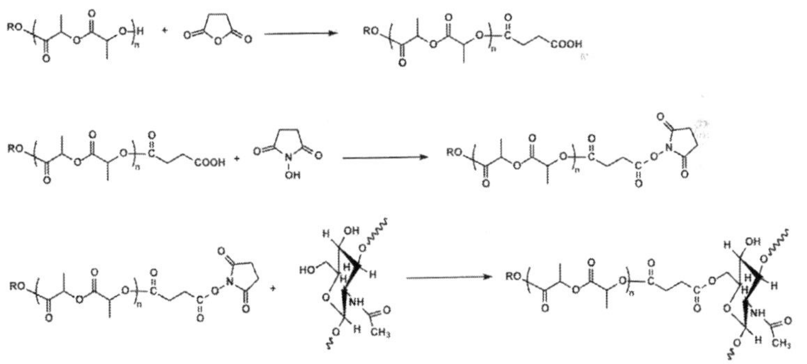

Figure 1.33 Grafting of polylactide onto polysaccharide via polylactide with *N*-succinimide end-group intermediate.

References

1. Gref, R., Domb, A., Quellec, P., Blunk, T., Müller, R. H., Verbavatz, J. M., Langer, R. (1995). *Adv Drug Delivery Rev,* **16**, 215–233.
2. Vert, M., Li, S., Garreau, H. (1995). *Macromol Symp,* **98**, 633–642.
3. Vert, M. (2000). *Macromol Symp,* **153**, 333–342.
4. Cohen, H., Levy, R. J., Gao, J., Fishbein, I., Kousaev, V., Sosnowski, S., Slomkowski, S., Golomb, G. (2000). *Gene Therapy,* **7**, 1896–1905.
5. Albertsson, A.-Ch., Varma, I. K. (2003). *Biomacromolecules,* **4**, 1466–1486.
6. Slomkowski, S. (2006). *Acta Pol Pharm,* **63**, 351–358.
7. Meng, F., Zhong, Z., Feijen, J. (2009). *Biomacromolecules,* **10**, 197–209.
8. Melchels, F. P. W., Feijen, J., Grijpma, D. W. (2009). *Biomaterials,* **30**, 3801–3809.
9. Slomkowski, S., Gosecki, M. (2011). *Curr Pharm Biotechnol,* **12**, 1823–1839.
10. Tyson, T., Malberg, S., Watz, V., Finne-Wistrand, A., Albertsson, A.-Ch. (2011). *Macromol Biosci,* **11**, 1432–1442.
11. Gref, R., Domb, A., Quellec, P., Blunk, T., Müller, R. H., Verbavatz, J. M., Langer, R. (2012). *Adv Drug Delivery Rev,* **64**, 316–326.
12. Chiellini, E. (2013). *Polimery,* **58**, 633–640.
13. Majerska, K., Duda, A. (2004). *J Am Chem Soc,* **126**, 1026–1027.
14. Shasteen, C., Choy, Y. B. (2011). *Biomed Eng Lett,* **1**, 163–167.
15. Duda, A. (2012). ROP of cyclic esters: mechanism of ionic and coordination processes, in Matyjaszewski, K., Möller, M. (eds.), *Polymer Science: A Comprehensive Reference,* vol. 4, Elsevier BV, Amsterdam, pp. 213–246.
16. Slomkowski, S., Penczek, S., Duda, A. (2014). *Polym Adv Technol,* **25**, 436–447.
17. Corneillie, S., Smet, M. (2015). *Polym Chem,* **6**, 850–867.
18. Evstropov, A. A., Lebedev, B. V., Kulagina, T. G., Lyudvig, E. B., Belenkaya, B. W. (1979). *Vysokomol Soedin, Ser A,* **21**, 2038–2044.
19. Hofman, A., Slomkowski, S., Penczek, S. (1984). *Makromol Chem,* **185**, 91–101.
20. Sosnowski, S., Duda, A., Slomkowski, S., Penczek, S. (1984). *Makromol Chem, Rapid Commun,* **5**, 551–557.

21. Sosnowski, S., Slomkowski, S., Penczek, S. (1993). *Macromolecules*, **26**, 5526–5527.

22. Dale, J., Schwartz, J. E. (1986). *Acta Chem Scand, Ser B*, **40**, 559–567.

23. Duda, A., Kubisa, P., Lapienis, G., Slomkowski, S. (2014). *Polimery*, **59**, 9–23.

24. Evstropov, A. A., Lebedev, B. V., Kiparisova, Y. G., Alekseyev, V. A., Stashina, G. A. (1980). *Vysokomol Soedin, Ser A*, **22**, 2450–2456.

25. Duda, A., Biela, T., Libiszowski, J., Penczek, S., Dubois, Ph., Mecerreyes, D., Jerome, R. (1998). *Polym Degrad Stab*, **59**, 215–222.

26. Duda, A., Penczek, S., Dubois, Ph., Mecerreyes, D., Jerome, R. (1996). *Macromol Chem Phys*, **197**, 1273–1283.

27. Ubaghs, L., Waringo, M., Keul, H., Hoecker, H. (2004). *Macromolecules*, **37**, 6755–6762.

28. Agarwal, S., Xie, X. (2003). *Macromolecules*, **36**, 3545–3549.

29. Sosnowski, S., Slomkowski, S., Penczek, S., Reibel, L. (1983). *Makromol Chem*, **184**, 2159–2171.

30. Hofman, A., Slomkowski, S., Penczek, S. (1987). *Makromol Chem Rapid Commun*, **8**, 387–391.

31. Dubois, Ph., Jerome, R., Teyssie, Ph. (1989). *Polym Bull*, **22**, 475–482.

32. Duda, A., Florjanczyk, Z., Hofman, A., Slomkowski, S., Penczek, S. (1990). *Macromolecules*, **23**, 1640–1646.

33. Kricheldorf, H. R., Scharnagl, N., Kreiser-Sanders, I. (1990). *Makromol Chem Macromol Symp*, **32**, 285–298.

34. Penczek, S., Duda, A., Slomkowski, S. (1992). *Makromol Chem Macromol Symp*, **54/55**, 31–40.

35. Baran, J., Duda, A., Kowalski, A., Szymanski, R., Penczek, S. (1997). *Macromol Symp*, **123**, 93–101.

36. Penczek, S., Duda, A., Szymanski, R. (1998). *Macromol Symp*, **132**, 441–449.

37. Penczek, S., Biela, T., Duda, A. (2000). *Macromol Rapid Commun*, **21**, 941–950.

38. Duda, A., Penczek, S. (2002). Mechanism of aliphatic polyester formation, in Steinbuechel, A., Doi, Y. (eds.), *Biopolymers*, Vol. 3b: *Polyesters II–Properties and Chemical Synthesis*, Wiley-VCH, Weinheim, pp. 371–430.

39. Duda, A., Kubisa, P., Lapienis G., Slomkowski, S. (2014). *Polimery*, **59**, 9–23.

40. Zhong, Z., Schneiderbauer, S., Dijkstra, P. J., Westerhausen, M., Feijen, J. (2001). *J Polym Environ*, **9**, 31–38.

41. Biela, T., Kowalski, A., Libiszowski, J., Duda, A., Penczek, S. (2006). *Macromol Symp*, **240**, 47–55.

42. Kowalski, A., Libiszowski, J., Duda, A., Penczek, S. (2000). *Macromolecules*, **33**, 1964–1971.

43. Tsuji, H. (2005). *Macromol Biosci*, **5**, 569–597.

44. Piao, L., Deng, M., Chen, X., Jiang, L., Jing, X. (2003). *Polymer*, **44**, 2331–2336.

45. Chisholm, M. H., Eilerts, N. W. (1996). *Chem Commun*, 853–854.

46. Chisholm, M. H., Eilerts, N. W., Huffman, J. C., Iyer, S. S., Pacold, M., Phomphrai, M. (2000). *J Am Chem Soc*, **122**, 11845–11854.

47. Chisholm, M. H., Gallucci, J., Phomphrai, K. (2003). *Chem Commun*, 48–49.

48. Nederberg, F., Connor, E. F., Möller, M., Glauser, T., Hedrick, J. L. (2001). *Angew Chem Int Ed*, **40**, 2712–2715.

49. Bourissou, D., Moebs-Sanchez, S., Martín-Vaca, B. (2007). *C R Chimie*, **10**, 775–794.

50. Kamber, N. E., Jeong, W., Waymouth, R. M., Pratt, R. C., Lohmeijer, B. G. G., Hedrick, J. L. (2007). *Chem Rev*, **107**, 5813–5840.

51. Kricheldorf, H. R. (2014). *Polycondensation History and New Results*, Springer, Heidelberg chap. 2, p. 9.

52. Gay-Lusac, J., Pelouze, J. (1933). *Ann Chem Pharm (Liebigs Ann Chem)*, **7**, 40.

53. Engelhardt, C. (1849). *Liebigs Ann Chem*, **70**, 241.

54. Knani, D., Gutman, A. L., Kohn, D. H. (1993). *J Polym Sci Part A: Polym Chem*, **31**, 1221–1232.

55. Uyama, H., Kobayashi, S. (1993). *Chem Lett*, 1149–1150.

56. MacDonald, R. T., Pulapura, S. K., Svirkin, Y. Y., Gross, R. A., Kaplan, D. L., Akkara, J., Swift, G., Wolk, S. (1995). *Macromolecules*, **28**, 73–78.

57. Uyama, H., Takeya, K., Hoshi, N., Kobayashi, S. (1995). *Macromolecules*, **28**, 7046–7050.

58. Namekawa, S., Uyama, H., Kobayashi, S. (1996). *Polym J*, **28**, 730–731.

59. Uyama, H., Kikuchi, H., Takeya, K., Kobayashi, S. (1996). *Acta Polym*, **47**, 357–360.

60. Henderson, L. A., Svirkin, Y. Y., Gross, R. A., Kaplan, D. L., Swift, G. (1996). *Macromolecules*, **29**, 7759–7766.

61. Bisht, K. S., Henderson, L. A., Gross, R. A., Kaplan, D. L., Swift, G. (1997). *Macromolecules*, **30**, 2705–2711.

62. Küllmer, K., Kikuchi, H., Uyama, H., Kobayashi, S. (1998). *Macromol Rapid Commun*, **19**, 127–130.

63. Kobayashi, S., Uyama, H., Namekawa, S. (1998). *Polym Degrad Stabil*, **59**, 195–201.

64. Kobayashi, S., Takeya, K., Suda, S., Uyama, H. (1998). *Macromol Chem Phys*, **199**, 1729–1736.

65. Kobayashi, S., Uyama, H., Namekawa, S., Hayakawa, H. (1998). *Macromolecules*, **31**, 5655–5659.

66. Kobayashi, S., Uyama, H. (1999). *Macromol Symp*, **144**, 237–246.

67. Matsumura, S., Ebata, H., Toshima, K. (2000). *Macromol Rapid Commun*, **21**, 860–863.

68. Kumar, A., Gross, R. A. (2000). *Biomacromolecules*, **1**, 133–138.

69. Suzuki, Y., Taguchi, S., Saito, T., Toshima, K., Matsumura, S., Doi, Y. (2001). *Biomacromolecules*, **2**, 541–544.

70. Kikuchi, H., Uyama, H., Kobayashi, S. (2002). *Polym J*, **34**, 835–840.

71. Suzuki, Y., Taguchi, S., Hisano, T., Toshima, K., Matsumura, S., Doi, Y. (2003). *Biomacromolecules*, **4**, 537–543.

72. Mei, Y., Kumar, A., Gross, R. (2003). *Macromolecules*, **36**, 5530–5536.

73. van der Mee, L., Helmich, F., de Bruijn, R., Vekemans, J., Palmans, A. R. A., Meijer, E. W. (2006). *Macromolecules*, **39**, 5021–5027.

74. Numata, K., Srivastava, R. K., Finne-Wistrand, A., Albertsson, A.-Ch., Doi, Y., Abe, H. (2007). *Biomacromolecules*, **8**, 3115–3125.

75. Kikuchi, H., Uyama, H., Kobayashi, S. (2000). *Macromolecules*, **33**, 8971–8975.

76. Kobayashi, S. (2006). *Macromol Symp*, **240**, 178–185.

77. Mahapatro, A., Kalra, B., Kumar, A., Gross, R. A. (2003). *Biomacromolecules*, **4**, 544–551.

78. Mahapatro, A., Kumar, A., Gross, R. A. (2004). *Biomacromolecules*, **5**, 62–68.

79. Slomkowski, S. (1984). *J Macromol Sci Part A: Polym Chem*, **21**, 1383–1404.

80. Slomkowski, S. (1985). *Makromol Chem*, **186**, 2581–2594.

81. Ito, K., Hashizuka, Y., Yamashita, Y. (1977). *Macromolecules*, **10**, 821–824.

82. Ito, K., Yamashita, Y. (1978). *Macromolecules*, **11**, 68–72.

83. Sosnowski, S., Slomkowski, S., Penczek, S. (1991). *Makromol Chem,* **192**, 1457–1465.

84. Sosnowski, S. (2008). *J Polym Sci, Part A, Polym Chem Ed,* **46**, 6978–6982.

85. Biela, T., Duda, A., Rode, K., Pasch, H. (2003). *Polymer,* **44**, 1851–1860.

86. Xu, F., Zheng, S.-Z., Lu, Y.-L. (2013). *J Polym Sci Part A: Polym Chem,* **51**, 4429–4439.

87. Brutman, J. P., Delgado, P. A., Hillmyer, M. A. (2014). *ACS Macro Lett,* **3**, 607–610.

88. Shao, J., Tang, Z., Sun, J., Li, G., Chen, X. (2014). *J Polym Sci, Part B: Polym Phys,* **52**, 1560–1567.

89. Jing, Z., Shi, X., Zhang, G., Qin, J. (2015). *Polym Adv Technol,* **26**, 223–233.

90. Danko, M., Libiszowski, J. L., Biela, T., Wolszczak, M., Duda, A. (2005). *J Polym Sci Part A: Polym Chem,* **43**, 4586–4599.

91. Adeli, M., Namazi, H., Du, F., Hönzke, S., Hedtrich, S., Keilitz, J., Haag, R. (2015). *RSC Adv,* **5**, 14958–14966.

92. Pitet, L. M., Hait, S. B., Lanyk, T. J., Knauss, D. M. (2007). *Macromolecules,* **40**, 2327–2334.

93. Zhao, R.-X., Li, L., Wang, B., Yang, W.-W., Chen, Y., He, X.-H., Cheng, F., Jiang, S.-Ch. (2012). *Polymer,* **53**, 719–727.

94. Gadzinowski, M., Sosnowski, S. (2003). *J Polym Sci, Part A: Polym Chem,* **41**, 3750–3760.

95. Slomkowski, S., Gadzinowski, M., Sosnowski, S., Radomska-Galant, I. (2005). *Polimery,* **50**, 546–554.

96. Slomkowski, S. (2007). Polyester nano and microparticles by polymerization and by self assembly of macromolecules, in *Nanoparticles for Pharmaceutical Applications,* Domb, J., Tabata, Y., Ravikumar, M. N. V., Farber, S. (eds.), American Scientific Publisher, Stevenson Ranch, Cal, chap. 16, pp. 288–303.

97. Slomkowski, S., Gadzinowski, M., Sosnowski, S., Radomska-Galant, I., Pucci, A., De Vita, C., Ciardelli, F. (2006). *J Nanosci Nanotech,* **6**, 3242–3251.

98. Slomkowski, S. (2010). *Macromol Symp,* **288**, 121–129.

99. Sosnowski, S., Slomkowski, S., Penczek, S., Florjanczyk, Z. (1991). *Makromol Chem,* **192**, 1457–1465.

100. Slomkowski, S., Gadzinowski, M., Sosnowski, S., de Vita, C., Pucci, A., Ciardelli, F., Jakubowski, W., Matyjaszewski, K. (2005). *Macromol Symp,* **226**, 239–252.

101. Samonina-Kosicka, J., DeRosa, Ch. A., Morris, W. A., Fan, Z., Fraser, C. L. (2014). *Macromolecules*, **47**, 3736–3746.

102. Hsiao, T.-S., Huang, P.-Ch., Lin, L.-Y., Yang, D.-J., Hong, J.-L. (2015). *Polym Chem*, **6**, 2264–2273.

103. Klok, H.-A., Becker, S., Schuch, F., Pakula, T., Mullen, K. (2003). *Macromol Biosci*, **3**, 729–741.

104. Dubois, Ph., Jerome, R., Teyssie, Ph. (1989). *Polym Bull*, **22**, 475–482.

105. Sosnowski, S., Gadzinowski, M., Slomkowski, S., Penczek, S. (1994). *J Bioact Compat Polym*, **9**, 345–366.

106. Barz, M., Armiñán, A., Canal, F., Wolf, F., Koynov, K., Frey, H., Zentel, R., Vicent, M. J. (2012). *J Control Release*, **163**, 63–74.

107. Yu, Y., Zou, J., Yu, L., Ji, W., Li, Y., Law, W.-Ch., Cheng, Ch. (2011). *Macromolecules*, **44**, 4793–4800.

108. Yu, Y., Chen, Ch.-K., Law, W.-Ch., Weinheimer, E., Sengupta, S., Prasad, P. N., Cheng, Ch. (2014). *Biomacromolecules*, **15**, 524–532.

109. Yu, Y., Chen, Ch.-K., Law, W.-Ch., Sun, Ha., Prasad, P. N., Cheng, Ch. (2015). *Polym Chem*, **6**, 953–961.

110. Berger, K., Gregorova, A. (2014). *J Appl Polym Sci*, **131**, article no. 41105.

111. Palumbo, F. S., Pitarresi, G., Mandracchia, D., Tripodo, G., Giammona, G. (2006). *Carbohydrate Polym*, **66**, 379–385.

112. Pavia, F. C., La Carrubba, V., Brucato, V., Palumbo, F. S., Giammona, G. (2014). *Mater Sci Eng C*, **41**, 301–308.

Chapter 2

Biodegradable Fatty Acid Polyesters

Konda Reddy Kunduru, Arijit Basu, and Abraham J. Domb
Institute for Drug Research, School of Pharmacy, Faculty of Medicine, Center for Nanoscience & Nanotechnology and The Alex Grass Center for Drug Design and Synthesis, The Hebrew University of Jerusalem 91120, Israel
avid@ekmd.huji.ac.il

Biodegradable polyesters synthesized from naturally occurring fatty acids have been used in drug delivery and temporary implantable devices. They provide control over flexibility, melting temperature, hydrophobicity, and pliability. Dicarboxylic acid containing fatty acids are used along with hydroxy acids to give proper control over the final structure. Synthesis, physical properties, pharmacokinetics, stability, toxicology, and biomedical applications of polyesters and poly(ester-anhydride)s are discussed.

2.1 Introduction

Vegetable oils are important renewable resource for synthesizing polyesters. Fatty acids are the main components of naturally occurring vegetable oils. They are available in pure form at

Handbook of Polyester Drug Delivery Systems
Edited by MNV Ravikumar

ISBN 978-981-4669-65-8 (Hardcover), 978-981-4669-66-5 (eBook)
www.panstanford.com

affordable cost [1, 2]. The fatty acids mainly used in the synthesis of biodegradable materials are stearic acid, oleic acid, linoleic acid, linolenic acid, erucic acid, veranolic acid, and ricinoleic acid (Fig. 2.1) [3, 4]. Other useful short chain fatty acids are azelic, sebacic, undecenoic, and brassylic acids. Azelic acid is usually prepared from oleic acid by ozonolysis. Alkaline pyrolysis of ricinoleic acid gives sebacic acid and pyrolysis gives undecenoic acid. Oxidation of erucic acid gives brassylic acid, a C_{13} chain dicarboxylic acid [5].

Polyesters have been extensively explored for biomedical uses. Polymer chemists are interested in these materials because they are environmentally friendly and biodegradable. Polyesters are prone to degradation at specific conditions; they also degrade in presence of naturally occurring microorganisms such as algae, fungi, and bacteria. Usage of polyesters has less effect on the environment, as they are nontoxic and biodegradable. Therefore, they are used widely in biomedical applications. Various polyesters have been reported in the literature, depending on the monomers used for synthesis [5]. We discuss the preparation and properties of two types of fatty acid based polymers, polyesters and poly(ester-anhydride).

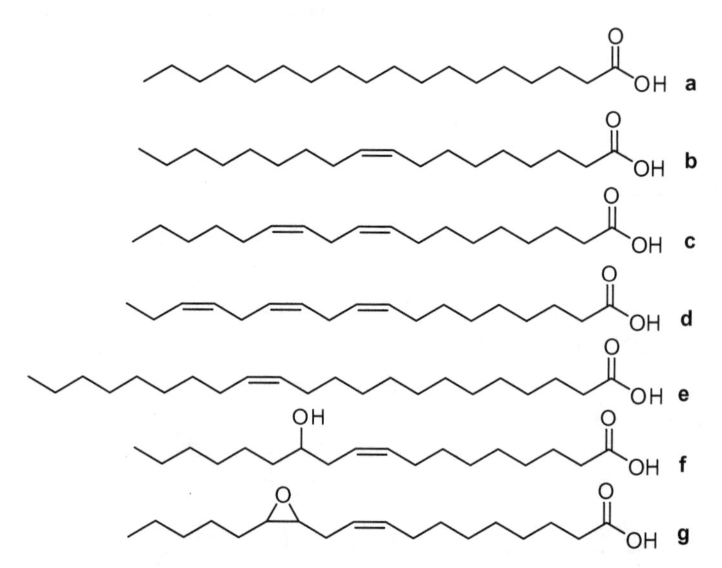

Figure 2.1 Major fatty acids as starting materials for the synthesis of biodegradable polymers: (a) stearic acid, (b) oleic acid, (c) linoleic acid, (d) linolenic acid, (e) erucic acid, (f) ricinoleic acid, and (g) vernolic acid.

2.2 Importance of Fatty Acids in Biodegradable Polymers

Biodegradability of polymers is important in drug delivery, as there is a need to remove this biomaterial once the drug is released from the polymer [6]. Different types of biodegradable polymers are studied for this purpose, such as polyesters, polyanhydrides, polyorthoesters, and polyphophazenes, on the basis of their chemical linkages in the polymer [7, 8]. The polymer which is used for the drug delivery application should contain the following characteristics [9–11]:

- The polymer must be biocompatible when implanted in the target organ.
- It must have enough hydrophobic nature, so that the drug is released in a predictable and controlled manner.
- It should be eliminated completely from the implantation site in a predictable manner.
- It must have a low melting point (<100°C) and must be soluble in common organic solvents.
- It should be flexible enough before and during degradation so that it does not crumble or fragment during use.
- It should be easy to manufacture at a reasonable cost.

Fatty acids from natural vegetable oils have all these properties, plus they provide hydrophobicity, flexibility, and injectability [9, 12]. Fatty acids are natural body components having readily hydrolysable bonds [12–16]. Synthetic methodologies for biodegradable polymers using vegetable oils and fatty acids have been reported. Fatty acids are either saturated or unsaturated in nature. Saturated fatty acids such as palmitic acid, lauric acid, or stearic acid are monofunctional carboxylic groups. Therefore, they have been used as chain terminators in the polymer synthesis [15, 17]. Unsaturated fatty acids such as oleic, linoleic, or linolenic acid contain double bonds along with carboxylic groups and they can be converted into fatty acid dimer and trimer having two or three carboxylic acid groups for further polymer synthesis. Ricinoleic acid is a bifunctional natural fatty acid with hydroxyl and carboxyl groups. It is also available at a cheaper price [18–20].

2.3 Synthesis of Fatty Acid-Based Biodegradable Polyesters

Polyesters can be synthesized either by polycondensation of hydroxy fatty acids or by ring-opening polymerization of cyclic esters or lactones. The monomers used for the preparation of polyesters are lactic, glycolic, hydroxybutyric, and hydroxycaproic acids. Lactic acid-based polyesters are safe and nontoxic in nature [21–23]. These polyesters are useful materials for controlled drug delivery. They are degraded to hydroxy acids, but are less sensitive to hydrolysis. Degradation is slower compared to polyanhydrides; ester bond is less sensitive to hydrolysis, making preparation, handling, and storage of this type of polymer easier.

Co-polyesters based on ricinoleic (RA) and lactic (LA) acids were prepared by thermal polycondensation method at 150°C. Polyesters containing 20% or more RA were liquid at room temperature. Transesterification of high molecular weight poly(lactic acid) (PLA) with RA resulted in oligomers. Further condensation at 150°C for 12 hours resulted in multiblock P(LA-RA) copolyesters (Fig. 2.2) [24]. A systematic study of the synthesis and stereocomplexation of L-lactic acid and ricinoleic acid-based copolyesters was reported. The relative degree of crystallinity of these copolyesters depends directly on PLA block size, which is the only difference between the corresponding polymers. P(L-LA-RA)s and enantiomeric D-PLA were mixed together in acetonitrile solution to form stereocomplexes. At least a block length of 10 lactic acid units are required to form a stereocomplex from P(L-LA-RA)s. The formed stereocomplexes exhibited higher melting temperature than the enantiomeric polymers, indicating stereocomplex formation. This stereocomplex can be used as a carrier for drugs and polymeric scaffolds for tissue engineering. They are low temperature melting biodegradable polymers with desired properties such as pliability, hydrophobicity, and softness [25].

Another study describes degradation and drug release from L-lactic acid and ricinoleic acid-based copolyesters. These copolyesters were synthesized by ring-opening polymerization,

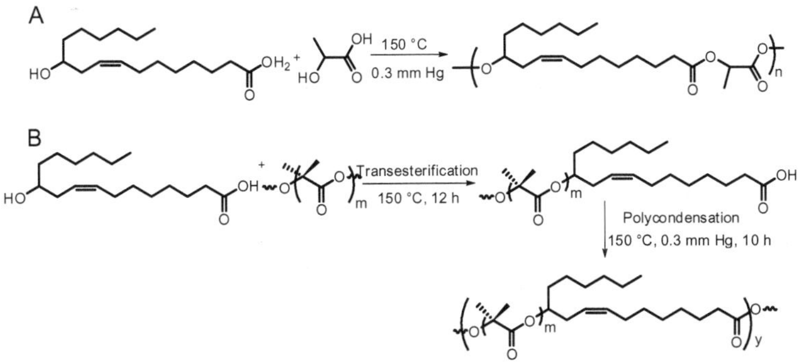

Figure 2.2 (A) Synthesis of poly(ricinoleic acid-lactic acid) condensation polymers. (B) Synthesis of poly(ricinoleic acid-lactic acid) by transesterification of polylactic acid and ricinoleic acid followed by re-polyesterification [24].

melt condensation, and transesterification of high molecular weight poly(lactic acid) (PLA) with ricinoleic acid, and repolymerization by condensation to yield random and block copolymers. The weight average molecular weights (M_w) of these polymers were between 3000 and 13000. All polymers exhibited zero-order weight loss, with a 20–40% loss after 60 days of incubation. Weight loss from the polymer is proportional to the lactic acid release into the degradation solution. A sudden decrease in molecular weight of the polymers was observed during the initial 20 days, showing a steady slow degradation pattern. During this slow degradation phase the number average molecular weight (M_n) found to be at 4000–2000 for at least another 40 days [26]. LA release from P(LA-RA), synthesized by condensation, is similar to LA release from pure PLA, because L-PLA is a relatively hydrophilic crystalline polymer, whereas P(LA-RA) 60:40 w/w is a relatively hydrophobic noncrystalline polymer. Hydrophobicity compensates for loss of crystallinity resulting in a similar degradation rate. LA release from P(LA-RA)s synthesized by transesterification and by ROP is also similar, due to the block-like nature of the structures [26].

We also reported synthesis, characterization, and polymerization of ricinoleic acid (RA) lactone. Lactones of RA were synthesized using dicyclohexylcarbodimide and (dimethylamino)pyridine as catalyst.

The conversion of the hydroxyl on C_{12} to an ester group shifted the C12 proton NMR signal from 3.6 to 2.7 ppm. FTIR spectrum showed the disappearance of hydroxyl broadening peak at 3400 cm^{-1} and an ester carbonyl peak appeared at 1730 cm^{-1} instead of the carboxylic acid peak at 1700 cm^{-1}. Mono- to hexalactones (1RM-6RM) obtained and polymerized with catalysts commonly used for ring-opening polymerization of lactones, under specific reaction conditions. Polymerization with Sn(Oct) was not successful for all the procedures and more reactive catalyst such as yttrium isopropoxide resulted in oligomers. Copolymerization with lactide (LA) by ROP using Sn(Oct) as catalyst yielded copolyesters with molecular weights in the range of 5000–16000 and melting temperatures of 100–130°C for copolymers containing 10–50% w/w ricinoleic acid residues (Fig 2.3) [27].

The polyester polyol was synthesized by the "core-first" method, which involves a polymerization of L-lactide by using a castor oil as multifunctional initiator using trifluoromethanesulfonicacid as a catalyst (Fig. 2.4). The band for carbonyl(C=O) of aliphatic ester splits into two, one at 1758 cm^{-1} for poly(L-lactide) chains, and the other one at 1745 cm^{-1} for castor oil from the FTIR spectrum. ^{1}H NMR spectrum for the lactide polymerization with castor oil hydroxyl groups, the peak related to C_{12} proton attached to the hydroxyl group (CH–OH) at 3.61 ppm disappeared, while a new peak at 4.32 ppm appeared which confirms the hydroxyl was connected to lactide chains. The weight average molecular weight of this polyester was in the range of 3000–5700. The thermal stability of these polyesters showed the dependence of weight loss on the arm length of the star-shaped polyesters. The polyesters have high crystallinity and a low melting point, according to DSC results, and are thus very promising for medical and pharmaceutical applications [28].

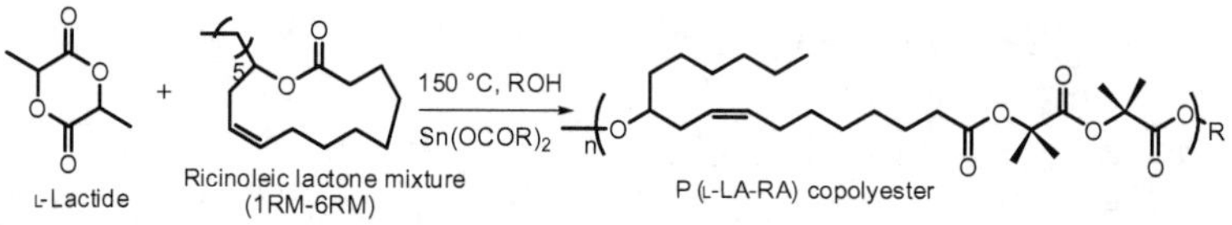

Figure 2.3 Ring-opening polymerization for the preparation of copolyester of L-lactide and ricinoleic acid lactone.

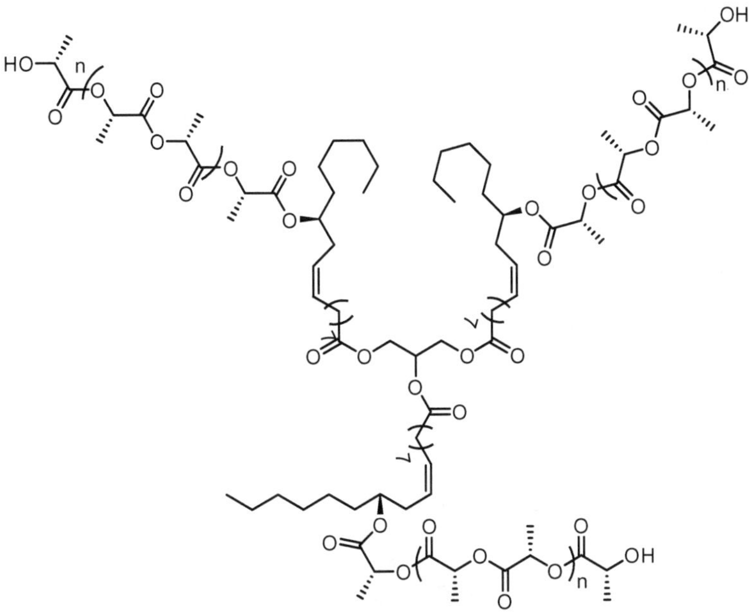

Figure 2.4 Structure of polyester of L-lactide and castor oil.

Poly(ester-anhydride)s were prepared by the melt polycondensation of diacid oligomers of poly(sebacic acid) (PSA) transesterified with ricinoleic acid (Fig. 2.5). The transesterification of PSA with ricinoleic acid to form oligomers was conducted via a melt bulk reaction between a high molecular weight PSA and ricinoleicacid. Polymers with weight-average molecular weights of 2000–60000 and melting temperatures of 24–77°C were obtained for PSA containing 20–90% (w/w) ricinoleic acid. NMR and IR analyses indicated the formation of ester bonds along the polyanhydride backbone. These biodegradable copolymers have potential use as drug carriers [29].

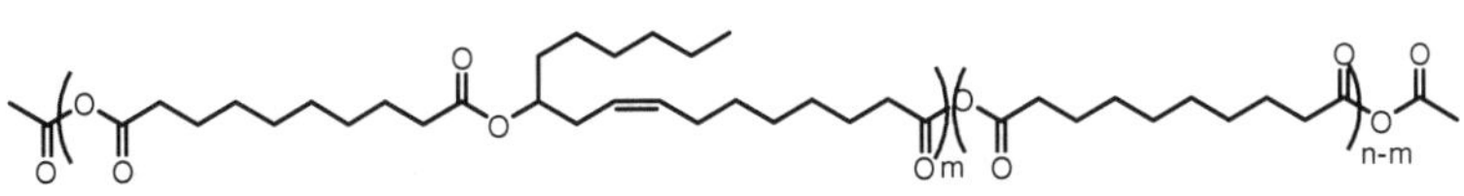

Figure 2.5 Structure of poly(sebacic acid-*co*-ricinoleic acid) [29].

Ricinoleic acid or its oligoesters and alkane dicarboxylic acids with C_6–C_{12} carbon chain lengths were used for the preparation of injectable poly(ester-anhydride)s by transesterification and repolymerization (Fig 2.6). Melting points decreased with the length of alkandicarboxylicacid from C_{12} to C_6 of poly(ester-anhydride) s. Poly(ester-anhydride)s prepared with dodecandioic, sebacic, suberic, adipic, and ricinoleic acids were injectable with a 21G needle. Poly(ester-anhydride)s prepared using ricinoleic acid oligomers produced higher melting polymers. Copolymers of ricinoleic acid with maleic or succinic anhydrides were also pasty injectable [30].

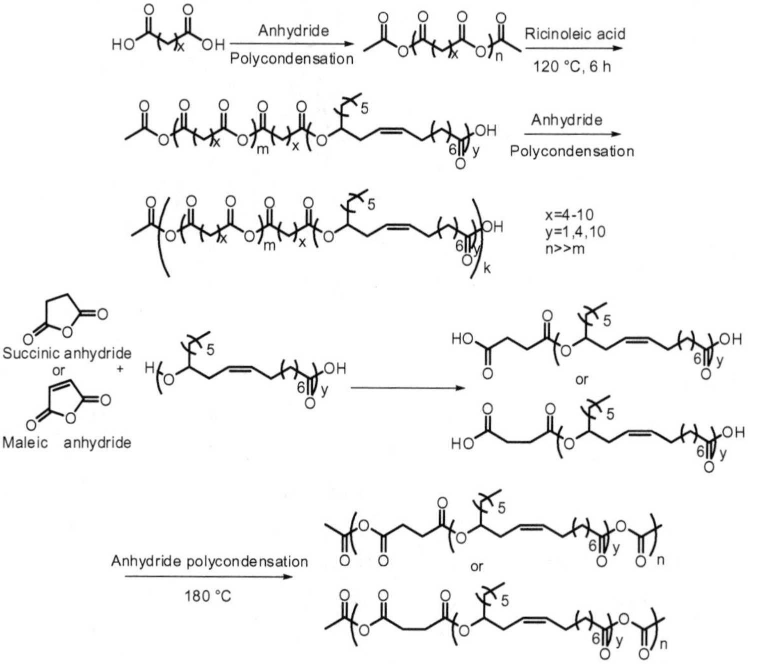

Figure 2.6 Synthesis of poly(ester-anhydride)s of ricinoleic acid and alkanedioic acids as well as maleic and succinic anhydrides [30].

Synthesis and *in vitro* analysis of poly(ester-anhydride) antimicrobial protection coatings composed of ricinoleic acid, sebacic acid, terephthalic acid, and isophthalic acid was conducted. Hydrolysis and controlled release were studied in the buffer phosphate and in water with different analytical methods. The mixture of the polymer with various fillers proved that poly(ester-anhydride)s were compatible with paint preparation [31].

Synthesis of aliphatic polyesters was conducted from adipic or sebacic acid and alkanediols, using inorganic acid catalyst. The monomer composition, reaction time, catalyst type, and reaction conditions were optimized to yield polyesters with weight average molecular weights of 23000 for adipic acid and 85000 for sebacic acid-based polyesters. The polymers melt at temperatures of 52–65°C and possess melt viscosity in the range of 5600–19400 cP. This route represents an alternative method for producing aliphatic polyesters for possible use in the preparation of degradable disposable medical supplies [32].

Oleic acid was epoxidized in to polymerizable monomer of epoxidized oleic acid followed by heating at 90–120°C obtained poly(oleic acid). The polymer was remained as a viscous liquid compared to monomer. The carboxyl group at 1701 cm^{-1} of oleic acid epoxide reduced after polymerization, while new peaks were observed at 1735 and 3462 cm^{-1}, suggesting the formation of ester group and hydroxyl groups (Fig. 2.7). This polyester may find an application in drug delivery [33].

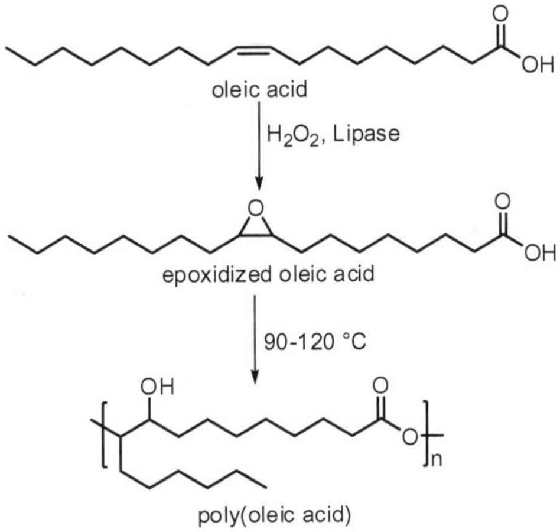

oleic acid

H_2O_2, Lipase

epoxidized oleic acid

90-120 °C

poly(oleic acid)

Figure 2.7 Synthesis of oleic acid polymers [33].

Poly(oleic diacid-*co*-glycerol) was synthesized from oleic diacid and glycerol as monomers with novozyme 435 or dibutyl tin oxide as catalyst. The resulting polyester with a molecular weight 6000 Da was obtained. The usage of enzyme catalyst did not form any cross link in the polyester and obtained a unique, soluble, hyperbranched material for biomedical applications (Fig. 2.8) [34].

Synthesis of a unique polyester based on the monomers glycerol, azelaic acid and succinic acid were prepared via catalyst free polyesterification. Thermal stability of the polymers increased with increasing number of azelaic acid units in the polyester. This class of polyesters can be further exploited by attaching biomolecules of proteins category (Fig. 2.9) [35].

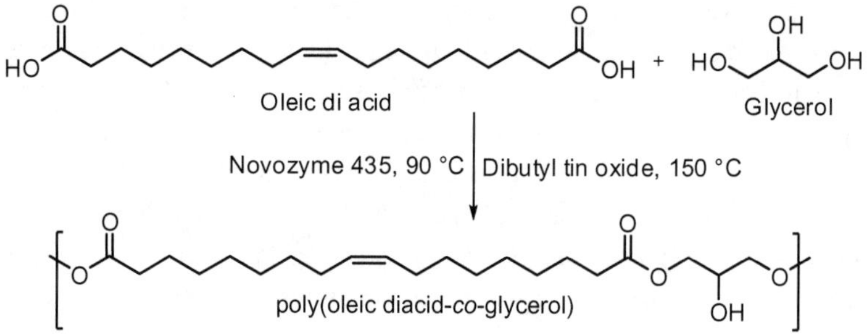

Figure 2.8 Synthesis of poly(oleic diacid-*co*-glycerol) [34].

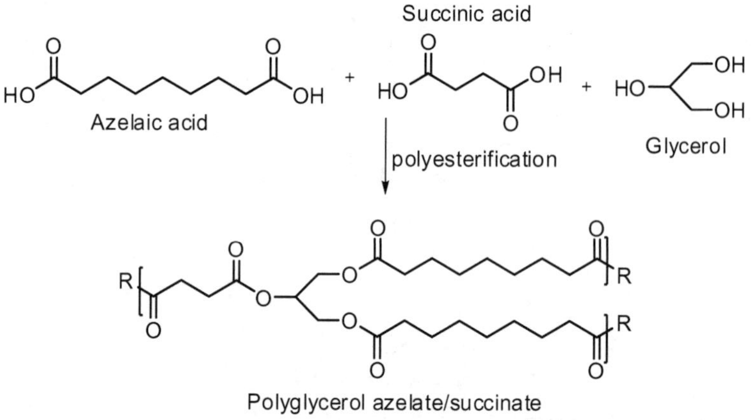

Figure 2.9 Polyesterification of azelaic acid, succinic acid and glycerol [35].

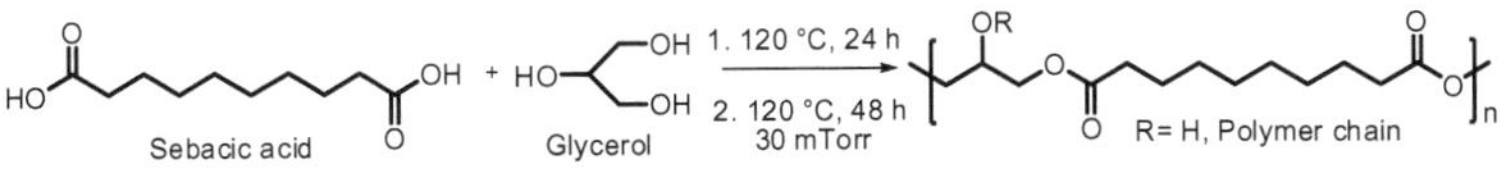

Figure **2.10** Chemical synthesis of poly(glycerol-sebacate) [36].

Poly(glycerol sebacate) (PGS) was prepared by polycondensation of glycerol and sebacic acid. This polyester is a transparent and almost colorless material. The resulting polymer had a small amount of crosslinkage and hydroxyl groups directly attached to the polymer backbone. FTIR spectrum showed an intense peak for carbonyl at 1740 cm^{-1} confirmed the ester bond (Fig. 2.10) [36].

Poly(glycerol sebacate)-*co*-lactic acid was synthesized by the catalyst and solvent free polycondensation of glycerol, sebacic acid and lactic acid in equal molar ratio at 110–120°C and for 2 to 6 days under nitrogen blanket. The formation of the polyester was confirmed from its FTIR absorption values at 1730 cm^{-1} for ester carbonyl group; the broad peak at 3460 cm^{-1} was due to hydrogen bonding hydroxyl groups [37].

2.4 Degradation and Elimination

Elimination of the polymer after its incorporation in the body is important for successful treatment. These are either removed through surgery or by degradation in the body followed by excretion as metabolites. Biodegradable and biocompatible polymer should be degraded in the biological system without causing any toxicity, and eliminated completely. Degradation is a process where polymers chains are cleaved, whereas erosion is the loss of the mass from the polymer matrix. Rate of degradation, swelling, porosity, and diffusion of oligomers and monomers from the polymer matrix, influences the erosion of the polymer [38–40]. Polyester PLGA undergoes bulk erosion, where the polymer mass is lost uniformly and the rate of erosion is dependent on the volume of the polymer instead of its surface area. In surface erosion, material loss will occur from the outside to the inside of the polymer matrix, and therefore the erosion rate depends on the surface area of the polymer instead of its volume. Surface erosion depends on the thickness of the polymer matrix. Thicker the matrix, longer the lifetime of the polymer disk [39, 40].

Copolyesters of ricinoleic acid lost weight uniformly during the degradation process of 60 days (20–40% loss) [26, 27]. Polymers lost ~10% of their weight for about the first 10 days. It was observed that lactic acid (LA)-based polymers lose up to 5% of LA immediately after the polymer is kept in aqueous solution. In case of block copolymers made of RA and LA segments, RA surrounded LA, causing small and gradual loss of LA due to the hydrophobic nature of RA. During the period of study, it was observed that the samples retain their original shape. Only partial disintegration of polymer was observed throughout the period of the study [10, 11].

In vitro degradation of poly(ester-anhydride) P(RA-SA) was studied by Krasko *et al.* with 50:50 ratios of polymer for a period of 2 months, and the polymers having a 70:30 ratio of polymer degraded in a period of more than 3 months. Anhydride bonds of the poly(ester-anhydride) were completely cleaved in a week's time and sebacic acid was released from the polymer. RA was not observed to release from the polymer during degradation, but alkaline hydrolysis showed RA release. It was also concluded that RA oligomers such as RA-RA, RA-SA, RA-RA-RA, and RA-RA-RA-RA were released instead of RA itself. Poly(ester-anhydride) samples degraded to their oligomers with an average molecular weight of 1000 Da in a buffer solution after approximately 1.5 months (Fig. 2.11). There was no difference in the SA:RA ratio for the initial 24 hours, after which a sudden fall in the SA content was observed because of its fast release. Degradation study for RA oligomers revealed a similar trend to the degradation studies of P(RA-SA). The degradation study of poly(ester-anhydride) progressed in two stages. Stage one lasted a week, during which the anhydride bonds were completely degraded and released SA units conjugated to both monomers with anhydride linkage. During the second stage the remaining oligoesters degraded into a shorter RA ester of dimers, trimers, and tetramers along with dimers of RASA [41].

The *in vivo* degradation and elimination of P(SA-RA) in 3:7 copolymer was studied after subcutaneous implantation in rats. IR spectroscopy revealed that the anhydride linkages of polymer were cleaved in a biological environment followed by a respective sharp decrease in the weight average molecular weight of the polymer and the initial rapid weight loss. The results obtained showed that the *in vivo* degradation pattern of P(SA-RA) 3:7 correlates well with

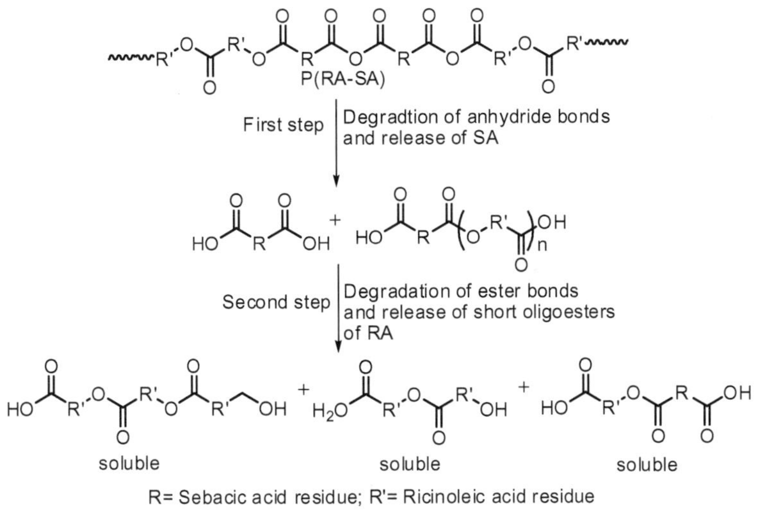

Figure 2.11 *In vitro* degradation of $P(SA\text{-}RA)_n$ in an aqueous medium [41].

that observed under *in vitro* conditions [41, 42]. The degradation pattern was the same as in *in vitro* studies for the polymer. The pattern of low molecular weight degradation products was also very similar. The findings from this study indicated that the degradation of P(SA-RA) 3:7 was determined by its chemical structure and not by the biological environment. The *in vitro* hydrolytic degradation of P(SA-RA) 3:7 was completely eroded over 4 days, resulting in 100% weight loss. On the other hand, it was observed that 42 days were required to hydrolyze 90% of the polymer and to eliminate from the implantation site under *in vivo* conditions (Fig. 2.12) [43].

Degradation of PGS both *in vitro* and *in vivo* was reported [36, 44]. According to one of the studies, PGS undergoes surface degradation and the main mechanism is the degradation of the ester linkages. Due to this slow loss of mechanical strength (tensile properties), relative to mass loss (per unit original area) occurs. Mass loss changes linearly with time, so detectable swelling and better retention of geometry were observed [36, 45, 46]. The authors found difficult to correlate *in vitro* and *in vivo* degradation behavior of PGS. When PGS was subcutaneously implanted in Sprague-Dawley rats, the polyester was completely absorbed without granulation or formation of scar tissue and also the implantation site was restored to its normal histological architecture within 60 days [36].

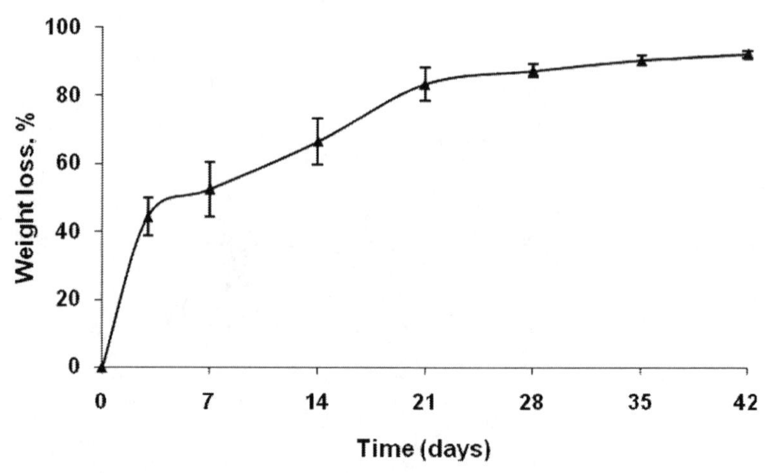

Figure 2.12 *In vivo* hydrolytic degradation and elimination from the implantation site of the P(SA-RA)3:7 copolymer monitored by weight loss. Polymer samples were implanted SC in the posteriolateral flank of SD rats. At each time point, the remaining polymer was excided, dried, and weighed. Data are expressed as the mean and standard deviation of six implants [43].

2.5 Stability of Fatty Acid-Based Polyesters

The stability of fatty acid-based polyesters or poly(ester-anhydride)s under different storage conditions has been reported. These polymers were hydrolytically unstable and prone to degradation. These polymer samples should be stored under moisture-free condition at –20°C [6, 47]. The stability of P(SA-RA) 3:7 poly(ester-anhydride) with or without gentamicin was studied at −17°C, 4°C, 25°C, and 37°C for a period of 6 months with respect to changes in the molecular weight and melting temperature. The molecular weights dropped to one third when the polymer was stored at 4°C for 30 days. For the formulations stored at −17°C, no change in the initial molecular weight was observed after 6 months. For the formulations stored at 37°C, no change in molecular weight was observed for 48 hours, followed by rapid degradation to short oligomers of 1000 Da within 2 weeks [27].

These polyesters should be sterilized before administration, as they are used as biodegradable implants. Exposure to ionization radiation is an effective method for the sterilization of moisture- and heat-sensitive polymers. Radiation of 2.5 Mrad is used for sterilization of medical devices [48, 49]. Ricinoleic acid-based copolyesters were exposed to γ-irradiation at a dose of 2.5 Mrad at room temperature. No change in molecular weight or crosslinking was observed. All the irradiated samples dissolved immediately at room temperature to form a clear solution, and also no differences were observed in the thermal and spectral analysis, color, and shape [27].

Similarly, RA-inserted PSA poly(ester-anhydride)s showed no significant change in molecular weight when irradiated. IR spectroscopy confirmed anhydride bonds of all the samples analyzed. There was no evidence of carboxylic acid peak at 1700 cm^{-1} for all the irradiated polymers; an anhydride peak was observed at 1813 cm^{-1}, and an ester peak at 1731 cm^{-1} indicated no hydrolysis of the polymers [29].

2.6 Biocompatibility

Polymers that do not elicit adverse/immunological reactions on tissue after implantation are biocompatible. Polymers excreted completely from the implanted site without causing any toxicity are called biodegradable. Generally toxicity arises from biomaterials by biologically active, but leachable, substances. They may cause thrombosis, or alter compatibility when the material degrades. Therefore, no single test provides all the information needed for safety assessment. Due to this reason, different tests are required to evaluate the toxicity of the biomaterials, and usually the test methods involve both *in vitro* and *in vivo* methods [50–58]. Generally, fatty acids are degraded by the β-oxidation pathway and excreted, as CO_2, completely from the body. Therefore, no deposition of polymer, oligomers, or monomers, which can cause adverse effects, should remain [47].

Biocompatibility of poly(ester-anhydride) P(SA-RA) 3:7 was evaluated after implantation in brain tissue of adult rats. The brain tissue from rats was collected on days 7, 14, and 21 for histological evaluation with respect to the host tissue response to poly(ester-anhydride) in brain parenchyma and a description of the micro architecture of the brain implant surface. One week after implantation of the polymer, there was no significant adverse reaction or inflammation. No macrophages, lymphocytes, or PMNL were observed at the implantation site. Fourteen days post implantation, a thick layer of astroglial cells along the brain–implant interface was observed. This was associated with capillary proliferation. Overall, no abnormalities or toxicities were observed in the brain tissue parenchyma, and also the cavity where the polymer was deposited remained unchanged until the end point of the experimental schedule [42, 59–61].

Castor oil and citric acid were prepared to obtain a viscous-branched and an elastomeric crosslinked material. These polymers degraded into natural components, metabolized, and eliminated after administration. The polymers showed good tolerability, stability, and long persistence *in vitro* and *in vivo*, with tissue biocompatibility. These polymers could be used in soft tissue augmentation (Fig. 2.13) [62].

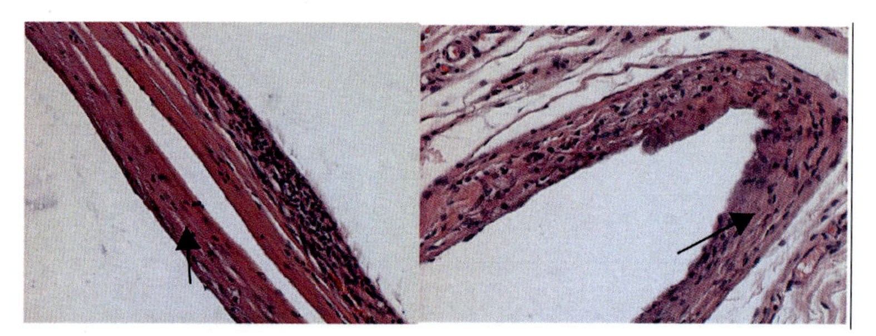

Figure 2.13 Histopathology of the implantation site. High-magnification (×20) view of the capsular reaction 6 months post-implantation of the P(CO:CIA) crosslinked:branched 1:2 w/w composition. There is excellent time-related progressive maturation of the capsule (arrow), with grade 1 evidence for mononuclear cell infiltration. The reaction is characterized by highly mature fibrotic capsular reaction surrounding a cavity. There was no to minimal (grade 0 to 1) evidence for mononuclear lymphocytic infiltration. The implanted material (i.e., filler) was not present within the cavity. No multinucleated giant cells reaction was noted [62].

Poly(glycerol sebacate)(PGS) was biocompatible. The NIH 3T3 fibroblast cells grown in the PGS sample showed normal morphology with enhanced growth rate compared to control. PGS was also tested for its biocompatibility *in vivo* by subcutaneous implantation in Sprague–Dawley rats, and it was found that the inflammatory response is the same as PLGA, and unlike PLGA, PGS induces very small fibrous capsule formation [36].

2.7 Biomedical Applications

Fatty acid-based polyesters and poly(ester-anhydride)s are used in the drug delivery for different categories of drugs and also for peptides, proteins, nucleic acids, and so on. These polymers are meant for the localized release of drugs in a controlled manner [47].

In vitro release of 5-flurouracil (5FU) and triamcinolone from P(LA-RA)s 60:40 prepared by different methods was studied. Drug release into phosphate buffer at pH 7.4 at 37°C from P(LA-RA)60:40 prepared by condensation of the acids was faster than from pasty P(PLA-RA) 60:40 synthesized by transesterification for both drugs. 5FU drug released faster than hydrophobic triamcinolone [26]. Cisplatin as a model drug for the release from the poly(ester-anhydride)s was studied in phosphate buffer. Cisplatin was released at an average rate of 0.45%/h for the first 73 hours from the pasty polymer. Drug release was much slower over a period of time. This is due to the increase in the release of sebacic acid from the polymer matrix, which in turn increases hydrophobicity in the surrounding medium. Polymers containing 30% ricinoleic acid released about 60% of the cisplatin within 100 hours, whereas the 50–60% ricinoleic acid-containing polymers showed only 15% release within 100 hours [29]. Gentamicin sulfate and triamcinolone were incorporated into the polymeric paste of poly(ester-anhydride)s at room temperature and injected into buffer phosphate solution at 37°C. A constant release of both drugs for over 30 days was obtained [30].

Tamoxifen citrate release studies were conducted in phosphate buffer saline at pH 7.4 from P(SA-RA) 70:30 w/w based implants with 10 and 20wt % drug loadings (Fig. 2.14). The drug released by 10 wt% drug-loaded P(SA-RA) 70:30 w/w implants after 30 days was found to be 42.36%, whereas the drug released by 20 wt% drug-loaded was about 62.60%. Both the 10 and 20 wt% drug-loaded polymers showed an initial burst release followed by a sustained-release effect. Unbound drug present on the surface of the polymer implants caused the initial burst effect. The lower drug release with 10 wt% drug loading is due to the fact that, because incorporated polymer concentration was higher than the drug loading, fewer microchannels formed in the implant, and fatty degradation products remained in the matrix and blocked further degradation. The 20 wt% drug-loaded polymer generated a thinner implant compared to 10 wt% drug-loaded implant. Consequently, 20 wt% drug-loaded implants showed faster release and the gradual higher penetration of water from the surface into the implant center, which leads to an increased rate of hydrolysis of the polymer [63, 64].

Tamsulosin hydrochloride (TAM) was incorporated in the P(SA-RA) (30:70 and 20:80) in 2–20% w/w and its release in buffer solution was monitored. The release rate of the drug from polymer was affected by drug content. The higher the drug content in the polymer matrix, the faster was the drug release. The drug

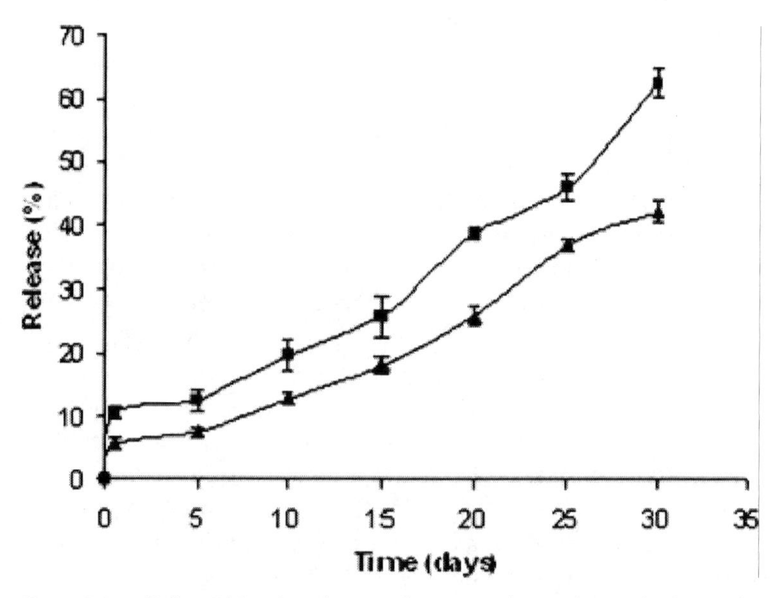

Figure 2.14 Release kinetics of tamoxifen citrate from P(SA-RA) 70:30 w/w implants in PBS (pH 7.4): (▲) 10 wt% drug-loaded implant and (■) 20 wt% drug-loaded implant [63].

formulation was found to be stable for a period of 6 months. The drug-loaded formulation became gel at the site of injection after it was injected into mice [65]. Poly(DL-lactic acid-*co*-castor oil) (60:40 w/w) based nanoparticles containing paclitaxel were prepared as a nano-size drug delivery system. It was confirmed from analysis that the drug was entrapped within the polymeric matrix. A prolonged period of drug release was observed from *in vitro* release studies [66]. Low molecular weight hydroxy fatty acid based copolymers were used for controlled drug release with the drugs methotrexate (hydrophobic) and 5-fluorouracil (hydrophilic). A negligible effect was observed for the release profiles of drug loaded at 5%, 10%, and 20% w/w of methotrexate. The entrapped drug was released in a controlled way from the polymers and had biocompatibility with the tissue [67].

Poly(ester-anhydride)s containing 30% or less of ricinoleic acid were liquid at body temperature having melting temperatures at 33°C. Because of this nature, these polymers were injectable at room temperature through a 23G needle by the administration of polymer formulations. The degradation studies of polymers were conducted in phosphate buffer at pH 7.4, and it was found that almost 80% of its weight was lost during 2 weeks of incubation. These liquid polymers solidified in buffer solution and decomposed completely after 8 weeks. The release of the drug cisplatin from both pasty and solid polymers was evaluated, and it was found that cisplatin release from the pasty polymers was different from solid formulations. Solid formulations released up to 80% of the drug in 3 weeks, whereas pasty polymers released complete drug in 5 weeks [68].

Gentamicin sulfate at 10–20% w/w was incorporated in P(SA-RA) paste of different molecular weights and the formulation was studied against *Staphylococcus aureus*. It was observed that 20% gentamicin had slower release profiles compared to formulations with 10% gentamicin. This is probably due to salt formation between gentamicin and the fatty degradation products of the polymer [69]. The toxicity evaluation of the formulations was examined by subcutaneous injection to rats. It was observed from the preliminary study that P(SA-RA) 3:7 loaded with 10–20% gentamicin sulfate could be used as an injectable biodegradable device for *in situ* treatment of osteomyelitis induced by *S. aureus* [61].

PGS was used in several soft tissue engineering applications such as retinal, nerve, vascular, myocardial repair, and also as adhesive sealant [70–75]. *Ex vivo* and *in vivo* studies indicated that PGS membranes are well tolerated in the subretinal space and also PGS membranes provided the selective apoptosis of the host photoreceptors without provoking inflammation of the tissue [76–78]. PGS-*co*-LA copolymers had promising sealant properties. The addition of lactic acid to PGS significantly improved the cytocompatibility of the polyester to that of pure PGS polyester. This copolymer can be applied easily at 45°C and solidifies quickly in a controlled manner when exposed to dry ice. It has a significantly higher adhesive strength compared to fibrin sealants. The primary cause of this enhancement in the adhesive behavior of PGS-*co*-LA is explained by the increase in the degree of polymerization and the greater presence of alcohol groups on the polymer chains [37]. PGS implants were prepared by adding 5-fluorouracil anticancer drug, and the drug release in PBS medium for 30 days was studied. The more the concentration of the drug in the implant, the faster is the release rate of the drug from the implant. After 7 days of study, the drug was released 100% from the implants, and this release pattern was similar to that of the Gliadel® wafer, which was commercially used for the treatment of recurrent glioblastoma and malignant gliomas [79]. Ciprofloxacin-HCl was encapsulated in the PGS polyester for the drug delivery. Ciprofloxacin was released from the PGS implant through osmosis and diffusion mechanisms. Therefore, because of their degradable nature, PGS-based delivery systems could be implanted at disease sites where device retrieval is restricted [80].

2.8 Conclusions and Future Perspectives

Polymers prepared from naturally occurring fatty acids are suitable for different biomedical applications. They are resistant to degradation and provide control over the release of incorporated drug molecules. A number of naturally occurring low-cost fatty acids such as oleic acid, erucic acid, and ricinoleic acid have been copolymerized with lactic acid, glycolic acid, and caprolactone to form polymer with desired and tunable properties.

A significant number of copolymer combinations is possible. This renders desirable control over the final physical or mechanical properties: pasty, rigid, liquid at room temperature, elastic, etc. Bifunctional acids like ricinoleic, lactic, or citric have been used as monomer. Rinoleic acid-based poly(ester-anhydride) or copolyesters provided pasty *in situ* forming injectable gels. In conclusion, these fatty acid-based polymers have remarkable prospects of being used in drug delivery and other implantation purposes. However, proper control over the synthesis of the polymers, block, random, or alternating, is warranted. A benchmark study is important to especially address these issues.

References

1. Biermann, U., Bornscheuer, U., Meier, M. A., Metzger, J. O., Schäfer, H. J. (2011). Oils and fats as renewable raw materials in chemistry, *Angew Chem Int Ed*, **50**, 3854–3871.
2. Miao, S., Wang, P., Su, Z., Zhang, S. (2014). Vegetable-oil-based polymers as future polymeric biomaterials, *Acta Biomater*, **10**, 1692–1704.
3. Meier, M. A., Metzger, J. O., Schubert, U. S. (2007). Plant oil renewable resources as green alternatives in polymer science, *Chem Soc Rev*, **36**, 1788–1802.
4. Xia, Y., Larock, R. C. (2010). Vegetable oil-based polymeric materials: synthesis, properties, and applications, *Green Chem*, **12**, 1893–1909.
5. Maisonneuve, L., Lebarbé, T., Grau, E., Cramail, H. (2013). Structure–properties relationship of fatty acid-based thermoplastics as synthetic polymer mimics, *Polym Chem*, **4**, 5472–5517.
6. Kumar, N., Langer, R. S., Domb, A. J. (2002). Polyanhydrides: an overview, *Adv Drug Deliv Rev*, **54**, 889–910.
7. Jain, J. P., Modi, S., Domb, A., Kumar, N. (2005). Role of polyanhydrides as localized drug carriers, *J Control Release*, **103**, 541–563.
8. Qiu, L. (2002). *In vivo* degradation and tissue compatibility of polyphosphazene blend films, *Sheng Wu Yi Xue Gong Cheng Xue Za Zhi*, **19**, 191–195.
9. Domb, A. J., Maniar, M. (1993). Absorbable biopolymers derived from dimer fatty acids, *J Polym Sci, Part A: Polym Chem*, **31**, 1275–1285.
10. Friend, D. R., Pangburn, S. (1987). Site-specific drug delivery, *Med Res Rev*, **7**, 53–106.

11. Weiner, A., Domb, A. (1994). Polymeric drug delivery systems for the eye. In *Polymeric Site-Specific Pharmacotherapy*, ed., Domb, A. J., Wiley, Chichester, 315–346.
12. Domb, A. J., Nudelman, R. (1995). Biodegradable polymers derived from natural fatty acids, *J Polym Sci, Part A: Polym Chem,* **33**, 717–725.
13. Sokolsky-Papkov, M., Shikanov, A., Kumar, N., Vaisman, B., Domb, A. J. (2008). Fatty acid based biodegradable polymers-synthesis and applications, *Bull Israel Chem Soc,* **23**, 12–17.
14. Desroches, M., Escouvois, M., Auvergne, R., Caillol, S., Boutevin, B. (2012). From vegetable oils to polyurethanes: synthetic routes to polyols and main industrial products, *Polym Rev,* **52**, 38–79.
15. Teomim, D., Domb, A. J. (1999). Fatty acid terminated polyanhydrides, *J Polym Sci, Part A: Polym Chem,* **37**, 3337–3344.
16. Teomim, D., Nyska, A., Domb, A. J. (1999). Ricinoleic acid-based biopolymers, *J Biomed Mater Res,* **45**, 258–267.
17. Teomim, D., Domb, A. J. (2001). Nonlinear fatty acid terminated polyanhydrides, *Biomacromolecules,* **2**, 37–44.
18. Shieh, L., Tamada, J., Chen, I., Pang, J., Domb, A., Langer, R. (1994). Erosion of a new family of biodegradable polyanhydrides, *J Biomed Mater Res,* **28**, 1465–1475.
19. Laurencin, C., Gerhart, T., Witschger, P., Satcher, R., Domb, A., Rosenberg, A., Hanff, P., Edsberg, L., Hayes, W., Langer, R. (1993). Bioerodible polyanhydrides for antibiotic drug delivery: *in vivo* osteomyelitis treatment in a rat model system, *J Orthop Res,* **11**, 256–262.
20. Shieh, L., Tamada, J., Tabata, Y., Domb, A., Langer, R. (1994). Drug release from a new family of biodegradable polyanhydrides, *J Control Release,* **29**, 73–82.
21. Kumar, N., Ravikumar, M. N., Domb, A. (2001). Biodegradable block copolymers, *Adv Drug Deliv Rev,* **53**, 23–44.
22. Penczek, S., Szymanski, R., Duda, A., Baran, J. (2003). Living polymerization of cyclic esters—a route to (bio)degradable polymers. Influence of chain transfer to polymer on livingness. In *Macromolecular Symposia,* Wiley Online Library, 261–270.
23. Seppälä, J. V., Helminen, A. O., Korhonen, H. (2004). Degradable polyesters through chain linking for packaging and biomedical applications, *Macromol Biosci,* **4**, 208–217.
24. Slivniak, R., Domb, A. J. (2005). Lactic acid and ricinoleic acid based copolyesters, *Macromolecules,* **38**, 5545–5553.

25. Slivniak, R., Langer, R., Domb, A. J. (2005). Lactic and ricinoleic acid based copolyesters stereocomplexation, *Macromolecules*, **38**, 5634–5639.

26. Slivniak, R., Ezra, A., Domb, A. J. (2006). Hydrolytic degradation and drug release of ricinoleic acid–lactic acid copolyesters, *Pharm Res*, **23**, 1306–1312.

27. Slivniak, R., Domb, A. J. (2006). Macrolactones and polyesters from ricinoleic acid, *Biomacromolecules*, **6**, 1679–1688.

28. Ristić, I. S., Marinović-Cincović, M., Cakić, S. M., Tanasić, L. M., Budinski-Simendić, J. K. (2013). Synthesis and properties of novel star-shaped polyesters based on L-lactide and castor oil, *Polym Bull*, **70**, 1723–1738.

29. Krasko, M. Y., Shikanov, A., Ezra, A., Domb, A. J. (2003). Poly (ester-anhydride)s prepared by the insertion of ricinoleic acid into poly(sebacic acid), *J Polym Sci, Part A: Polym Chem*, **41**, 1059–1069.

30. Krasko, M. Y., Domb, A. J. (2007). Pasty injectable biodegradable polymers derived from natural acids, *J Biomed Mater Res*, **83**, 1138–1145.

31. Fay, F., Linossier, I., Langlois, V., Vallee-Rehel, K., Krasko, M. Y., Domb, A. J. (2007). Protecting biodegradable coatings releasing antimicrobial agents, *J Appl Polym Sci*, **106**, 3768–3777.

32. Sokolsky-Papkov, M., Langer, R., Domb, A. J. (2011). Synthesis of aliphatic polyesters by polycondensation using inorganic acid as catalyst, *Polym Adv Technol*, **22**, 502–511.

33. Miao, S., Zhang, S., Su, Z., Wang, P. (2008). Chemoenzymatic synthesis of oleic acid-based polyesters for use as highly stable biomaterials, *J Polym Sci, Part A: Polym Chem*, **46**, 4243–4248.

34. Yang, Y., Lu, W., Cai, J., Hou, Y., Ouyang, S., Xie, W., Gross, R. A. (2011). Poly(oleic diacid-*co*-glycerol): comparison of polymer structure resulting from chemical and lipase catalysis, *Macromolecules*, **44**, 1977–1985.

35. Baharu, M. N., Kadhum, A. A. H., Al-Amiery, A. A., Mohamad, A. B. (2014). Synthesis and characterization of polyesters derived from glycerol, azelaic acid, and succinic acid, *Green Chem Lett Rev*, **8**, 31–38.

36. Wang, Y., Ameer, G. A., Sheppard, B. J., Langer, R. (2002). A tough biodegradable elastomer, *Nat Biotechnol*, **20**, 602–606.

37. Chen, Q., Liang, S., Thouas, G. A. (2011). Synthesis and characterisation of poly(glycerol sebacate)-*co*-lactic acid as surgical sealants, *Soft Matter*, **7**, 6484–6492.

38. Tamada, J., Langer, R. (1993). Erosion kinetics of hydrolytically degradable polymers, *Proc Natl Acad Sci U S A,* **90**, 552–556.

39. Göpferich, A. (1997). Erosion of composite polymer matrices, *Biomaterials,* **18**, 397–403.

40. Gopferich, A., Langer, R. (1993). Modeling of polymer erosion, *Macromolecules,* **26**, 4105–4112.

41. Krasko, M. Y., Domb, A. J. (2005). Hydrolytic degradation of ricinoleic-sebacicester-anhydride copolymers, *Biomacromolecules,* **6**, 1877–1884.

42. Shikanov, A., Vaisman, B., Krasko, M. Y., Nyska, A., Domb, A. J. (2004). Poly(sebacic acid-co-ricinoleic acid) biodegradable carrier for paclitaxel: *in vitro* release and *in vivo* toxicity, *J Biomed Mater Res A,* **69**, 47–54.

43. Vaisman, B., Ickowicz, D. E., Abtew, E., Haim-Zada, M., Shikanov, A., Domb, A. J. (2013). *In vivo* degradation and elimination of injectable ricinoleic acidbased poly (ester-anhydride), *Biomacromolecules,* **14**, 1465–1473.

44. Chen, Q.-Z., Ishii, H., Thouas, G. A., Lyon, A. R., Wright, J. S., Blaker, J. J., Chrzanowski, W., Boccaccini, A. R., Ali, N. N., Knowles, J. C., Harding, S. E. (2010). An elastomeric patch derived from poly(glycerol sebacate) for delivery of embryonic stem cells to the heart, *Biomaterials,* **31**, 3885–3893.

45. Jaafar, I., Ammar, M., Jedlicka, S., Pearson, R., Coulter, J. (2010). Spectroscopic evaluation, thermal, and thermomechanical characterization of poly(glycerol-sebacate) with variations in curing temperatures and durations, *J Mater Sci,* **45**, 2525–2529.

46. Wang, Y., Kim, Y. M., Langer, R. (2003). *In vivo* degradation characteristics of poly(glycerol sebacate), *J Biomed Mater Res A,* **66**, 192–197.

47. Jain, J. P., Sokolsky, M., Kumar, N., Domb, A. (2008). Fatty acid based biodegradable polymer, *Polym Rev,* **48**, 156–191.

48. Helliwell, P. (1966). A review of sterilization and disinfection, *Proc R Soc Med,* **59**, 76–77.

49. Deng, J.-S., Li, L., Stephens, D., Tian, Y., Harris, F. W., Cheng, S. Z. (2002). Effect of γ-radiation on a polyanhydride implant containing gentamicin sulfate, *Int J Pharm,* **232**, 1–10.

50. Říhová, B. (1996). Biocompatibility of biomaterials: hemocompatibility, immunocompatiblity and biocompatibility of solid polymeric materials and soluble targetable polymeric carriers, *Adv Drug Deliv Rev,* **21**, 157–176.

51. Fournier, E., Passirani, C., Montero-Menei, C., Benoit, J. (2003). Biocompatibility of implantable synthetic polymeric drug carriers: focus on brain biocompatibility, *Biomaterials*, **24**, 3311–3331.
52. Kou, J. H., Emmett, C., Shen, P., Aswani, S., Iwamoto, T., Vaghefi, F., Cain, G., Sanders, L. (1997). Bioerosion and biocompatibility of poly (D,L-lactic-*co*glycolic acid) implants in brain, *J Control Release*, **43**, 123–130.
53. Lee, J., Jallo, G. I., Penno, M. B., Gabrielson, K. L., Young, G. D., Johnson, R. M., Gillis, E. M., Rampersaud, C., Carson, B. S., Guarnieri, M. , (2006). Intracranial drug-delivery scaffolds: biocompatibility evaluation of sucrose acetate isobutyrate gels, *Toxicol Appl Pharmacol*, **215**, 64–70.
54. Tamargo, R. J., Myseros, J. S., Epstein, J. I., Yang, M. B., Chasin, M., Brem, H. (1993). Interstitial chemotherapy of the 9L gliosarcoma: controlled release polymers for drug delivery in the brain, *Cancer Res*, **53**, 329–333.
55. Walter, K. A., Cahan, M. A., Gur, A., Tyler, B., Hilton, J., Colvin, O. M., Burger, P. C., Domb, A., Brem, H. (1994). Interstitial taxol delivered from a biodegradable polymer implant against experimental malignant glioma, *Cancer Res*, **54**, 2207–2212.
56. Brem, H., Domb, A., Lenartz, D., Dureza, C., Olivi, A., Epstein, J. I. (1992). Brain biocompatibility of a biodegradable controlled release polymer consisting of anhydride copolymer of fatty acid dimer and sebacic acid, *J Control Release*, **19**, 325–329.
57. Tamargo, R. J., Epstein, J. I., Reinhard, C. S., Chasin, M., Brem, H. (1989). Brain biocompatibility of a biodegradable, controlled-release polymer in rats, *J Biomed Mater Res*, **23**, 253–266.
58. Mokrý, J., Karbanová, J., Lukáš, J., Palečková, V., Dvořánková, B. (2000). Biocompatibility of HEMA copolymers designed for treatment of CNS diseases with polymer-encapsulated cells, *Biotechnol Prog*, **16**, 897–904.
59. Vaisman, B., Motiei, M., Nyska, A., Domb, A. J. (2010). Biocompatibility and safety evaluation of a ricinoleic acid-based poly(ester-anhydride) copolymer after implantation in rats, *J Biomed Mater Res*, **92**, 419–431.
60. Shikanov, A., Ezra, A., Domb, A. J. (2005). Poly(sebacic acid-*co*-ricinoleic acid) biodegradable carrier for paclitaxel—effect of additives, *J Control Release*, **105**, 52–67.
61. Krasko, M. Y., Golenser, J., Nyska, A., Nyska, M., Brin, Y. S., Domb, A. J. (2007). Gentamicin extended release from an injectable polymeric implant, *J Control Release*, **117**, 90–96.

62. Ickowicz, D. E., Haim-Zada, M., Abbas, R., Touitou, D., Nyska, A., Golovanevski, L., Weiniger, C. F., Katzhendler, J., Domb, A. J. (2014). Castor oil–citric acid copolyester for tissue augmentation, *Polym Adv Technol*, **25**, 1323–1328.

63. Hiremath, J., Kusum Devi, V., Devi, K., Domb, A. (2008). Biodegradable poly(sebacic acid-co-ricinoleic-ester-anhydride) tamoxifen citrate implants: preparation and *in vitro* characterization, *J Appl Polym Sci*, **107**, 2745–2754.

64. Akbari, H., D'Emanuele, A., Attwood, D. (1998). Effect of geometry on the erosion characteristics of polyanhydride matrices, *Int J Pharm*, **160**, 83–89.

65. Havivi, E., Farber, S., Domb, A. J., (2011). Poly(sebacic acid-co-ricinoleic acid) biodegradable carrier for delivery of tamsulosin hydrochloride, *Polym Adv Technol*, **22**, 114–118.

66. Hiremath, J., Rajeshkumar, A., Ickowicz, D., Domb, A. , (2013). Preparation and *in vitro* characterization of paclitaxel containing poly (lactic acid cocastor oil)-based nanodispersions, *J Drug Deliv Sci Technol*, **23**, 439–444.

67. Jain, J. P., Modi, S., Kumar, N. (2008). Hydroxy fatty acid based polyanhydride as drug delivery system: synthesis, characterization, *in vitro* degradation, drug release, and biocompatibility, *J Biomed Mater Res*, **84**, 740–752.

68. Shikanov, A., Domb, A. J. (2006). Poly(sebacic acid-*co*-ricinoleic acid) biodegradable injectable *in situ* gelling polymer, *Biomacromolecules*, **7**, 288–296.

69. Li, L. C., Deng, J., Stephens, D. (2002). Polyanhydride implant for antibiotic delivery—from the bench to the clinic, *Adv Drug Deliv Rev*, **54**, 963–986.

70. Redenti, S., Neeley, W. L., Rompani, S., Saigal, S., Yang, J., Klassen, H., Langer, R., Young, M. J. (2009). Engineering retinal progenitor cell and scrollable poly(glycerol-sebacate) composites for expansion and subretinal transplantation, *Biomaterials*, **30**, 3405–3414.

71. Sundback, C. A., Shyu, J. Y., Wang, Y., Faquin, W. C., Langer, R. S., Vacanti, J. P., Hadlock, T. A. (2005). Biocompatibility analysis of poly(glycerol sebacate) as a nerve guide material, *Biomaterials*, **26**, 5454–5464.

72. Motlagh, D., Yang, J., Lui, K. Y., Webb, A. R., Ameer, G. A. (2006). Hemocompatibility evaluation of poly(glycerol-sebacate) *in vitro* for vascular tissue engineering, *Biomaterials*, **27**, 4315–4324.

73. Chen, Q., Jin, L., Cook, W. D., Mohn, D., Lagerqvist, E. L., Elliott, D. A., Haynes, J. M., Boyd, N., Stark, W. J., Pouton, C. W., Stanley, E. G., Elefanty,

A. G. (2010). Elastomeric nanocomposites as cell delivery vehicles and cardiac support devices, *Soft Matter*, **6**, 4715–4726.

74. Lligadas, G., Ronda, J. C., Galià, M., Cádiz, V. , (2013). Renewable polymeric materials from vegetable oils: a perspective, *Mater Today*, **16**, 337–343.
75. Rai, R., Tallawi, M., Grigore, A., Boccaccini, A. R. (2012). Synthesis, properties and biomedical applications of poly(glycerol sebacate) (PGS): a review, *Prog Polym Sci*, **37**, 1051–1078.
76. Pritchard, C. D., Arnér, K. M., Langer, R. S., Ghosh, F. K. (2010). Retinal transplantation using surface modified poly(glycerol-*co*-sebacic acid) membranes, *Biomaterials*, **31**, 7978–7984.
77. Pritchard, C. D., Arnér, K. M., Neal, R. A., Neeley, W. L., Bojo, P., Bachelder, E., Holz, J., Watson, N., Botchwey, E. A., Langer, R. S., Ghosh, F. K. (2010). The use of surface modified poly(glycerol-*co*-sebacic acid) in retinal transplantation, *Biomaterials*, **31**, 2153–2162.
78. Ghosh, F., Neeley, W. L., Arnér, K., Langer, R. (2011). Selective removal of photoreceptor cells *in vivo* using the biodegradable elastomer poly(glycerol sebacate), *Tissue Eng Part A*, **17**, 1675–1682.
79. Sun, Z.-J., Chen, C., Sun, M.-Z., Ai, C.-H., Lu, X.-L., Zheng, Y.-F., Yang, B.-F., Dong, D.-L. (2009). The application of poly(glycerol–sebacate) as biodegradable drug carrier, *Biomaterials*, **30**, 5209–5214.
80. Tobias, I. S., Lee, H., Engelmayr Jr, G. C., Macaya, D., Bettinger, C. J., Cima, M. J. (2010). Zero-order controlled release of ciprofloxacin-HCl from a reservoir-based, bioresorbable and elastomeric device, *J Control Release*, **146**, 356–362.

Chapter 3

Enzymatically Synthesized Polyesters for Drug Delivery

Shengfan Lu,[a,b] Jianxun Ding,[a] Jinjin Chen,[a] Wei Wang,[b] Xiuli Zhuang,[a] and Xuesi Chen[a]
[a]*Key Laboratory of Polymer Ecomaterials, Changchun Institute of Applied Chemistry, Chinese Academy of Sciences, Changchun 130022, P. R. China*
[b]*Laboratory of Functional Polymers, Changchun University of Science and Technology, Changchun 130022, P. R. China*
jxding@ciac.ac.cn

Nowadays, the enzymatically synthesized polyesters (ESPs) have become a promising type of environment-friendly biomaterials, mainly because they exhibit highly efficient syntheses, excellent biocompatibility and biodegradability, and adjustable chemical reaction activities and physical properties. ESPs have been widely applied in various biomedical fields, including drug delivery, gene transfection, tissue engineering, and so forth. In this chapter, the main preparation strategies of the ESPs are summarized, the applications in the field of drug delivery are highlighted, and the prospects of future developments are discussed.

Handbook of Polyester Drug Delivery Systems
Edited by MNV Ravikumar

ISBN 978-981-4669-65-8 (Hardcover), 978-981-4669-66-5 (eBook)
www.panstanford.com

3.1 Introduction

Polyesters have been widely used in a variety of biomedical fields, especially as the matrices of drug delivery systems, benefited from excellent advantages such as good biocompatibility and biodegradability [1–8]. Although the traditional syntheses of polyesters, including polylactide (PLA) [9], poly(ε-caprolactone) (PCL) [10], poly(ethylene terephthalate) (PET) [11], poly(butylene succinate) (PBS) [12], and so on, catalyzed by organometallic compounds are well developed, there are still many obstructions for their wide applications in different biomedical fields. For example, most of the organometallic catalysts are toxic and hard to be cleared after polymerization [13]. The reaction conditions are always strict, such as high temperature (e.g., ~110–280°C) and absolute dryness [6, 14]. Meanwhile, the high reaction temperature is not suitable for the thermally or chemically unstable monomers and functional groups, including siloxane, epoxy, vinyl moieties, and so on. Furthermore, traditional chemical catalysts generally lack selectivity, and the designed functional polyesters are hard to be produced. The functional side or terminal groups, or subsequently conjugated bioactive agents, are important to meet the special requirements for further biomedical applications. Thus, the discovery of advanced and versatile catalysts to catalyze the efficient polymerization of polyesters under mild conditions with chemo-, regio-, and enantio-selectivity is of great significance [15].

In the 1980s, Klibanov and coworkers first used porcine pancreatic lipase (PPL) as a nearly non-aqueous phase catalyst to catalyze the transesterification reaction between tributyrin and various primary and secondary alcohols [16]. Subsequently, the researchers have paid much attention to the enzyme-catalyzed organic reactions [17]. Enzymes are biological catalysts with nontoxicity and can be applied to prepare the relatively specific products with chemo- and stereoselectivity [18]. There are many advantages for using enzymes as catalysts in comparison with the traditional ones, such as (i) mild reaction conditions, (ii) nontoxicity, (iii) high selectivity, and (iv) recyclability [19]. More attractively, the enzyme catalysts have moved from laboratory research to industrial applications in the 21st century [20]. With these excellent characteristics, enzyme catalysts have also been developed for the syntheses of polyesters

[21], and more types of polyesters synthesized by enzyme catalyst have been used in the biomedical fields, owing to the solutions of the toxicity and the lack of functional side or terminal groups for intelligence or further modification with bioactive ligands.

There are two major modes of the enzyme-catalyzed polymerization of polyesters: enzymatic ring-opening polymerization (ROP) and polycondensation (PC) [22]. Lipases are the mostly used enzymes in the syntheses of polyesters, such as PPL from mammalians [23]; *Aspergillus niger* lipase (ANL), *Candida antartica* lipase (CAL), *Candida cylindracea* lipase (CCL), *Candida rugosa* lipase (CRL), *Mucor javanicus* lipase (MJL), *Penicillium roqueforti* lipase (PRL), *Rhizopus delemar* lipase (RDL), *Rhizopus japonicas* lipase (RJL), and *Rhizomucor miehei* lipase (RML) from fungi [24, 25]; and *Pseudomonas aeruginosa* lipase (PAL), *Pseudomonas cepacia* lipase (PCL_E), *Pseudomonas fluorescens* lipase (PFL), and *Yarrowia lipolytica* lipase (YLL) from bacteria [25–27]. In addition, some modified and enzymatically synthesized polyesters (ESPs) have also been prepared by decorating the side or end groups post-polymerization [28]. Furthermore, ESPs with good biocompatibility and biodegradability can be used as carriers, which can transport drugs into the diseased cells without raising worries about the toxicities and residues of matrices [29]. This chapter summarizes the developments of the enzyme-catalyzed polymerization of polyesters over the past 30 years, highlights the applications of the ESPs in drug delivery, and demonstrates the possible advancements.

3.2 Enzymatic Syntheses of Polyesters

As mentioned before, there are two major strategies for the enzyme-catalyzed polymerization of polyesters: the enzymatic ROP and PC [22]. The enzymatic PC through the reaction between a carboxylic acid or its ester and an alcohol group is complicated and requires removing small molecules, such as ethanol, methanol, and water, during the polymerization [30]. The enzymatic ROP of cyclic monomers, such as lactones, is relatively simple and doesn't involve the removal of any small molecules. Early on, the enzymatic synthesis reactions of polyesters mainly focuses on the ROP mode [31]. However, the cyclic monomers needed for the enzymatic ROP are not only difficult to be synthesized but also costly. In contrast, the monomers suitable for the enzymatic PC are easier to be obtained [19]. So, the enzymatic PC has become the focus of ESP research in recent years [15]. Both the enzyme-catalyzed ROP and PC for the preparation of the ESPs have broad application prospects in the category of biological materials.

3.2.1 Enzymatic ROP

As mentioned above, the enzymatic ROP is relatively simple and doesn't need the removal of by-products. As a result, the early enzymatic polymerization of polyesters mainly focuses on the approach of ROP [13, 32]. It was a breakthrough discovered by Gutman's group and Kobayashi's group independently in the enzymatic ROP of lactones in 1993 for both [33, 34]. Since then, lactones have been extensively studied as monomers. In general, a wide range of enzymatic catalysts can be usually used for the syntheses of polyesters by ROP, which can be recycled [35, 36]. It can also provide versatile synthetic tools for the enantioselective polymerization of polyesters [37]. The ROP of typical lactones that are catalyzed by different enzymes are listed in Table 3.1 [25, 38].

As listed in Fig. 3.1, there are many kinds of lactones and other cyclic monomers for the enzymatic ROP to synthesize various polyesters [31]. A variety of monomers with different lactone rings have diverse reaction conditions. The process of the enzymatic ROP is shown in Fig. 3.2 [21, 29, 39–42]. Of course, the reaction solvents are also very important to conduct green chemistry [43]. The lipase-catalyzed ROP is normally carried out in bulk or in an organic solvent, such as *n*-heptane, toluene, 1,4-dioxane, and diisopropyl ether. The organic solvents are employed to increase the conversion of monomers. For example, Barrera-Rivera *et al.* used YLL to synthetize PCL in *n*-heptane with 100% monomer conversion [44]. Moreover, the typical environment-friendly solvents, such as water, supercritical carbon dioxide ($scCO_2$), and ionic liquids, have also been often used. The first example using water as a solvent is the lipase-catalyzed ROP of five lactone monomers, i.e., ε-caprolactone (ε-CL), 8-octanolide (9-membered lactone, OL), 11-undecanolide (12-membered lactone, UDL), 12-dodecanolide (13-membered lactone, DDL), and 15-pentadecanolide (16-membered lactone, PDL) [45].

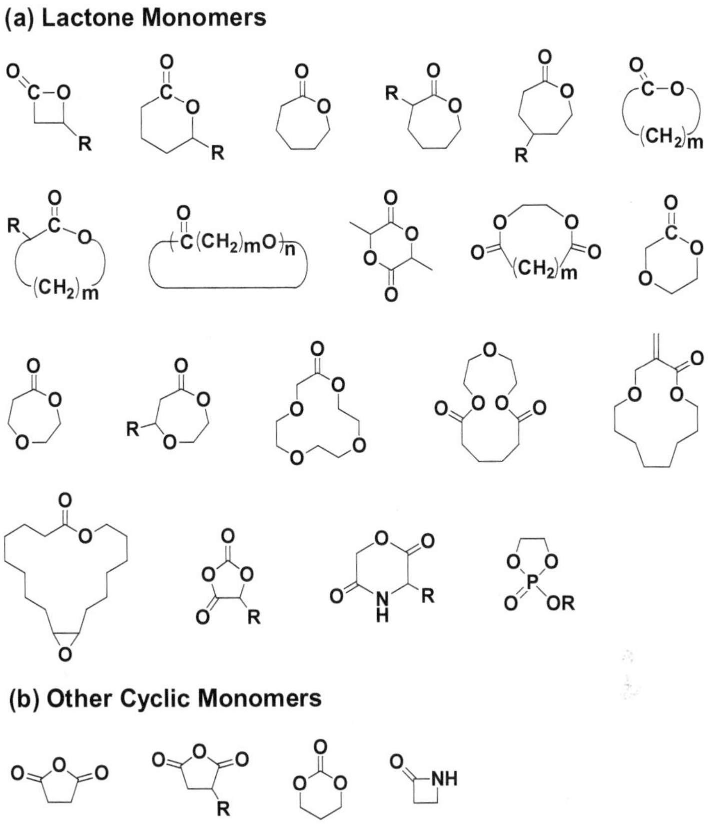

(a) Lactone Monomers

(b) Other Cyclic Monomers

Figure 3.1 Typical cyclic monomers for enzyme-catalyzed ROP [31].

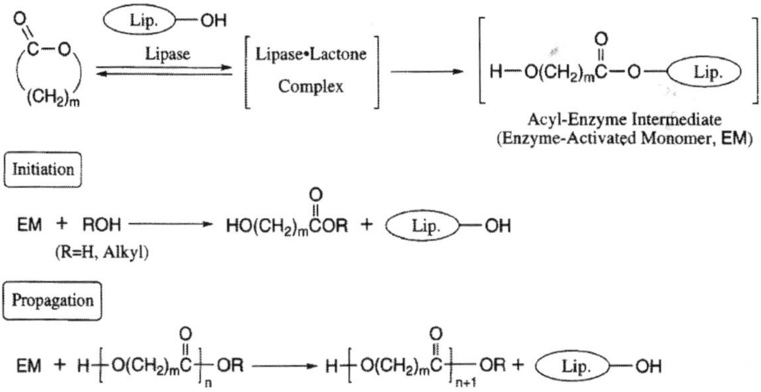

Lip.—OH
Lipase
Lipase•Lactone Complex
Acyl-Enzyme Intermediate (Enzyme-Activated Monomer, EM)

Initiation

$$EM + ROH \longrightarrow HO(CH_2)_mCOR + \text{Lip.—OH}$$

(R=H, Alkyl)

Propagation

$$EM + H{+}O(CH_2)_mC{+}_nOR \longrightarrow H{+}O(CH_2)_mC{+}_{n+1}OR + \text{Lip.—OH}$$

Figure 3.2 Schematic mechanism of lipase-catalyzed ROP of lactones [42].

Table 3.1 Typical types of enzymatically catalyzed ROP

Cyclic monomers	Enzymes from mammalians	Enzymes from fungi	Enzymes from bacteria
β-Propiolactone	PPL	–	PCL_E
α-Methyl-β-Propiolactone	PPL	CCL	–
δ-Valerolactone	PPL	CCL	PCL_E, PFL
γ-Valerolactone	PPL	RJL	PCL_E
α-Methyl-γ-Valerolactone	–	CAL	–
ε-Caprolactone	PPL	ANL, CAL, CCL, RDL	PAL, PCL_E, PFL
γ-Caprolactone	–	–	PCL_E
α-Methyl-γ-Caprolactone	–	CAL	–
Octalactones	–	CAL, CCL	PAL, PCL_E, PFL
UDL (11-Undecanolide, 12-Membered Lactone)	PPL	CCL	PCL_E, PFL
DDL (13-Dodecanolide, 12-Membered Lactone)	PPL	ANL, CCL	PAL, PCL_E, PFL
PDL (15-Pentadecanolide, 16-Membered Lactone)	PPL	ANL, CAL, CCL, CRL, MJL, PRL, RJL	PCL_E, PFL
HDL (16-Hexadecanolide, 17-Membered Lactone)	PPL	CCL	PCL_E, PFL

Source: From Ref. 25.

3.2.2 Enzymatic PC

The enzymatic PC can be used in the syntheses of polyesters through the esterification of dicarboxylic acids and glycols, or the transesterifications of carboxylic acids and esters, alcohols and esters, and esters and esters, as shown in Fig. 3.3 [15, 46]. Compared with the enzymatic ROP, the enzymatic PC has obvious advantages in the monomer synthesis and selection. However, it produces water and other small molecules during the reaction. Thus, the removal of by-products is necessary to ensure the obtained ESP without secondary pollution. The main approaches to remove the small molecule by-products are adsorption by molecular sieves and

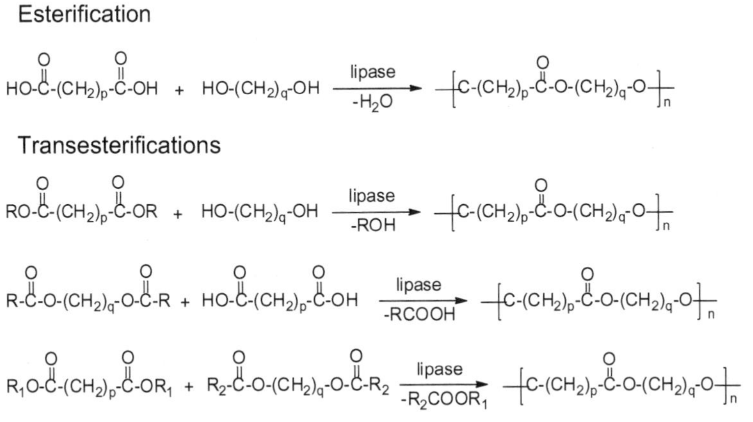

Esterification

$$HO\text{-}\overset{O}{\overset{\|}{C}}\text{-}(CH_2)_p\text{-}\overset{O}{\overset{\|}{C}}\text{-}OH + HO\text{-}(CH_2)_q\text{-}OH \xrightarrow[-H_2O]{\text{lipase}} \left[C\text{-}(CH_2)_p\text{-}\overset{O}{\overset{\|}{C}}\text{-}O\text{-}(CH_2)_q\text{-}O \right]_n$$

Transesterifications

$$RO\text{-}\overset{O}{\overset{\|}{C}}\text{-}(CH_2)_p\text{-}\overset{O}{\overset{\|}{C}}\text{-}OR + HO\text{-}(CH_2)_q\text{-}OH \xrightarrow[-ROH]{\text{lipase}} \left[C\text{-}(CH_2)_p\text{-}\overset{O}{\overset{\|}{C}}\text{-}O\text{-}(CH_2)_q\text{-}O \right]_n$$

$$R\text{-}\overset{O}{\overset{\|}{C}}\text{-}O\text{-}(CH_2)_q\text{-}O\text{-}\overset{O}{\overset{\|}{C}}\text{-}R + HO\text{-}\overset{O}{\overset{\|}{C}}\text{-}(CH_2)_p\text{-}\overset{O}{\overset{\|}{C}}\text{-}OH \xrightarrow[-RCOOH]{\text{lipase}} \left[C\text{-}(CH_2)_p\text{-}\overset{O}{\overset{\|}{C}}\text{-}O\text{-}(CH_2)_q\text{-}O \right]_n$$

$$R_1O\text{-}\overset{O}{\overset{\|}{C}}\text{-}(CH_2)_p\text{-}\overset{O}{\overset{\|}{C}}\text{-}OR_1 + R_2\text{-}\overset{O}{\overset{\|}{C}}\text{-}O\text{-}(CH_2)_q\text{-}O\text{-}\overset{O}{\overset{\|}{C}}\text{-}R_2 \xrightarrow[-R_2COOR_1]{\text{lipase}} \left[C\text{-}(CH_2)_p\text{-}\overset{O}{\overset{\|}{C}}\text{-}O\text{-}(CH_2)_q\text{-}O \right]_n$$

Figure 3.3 Four condensation reactions for syntheses of esters [45].

volatilization under reduced pressure [30]. The molecular weight and end group can be controlled by the following factors: (i) the water content in the reaction system, (ii) the ratio of enzyme and substrate, (iii) the monomer percentage of all the materials, and (iv) the reaction temperature and medium. In addition, research on the enzymatic PC mainly focuses on the activation of monomers to improve the rate of acyl transfer reaction [47]. The activation of monomers mainly refers to the esterification of carboxylic acid, such as alkylation [19, 46]. For example, poly(butylene succinate) was efficiently synthesized by Gross and coworkers using diethyl succinate and 1,4-butanediol as monomers and CAL as a catalyst [48]. The polyester exhibits the weight-average molecular weight (M_w) of 38,000 g mol^{-1} and polydispersity index (PDI) of 1.39.

3.3 Postpolymerization Functionalization of ESPs

ESPs exhibit greatly promising applications in the field of medicine owing to their good biocompatibility and biodegradability. However, the ESPs produced directly by the enzymatic ROP or PC receive a lot of restrictions, such as lack of some functional groups with responsiveness, bioactivity, or reactivity. To address the problems, functionalized monomers are prepared and used to synthesize

ESPs for further functionalization. The modification of part chains formed by the functionalized monomers is applied in order to add the function groups, such as hydroxyl, carboxyl, and alkenyl ones in the side chains, to the ESPs. The modification of their side chains or terminal chains can obtain some necessary characteristics for biomedical applications, such as amphiphilicity.

As a typical example, a novel method for the modification of polyglobalide was conducted after the enzymatic ROP of globalide [49]. 6-Mercapto-1-hexanol (MH), butyl-3-mercapto propionate (BMP), and *N*-acetylcysteamine (nACA) were conjugated to the side chain via thiol-ene "click" chemistry, individually. Since thiol-ene "click" chemistry is efficient and facile, it has been widely used in the functionalization of various polymers [50]. The introduction of primary alcohol-terminated MH and nACA, which both potentially enabled the further modification of polymer via the esterification or amidation of the deacetylated amine, respectively, to introduce functional amino acid groups, endowed the ESP with chemical reaction activities for further modification and various physical properties, such as hydrophilicity. In addition, Cheong's group prepared the polyester that exhibited the side hydroxyl group through the enzymatic PC of glycerol and divinyl ester using Novozyme-435 (i.e., CAL-B) as a catalyst. The side hydroxyl group in the ESP could be used for covalent drug conjugate of xanthorrhizol through a biodegradable linker (Fig. 3.4) [51].

3.4 ESP-Based Drug Delivery Platforms

The controlled and targeted release of drugs *in vivo* is one of the hot topic studies currently in the interdisciplinary fields of chemistry, pharmacy, and medicine [52–54]. As we know, traditional small-molecule drugs (SMDs) have certain side effects. The quick and accurate accumulation in lesion site and the long-term sustained release of drugs in a safe range are two important issues regarding today's drug delivery. Nowadays, various kinds of polymeric drug delivery systems, such as polymer–drug conjugates, micelles, vesicles, and nanogels, are being exploited for controlled drug release. These nanosized platforms can overcome the dose limitations of SMDs, i.e., the limitation of the maximal dose in practice due to the serious toxicities and side effects [55]. In addition, the systems

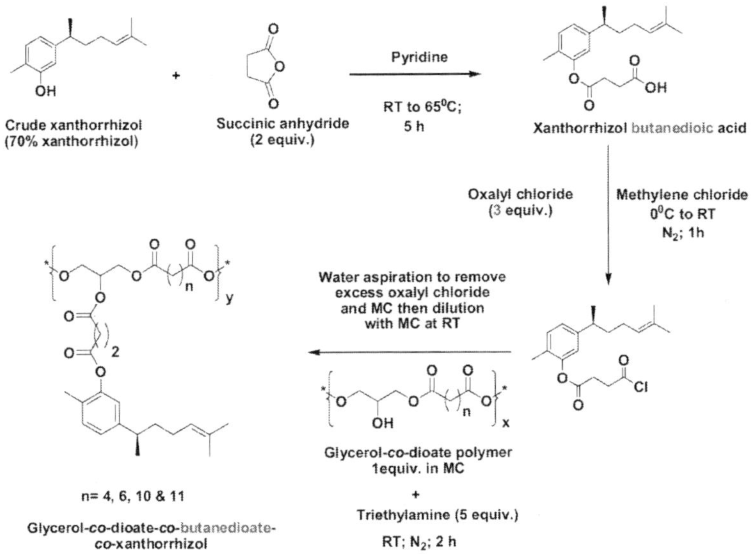

Crude xanthorrhizol (70% xanthorrhizol)
Succinic anhydride (2 equiv.)
Pyridine
RT to 65°C; 5 h
Xanthorrhizol butanedioic acid
Oxalyl chloride (3 equiv.)
Methylene chloride 0°C to RT N_2; 1h
Water aspiration to remove excess oxalyl chloride and MC then dilution with MC at RT
Glycerol-*co*-dioate polymer 1equiv. in MC
+
Triethylamine (5 equiv.)
RT; N_2; 2 h
n= 4, 6, 10 & 11
Glycerol-*co*-dioate-co-butanedioate-*co*-xanthorrhizol

Figure 3.4 Synthesis of poly((glycerol-*g*-succinate-xanthorrhizol)-*co*-dioate) [51].

have many other advantages: (i) improving the selectivity of drugs, (ii) increasing the action duration of drugs, and (iii) reducing the toxicity of drugs. ESPs do not involve the use of metal catalysts in the synthesis process [41]. So, ESPs should be one kind of ideal materials for drug delivery.

3.4.1 ESP–Drug Conjugates

Polymer conjugates are a well-known and widely used method to improve the treatment efficacy of SMDs since 1975 [5, 56–62]. They generally have higher stability, water solubility, biocompatibility, and specific targeting to some tissues or organs [63]. Polyesters are one of the most important materials for bonding drugs. As typical examples, two kinds of linear polyester–ketoprofen conjugates were synthesized through the enzyme-catalyzed PC by Yu, Wang, and coworkers [64, 65]. First, ketoprofen, an important 2-arylpropionic acid carboxylic nonsteroidal anti-inflammatory drug (NSAID), was modified to obtain the prodrug-contained monomers. Then, the polyester–ketoprofen prodrugs with different chemical structures and molecular weights were prepared by the PCs of ketoprofen glycerol ester (i.e., 1-*O*-ketoprofen glycerol ester), PEG, and divinyl sebacate [64], and PEG and 2-ketoprofen malic acid (thiomalic acid or aspartic acid) dimethyl ester catalyzed by CAL-B for both [65]. The bonded ketoprofen released from the polyester prodrugs slowly in the gastric acidic environment, and exhibited accelerated release in the intestinal alkaline condition. The results indicated selective drug delivery through containing ketoprofen in polyester backbone, which reduced toxicity and upregulated bioavailability. Similarly, as shown in Fig. 3.5, a series of biodegradable polyester prodrugs composed of three kinds of NSAID branches and poly(amide-*co*-ester) backbone were synthesized by the CAL-B-catalyzed PC of the profen-containing diol monomers and diesters by Lin's group [66]. The ESPs had relatively high drug-loading contents (DLCs) of 44.7–59.7 wt% because every repeat unit contained one drug molecule. The ESPs could release the drug effectively under physiological conditions with enzyme (e.g., CAL-B), which indicated that the obtained prodrugs could be a promising candidate for extending pharmacological efficacies by delaying drug release. In addition, the poly(amide-*co*-ester) prodrugs of NSAIDs could be further functionalized with PEG, targeting groups, and quaternary ammonium compounds. The supplemented agents endowed the ESPs with some special properties, such as amphiphilicity, targetability, and antibacterial activity.

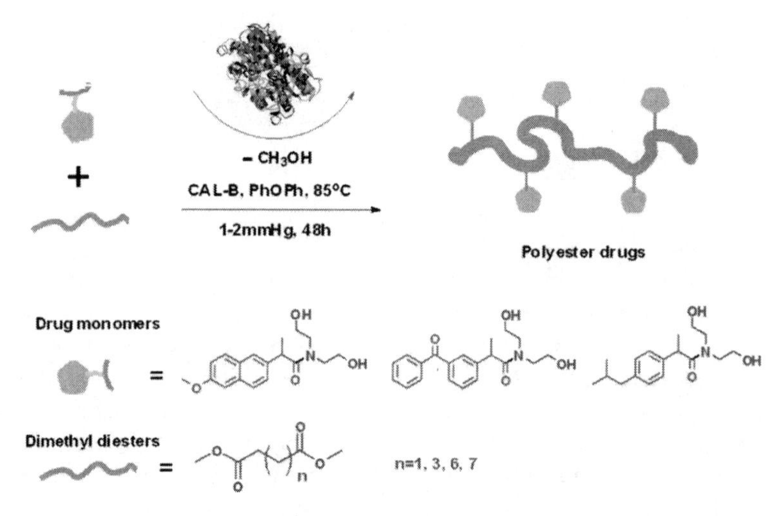

Figure 3.5 CAL-B catalyzed condensation polymerization of *N,N*-bis(2-hydroxyethyl)-2-ary propanamide with diesters (PhOPh = diphenyl ether) [66].

3.4.2 Drug-Encapsulated ESP-Based Platforms

The encapsulation strategies are pivotal for the delivery of fragile, poorly soluble, and/or toxic compounds [67]. They can make a stronger therapeutic effect and minimized side effects via the selective and sustained release of the loaded drug in particles. For these reasons, some novel methods of encapsulation are developed by using advanced materials and various types of drug–carrier interactions [68].

The syntheses of poly(ω-pentadecalactone-*co*-*p*-dioxanone) (poly(PDL-*co*-DO)) copolyesters through the enzymatic ROP of ω-pentadecalactone (PDL) and *p*-dioxanone (DO) with CAL-B as the catalyst were reported by Gross, Scandola, and coworkers [69]. Three kinds of copolyesters with different hydrophilicity were prepared by changing the monomer ratio. Doxorubicin (DOX) is a commercial drug used in the clinical treatment of various cancers, like leukemia, lymphomas, breast carcinoma, and many other solid tumors [70–76]. However, it is limited by its low tolerance because of its high cardiotoxicity, such as irreversible cardiomyopathy and congestive heart failure [77]. In this work, DOX was loaded into the poly(PDL-*co*-DO) nanoparticles from two types of copolyesters containing 42 and 69 mol% of DO unit with the DLCs of 2.1 and 1.3 wt%, respectively. The diameters of the two DOX-loaded nanoparticles from scanning electron microscopy (SEM) were 235 ± 47 and 207 ± 44 nm, respectively. In addition, the poly(PDL-*co*-DO) copolymer with 69 mol% of DO segment were capable of encapsulating a polynucleotide (e.g., siRNA) apart from a typical antitumor drug (e.g., DOX). The drug-loaded poly(PDL-*co*-DO) nanoparticles exhibited controlled and continuous release of drugs over an extensive period of time (i.e., 20–60 days). In particular, the luciferase-targeted siRNA (siLUC)-loaded poly(PDL-*co*-DO) nanoparticle was active in inhibiting the luciferase gene expression in LUC-RKO cells (a human colon carcinoma cell line).

Liu, Jiang, and colleagues reported a series of DOX-encapsulated nanoparticles fabricated from poly(butylene-*co*-sebacate-*co*-glycolate) (PBSG) copolymers by a single emulsification-solvent evaporation process [78]. The PBSG polyesters were synthesized via the CAL-B-catalyzed PC of 1,4-butanediol (BD), ethyl glycolate (EGA), and diethyl sebacate (DES). The average hydrodynamic diameters (D_hs) of the DOX-loaded PBSG nanoparticles were between about 245 and 280 nm with narrow distribution. The zeta potentials of the DOX-encapsulated nanoparticles were slightly positive, ranging from around +6.5 to +15.5 mV. The DLCs of the DOX-loaded PBSG nanoparticles were in the range between ~45 and 65 μg mg^{-1}. The DOX-encapsulated PBSG nanoparticles exhibited a slow and sustained drug release profile in phosphate-buffered saline (PBS) at 37°C over an extended period of time (i.e., 60 days), owing to the gradual hydrolytic degradation. In addition, the systems showed highly efficient internalization toward HeLa cells (a human cervical cancer cell line, up to 95%) after co-incubation for 4 hours, and exhibited upregulated cytotoxicities than free DOX in low drug concentrations (≤ 0.5 μM).

Watterson, Parmar, Kumar, and colleagues designed and prepared a series of amphiphilic copolymers composed of PEG600 and ESPs with various pendant functional groups, such as hydroxyl and carboxyl groups (Fig. 3.6) [79]. The copolymers self-assembled into micelles in aqueous media with D_hs in the range of 20–50 nm. These self-organized micelles were highly efficient drug delivery vehicles for both hydrophobic and partially hydrophilic drugs, both transdermally and orally, as they have the ability to encapsulate guest molecules during self-assembly. The micelles loaded with anti-inflammatory agents (i.e., aspirin and naproxen) resulted in significant reduction in inflammation after being applied topically *in vivo*. In detail, the reduction percentage in inflammation after the treatment using micelle containing aspirin or naproxen was 62% or 64% compared to that of the group without treatment, respectively.

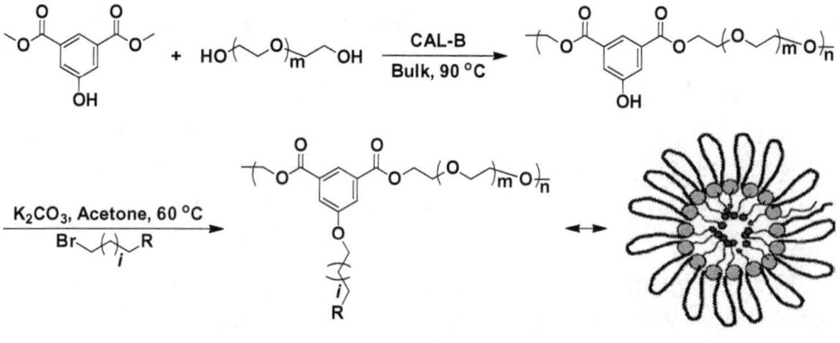

CAL-B, Bulk, 90 °C

K_2CO_3, Acetone, 60 °C

(1) R = CH_3, *i* = 7; (2) R = $^{20}CH_3$, *i* = 7; (3) R = OH, *i* = 10; (4) R = OH, *i* = 11; (5) R = COOH, *i* = 8; (6) R = COOH, *i* = 9

Figure 3.6 Chemoenzymatic syntheses of functionalized amphiphilic polyesters and their self-assemblies in aqueous medium to form micelles [79].

Poly(glycerol adipate-*co*-ω-pentadecalactone) (P(GA-*co*-PDL)) was enzymatically synthesized by a combination of enzymatic ROP and PC [80]. The microparticles were produced by spray drying directly from double emulsion with and without 0.5–1.5 wt% of dispersibility enhancers (e.g., L-arginine and L-leucine) using sodium fluorescein (SF) as a model hydrophilic drug. The incorporation of L-leucine (i.e., 1.5 wt%) reduced the burst release (i.e., 24.04 ± 3.87%) of SF compared to the unmodified formulation (i.e., 41.87 ± 2.46%). The spray-dried microparticles of P(GA-*co*-PDL) with 1.5 wt% of L-leucine showed reduced toxicity compared with that of PLGA at concentrations up to 5.0 mg mL^{-1} toward human bronchial epithelial 16HBE14o– cell line after co-incubation for 72 hours.

3.5 Conclusions and Outlook

ESPs with high specificity and adjustable features can be facilely and "green" synthesized through the enzymatic ROP or PC. In addition, it solves the toxicity and biological incompatibility caused by the traditional toxic chemical catalysts in the synthesized polyesters. A variety of chemical, physical, and biological properties, such as chemical reaction activities, solubility, and bioactivities, can be facilely pre-designed and regulated for ESPs. SMDs can be combined with ESPs through chemical conjugation or physical encapsulation, which reduce the side effects and upregulate the efficacy and bioavailability. Apparently, the applications of ESPs are manifold and multilevel. So ESPs will certainly be a common science and technology and be widely applied in the biomedical fields, such as drug delivery, in future developing process.

References

1. Zhang, L., Xiong, C., Deng, X. (1995). Biodegradable polyester blends for biomedical application. *J Appl Polym Sci*, **56**, 103–112.
2. Jiao, Y.-P., Cui, F.-Z. (2007). Surface modification of polyester biomaterials for tissue engineering. *Biomed Mater*, **2**, R24.
3. Lim, Y.-B., Han, S.-O., Kong, H.-U., Lee, Y., *et al.* (2000). Biodegradable polyester, poly[α-(4-aminobutyl)-L-glycolic acid], as a non-toxic gene carrier. *Pharm Res*, **17**, 811–816.

4. Li, D., Ding, J. X., Tang, Z. H., Sun, H., *et al.* (2012). *In vitro* evaluation of anticancer nanomedicines based on doxorubicin and amphiphilic Y-shaped copolymers. *Int J Nanomedicine*, **7**, 2687–2697.
5. Ding, J., Li, D., Zhuang, X., Chen, X. (2013). Self-assemblies of pH-activatable PEGylated multiarm poly(lactic acid-*co*-glycolic acid)-doxorubicin prodrugs with improved long-term antitumor efficacies. *Macromol Biosci*, **13**, 1300–1307.
6. Li, D., Sun, H., Ding, J., Tang, Z., *et al.* (2013). Polymeric topology and composition constrained polyether-polyester micelles for directional antitumor drug delivery. *Acta Biomater*, **9**, 8875–8884.
7. Wang, J., Li, D., Li, T., Ding, J., *et al.* (2015). Gelatin tight-coated poly(lactide-*co*-glycolide) scaffold incorporating rhBMP-2 for bone tissue engineering. *Materials*, **8**, 1009–1026.
8. Wang, J., Xu, W., Ding, J., Lu, S., *et al.* (2015). Cholesterol-enhanced polylactide-based stereocomplex micelle for effective delivery of doxorubicin. *Materials*, **8**, 216–230.
9. Kulkarni, R. K., Moore, E. G., Hegyeli, A. F., Leonard, F. (1971). Biodegradable poly(lactic acid) polymers. *J Biomed Mater Res*, **5**, 169–181.
10. Wang, L., Ma, W., Gross, R. A., McCarthy, S. P. (1998). Reactive compatibilization of biodegradable blends of poly(lactic acid) and poly(ε-caprolactone). *Polym Degrad Stab*, **59**, 161–168.
11. Srinivasan, R., Mayne-Banton, V. Self-developing photoetching of poly(ethylene terephthalate) films by far-ultraviolet excimer laser radiation. *Appl Phys Lett* 1982, **41**, 576–578.
12. Shibata, M., Inoue, Y., Miyoshi, M. (2006). Mechanical properties, morphology, and crystallization behavior of blends of poly(L-lactide) with poly(butylene succinate-*co*-L-lactate) and poly(butylene succinate). *Polymer*, **47**, 3557–3564.
13. Mespouille, L., Coulembier, O., Kawalec, M., Dove, A. P., Dubois, P. (2014). Implementation of metal-free ring-opening polymerization in the preparation of aliphatic polycarbonate materials. *Prog Polym Sci*, **39**, 1144–1164.
14. Ajellal, N., Carpentier, J.-F., Guillaume, C., Guillaume, S. M., *et al.* (2010). Metal-catalyzed immortal ring-opening polymerization of lactones, lactides and cyclic carbonates. *Dalton Trans*, **39**, 8363–8376.
15. Yu, Y., Wu, D., Liu, C., Zhao, Z., *et al.* (2012). Lipase/esterase-catalyzed synthesis of aliphatic polyesters via polycondensation: a review. *Process Biochem*, **47**, 1027–1036.

16. Zaks, A., Klibanov, A. M. (1984). Enzymatic catalysis in organic media at 100-degrees-C. *Science*, **224**, 1249–1251.

17. Seo, K. S., Castano, M., Casiano, M., Wesdemiotis, C., *et al.* (2014). Enzyme-catalyzed quantitative chain-end functionalization of poly(ethylene glycol)s under solventless conditions. *RSC Adv*, **4**, 1683–1688.

18. Le-Huu, P., Heidt, T., Claasen, B., Laschat, S., Urlacher, V. B. (20015). Chemo-, regio-, and stereoselective oxidation of the monocyclic diterpenoid *β*-cembrenediol by P450 BM3. *ACS Catal*, **5**, 1772–1780.

19. Gross, R. A., Ganesh, M., Lu, W. (2010). Enzyme-catalysis breathes new life into polyester condensation polymerizations. *Trends Biotechnol*, **28**, 435–443.

20. Schmid, A., Dordick, J. S., Hauer, B., Kiener, A., *et al.* (2001). Industrial biocatalysis today and tomorrow. *Nature*, **409**, 258–268.

21. Kobayashi, S. (1999). Enzymatic polymerization: a new method of polymer synthesis. *J Polym Sci A: Polym Chem*, **37**, 3041–3056.

22. Kobayashi, S. (2009). Recent developments in lipase-catalyzed synthesis of polyesters. *Macromol Rapid Commun*, **30**, 237–266.

23. Kikuchi, H., Uyama, H., Kobayashi, S. (2000). Lipase-catalyzed enantioselective copolymerization of substituted lactones to optically active polyesters. *Macromolecules*, **33**, 8971–8975.

24. Treichel, H., de Oliveira, D., Mazutti, M., Di Luccio, M., Oliveira, J. V. (2010). A review on microbial lipases production. *Food Bioprocess Technol*, **3**, 182–196.

25. Deng, M.-M., Yu, J.-G. (2007). Enzyme catalyzed synthesis of aliphatic polyesters. *Polym Bull—Beijing*, **3**, 61.

26. Cameron, D. J. A., Shaver, M. P. (2011). Aliphatic polyester polymer stars: synthesis, properties and applications in biomedicine and nanotechnology. *Chem Soc Rev*, **40**, 1761–1776.

27. Piotrowska, U., Sobczak, M. (2014). Enzymatic polymerization of cyclic monomers in ionic liquids as a prospective synthesis method for polyesters used in drug delivery systems. *Molecules*, **20**, 1–23.

28. Seyednejad, H., Ghassemi, A. H., van Nostrum, C. F., Vermonden, T., Hennink, W. E. (2011). Functional aliphatic polyesters for biomedical and pharmaceutical applications. *J Control Release*, **152**, 168–176.

29. Jerome, C., Lecomte, P. (2008). Recent advances in the synthesis of aliphatic polyesters by ring-opening polymerization. *Adv Drug Deliv Rev*, **60**, 1056–1076.

30. Binns, F., Roberts, S. M., Taylor, A., Williams, C. F. (1993). Enzymatic polymerization of an unactivated diol diacid system.*J Chem Soc, Perkin Trans 1*, 899–904.

31. Yang, Y., Yu, Y., Zhang, Y., Liu, C., *et al.* (2011). Lipase/esterase-catalyzed ring-opening polymerization: a green polyester synthesis technique. *Process Biochem*, **46**, 1900–1908.

32. Castano, M., Zheng, J., Puskas, J. E., Becker, M. L. (2014). Enzyme-catalyzed ring-opening polymerization of [varepsilon]-caprolactone using alkyne functionalized initiators. *Polym Chem*, **5**, 1891–1896.

33. Knani, D., Gutman, A. L., Kohn, D. H. (1993). Enzymatic polyesterification in organic media. Enzyme-catalyzed synthesis of linear polyesters. I. Condensation polymerization of linear hydroxyesters. II. Ring-opening polymerization of ε-caprolactone. *J Polym Sci A: Polym Chem*, **31**, 1221–1232.

34. Uyama, H., Kobayashi, S. (1993). Enzymatic ring-opening polymerization of lactones catalyzed by lipase. *Chem Lett*, **7**, 1149–1150.

35. Miletić, N., Nastasović, A., Loos, K. (2012). Immobilization of biocatalysts for enzymatic polymerizations: possibilities, advantages, applications. *Bioresour Technol*, **115**, 126–135.

36. Idris, A., Bukhari, A. (2012). Immobilized *Candida antarctica* lipase B: hydration, stripping off and application in ring opening polyester synthesis. *Biotechnol Adv*, **30**, 550–563.

37. Robert, C., Thomas, C. M. (2013). Tandem catalysis: a new approach to polymers. *Chem Soc Rev*, **42**, 9392–9402.

38. Gross, R. A., Kumar, A., Kalra, B. (2001). Polymer synthesis by *in vitro* enzyme catalysis. *Chem Rev*, **101**, 2097–2124.

39. Macdonald, R. T., Pulapura, S. K., Svirkin, Y. Y., Gross, R. A., *et al.* (1995). Enzyme-catalyzed epsilon-caprolactone ring-opening polymerization. *Macromolecules*, **28**, 73–78.

40. Uyama, H., Takeya, K., Hoshi, N., Kobayashi, S. (1995). Lipase-catalyzed ring-opening polymerization of 12-dodecanolide. *Macromolecules*, **28**, 7046–7050.

41. Kobayashi, S., Uyama, H., Kimura, S. (2001). Enzymatic polymerization. *Chem Rev*, **101**, 3793–3818.

42. Namekawa, S., Suda, S., Uyama, H., Kobayashi, S. (1999). Lipase-catalyzed ring-opening polymerization of lactones to polyesters and its mechanistic aspects. *Int J Biol Macromol*, **25**, 145–151.

43. Warner, J. C. (2012). Green chemistry: theory and practice. *Abstr Pap Am Chem Soc*, **244**.

44. Barrera-Rivera, K. A., Flores-Carreon, A., Martinez-Richa, A. (2008). Enzymatic ring-opening polymerization of epsilon-caprolactone by a new lipase from *Yarrowia lipolytica*. *J Appl Polym Sci*, **109**, 708–719.

45. Kobayashi, S. (2010). Lipase-catalyzed polyester synthesis: a green polymer chemistry. *Proc Jpn Acad Series B: Phys Biol Sci*, **86**, 338–365.

46. Warwel, S., Demes, C., Steinke, G. (2001). Polyesters by lipase-catalyzed polycondensation of unsaturated and epoxidized long-chain alpha,omega-dicarboxylic acid methyl esters with diols. *J Polym Sci A: Polym Chem*, **39**, 1601–1609.

47. Uyama, H., Kobayashi, S. (2006). Enzymatic synthesis of polyesters via polycondensation, in Kobayashi, S., Ritter, H., Kaplan, D. (eds.), *Enzyme-Catalyzed Synthesis of Polymers*, Springer, Berlin/Heidelberg, pp. 133–158.

48. Azim, H., Dekhterman, A., Jiang, Z., Gross, R. A. (2006). *Candida antarctica* lipase b-catalyzed synthesis of poly(butylene succinate): shorter chain building blocks also work. *Biomacromolecules*, **7**, 3093–3097.

49. Ates, Z., Thornton, P. D., Heise, A. (2011). Side-chain functionalisation of unsaturated polyesters from ring-opening polymerisation of macrolactones by thiol-ene click chemistry. *Polym Chem*, **2**, 309–312.

50. Hoyle, C. E., Bowman, C. N. (2010). Thiol–ene click chemistry. *Angew Chem Int Ed*, **49**, 1540–1573.

51. Shafioul, A. S. M., Pyo, J. I., Kim, K. S., Cheong, C. S. (2012). Synthesis of poly (glycerol-*co*-dioate-*co*-butanedioate-*co*-xanthorrhizol) ester and a study of chain length effect on pendant group loading. *J Mol Catal B-Enzym*, **84**, 198–204.

52. Liao, L., Liu, J., Dreaden, E. C., Morton, S. W., *et al.* (2014). A convergent synthetic platform for single-nanoparticle combination cancer therapy: ratiometric loading and controlled release of cisplatin, doxorubicin, and camptothecin. *J Am Chem Soc*, **136**, 5896–5899.

53. Kanapathipillai, M., Brock, A., Ingber, D. E. (2014). Nanoparticle targeting of anti-cancer drugs that alter intracellular signaling or influence the tumor microenvironment. *Adv Drug Deliv Rev*, **79–80**, 107–118.

54. Ding, J., Xu, W., Zhang, Y., Sun, D., *et al.* (2013). Self-reinforced endocytoses of smart polypeptide nanogels for "on-demand" drug delivery. *J Control Release*, **172**, 444–455.

55. Maeda, H., Wu, J., Sawa, T., Matsumura, Y., Hori, K. (2000). Tumor vascular permeability and the EPR effect in macromolecular therapeutics: a review. *J Control Release*, **65**, 271–284.

56. Sun, D., Ding, J., Xiao, C., Chen, J., *et al.* (2014). Preclinical evaluation of antitumor activity of acid-sensitive PEGylated doxorubicin. *ACS Appl Mater Interfaces*, **6**, 21202–21214.

57. Xu, W., Ding, J., Xiao, C., Li, L., *et al.* (2015). Versatile preparation of intracellular-acidity-sensitive oxime-linked polysaccharide-doxorubicin conjugate for malignancy therapeutic. *Biomaterials*, **54**, 72–86.

58. Xu, W., Ding, J., Li, L., Xiao, C., *et al.* (2015). Acid-labile boronate-bridged dextran-bortezomib conjugate with up-regulated hypoxic tumor suppression. *Chem Commun*, **51**, 6812–6815.

59. Zhang, J. -C., Ding, J. -X., Xiao, C. -S., He, C. -L., *et al.* (2012). Synthesis and characterization of tumor-acidity-sensitive poly(L-lysine)-doxorubicin conjugates. *Chem J Chinese Univ*, **33**, 2809–2815. [Chinese]

60. Zhang, Y., Xiao, C., Li, M., Chen, J., *et al.* (2013). Co-delivery of 10-hydroxycamptothecin with doxorubicin conjugated prodrugs for enhanced anticancer efficacy. *Macromol Biosci*, **13**, 584–594.

61. Zhang, Y., Xiao, C., Li, M., Ding, J., *et al.* (2014). Core-cross-linked micellar nanoparticles from a linear-dendritic prodrug for dual-responsive drug delivery. *Polym Chem*, **5**, 2801–2808.

62. Ringsdorf, H. (1975). Structure and properties of pharmacologically active polymers. *J Polym Sci C: Polym Symp*, **51**, 135–153.

63. Pasut, G., Veronese, F. M. (2007). Polymer-drug conjugation, recent achievements and general strategies. *Prog Polym Sci*, **32**, 933–961.

64. Wang, H.-Y., Zhang, W.-W., Wang, N., Li, C., *et al.* (2010). Biocatalytic synthesis and *in vitro* release of biodegradable linear polyesters with pendant ketoprofen. *Biomacromolecules*, **11**, 3290–3293.

65. Wu, W.-X., Wang, H.-Y., Wang, N., Zhang, W.-W., *et al.* (2013). Enzymatic synthesis and characterization of thermosensitive polyester with pendent ketoprofen. *Polymers*, **5**, 1158–1168.

66. Qian, X., Wu, Q., Xu, F., Lin, X. (2013). Lipase-catalyzed synthesis of polymeric prodrugs of nonsteroidal anti-inflammatory drugs. *J Appl Polym Sci*, **128**, 3271–3279.

67. Ding, J., Chen, L., Xiao, C., Chen, L., *et al.* (2014). Noncovalent interaction-assisted polymeric micelles for controlled drug delivery. *Chem Commun*, **50**, 11274–11290.

68. Kita, K., Dittrich, C. (2011). Drug delivery vehicles with improved encapsulation efficiency: taking advantage of specific drug-carrier interactions. *Expert Opin Drug Deliv*, **8**, 329–342.

69. Liu, J., Jiang, Z., Zhang, S., Liu, C., *et al.* (2011). Biodegradation, biocompatibility, and drug delivery in poly (omega-pentadecalactone-*co*-p-dioxanone) copolyesters. *Biomaterials*, **32**, 6646–6654.

70. Wang, W., Ding, J., Xiao, C., Tang, Z., *et al.* (2011). Synthesis of amphiphilic alternating polyesters with oligo(ethylene glycol) side chains and potential use for sustained release drug delivery. *Biomacromolecules*, **12**, 2466–2474.

71. Ding, J., Shi, F., Li, D., Chen, L., *et al.* (2013). Enhanced endocytosis of acid-sensitive doxorubicin derivatives with intelligent nanogel for improved security and efficacy. *Biomater Sci*, **1**, 633–646.

72. Ding, J., Xiao, C., Yan, L., Tang, Z., *et al.* (2011). pH and dual redox responsive nanogel based on poly(L-glutamic acid) as potential intracellular drug carrier. *J Control Release*, **152**, E11–E13.

73. Shi, F., Ding, J., Xiao, C., Zhuang, X., *et al.* (2012). Intracellular microenvironment responsive PEGylated polypeptide nanogels with ionizable cores for efficient doxorubicin loading and triggered release. *J Mater Chem*, **22**, 14168–14179.

74. Ding, J., Chen, J., Li, D., Xiao, C., *et al.* (2013). Biocompatible reduction-responsive polypeptide micelles as nanocarriers for enhanced chemotherapy efficacy *in vitro*. *J Mater Chem B*, **1**, 69–81.

75. Ding, J., He, C., Xiao, C., Chen, J., *et al.* (2012). pH-responsive drug delivery systems based on clickable poly(L-glutamic acid)-grafted comb copolymers. *Macromol Res*, **20**, 292–301.

76. Ding, J., Shi, F., Xiao, C., Lin, L., *et al.* (2011). One-step preparation of reduction-responsive poly(ethylene glycol)-poly (amino acid)s nanogels as efficient intracellular drug delivery platforms. *Polym Chem*, **2**, 2857–2864.

77. Wang, A. Z., Langer, R., Farokhzad, O. C. (2012). Nanoparticle delivery of cancer drugs. *Annu Rev Med*, **63**, 185–198.

78. Yang, Z., Zhang, X., Luo, X., Jiang, Q., *et al.* (2013). Enzymatic synthesis of poly(butylene-*co*-sebacate-*co*-glycolate) copolyesters and evaluation of the copolymer nanoparticles as biodegradable carriers for doxorubicin delivery. *Macromolecules*, **46**, 1743–1753.

79. Kumar, R., Chen, M. H., Parmar, V. S., Samuelson, L. A., *et al.* (2004). Supramolecular assemblies based on copolymers of PEG600 and

functionalized aromatic diesters for drug delivery applications. *J Am Chem Soc*, **126**, 10640–10644.

80. Tawfeek, H., Khidr, S., Samy, E., Ahmed, S., *et al.* (2011). Poly(glycerol adipate-*co*-omega-pentadecalactone) spray-dried microparticles as sustained release carriers for pulmonary delivery. *Pharm Res*, **28**, 2086–2097.

Chapter 4

Overview of Methods of Making Polyester Nano- and Microparticulate Systems for Drug Delivery

Nadim Ahamad,* Dadi A. Srinivasa Rao,* and Dhirendra S. Katti
Department of Biological Sciences and Bioengineering, Indian Institute of Technology, Kanpur 208016, Uttar Pradesh, India
dsk@iitk.ac.in

4.1 Introduction

During the past two decades the development of drug delivery systems (DDSs) has been a revolutionary cue for the advancement of conventional approaches of drug administration in medicine and thus has opened new frontiers such as micromedicine and nanomedicine. These frontiers encompass therapeutic, diagnostic, as well as theranostic approaches and have witnessed seminal growth in the recent past [1]. Micro-/nanotechnology that enables the fabrication of fine and ultrafine structures (e.g., particles, fibers, wires, etc., on

*These authors contributed equally.

Handbook of Polyester Drug Delivery Systems
Edited by MNV Ravikumar

ISBN 978-981-4669-65-8 (Hardcover), 978-981-4669-66-5 (eBook)
www.panstanford.com

the nanometer scale) using diverse materials has played a pivotal role in the development of these frontier areas. The success of micro-/nanomedicine is governed to a great extent by the availability of polymeric material which can be used for human applications. Till date, multiple polymeric materials have been approved by the United States Food and Drug Administration (US FDA) [2] for their application as DDSs, which largely include polyesters, polyacrylates, polyanhydrides, polyamides, and polycarbonates. Among these, polyesters represent a major class where multiple polyester-based DDSs have been approved by the US FDA for application in humans [3, 4], including polylactic acid (PLA), polyglycolic acid (PGA), polylactic-*co*-glycolic acid (PLGA), polydioxanone, polycaprolactone (PCL), and poly(trimethyl carbonate) [3, 5, 67]. Further, the potential of polyester-based polymers as versatile DDSs has been demonstrated by their ability to encapsulate diverse therapeutic molecules/cargo (e.g., synthetic drugs, proteins, or deoxyribonucleic acid [DNA]) inside the polymer matrix [8, 9]. In contrast to the administration of a free drug, encapsulation of a drug inside polyester-based particulate DDSs not only ensures its protection in hostile environments and desirable release kinetics, but can also improve its bioavailability and thus facilitate high payload delivery [2]. Currently there are more than 10 polyester-based DDSs available commercially in the market [9]. For example, two of the PLGA-based DDSs are Zoladex® (PLGA/goserelin acetate), which is used for the treatment of prostate cancer, and Lupron Depot® (PLGA/leuprolide acetate), which is used for the treatment of prostate cancer and endometriosis [10]. Due to their desirable properties, polyester-based materials show high potential for the development of future DDSs employing DNA, small interfering ribonucleic acid (siRNA), messenger ribonucleic acid (mRNA), aptamer, and other novel synthetic/peptide-based therapeutic molecules. While existing polyester-based DDSs have shown good success, innovations in polymer chemistry combined with the discovery of novel therapeutic molecules are continuously fueling the development of newer biodegradable, biocompatible, tissue-specific, stimuli-responsive (pH, temperature, pressure, light, ionic concentration, and catalytic based) polyester-based DDSs [11].

The success of DDSs is governed not only by the class of material chosen, but also the method of fabrication used. Interestingly, polyesters are a class of polymers for which a large number of existing methods can be used for the fabrication of particles with desired physical properties (e.g., size, shape, encapsulation efficiency, and drug release kinetics). From the initial development of emulsion-based methods for fabrication of polymeric micro-/nanoparticles, there have been significant advancements in technology which are focused on providing a fine control on the dispersity of particles, the type of cargo that can be encapsulated (hydrophilic or lipophilic), percent yield, shelf life of the encapsulants, ease of fabrication, cost-effectiveness, and environmental toxicity. Some of the examples of the techniques for the fabrication of polymeric micro-/nanoparticles include nanoprecipitation, electrospraying, micro-/nanoemulsion-based techniques, layer-by-layer fabrication, microfluidics, nanoimprinting, and lithography [2, 12–15]. The rationale for selection of a method most often is governed by the end application of the DDSs while accounting for the properties of the particles that can be controlled by that particular technique. Therefore, in order to choose an appropriate technique for the fabrication of polyester-based particulate DDSs, one needs to have a proper understanding of existing methods for fabrication of particles.

This chapter summarizes the existing methods for the synthesis of polyester-based micro-/nanoparticles along with their merits and demerits. Additionally, the scalability and versatility of the method which determines the translation potential of a formulation has also been given some consideration. Further, drying and storage of particles is another important aspect that has to be considered during the development of a successful DDS. With many of the DDSs having protein or nucleic acid as the therapeutic molecule, drying or storage under normal conditions remains a challenge, as the therapeutic molecules can undergo denaturation and consequently result in loss of activity. Hence, in this chapter, due consideration has also been given to the standard methods of drying and storage of particles. Finally, the challenges associated with fabrication of polyester-based particulate DDSs have been briefly covered toward the end of the chapter.

4.2 Methods for Synthesis of Micro-/Nanoparticles

4.2.1 Conventional Methods for Micro-/Nanoparticle Synthesis

4.2.1.1 One-step fabrication methods

4.2.1.1.1 *Nanoprecipitation*

The technique of nanoprecipitation is based on the principle of interfacial deposition of a polymer due to displacement of its solvent by a nonsolvent. In this method, a polyester solution prepared in an organic solvent is added to a nonsolvent (an aqueous solvent containing surfactant) of that polyester. Upon addition, the organic solvent spontaneously diffuses into the aqueous phase, resulting in precipitation of the polyester as fine particles (nanoparticles). The organic solvent is then removed by evaporation to obtain polyester nanoparticles [16, 17]. During this process, the surfactant gets adsorbed onto the interface of nanoparticles, thus preventing their aggregation. A primary requirement for preparation of nanoparticles by nanoprecipitation method is miscibility of the organic solvent with the aqueous solvent system. Hence, acetone which is miscible with water and has the property to solubilize hydrophobic polyesters has been widely used as the organic solvent of choice in nanoprecipitation [17].

In the nanoprecipitation method, processing parameters such as polymer concentration, surfactant concentration, phase volume ratio of aqueous to non-aqueous solvents, temperature, time and pressure used during evaporation govern the size, morphology, polydispersity, drug loading, and physicochemical stability of the fabricated nanoparticles. Increase in polymer concentration or reduction in phase volume ratios of organic to aqueous phase leads to the formation of larger particles [16]. The concentration of surfactant defines the physical stability of the colloidal dispersion thereby influencing the size of particles formed. Use of suboptimal concentration of surfactant causes coalescence of particles, which in turn causes an increase in particle size. Chemical stability of the polymer can be affected by the temperature used during evaporation of organic solvent. This can be of significance as polyesters are susceptible to degradation at higher temperatures. Hence, an optimum temperature should be selected on the basis of glass transition temperature (T_g), melting temperature (T_m), and degradation temperature of the polyester being used for the preparation of nanoparticles.

Nanoprecipitation is widely used for the fabrication of drug-loaded nanoparticulate DDSs. During the preparation of drug-loaded nanoparticles, the drug and polyester are usually dissolved in an organic solvent. Since most of the drugs are hydrophobic in nature, they are soluble in organic solvents, and hence can be easily encapsulated into particles during nanoprecipitation, thereby providing a high loading efficiency [18]. However, the loading efficiency of hydrophilic drugs is relatively lower as hydrophilic/water-soluble drugs readily diffuse out into the aqueous phase during preparation. The drug loading efficiency of a drug can be affected by formulation parameters such as pH of aqueous phase and inclusion of excipients. These parameters were studied by Govender *et al.* to enhance the entrapment efficiency of a water-soluble drug (procaine hydrochloride) into PLGA 50:50 nanoparticles. Their results indicated that change in the pH of aqueous phase from 5.8 to 9.3 improved the yield of nanoparticle from 65.1% to 93.4%, drug content from 0.3% to 1.3% w/w, and drug entrapment from 11.0% to 58.2% [19]. The method of nanoprecipitation is also suitable for the preparation of nanoparticles with desired release kinetics when produced in lower quantities. Chorny *et al.* studied the influence of different formulation parameters such as concentration of polyester (PLA), ratio of organic solvent to nonsolvent (aqueous solvent), and surfactant concentration on nanoparticles size, drug recovery, and drug release kinetics of PLA nanoparticles prepared by the nanoprecipitation method. Their results demonstrated that particle size is mainly influenced by the amounts of polymer and its nonsolvent; further, drug release is affected by PLA concentration, while the stability and yield is affected by drug to polymer ratio [18]. Hence, by choosing optimal concentrations of the polyester, surfactant, and ratio of non-aqueous to aqueous phase, particulate delivery systems with high drug loading can be prepared using the nanoprecipitation method at the laboratory scale. However, large-scale production of nanoparticles using the nanoprecipitation method still remains a challenge.

4.2.1.1.2 *Electrospraying*

Fundamentally, this technique is similar to the electro-spray ionization (ESI) process used in mass spectrometry and involves spraying of a polymeric solution under the influence of high voltage in order to fabricate micro-/nanoparticles [20–24]. The experimental setup consists of a syringe pump, a high-voltage power supply unit, and a grounded collector (Fig. 4.1). The syringe pump is programmed to enable smooth flow of the polyester solution at desired flow rate through a metallic needle. The needle is attached to a high-voltage power source and hence acts as an anionic electrode. Under the influence of repulsive forces caused due to the high charge on the polyester solution, the polyester solution protrudes out of the needle tip to form a cone like structure called the Taylor cone [21, 25]. Depending on the process and solution parameters used, the formed Taylor cone under the influence of the high voltage can lead to the formation of distinct morphological structures such as thin films, particles, beaded fibers, or smooth fibers upon evaporation of solvent. The various parameters that influence the size and morphology of electrosprayed polyesters can be classified into three categories: (i) solution parameters, such as concentration, viscosity, surface tension, boiling point of the solvent, and conductivity of the solution; (ii) process parameters, such as applied voltage, needle gauge, flow rate, distance between needle, and collector; and (iii) ambient parameters, such as temperature and humidity of processing environment [24, 26–36]. Hence, these parameters can be optimized to fabricate polymeric particles of desired size and shape [24]. Mahaling and Katti studied the influence of polymer concentration and solvent type on the structure/morphology of electrosprayed poly[(R)-3-hydroxybutyric acid] solution. They demonstrated that change in polymer concentration and type of solvent can lead to distinct morphologies (such as porous thin films, microparticles, beaded fibers, and smooth fibers) of the electrosprayed/electrospun polymeric solutions [37].

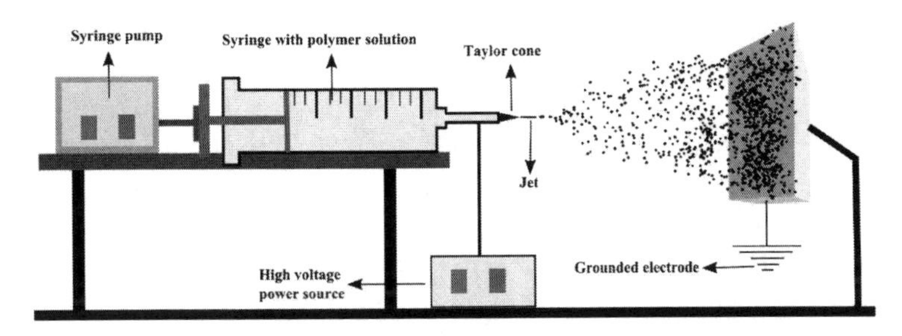

Figure 4.1 Schematic representation of the process of electrospraying for the fabrication of micro-/nanoparticles.

In addition to polymer concentration, presence or absence of surfactant(s) in the polyester solution can influence the morphology of electrosprayed polymers. It has been shown that blending of PLGA 85:15 with Pluronics® (F127, P123, and L121) altered the morphology of electrosprayed microparticles from doughnut shape (with PLGA 85:15 alone) to spherical shape (upon blending with Pluronics®) [36]. The process of electrospraying has been employed to fabricate polyester-based micro-/nanoparticulate DDSs with the possibility of entrapping both hydrophobic and hydrophilic drugs. Hydrophobic drug entrapped polyester particles can be prepared by electrospraying of a low boiling point organic solution of drug and polymer [38], whereas hydrophilic drug encapsulated particles have been prepared by emulsifying an aqueous solution of a drug in a non-aqueous polymer solution and subsequently electrospraying the resultant emulsion [39]. Hydrophilic protein molecules such as bovine serum albumin (BSA) can also be encapsulated in particles using the electrospraying technique by optimization of process parameters. Xu and Hanna comprehensively studied different processing variables such as organic/aqueous phase volume ratio, protein/polyester weight ratio, viscosity, electrical conductivity and surface tension on size, morphology, yield, and entrapment efficiency of fabricated particles using the system of BSA-loaded PLA microparticles. They observed that entrapment efficiency of BSA increased with increase in organic/aqueous phase ratio and decreased with increase in BSA/PLA ratio [39]. In another study, Xie and Wang studied the structure and bio-activity of electrosprayed BSA-loaded PLGA 50:50 microparticles. They demonstrated that BSA can be encapsulated in PLGA particles without altering its structure and function using the electrospraying technique [40]. The aforementioned studies demonstrate that both hydrophobic and hydrophilic drugs and protein molecules can be loaded in polyester particles fabricated using the process of electrospraying.

While the electrospraying process is a facile technique for the fabrication of polyester-based micro-/nanoparticles, it is not amenable to the fabrication of particles at a large scale. Since a single electrospraying apparatus does not lend itself to large-scale production of particulate DDSs, a multi-nozzle electrospray (MES) system has been developed to facilitate production at a relatively larger scale [41]. However, this process can lead to aggregation or irreversible fusion of particles due to poor evaporation of the solvent during a continuous electrospraying process. Hence, surfactant/protective colloid solutions have been employed to overcome this limitation in a continuous electrospraying process. Almería *et al.* demonstrated the use of a multiplexed electrospray process for single-step synthesis of amphiphilic drug entrapped PLGA microparticles with high entrapment efficiency. For this process, polyvinyl alcohol (PVA) solution was used as a surfactant for the collection of particles [41]. Hence, the aforementioned studies demonstrate that electrospraying is a facile technique that can be used for the fabrication of hydrophobic and/or hydrophilic drug loaded micro-/nanoparticulate DDSs at a small as well as relatively larger scale.

4.2.1.1.3 *High-pressure homogenization*

High-pressure homogenization (HPH) is one of the widely used methods for large-scale production of nanosuspensions and nanoparticles. Size reduction of suspensions and emulsions to obtain a very small size of the dispersed phase is performed using a high-pressure homogenizer also known as microfluidiser® [42]. In this process, the application of high pressure causes a reduction in the size of the dispersed phase by the cumulative effect of cavitation forces, shear forces, and collision between emulsion droplets. The size and polydispersity of colloidal dispersions prepared by HPH can be controlled using processing parameters such as polymer concentration, surfactant concentration, total volume of solution, volume ratio of aqueous to non-aqueous phase, applied pressure, number of homogenization cycles, and operating temperature. Lamprecht *et al.* optimized various processing parameters during the preparation of PLGA 50:50 and PCL nanoparticles by HPH and emulsification process. They observed that the extent of reduction in the size of emulsion droplet is directly proportional to the applied pressure and the number of homogenization cycles used [42]. HPH is widely used for large-scale fabrication of emulsion-based dispersions and particulate DDSs. Further, it is also used for reduction in size of emulsion droplets for the preparation of nanoparticles by emulsion and solvent evaporation method. Hence, HPH is an easy, relatively inexpensive, and reproducible process for the large-scale production of polyester-based particulate DDSs such as nanoparticles, nanocrystals, and nanoemulsions.

4.2.1.2 Two-step fabrication

4.2.1.2.1 *Emulsion solvent evaporation*

Fabrication of micro-/nanoparticles from preformed polymers (including polyesters) using the emulsion solvent evaporation method has been accepted as one of the earliest and most commonly used methods [43]. Briefly, the polyester of interest is initially dissolved in a volatile organic solvent (e.g., chloroform or dichloromethane) followed by emulsification (oil-in-water emulsion or O/W emulsion) using homogenization/ultrasonication into an excess volume of immiscible aqueous solution containing surfactant (e.g., polyvinyl alcohol (PVA) or sodium dodecyl sulfate [SDS]). The emulsion is then evaporated under ambient conditions resulting in condensation of the polyester in the form of micro-/nanoparticles [44–46]. Desgouilles *et al.* demonstrated that it requires approximately 70–80 minutes for complete evaporation of ethyl acetate from 50 ml of emulsion under controlled conditions. In the same study the authors also observed that the first phase of evaporation is crucial and results in up to 90% solvent evaporation that eventually results in decrease in the size of particles. However, during the second phase of evaporation, wherein the remaining fraction of solvent is evaporated, an increment in size and polydispersity of particles was observed [47]. The key parameters that influence the diameter of particles are concentration of polyester solution, phase volume ratio of non-aqueous to aqueous phase, concentration of surfactant, and the intensity and duration of applied mechanical energy (homogenization or ultrasonication) [48]. Emulsion solvent evaporation is a versatile method for fabrication of micro-/nanoparticles and has been widely explored for a large variety of polyesters including PCL, PLA, and PLGA [49, 50] or their PEGylated copolymers, including polyethylene glycol (PEG)-PLA [51] and PEG-PLGA [50, 52]. This method was originally developed for encapsulation of lipophilic drugs as hydrophilic drugs (e.g., protein) would diffuse from the polyester droplet into the external aqueous phase that resulted in poor encapsulation efficiency. This problem has been largely overcome by using a double emulsion, i.e., water-in-oil-in-water (W/O/W) method, where the hydrophilic drug dissolved in aqueous solution (W1) is first emulsified in the polyester solution (O phase) and this primary emulsion is further emulsified in a secondary aqueous phase (W2) containing surfactant. The resultant double emulsion (W1/O/W2) is processed further similar to the single emulsion method [53]. Overall, the emulsion solvent evaporation method is a simple and widely accepted method for fabrication of polyester-based spherical micro-/nanoparticles with relatively higher drug entrapment efficiency.

4.2.1.2.2 *Emulsion solvent diffusion method*

This method fundamentally differs from the emulsion solvent evaporation method in terms of the mechanism of particle synthesis. In contrast to solvent evaporation method where particle formation is triggered by polymer condensation, in solvent diffusion method particles are formed purely by the diffusion phenomenon [54]. Unlike the solvent evaporation method, in this method the polymer is initially dissolved in a partially water-miscible organic solvent, e.g., benzyl alcohol or propylene carbonate (pre-saturated with water), [55] followed by emulsification in water-saturated organic solvent to produce an O/W emulsion. The emulsion containing dispersed droplets is diluted with excess volume of aqueous phase containing stabilizer. The organic solvent immediately starts diffusing (extraction) from polyester-solvent droplets (discontinuous phase) toward the external aqueous phase (continuous phase), which eventually results in the formation of spherical polyester micro-/nanoparticles from the emulsion droplets. Hence, this method is also known as the emulsification solvent displacement method (ESDM). In contrast to the solvent evaporation method, extraction of solvent from the droplets in the diffusion method is relatively faster and commences within milliseconds, resulting in a remarkable drop in particles size [56]. Finally, the remaining organic solvent from the aqueous phase is removed using filtration or evaporation. Generally, this method can produce particles having diameter in the range of 150 nm to a few micrometers. Further, the particle size can be controlled by modulating parameters like miscibility of organic solvent with water, stirring rate, and type and concentration of stabilizer. Generally, higher the degree of condensation of polyester inside the emulsion droplet due to solvent diffusion [57, 58] or higher the miscibility of organic solvent with water [59], smaller is the size of the particles produced.

The drug encapsulation efficiency obtained using this method is relatively higher when compared to the conventional solvent evaporation method. This method is used to fabricate micro-/nanoparticles with spherical morphology, but there are studies which show that by varying the oil to aqueous phase ratio micro-/nanoparticles with nonspherical morphology (like nanocapsules) can be fabricated [56, 60]. Overall, this method is a simple, efficient, and cost-effective method that can produce polyester particles with a narrow size distribution with minimal batch-to-batch variations.

4.2.1.2.3 *Emulsion reverse salting-out*

Being a derivative of the emulsion solvent diffusion method, the emulsion reverse salting out method has received significant attention for the fabrication of polyester-based particles with desired properties. The composition of the emulsion is the key factor which distinguishes it from the conventional emulsion methods. Briefly, in the emulsion salting-out method, the polyester is initially dissolved in an organic solvent like acetone, which is totally miscible with water, followed by emulsification in a salt-saturated solution containing magnesium chloride, or calcium chloride, or a non-electrolyte solution containing sucrose [61]. The presence of the salt in high concentrations in the emulsion prevents acetone from mixing with water, and thereby promotes the formation of small emulsion droplets dispersed in the saturated salt solution [62]. Moreover, the saturated salt solution retains water for its own solubilization, and thus changes the miscibility of water in the other solvent, i.e., acetone. Further, the generated microemulsion is diluted with an excess volume of water that eventually decreases the salt concentration, thereby inducing solvent diffusion by the reverse salting-out principle. Diffusion of solvent from the microemulsion droplets into the external continuous phase of salt solution triggers polymer precipitation and hence formation of micro-/nanoparticles. After the micro-/nanoparticles are formed, the salting-out agents (electrolytes/non-electrolytes) and the solvent are removed using the cross-flow filtration process. The choice of an appropriate salting-out agent is critical for fabrication of micro-/nanoparticles with controlled size, shape, and drug entrapment efficiency. Moreover, the procedure used in this method is compatible with sensitive biological molecules like proteins and does not cause much stress due to organic solvents. However, one of the major limitations associated with this method is that it is not suitable for encapsulation of hydrophilic molecules [63–65]. In addition to this, the cumbersome washing procedure required to remove salts and solvents from the particles can result in poor yield of micro-/nanoparticles. Finally, the key parameters that govern the properties of micro-/nanoparticles in this method are polymer concentration, intensity and duration of stirring, solvent type, and nature of the salting-out agents. [62, 66] In summary, this method is relatively less cost-effective and more challenging for large-scale production of particles.

4.2.1.3 Multistep fabrication

4.2.1.3.1 *Layer-by-layer fabrication*

Fabrication of particulate DDSs using layer-by-layer (LbL) technique for drug delivery applications is a relatively recent technique that involves the assembling of two interacting polymers onto a core material in a layered manner. The core material can either be sacrificial or nonsacrificial. A sacrificial core is removed before drug loading, whereas, a nonsacrificial core is permanent and remains as part of the DDS. The selection of the core material is based on the type and nature of the polymers being employed for LbL assembly. This technique was first described by Decher *et al.* for the fabrication of ultrathin multilayer films [67]. Later on, the technique was explored in the field of drug delivery by Caruso *et al.* for making particulate DDSs intended for spatial and temporal targeting applications [68]. Although this method is not employed exclusively for the preparation of polyester-based particulate systems, the underlying principle is used for the surface modification and/or surface functionalization of polyester particle-based DDSs. Since many polyesters such as PLGA and PLA are biocompatible, biodegradable, and provide sustained release of entrapped drugs, particulate DDSs fabricated using these polymers are used as templates (nonsacrificial core), which are further coated with interacting polymers and finally surface functionalized for targeted drug delivery applications. During this process, the core material (polyester particle) is coated with multiple layers of oppositely charged polymers (such as chitosan and alginate) through electrostatic interactions. Such LbL coatings of chitosan/alginate onto PLGA nanoparticles help in both reducing nonspecific protein interaction with PLGA nanoparticles (antifouling property) and easy attachment of active targeting moieties for targeted drug delivery applications [69]. The polymers deposited onto the polyester core material offer a wide range of chemical functionalities which can further be used for enabling crosslinking reactions such as click chemistry, carbodiimide chemistry, disulfide linkages, etc. LbL capsules have also been used for active targeting applications by conjugating antibodies or other active targeting molecules [70–72]. Due to their negative charge, preformed drug-loaded polyester-based PLA nanoparticles can be used for the coating of positively charged polymers in a layered manner for sustained drug delivery applications. Jiao *et al.* demonstrated this by using PLA-nanoparticles and poly(ethylene imine) (PEI) polymer-based LbL thin-film assemblies for sustained drug delivery applications [73]. Hence, the LbL technique is employed for the fabrication of targeted DDSs using preformed polyester-based particles as templates. However, this technique is not suitable for the large-scale production of particles due to the involvement of multiple steps.

4.2.1.3.2 *Rapid expansion of supercritical solution*

Synthesis of drug-loaded polymeric micro-/nanoparticles using the supercritical fluid technology (SFT) has received a lot of interest in the past three decades [74]. In SFT, particles are fabricated using supercritical fluids which are pure and nontoxic and hence are considered to be environmentally friendly [74, 75]. Supercritical fluids are substances that remain above their critical temperature and pressure where no distinct liquid or gaseous state exists. Supercritical fluids can diffuse through solids like gases or can dissolve compounds like liquid and hence can be used as a potential substitute for organic solvents during the synthesis of polyester micro-/nanoparticles. Supercritical fluids have liquid like density in supercritical phase and, therefore, are also known as "dense gases" [76]. Supercritical carbon dioxide ($scCO_2$) and supercritical water (scH_2O) are the most commonly preferred supercritical fluids (solvents).

There are various techniques available for the synthesis of polyester particles using SFT, including static supercritical fluid process (SSF), supercritical antisolvent process (SAS), particles from gas-saturated solutions (PGSS), gas antisolvent process (GAS), and depressurization of an expanded liquid organic solution (DELOS). However, rapid expansion of supercritical solution (RESS) has been the most popular for providing improved control, flexibility, and operational ease for the fabrication of polyester particles. The RESS technology was originally developed during modeling and analysis of flow pattern and nucleation processes of liquids [77, 78]. The technique is based on the principle of co-precipitation of a drug–polymer mixture using a supercritical nonsolvent (antisolvent) like $scCO_2$. During the process of particle synthesis the supercritical fluid (having drug–polyester mixture) is first saturated with a solid substrate followed by depressurization of solution through a heated nozzle into a low-pressure compartment where rapid nucleation of substrate results in formation of small particles. The produced particles are finally collected from a gaseous stream [79]. Therefore, this method is also known as supercritical nucleation [77]. The setup for the technique is designed with three distinct compartments: first, a high-pressure stainless steel cell, followed by a syringe pump with a fine heated nozzle and a third compartment as expansion unit, where scCO2 undergoes expansion, resulting in nucleation of the polyester mixture. The temperature, pressure drop, impact distance from nozzle to expansion unit, and nozzle geometry are the key parameters which govern the properties of the fabricated particles

[77, 80–82]. Generally, for small-scale production a single nozzle can be used; however, translation to a bulk production volume requires upgradation to a multi-nozzle system or porous sintered disk. The particles collected from the gaseous phase are directly ready to be used without any further purification or processing, which makes it advantageous for the synthesis of particles involving fragile/sensitive molecules (e.g., DNA or protein).

One of the potential limitations of RESS technology is the limited solubility of polyester-based polymeric systems and drug molecules in supercritical fluids. While the use of a co-solvent can help in improving solubility, it may not be feasible for many of the polymers. Moreover, aggregation of droplets during free jet expansion due to electrostatic interactions between the solution and needle can result in the alternation of the shape of particles from spherical to needle. RESS technology is also associated with the shortcoming of collecting particles directly from the gaseous phase. To a large extent, this limitation has been overcome by the recent development of a variant of RESS known as *rapid expansion of a supercritical solution into a liquid solvent* (RESOLV). In this technique particles can be directly collected into a liquid phase without any further purification or processing [83]. Therefore, the RESS method shows potential to be used for production of particles at the industrial scale at a relatively lower cost.

4.2.2 Advanced Methods for Particles Synthesis

4.2.2.1 Microfluidics

Emulsion-based conventional methods for the fabrication of micro-/nanoparticles, though robust, are associated with the limitation of heterogeneous nature of emulsion droplet. The heterogeneity in emulsion droplets results in the fabrication of non-uniform particles with varying physical properties (morphology, drug encapsulation efficiency, and non-uniform release profile) [2]. This shortcoming associated with conventional emulsion-based methods has largely been overcome with the development of microfluidics. Microfluidics is a relatively new technique for the fabrication of near monodisperse complex particulate DDSs, including polyester-based polymeric particles [84]. Microfluidics is an interdisciplinary approach for handling and manipulating fluids at a very small scale (nano- to picoliter volume) under confined space with the help of an ultrafine microfluidic device (microchannels or microcapillaries) [85]. This technology has been successfully used for fabrication of inkjet printheads, DNA chips, and lab-on-a-chip devices [86]. A microfluidic device represents the functional unit of a microfluidic system which enables rapid thermodynamic mixing of two or more solutions together (up to milliseconds) and thus provides a homogenous microenvironment for particle formation. The design of microfluidic devices can be categorized into three distinct groups: (i) flow focusing, (ii) T junction, and (iii) concentric capillaries [2] (Fig. 4.2). Microfluidic devices can be fabricated using glass, polydimethylsiloxane (PDMS), silicon, polyurethane, polycarbonate, or stainless steel using microfabrication techniques

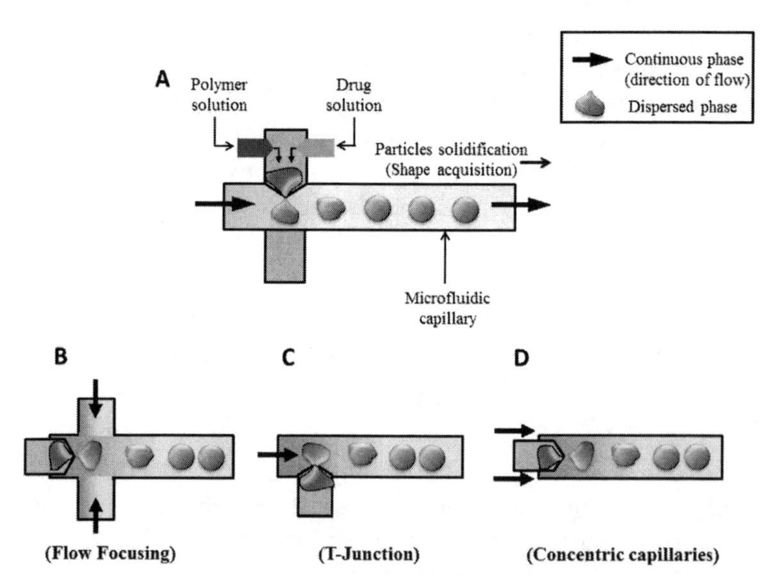

Figure 4.2 (A) Schematic showing the process of fabrication of polymeric particles by the *microfluidics technique*. Particles are formed inside the microfluidic capillary wherein the polymer–drug solution is mixed with continuous aqueous phase flowing under pressure. Schematic representation of the three most common variants of microfluidic devices are (B) flow focusing, (C) T junction, and (D) concentric capillary system. [Adapted from Zhang Y, Chan HF, Leong KW, *Adv. Drug Delivery Rev.*, **65**(1), 104–120 (2013).]

like micromilling, micromachining, and lithography. The synthesis of particles within a microfluidic device is facilitated by the competition between shear stress imposed by the flow of a continuous phase and the interfacial force [87]. Depending on the type of material used, emulsion droplets in the microfluidic device can be solidified using condensation, ionic crosslinking, radical polymerization, solvent evaporation, or diffusion method [85].

The processing and manipulation of a solution at a very small scale not only improves accuracy of reaction but also consumes significantly lesser amounts of reagents and energy. Moreover, microfluidics also offers flexibility for multistep reaction design, low-cost synthesis, portability for *in situ* usage, and efficiency in reaction, which makes it a versatile tool for online quality control [2, 88].

In the field of fabrication of particle-based DDSs, microfluidics has received significant attention in the past decade. Microfluidics has been widely explored for fabrication of particles with desired physical properties like size, shape, entrapment efficiency, and release properties of therapeutic molecules. This technique imparts fine control on size of particles in the range 10 nm to a few micrometers. In the micro-droplet-based microfluidics technique (Fig. 4.2) the droplet size is a function of flow rate of continuous phase and dispersed phase. Karnik *et al.* demonstrated the influence of flow rate of solvent stream on the size of particle. In their study, when a PLGA-PEG solution (50 mg/ml in acetonitrile [discontinuous phase]) was allowed to mix with a binary mixture of acetonitrile–water (continuous phase) at a flow rate of 0.5 and 10 μL/min for acetonitrile and water, respectively, the size of PLGA particles obtained was 10–50 nm. However, when the flow rate of acetonitrile was increased to 1.0 μL/min while keeping the flow rate for water constant (10 μL/min) (varying the mixing time of both phases by varying the mixing ratio of binary solvent), a reduction in the upper limit of particles size was observed (23–29 nm) [84, 88].

Microfluidics is not limited to the fabrication of polyester-based nanoparticles, particles in micrometer size range (microsphere/microparticles) have also been reported using PLA (6.0 to 50 μm) [89] and PLGA (up to 30 μm) [90, 91]. In addition to the regular spherical geometry of particles, microfluidics can further be tuned to fabricate particles with core–shell architecture [92] or Janus particles, which have the ability to encapsulate two drugs having distinct solubility in water and organic solvent [93]. Moreover, the entrapment efficiency of drugs can be improved to up to 90% or greater using the microfluidics-based technology (entrapment efficiency using conventional methods is ~50–90%) [94, 95]. While microfluidics offers multiple advancements over conventional methods, it is associated with some shortcomings too. The use of narrow bore microfluidics device (microcapillary) in microfluidics technique creates one of the major problems. This provides low throughput (a few milliliters per hour) and hence cannot produce more than a few grams of particles per hour, thus impeding its industrial application for bulk fabrication. This shortcoming can be overcome to some extent by using a multichannel microfluidics device. Another limitation associated with microfluidics is the difficulty of performing downstream operations, like extraction, washing, and purification of particles, at a small scale. Small dimensions of microfluidic devices also make them prone to blockage or clogging. Hence, microfluidics is a fascinating technique for fabrication of near monodispersed micro-/nanoparticles and shows potential to be used for bulk fabrication of polyester-based micro-/nanoparticles for drug delivery applications.

4.2.2.2 Particle replication in non-wetting template (PRINT)

Particle replication in non-wetting template (PRINT) is a technique that can be applied to a diverse range of materials for the fabrication of particles with desired size and shape. PRINT enables the fabrication of particles with relatively higher degree of monodispersity both in terms of size and shape. This technique was developed by DeSimone and coworkers in 2005 as a method for the fabrication of monodisperse shape-specific nanomaterials [96]. PRINT is a continuous, roll-to-roll, high-resolution technique inspired by microelectronics for the reproducible fabrication of monodisperse particles. Unlike other top-down approaches (milling, grinding, and microemulsions) for the fabrication of nanomaterials, PRINT technology has the ability to provide fine control on the size and shape of particles [97]. PRINT can be used for fabrication of micro-/nanoparticles with a wide range of diameters from 80 nm to 20 μm using a variety of materials, including PEG, proteins, and polyesters such as PLGA [98].

PRINT is fundamentally based on the principle of soft lithography, wherein a mold (also known as stamp) with defined patterns/features is developed and is used to transfer the pattern in the form of material on a substrate [99]. Briefly, a pre-polymer solution is poured over a substrate and molded with the help of a featured stamp (device patterned with wells/groves) by the application of a small amount of pressure. The polymer solution occupies spaces between the features and does not deposit in the form of a scum layer as produced in soft lithography. The low surface energy of the mold, which is crucial for the efficacy of PRINT, prevents the overflow of solution to non-cavity regions and eventually leads to fabrication of discrete particles. For polymers dissolved in volatile organic solvents, an additional step for phase transition or solvent evaporation is required, which induces the solidification of emulsion droplets inside cavities of the mold.

The non-wettability and resistance to swelling combined with low surface energy are the key properties of a mold used for PRINT. In contrast to the conventional PDMS polymer which undergoes swelling when exposed to nonpolar organic solvents (e.g., toluene, hexane), fluoropolymers like perfluorinated polyether (PFPE) have been developed as inert materials, which are non-wetting and exhibit excellent swelling resistance against organic solvents [96]. Moreover, PFPE enables the fabrication of particles with lesser than 20 nm features with high fidelity [100, 101]. In addition to this, PFPE also exhibits extremely low surface energy, low modulus, high gas permeability, and low toxicity.

There are a large number of techniques available for the fabrication (patterning) of micro-patterned molds. However, micro-contact printing (μCP), replica molding (REM), and micro-transfer molding are widely accepted techniques [102]. The designing of the micro-/nanopattern is created using Autodesk AutoCAD (computer aided design) tool, which is further printed using high-resolution printers (at least 50–60 dpi with dot size of 5 μm) in the form of a "mask." The mask represents the intermediate unit which is further used to generate a "master" (or solid replica) using photolithography or electron beam lithography (EBL) techniques. Finally, these masters are subjected to the casting of a liquid precursor in the form of complementary molds which is used as the final device for PRINT [102].

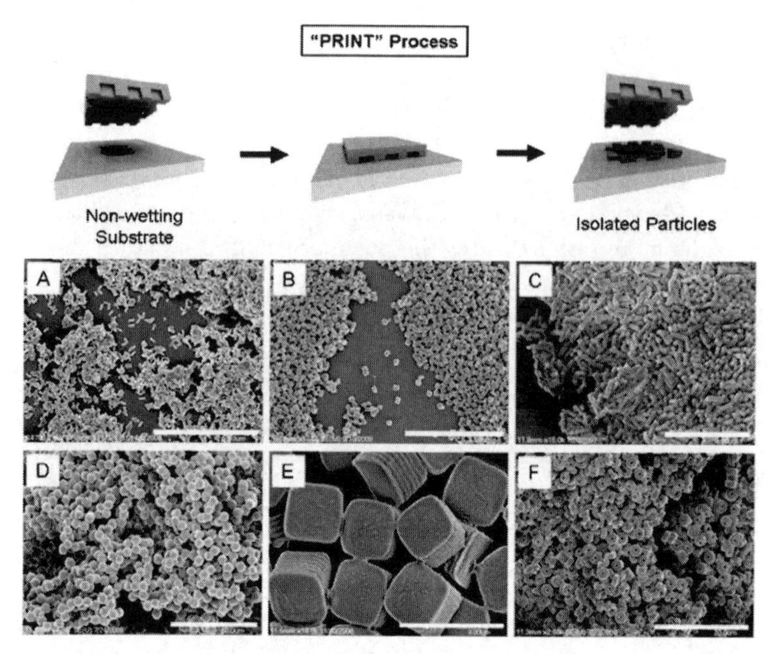

Figure 4.3 Schematic showing the fabrication of micro-/nanostructures using the particles replication in non-wetting template (PRINT) process (top). (Image reproduced with permission from Rolland J P, Maynor BW, Euliss L E, *et al.*, *J. Am. Chem. Soc.*, 2005, 127:10096–10100. Copyright 2005, American Chemical Society.) A–F are scanning electron micrographs of the PLGA micro-/nanoparticles of varying size and shapes (A, B and C are cylindrical particles with varying dimensions; D, roughly spherical; E, cubical with ridges; and F, particles with center fenestration) fabricated using the PRINT process. Scale bars: (A) 5 μm, (B) 4 μm, (C) 3 μm, (D) 10 μm, (E) 3 μm, and (F) 20 μm. (Image reproduced with permission from Enlow EM, Luft JC, Napier ME, DeSimone JM, *Nano Lett.*, **11**(2), 808–813 (2011). Copyright 2011, American Chemical Society.)

One of the remarkable applications of PRINT is generating particles with controlled shape wherein the particles with diverse shapes ranging from cylindrical, cubical, spherical, and rod shape have been reported [96, 98]. Gratton *et al.* intended to study the effect of particles design on cellular internalization pathway (endocytosis) using HeLa cells. They generated a library of PEG-based particles of different sizes (2–5 μm), shapes (spherical, cubic, and cylindrical) and surface charge (zeta potential from +34.8 to –33.7 mV) by employing the fascinating PRINT technology [103]. Similarly, in another study Warefta *et al.* demonstrated the potential of PRINT technology to fabricate lipid coated-PLGA particles encapsulating siRNA inside PLGA (32–46% encapsulation efficiency) [104]. For drug delivery applications the loading and entrapment efficiency of cargo in polymeric nanoparticles using PRINT is comparable to that obtained using microfluidics. Elizabeth *et al.* demonstrated a drug loading of up to 40% (w/w) for docetaxel and entrapment efficiency >90% inside PLGA particles fabricated using PRINT technology [105]. Overall, the PRINT technique can be an effective technique for the bulk production of polyester-based micro-/nanoparticles of specific shape.

4.3 Scalability for Large-Scale Production

Polyester-based polymeric particles are versatile and have been widely explored for therapeutic as well as diagnostic applications. A majority of polyester-based polymers are biodegradable and biocompatible, and hence used as DDSs. One of the major obstacles in the introduction of nanoparticle-based therapies (nanomedicine) into pharmaceutical markets or clinic is the bulk production of particles. Unlike microparticles, the bulk production of nanoparticles is relatively more challenging. A majority of the methods discussed in this chapter show promising results during production at bench scale; however, the properties of the polyester particles can be altered when produced at the pilot or large scale. The various formulation aspects for the preparation of polyester-based micro-/nanoparticles are briefly summarized in Table 4.1. Additionally, collection and purification of nanoparticles from suspensions requires large centrifugal force, which can be provided by ultracentrifugation, but it limits the usage of larger volumes and also increases the cost of production significantly. Furthermore, the ingredients of the formulation also have impact on its translation potential. For example, use of chloroform (toxic/anesthetic agent) for dissolving polymers while performing solvent evaporation method is acceptable at smaller scales (bench scale), but when translated to the bulk scale, it can result in enhanced exposure and, thus, toxicity. Therefore, when the acceptability of a particulate system (or method of preparation) by regulatory agencies (e.g., FDA) is considered, a formulation with residual toxic components or additives like organic solvents or surfactants is highly undesirable.

Generally, the following points have to be considered for large-scale production of any particle-based DDSs: (i) technical challenges for translation of the method from bench to large scale (e.g., multi-parameter process or long response time), (ii) toxicological profile of ingredients (e.g., organic solvent or surfactants) used, (iii) quality assurance of the formulation, (iv) production rate, (v) cost-effectiveness, and (vi) batch-to-batch variations [106]. Methods including electrospraying, HPH, emulsion solvent evaporation, RESS, microfluidics, and PRINT have been demonstrated as potential methods for bulk production of micro-/nanoparticles.

Among all the known technique used for the fabrication of polyester-based particulate DDSs, high-pressure homogenization (HPH) and microfluidics are amenable to large-scale production [107]. HPH is generally used in industries for processing of large quantities of emulsions like milk. Special homogenization setups have been designed which can homogenize feed emulsion volumes of up to 60 liters (e.g., LAB60 setup, capacity up to 60 l/h). This homogenizer operates at pressures between 100 and 700 bars at room temperature and has been used for the fabrication of solid lipid nanoparticles [107, 108]. However, the application of the HPH technique for production of polyester-based particles at a larger scale has not been widely acclaimed. In addition to this, polyester-based emulsions tend to get agglomerated when the polymer concentration is increased beyond a certain value. Therefore, emulsion-based methods are heterogeneous and are challenging for effective scale-up [109, 110].

Similar to HPH, microfluidics serves as another method that can be used to fabricate polyester-based micro-/nanoparticles at larger scales. Unlike, conventional microfluidics where a single jet droplet generator is used, for scale-up purposes a multi-jet (up to 128 jets) droplet generator has been reported to produce large amounts of nanoparticles per hour (up to 0.3 kg/h) [111]. Therefore, it can be concluded that only some of the discussed methods can be used for bulk production of polyester-based microparticles (e.g., HPH or emulsion-based methods) or micro-/nanoparticles (e.g., microfluidics or PRINT technology) at industrial scale.

4.4 Recovery and Purification Methods

After the fabrication of polyester particles, it is necessary to purify them by the removal of impurities such as adhered surfactants and residual organic solvents. The processes that are commonly employed for the recovery and purification of polyester particle dispersions are evaporation under reduced pressure, dialysis, filtration, and centrifugation [14]. These recovery and purification processes are not specific to polyester particles and are generic to all polymeric particulate DDSs.

4.4.1 Evaporation under Reduced Pressure

During the fabrication of polyester-based particles, wet dispersions are subjected to evaporation under vacuum to remove volatile organic solvents. The rigid particles formed are often associated with minute quantities of residual organic solvent, which is removed using evaporation under reduced pressure. This is because particle associated volatile organic impurities are difficult to remove completely during the process of preparation. Hence, the residual solvents are removed during the process of purification by evaporating the particle dispersions under vacuum. Generally, low-temperature conditions are preferred for purification because most polymers are susceptible to degradation at elevated temperatures. As the boiling point of a solvent depends on atmospheric pressure, application of high vacuum helps in reducing the boiling point of solvents and thereby facilitates removal of undesired solvents at low temperature. The extent of heating is determined by the thermal stability of the drug/polymer and the boiling point of solvents being removed during the process [112]. Overall, the process of evaporation under reduced pressure is widely employed during preparation and purification of particles by methods such as nanoprecipitation, emulsion solvent diffusion, and emulsion reverse salting-out.

4.4.2 Dialysis

In dialysis, separation and purification of particle dispersions can be achieved on the basis of diffusion of molecules through a semipermeable membrane having desired pore size (molecular weight cut-off range). The particulate dispersions that need to be purified are packed into a dialysis membrane with desired pore size and are then immersed in the dialyzing medium with continuous stirring. This leads to concentration gradient based diffusion of particle associated impurities into the dialyzing medium. The dialyzing medium is periodically replaced with a fresh medium until the desired level of purification is achieved [113]. While the process of dialysis is routinely used for the purification of particulate dispersions at the laboratory scale, it may not be desirable for large-scale separation and purification of particulate systems.

4.4.3 Filtration

The unit operation of filtration is used for the separation of solids from liquids, i.e., separation of particles from colloidal dispersions. In this process, particles are separated from their dispersions by passing the dispersion through a filter (membrane having pores of defined size range). This method has been reported to be rapid and efficient when compared to dialysis for the separation of particles [114]. However, caking is a common problem associated with this method that can arise due to fusion of particles. This limitation can be overcome by using other types of filtration such as cross-flow filtration, diafiltration, and gel filtration techniques [114, 115]. Owing to its simplicity and low cost, the process of filtration has been widely used for large-scale separation of particles from their dispersions.

4.4.4 Centrifugation

In centrifugation, particulate dispersions are subjected to high gravitational force in order to sediment particles. Upon sedimentation of particles, the supernatant-containing impurities can be easily removed by decantation, thereby enabling separation and purification of particles. This process can be repeated until the desired level of purification is achieved. Although centrifugation is commonly used for the separation and purification of particulate DDSs at laboratory scale, it is associated with the drawback of causing fusion of particles when high gravitational force is applied. Need of high centrifugal force combined with the difficulty of handling large volumes limits the use of centrifugation for industrial-scale separation of nanoparticles.

4.5 Drying and Storage of Particles

4.5.1 Drying of Particles

Since majority of the polyester-based delivery systems discussed in above sections were prepared in aqueous solvents, they must be processed to completely remove bound and unbound water as polyesters are susceptible to hydrolysis in the presence of water vapor/aqueous solvents. Hence, the particle dispersions obtained after purification are subjected to drying in order to obtain dry particles (in powder form). The most commonly used techniques for the drying of particulate delivery systems are freeze drying and spray drying.

4.5.1.1 Freeze drying

Freeze drying, also known as lyophilization, is a commonly used technique in pharmaceutical industry to remove bound and unbound water from thermolabile therapeutic agents. Since drug loaded polyester-based particles are susceptible to degradation at higher temperatures, removal of water at subzero temperatures is performed using lyophilization. During lyophilization, vacuum is applied to the system in order to reach the triple point of water where sublimation occurs. At this point, water (in the form of ice) associated with particles sublimes to yield dried particles. This is one of the most widely used methods for the drying of particulate delivery systems. However, particles are subjected to mechanical stress during the process due to crystallization of ice. This can lead to agglomeration, which eventually affects the polydispersity of the formulation. This problem can be partially overcome by optimizing lyophilization conditions such as extent of applied vacuum and duration of lyophilization. Another commonly used approach to overcome this problem is to utilize cryoprotectants (lyoprotectants).

Polyhydroxyl compounds such as trehalose, mannose, sucrose, glucose, lactose, maltose and mannitol (sugars), PEG, or glycerol are commonly used cryoprotectants. These compounds improve resistance of the particles to stresses during freezing by formation of hydrogen bonds between cryoprotectant and the polar groups of the nanoparticles [116, 117]. The ratio of nanoparticles to cryoprotectant concentration plays an important role in stabilizing nanoparticles during freeze drying as reported by Jaeghere *et al.* and Saez *et al.* Jaeghere *et al.* studied the effect of different concentrations of trehalose on the stability of PLA nanoparticles during lyophilization and reported that the optimum ratio of PLA to trehalose for the protection of PLA nanoparticles was 1:1 [118]. Saez *et al.* conducted an experiment with different concentrations of glucose, sucrose, and trehalose to preserve the properties of PLGA and PCL nanoparticles during lyophilization and observed that 20% sucrose and 20% glucose had an acceptable cryoprotective effect on PLGA and PCL nanoparticles, respectively [116, 119]. PVA, which is used as a surfactant during nanoparticle preparation, can also act as a cryoprotectant while maintaining the size of nanoparticles during lyophilization. These studies indicate that concentration of cryoprotectants used during lyophilization governs particle stability. Hence, a compatible cryoprotectant (poly hydroxyl compounds) with appropriate concentration should be selected for lyophilization of polyester-based particulate DDSs.

4.5.1.2 Spray drying

Spray drying is a widely used technique for the drying of particle suspensions. During spray drying, an atomizer sprays particle dispersions as small droplets into a preheated chamber where the solvent is evaporated to result in the formation of dry free-flowing particles. This method is capable of drying a large volume of suspensions and hence is commonly used in the industrial setting. The boiling point of the solvent determines the temperature to be used during the spray-drying process. Hence, suspensions prepared in low boiling point solvents can be spray dried at low temperatures. However, this method is not suitable for thermolabile substances and high boiling point solvents.

4.5.2 Storage of particles

The storage conditions for polyester-based particulate delivery systems are specified based on the stability of polymers and cargo under environmental conditions such as light, temperature, humidity, etc. Storage conditions depend not only on the properties of the polymer but also on the nature of the drug being entrapped/encapsulated. In order to preserve particle size, shape (morphology), stability, and therapeutic effect, appropriate storage conditions should be selected. It has been reported that exposure to light adversely affects the structure of polyesters by causing breakage of ester bonds [120]. Such photo-induced degradation can lead to alteration in drug release kinetics, which ultimately affects the therapeutic effect [121, 122]. Hence, amber-colored containers are recommended for storage of polyester-based particulate DDSs. In addition to light, temperature is also known to play an important role in maintaining the physical stability of polyester-based particles during storage. The T_g of most polyesters is above physiological temperature, which indicates their glassy nature at 37°C. Hence, they should be stored below 37°C to preserve their physical state (glassy state) [122]. In addition to light and temperature, humidity of the storage environment also affects stability of polyesters as they are susceptible to hydrolysis in the presence of water vapor/aqueous solvents.

4.6 Challenges and Future Prospects

In spite of having a number of methods for fabrication of polyester-based micro-/nanoparticles, only a few of them can be considered for bulk fabrication. This is because the success of a particle-based pharmaceutical product depends on the uniformity by which the formulation with desired properties can be prepared. The different methods available for fabrication of polyester-based micro-/nanoparticles vary in terms of reproducibility, applicability to a class of polymers, type of cargo which can be entrapped, physical properties of particles (e.g., size and shape), drug loading/entrapment efficiency, and finally feasibility for bulk production (Table 4.1). With some of the existing methods (such as electrospraying, solvent evaporation-based methods, HPH, and RESS), it is feasible to fabricate particles in the micrometer size range; however, the properties of the particles may get altered when the same method is used for fabrication of particles in the nanometer size range. This signifies that the methods for fabrication of micro-/nanoparticles have to be optimized for individual application. Additionally, in order to maintain the final yield of particles, collection and purification of nanoparticles from suspension needs to be performed under high centrifugal force using ultracentrifugation, which increases the cost of production significantly. Hence, it remains a challenge to develop a single method which can be used for multiple applications. Therefore, the fabrication method used for every polyester-based particulate DDS needs to be optimized for each specific application.

Microfluidics has emerged as a fascinating technique for the production of polyester-based nanoparticles as small as 10 nm with high degree of monodispersity. However, improper mixing of polymer solution in microcapillary or fluctuation in flow rates may result in altered properties of particles. Recently, Guzowski *et al.* have reported the use of automatic valves (automated active droplet producing system) to provide a fine control on the flow rate in order to produce monodispersed nanoparticles [123]. Another challenge in microfluidics which is associated with the sorting of particles after fabrication has been overcome by combining microfluidics with other techniques like sedimentation [124] and hydrodynamic filtration [125]. The microfluidics technique is amenable to further improvements by implementing enhanced automation programing, feasibility for scale-up production, and incorporation of coupled co-screening on the chip.

After microfluidics, PRINT technology offers advantage for controlled synthesis of polyester-based micro-/nanoparticles. The advantage of PRINT to fabricate micro-/nanoparticles with desired shape distinguishes it from the rest of the techniques. However, unlike the microemulsion-based methods, it is challenging to encapsulate amphiphilic drugs at higher entrapment efficiency for polyester-based materials using the PRINT technology. Finally, it is surmised that the method for fabrication of polyester-based micro-/nanoparticles is crucial for the success of a polyester-based DDSs.

Table 4.1 Comparison between various methods used for fabrication of polyester-based micro-/nanoparticles

SN	Method	Size range and shape	Advantage(s)	Potential limitation(s)
1	Nanoprecipitation	~60 nm–1 μm; Spherical	Reproducible method More suitable for encapsulation of hydrophobic cargo	Miscibility of organic and aqueous solvents is a prerequisite for this method, which limits the use of wide varieties of polyesters The temperature used for evaporation should be lower than the glass transition temperature (T_g) and the melting temperature (T_m) of the polymer Entrapment efficiency of hydrophilic drugs is low Not amenable to large-scale production
2	Electrospraying	~500 nm–10 μm; Spherical and non-spherical shape (doughnut)	Reproducible method Both hydrophilic and hydrophobic cargo can be encapsulated Amenable to large-scale production	Demands low boiling point solvents Change in ambient conditions (e.g., temperature and humidity) can alter the results significantly
3	High-pressure homogenization (HPH)	~65 nm–550 nm; Spherical	Reproducible method More suitable for encapsulation of hydrophobic cargo Amenable to large-scale production	Poor entrapment efficiency of hydrophilic cargo

(*Continued*)

Table 4.1 *(Continued)*

SN	Method	Size range and shape	Advantage(s)	Potential limitation(s)
4	Emulsion solvent evaporation method	~150 nm–100 μm; Spherical	Reproducible method Both hydrophilic and hydrophobic cargo can be encapsulated Amenable to large-scale production	Relatively lesser entrapment efficiency of hydrophilic cargo
5	Emulsion solvent diffusion method	~150 nm to few microns; Spherical	Reproducible method Both hydrophilic and hydrophobic cargo can be encapsulated	Relatively poor entrapment efficiency of hydrophilic cargo, as the cargo may diffuse into external aqueous phase Not amenable to large-scale production
6	Emulsion reverse salting-out	~100 nm–1000 nm; Spherical	Largely used for entrapment of hydrophobic cargos, but hydrophilic cargo can be loaded with poor encapsulation efficiency	Relatively less reproducibility Less suitable for entrapment of hydrophilic cargo Removal of salts and solvents from nanoparticles is tedious and can result in poor yield Not amenable to large-scale production
7	Layer-by-layer fabrication	~500 nm–100 μm; Spherical and non-spherical (depending on the shape of template)	Reproducible method Both hydrophilic and hydrophobic cargo can be encapsulated Can be used for surface modification of particles based on the nature of coating material	Not amenable to large-scale production

SN	Method	Size range and shape	Advantage(s)	Potential limitation(s)
8	Rapid expansion of supercritical solution (RESS)	~10 nm–50 μm; Spherical and non-spherical (needle)	Reproducible method Both hydrophilic and hydrophobic cargo can be encapsulated Amenable to large-scale production	Limited solubility of polyester-based polymeric systems and drug molecules in supercritical fluids Difficulty with collection of particles directly from gaseous stream
9	Microfluidics	~10 nm–50μm; Spherical (includes core–shell)	High reproducibility Both hydrophilic and hydrophobic cargo can be encapsulated Amenable to large-scale production	Clogging of microfluidic capillary with high polymer concentration Difficulty with downstream operations
10	Particle replication in non-wetting template (PRINT)	~80 nm–20 μm; High control on shape (cylindrical, cubical, spherical, and rod shape)	High reproducibility Both hydrophilic and hydrophobic cargo can be encapsulated Amenable to large-scale production	Wettability of template may cause formation of scum layer and non-uniform shape of particles

Although a large number of methods have been developed, only a few of them meet the many requirements of application as DDSs. Developing a versatile and robust method for the fabrication of micro-/nanoparticles is still a challenge and has been continuously attempted by researchers either by proposing a new method or by incorporating significant modification to existing ones. Similarly, design/modification of new polyester-based polymers using greener approaches is another important area of research in polymer chemistry. Together with the development of new polyesters, the development of new methods may bring new perspectives in the area of fabrication of next-generation DDSs.

4.7 Summary

Polyester-based DDSs are versatile and known to play a pivotal role in healthcare and medicine. Similar to the liposome-based DDSs which are FDA approved for certain application [126–129], polyester-based DDSs (e.g., PLGA) too have been FDA approved for specific applications and serve as a potential alternative to lipids owing to their improved stability under *in vivo* conditions [8, 12]. Currently, there are multiple polyester particle-based products available in the market such as Zoladex®, Lupron Depot®, Atridox®, Nutropin Depot®, Ozurdex®, and Eligard® [2, 10]. Polyester is a class of polymers where most of the existing methods can be used for the fabrication of particles with different sizes and shapes. The methods discussed in this chapter enable fabrication of polyester-based particles with desired properties (size, morphology, monodispersity, drug encapsulation efficiency, and percentage yield) to facilitate their use in a specific application (passive or active targeting). The choice of an appropriate method depends on the end application of the particle system. For example, some of the methods provide good entrapment efficiency, whereas they may not be effective in controlling the dispersity or final yield of particles (e.g., some of the emulsion-based methods). Similarly, electrospraying is a simple and facile method for fabrication of drug-loaded particles in one step; however, the application of electric field can impact the structure of sensitive bioactive molecules used and may at times result in loss of function. On the other hand, the microfluidics technique imparts good control on dispersity (almost monodisperse particles), entrapment efficiently, particle yield, and collection. However, microfluidics may not be as amenable to large-scale production as some of the other techniques (Table 4.1).

Translational potential is yet another factor that has to be given due consideration while selecting a method for fabrication of particles. Most of the methods show promising results when conducted at bench scale (microemulsion-based methods, nanoprecipitation, HPH, and RESS). However, the properties of the particles are not consistent at a larger scale. Presence of any critical step like requirement of ultracentrifugation for purification of nanoparticles (microemulsion-based methods, nanoprecipitation, and RESS) or need to use saturated salt concentration (use of $CaCl_2$ or $MgCl_2$ in reverse salting-out method) is acceptable at bench scale, but may increase the cost of production significantly at industrial scale.

Interestingly, most of the existing methods for fabrication of micro-/nanoparticles are applicable to polyester-based polymers. In addition to this, polyesters-based polymers offer the flexibility to encapsulate a wide range of cargo/drugs with diverse physicochemical properties. Finally, innovations in the designing, novel modification of polyester-based polymers, and the method for fabrication of micro-/nanoparticles are continuously fueling the development of newer DDSs.

Acknowledgements

NA and DASR acknowledge University Grants Commission Council for Scientific and Industrial Research (UGC-CSIR) and Ministry of Human Resource Development (MHRD), India, respectively, for their research fellowship. DSK acknowledges IIT Kanpur, Department of Biotechnology (DBT) India, Department of Science and Technology (DST) India, DST-Nanomission India, and Science and Engineering Research Board (SERB) India for the research support received.

References

1. Xie, J., Lee, S., and Chen, X. (2010). Nanoparticle-based theranostic agents. *Adv Drug Deliv Rev*, **62**, 1064–1079.

2. Zhang, Y., Chan, H. F., and Leong, K. W. (2013). Advanced materials and processing for drug delivery: the past and the future. *Adv Drug Deliv Rev,* **65**, 104–120.
3. Patrick, P. D., Rahul, C. M., Angie, G. H., and Thanoo, B. C. (1993). In *Polymeric Delivery Systems,* Vol. **520**, 53–79 (American Chemical Society).
4. Ulery, B. D., Nair, L. S., and Laurencin, C. T. (2011). Biomedical applications of biodegradable polymers. *J Polym Sci B Polym Phys,* **49**, 832–864.
5. Anderson, J. M. and Shive, M. S. (1997). Biodegradation and biocompatibility of PLA and PLGA microspheres. *Adv Drug Deliv Rev,* **28**, 5–24.
6. Jameela, S. R. *et al.* (1996). Poly(ε-caprolactone) microspheres as a vaccine carrier. *Curr Sci,* **70**, 669–671.
7. Treiser, M. Abramson, S., Langer, R., and Kohn, J. (2013). In *Biomaterials Science* (3rd Edition). (eds. Ratner, B. D., Hoffman, A. S., Schoen, F. J., and Lemons, J. E.) 179–195 (Academic Press).
8. Jain, R., Shah, N. H., Malick, A. W., and Rhodes, C. T. (1998). Controlled drug delivery by biodegradable poly(ester) devices: different preparative approaches. *Drug Dev Ind Pharm,* **24**, 703–727.
9. Allen, T. M. and Cullis, P. R. (2004). Drug delivery systems: entering the mainstream. *Science,* **303**, 1818–1822.
10. US Food and Drug Administration Website. (http://www.accessdata.fda.gov/scripts/cder/drugsatfda/).
11. Roy, D., Cambre, J. N., and Sumerlin, B. S. (2010). Future perspectives and recent advances in stimuli-responsive materials. *Prog Polym Sci,* **35**, 278–301.
12. Jain, R. A. (2000). The manufacturing techniques of various drug loaded biodegradable poly(lactide-*co*-glycolide) (PLGA) devices. *Biomaterials,* **21**, 2475–2490.
13. Rao, J. P. and Geckeler, K. E. (2011). Polymer nanoparticles: preparation techniques and size-control parameters. *Prog Polym Sci,* **36**, 887–913.
14. Vauthier, C. and Bouchemal, K. (2009). Methods for the preparation and manufacture of polymeric nanoparticles. *Pharm Res,* **26**, 1025–1058.
15. Chakraborty, S., Liao, I. C., Adler, A., and Leong, K. W. (2009). Electrohydrodynamics: a facile technique to fabricate drug delivery systems. *Adv Drug Deliv Rev,* **61**, 1043–1054.

16. Stainmesse, S., Orecchioni, A.-M., Nakache, E., Puisieux, F., and Fessi, H. (1995). Formation and stabilization of a biodegradable polymeric colloidal suspension of nanoparticles. *Colloid Polym Sci,* **273**, 505–511.

17. Fessi, H., Puisieux, F., Devissaguet, J. P., Ammoury, N., and Benita, S. (1989). Nanocapsule formation by interfacial polymer deposition following solvent displacement. *Int J Pharm,* **55**, R1–R4.

18. Chorny, M., Fishbein, I., Danenberg, H. D., and Golomb, G. (2002). Lipophilic drug loaded nanospheres prepared by nanoprecipitation: effect of formulation variables on size, drug recovery and release kinetics. *J Control Release,* **83**, 389–400.

19. Govender, T., Stolnik, S., Garnett, M. C., Illum, L., and Davis, S. S. (1999). PLGA nanoparticles prepared by nanoprecipitation: drug loading and release studies of a water soluble drug. *J Control Release,* **57**, 171–185.

20. Fenn, J., Mann, M., Meng, C., Wong, S., and Whitehouse, C. (1989). Electrospray ionization for mass spectrometry of large biomolecules. *Science,* **246**, 64–71.

21. Doshi, J. and Reneker, D. H. (1995). Electrospinning process and applications of electrospun fibers. *J Electrostat,* **35**, 151–160.

22. Drozin, V. G. (1995). The electrical dispersion of liquids as aerosols. *J Colloid Sci,* **10**, 158–164.

23. Vonnegut, B. and Neubauer, R. L. (1952). Production of monodisperse liquid particles by electrical atomization. *J Colloid Sci,* **7**, 616–622.

24. Arya, N., Chakraborty, S., Dube, N., and Katti, D. S. (2009). Electrospraying: a facile technique for synthesis of chitosan-based micro/nanospheres for drug delivery applications. *J Biomed Mater Res B,* **88**, 17–31.

25. Taylor, G. (1969). Electrically driven jets. *Proc R Soc A,* **313**, 453–475.

26. Wang, C. *et al.* (2009). Correlation between processing parameters and microstructure of electrospun poly (D,L-lactic acid) nanofibers. *Polymer,* **50**, 6100–6110.

27. Almería, B., Deng, W., Fahmy, T. M., and Gomez, A. (2010). Controlling the morphology of electrospray-generated PLGA microparticles for drug delivery. *J Colloid Interface Sci,* **343**, 125–133.

28. Park, C. H. and Lee, J. (2009). Electrosprayed polymer particles: effect of the solvent properties. *J Appl Polym Sci,* **114**, 430–437.

29. Casper, C. L., Stephens, J. S., Tassi, N. G., Chase, D. B., and Rabolt, J. F. (2004). Controlling surface morphology of electrospun polystyrene fibers: effect of humidity and molecular weight in the electrospinning process. *Macromolecules,* **37**, 573–578.

30. De Vrieze, S. *et al.* (2009). The effect of temperature and humidity on electrospinning. *J Mater Sci,* **44**, 1357–1362.
31. Giller, C. B., Chase, D. B., Rabolt, J. F., and Snively, C. M. (2010). Effect of solvent evaporation rate on the crystalline state of electrospun nylon 6. *Polymer,* **51**, 4225–4230.
32. Koski, A., Yim, K., and Shivkumar, S. (2004). Effect of molecular weight on fibrous PVA produced by electrospinning. *Mater Lett,* **58**, 493–497.
33. Wannatong, L., Sirivat, A., and Supaphol, P. (2004). Effects of solvents on electrospun polymeric fibers: preliminary study on polystyrene. *Polym Int,* **53**, 1851–1859.
34. Deitzel, J., Kleinmeyer, J., Harris, D., and Beck Tan, N. (2001). The effect of processing variables on the morphology of electrospun nanofibers and textiles. *Polymer,* **42**, 261–272.
35. Jain, A. K., Sood, V., Bora, M., Vasita, R., and Katti, D. S. (2014). Electrosprayed inulin microparticles for microbiota triggered targeting of colon. *Carbohydr Polym,* **112**, 225–234.
36. Seth, A. and Katti, D. S. (2012). A one-step electrospray-based technique for modulating morphology and surface properties of poly(lactide-*co*glycolide) microparticles using Pluronics®. *Int J Nanomed,* **7**, 5129–5136.
37. Mahaling, B. and Katti, D. S. (2014). Fabrication of micro-structures of poly [(R)-3-hydroxybutyric acid] by electro-spraying/-spinning: understanding the influence of polymer concentration and solvent type. *J Mater Sci,* **49**, 4246–4260.
38. Bohr, A., Kristensen, J., Dyas, M., Edirisinghe, M., and Stride, E. (2012). Release profile and characteristics of electrosprayed particles for oral delivery of a practically insoluble drug. *J R Soc Interface,* **9**, 2437–2449.
39. Xu, Y. and Hanna, M. A. (2006). Electrospray encapsulation of water-soluble protein with polylactide: effects of formulations on morphology, encapsulation efficiency and release profile of particles. *Int J Pharm,* **320**, 30–36.
40. Xie, J. and Wang, C. H. (2007). Encapsulation of proteins in biodegradable polymeric microparticles using electrospray in the Taylor cone-jet mode. *Biotechnol Bioeng,* **97**, 1278–1290.
41. Almería, B., Fahmy, T. M., and Gomez, A. (2011). A multiplexed electrospray process for single-step synthesis of stabilized polymer particles for drug delivery. *J Control Release,* **154**, 203–210.
42. Lamprecht, A. *et al.* (1999). Biodegradable monodispersed nanoparticles prepared by pressure homogenization-emulsification. *Int J Pharm,* **184**, 97–105.

43. Vanderhoff, J. W., El-Aasser, M. S., and Ugelstad, J. (1979). Polymer emulsification process. US Patent 4 177177.

44. Anton, N., Benoit, J. P., and Saulnier, P. (2008). Design and production of nanoparticles formulated from nano-emulsion templates: a review. *J Control Release,* **128**, 185–199.

45. Vauthier, C., Fattal, E., and Labarre, D. (2004). From polymer chemistry and physicochemistry to nanoparticulate drug carrier design and applications. In *Biomaterial Handbook-Advanced Applications of Basic Sciences and Bioengineering,* 563–598 (Marcel Dekker).

46. Allemann, E., Gurny, R., and Doelker, E. (1993). Drug-loaded nanoparticles-preparation methods and drug targeting issues. *Eur J Pharm Biopharm,* **39**, 173–191.

47. Desgouilles, S. *et al.* (2003). The design of nanoparticles obtained by solvent evaporation: a comprehensive study. *Langmuir,* **19**, 9504–9510.

48. Bilati, U., Allémann, E., and Doelker, E. (2003). Sonication parameters for the preparation of biodegradable nanocapsules of controlled size by the double emulsion method. *Pharm Dev Technol,* **8**, 1–9.

49. Mundargi, R. C., Babu, V. R. Rangaswamy, V., Patel, P., and Aminabhavi, T. M. (2008). Nano/micro technologies for delivering macromolecular therapeutics using poly(D,L-lactide-*co*-glycolide) and its derivatives. *J Control Release,* **125**, 193–209.

50. Gref, R. *et al.* (1994). Biodegradable long-circulating polymeric nanospheres. *Science,* **263**, 1600–1603.

51. Bazile, D. *et al.* (1995). Stealth Me.PEG-PLA nanoparticles avoid uptake by the mononuclear phagocytes system. *J Pharm Sci,* **84**, 493–498.

52. Avgoustakis, K. (2004). Pegylated poly(lactide) and poly(lactide-*co*-glycolide) nanoparticles: preparation, properties and possible applications in drug delivery. *Curr Drug Deliv,* **1**, 321–333.

53. Feczko, T., Toth, J., Dósa, G., and Gyenis, J. (2011). Optimization of protein encapsulation in PLGA nanoparticles. *Chem Eng Process,* **50**, 757–765.

54. Quintanar-Guerrero, D., Allémann, E., Doelker, E., and Fessi, H. (1997). A mechanistic study of the formation of polymer nanoparticles by the emulsification-diffusion technique. *Colloid Polym Sci,* **275**, 640–647.

55. Leroux, J. C., Allemann, E., Doelker, E., and Gurny, R. (1995). New approach for the preparation of nanoparticles by an emulsification-diffusion method. *Eur J Pharm Biopharm,* **41**, 14–18.

56. Moinard-Chécot, D., Chevalier, Y., Briançon, S., Beney, L., and Fessi, H. (2008). Mechanism of nanocapsules formation by the emulsion-diffusion process. *J Colloid Interface Sci,* **317**, 458–468.

57. Quintanar-Guerrero, D., Fessi, H., Allémann, E., and Doelker, E. (1996). Influence of stabilizing agents and preparative variables on the formation of poly(D,L-lactic acid) nanoparticles by an emulsification-diffusion technique. *Int J Pharm,* **143**, 133–141.

58. Quintanar-Guerrero, D., Ganem-Quintanar, A., Allémann, E., Fessi, H., and Doelker, E. (1998). Influence of the stabilizer coating layer on the purification and freeze-drying of poly(D,L-lactic acid) nanoparticles prepared by an emulsion-diffusion technique. *J Microencapsulation,* **15**, 107–119.

59. Quintanar-Guerrero, D., Tamayo-Esquivel, D., Ganem-Quintanar, A., Allémann, E., and Doelker, E. (2005). Adaptation and optimization of the emulsification-diffusion technique to prepare lipidic nanospheres. *Eur J Pharm Sci,* **26**, 211–218.

60. Quintanar-Guerrero, D., Allémann, E., Doelker, E., and Fessi, H. (1998). Preparation and characterization of nanocapsules from preformed polymers by a new process based on emulsification-diffusion technique. *Pharm Res,* **15**, 1056–1062.

61. Ibrahim, H., Bindschaedler, C., Doelker, E., Buri, P., and Gurny, R. (1992). Aqueous nanodispersions prepared by a salting-out process. *Int J Pharm,* **87**, 239–246.

62. Allemann, E., Gurny, R., and Doelker, E. (1992). Preparation of aqueous polymeric nanodispersions by a reversible salting-out process: influence of process parameters on particle size. *Int J Pharm,* **87**, 247–253.

63. Allemann, E., Leroux, J. C., Gurny, R., and Doelker, E. (1993). *In vitro* extendedrelease properties of drug-loaded poly(D,L-lactic acid) nanoparticles produced by a salting-out procedure. *Pharm Res,* **10**, 1732–1737.

64. Allemann, E., Doelker, E., and Gurny, R. (1993). Drug loaded poly(lactic acid) nanoparticles produced by a reversible salting-out process: purification of an injectable dosage form. *Eur J Pharm Biopharm,* **39**, 13–18.

65. Zhang, Z., Grijpma, D. W., and Feijen, J. (2006). Poly(trimethylene carbonate) and monomethoxy poly(ethylene glycol)-block-poly(trimethylene carbonate) nanoparticles for the controlled release of dexamethasone. *J Control Release,* **111**, 263–270.

66. Konan, Y. N., Gurny, R., and Allémann, E. (2002). Preparation and characterization of sterile and freeze-dried sub-200 nm nanoparticles. *Int J Pharm,* **233**, 239–252.

67. Decher, G., Hong, J., and Schmitt, J. (1992). Buildup of ultrathin multilayer films by a self-assembly process: III. Consecutively alternating adsorption of anionic and cationic polyelectrolytes on charged surfaces. *Thin Solid Films,* **210**, 831–835.

68. Ariga, K., Lvov, Y. M., Kawakami, K., Ji, Q., and Hill, J. P. (2011). Layer-by-layer self-assembled shells for drug delivery. *Adv Drug Deliv Rev,* **63**, 762–771.

69. Zhou, J. *et al.* (2010). Layer by layer chitosan/alginate coatings on poly (lactide*co*- glycolide) nanoparticles for antifouling protection and folic acid binding to achieve selective cell targeting. *J Colloid Interface Sci,* **345**, 241–247.

70. Cortez, C. *et al.* (2006). Targeting and uptake of multilayered particles to colorectal cancer cells. *Adv Mater,* **18**, 1998–2003.

71. Johnston, A. P. *et al.* (2012). Targeting cancer cells: controlling the binding and internalization of antibody-functionalized capsules. *ACS Nano,* **6**(8), 6667–6674.

72. Sivakumar, S. *et al.* (2009). Degradable, surfactant-free, monodisperse polymer-encapsulated emulsions as anticancer drug carriers. *Adv Mater,* **21**, 1820–1824.

73. Jiao, Y. H. *et al.* (2010). Layer-by-layer assembly of poly(lactic acid) nanoparticles: a facile way to fabricate films for model drug delivery. *Langmuir,* **26**, 8270–8273.

74. Byrappa, K., Ohara, S., and Adschiri, T. (2008). Nanoparticles synthesis using supercritical fluid technology - towards biomedical applications. *Adv Drug Deliv Rev,* **60**, 299–327.

75. Deshpande, P. B. *et al.* (2011). Supercritical fluid technology: concepts and pharmaceutical applications. *PDA J Pharm Sci Technol,* **65**, 333–344.

76. Nalawade, S. P., Picchioni, F., and Janssen, L. P. B. M. (2006). Supercritical carbon dioxide as a green solvent for processing polymer melts: processing aspects and applications. *Prog Polym Sci,* **31**, 19–43.

77. Reverchon, E. and Adami, R. (2006). Nanomaterials and supercritical fluids. *J Supercrit Fluids,* **37**, 1–22.

78. Leduc, P. R. *et al.* (2007). Towards an *in vivo* biologically inspired nanofactory. *Nat Nanotechnol,* **2**, 3–7.

79. Weber, M. and Thies, M. C. (2002). Understanding the RESS process, in *Supercritical Fluid Technology in Materials Science and Engineering: Syntheses, Properties, and Applications,* 387–427.

80. Jung, J. and Perrut, M. (2001). Particle design using supercritical fluids: Literature and patent survey. *J Supercrit Fluids,* **20**, 179–219.

81. Reverchon, E. and Pallado, P. (1996). Hydrodynamic modeling of the RESS process. *J Supercrit Fluids,* **9**, 216–221.

82. Date, A. A. and Patravale, V. B. (2004). Current strategies for engineering drug nanoparticles. *Curr Opin Colloid Interface Sci,* **9**, 222–235.

83. Sun, Y. P. and Rollins, H. W. (1998). Preparation of polymer-protected semiconductor nanoparticles through the rapid expansion of supercritical fluid solution. *Chem Phys Lett,* **288**, 585–588.

84. Karnik, R. *et al.* (2008). Microfluidic platform for controlled synthesis of polymeric nanoparticles. *Nano Lett,* **8**, 2906–2912.

85. Khan, I. U., Serra, C. A., Anton, N., and Vandamme, T. (2013). Continuous-flow encapsulation of ketoprofen in copolymer microbeads via co-axial microfluidic device: influence of operating and material parameters on drug carrier properties. *Int J Pharm,* **441**, 809–817.

86. Dittrich, P. S. and Manz, A. (2006). Lab-on-a-chip: microfluidics in drug discovery. *Nat Rev Drug Discovery,* **5**, 210–218.

87. Dendukuri, D. and Doyle, P. S. (2009). The synthesis and assembly of polymeric microparticles using microfluidics. *Adv Mater,* **21**, 4071–4086.

88. Wang, J. T., Wang, J., and Han, J. J. (2011). Fabrication of advanced particles and particle-based materials assisted by droplet-based microfluidics. *Small,* **7**, 1728–1754.

89. Watanabe, T., Ono, T., and Kimura, Y. (2011). Continuous fabrication of monodisperse polylactide microspheres by droplet-to-particle technology using microfluidic emulsification and emulsion-solvent diffusion. *Soft Matter,* **7**, 9894–9897.

90. Hung, L. H., Teh, S. Y., Jester, J., and Lee, A. P. (2010). PLGA micro/nanosphere synthesis by droplet microfluidic solvent evaporation and extraction approaches. *Lab Chip,* **10**, 1820–1825.

91. Xu, Q. *et al.* (2009). Preparation of monodisperse biodegradable polymer microparticles using a microfluidic flow-focusing device for controlled drug delivery. *Small,* **5**, 1575–1581.

92. Lensen, D., Van Breukelen, K., Vriezema, D. M., and Van Hest, J. C. M. (2010). Preparation of biodegradable liquid core PLLA microcapsules and hollow PLLA microcapsules using microfluidics. *Macromol Biosci,* **10**, 475–480.

93. Nie, Z., Li, W., Seo, M., Xu, S., and Kumacheva, E. (2006). Janus and ternary particles generated by microfluidic synthesis: design, synthesis, and self-assembly. *J Am Chem Soc,* **128**, 9408–9412.

94. He, T. *et al.* (2011). A modified microfluidic chip for fabrication of paclitaxelloaded poly(L-lactic acid) microspheres. *Microfluid Nanofluid,* **10**, 1289–1298.

95. Ho, C. C., Keller, A., Odell, J. A., and Ottewill, R. H. (1993). Preparation of monodisperse ellipsoidal polystyrene particles. *Colloid Polym Sci,* **271**, 469–479.

96. Rolland, J. P. *et al.* (2005). Direct fabrication and harvesting of monodisperse, shape-specific nanobiomaterials. *J Am Chem Soc,* **127**, 10096–10100.

97. Canelas, D. A., Herlihy, K. P., and DeSimone, J. M. (2009). Top-down particle fabrication: control of size and shape for diagnostic imaging and drug delivery. *Wiley Interdiscip Rev Nanomed Nanobiotechnol,* **1**, 391–404.

98. Petras, R. A., Ropp, P. A., and DeSimone, J. M. (2008). Reductively labile PRINT particles for the delivery of doxorubicin to HeLa cells. *J Am Chem Soc,* **130**, 5008–5009.

99. Xia, Y. and Whitesides, G. M. (1998). Soft lithography. *Angew Chem Int Ed,* **37**, 550–575.

100. Rolland, J. P., Hagberg, E. C., Denison, G. M., Carter, K. R., and De Simone, J. M. (2004). High-resolution soft lithography: enabling materials for nanotechnologies. *Angew Chem Int Ed,* **43**, 5796–5799.

101. Williams, S. S. *et al.* (2010). High-resolution PFPE-based molding techniques for nanofabrication of high-pattern density, sub-20 nm features: a fundamental materials approach. *Nano Lett,* **10**, 1421–1428.

102. Qin, D., Xia, Y., and Whitesides, G. M. (2010). Soft lithography for micro- and nanoscale patterning. *Nat Protoc,* **5**, 491–502.

103. Gratton, S. E. A. *et al.* (2008). The effect of particle design on cellular internalization pathways. *Proc Natl Acad Sci U S A,* **105**, 11613–11618.

104. Hasan, W. *et al.* (2011). Delivery of multiple siRNAs using lipid-coated PLGA nanoparticles for treatment of prostate cancer. *Nano Lett,* **12**, 287–292.

105. Enlow, E. M., Luft, J. C., Napier, M. E., and DeSimone, J. M. (2011). Potent engineered PLGA nanoparticles by virtue of exceptionally high chemotherapeutic loadings. *Nano Lett,* **11**, 808–813.

106. Wise, D. L. (2000). Large scale production of solid lipid nanoparticels (SLN) and Nanosuspensions (DIssoCubes). In *Handbook of Pharmaceutical Controlled Release Technology, 1st Ed.* Chap. 18, 359–381 (CBS Publisher).
107. Müller, R. H., Benita, S., and Bohm, B. (eds.) (1998). Nanoemulsions. In *Emulsions and Nanosuspensions for the Formulation of Poorly Soluble Drugs,* Chap. 9, 166–170 (Medpharm).
108. Jenning, V., Lippacher, A., and Gohla, S. H. (2002). Medium scale production of solid lipid nanoparticles (SLN) by high pressure homogenization. *J Microencapsulation,* **19**, 1–10.
109. Murakami, H., Kobayashi, M., Takeuchi, H., and Kawashima, Y. (1999). Preparation of poly(D,L-lactide-*co*-glycolide) nanoparticles by modified spontaneous emulsification solvent diffusion method. *Int J Pharm,* **187**, 143–152.
110. Vauthier, C. and Bouchemal, K. (2011). In *Intracellular Delivery,* Vol. **5**, 433–456 (Springer Netherlands).
111. Nisisako, T. and Torii, T. (2008). Microfluidic large-scale integration on a chip for mass production of monodisperse droplets and particles. *Lab Chip,* **8**, 287–293.
112. Szoka, F. and Papahadjopoulos, D. (1978). Procedure for preparation of liposomes with large internal aqueous space and high capture by reverse-phase evaporation. *Proc Natl Acad Sci U S A,* **75**, 4194–4198.
113. Nehilla, B. J., Bergkvist, M., Popat, K. C., and Desai, T. A. (2008). Purified and surfactant-free coenzyme Q10-loaded biodegradable nanoparticles. *Int J Pharm,* **348**, 107–114.
114. Sweeney, S. F., Woehrle, G. H., and Hutchison, J. E. (2006). Rapid purification and size separation of gold nanoparticles via diafiltration. *J Am Chem Soc,* **128**, 3190–3197.
115. Dalwadi, G., Benson, H. A., and Chen, Y. (2005). Comparison of diafiltration and tangential flow filtration for purification of nanoparticle suspensions. *Pharm Res,* **22**, 2152–2162.
116. Allison, S. D. *et al.* (1998). Effects of drying methods and additives on structure and function of actin: mechanisms of dehydration-induced damage and its inhibition. *Arch Biochem Biophys,* **358**, 171–181.
117. Abdelwahed, W., Degobert, G., and Fessi, H. (2006). A pilot study of freeze drying of poly (epsilon-caprolactone) nanocapsules stabilized by poly (vinyl alcohol): formulation and process optimization. *Int J Pharm,* **309**, 178–188.

118. De Jaeghere, F. *et al.* (1999). Formulation and lyoprotection of poly(lactic acid*co*- ethylene oxide) nanoparticles: influence on physical stability and *in vitro* cell uptake. *Pharm Res,* **16**, 859–866.

119. Saez, A., Guzman, M., Molpeceres, J., and Aberturas, M. (2000). Freeze-drying of polycaprolactone and poly (D,L-lactic-glycolic) nanoparticles induce minor particle size changes affecting the oral pharmacokinetics of loaded drugs. *Eur J Pharm Biopharm,* **50**, 379–387.

120. Yixiang, D., Yong, T., Liao, S., Chan, C. K., and Ramakrishna, S. (2008). Degradation of electrospun nanofiber scaffold by short wave length ultraviolet radiation treatment and its potential applications in tissue engineering. *Tissue Eng A,* **14**, 1321–1329.

121. Loo, S. C., Tan, Z. Y., Chow, Y. J., and Lin, S. L. (2010). Drug release from irradiated PLGA and PLLA multi-layered films. *J Pharm Sci,* **99**, 3060–3071.

122. Makadia, H. K. and Siegel, S. J. (2011). Poly lactic-*co*-glycolic acid (PLGA) as biodegradable controlled drug delivery carrier. *Polymers (Basel),* **3**, 1377–1397.

123. Guzowski, J., Korczyk, P. M., Jakiela, S., and Garstecki, P. (2011). Automated high-throughput generation of droplets. *Lab Chip,* **11**, 3593–3595.

124. Huh, D. *et al.* (2007). Gravity-driven microfluidic particle sorting device with hydrodynamic separation amplification. *Anal Chem,* **79**, 1369–1376.

125. Yamada, M. and Seki, M. (2005). Hydrodynamic filtration for on-chip particle concentration and classification utilizing microfluidics. *Lab Chip,* **5**, 1233–1239.

126. Batist, G. *et al.* (2001). Reduced cardiotoxicity and preserved antitumor efficacy of liposome-encapsulated doxorubicin and cyclophosphamide compared with conventional doxorubicin and cyclophosphamide in a randomized, multicenter trial of metastatic breast cancer. *J Clin Oncol,* **19**, 1444–1454.

127. Glantz, M. J. *et al.* (1999). Randomized trial of a slow-release versus a standard formulation of cytarabine for the intrathecal treatment of lymphomatous meningitis. *J Clin Oncol,* **17**, 3110–3116.

128. Glantz, M. J. *et al.* (1999). A randomized controlled trial comparing intrathecal sustained-release cytarabine (DepoCyt) to intrathecal methotrexate in patients with neoplastic meningitis from solid tumors. *Clin Cancer Res,* **5**, 3394–3402.

129. Adler-Moore, J. (1994). AmBisome targeting to fungal infections. *Bone Marrow Transplant,* **14**, S3–S7.

Chapter 5

Electrospun Biodegradable Polyester Micro-/Nanofibers for Drug Delivery and Their Clinical Applications

Xin Zhao,[a,b] Divia Hobson,[b] Zhi Yuan (William) Lin,[b] and Wenguo Cui[a,b]

[a]*Department of Orthopedics, The First Affiliated Hospital of Soochow University, Orthopedic Institute, Soochow University, 708 Renmin Road, Suzhou, Jiangsu 215006, P.R. China.*

[b]*School of Engineering and Applied Sciences, Harvard University, Cambridge, MA 02138, USA*

wgcui80@hotmail.com

Electrospun biodegradable micro-/nanopolyester fibers have attracted increasing interest for the designing of site-specific controlled drug delivery system in the body. Currently, different types of drugs, hydrophobic or hydrophilic drugs, growth factors, or genes have been entrapped within the interior or physically immobilized on the surfaces of these fibers by versatile drug encapsulation methods, including blending, coaxial process, emulsion, and surface modification. The advance in creating drug-loaded micro-/nanofib-

Handbook of Polyester Drug Delivery Systems
Edited by MNV Ravikumar

ISBN 978-981-4669-65-8 (Hardcover), 978-981-4669-66-5 (eBook)
www.panstanford.com

ers have made them applicable in numerous clinical applications, including wound dressing, cancer therapy, adhesion barrier, as well as regeneration of human soft and hard tissues. This book chapter reviews micro-/nanofibers associated with different drugs and their clinical applications.

5.1 Introduction

For the past few decades, biodegradable micro-/nanofibers on the order of a few nanometers to several micrometers have attracted significant research interest due to their applicability to the developments of targeted drug delivery devices which maximize therapeutic effects and minimize side effects [1]. The wide application of these micro-/nanofibers has also been attributed to their remarkable characteristics, including high surface area-to-volume ratio and interconnected porosity with tunable pore size [1]. The drug diffusion system is consequently more efficient as diffusion of the drug is encouraged by the high surface area of small fibers and three-dimensional (3D) open porous structure [2]. Moreover, the degradation products of the polymeric materials can be diffused through the pores and will not accumulate at the implantation site. Other advantages of the biodegradable micro-/nanofibers, including the potential for effective surface functionalization, modifiable surface morphology, and structural similarity to the extracellular matrix (ECM), have extended their clinical applications from drug delivery to tissue engineering [3].

Linear aliphatic polyesters, such as poly(lactic acid) (PLA), poly(ε-caprolactone) (PCL), poly(lactide-*co*-caprolactone) (PLCL), and poly(lactic-*co*-glycolic acid) (PLGA), are the most frequently used polymers to fabricate into micro-/nanofibers for drug delivery applications [3, 4]. These biodegradable polyester micro-/nanofibers allow for easy tailoring of their mechanical, architectural, degradation, and biological properties dependent on the molecular structure, morphology and molecular weight [3, 4]. In addition, they can degrade *in vivo* into nontoxic end products [5]. Moreover, they can be easily blended with natural polymers like collagen, gelatin, hyaluronic acid (HA), chitosan, and silk fibroin, or with other synthetic polymers such as water-soluble poly(*N*-vinylpyrrolidone) (PVP), poly(ethylene glycol) (PEG), polyvinyl alcohol (PVA), and polyethylene oxide (PEO) to further adapt their properties (e.g., hydrophobicity, hydrophilicity) to meet various needs [3, 4].

To produce biodegradable polyester micro-/nanofibers, various approaches, including electrospinning, self-assembly, and phase separation, have been used. Of these methods, electrospinning is the most widely used and preferred due to its comprehensibility, cost-effectiveness, potential for greater scale, versatility, and ability to spin a large variety of polymers [3]. Via varying solution parameters, like the viscosity, molecular weight and polymer concentration in addition to process parameters such as the applied voltage, distance from tip to collector and conductivity, the fiber morphology can be manipulated to attain desired properties for a broad spectrum of applications [6]. A schematic diagram of the electrospinning process to produce polymer fibers is depicted in Fig. 5.1. There are three fundamental components of the electrospinning process: a high- voltage supplier, a syringe with a needle of small diameter (i.e., a spinneret), and a metal collecting plate. An electrically charged jet of polymer solution is created through the application of a high voltage. The solution is released out of the pipette and solidifies after solvent evaporation before accumulating as a web of small fibers ranging from nanometers to micrometers on the collection screen.

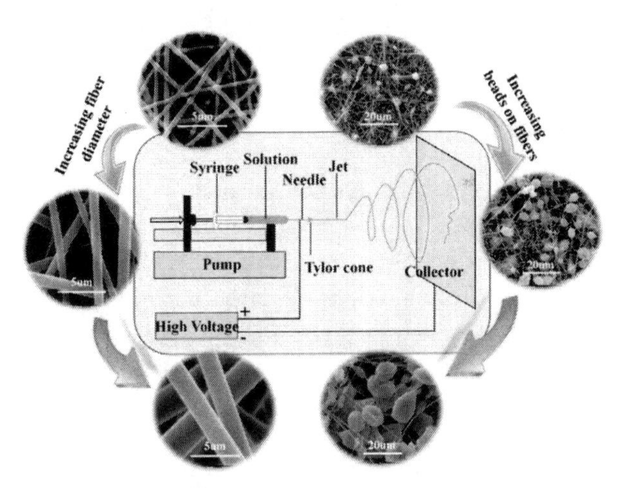

Figure 5.1 The electrospinning process of fibers of varying diameters.

Electrospinning allows drugs to be directly encapsulated into the electrospun fibers. Encapsulated drugs can be released from the electrospun fibers via diffusion and/ or polymer degradation [5]. Drug release behavior is thus highly dependent on the extent of drug molecule dispersion in the electrospun fibers, the physical and chemical properties of the drug and the polymer, as well as the morphology of the fibers. By proper selection of an electrospinning technique or a drug–polymer–solvent system, the drug molecule dispersion in the electrospun fibers and the fiber morphology can be modulated and thus the drug encapsulation efficiency and the drug release rates [7]. Various types of drugs can be encapsulated in the electrospun fibers, endowing the fibers with different functions and broadening their clinical applications as wound dressings, in cancer therapy, as physical barriers to adhesion formation, and for other tissue engineering applications [8].

In this chapter, we reviewed the drug incorporation techniques, various types of drugs delivered, and the clinical application of the drug-loaded electrospun polyester micro-/nanofibers.

5.2 Drug Incorporation Techniques

Various techniques have been used to desirably incorporate the therapeutic agents into the electrospun fibers, including blending, coaxial process, emulsion, surface modification, etc.

5.2.1 Blending

In blending electrospinning, drug encapsulation is achieved through a one-phase electrospinning approach as the drug is dissolved or dispersed in the polymer solution (Fig. 5.2A). The interaction of the polymer with the drug as well as the physicochemical properties of the polymer act as factors in determining the efficiency of drug encapsulation, dispersion of the drug in the fibers, and drug release kinetics. Isolated distribution of the drug inside the solution can be caused by insufficient solubility of the drug in the polymer, where the drug molecules could migrate to or near the fiber surface during the electrospinning process, resulting in a large initial burst release [7]. Therefore, it is necessary to match the hydrophobic–hydrophilic properties of the drug and the polymer when using blending electrospinning. Owing to the hydrophobic nature of polyesters, lipophilic drugs such as rifampicin and paclitaxel can usually be easily dissolved in the polyester solutions and released in a controlled manner. For hydrophilic drugs such as doxorubicin hydrochloride (DOX) [9–12], hydrophilic polymers such as gelatin [9], PEG [11], PVA [12], or amphiphilic copolymers like PEG-b-PLA [10] can be added to the polyesters to enhance the efficiency of drug loading and control the drug release rates.

Blending electrospinning generates fibers with a single layer structure. In order to produce fibers with a core–shell morphology to protect easily denatured biological agents, growth factors (GFs), and even genes dissolved in the fiber cores [13–15], the coaxial process or emulsion electrospinning methods can be employed (see Fig. 5.2) [16, 17].

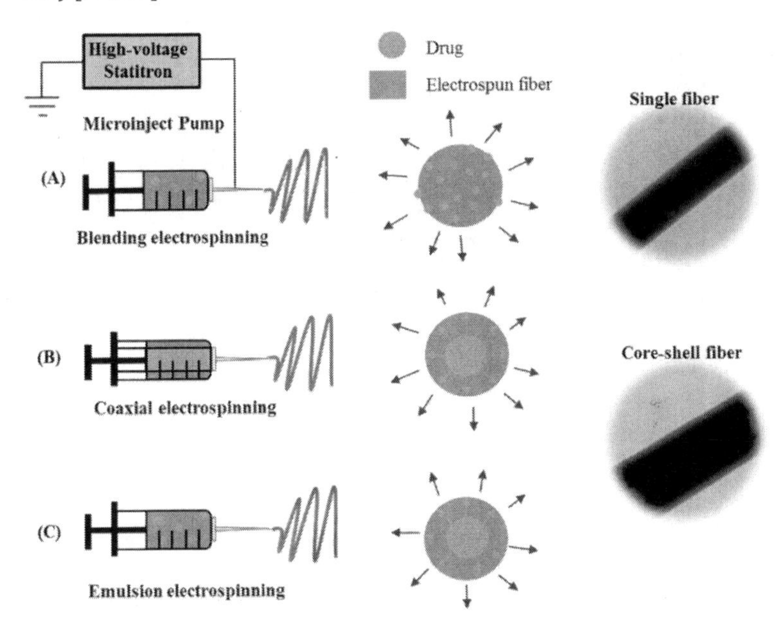

Figure 5.2 Drug incorporation techniques: (A) blending, (B) coaxial process, and (C) emulsion electrospinning.

5.2.2 Coaxial Process

The coaxial electrospinning process results in enhanced biomolecule functionality by having the biomolecule solution compose the inner jet while being co-electrospun with a polymer solution forming the outer jet (Fig. 5.2B) [18]. By this method, the core ingredient is shielded by the shell polymer to avoid direct contact to the biological environment. The shell polymer also improves the sustained release of the therapeutic agents [19]. The coaxial electrospinning process allows for the bioavailability of unstable biological molecules to be maintained, making it advantageous over blending electrospinning. Apart from the core-seeded bioactive molecules, the shell can be occupied with other types of pharmaceutical compounds for different applications [15, 20].

5.2.3 Emulsion Electrospinning

In the emulsion electrospinning process, the oil phase is created by the emulsion of the drug or aqueous protein solution in the polymer solution and then electrospun [21]. If the drug used has a sufficiently low molecular weight, the biomolecule-loaded phase can be distributed within the fiber or a core–shell fibrous structure could be configured as macromolecules amalgamate in the aqueous phase [21, 22] (Fig. 5.2C). The prospective benefit of emulsion electrospinning over the traditional blending technique is the eliminated need for a common solvent as the drug and the polymer are dissolved in applicable solvents. Consequently, a number of combinations of hydrophilic drugs and hydrophobic polymers can be employed while maintaining minimal drug contact with the organic solvent during processing [21].

5.2.4 Surface Modification

Surface modification is a promising method to introduce biofunctionality into micro-/nanofibers by conjugating therapeutic agents to the fiber surfaces, resulting in structural and biochemical similarity to the native tissue. Moreover, this strategy is typically applied to resolve the issues of large initial burst release and short release time as the biomolecules are surface-immobilized [23]. Conversely, the target biomolecule is scarcely released after chemical immobilization on the surface and therefore more suited for delivery of genes or GFs which need to be available for a long time to achieve certain therapeutic effect [24–26].

5.2.5 Other Electrospinning Techniques

Unlike the techniques described above, the tri-/multiaxial spinneret has been designed to fabricate micro-/nanofibers with two to five channels. For example, a technique that involved the simultaneous jet electrospinning of several channels of dissimilar fluids was designed by Zhao and co-researchers to produce microtubes with a biomimicking hierarchical multichamber structure [27, 28]. Moreover, composite fibrous meshes were fabricated from a variety of polymers and drugs by the two-stream electrospinning method, by which two drug-loaded polymers, possessing individual key functionalities, were concurrently electrospun. This achieved distinct fiber populations possessing collective properties [29].

5.3 Types of Drugs Released

Thanks to the versatile drug encapsulating techniques, different types of drugs, including hydrophobic (e.g., paclitaxel, rifampin), hydrophilic (e.g., DOX, tetracycline hydrochloride), and biomacromolecules such as proteins and DNA, can be delivered via electrospun polyester fibers.

5.3.1 Hydrophobic Drugs

Hydrophobic drugs and polyesters are usually soluble in the same solvents (e.g., dichloromethane) and can be dissolved in the polymer and electrospun. In cases where the drug (e.g., ginsenoside-Rg3 [Rg3]) and the polymer require different solvents to be dissolved, a relatively small quantity of solvent (e.g., hexafluoro-2-propanol, HFIP) may be utilized to solubilize the drug prior to its inclusion in the polymer solution. Many hydrophobic drugs have been delivered using electrospun polyester fibers, including antifever paracetamol and anti-inflammatory ibuprofen (IBU) to name a few. For example, Cui *et al.* investigated the use of electrospun poly(L-lactic acid) (PLLA) fibers inoculated with antifever paracetamol as drug delivery vehicles. It was found that the drug release rates can be controlled by the fiber diameters, drug contents and fiber degradation [30]. The authors further modified the PLLA fibers with acetal groups which endowed the fibers with acid responsive property, i.e., the fiber degradation and the drug release responded to the local pH variations [31]. Additionally, Hu *et al.* fabricated electrospun composite PLLA fibers consisting of IBU-loaded mesoporous silica nanoparticles (MSNs), which could increase the drug-loading capacities. This design extended the drug diffusive route, from MSNs to PLLA and from PLLA to environment, with drug release period as long as 70 days [32]. These authors then implanted these fibrous membranes into chickens to study their effect on inflammation and peritendinous adhesions [33]. It was found that the PLLA-MSN-IBU fibrous membranes had improved anti-inflammatory and anti-adhesion effects as compared to the PLLA-IBU fibrous membrane. More recently, Zhao *et al.* developed acid-responsive electrospun PLLA and PLGA fibers to incorporate sodium bicarbonate ($NaHCO_3$) and IBU via different blending electrospinning techniques. The electrospun fibers demonstrated rapid acid responsiveness and controlled drug release dependent on the concentrations of $NaHCO_3$ [34, 35].

5.3.2 Hydrophilic Drugs

Hydrophilic drugs and polyesters are usually not soluble in the same solvent. In this case, they can be loaded in certain drug carriers like MSNs which can be distributed in the polymer solution and subsequently fabricated by blending electrospinning. In another technique for resolving drug and polymer insolubility in a common solvent, the drug and polymer are dissolved in two immiscible solvents so that the two solutions can be kept in separate capillaries and coaxially electrospun or blended to produce an emulsion that can be electrospun. Hydrophilic drugs delivered include antitumor DOX and antimalaria chloroquine (CQ).

Qiu *et al.* used MSNs as devices to carry anticancer DOX, and the DOX-loaded MSNs were subsequently incorporated into PLLA nanofibers via electrospinning. It was found that the electrospun composite fibers possessed a high loading capacity of DOX and that drugs can be released in a controlled and prolonged manner dependent on the drug or MSN concentrations. Moreover, the composite fibers exhibited higher and longer *in vitro* antitumor efficacy compared to their MSN-free counterparts [36]. More recently, Zhou *et al.* employed the microsol-electrospinning (emulsion electrospinning) technique to encapsulate antimalaria CQ-loaded HA sol nanoparticles inside PLLA fibers. CQ was loaded into HA sol nanoparticles to preserve its activity by preventing diffusion of the organic solvent. By stretching the soft HA sol particles, composite fibers with core–shell structure were produced. *In vitro* drug release study revealed that the drug can be released for more than 40 days and the drug release rates could be fine-tuned by changing the microsol particle concentration or the amount of loaded drug [37].

5.3.3 Growth Factor Delivery

Growth factors (GFs) readily lose their activity during the chemical or physical process; therefore, the preservation of protein activity is critical for successful GF delivery [38]. Typically, in order to preserve GF activity, a solvent that is immiscible with the solvent used to dissolve the polymer is selected to dissolve the GFs in order for the solutions to be electrospun coaxially, or blended to produce an emulsion that can be electrospun. Moreover, GFs can be grafted on the surface of the electrospun fibers through fiber immersion in a drug solution. There are prospective advantages in the integration of GFs into ECM-mimicking scaffolds, particularly for tissue engineering applications, e.g., when damaged tissues lack the required potential for regeneration [39]. GFs incorporated include nerve GF (NGF), fibroblast GF (FGF), vascular endothelial GF (VEGF), etc.

For example, Mottaghitalab *et al.* evaluated attachment and proliferation of human neuroblastoma and human glioblastoma-astrocytoma cell lines in NGF surface-conjugated chitosan/PVA scaffolds. However, using this system, release of NGF was mainly from localized NGF on the surface and consequently could only last for 10 days [26]. Production of core–shell fibers using emulsion or coaxial electrospinning could be a rescue in preservation of bioactivity of GFs and extension of drug release. For instance, Yang *et al.* fabricated core–shell structured ultrafine PLLA fibers as carriers of a model protein bovine serum albumin (BSA) using emulsion electrospinning. The electrospun fibers demonstrated high structural integrity of core–shell fiber and the BSA could be released for a couple of weeks [40]. Zhao *et al.* reported on a basic FGF (bFGF)-loaded electrospun PLGA fibrous membrane for repairing rotator cuff tear (RCT). Ultrafine fibers with a core–sheath structure fabricated using emulsion electrospinning had secured the bioactivity of bFGF for 3 weeks. *In vivo* results demonstrated enhanced tendon-bone healing for 4 weeks [41]. Liao and Leong detailed the use of nanofibers as vehicles for the controlled delivery of VEGFC, VEGFA, and platelet-derived GF BB. The delivery of these GFs was intended to cultivate the local lymphatic and vascular systems. A sustained release period of 10–14 days for GFs from core–shell polyurethane nanofibers was achieved while bioactivity was observed to be analogous to fresh GFs used for the treatment of hemophilia. Additionally, the *in vivo* results aligned well with the outcomes of the *in vitro* testing [42].

5.3.4 DNA and siRNA Delivery

Target genes perform differently from GFs as they assimilate into the host genome of endogenous cells. Tissue formation is augmented by the transformation of transfected cells into bio-activated actors. Consequently, successful gene delivery through fibers requires that the active gene can be released from the fiber to allow integration into the host genome. Prior to this integration, the target gene is packed within vectors to prevent the target genes from being taken up by surrounding cells [43]. Conversely, vectors can contribute to the most significant impediment in gene transfection by transporting genes through the lipid bilayer of the cell membrane. Like GF delivery, coaxial and emulsion electrospinning and surface modification are the preferred techniques to incorporate genes into polyester fibers.

Many different approaches have been tried in the area of tissue engineering to produce nanofibrous scaffolds functionalized with DNA for the purpose of gene delivery [18, 19, 44]. Blending DNA in electrospinning solution was unsuccessful as poor encapsulation and transfection efficiency were observed. [44]. The incorporation of particles loaded with DNA into nanofibers [45], core–shell nanofibers [18, 19], and surface modification [46, 47] were strategies executed against this low transfection efficiency. For instance, Nie *et al.* produced PLGA/hydroxylapatite (HAp) composite scaffolds encapsulated with DNA/chitosan nanoparticles and found that these scaffolds had greater cell attachment and viability as well as desirable DNA transfection efficiency through cell culture experiments. These observations showed the potential for use of DNA/chitosan nanoparticles encapsulated PLGA/HAp composite scaffolds in bone regeneration [45]. Saraf *et al.* fabricated fibrous mesh scaffolds by means of coaxial electrospinning and detailed the encapsulation of nonviral gene delivery vectors as well as release over a period up to 60 days. Several fiber mesh scaffolds were fabricated with a core housing plasmid DNA (pDNA) and the sheath of coaxial fiber containing poly(ethylenimine)-HA (PEI-HA), the nonviral gene delivery vector [19]. Successful cell transfection of released complexes of pDNA with PEI-HA, were observed and found to promote the expression of enhanced green fluorescent protein (EGFP) in model fibroblast-like cells *in vitro* over 60 days.

siRNA, as a class of bioactive macromolecules, has been recognized as capable of muting particular protein expression. This phenomenon proves useful under certain circumstances, for example, when the secretion of inhibitory factors precludes the tissue-repair process, or in cancer treatment, where tumor growth is escalated by certain genes [48]. The first report on the encapsulation of siRNA within PCL nanofibers was published by Cao *et al.* Sustained bioactivity of glyceraldehyde-3-phosphate dehydrogenase siRNA throughout the electrospinning process resulted in a repression efficiency of 61–81% while transfection reagent TransIT-TKOTM (Mirus Bio LLC, Madison, WI, USA) was present in human embryonic kidney 293 cells [11]. On the other hand, the siRNA release rate was observed to be significantly slow owing to the hydrophobicity and degradation rate of PCL nanofibers. The encapsulation of siRNA and transfection reagent TKO complexes into poly(caprolactone-*co*-ethyl ethylene phosphate) was performed as a means of achieving faster siRNA release and ameliorating gene-silencing efficiency. This approach proved successful in increasing the siRNA release rate. Moreover, substantial gene silencing was exhibited by mouse fibroblast NIH 3T3 cells seeded on the scaffolds although no transfection reagent was present [6]. Moreover, the integration of siRNA-loaded nanoparticles into nanofibers was explored as a means of accomplishing matrix metalloproteinase-responsive release [49, 50].

5.3.5 Other Types of Drugs

Other types of drugs, e.g., nanoparticles, can also be incorporated in the electrospun fibers. For example, Liu *et al.* evaluated the potential of silver nanoparticles (AgNPs)-loaded PLLA electrospun fibrous membranes to prevent infection and adhesion formation. It was found that the Ag ion release rates could be fine-tuned by its concentration. More importantly, the AgNP-loaded PLLA fibrous membranes had a significant effect on prevention of cell adhesion and proliferation without detrimental effects on cells. The fibrous membranes also demonstrated antibacterial activity against Gram-positive *Staphylococcus aureus* and *Staphylococcus epidermidis* as well as Gram-negative *Pseudomonas aeruginosa* [51].

The implementation of many unique drug encapsulation techniques has enabled a variety of drug types to be delivered through electrospun polyester fibers with controllable release profiles. Fiber modification and various blend electrospinning techniques have been utilized to enable acid responsiveness in fibers so that degradation and hydrophobic drug release can respond to local pH environments [30, 33, 34]. The use of specific drug carriers like MSNs to increase drug loading capacities, extend drug diffusive routes and prolong drug release has also been reported in studies for delivery of both hydrophobic and hydrophilic drugs [31, 35]. Core-shell fiber production fabricated by emulsion and coaxial electrospinning has additionally been proven successful for the preservation of hydrophilic drug activity, GF, and gene bioactivity and extension of drug release [36, 39, 40]. This novel process allows the integration of water soluble drugs in hydrophobic or amphiphilic polymer fibers. Other methods to incorporate GFs or other bioactive agents involve surface grafting of the reagents on the surface of electrospun fibers through fiber immersion in a drug solution. In terms of other drug types such as nanoparticles, the controlled release of nanoparticles has also been demonstrated through (AgNPs)-loaded PLLA electrospun fibers with tunable release rates by varying the nanoparticle concentration [51].

5.4 Clinical Applications

5.4.1 Wound Dressings

Wound dressings prevent exogenous microorganisms from entering the wound and absorb exudate while improving overall aesthetic value [52]. Of late, electrospun micro-/nanofibers demonstrated great suitability for application in wound dressings due to their unique characteristics. For example, exceptionally high surface area allows the electrospun fibers to efficiently absorb exudates and regulate wound moisture [12, 52]. High porosity of the fibrous membrane leads to significant air permeability for cell respiration while the small pore size can prevent bacterial infections. Furthermore, fibrous dressings have great potential to be functionalized with numerous bioactive molecules. Finally, from an aesthetic perspective, fibers are advantageous as they allow the normal skin cell proliferation to occur, leading to scar-free regeneration [53, 54]. As naked polymeric fibers may not completely prevent infections, or the healed skin may suffer from large area of scar formation, bioactive dressings containing antibiotics or anti-scar reagents are usually required at the beginning of wound healing [12].

For example, Sun *et al.* developed PLGA fibers surface-modified with bFGF. Augmented cell adhesion and viability was determined in bFGF-gratfed PLGA electrospun fibrous scaffolds through *in vitro* testing. When the fibers were implanted, they were able to shorten the wound healing time, accelerate epithelialization and promote skin remodeling [30]. Anti-inflammatory drugs can also be included in the polymer fibers to inhibit inflammation and scar formation. For example, Yuan *et al.* designed an IBU-loaded PLLA fibrous scaffold to prevent excessive inflammation and to achieve scarless healing [55]. *In vivo* tests demonstrated that animals treated with IBU-loaded PLLA fibrous scaffolds exhibited reduced inflammation, accelerated healing process and regulated collagen deposition, leading to decreased scar area and scarless healing. Moreover, loading anti-scar formation drug, e.g., Rg3, allowed the fibrous membranes to assist wound healing and reduce hypertrophic scar formation (see example in Fig. 5.3) [56–59]. Rg3 release from these materials

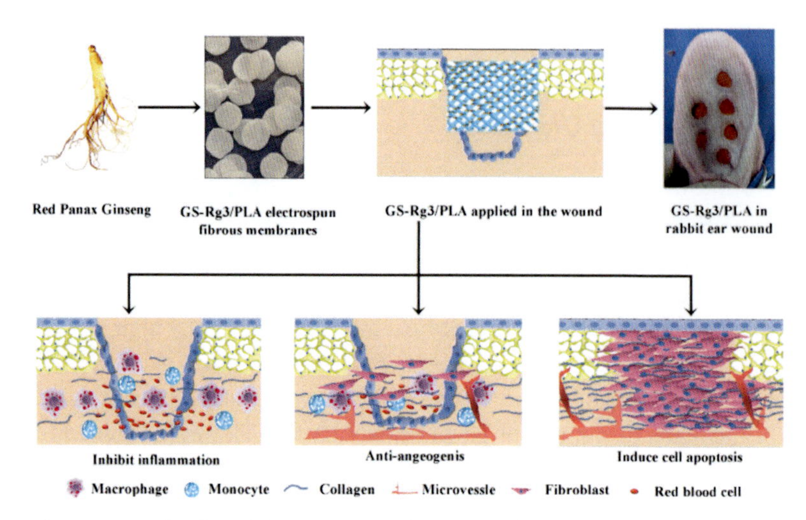

Figure 5.3 Implantable ginsenoside-Rg3-loaded electrospun fibrous membranes encouraging skin regeneration and inhibiting hypertrophic scarring [57].

such as PLGA, could be controlled via its concentration. Additionally, when the materials were used to cover a wounded area, the wounds could be healed and complete re-epithelialization was observed in 4 weeks with reduced scar formation. These polyester materials can also be modified using chitosan [56], HA [58], etc., to improve their hydrophilicity and cell adhesion properties.

5.4.2 Cancer Therapy

Strategies for developing successful postsurgical drug delivery have targeted prolonged function, maximal efficiency and minimal unwanted effect to healthy tissue [60]. Solid tumor sites can be covered with electrospun fabrics containing anticancer drugs: providing high local dosage with small amounts of the drug while reducing the need for frequent administration and ultimately resulting in patient convenience.

An early study by Zeng *et al.* examined the encapsulation of anticancer drugs, including hydrophobic paclitaxel and doxorubin base and hydrophilic DOX in electrospun PLLA fibers. Hydrophobic paclitaxel and doxorubin base were found to produce a constant drug release profile whereas hydrophilic DOX exhibited an obvious burst release [7]. Recently, the inhibitory effect against cervical carcinoma was examined *in vivo* through released dichloroacetate from PLLA electrospun mats [61]. Within a 3-week period, half of the tumor-bearing mice recovered and reduction in tumor mass and volume was observed.

Natural therapeutic agents with established anticancer properties are of significant interest for cancer therapy as they commonly have fewer side effects than synthetic anticancer drugs [62, 63]. In one study, Shao *et al.* fabricated green tea polyphenol (GTP)-loaded nanofibers that retained the chemical structure of bioactive substances prior to release into the medium. GTP was first noncovalently adsorbed on the surface of multiwall carbon nanotubes (MWCNTs), and subsequently incorporated into PCL nanofibers. This technique resulted in a significantly lower burst release of GTP over the initial two-day period. Retardation of the proliferation of human hepatocellular carcinoma cells (Hep G2) was observed while cytotoxicity to non-tumor osteoblast cells was low [63]. An additional class of materials with confirmed antitumor properties is inorganic compounds such as cisplatin; however, the difficulty in utilizing these compounds is their short half-life in the biological environment. The incorporation of titanocene dichloride in PLLA nanofibers was used as an approach for increasing the efficiency of inorganic anticancer drug and resulted in the suppression of lung tumor cells [64]. The hydrophobicity, poor solubility, and instability of antitumor agents pose challenges for successful sustained release. To overcome these challenges, PLLA-PEG electrospun nanofibers were loaded with an insoluble anticancer drug, hydroxycamptothecin (HCPT), utilizing 2-hydroxypropyl-β-cyclodextrin (HPCD) as a solubilizer [65]. *In vitro* testing demonstrated that HCPT-loaded electrospun fibers had substantially greater inhibition of human mammary gland MCF-7 cancer cells than the free drug over the initial 72 hour incubation period.

The emulsion electrospinning method was explored in the presence of HPCD to address the problem of burst release by producing core–shell nanofibers. Compared to the original profile from blended electrospun fibers, constant release was achieved as a result of the formation of preferential HCPT/HPCD complexes. Furthermore, the inhibitory activity of core–shell HCPT-loaded fibers opposing Hep G2 was found to be 0.2 times higher than HCPT free fibers over a 72 hour incubation period [2]. Huang *et al.* encapsulated iron oxide nanoparticles (IONPs) in electrospun polystyrene (PS) fiber webs, which enables substantial heating of the environment with the application of an alternating magnetic field as an anticancer tactic. The collagen-functionalized fiber surface allowed high attachment of human SKOV-3 ovarian cancer cells to the fiber. This enabled all fiber-associated cancer cells to be killed with the application of an alternating magnetic field to the fiber webs after a 10 minute duration [66].

5.4.3 Adhesion Barrier

Adhesion formation after tendon injury is a major clinical issue [67]. It not only causes gliding dysfunction and pain but also requires complicated reoperative surgery. The biomimetic electrospun fibers can be used to prevent adhesion formation during tendon healing by inhibiting the inflammatory reaction or isolating invasion of exogenous cells from the surrounding tendon sheath.

For example, Liu *et al.* employed a combination of sequential and microgel electrospinning techniques to fabricate a biomimetic bilayer sheath membrane with PCL fibrous membrane being the outer layer and HA microsol PCL (HA/PCL) fibrous membrane being the inner layer [68]. Experimental results proved the anti-adhesion property of the outer layer by showing lower cell proliferation on its surface compared to tissue culture plates. Results using a chicken model have demonstrated reduced peritendinous adhesions as well as improved tendon gliding by the application of this sheath membrane. Via incorporation with certain drugs, e.g., IBU, additional properties can be added to the electrospun fibers. That is, the IBU loaded electrospun fibers can prevent peritendinous adhesions while being against inflammation [33, 69]. By incorporating with certain bioactive reagents, such as bFGF to enhance cell behaviors, including adhesion and proliferation, the electrospun fibers can work simultaneously as physical barrier to prevent tendon adhesion and tissue engineering scaffolds to support tendon healing. For instance, Liu *et al.* encapsulated dextran glassy nanoparticles (DGNs) loaded with bFGF into a PLLA copolymer fiber by emulsion electrospinning (Fig. 5.4). It was found that the bFGF can be well preserved and could be released for a month. *In vivo* results demonstrated that the bFGF loaded PLLA fibers could effectively prevent peritendinous adhesion to promote intrinsic tendon healing [70]. Moreover, bioactive reagents such as celecoxib have been added in the electrospun fibers to further inhibit adhesion formation via reducing collagen I and collagen III expressions, inflammation reaction, and fibroblast proliferation [67].

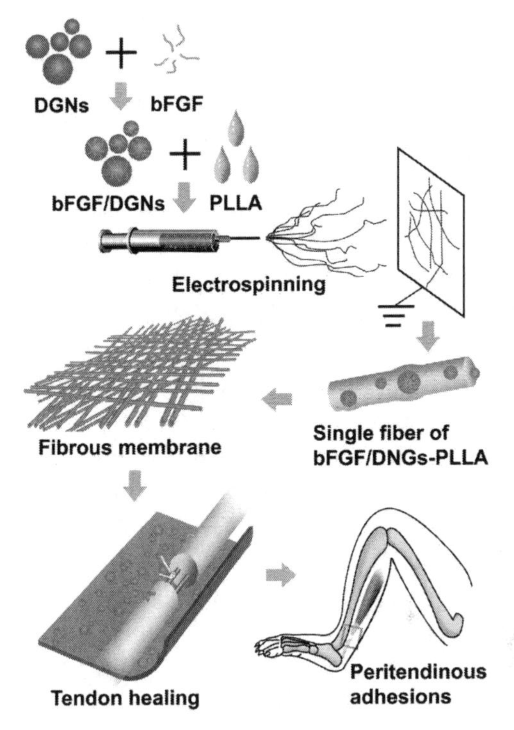

Figure 5.4 bFGF/DGN-loaded PLLA fibrous membrane was electrospun following the incorporation of bFGF into dextran glassy nanoparticles (DGNs) by a freeze-induced particle-forming process. The fabricated bFGF/DGN-loaded PLLA fibrous membrane was able to maintain the biological activity of bFGF, release bFGF sustainably, stimulate intrinsic tendon healing, and inhibit extrinsic healing. Modified from [70].

5.4.4 Tissue Engineering

Owing to the versatility of electrospinning, i.e., the ability to spin various materials, fibers with large surface area that mimic the natural ECM in terms of scale and structure have been produced with a broad spectrum of bioactive reagents and loaded to engineer various tissues, including vasculature, bone, and tendon. [71].

5.4.4.1 Vascular tissue engineering

Electrospun vascular grafts have been studied to solve issues during vascular regeneration like thrombus formation, occlusion, intimal hyperplasia and vasospasm [72–76]. Surface modification presents a potential approach to tackle this challenge. Researchers have enhanced endothelialization process by surface modification of electrospun nanofibers to attract endothelial cells (ECs) as they exhibit antithrombotic properties [77]. For instance, rapid endothelialization was accomplished through fiber surface modification with EC capturing ligands [78]. The development of small-diameter PCL nanofibrous vascular grafts coated with Nap-FFGRGD, an arginine-glycine-aspartic acid (RGD)-containing molecule, has been reported [79]. Self-assembly of the RGD molecule and the hydrophobic naphthalene groups to form a RGD layer on the hydrophobic surface was found. *In vivo* tests involving the implantation of PCL grafts and RGD-PCL grafts in rabbit carotid arteries for 2 and 4 week periods demonstrated confluent EC with alignment similar to that of the native vessel. In contrast, the EC on the PCL graft had comparatively random orientation. The endothelialization rates for RGD–PCL grafts (27.2% ± 11.5% and 51.1% ± 6.4% at 2 and 4 weeks, respectively) were much faster than that of the PCL grafts (1.8% ± 1.1% and 11.5% ± 3.2% at 2 and 4 weeks, respectively). In another study, a polypeptide named hirudin was conjugated to the surface of PLLA fibers using an intermediate linker of PEG (PEG) [80]. Properties of Hirudin and PEG are blood anticoagulation and ability to diminish platelet aggregation, respectively. Moreover, Hirudin-PEG modified polyurethane or PCL grafts demonstrated entire endothelial coverage after being implanted for 1 and 6 month periods. The ECs had morphologically likeness to those in the native vasculature and exhibited alignment in the direction of blood flow. With the exception of polypeptides [81, 82], the following proteins have been employed to modify graft surface and contribute to EC growth: collagen [83, 84], fibronectin [85], gelatin [86], and hydrophobin [87]. Moreover, paclitaxel or rapamycin loaded PCL fibers have been used as membrane to cover cardiac stent to reduce inflammation and scar formation after implantation with no detrimental effect on the normal function of cardiac stent [72–74]. Additionally, highly flexible and rapidly degradable papaverine-loaded electrospun PEG/PLLA fibrous membranes were developed to prevent vasospasm and repair vascular tissue when being wrapped around vascular suturing (Fig. 5.5) [88].

In addition, inhibition of thrombosis in tissue engineered blood vessels is proposed by rapid endothelization along the lumen of grafts and subsequent proliferation of vascular smooth muscle cells (VSMCs) on the outside. Zhang *et al.* developed two modified coaxial electrospinning techniques to fabricate double-layered membranes of a chitosan hydrogel/PEG-b-poly(L-lactide-*co*-caprolactone)

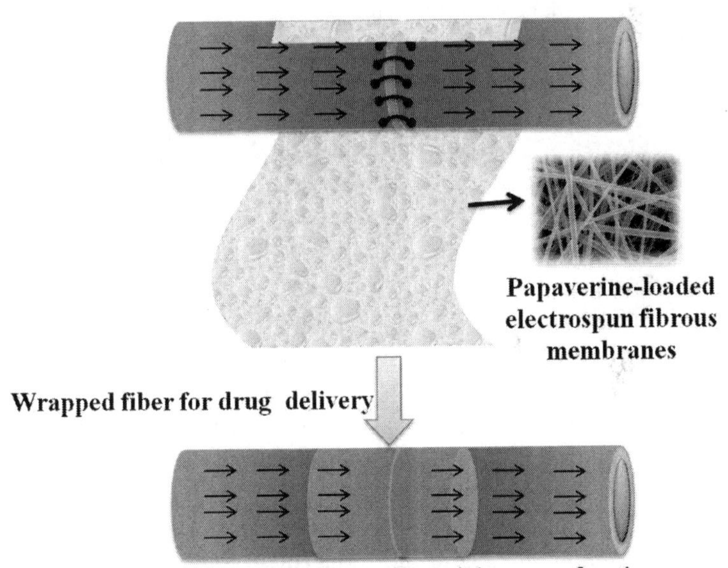

Figure 5.5 An illustration showing topical application of electrospun fibrous membranes around vascular suturing for preventing vasospasm and repairing vascular tissue. Modified from [88].

(PELCL) inner membrane loaded with VEGF and an emulsion/PELCL outer membrane loaded with platelet-derived growth factor (PDGF). VEGF and PDGF were encapsulated in order to regulate the proliferation of vascular endothelial cells (VECs) and VSMCs. Cell culture results demonstrated that VECs rapidly adhered and proliferated in the first 6 days and VSMCs had an accelerated growth rate in relation to other samples after the 6 day period. Furthermore, *in vivo* implantation of the double-layered membranes in a rabbit carotid artery model over 4 weeks resulted in VEC adhesion on the lumen and VSMCs on the outer layer absent of thrombosis. This novel strategy is promising as a substitute for small diameter vascular regeneration and addresses issues of burst and restenosis in current grafts [25].

5.4.4.2 Bone tissue engineering

Since ECM of bone is mainly composed of collagen (organic component) and HAp (inorganic component), the loading of calcium phosphate nanoparticles, e.g., tricalcium phosphate (TCP), dicalcium silicate or HAp in electrospun polymer fibers stimulates the biomineralization process and was found to enhance bone regeneration. As a result these loaded fibers can be used to replace diseased or damaged bones [89–91]. Improved growth of human fetal osteoblasts was stimulated by the synergistic effects of collagen, including the provision of extra cell recognition sites, and HAp, acting as a chelating agent for mineralization. Furthermore, 57% higher mineral deposition was observed on PLLA/collagen/HAp nanofibers than the PLLA/HAp nanofibers [92]. Another ceramic material widely used in bone-related biomedical applications is bioactive glass. Silica-based bioactive glass, in particular, is advantageous over HAp due to its greater bone-bonding ability [93] and silicon content; as silicon is a vital component for *in vivo* bone formation [94–96]. The bioactivity of PCL fibers was improved with the addition of nanofibrous bioactive glass of composition $70SiO_2\_25CaO\_5P_2O_5$. Upon immersion in a simulated body fluid (SBF), rapid formation of an apatite layer was observed on the fiber surface. Moreover, it was determined that greater osteoblastic cell attachment (MC3T3-E1) occurred on the nanocomposite membrane compared to the pure PCL membrane [97]. These results are promising for future applications in the field of bone regeneration and establish bioactive glass fibers as osteogenic stimulants. *In vivo* studies conducted in Sprague–Dawley albino rats further proved tremendous biocompatibility of PCL matrix integrated nanofibrous bioactive glass [98]. Apart from chelating agents for mineralization, GFs (e.g., bone morphorgenic protein-2) or chemical reagents like dexamethasone have also been added into the electrospun fibers to stimulate osteogenic differentiation [99].

5.4.4.3 Tendon tissue engineering

Applications of electrospinning in tissue engineering pertaining specifically to connective tissues, tendons and ligaments have been established. For example, electrospun polyester such as PLGA, PCL fibers have been used for augmentation of RCT repair with increased area of fibrocartilage and improved collagen organization (see example in Fig. 5.6) [41, 100, 101]. In addition, this strategy has demonstrated success in repairing tendon with no postsurgical peritendinous adhesion formation and tendon sheath infection [51]. Recent advancement in tendon/ligament tissue engineering has involved the use of hybrid scaffolds that combine electrospun fibers with knitted structures [102, 103]. The mechanical properties of knitted structures as well as the topographical characteristics of electrospun fibers were utilized in these scaffolds. These properties fulfill the mechanical prerequisite for tendon/ligament grafting and support cell attachment, proliferation, and differentiation. The silk-knitted structure coated with poly(L-lactic-*co*-ε-caprolactone) (PLCL) microfibers forming a composite scaffold was found to have an elastic modulus of 150 MPa, comparable to the native tendon and ligament modulus of 50–100 MPa. Moreover, the seeding efficacy of rat mesenchymal stem cells (MSCs) was greater on the hybrid scaffold than on the uncoated knitted structure. After 1 week of cell culture, collagen types I and III were found to be present in the tendon/ligament tissue through immunostaining [102]. In another study, bFGF was loaded into degummed knitted silk microfibrous scaffolds coated with PLGA nanofibers. Tenogenic differentiation of stem cells into tendon/ligament fibroblasts was stimulated by this GF, which can be integrated into fibers by either the blending or coaxial electrospinning methods [103]. Both methods demonstrate sustained release from the fibers as well as increased collagen production and improved fibroblast differentiation [104].

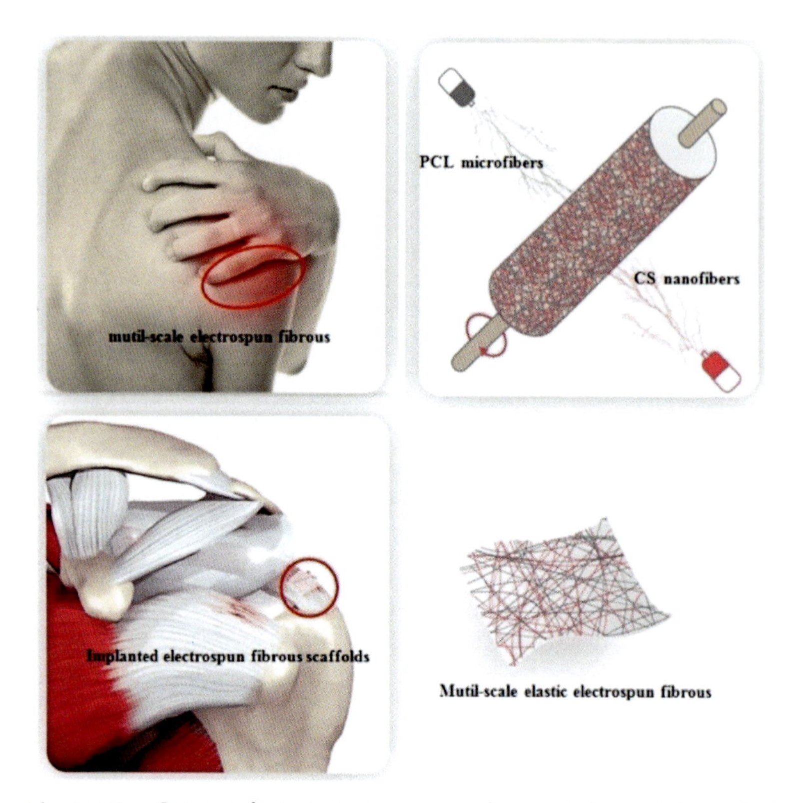

Figure 5.6 Stagger-electrospinning process by a rotating metal mandrel produces a dual scaffolding system of PCL microfibers interspersed with chitosan (CS) nanofibers. Modified from [101].

In another study, Peach *et al.* fabricated electrospun PCL matrices surface modified with polyphosphazene for the purpose of tendon tissue engineering and augmentation. PCL fiber matrices functionalized with poly[(ethyl alanato)1(*p*-methyl phenoxy)1] phosphazene (PNEA-mPh) were compared with smooth PCL fibers and found to have a more roughened hydrophilic surface that resulted in enhanced cell adhesion and superior cell-construct infiltration. Functionalized matrices also achieved more prominent tenogenic differentiation, having greater tenomodulin expression and superior phenotypic maturity [20].

More recently, Manning *et al.* developed a scaffold for delivery of GFs and cells in a surgically manageable form for tendon repair. PDGF and adipose-derived mesenchymal stem cells (ADSCs) were incorporated into a hydrogel heparin/fibrin-based delivery system and subsequently layered with an electrospun nanofiber PLGA backbone for structural integrity during surgical handling and tendon implantation. The heparin/fibrin drug delivery system permitted the controlled simultaneous delivery of PDGF and ADSCs. *In vitro* studies demonstrated cell viability and sustained GF release were achieved. *In vivo* study in a large animal tendon model further verified the viability of cells in the tendon repair environment as well as the clinical relevancy of the approach. In addition, viable ADSCs were detected at the repair 9 days post operation as well as increased total DNA in the ADSC treated tendons. This novel layered scaffold has demonstrated the potential for ameliorated tendon healing through the concurrent delivery of cells and GFs in model suitable for surgical handling [105].

5.4.4.4 Skin tissue engineering

In skin tissue regeneration, micro-/nanofibers have shown great potential to mimic skin ECM in both morphology and composition, and thus may be promising tissue engineering scaffolds for skin substitutes [48]. A number of synthetic polymers have been electrospun and assessed for wound healing applications, including polyurethane, poly(3-hydroxybutyrate-*co*-3-hydroxyvalerate) (PHBV), PCL, PLGA, and PLLA. As an example, electrospun PHBV fibers have been demonstrated as a suitable skin scaffold due to ability to support human fibroblast and keratinocyte adhesion and proliferation in addition to oxygen permeability and biocompatibility. Similarly, PCL-blended collagen nanofibrous membranes have been determined to have the potential applications as dermal substitutes for the treatment of skin defects and burn injuries due to favorable attachment and proliferation of fibroblasts [29]. Moreover, electrospun synthetic polymers are suitable as skin scaffolds due to their ability to encapsulate wound healing mediators and GFs. For example, PCL-PEG-PCL amine terminated block copolymers were electrospun to fabricate nanofibers with surface functionalized with epidermal growth factor (EGF) [21]. Efficacy of this novel EGF containing nanofiber model was established in diabetic animals with dorsal full-thickness wounds as treatment of EGF-loaded nanofibers resulted in improved wound healing and closure than the control groups with no EGF.

Another strategy to promote skin regeneration involved the encapsulation of bFGF through emulsion electrospinning to into ultrafine PELA fibers with core-sheath structure. Skin regeneration was evaluated *in vivo* as dorsal wounds in diabetic rats were covered with bFGF-loaded scaffolds and found to have a significantly higher wound-healing rate as well as full re-epithelialization as compared to the control with no bFGF. In addition, higher density and mature capillary vessels were produced in the 2 weeks following bFGF-loaded fiber application. bFGF was observed to have a low initial burst release of 14.0% ± 2.2% while subsequent gradual release, for approximately 4 weeks, improved collagen deposition and ECM remodeling resulting in greater simulation of the native tissue [53].

An additional study fabricated core–shell electrospun gelatin/PLLCL fibers encapsulated with multiple epidermal induction factors (EIF), including EGF, insulin, hydrocortisone, and retinoic acid in order to allow sustained release and protect healing mediators from hydrolytic enzyme attack. Burst release was not observed from the core-shell nanofibers while a 44.9% burst release from EIF blend electrospun nanofibers was detected over 15 days. Furthermore, evaluation after 15 days of cell culturing ADSCs demonstrated increased ADSC proliferation and a higher percentage of ADSCs differentiating to epidermal lineages on the EIF core–shell nanofibers as compared to blend electrospun fibers as a result of sustained EIF release. These findings demonstrated the potential for EIF encapsulated core–shell nanofibers as a tissue engineered graft for skin regeneration [106].

5.5 Conclusion and Future Prospects

Over the past decade, biomimetic micro-/nanofibers have emerged as powerful tools for constructing ECM-mimicking scaffolds for drug delivery and tissue regeneration applications. Recent advances in electrospinning techniques have allowed for the development of fibrous scaffolds with controllable compositions, structures and surface chemistries. Relying on the utilization of different fabrication techniques, various bioactive compounds can be incorporated within the fibrous scaffolds by blending, coaxial process, emulsion and surface modification. A broad spectrum of drugs with various activities such as anti-inflammation, antitumor, or bioactive molecules such as GFs or genes to aid skin, bone or tendon healing have been incorporated in the fibers and released in a controlled manner, potentially creating a more functional and biomimetic environment for clinical applications, including wound dressings, cancer treatment, adhesion barriers, and other tissue engineering applications.

Despite heady progress, several technical hurdles remain to be addressed for moving these promising techniques from bench top to bed side. A major hurdle encountered in the current electrospinning technique is the limited ability to control architectural properties like pore size and scaffold porosity, and seen as a likely cause of poor cell infiltration and failure in tissue engineering. Additionally, current knowledge on the role of the material composition, fiber arrangement, structural geometry, surface properties and release profiles of various fibrous scaffolds in the modulation of cellular responses and tissue morphogenesis is still incomplete. A thorough understanding of cell–fibrous matrix interactions would yield a better design of biomimetic fibrous scaffolds for drug delivery and tissue engineering applications. Improvement of the existing techniques and combination of novel techniques in the fiber fabrication area would greatly accelerate the progress in the burgeoning field of tissue engineering [4].

Acknowledgements

This work was supported by the National Natural Science Foundation of China (51373112 and 51003058) and China Scholarship Council (CSC).

References

1. Soppimath, K. S., Aminabhavi, T. M., Kulkarni, A. R., and Rudzinski, W. E. (2001). Biodegradable polymeric nanoparticles as drug delivery devices. *J Control Release,* **70**, 1–20.
2. Luo, X., Xie, C., Wang, H., Liu, C., Yan, S., and Li, X. (2012). Antitumor activities of emulsion electrospun fibers with core loading of hydroxycamptothecin via intratumoral implantation. *Int J Pharm,* **425**, 19–28.

3. Coolen, N. A., Vlig, M., van den Bogaerdt, A. J., Middelkoop, E., and Ulrich, M. M. (2008). Development of an *in vitro* burn wound model. *Wound Repair Regen*, **16**, 559–567.
4. He, C., Nie, W., and Feng, W. (2014). Engineering of biomimetic nanofibrous matrices for drug delivery and tissue engineering. *J Mater Chem B*, **2**, 7828–7848.
5. Sill, T. J. and von Recum, H. A. (2008). Electrospinning: applications in drug delivery and tissue engineering. *Biomaterials*, **29**, 1989–2006.
6. Bhardwaj, N. and Kundu, S. C. (2010). Electrospinning: a fascinating fiber fabrication technique. *Biotechnol Adv*, **28**, 325–347.
7. Dahlin, R. L., Kasper, F. K., and Mikos, A. G. (2011). Polymeric nanofibers in tissue engineering. *Tissue Eng Part B Rev*, **17**, 349–364.
8. Torres Vargas, E. A., do Vale Baracho, N.C., de Brito, J., and de Queiroz, A. A. A. (2010). Hyperbranched polyglycerol electrospun nanofibers for wound dressing applications. *Acta Biomater*, **6**, 1069–1078.
9. Hu, Q., Li, B., Wang, M., and Shen, J. (2004). Preparation and characterization of biodegradable chitosan/hydroxyapatite nanocomposite rods via *in situ* hybridization: a potential material as internal fixation of bone fracture. *Biomaterials*, **25**, 779–785.
10. Zaulyanov, L. and Kirsner, R. S. (2007). A review of a bi-layered living cell treatment (Apligraf®) in the treatment of venous leg ulcers and diabetic foot ulcers. *Clin Interv Aging*, **2**, 93–98.
11. Taira, B. R., Singer, A. J., McClain, S. A., Lin, F., Rooney, J., Zimmerman, T., *et al.* (2009). Rosiglitazone, a PPAR-gamma ligand, reduces burn progression in rats. *J Burn Care Res*, **30**, 499–504.
12. Martínez-Santamaría, L., Conti, C. J., Llames, S., García, E., Retamosa, L., Holguín, A., *et al.* (2013). The regenerative potential of fibroblasts in a new diabetes-induced delayed humanised wound healing model. *Exp Dermatol*, **22**, 195–201.
13. Liu, X., Liu, S., Liu, S., and Cui, W. (2014). Evaluation of oriented electrospun fibers for periosteal flap regeneration in biomimetic triphasic osteochondral implant. *J Biomed Mater Res B Appl Biomater*, **102**, 1407–1414.
14. Sombatmankhong, K., Sanchavanakit, N., Pavasant, P., and Supaphol, P. (2007). Bone scaffolds from electrospun fiber mats of poly(3-hydroxybutyrate), poly(3-hydroxybutyrate-*co*-3-hydroxyvalerate) and their blend. *Polymer*, **48**, 1419–1427.
15. Zomer Volpato, F., Almodovar, J., Erickson, K., Popat, K. C., Migliaresi, C., and Kipper, M. J. (2012). Preservation of FGF-2 bioactivity using hepa-

rin-based nanoparticles, and their delivery from electrospun chitosan fibers. *Acta Biomater,* **8**, 1551–1559.

16. Zeng, J., Xu, X., Chen, X., Liang, Q., Bian, X., Yang, L., *et al.* (2003). Biodegradable electrospun fibers for drug delivery. *J Control Release,* **92**, 227–231.
17. Sinha-Ray, S., Pelot, D. D., Zhou, Z. P., Rahman, A., Wu, X. F., and Yarin, A. L. (2012). Encapsulation of self-healing materials by coelectrospinning, emulsion electrospinning, solution blowing and intercalation. *J Mater Chem,* **22**, 9138–9146.
18. James, R., Kumbar, S. G., Laurencin, C. T., Balian, G., and Chhabra, A. B. (2011). Tendon tissue engineering: Adipose 1 derived stem cell and GDF-5 mediated regeneration using electrospun matrix systems. *Biomed Mater,* **6**, 025011.
19. Harding, K. G., Morris, H. L., and Patel, G. K. (2002). Science, medicine and the future: healing chronic wounds. *BMJ,* **324**, 160–163.
20. Peach, M. S., James, R., Toti, U. S., Deng, M., Morozowich, N. L., Allcock, H. R., *et al.* (2012). Polyphosphazene functionalized polyester fiber matrices for tendon tissue engineering: *in vitro* evaluation with human mesenchymal stem cells. *Biomed Mater,* **7**, 045016.
21. Choi, J. S., Leong, K. W., and Yoo, H. S. (2008). *In vivo* wound healing of diabetic ulcers using electrospun nanofibers immobilized with human epidermal growth factor (EGF). *Biomaterials,* **29**, 587–596.
22. Lannutti, J., Reneker, D., Ma, T., Tomasko, D., and Farson, D. (2007). Electrospinning for tissue engineering scaffolds. *Mater Sci Eng, C,* **27**, 504–509.
23. Agarwal, S., Wendorff, J. H., and Greiner, A. (2008). Use of electrospinning technique for biomedical applications. *Polymer,* **49**, 5603–5621.
24. Davidson, J. M. (1998). Animal models for wound repair. *Arch Dermatol Res,* **290**, S1–S11.
25. Zhang, H., Jia, X., Han, F., Zhao, J., Zhao, Y., Fan, Y., *et al.* (2013). Dual-delivery of VEGF and PDGF by double-layered electrospun membranes for blood vessel regeneration. *Biomaterials,* **34**, 2202–2212.
26. Nichol, J. W., Koshy, S. T., Bae, H., Hwang, C. M., Yamanlar, S., and Khademhosseini, A. (2010). Cell-laden microengineered gelatin methacrylate hydrogels. *Biomaterials,* **31**, 5536–5544.
27. Zhao, Y., Cao, X., and Jiang, L. (2007). Bio-mimic multichannel microtubes by a facile method. *J Am Chem Soc,* **129**, 764–765.

28. Zhao, T., Liu, Z., Nakata, K., Nishimoto, S., Murakami, T., Zhao, Y., *et al.* (2010). Multichannel TiO_2 hollow fibers with enhanced photocatalytic activity. *J Mater Chem,* **20**, 5095–5099.
29. Sundaramurthi, D., Krishnan, U. M., and Sethuraman, S. (2014). Electrospun nanofibers as scaffolds for skin tissue engineering. *Polym Rev,* **54**, 348–376.
30. Cui, W., Li, X., Xie, C., Zhuang, H., Zhou, S., and Weng, J. (2010). Hydroxyapatite nucleation and growth mechanism on electrospun fibers functionalized with different chemical groups and their combinations. *Biomaterials,* **31**, 4620–4629.
31. Cui, W., Qi, M., Li, X., Huang, S., Zhou, S., and Weng, J. (2008). Electrospun fibers of acid-labile biodegradable polymers with acetal groups as potential drug carriers. *Int J Pharm,* **361**, 47–55.
32. Hu, C. and Cui, W. (2012). Hierarchical structure of electrospun composite fibers for long-term controlled drug release carriers. *Adv Healthc Mater,* **1**, 809–814.
33. Hu, C., Liu, S., Zhang, Y., Li, B., Yang, H., Fan, C., *et al.* (2013). Long-term drug release from electrospun fibers for *in vivo* inflammation prevention in the prevention of peritendinous adhesions. *Acta Biomater,* **9**, 7381–7388.
34. Zhao, J. Liu, S., Li, B., Yang, H., Fan, C., and Cui, W. (2013). Stable acid-responsive electrospun biodegradable fibers as drug carriers and cell scaffolds. *Macromol Biosci,* **13**, 885–892.
35. Zhao, J. and Cui, W. (2014). Fabrication of acid-responsive electrospun fibers via doping sodium bicarbonate for quick releasing drug. *Nanosci Nanotechnol Lett,* **6**, 339–345.
36. Qiu, K., He, C., Feng, W., Wang, W., Zhou, X., Yin, Z., *et al.* (2013). Doxorubicin-loaded electrospun poly(L-lactic acid)/mesoporous silica nanoparticles composite nanofibers for potential postsurgical cancer treatment. *J Mater Chem B,* **1**, 4601–4611.
37. Zhou, L., Zhu, C., Edmonds, L., Yang, H., Cui, W., and Li, B. (2014). Microsol-electrospinning for controlled loading and release of water-soluble drugs in microfibrous membranes. *RSC Adv,* **4**, 43220–43226.
38. Ji, W., Sun, Y., Yang, F., van den Beucken, J. J., Fan, M., Chen, Z., *et al.* (2011). Bioactive electrospun scaffolds delivering growth factors and genes for tissue engineering applications. *Pharm Res,* **28**, 1259–1272.
39. Tabata, Y. (2000). The importance of drug delivery systems in tissue engineering. *Pharm Sci Technol Today,* **3**, 80–89.
40. Yang, Y., Li, X., Cui, W., Zhou, S., Tan, R., and Wang, C. (2008). Structural stability and release profiles of proteins from core-shell poly (DL-lac-

tide) ultrafine fibers prepared by emulsion electrospinning. *J Biomed Mater Res A,* **86**, 374–385.

41. Eming, S. A., Brachvogel, B., Odorisio, T., and Koch, M. (2007). Regulation of angiogenesis: wound healing as a model. *Prog Histochem Cytochem,* **42**, 115–170.
42. Liao, I. C. and Leong, K. W. Efficacy of engineered FVIII-producing skeletal muscle enhanced by growth factor-releasing co-axial electrospun fibers. *Biomaterials,* **32**, 1669–1677.
43. Roy, K., Wang, D., Hedley, M. L., and Barman, S. P. (2003). Gene delivery with *in situ* crosslinking polymer networks generates long-term systemic protein expression. *Mol Ther,* **7**, 401–408.
44. Luu, Y. K., Kim, K., Hsiao, B. S., Chu, B., and Hadjiargyrou, M. (2003). Development of a nanostructured DNA delivery scaffold via electrospinning of PLGA and PLA-PEG block copolymers. *J Control Release,* **89**, 341–353.
45. Nie, H. and Wang, C. H. (2007). Fabrication and characterization of PLGA/HAp composite scaffolds for delivery of BMP-2 plasmid DNA. *J Control Release,* **120**, 111–121.
46. Kim, H. S. and Yoo, H. S. (2010). MMPs-responsive release of DNA from electrospun nanofibrous matrix for local gene therapy: *in vitro* and *in vivo* evaluation. *J Control Release,* **145**, 264–271.
47. Zou, B., Liu, Y., Luo, X., Chen, F., Guo, X., and Li, X. (2012). Electrospun fibrous scaffolds with continuous gradations in mineral contents and biological cues for manipulating cellular behaviors. *Acta Biomater,* **8**, 1576–1585.
48. Meinel, A. J., Germershaus, O., Luhmann, T., Merkle, H. P., and Meinel, L. (2012). Electrospun matrices for localized drug delivery: current technologies and selected biomedical applications. *Eur J Pharm Biopharm,* **81**, 1–13.
49. Kim, H. S. and Yoo, H. S. (2013). Matrix metalloproteinase-inspired suicidal treatments of diabetic ulcers with siRNA-decorated nanofibrous meshes. *Gene Ther,* **20**, 378–385.
50. Chen, M., Gao, S., Dong, M., Song, J., Yang, C., Howard, K. A., *et al.* Chitosan/siRNA nanoparticles encapsulated in PLGA nanofibers for siRNA delivery. *ACS Nano,* **6**, 4835–4844.
51. Liu, S., Zhao, J., Ruan, H., Wang, W., Wu, T., Cui, W., *et al.* (2013). Antibacterial and anti-adhesion effects of the silver nanoparticles-loaded poly(L-lactide) fibrous membrane. *Mater Sci Eng, C,* **33**, 1176–1182.

52. Khil, M. S., Cha, D. I., Kim, H. Y., Kim, I. S., and Bhattarai, N. (2003). Electrospun nanofibrous polyurethane membrane as wound dressing. *J Biomed Mater Res B Appl Biomater,* **67**, 675–679.

53. Zhang, H., Jia, X., Han, F., Zhao, J., Zhao, Y., Fan, Y., *et al.* (2013). Dual-delivery of VEGF and PDGF by double-layered electrospun membranes for blood vessel regeneration. *Biomaterials,* **34**, 2202–2212.

54. Boateng, J. S., Matthews, K. H., Stevens, H. N., and Eccleston, G. M. (2008). Wound healing dressings and drug delivery systems: a review. *J Pharm Sci,* **97**, 2892–2923.

55. Yuan, Z., Zhao, J., Chen, Y., Yang, Z., Cui, W., and Zheng, Q. (2014). Regulating inflammation using acid-responsive electrospun fibrous scaffolds for skin scarless healing. *Mediators Inflamm,* 2014–2011.

56. Sun, X., Cheng, L., Zhu, W., Hu, C., Jin, R., Sun, B., *et al.* (2014). Use of ginsenoside Rg3-loaded electrospun PLGA fibrous membranes as wound cover induces healing and inhibits hypertrophic scar formation of the skin. *Colloids Surf B Biointerfaces,* **115**, 61–70.

57. Cheng, L., Sun, X., Hu, C., Jin, R., Sun, B., Shi, Y., *et al.* (2013). *In vivo* inhibition of hypertrophic scars by implantable ginsenoside-Rg3-loaded electrospun fibrous membranes. *Acta Biomater,* **9**, 9461–9473.

58. Cheng, L., Sun, X., Li, B., Hu, C., Yang, H., Zhang, Y., *et al.* (2013). Electrospun Ginsenoside Rg3/poly(lactic-*co*-glycolic acid) fibers coated with hyaluronic acid for repairing and inhibiting hypertrophic scars. *J Mater Chem B,* **1**, 4428–4437.

59. Cui, W., Cheng, L., Hu, C., Li, H., Zhang, Y., and Chang, J. (2013). Electrospun poly(L-lactide) fiber with ginsenoside rg3 for inhibiting scar hyperplasia of skin. *PLoS One,* **8**, e68771.

60. Pradilla, G., Wang, P. P., Gabikian, P., Li, K., Magee, C. A., Walter, K. A., *et al.* (2006). Local intracerebral administration of Paclitaxel with the paclimer delivery system: toxicity study in a canine model. *J Neurooncol,* **76**, 131–138.

61. Liu, D., Liu, S., Jing, X., Li, X., Li, W., and Huang, Y. (2012). Necrosis of cervical carcinoma by dichloroacetate released from electrospun polylactide mats. *Biomaterials,* **33**, 4362–4369.

62. Suwantong, O., Opanasopit, P., Ruktanonchai, U., and Supaphol, P. (2007). Electrospun cellulose acetate fiber mats containing curcumin and release characteristic of the herbal substance. *Polymer,* **48**, 7546–7557.

63. Shao, S., Li, L., Yang, G., Li, J., Luo, C., Gong, T., *et al.* (2011). Controlled green tea polyphenols release from electrospun PCL/MWCNTs composite nanofibers. *Int J Pharm,* **421**, 310–320.

64. Chen, P., Wu, Q. S., Ding, Y. P., Chu, M., Huang, Z. M., and Hu, W. (2010). A controlled release system of titanocene dichloride by electrospun fiber and its antitumor activity *in vitro*. *Eur J Pharm Biopharm*, **76**, 413–420.

65. Xie, C., Li, X., Luo, X., Yang, Y., Cui, W., Zou, J., *et al.* (2010). Release modulation and cytotoxicity of hydroxycamptothecin-loaded electrospun fibers with 2-hydroxypropyl-β-cyclodextrin inoculations. *Int J Pharm*, **391**, 55–64.

66. Huang, C., Soenen, S. J., Rejman, J., Trekker, J., Chengxun, L., Lagae, L., *et al.* (2012). Magnetic electrospun fibers for cancer therapy. *Adv Funct Mater*, **22**, 2479–2486.

67. Jiang, S., Zhao, X., Chen, S., Pan, G., Song, J., He, N., *et al.* (2014). Down-regulating ERK1/2 and SMAD2/3 phosphorylation by physical barrier of celecoxib-loaded electrospun fibrous membranes prevents tendon adhesions. *Biomaterials*, **35**, 9920–9929.

68. Liu, S., Zhao, J., Ruan, H., Tang, T., Liu, G., Yu, D., *et al.* (2012). Biomimetic sheath membrane via electrospinning for antiadhesion of repaired tendon. *Biomacromolecules*, **13**, 3611–3619.

69. Liu, S., Hu, C., Li, F., Li, X. J., Cui, W., and Fan, C. (2013). Prevention of peritendinous adhesions with electrospun ibuprofen-loaded poly(L-lactic acid)-polyethylene glycol fibrous membranes. *Tissue Eng Part A*, **19**, 529–537.

70. Liu, S., Qin, M., Hu, C., Wu, F., Cui, W., Jin, T., *et al.* (2013). Tendon healing and anti-adhesion properties of electrospun fibrous membranes containing bFGF loaded nanoparticles. *Biomaterials*, **34**, 4690–4701.

71. Goh, Y-F., Shakir, I., and Hussain, R. (2013). Electrospun fibers for tissue engineering, drug delivery, and wound dressing. *J Mater Sci*, **48**, 3027–3054.

72. Zhu, Y. Q., Hu, C. M., Li, B., Yang, H. L., Cheng, Y. S., and Cui, W. G. (2013). A high flexible paclitaxel-loaded poly(ε-caprolactone) electrospun fibrous membrane covered stents for benign cardia stricture. *Acta Biomater*, **9**, 8328–8336.

73. Zhu, Y. Q., Cui, W. G., Cheng, Y. S., Chang, J., Chen, N. W., Yan, L., *et al.* (2013). Biodegradable rapamycin-eluting nano-fiber membrane-covered metal stent placement to reduce fibroblast proliferation in experimental stricture in a canine model. *Endoscopy*, **45**, 458–468.

74. Soppimath, K. S., Aminabhavi, T. M., Kulkarni, A. R., and Rudzinski, W. E. (2001). Biodegradable polymeric nanoparticles as drug delivery devices. *J Control Release*, **70**, 1–20.

75. Yu, J., Wang, A., Tang, Z., Henry, J., Li-Ping Lee, B., Zhu, Y., *et al.* (2012). The effect of stromal cell-derived factor-1alpha/heparin coating of

biodegradable vascular grafts on the recruitment of both endothelial and smooth muscle progenitor cells for accelerated regeneration. *Biomaterials,* **33**, 8062–8074.

76. Browning, M. B., Dempsey, D., Guiza, V., Becerra, S., Rivera, J., Russell, B., *et al.* (2012). Multilayer vascular grafts based on collagen-mimetic proteins. *Acta Biomater,* **8**, 1010–1021.
77. Michiels, C. (2003). Endothelial cell functions. *J Cell Physiol,* **196**, 430–443.
78. de Mel, A., Jell, G., Stevens, M. M., and Seifalian, A. M. (2008). Biofunctionalization of biomaterials for accelerated *in situ* endothelialization: a review. *Biomacromolecules,* **9**, 2969–2979.
79. Zheng, W., Wang, Z., Song, L., Zhao, Q., Zhang, J., Li, D., *et al.* (2012). Endothelialization and patency of RGD-functionalized vascular grafts in a rabbit carotid artery model. *Biomaterials,* **33**, 2880–91.
80. Hashi, C. K., Derugin, N., Janairo, R. R., Lee, R., Schultz, D., Lotz, J., *et al.* (2010). Antithrombogenic modification of small-diameter microfibrous vascular grafts. *Arterioscler Thromb Vasc Biol,* **30**, 1621–1627.
81. Blit, P. H., Battiston, K. G., Yang, M., Paul Santerre J., and Woodhouse, K. A. (2012). Electrospun elastin-like polypeptide enriched polyurethanes and their interactions with vascular smooth muscle cells. *Acta Biomater,* **8**, 2493–2503.
82. Wise, S. G., Byrom, M. J., Waterhouse, A., Bannon, P. G., Weiss, A. S., and Ng, M. K. (2011). A multilayered synthetic human elastin/polycaprolactone hybrid vascular graft with tailored mechanical properties. *Acta Biomater,* **7**, 295–303.
83. He, W., Ma, Z., Teo, W. E., Dong, Y. X., Robless, P. A., Lim, T. C., *et al.* (2009). Tubular nanofiber scaffolds for tissue engineered small-diameter vascular grafts. *J Biomed Mater Res A,* **90**, 205–216.
84. He, W., Ma, Z., Yong, T., Teo, W. E., and Ramakrishna, S. (2005). Fabrication of collagen-coated biodegradable polymer nanofiber mesh and its potential for endothelial cells growth. *Biomaterials,* **26**, 7606–7615.
85. Zhu, Y., Leong, M. F., Ong, W. F., Chan-Park, M. B., and Chian, K. S. (2007). Esophageal epithelium regeneration on fibronectin grafted poly(L-lactide-*co*-caprolactone) (PLLC) nanofiber scaffold. *Biomaterials,* **28**, 861–868.
86. Ma, Z., Kotaki, M., Yong, T., He, W., and Ramakrishna, S. (2005). Surface engineering of electrospun polyethylene terephthalate (PET) nanofibers towards development of a new material for blood vessel engineering. *Biomaterials,* **26**, 2527–2536.

87. Zhang, M., Wang, Z., Wang, Z., Feng, S., Xu, H., Zhao, Q., *et al.* (2011). Immobilization of anti-CD31 antibody on electrospun poly(varepsilon-caprolactone) scaffolds through hydrophobins for specific adhesion of endothelial cells. *Colloids Surf B Biointerfaces*, **85**, 32–39.

88. Zhu, W., Liu, S., Zhao, J., Liu, S., Jiang, S., Li, B., *et al.* (2014). Highly flexible and rapidly degradable papaverine-loaded electrospun fibrous membranes for preventing vasospasm and repairing vascular tissue. *Acta Biomater,* **10**, 3018–3028.

89. Beldon, P. (2003). Skin grafts 1: theory, procedure and management of graft sites in the community. *Br J Community Nurs*, **8**, S8, S10–S12, S14 passim.

90. Zonari, A., Cerqueira, M. T., Novikoff, S., Goes, A. M., Marques, A. P., Correlo V. M., *et al.* (2014). Poly(hydroxybutyrate-*co*-hydroxyvalerate) Bilayer Skin Tissue Engineering Constructs with Improved Epidermal Rearrangement. *Macromol Biosci,* **14**, 977–990.

91. Cui, W., Li, X., Xie, C., Zhuang, H., Zhou, S., and Weng, J. (2010). Hydroxyapatite nucleation and growth mechanism on electrospun fibers functionalized with different chemical groups and their combinations. *Biomaterials,* **31**, 4620–4629.

92. Prabhakaran, M. P., Venugopal, J., and Ramakrishna, S. (2009). Electrospun nanostructured scaffolds for bone tissue engineering. *Acta Biomater,* **5**, 2884–2893.

93. So, K., Fujibayashi, S., Neo, M., Anan, Y., Ogawa, T., Kokubo, T., *et al.* (2006). Accelerated degradation and improved bone-bonding ability of hydroxyapatite ceramics by the addition of glass. *Biomaterials,* **27**, 4738–4744.

94. Jugdaohsingh, R., Tucker, K. L., Qiao, N., Cupples, L. A., Kiel, D. P., and Powell, J. J. (2004). Dietary silicon intake is positively associated with bone mineral density in men and premenopausal women of the Framingham Offspring cohort. *J Bone Miner Res*, **19**, 297–307.

95. Carlisle, E. M. (1972). Silicon: an essential element for the chick. *Science,* **178**, 619–621.

96. Seaborn, C. D. and Nielsen, F. H. (2002). Silicon deprivation and arginine and cystine supplementation affect bone collagen and bone and plasma trace mineral concentrations in rats. *J Trace Elem Exp Med*, **15**, 113–122.

97. Lee, H. H., Yu, H. S., Jang, J. H., and Kim, H. W. (2008). Bioactivity improvement of poly(epsilon-caprolactone) membrane with the addition of nanofibrous bioactive glass. *Acta Biomater,* **4**, 622–629.

98. Jo, J. H., Lee, E. J., Shin, D. S., Kim, H. E., Kim, H. W., Koh, Y. H., *et al.* (2009). *In vitro/in vivo* biocompatibility and mechanical properties of bioactive glass nanofiber and poly(epsilon-caprolactone) composite materials. *J Biomed Mater Res B Appl Biomater,* **91**, 213–220.

99. Su, Y., Su, Q., Liu, W., Lim, M., Venugopal, J. R., Mo, X., *et al.* (2012). Controlled release of bone morphogenetic protein 2 and dexamethasone loaded in core-shell PLLACL-collagen fibers for use in bone tissue engineering. *Acta Biomater,* **8**, 763–771.

100. Zhao, S., Xie, X., Pan, G., Shen, P., Zhao, J., and Cui, W. (2015). Healing improvement after rotator cuff repair using gelatin-grafted poly(L-lactide) electrospun fibrous membranes. *J Surg Res,* **193**, 33–42.

101. Zhao, S., Zhao, X., Dong, S., Yu, J., Pan, G., Zhang, Y., *et al.* (2015). Hierarchical, stretchable and stiff fibrous biotemplate engineered using stagger-electrospinning for augmentation of rotator cuff tendon-healing. *J Mater Chem B,* **3**, 990–1000.

102. Vaquette, C., Kahn, C., Frochot, C., Nouvel, C., Six, J. L., De Isla, N., *et al.* (2010). Aligned poly(L-lactic-*co*-ε-caprolactone) electrospun microfibers and knitted structure: a novel composite scaffold for ligament tissue engineering. *J Biomed Mater Res A,* **94**, 1270–1282.

103. Sahoo, S., Toh, S. L., and Goh, J. C. H. (2010). A bFGF-releasing silk/PLGA-based biohybrid scaffold for ligament/tendon tissue engineering using mesenchymal progenitor cells. *Biomaterials,* **31**, 2990–2998.

104. Sahoo, S., Ang, L. T., Goh, J. C., and Toh, S. L. (2010). Growth factor delivery through electrospun nanofibers in scaffolds for tissue engineering applications. *J Biomed Mater Res A,* **93**, 1539–1550.

105. Manning, C. N., Schwartz, A. G., Liu, W., Xie, J., Havlioglu, N., Sakiyama-Elbert, S. E., *et al.* (2013). Controlled delivery of mesenchymal stem cells and growth factors using a nanofiber scaffold for tendon repair. *Acta Biomater,* **9**, 6905–6914.

106. Jin, G., Prabhakaran, M. P., Kai, D., and Ramakrishna, S. (2013). Controlled release of multiple epidermal induction factors through core-shell nanofibers for skin regeneration. *Eur J Pharm Biopharm,* **85**, 689–698.

Chapter 6

Overview of Polylactide-*co*-Glycolide Drug Delivery Systems

Hongkee Sah
College of Pharmacy, Ewha Womans University, 52 Ewhayeodae-gil, Seodaemun-gu, Seoul 120-750, Republic of Korea
hsah@ewha.ac.kr

6.1 Introduction

Poly(lactide-*co*-glycolide) and its related polymers such as poly(lactic-*co*-glycolic acid), poly(lactide), and poly(lactic acid) are FDA-approved inactive ingredients for use in drug products, medical devices, and combination products. From now on, all these polymers will simply be abbreviated as PLGA in the text. PLGA has not only a proven history of safety but also a plethora of clinical experiences. Depending on polymer characteristics, PLGA displays a wide range of degradation rates and mechanical strengths [1]. PLGA has been long used as stitches, suture, pins, and screws for surgical and orthopedic applications. For example, Vicryl (Ethicon Inc.) is a surgical suture composed of 90% glycolide and 10% L-lactide, which is completely dissolved between 56 and 70 days. SmartPin (Linvatec Biomaterials

Handbook of Polyester Drug Delivery Systems
Edited by MNV Ravikumar

ISBN 978-981-4669-65-8 (Hardcover), 978-981-4669-66-5 (eBook)
www.panstanford.com

Ltd.) composed of poly-96L/4D-lactide copolymer is an example of bone fixation pins.

Advances in pharmaceutical technology have made it possible to fabricate PLGA into new dosage forms with varying drug release rates and degradation profiles. An injectable microsphere dosage form delivering luteinizing hormone releasing hormone (LHRH) peptides was one of the first generation of pioneering extended release depot products [2, 3]. Its success made in 1980s was followed by intensive research in the product development making use of innovative PLGA technologies. As result, versatile dosage forms have appeared at all scales of implants, gels, micro- and nanofibers, micelles, and nanoparticles, to name a few. There are also numerous studies reporting out-of-the-box PLGA-based composites.

There are other classes of commercial PLGA products which demonstrate great potentials for tissue engineering and regenerative medicine. PLGA is fabricated into various three-dimensional architectures to function as scaffolds for target tissues [4, 5]. Once the regenerative process primes, scaffolds dissolve away at a preprogrammed rate. Scaffolds can be shaped into microparticles, rods, nanofibers, meshes, foams, thin films, and any target structure. Finally, PLGA materials are also being applied for developing animal drug products, absorbable wrinkle fillers, and combination products such as drug-eluting stents and drug-releasing embolic microspheres. This chapter is aimed at providing brief overviews on representative PLGA-based dosage forms in commercial products and in future product development. Orthopedic applications are disregarded in this review. Also, preparation techniques and methodologies are not discussed in this chapter but dealt elsewhere. The contents described in this chapter would help broaden dimensions of PLGA applications in drug products, medical devices, and combination products.

6.2 Microspheres

6.2.1 Long-Acting Release (LAR) Microspheres

PLGA microspheres have found versatile applications in drug delivery, diagnostics, theranostics, and analytical testing. Our discussion in this section is primarily focused on their application in drug delivery. Currently, there are several commercial products of PLGA microspheres in the marketplace (Table 6.1). They are usually administered intramuscularly, subcutaneously, or subgingivally into patients through syringe needles. The first once-monthly microsphere formulation of leuprorelin was the Lupron Depot that received the US FDA approval in 1989. Since then, similar PLGA microsphere products have been developed worldwide. Ovarian and testicular steroidogenesis can be suppressed by the chronic, continual administration of peptide agonists (e.g., leuprorelin, goserelin, buserelin, or nafarelin) of the natural gonadotropin releasing hormone (GnRH, LHRH). At present, Lupron Depot (leuprolide acetate for depot suspension) 7.5 mg for 1 month, 22.5 mg for 3 months, 30 mg for 4 months, and 45 mg for 6 months are available for the palliative treatment of advanced prostate cancer. Similar LHRH agonist microsphere products include Pamorelin LA, Decapeptyl, and Trelstar. All the PLGA microspheres contain triptorelin as an active ingredient. They are sold under names of Trelstar in North America, while their brand names in Europe and Latin America are Decapeptyl and Pamorelin. Its brand name in India is Pamorelin. Typical release patterns of the peptide agonists from these microspheres are characterized by an initial burst followed by steady and sustained release.

Risperdal Consta is an extended release PLGA microsphere formulation of risperidone in strengths of 25, 37.5, and 50 mg. Risperdal Consta is administered every 2 weeks to the upper arm (deltoid muscle) or buttock (gluteal muscle) of patients to treat schizophrenia or bipolar I disorder [6]. Upon administration, this product provides negligible initial drug release followed by a lag time of 3 weeks, sustained drug release maintaining from 4 to 6 weeks, and diminishment of drug release by week 7. Because of these features, the intramuscular injection of this product every 2 weeks brings about therapeutic plasma drug concentrations.

Sandostatin LAR Depot is an injectable suspension of PLGA microspheres loaded with octreotide acetate [7]. Following a deep subcutaneous injection, it provides the sustained release of octreotide acetate that is biologically similar to the natural somatostatin. This product controls effectively both growth hormone and insulin growth factor 1, so as to reduce their blood concentrations to normal. Sandostatin LAR Depot is used to treat acromegalic patients. It is also indicated for long-term treatment of the severe diarrhea and flushing episodes associated with metastatic carcinoid tumors. Similarly, Somatuline Depot is also a LAR microsphere formulation of lanreotide acetate which is biologically similar to somatostatin.

Table 6.1 Commercial long-acting release PLGA microsphere products

Brand name	Company	API	Dose interval	Major indication
Lupron Depot	Abbvie	leuprolide acetate	1, 3, 4, 6 months	prostate cancer, endometriosis, precocious puberty
Trenantoe Depot	Takeda	leuprolide acetate	1, 3, 4, 6 months	prostate cancer, endometriosis, precocious puberty
Decapeptyl Depot	Debiopharm	triptorelin	1, 3, 6 months	prostate cancer
Trelstar	Actavis	triptorelin pamoate	1, 3, 6 months	prostate cancer
Pamorelin LA	Dr. Reddy's	triptorelin	6 months	prostate cancer
Risperdal Consta	Jassen	risperidone	2 weeks	schizophrenia, bipolar I disorder.
Sandostatin LAR Depot	Sandoz	octreotide acetate	4 weeks	acromegaly, carcinoid syndrome (e.g., severe diarrhea)
Somatuline LA	Ipsen	lanreotide acetate	2 weeks	acromegaly, thyrotropic adenomas, carcinoid tumors
Vivitrol	Alkermes	naltrexone HCl	4 weeks	opioid or alcohol dependence
Bydureon	AstraZeneca	exenatide	1 week	type 2 diabetes mellitus
Arestin	OraPharma	minocycline HCl	3 weeks	periodontitis

Naltrexone is a drug of choice for the treatments of opioid dependence and alcohol addiction. Oral naltrexone tablet was approved in 1984. When given orally, naltrexone is subject to extensive first pass metabolism. In addition, the issue of poor patient compliance is one of major challenges to the success of naltrexone treatment. A PLGA microsphere formulation of naltrexone is Vivitrol which is administered by gluteal intramuscular injection. This product provides desired naltrexone plasma concentrations for 30 days [8]. The gradual PLGA hydrolysis, particularly 2 to 4 weeks after injection, contributes to sustained drug release and its prolonged absorption into the systemic circulation.

Bydureon is the most recently approved PLGA microsphere formulation which provides the sustained release of exenatide [9]. It is given subcutaneously once every 7 days to adults with type 2 diabetes mellitus. This microsphere product is the first non-insulin, once-weekly injectable formulation. After the initial release of exenatide, the microspheres slowly breakdown and allows the extended release of exenatide into the body for 7 days.

Arestin is a microsphere product encapsulating minocycline hydrochloride, a tetracycline derivative. It was approved by the US FDA in 2001. Each unit dose delivers 1 mg minocycline encapsulated in approximately 3 mg PLGA. This product, when administered subgingivally into periodontal pockets, slowly releases the antibiotic for up to 3 weeks. Arestin works well for pockets that are 5–6 mm. Such local antibiotics delivery helps minimize total exposure of the body to antibiotics, reduce associated risks for systemic complications and improve patient compliance.

Duralease (Merial) and Celerin (PR Pharmaceuticals, Inc.) were PLGA microsphere products of estradiol benzoate to be injected under the skin of the ears of heifer and suckling calves. These products were designed to increase the rate of weight gain and to improve feed efficiency. On the basis of a business strategic decision, however, PR Pharmaceuticals, Inc., stopped their manufacturing and marketing. Accordingly, the FDA withdrew the approval of the NADAs (New Animal Drug Applications) in 2007 [10].

As shown above, there have been great successes in the commercialization of PLGA microsphere products for small molecules and peptides. PLGA microsphere formulations have also been explored as depot systems for macromolecules such as polypeptides, proteins, antigens, DNA, and RNA. There are a number of challenges in terms of formulation, manufacturing process, and their stability during *in vivo* delivery [11–14]. Innovative strategies for overcoming relevant obstacles would help commercialize PLGA microsphere products for macromolecules delivery.

6.2.2 Embolic Drug-Eluting Microspheres

Embolization is useful to achieve tumor regression by blocking terminal blood vessels and depleting a target tumor of nutrient supply. On the basis of this rationale, embolic microspheres have been explored for the treatments of uterine fibroids, hypervascular tumors, internal bleeding (e.g., gastrointestinal, urinary, renal, and varicose bleeding), or arteriovenous malformations. Embospheres prepared using tris-acryl gelatin are available in six size ranges (40~1,200 μm), and physicians can select adequate microspheres to block the microvasculature of a target tumor. Other types of nonbiodegradable microspheres which are under clinical trials for chemoembolization therapy are Hepaspheres/QuadraSpheres (Merit, USA) and DC Bead microspheres (Biocompatibles Ltd.). Chitosan microspheres, alginate microspheres, and PLGA-based microspheres are also being investigated for their potential application in chemoembolization. In particular, PLGA microspheres can carry various drugs (e.g., angiogenesis inhibitor, doxorubicin, cisplatin, mitomycin, norcantharidin, sunitinib, and sorafenib tosylate), thereby providing not only embolizing effects but also sustained local drug delivery [15–18]. PLGA microspheres with the capability of long-acting drug release are likely to be used chemoembolization for the treatment of many types of unresectable tumors.

6.2.3 Aesthetic Microspheres

Cosmetic wrinkle fillers are used to restore smile lines, frown lines, crows' feet, nasolabial folds, and forehead furrows. Examples of absorbable wrinkle fillers are collagen, hyaluronic acid, calcium hydroxyapatite, and poly-L-lactic acid (Table 6.2). Because they undergo degradation in the body, their function is temporary. Botox is not a wrinkle filler: it keeps muscles from tightening, thereby helping minimize the appearance of wrinkles. Microparticulated polymethylmethacrylate is the only non-degradable wrinkle filler approved for the correction of nasolabial folds.

Table 6.2 Comparison of various commercial wrinkle fillers in the market

Brand name	Company	Filler material	Major indications
Restylane silk	Valeant/ Medicis	Hyaluronic acid with lidocaine	Perioral rhytids (wrinkles around the lips)
Juvederm Volumna XC	Allergan	Hyaluronic acid with lidocaine	Age-related volume deficit in the midface
Sculptra aesthetic	Sanofi Aventis	Poly-L-lactic acid	Nasolabial fold contour deficiencies and other facial wrinkles
Evolence collagen filler	Colibar Lifescinece	Collagen	Facial wrinkles and folds (e.g., nasolabial fold)
Elevess	Anika Therapeutics	Hyaluronic acid with lidocaine	Facial wrinkles and folds (e.g., nasolabial fold)
Radiesse	Bioform Medical	Hydroxylapatite	Facial fat loss, facial wrinkles and folds (e.g., nasolabial fold)
Artefill	Suneva Medical	Polymethyl-methacrylate, collagen	Facial tissue around the mouth (e.g., nasolabial folds)

Sculptra (Sanofi-Aventis, Dermis Laboratories) is a commercial biodegradable wrinkle filler that is made of PLGA (specifically, poly-L-lactic acid) [19]. The polymer is composed of nontoxic lactic acid that is a normal metabolite found in the body. As a result, it is highly biocompatible, biodegradable and does not require an allergy test. This product is used to fill out facial contours and to correct defects such as wrinkles, nasolabial folds and furrows. The administration regime of the microsphere product consists of 4 injection sessions that are scheduled above 3 weeks apart. The 2-year durability of Sculptra is much longer than the durability observed with other biodegradable fillers such as collagen and hyaluronic acid. Another interesting feature of Sculptra is that its full cosmetic effect develops 6 to 8 weeks after injection under the skin. The reason is that the polyester polymer induces the synthesis and deposition of the body's own collagen. These events take several weeks to occur. Namely, the effect of Sculptra originates not only from the mechanical filling effect but also the deposition of natural collagen on the microspheres. To maximize the deposition of collagen around the microspheres, their average size is tailored to 50 μm. After the polymer degradation is complete, much of the deposited collagen remains in the injection site until it is completely degraded. This is why the effect of Sculptra lasts almost 2 years.

6.3 Implants

6.3.1 Subdermal Implants

Zoladex 3.6 mg (AstraZeneca) is a pellet-type PLGA implant. When the implant is injected under the skin of the abdomen, it slowly dissolves to release goserelin acetate for 4 weeks. Zoladex 10.8 mg delivers the GnRH agonist for 12 weeks. As described earlier, clinical inhibition of gonadotropin release leads to the subsequent reduction of serum testosterone or estradiol to castration level. The products are indicated for the palliative treatment of advanced carcinoma of the prostate and for the management of endometriosis. Similar rod-shaped implants delivering a GnRH agonist are Profact Depot and Suprefact Depot sold by Sanofi-Aventis. In fact, both implants are the same product containing buserelin acetate. These products are subcutaneously injected into lateral abdominal wall every 2 or 3 months. They are packaged in a sterile ready-to-use disposable applicator with a needle (internal needle diameter of 1.4 mm). Suprefact Depot 2 months consists of two identical rods, whereas Suprefact Depot 3 months consists of three identical rods.

6.3.2 Intravitreal Implant

Ozurdex (Allergan Pharmaceuticals) is an intravitreal implant containing 0.7 mg dexamethasone in a rod-shaped PLGA implant. The product is approximately 0.46 mm in diameter and 6 mm in length. Dexamethasone is a potent anti-inflammatory agent that inhibits multiple inflammatory cytokines, thereby suppressing edema, fibrin deposition, capillary leakage, and migration of inflammatory cells. The implant is formed from a mixture of PLGAs with hydrophilic carboxyl end groups and hydrophobic ester end groups. Their molar ratio of lactide:glycolide ratio is 50:50. A patent discloses that the implant is produced by an extrusion process, which is known as the Novadur drug delivery technology [20]. Briefly, micronized PLGAs and dexamethasone powders are thoroughly blended. This mixture is fed into a DACA Microcompounder-Extruder with its temperature preset at 115°C. The extruding filament is cut into a target implant size. Detailed single and double extrusion processes are disclosed in the patent. Ozurdex is indicated for the treatment of non-infectious uveitis affecting the posterior segment of the eye and macular edema following retinal vein occlusion. Being placed into the vitreous gel of the eye, this slowly degrading implant releases dexamethasone to the vitreous, thereby making its concentration equivalent to 0.05 μg/ml within 48 hours and to maintain its concentration at 0.03 μg/ml for at least 3 weeks [21].

6.4 *In situ* Formation of a Solid Implant from a Flowable Liquid

Atrigel is a PLGA drug delivery system to be injected as a liquid. This technology was first developed by Atrix Laboratories but is now a registered trademark of Tolmar Therapeutics, Inc. Upon injection of the formulation to a target site, the solvent dissolving PLGA is drained off with the body fluids. This results in the solidification of PLGA and encapsulation of a drug into the solid mass. As PLGA degrades, the drug is slowly released. One of the original patents describing the Atrigel drug delivery system was made public in 2000 [22]. A major advantage of this technology is that larger amounts of drugs can be safely incorporated into an implant by a simple one-step process. The solvents used in the practice of the Atrigel technology are polar and water-miscible. They should also possess strong solvation power on various kinds of PLGAs. Representative examples include *N*-methyl-2-pyrrolidone (NMP), 2-pyrrolidone, dimethylformamide, dimethyl sulfoxide, propylene carbonate, caprolactam, triacetin, glycofurol, or any combination thereof. Among them, the most preferred solvent is NMP, which is used for the Atrigel technology–based commercial products such as Eligard and Atridox.

Eligard (Sanofi-Aventis), which is indicated for the palliative treatment of advanced prostate cancer, is supplied in two separate syringes [23]. One syringe contains PLGA dissolved in NMP, and the other one contains leuprolide acetate. Their contents are mixed together before subcutaneous administration to patients. When the water-miscible solvent dissipates into the body fluids, a solid depot appears at the injection site. Depending on the product formulation, a single dose of Eligard provides the sustained release of leuprolide acetate over 1~6 months (Table 6.3). Critical formulation parameters are the type & concentration of PLGA (lactide:glycolide molar ratio, molecular weight, and end groups), the solvent amount, and drug payload. It may be worthwhile to mention the nature of PLGA end groups. Depending on the method of polymerization reaction, the terminal groups of PLGA can be hydroxyl, carboxyl, or ester. For example, polycondensation of lactic acid and/or glycolic acid leads to the formation of PLGA with hydroxyl and carboxyl end groups. PLGA with the same terminal end groups is prepared by ring-opening polymerization of dimers (e.g., lactide and/or glycolide) with water. By contrast, ring-opening of the dimers with a mono functional alcohol (e.g, methanol, ethanol, or 1-dodecanol) results in the formation of PLGA with one hydroxyl group and one ester end group. Furthermore, ring-opening polymerization of the dimers with diol, triol, tetraol, or pentaol provides PLGA with hydroxyl end groups. PLGA with hexanediol shown in Table 6.3 might have been synthesized by ring-opening polymerization of dimers with hexane-1,6-diol.

Atridox also makes use of the Atrigel delivery system for the subgingival controlled release of doxycycline hyclate. The combined treatment of scaling, root planning and local antibiotics is regarded as the best to treat recurring periodontal infection. Like Eligard, this product is composed of a two-syringe mixing system: one syringe contains a flowable polymeric solution (36.7% PLGA dissolved in 63.3% NMP), and the other syringe contains doxycycline hyclate. When its mixture is administered subgingivally or placed in pockets between the tooth and gum, it is solidified into an implant that can release the drug for up to 3 weeks. Many product developments based on the Atrigel drug delivery system are underway. A good example is a buprenorphine formulation that enables the delivery of buprenorphine over the periods of 14 days to about 3 months [24].

Table 6.3 The composition of constituted Eligard® formulations

Product name	**Eligard® 7.5 mg**	**Eligard® 22.5 mg**	**Eligard® 30 mg**	**Eligard® 45 mg**
PLGA end groups	carboxyl	hexanediol	hexanediol	hexanediol
Lactide:glycolide	50:50	75:25	75:25	85:15
PLGA used (mg)	82.5	158.6	211.5	165.0
NMP used (mg)	160.0	193.9	258.5	165.0
Leuprolide base (mg)	7	21	28	42
Administered dose (mg)	250	375	500	375
Injection volume (ml)	0.25	0.375	0.5	0.375
Release duration (month)	1	3	4	6

Using the technology of the *in situ* implant formation, Merial introduced Longrange for the extended release of eprinomectin. This product is the first sustained-release injectable cattle dewormer that offers season-long parasite control for 100 to 150 days, depending on parasite species [25]. This product is a ready-to-use injectable solution consisting of eprinomectin, NMP, triacetin, and PLGA. When injected subcutaneously into the shoulder of cattle, the solvent diffusion into the body fluids triggers the formation of a PLGA implant to release eprinomectin in a sustained manner. Following the injection, an initial burst release makes the plasma drug concentration reach a peak level quickly. Its concentration then starts to decrease steadily, but beginning about 70 days, the drug concentration increases due to PLGA degradation. The company's report on pharmacokinetic studies of Longrange in cattle demonstrated that a single subcutaneous injection of the product achieved nematicial plasma concentrations above 0.5 ng/ml for 150 days. As opposed to conventional parasite control products requiring 3 to 4 treatments during the gazing season, this product provides effective parasite control by a single injection.

6.5 PLGA-Based Micelles

Amphiphilic block copolymers such as PEG-PLGA can self-assemble into micelles in an aqueous phase [26]. They are usually in the size range of 10 to 100 nm and have been extensively investigated for anticancer therapy. In a polyethylene glycol (PEG)-PLGA micellar system, PEG functions as a hydrophilic corona, while PLGA serves as a hydrophobic core in which a hydrophobic drug is dissolved. PEG-PLGA micelles tend to display longer blood residence time than plain PLGA nanoparticles do. This results in the maximization of the benefit of the so-called enhanced permeability and retention (EPR) effect. Genexol-PM (Samyang Corp., Korea) is a product based on the PEG-PLGA micellar technology. The specific material used for micelle formation is methoxypoly(ethylene glycol)-block-poly(D,L-lactide). Paclitaxel is dissolved in the core of micelles without the use of toxic solubilizers such as Cremophor-EL. This product received the Korea FDA approval for the treatment of breast cancer and non-small cell lung cancer [27–29]. It not only increases drug accumulation in tumor but also allows the use of a higher dose of paclitaxel than does a conventional Cremophor EL-based formulation. Instead of PEG, different polymers are conjugated to PLGA, in order to make various types of polymeric micelles. Examples of relevant block copolymers include polyethyleneimine (PEI)-PLGA, poly-L-lysine (PLL)-PLGA, poly-L-cysteine-PLGA, and siRNA-PLGA. PEI-PLGA or PLL-PLGA micelles are sometimes complexed with polyanions such as plasmid DNA. These polymeric micelle-based polyplexes have been used as nonviral gene delivery vectors, to replace typical PEI-DNA complexes. A polymeric micellar delivery system can be further functionalized in many ways. For example, targeting ligands can be attached to PEG to direct PEG-PLGA micelles toward targeting tissues. Also, incorporation of an imaging agent enables tracking micelles *in vivo* for biodistribution studies. Moreover, pH-, thermo-, ultrasound-, or light-sensitive block copolymers are being developed to control micelle breakdown and to trigger drug release at specific tissues or cell compartments [30].

6.6 PLGA-Based Nanoparticles

PLGA nanoparticles can dissolve hydrophobic drugs, provide their long-acting release, protect them from degradation, benefit from the EPR effect, and facilitate intracellular drug delivery. Therefore, injectable PLGA nanoparticles have been used for various clinical applications [31–33]. However, clinical translation of plain PLGA nanoparticles have been plagued by several problems—rapid clearance from the systemic circulation, incapability of recognizing different cell types, inherent negative surface charge, inability to pass through the blood–brain barrier, and the like. Plain PLGA nanoparticles have evolved into various nanoparticulate systems, in order to achieve a targeted functionality and to improve *in vivo* performance. Representative approaches include surface modification of bare PLGA nanoparticles by ligand conjugation, PEGylation, lipid adsorption, surfactant coating, polyion pairing, and encasement by cells or artificial lipid membrane. As illustrated in Fig. 6.1, these PLGA nanoparticles can also act as a promising carrier for multi-drug delivery that can achieve sustained delivery of multiple drugs in a programmed order from a single entity [34]. Discussed in the following section are some representative classes of functional PLGA nanoparticles. The first generation of functional PLGA nanoparticles is PEGylated PLGA nanoparticles that are made of PEG-PLGA copolymers. They are different from PEG-PLGA micelles in terms of size and physicochemical properties of PEG-PLGA constituents. PEGylated PLGA nanoparticles can retain the stealth benefit of PEG, thereby displaying extended systemic circulation time. However, there are some cautions on the undesirable effects of PEGyation: the body develops immune responses toward repeated administration of PEGylated nanoparticles, and the hydrophilic PEG corona interferes with the interactions between nanoparticles and cells [35].

A second major class of functional PLGA nanoparticles is core–shell-type hybrid nanoparticles [36, 37]. They are primarily composed of PLGA nanoparticles, a lipid layer or a membrane encasing PLGA nanoparticles, and a hydrophilic stealth component such as PEG. In fact, PEG can be conjugated to either PLGA or a lipidic material such as distearoyl phosphatidylethanolamine. In general, high drug encapsulation efficiencies are attained with core–shell-

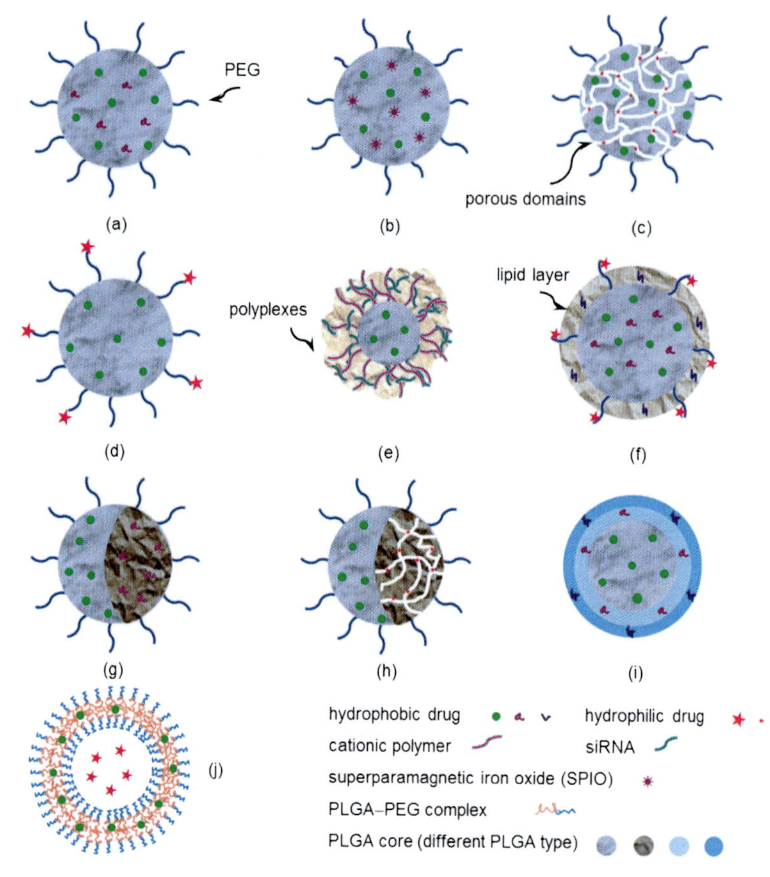

Figure 6.1 Illustration of various kinds of PLGA-based nanoparticulate systems for multi-drug delivery: (a) typical nanoparticles, (b) theranostic nanoparticles, (c) porous nanoparticles with both hydrophobic and hydrophilic drugs, (d) drug-spacer-PLGA nanoparticles with a hydrophobic drug in its core, (e) polyplex nanoparticles, (f) core–shell-type lipid-PLGA hybrid nanoparticles, (g, h) Janus nanoparticles, (i) onion-type nanoparticles, and (j) polymersomes. Copyright © 2015 Sah and Sah, publisher and licensee Hindawi Publishing Corporation.

type hybrid nanoparticles [36]. It is also possible to accommodate multiple drugs, thereby enabling drug cocktail therapy and temporal drug release. The so-called nanocell belongs to core–shell-type hybrid nanoparticles, because a nuclear PLGA nanoparticle is encased by an extranuclear lecithin/PEGylated lipid envelope. Instead of a lipid envelope, an erythrocyte membrane was also used to encapsulate PLGA nanoparticles. For example, red blood cells devoid of cytoplasmic contents were mixed with preformed PLGA nanoparticles, and the mixture was extruded together to prepare biomimetic cell-PLGA hybrid nanoparticles [38].

In addition to core–shell-type hybrid nanoparticles, other types of PLGA-based nanoparticles make it possible to accommodate multiple drugs with dissimilar water solubility and/or different mechanisms of action. For example, dual drugs with fundamentally different water solubility (e.g., paclitaxel and doxorubicin HCl) were nanoencapsulated into Janus PLGA nanoparticles with bi-compartmental morphology [39, 40]. Janus nanoparticles can also provide the sustained release of multiple drugs in a specific order. In addition, onion-like PLGA nanoparticles with multiple drug layers can be prepared by electrohydrodynamic atomization that breaks a liquid jet into tiny droplets by use of electrical forces. For example, a coaxial electrohydrodynamic atomization system using 3 different needles was reported to produce triple-layered nanoparticles consisting of PLGA and two other polymers [41].

Polymersomes are bilayer vesicles made of amphiphilic polymers such as PEG-PLGA [42, 43]. Compared to liposomes, polymersomes can contain larger amounts of PEG and higher drug payloads. Depending on the property of PEG-PLGA, polymersomes show great aqueous stability without drug leaching. Both hydrophilic and hydrophobic drugs can be loaded together into polymersomes, making it possible to practice a drug cocktail therapy. The duration of drug release can be tailored by adjusting the characteristics of PEG-PLGA.

As discussed so far, PLGA-based nanoparticles have evolved into many promising drug carrier platforms such as polymeric micelles, polyplexes, Janus nanoparticles, and biomimetic nanocarriers such as core–shell-type hybrid nanoparticles, nanocells, and polymersomes [44]. These PLGA-based nanoparticulate carriers are further functionalized by targeting ligands including monoclonal antibodies, peptides, proteins, aptamers, cell-recognizing moieties, and diagnostic agents. Such derivatization is often performed on PEG-PLGA block copolymers with different end groups. Amine-, aldehyde-, carboxyl-, maleimide-, succinimidyl ester-, and sulfhydryl-functionalized PEG-PLGA polymers are often used to functionalize their surface with diverse ligands [44]. It is expected that these functional PLGA-based nanoparticles would enable drug targeting, temporal drug release, multiple drug cocktail therapy, and theranostic applications.

6.7 Biodegradable Composite Scaffolds

A number of natural and synthetic materials have been engineered into biocompatible scaffolds for bone repair and replacement. Bone grafting scaffolds are usually formed from composites of one or more types of materials. PLGA is a popular base material consisting of many bone grafting scaffolds. PLGA contributes to providing adequate mechanical strength to support dimensional integrity required as scaffolds [45]. On the other hand, calcium phosphates (e.g., β-tricalcium phosphate) have excellent biocompatibility, binding affinity to the bone, and osteoconductivity. Based on these features, PLGA-calcium phosphate composites have been developed for dental and orthopedic applications. OsteoScaf (Texas Innovative Medical Devices) is an example of PLGA-based bone grafting materials. It is indicated for intrabony periodontal osseous defects, augmentation of bony defects of the alveolar ridge, filling of tooth extraction sochets, and sinus elevation grafting. This product is a three-phase composite scaffold consisting of one PLGA phase and two osteoclast-resorbable calcium phosphate phases. The scaffold is prepared by a process of particle leaching and phase inversion from PLGA and two calcium phosphates [46, 47]. OsteoScaf is available as a particulate, cylinder or block. This product has highly interconnected macroporous structure of larger than 80% porosity and 0.25~1.2 mm pore size. These quality attributes are similar to those of the human cancellous bone. Both PLGA and calcium phosphate are bioresorbed over time during the healing process. The scaffold can also be used as a carrier template for drugs and cells [48, 49].

Guidor easy-graft (Sunstar Americas, Inc., Degradable Solutions AG) is an alloplastic bone grafting product which is indicated for the treatment of intraoral and maxillofacial osseous defects. The macroporous dental scaffold consists of two components, that is, a syringe containing PLGA-coated β-tricalcium phosphate granules (500~1,000 μm) and an ampoule containing the mixture of NMP and water. Pouring the solvent mixture inside the syringe and their mixing causes the formation of a putty-like mass. This can be dispensed and shaped into a defective region in the bone. This happens due to the partial solvation of PLGA by NMP. The solvent diffusion into the surrounding body fluids (e.g., the blood or saliva) leaves the putty-like material into a hardened, porous scaffold in minutes. In other words, the *in situ* extraction of the solvent results in the solidification of PLGA and subsequent aggregation of β-tricalcium phosphate granules. Its porosity contributes to the absorption of the blood and the improvement of osteoconductivity, thereby accelerating the healing process.

Finally, the BD™ OPLA (Open-Cell PolyLactic acid) scaffold is a sponge-like scaffold synthesized from polylactic acid. Its dimension is 5 mm × 3 mm × 0.039 cm^3 (diameter × height × volume), while average pore size is 100~200 μm. Its morphology is featured with a spherical facetted architecture which promotes the growth and differentiation of various cells [50, 51].

6.8 Nanofibers

PLGA nanofibers are biomimetic solid forms which can be engineered into varying three-dimensional structure. Figure 6.2 shows a typical scanning electron microscope image of electrospun PLGA nanofibers [52]. In fact, the native extracellular microenvironment of tissue is a porous network of fibrous proteins which constitute the three-dimensional architecture to a tissue. The electrospun structure made of PLGA nanofibers shares many similar features with the extracellular matrix (ECM) of natural tissues, in terms of porosity, mechanical properties, and morphology [53, 54]. Additionally, PLGA nanofibers are completely absorbable and have high surface to volume ratios. Because of these advantages, PLGA nanofibers have received wide interests in the areas of drug delivery and tissue engineering scaffolds [55].

Figure 6.2 A SEM micrgraph of electrospun PLGA nanofibers (5000× magnification, the bar size is 6 μm). Copyright © 2011 Seil and Webster, publisher and licensee Dove Medical Press Ltd.

Polymeric nanofibers can be produced by various methods such as melt spinning, electrospinning, and melt blowing, to name a few. Most PLGA nanofibers are formed *ex situ* by electrospinning which requires a high voltage equipment. An electric field is applied to a droplet of polymer melt or solution on the tip of a nozzle, so as to produce polymeric fibers [56, 57]. Depending on the application purpose, electrospun nanofibers can be fabricated into mats, plates, meshes, foams, membranes, and tubes. For example, nanofibrous PLGA meshes containing bioactive agents can serve as excellent templates for cell adhesion, proliferation and differentiation. Many chemicals, proteins, DNA, RNA, and even cells can also be encapsulated into PLGA nanofiber scaffolds [58]. When polymeric nanofibers are manufactured as a tubular shape, they can be employed to create vascularized tissue [59]. Additionally, they can be fabricated into other three-dimensional architectures to be fitted for bones and cartilage regenerations. Nanofibrous foam and mat can be applied on the surface of the skin, to encourage the generation of normal skin growth and to protect the wound from bacterial infiltration. Their high surface area is advantageous for fluid absorption and drug delivery. PLGA nanofibers can also be deposited as a thin porous film onto a hard prosthetic implant device.

As an alternative to the popular electrospinning process, a solution blow spinning technique has been proposed to produce PLGA nanofibers [60]. This solution blow spinning requires only a simple apparatus, a high-pressure gas source, and a concentrated polymer solution in a volatile solvent. The solution blow spinning process does not require a high voltage equipment with an electrically conductive collector. It can be used to coat any surface type of materials. Using this technique, Behrens *et al.* were able to generate PLGA nanofiber mats on any surface *in situ*, by use of a commercial airbrush and compressed CO_2 [61]. Their nanofiber mat had a median nanofiber diameter of 377 nm, which is similar to the natural fibrin fiber diameter (~376 nm). This PLGA nanofiber system seems to have great potentials as a surgical hemostatic sealant and scaffold for tissue repair.

6.9 Drug-Eluting Metallic Stent (DES)

The surface of metallic stents is usually coated with polymers, in order to improve their biocompatibility, to reduce inflammatory response, and to retain the capability of drug elution [62]. There are several FDA-approved DESs which release everolimus, paclitaxel, sirolimus, or zotarolimus. These DESs are used for patients with angina pectoris or silent ischemia and greater than 50% stenosis of one or more coronary arteries. Early generations of DES used non-biodegradable polymers to manipulate drug release. For example, the surface of Cypher (Cordis) is coated with a mixture of poly(ethylene-*co*-vinyl acetate) and poly(butylmethacrylate) containing sirolimus. The surface of Taxus (Boston Scientific) is coated by paclitaxel-containing poly(styrene-*b*-isobutylene-*b*-styrene). In comparison, a multitude of DESs have opted to use biodegradable polymers. Among them, PLGA is one of the most preferred polymers, due to its long history of safety, biocompatibility, and regulatory approvals in use of drug products and medical devices. Examples of PLGA-coated stents designed to release sirolimus are Sparrow NiTi, Nobori, Excel, and

Cura [63–66]. DESs coated by paclitaxel-containing PLGA are Conor Medstent (Conor Medsystems) and Infinnium (Sahajanand Medical Technologies) [67]. Coating strategy is an important parameter to affect the stent performance [68]. For example, an electrodeposition coating technique was utilized to coat the surface of a metallic stent with polymeric nanoparticles [69]. Compared to conventional DESs, this nanoparticle-eluting stent was shown to display interesting features in regard to vascular compatibility and drug release pattern. So far, numerous studies have demonstrated promising results of biodegradable polymer-coated DES, but there are still many challenges to product development in regards to thrombosis, inflammation, coating strategies, and drug release profiles. Optimization of coating polymers, polymer-drug formulations, and coating techniques would contribute to the better performance of DESs in many applications such as the coronary artery disease.

All DESs described so far leave metals in the body. By contrast, Abbott Laboratories has been developing a totally dissolvable and metal-free stent made of only PLGA (more specifically, poly-L-lactide and poly-D,L-lactide that can elute everolimus). The PLGA stent is named as Absorb. This stent supports the revascularization and restoration of the blood vessel until the artery can remain open on its own. After then, it completely dissolves in the artery by 3 years. Absorb can be considered a bioresorbable vascular scaffold in the sense that its temporary skeleton completely dissolves away in the body. Massive clinical trials of comparing Absorb's safety and efficacy with those of the metallic everolimus-eluting Xience stent have been underway toward patients with coronary artery disease. Similar technologies are being applied for the development of Elixir Medical's DESolve and Boston Scientific's Synergy stents. From regulatory perspective, all these bioresorbable PLGA stents are considered experimental and investigational in the United States, because their effectiveness and safety are not approved by the FDA. However, Elixir Medical's DESolve has been launched in overseas markets. In summary, innovative stent types are being developed to overcome the issues of conventional metallic stents. A dissolvable, vascular PLGA scaffold type is an important class of emerging new stents.

6.10 Summary

As discussed so far, PLGA polymers have a number of advantages as drug carriers, scaffolds for tissue engineering and regenerative medicine, orthopedic devices, and surgical materials. Innovative formulations and manufacturing techniques have been developed for their clinical translation and the expansion of clinical opportunities for PLGA applications. In particular, representative dosage forms attracting strong interest for product research and development are microspheres, ready-to-use solid implants, *in situ* implants/gels, nanofibers, micelles, polyplexes, polymersomes, nanoparticles, stents, and scaffolds with varying morphology. These dosage forms will enable unrestricted PLGA applications in controlled drug release, multi-drug delivery, tissue engineering, vascular engineering, orthopedics, and combination products.

Acknowledgements

This study was supported by the Korea SGER Program through the National Research Foundation of Korea (NRF) funded by the Ministry of Education, Science and Technology (NRF-2014R1A1A2A16054899).

References

1. Houchin, M. L. and Topp, E. M. (2009). Physical properties of PLGA films during polymer degradation, *J Appl Polym Sci,* **114**, 2848–2854.
2. Sanders, L. M., McRae G. I., Vitale, K. M., and Kell, B. A. (1985). Controlled delivery of an LHRH analogue from biodegradable injectable microspheres, *J Control Release,* **2**, 187–195.
3. Couvreur, P. and Puisieux, F. (1993). Nano- and microparticles for the delivery of polypeptides and proteins, *Adv Drug Deliv Rev,* **10**, 141–162.
4. Pan, Z. and Ding, J. (2012). Poly(lactide-*co*-glycolide) porous scaffolds for tissue engineering and regenerative medicine, *Interface Focus,* **2**, 366–377.
5. Félix Lanao, R. P., Jonker A. M., Wolke, J. G., *et al.* (2013). Physicochemical properties and applications of poly(lactic-*co*-glycolic acid) for use in bone regeneration, *Tissue Eng Part B Rev,* **19**, 380–390.

6. Manchanda, R., Chue, P., Malla, A., *et al.* (2013). Long-acting injectable antipsychotics: evidence of effectiveness and use, *Can J Psychiatry*, **58**, 5S–13S.

7. Lancranjan, I., Bruns, C., Grass, P., *et al.* (1996). Sandostatin LAR: a promising therapeutic tool in the management of acromegalic patients, *Metabolism*, **45**, 67–71.

8. Ciraulo, D. A., Dong, Q., Silverman, B. L., Gastfriend, D. R., and Pettinati, H. M. (2008). Early treatment response in alcohol dependence with extended-release naltrexone, *J Clin Psychiatry*, **69**, 190–195.

9. Ryan, G. J., Moniri, N. H., and Smiley, D. D. (2013). Clinical effects of once-weekly exenatide for the treatment of type 2 diabetes mellitus, *Am J Health Syst Pharm*, **70**, 1123–1131.

10. US Department of Health and Human Services. (2007). *Federal Register*, **72**, 19665.

11. Mao, S., Guo C., Shi Y., and Li, L. C. (2012). Recent advances in polymeric microspheres for parenteral drug delivery–part 1, *Expert Opin Drug Deliv*, **9**, 1161–1176.

12. Schwendeman, S. P., Shah, R. B., Bailey, B. A., and Schwendeman, A. S. (2014). Injectable controlled release depots for large molecules, *J Control Release*, **190**, 240–253.

13. Wang, L., Liu, Y., Zhang, W., *et al.* (2013). Microspheres and microcapsules for protein delivery: strategies of drug activity retention, *Curr Pharm Des*, **19**, 6340–6352.

14. Tamber, H., Johansen, P., Merkle, H. P., and Gander B. (2005). Formulation aspects of biodegradable polymeric microspheres for antigen delivery, *Adv Drug Deliv Rev*, **57**, 357–376.

15. Liu, X., Heng Paul W. S., Li, Q., and Chan, L. W. (2006). Novel polymeric microspheres containing norcantharidin for chemoembolization, *J Control Release*, **116**, 35–41.

16. Kettenbach, J., Stadler, A., Katzler, I. V., *et al.* (2008). Drug-loaded microspheres for the treatment of liver cancer: review of current results, *Cardiovasc Intervent Radiol*, **31**, 468–476.

17. Wang, Y., Benzina, A., Molin, D. G., *et al.* (2015). Preparation and structure of drug-carrying biodegradable microspheres designed for transarterial chemoembolization therapy, *J Biomater Sci Polym Ed*, **26**, 77–91.

18. Qian, J., Truebenbach, J., Graepler, F., *et al.* (2003). Application of poly-lactide-*co*-glycolide-microspheres in the transarterial

chemoembolization in an animal model of hepatocellular carcinoma, *World J Gastroenterol,* **9**, 94–98.

19. Sherman, R. N. (2006). Sculptra: the new three-dimensional filler, *Clin Plast Surg,* **33**, 539–550.

20. Shiah, J. G., Bhagat, R., Blanda, W. M., *et al.* (2011). Ocular implant made by a double extrusion process, *US Patent* 8,034,366.

21. Wong, V. G., and Hu, M. W. L. (2011). Implants and methods for treating inflammation-mediated conditions to the eye, *US Patent* 8,063,031.

22. Chandrashekar, B. L., Zhou, M., Jarr, E. M., and Dunn, R. L. (2000). Controlled release liquid delivery compositions with low initial drug burst, *US Patent* 6,143,314.

23. Dunn, R. L., Garrett, J. S., Ravivarapu, H., and Chandrashekar, B. L. (2013). Polymeric delivery formulations of leuprolide with improved efficacy, *US Patent* 8,486,455.

24. Norton, R. L., Watkins, A., and Zhou, M. (2014). Injectable flowable composition comprising buprenorphine, *US Patent* 8,921,387.

25. Kunkle, B. N., Williams, J. C., Johnson, E. G., *et al.* (2013). Persistent efficacy and production benefits following use of extended-release injectable eprinomectin in grazing beef cattle under field conditions, *Vet Parasitol,* **192**, 332–337.

26. Ashjari, M., Khoee, S., Mahdavian, A. R., and Rahmatolahzadeh, R. (2012). Self-assembled nanomicelles using PLGA-PEG amphiphilic block copolymer for insulin delivery: a physicochemical investigation and determination of CMC values, *J Mater Sci Mater Med,* **23**, 943–953.

27. Davis, M. E., Chen, Z. G., and Shin, D. M. (2008). Nanoparticle therapeutics: an emerging treatment modality for cancer, *Nat Rev Drug Discov,* **7**, 771–782.

28. Kim, T. Y., Kim, D. W., Chung, J. Y., *et al.* (2004). Phase I and pharmacokinetic study of Genexol-PM, a Cremophor-free, polymeric micelle-formulated paclitaxel, in patients with advanced malignancies, *Clin Cancer Res,* **10**, 3708–3716.

29. Kim, D. W., Kim, S. Y., Kim, H. K., *et al.* (2007). Multicenter phase II trial of Genexol-PM, a novel Cremophor-free, polymeric micelle formulation of paclitaxel, with cisplatin in patients with advanced non-small-cell lung cancer, *Ann Oncol,* **18**, 2009–2014.

30. Oerlemans, C., Bult, W., and Bos, M. (2010). Polymeric micelles in anticancer therapy: targeting, imaging and triggered release, *Pharm Res,* **27**, 2569–2589.

31. Danhier, F., Ansorena, E., Silva, J. M., *et al.* (2012). PLGA-based nanoparticles: an overview of biomedical applications, *J Control Release,* **161**, 505–522.
32. Mohammadi-Samani, S. and Taghipour, B. (2015). PLGA micro and nanoparticles in delivery of peptides and proteins; problems and approaches, *Pharm Dev Technol,* **20**, 385–393.
33. Grottkau, B. E., Cai, X., Wang, J., Yang, X., and Lin, Y. (2013). Polymeric nanoparticles for a drug delivery system, *Curr Drug Metab,* **14**, 840–846.
34. Sah, E. and Sah, H. (2015). Recent trend in preparation of poly(lactide-*co*-glycolide) nanoparticles by mixing polymeric organic solution with antisolvent, *J Nanomater,* **2015**, article ID 794601.
35. Amoozgar, Z. and Yeo, Y. (2012). Recent advances in stealth coating of nanoparticle drug delivery systems. *Wiley Interdiscip Rev Nanomed Nanobiotechnol,* **4**, 219–233.
36. Mandal, B., Bhattacharjee, H., Mittal, N., *et al.* (2013). Core-shell-type lipid-polymer hybrid nanoparticles as a drug delivery platform, *Nanomedicine,* **9**, 474–491.
37. Tan, S., Li, X., Guo, Y., and Zhang, Z. (2013). Lipid-enveloped hybrid nanoparticles for drug delivery, *Nanoscale,* **5**, 860–872.
38. Hu, C. M., Zhang, L., Aryal, S., *et al.* (2011). Erythrocyte membrane-camouflaged polymeric nanoparticles as a biomimetic delivery platform, *Proc Natl Acad Sci U S A,* **108**, 10980–10985.
39. Xie, H., She, Z. G., Wang, S., Sharma, G., and Smith, J. W. (2012). One-step fabrication of polymeric janus nanoparticles for drug delivery, *Langmuir,* **28**, 4459–4463.
40. Wei, Z. N., Li, W., Seo, M., Xu, S., and Kumacheva, E. (2006). Janus and ternary particles generated by microfluidic synthesis: design, synthesis, and self-assembly, *J Am Chem Soc,* **128**, 9408–9412.
41. Labbaf, S., Deb, S., Cama, G., Stride, E., and Edirisinghe, M. (2013). Preparation of multicompartment sub-micron particles using a triple-needle electrohydrodynamic device, *J Colloid Interface Sci,* **409**, 245–254.
42. Yu, Y., Pang, Z., Lu, W., *et al.* (2012). Self-assembled polymersomes conjugated with lactoferrin as novel drug carrier for brain delivery, *Pharm Res,* **29**, 83–96.
43. Discher, D. E., Ortiz, V., Srinivas, G., *et al.* (2007). Emerging application of polymersomes in delivery: from molecular dynamics to shrinkage of tumors, *Polym Sci,* **32**, 838–857.

44. Sah, H., Thoma, L. A., Desu, H. R., Sah, E., and Wood, G. C. (2013). Concepts and practices used to develop functional PLGA-based nanoparticulate systems, *Int J Nanomedicine,* **8**, 747–765.

45. Middleton, J. C. and Tipton, A. J. (2000). Synthetic biodegradable polymers as orthopedic devices. *Biomaterials,* **21**, 2335–2346.

46. Lickorish, D., Guan, L., and Davies, J. E. (2007). A three-phase, fully resorbable, polyester/calcium phosphate scaffold for bone tissue engineering: evolution of scaffold design, *Biomaterials,* **28**, 1495–1502.

47. Guan, L. and Davies, J. E. (2004). Preparation and characterization of a highly macroporous biodegradable composite tissue engineering scaffold, *J Biomed Mater Res A,* **71**, 480–487.

48. Davies, J. E., Matta, R., Mendesm, V. C., and Perri de Carvalho, P. S. (2010). Development, characterization and clinical use of a biodegradable composite scaffold for bone engineering in oro-maxillo-facial surgery, *Organogenesis,* **6**, 161–166.

49. Beltzer, C., Hägele, J., Wilke, A., *et al.* (2013). Monitoring degradation process of PLGA/Cap scaffolds seeded with mesenchymal stem cells in a critical-sized defect in the rabbit femur using Raman spectroscopy, *J Bone Marrow Res,* **1**, 133.

50. Gotlieb, E. L., Murray, P. E., Namerow, K. N., Kuttler, S., and Garcia-Godoy, F. (2008). An ultrastructural investigation of tissue-engineered pulp constructs implanted within endodontically treated teeth, *J Am Dent Assoc,* **139**, 457–465.

51. Alexander, D., Hoffmann, J., and Munz, A. (2008). Analysis of OPLA scaffolds for bone engineering constructs using human jaw periosteal cells, *J Mater Sci Mater Med,* **19**, 965–974.

52. Seil, J. T. and Webster, T. J. (2011). Spray deposition of live cells throughout the electrospinning process produces nanofibrous three-dimensional tissue scaffolds, *Int J Nanomedicine,* **6**, 1095–1099.

53. Li, W. J., Laurencin, C. T., Caterson, E. J., Tuan, R. S., and Ko, F. K. (2002). Electrospun nanofibrous structure: A novel scaffold for tissue engineering, *J Biomed Mater Res,* **60**, 613–621.

54. Ma, Z., Kotaki, M., Inai, R., and Ramakrishna, S. (2005). Potential of nanofiber matrix as tissue-engineering scaffolds, *Tissue Eng,* **11**, 101–109.

55. Kanani, A. G. and Bahrami, S. H. (2010). Review on electrospun nanofibers scaffold and biomedical applications, *Trends Biomater Artif Organs,* **24**, 93–115.

56. Agarwal, S., Wendorff, J. H., and Greiner, A. (2008). Use of electrospinning technique for biomecial applications, *Polymer*, **49**, 5603–5621.

57. Lu, W., Sun, J., and Jiang X. (2014). Recent advances in electrospinning technology and biomedical applications of electrospun fibers, *J Mater Chem B*, **2**, 2369–2380.

58. Kim, H. S. and Yoo, H. S. (2014). Therapeutic application of electrospun nanofibrous meshes, *Nanomedicine (Lond)*, **9**, 517–533.

59. Novosel, E. C., Kleinhans, C., and Kluger, P. J. (2011). Vascularization is the key challenge in tissue engineering. *Adv Drug Deliv Rev*, **63**, 300–311.

60. Medeiros, E. S., Glenn, G. M., Klamczynski, A. P., Orts, W. J., and Mattoso, L. H. C. (2009). Solution blow spinning: a new method to produce micro- and nanofibers from polymer solutions, *J Appl Polym Sci*, **113**, 2322–2330.

61. Behrens, A. M., Casey, B. J., Sikorski, M. J., *et al.* (2014). *In situ* deposition of PLGA nanofibers via solution blow spinning, *ACS Macro Lett*, **3**, 249–254.

62. Ma, X., Wu, T., and Robich, M. P. (2012). Drug-eluting stent coatings, *Interv Cardiol*, **4**, 73–83.

63. Abizaid, A. C., De Ribamar Costa Junior, J., Whitbourn, R. J., and Chang, J. C. (2007). The CardioMind coronary stent delivery system: stent delivery on a 0.014″ guidewire platform, *EuroIntery*, **3**, 154–157.

64. Martin, D. M. and Boyle, F. J. (2011). Drug-eluting stents for coronary artery disease, *Med Eng Phys*, **33**, 148–163.

65. Ge, J., Qian, J., Wang, X., *et al.* (2007). Effectiveness and safety of the sirolimus-eluting stents coated with bioabsorbable polymer coating in human coronary arteries, *Catheter Cardiovasc Interv*, **69**, 198–202.

66. Lee, C. H., Lim, J., Low, A., *et al.* (2007). Sirolimus-eluting, bioabsorbable polymer-coated constant stent (Cura™) in acute ST-elevation myocardial infarction: a clinical and angiographic study (CURAMI Registry), *J Invas Cardiol*, **19**, 182–185.

67. Vranckx, P., Serruys, P., Gambhir, S., *et al.* (2006). Biodegradable-polymer-based, paclitaxel-eluting Infinnium™ stent: 9-month clinical and angiographic follow-up results from the SIMPLE II prospective multicenter registry study, *EuroIntery*, **2**, 310–317.

68. Niu, X., Yang, C., Chen, D., *et al.* (2014). Impact of drug-eluting stents with different coating strategies on stent thrombosis: A meta-analysis of 19 randomized trials, *Cardiol J*, **21**, 557–568.

69. Nakano, K., Egashira, K., Masuda, S., *et al.* (2009). Formulation of nanoparticle-eluting stents by a cationic electrodeposition coating technology: efficient nano-drug delivery via bioabsorbable polymeric nanoparticle-eluting stents in porcine coronary arteries. *JACC Cardiovasc Interv*, **2**, 277–283.

Chapter 7

Overview of Polycaprolactone-Based Drug Delivery System

Aurelio Salerno
Institute of Materials Science of Barcelona, Spanish National Research Council, Campus de la UAB, Bellaterra, 08193, Spain
Centre for Advanced Biomaterials for Health Care@CRIB, Istituto Italiano di Tecnologia, Largo Barsanti e Matteucci 53, 80125, Naples, Italy
asalerno@icmab.es, asalerno@unina.it

Biodegradable polymers are essential components of drug delivery systems as they may provide many opportunities to improve the therapeutic efficacy of bioactive molecules for tissue engineering and health care applications. Polycaprolactone (PCL) is one of the most used and interesting polymer for drug delivery. This is because PCL is highly biocompatible with cells and biological tissues and can be processed into appropriate forms for drug delivery purposes. The main aim of this chapter is to provide the reader with an overview of the drug delivery application of PCL as well as its blends and composites. The use of different PCL formulations, namely blends and composites, nano- and microparticles, fibers, and porous scaffolds, in tissue engineering and drug delivery are described.

Handbook of Polyester Drug Delivery Systems
Edited by MNV Ravikumar

ISBN 978-981-4669-65-8 (Hardcover), 978-981-4669-66-5 (eBook)
www.panstanford.com

7.1 Introduction

Tissue engineering (TE) and health care (HC) are pioneering and continuously evolving research fields aiming to develop therapeutic treatments to fight disease and to repair/regenerate biological tissues and organs. Current approaches of TE and HC are highly multidisciplinary and complex, as they require the integration of emerging knowledge in the physical and life sciences with frontier engineering and clinical medicine. In particular, in biomaterials and molecular cues/drugs are two main components of TE and HC strategies as they are devoted to stimulate and drive the intrinsic regenerative capacities of cells and tissues [1].

To date, the integration between biomaterials and drugs into multifunctional biomedical devices has enabled the regeneration of several tissues in the laboratory, and laboratory-grown organs are continuously producing novel implants. Nevertheless, many obstacles and challenges remain for the successful clinical implementation of biomedical devices, mainly the difficulty of replicating the complex hierarchical architecture of tissues, such as bone and dermis at subcellular, cellular, and tissue levels. Overcoming these limitations can probably be achieved by designing novel multifunctional biomaterials provided by a suitable three-dimensional structure and, concomitantly, capable of coordinating, both spatially and temporally, the delivering of biological signals to cells [2–4]. This last issue is especially critical as simple bolus administration of drugs and bioactive molecules cannot be effective because of their uncontrollable diffusion rate and their concomitant enzymatic digestion or deactivation when in contact to biological environments [3]. Conversely, drug delivery platforms can prevent drug inactivation during the whole release duration and can provide local delivery and prolonged exposition of the bioactive molecules to the target sites [3].

Polypeptide growth factors are powerful regulators of biological function. They modulate many cellular functions, including migration, proliferation, differentiation, and survival [2, 5, 6]. Therefore, additional direction over cell fate, beyond control of biomaterials chemistry, can be achieved through the spatial and temporal controlled incorporation and release of these molecular cues. For example, porous synthetic scaffolds made of polycaprolactone (PCL) have been fabricated for the controlled release of grow factors able to recruit mesenchymal stem cells from the host [7]. These scaffolds limited scar tissue formation when implanted *in vivo* and enhanced cartilage regeneration. Different types of polymeric scaffolds have been also developed for the controlled delivery of growth factors, such as vascular endothelial growth factor (VEGF), to stimulate implant vascularization and blood vessel formation [8].

PCL is a hydrophobic, semicrystalline biodegradable polymer synthesized by ring-opening polymerization of ε-caprolactone (Fig. 7.1) [9]. Owing to its biocompatibility and biodegradability, PCL formulations are currently used for controlled drug delivery in TE and HC applications. Some examples are soft and hard tissue compatible biodegradable PCL films [10, 11] as well as PCL particles and porous scaffolds for regeneration of bone [12–14], skin [15], nerve [16], and blood vessel [17].

The goal of this chapter is to provide to the reader an overview of current biomedical use of PCL for drug delivery in TE and HC. In particular, Section 7.2 describes the most important chemical and physical properties of PCL which make them a suitable candidate material for biomedical applications. Section 7.3 addresses the different aspects related to the processing of PCL for the design and fabrication of biomedical devices for drug delivery and TE. Finally, Section 7.4 presents some concluding remarks.

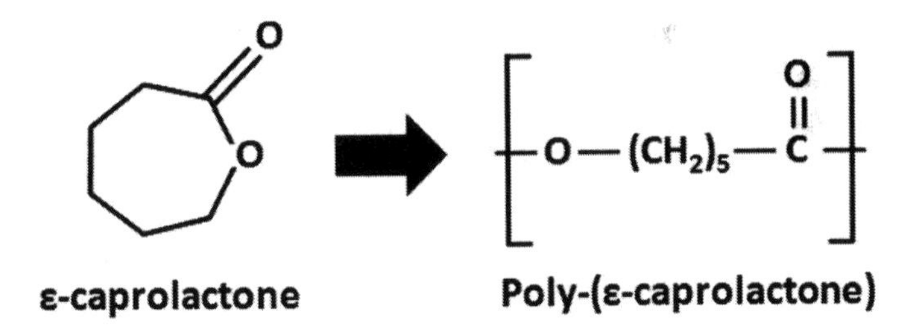

Figure 7.1 Scheme of the synthesis of PCL by ring-opening polymerization of ε-caprolactone.

7.2 Overview of PCL Properties Relevant for TE Applications

The use of PCL as biomaterials for TE and HC has grown significantly in the past decade [18]. In fact, PCL is a linear aliphatic polyester approved by the US Food and Drug Administration as, for example, drug delivery device, suture, or filler for skull defects [18, 19]. The glass transition temperature and melting range of PCL are –60°C and 59–64°C, respectively. However, the melting range depends strongly on the crystalline properties of PCL, which in turn is dictated by the molecular weight (normally varying from 10 to 42.5 kDa) and, to some extent, by the fabrication process [20, 21]. The degree of crystallinity of PCL, which affects important properties, such as polymer mechanical strength and degradation, can vary from 30% to 50 % [22] up to 60–80% [23–25]. As reflected in the typical X-ray diffraction spectra of PCL, this polymer has two distinct diffraction peaks at 2θ = 21.58° and 2θ = 23.88°, corresponding to the (110) and (200) planes, respectively, and indicating the presence of an orthorhombic unit cell [26].

The bulk compressive and tensile mechanical properties of PCL have been reported in the literature by both manufacturers and researchers. Ang *et al.* studied the effect of degradation on the mechanical properties of PCL (MW = 10–20 kDa) processed by melt molding and reported a compressive modulus equals to 407 MPa for the undegraded sample [27]. Several groups have also reported the tensile mechanical properties of bulk PCL processed by conventional methods and an overview of the obtained values can be found in [28]. As example, Engelberg and Kohn reported a tensile modulus of 400 MPa and tensile strength of 16 MPa for compression-molded PCL (MW = 42.5 kDa) [29]. Compared to other aliphatic linear polyesters, such as polyglycolic acid (PGA) and polylactic acid (PLA), PCL is provided by lower stiffness while higher strain-at-break and, therefore, is regarded as a ductile material. These differences in the mechanical properties of PCL in comparison with PLA and PGA are ascribable mainly to the differences in the chemical structure of the polymers which, in turn, affect their thermal behavior. It is noteworthy that PCL has significantly lower glass transition and melting temperatures if compared to both PLA and PGA photopolymers [29].

The biodegradability of PCL is dictated by the presence of hydrolytically unstable aliphatic-ester linkage in its chain structure (Fig. 7.1). In particular, literature investigations reported that PCL degradation often proceed via both bulk and surface erosion, resulting in the formation of 5-hydroxyhexanoic acids (caprice acids) products [27]. In particular, the degradation starts with the diffusion of the soaking medium in the amorphous regions of PCL and subsequently, the PCL chains fragmented into smaller ones and the degradation is auto-catalyzed by carbonyl end groups of fragmented polymeric chain [30, 31]. The weight loss usually starts after 4–6 months, which is the minimum time required to observe the formation of PCL fragments small enough to diffuse through the polymeric matrix into the soaking medium [30]. If compared to PLA and PGA photopolymers, the overall PCL degradation is significantly lower as its take place in 2–4 years, as a consequence of the enhanced surface hydrophilicity and crystallinity of PCL [18, 30]. Anyway, the degradation rate of PCL and its weight loss kinetic were reported to be adequate for matching the rate of bone formation *in vivo* [18]. Most importantly the presence of enzymes, such as lipase, into the degradation medium can accelerate significantly PCL degradation, as demonstrated by the formation of small oligomers and monomers only after 3–4 weeks of incubation [31, 32]. Regarding the degradation rate of PCL *in vivo*, it has been reported that this process occurs in two stages: first, the non-enzymatic hydrolytic cleavage of ester groups and subsequently, when the polymer is almost crystalline and provided of low molecular weight chains, degradation progresses by digestion by macrophages, giant cells and fibroblasts [18].

The biocompatibility is defined as the ability of a material to perform with an appropriate host response in a specific application [33]. In general, the biocompatibility of biodegradable polymers, such as aliphatic polyesters, is directly correlated to the mechanisms of their biodegradation. *In vitro* biocompatibility is evaluated through cell culture systems and is often used to carry out a first screening of the biological response of a biomaterial. This approach can be used, for instance, to assess the effect of chemical, topographical and structural properties of materials on cell adhesion, proliferation, migration and differentiation [34–36]. There is a large literature investigation on the *in vitro* biocompatibility performance of PCL.

Serrano and coworkers prepared PCL membranes with either smooth or rough surface for the *in vitro* culture of fibroblasts for developing vascular grafts [10]. The results of their study proved the ability of PCL to promote the adhesion, growth, viability, morphology, and mitochondrial activity of cells. In another work, Porter and coworkers developed a solvent-free template synthesis technique useful to fabricate PCL nanowire surfaces as a building block for the development of three-dimensional bone scaffolds [37]. The *in vitro* study revealed that the topographical feature of the surface of PCL samples may also affect cell response. Indeed, the authors reported that PCL nanowire surfaces enhanced mesenchymal stem cells adhesion and viability as compared with smooth PCL and also accelerated the production of bone extracellular matrix. *In vitro* cell culture onto porous PCL scaffolds has been also deeply investigated as model system for the regeneration of complex tissues. Some examples are the works by Salerno and coworkers which investigated the effect of the pore structure of PCL scaffold on bone regeneration [14, 38] as well as the works by Izquierdo *et al.* [24] and by Prabhakaran *et al.* [39] for *in vitro* cartilage and nerve regeneration, respectively. Despite the *in vitro* biological performance, *in vivo* assessment of the peri-implant and host responses, mainly immunogenic, carcinogenic and thrombogenic responses is the ultimate essential way to demonstrate the biocompatibility of a biomaterial for a specific application. The complexity of these host responses is a result of a cascade of processes involving multiple interdependent material-tissue interactions [18]. For example, the release of acidic degradation products from bioresorbable polymers and implants is responsible for the observed inflammatory reactions which, in turn, also depend on the site of implantation. Indeed, the capacity of the surrounding tissues to eliminate the degradation products is correlated to the intensity and duration of the inflammatory response [18]. The application of PCL for the fabrication of biodegradable implants for TE is widespread. Williams and coworkers have designed and fabricated porous PCL scaffolds by combining computational design and rapid prototyping fabrication [40]. The scaffolds were provided of porosity and mechanical properties adequate for bone regeneration and, after *in vivo* implantation, showed correct new tissue formation. Fechek *et al.* tested the possible use of PCL scaffolds prepared by means of porogen leaching technique as platform for the chondrogenic differentiation of embryonic stem cells and ultimately, *in vivo* cartilage regeneration [41]. The feasibility of PCL to be used as scaffold material for *in vivo* TE has been also demonstrated for application such as wound dressing [42], vascular graft substitution [43] and nerve regeneration [44]. In conclusion, all of these *in vitro* and *in vivo* studies demonstrated that PCL may be an excellent candidate material for a large variety of TE applications.

7.3 Processing of PCL for the Fabrication of Drug Delivery Devices

The use of PCL in drug delivery for TE and HC is widespread. This is because, as previously discussed, PCL is biocompatible and biodegradable and is suitable for controlled drug delivery due to a high permeability to many drugs [30]. Biodegradation of PCL is slow in comparison to other polymers, such PLA and PGA, so it is most suitable for long-term delivery (up to 1–2 years). However, the design of the appropriate PCL formulation is essential for the success of the drug delivery device. As shown in Fig. 7.2, several approaches have been developed for the fabrication of PCL platforms for TE and HC purposes.

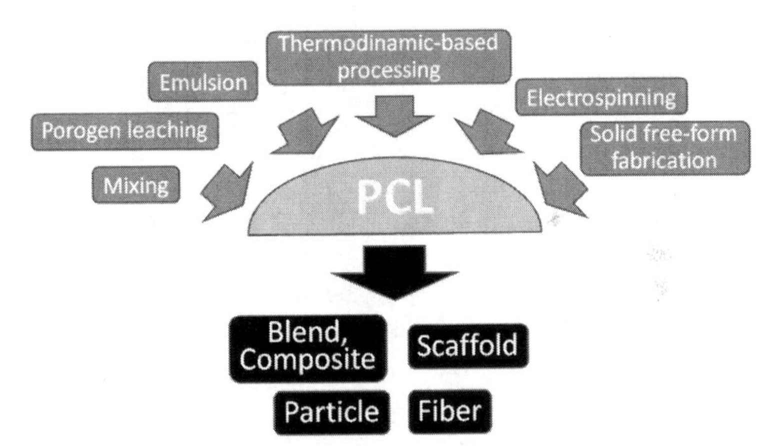

Figure 7.2 Scheme of the processing techniques that have been developed and can be conceivably used to fabricate PCL samples in form of blends/composites, nano- and microparticles, fiber, and porous scaffolds.

Blending two or more polymers or dispersing an inorganic powder phase into a polymeric matrix are well-used approaches whenever it is required to control properties by using conventional technology at relatively low cost. These multi-phase materials can be created with the purpose of improving physical properties, such as elastic modulus and yield strength, for application such as bone regeneration [45] or to modulate the degradation rate and the hydrophilic properties of PCL [46–50]. For example, by blending PCL and natural polymers, such as proteins and saccharides, it was possible to prepare multi-phase materials with enhanced wettability, faster degradation rate and able to improve cell differentiation. In particular, blends of PCL and zein, a vegetal protein extracted from corn, were used for preparing multi-phase materials for bone TE (Fig. 7.3) [46, 47]. Indeed, when blended with PCL, the natural polymer promoted the absorption of the biological medium and the material start to degrade after only few days of soaking [46, 47]. Furthermore, the higher hydrophilicity and peculiar topographical properties of the multi-phase material resulted in the enhancement of *in vitro* stem cells and osteoblasts adhesion and differentiation [46, 47]. Blends of PCL with other polyesthers, such as PLA was also reported to control the degradation behavior and the tensile properties of the sample [49, 50]. Indeed, the addition of PLA may allow fabricating PCL-based materials with enhanced stiffness and strength and accelerated degradation in physiological medium [49, 50]. PCL-PLA blends with incorporated calcium phosphates were also developed for guided bone regeneration [51].

Polymer blending also enabled the control and optimization of the drug releasing capability of PCL [45, 52, 53]. For example, Suárez-González *et al.* have recently developed mineral coatings on PCL scaffolds to serve as templates for growth factor binding and release [52]. Mineral coatings were formed using a biomimetic approach that consisted in the incubation of scaffolds in modified simulated body fluids. The results of their work demonstrated that vascular endothelial growth factor and bone morphogenetic protein were bound with efficiencies up to 90% to mineral-coated PCL scaffolds. Furthermore, the release kinetics of these compounds can be efficiently controlled by modulating the solubility of the mineral coating [52]. In another work, researchers used polyethylene oxide (PEO) to prepare PEO/PCL blends for the production of monolithic matrices for oral drug delivery [54]. Several batches of matrix material were prepared with carvedilol used as the active pharmaceutical ingredient. The matrices were prepared by extrusion and the effect of screw speed and barrel temperature on the drug delivery properties of devices was investigated. Most importantly, dissolution testing showed that increasing PEO concentration resulted in the acceleration of drug release [54].

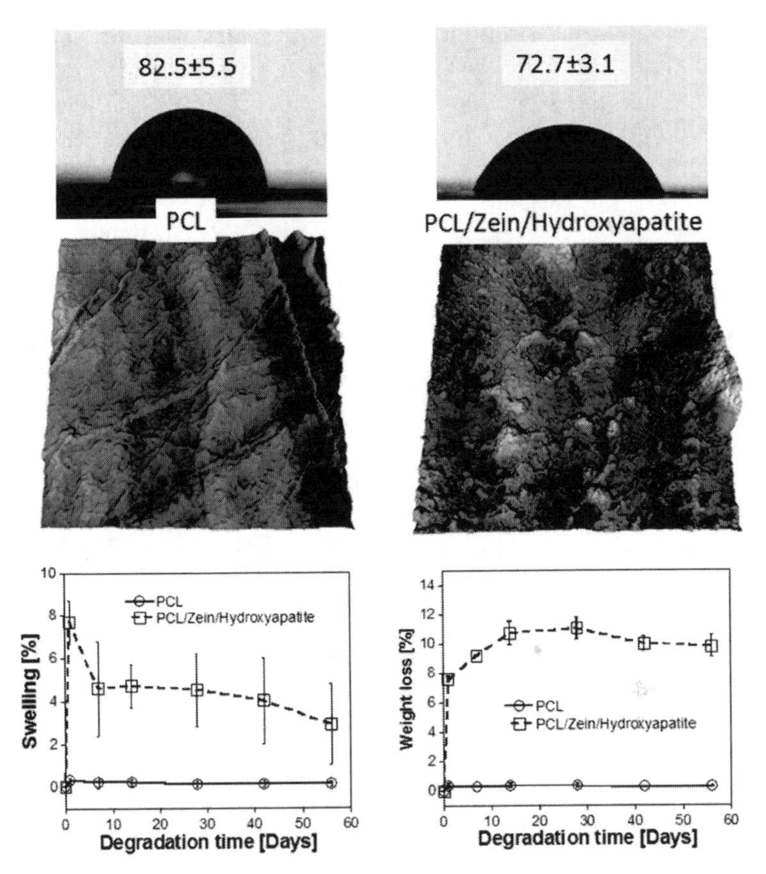

Figure 7.3 Effect of composition of PCL and PCL/zein/hydroxyapatite materials on wettability, topography, and *in vitro* degradation in phosphate buffer saline solution. Adapted from [46] and [47].

In recent years, an increasing interest has raised from both the industry and the research community in the development of novel nano and microparticles constituted by PCL-based materials suitable for drug delivery applications. Indeed, the PCL nano- and micro-carriers can effectively achieve the required therapeutic concentration of the drug at the target site for a desired period of time and limiting undesirable effects on other organs while circulating in the body [55]. Furthermore, the evolution of micro and nanotechnology processes has enabled the extensive production of particles of different sizes, shapes and structures. For instance, PCL nanoparticles, with spherical shape and uniform size distribution (250–300 nm), were prepared by using solvent displacement method for tamoxifen encapsulation and delivery [55]. Magnetic PCL nanoparticles containing the anti-cancer drugs cisplatin and gemcitabine were prepared by the o/w emulsion method and used as magnetically responsive drug carrier for cancer treatment [56].

PCL is an excellent candidate material also for drug microencapsulation as PCL microparticles can be prepared following different approaches, such as emulsion [31, 57], phase separation [25], and supercritical antisolvent precipitation [58], enabling high versatility in terms of the design of the structural characteristic and drug-releasing rate. Figure 7.4 shows a simple and versatile approach for the fabrication of porous PCL microparticles with controlled morphology and size distribution [25]. The process involves the dissolution of PCL in ethyl lactate, followed by the cooling of the solution to induce the phase separation and ultimately, the coagulation of the polymeric particles in water.

As shown in the SEM image of Fig. 7.4A, the microparticles are provided of a porous surface with lamellar morphology and pore sizes of the order of a few microns induced by the spherulitic crystallization of the polymer during solution cooling. Furthermore, as shown in Fig. 7.4B, by combining thermally induced phase separation and gelatin particles leaching techniques it has been possible the fabrication of multiscaled porous PCL microparticles with potential application as cell carries for TE [25]. The drug release kinetic is strongly affected by the crystalline structure of PCL microparticles. Indeed, it was reported that the release of antispasmodic drug papaverine from PCL microparticles was controlled by drug diffusion through the amorphous region of the polymer matrix, not by polymer erosion [59]. In particular, the increase of PCL molecular weight resulted in the decrease of the overall fraction of crystallinity and, in turn, resulted in a lower drug release rate. Moreover, PCL microparticles with different crystal structure while similar crystallinity fraction were prepared by applying different thermal histories. As a direct consequence, the samples annealed at a higher temperature were provided of larger crystallites and higher crystal perfection and resulted in more sustained release pattern [59]. In another work, microparticles of PCL and poly(hydroxybutyrate-cohydroxyvalerate) blend were prepared for the encapsulation and sustained release of two anti-inflammatory drugs, namely diclofenac and indomethacin [60]. The mechanism of release was mainly controlled by the drug diffusion and the drug release profiles were modulated by varying the PCL concentration in the blend. In particular, by increasing the amount of PCL, from 0 up to 50% in weight, the porosity of the microparticles increased and consequently, the release rates of diclofenac and indomethacin from microparticles were accelerated.

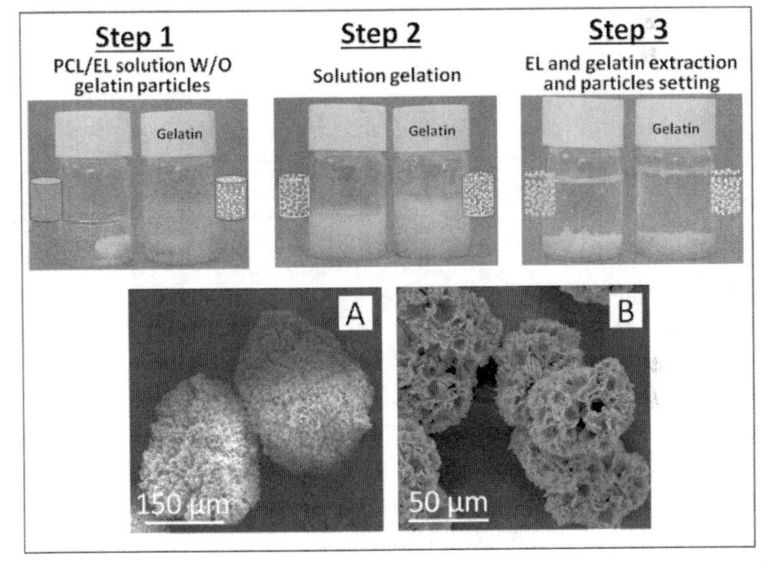

Figure 7.4 Scheme of the preparation of PCL microparticles by means of thermal induced phase separation technique from solution in ethyl lactate as green solvent. Morphology of PCL particles prepared (A) without and (B) with the addition of gelatin spheres as leachable porogen. Adapted from [25].

Biocompatible, biodegradable, and mechanically stable PCL fibers have found extensive use in the biomedical field as suture and fixation materials [61, 62]. Furthermore, as exhaustively reviewed by Cipitria and coworkers [21], with the advent of electrospinning technique in the mid-1990s, the fabrication of nano- and microfibrous nonwoven PCL maths were largely investigated as scaffolds for TE applications. This is because electrospinning is a relatively simple and low-cost technique enabling the fabrication of nanostructured fibrous materials mimicking the morphology and structure of the extracellular matrix of biological tissues [63]. A basic electrospinning system usually consists of a high-voltage power supply, a spinneret, and a grounded collecting plate. The polymer, in solution or in melt state, is fed through the spinneret under an external electric field which induces the formation of a tiny jet. The jet propagates toward the collecting plate while the fiber solidifies due to the evaporation of the solvent or the decrease of the temperature. As a result, a nonwoven fibrous matrix with a large surface area-to-volume ratio and a small pore size is achieved [63, 64]. Compared with other drug releasing forms, electrospun PCL nanofibers have the advantage of the control of the drug release by the modulation of fibers' morphology, porosity, and composition [65]. Furthermore, the very small diameter of the nanofibers can provide short diffusion passage length while their high surface area is helpful to achieve a controlled mass transfer and efficient drug release [66]. The electrospinning of PCL was used, among other, for the controlled release of heparin [17], ketoprofen [67], triclosan [68] and the antibiotic Biteral® [62]. Blends of PCL and PEO were also prepared via electrospinning technique for the controlled release of lysozyme [65]. The release rate of lysozyme from the samples was finely modulated by varying the composition of the blend and in particular, drug release was promoted by increasing the amount of PEO [65]. Similarly, electrospinning of blends of PCL and PLA was used to accelerate the release rate of the antimicrobial triclosal [68]. Most interestingly, coaxial electrospinning was developed for the fabrication of drug-loaded PCL nanofibers in the form of core–shell structure, which have potential applications in functional dressing for wound healing [66] and drug releasing scaffold [69]. The main advantage of this approach is that the drug is completely confined by the external polymeric shell and therefore, the burst release of the encapsulated drugs as well as its deterioration during the fabrication process can be avoided.

A clinically effective TE approach for the repair/regeneration of complex three dimensional tissues, such as bone and cartilage, requires a porous scaffold able to support cell migration, proliferation, differentiation, and survival. There are several techniques suitable for fabricating porous PCL scaffolds. Figures 7.5 and 7.6 show scanning electron microscopy images evidencing the morphology of porous PCL scaffolds prepared by gas foaming and phase separation, respectively. Gas foaming technique is based on the high-pressure solubilization of a nontoxic blowing agent, mainly CO_2, N_2, or a mixture of both, within the polymer. The system is then brought into a supersaturated state either by increasing temperature or by reducing pressure, with the consequent nucleation and growth of gas bubbles. Finally, the decrease of the temperature and the crystallization of the polymeric matrix stabilize the pore structure [23, 70, 71]. The gas foaming technique allows a fine control over the porous network of the scaffolds by the selection of the proper operating conditions, mainly the blowing agent type and concentration, the foaming temperature and the pressure drop profile [70]. Furthermore, composites of PCL and calcium phosphate nano- and micro-particles can be processed by foaming technique for the preparation of porous scaffolds for bone regeneration [71, 72].

As shown in Figs. 7.5A–C, the combination of gas foaming and porogen leaching techniques enabled the fabrication of PCL scaffolds provided of highly interconnected porosity and controlled morphology. In particular, by blending PCL with small NaCl particles (5 μm mean size) it was possible to prepare highly interconnected scaffolds with homogeneous morphology and controlled porosity and pore size (Fig. 7.5A) [73]. These aspects are very important as the control of pore size and interconnectivity is strongly required for the transport of fluids and nutrients in the entire scaffold. A completely different morphology was obtained by blending PCL with larger NaCl particles (300–500 μm size range). Indeed, for this system the gas foaming induced the formation of small pores (10–50 μm) in the walls of the large cubic-shaped pores created by the leaching of NaCl particles (Fig. 7.5B) [74].

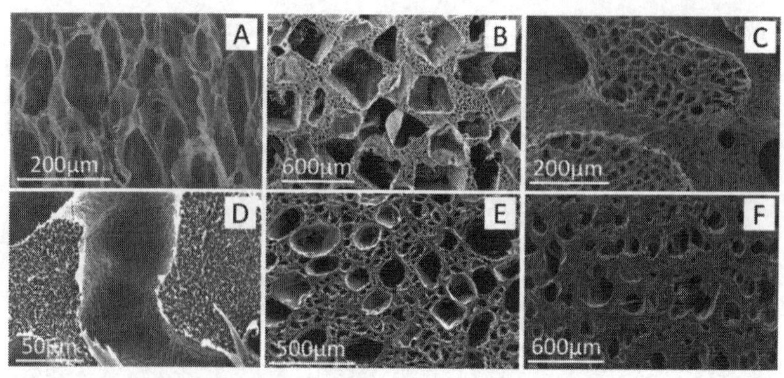

Figure 7.5 Morphology of porous PCL scaffolds prepared by means of gas foaming-based processes: (A) porous PCL scaffold prepared by combining gas foaming and NaCl leaching and by using NaCl particles with 5 µm mean size (adapted from [74]); (B) porous PCL scaffold prepared by combining gas foaming and NaCl leaching and by using NaCl particles with 300–500 µm mean size (adapted from [75]); (C) porous PCL scaffold prepared by combining gas foaming and gelatin leaching from co-continuous blends (adapted from [14, 38]); (D) porous PCL scaffold prepared by low temperature supercritical CO_2 foaming and by using ethyl lactate as co-plasticizer (adapted from [76]); (E) porous PCL scaffold prepared by low temperature supercritical CO_2 foaming, by using ethyl acetate as co-plasticizer and by applying a double step of depressurization (adapted from [73]); (F) porous PCL/zein scaffold prepared by supercritical CO_2 foaming (adapted from [77]).

Porous PCL scaffold with a double-scale pore size distribution can be also prepared by gas foaming of co-continuous blends of PCL and thermoplastic gelatin. In this case, the scaffolds were provided of large and elongated pores crossing the entire scaffold section and surrounded by smaller pores formed during foaming (Fig. 7.5C) [14, 39]. Another interesting foaming process suitable for the fabrication of PCL scaffolds with complex morphology and pore structure is based on the use of supercritical CO_2 ($scCO_2$). This process takes advantage of the capability of $scCO_2$ to induce the melting of PCL even at temperatures lower than 40°C and when the pressure rise values higher than 10 MPa [23, 71, 72]. As a direct consequence, PCL foaming can be carried out when the polymer is in the so-called solid state. The solid-state foaming of PCL can be useful to fabricate porous scaffolds with pores in the micrometric size scale and characterized by nanoscale fibrous pore wall morphology, as that showed in Fig. 7.5D [75]. This process can be improved further by adding proper plasticizers to $scCO_2$, such as in the case of ethyl lactate and ethyl acetate (Fig. 7.5E) [72]. Indeed, even the addition of small amount of plasticizers, up to 0.2 molar % with respect to $scCO_2$, can enhance the melting and subsequent foaming of PCL. Finally, as also previously shown, by blending PCL and natural polymers, such as zein, it was possible to prepare porous multi-phase scaffolds with interconnected porosity and accelerated degradation for bone TE [76].

The phase separation is a well-known technique which has been applied to PCL-based systems for the preparation of porous scaffolds. This process involves the preparation of a homogeneous solution of the polymer followed by the exposure of the solution to a nonsolvent or by the decrease of the temperature. As a direct consequence the solution separates into polymer-rich and polymer-lean phases. Later, the removal of the solvent within the polymer-lean phase by solvent evaporation, sublimation or solvent/nonsolvent exchange, allows achieving an interconnected porous network within the crystallized polymer-rich phase. In Fig. 7.6 is shown the morphology of PCL scaffolds obtained by thermally induced phase separation and starting from solution in ethyl lactate [77]. Figure 7.6 reports the morphology of porous PCL scaffolds prepared by phase separation (Fig. 7.6A,B) and phase separation combined with NaCl leaching (Fig. 7.6C,D). As shown, the scaffold structure is formed by the aggregation of porous microparticles, while the addition of NaCl induced the formation of an interconnected network of pores of several hundred of micrometers in size, replicating the size of the particulate porogen (Fig. 7.6C). Furthermore, the close observation of the sample surface, reported in Fig. 7.6B,D, highlights the lamellar structure of the polymer. The peculiar structure of PCL scaffolds consisting of aggregated microparticles depends on the fact that, during cooling, the system crosses the crystallization temperature and the polymer-rich domains crystallize forming spherulites which entrap the polymer-lean domains. Spherulites grow forming porous microparticles which finally sintered together when they reached impingement because of the presence of residual solvent [77].

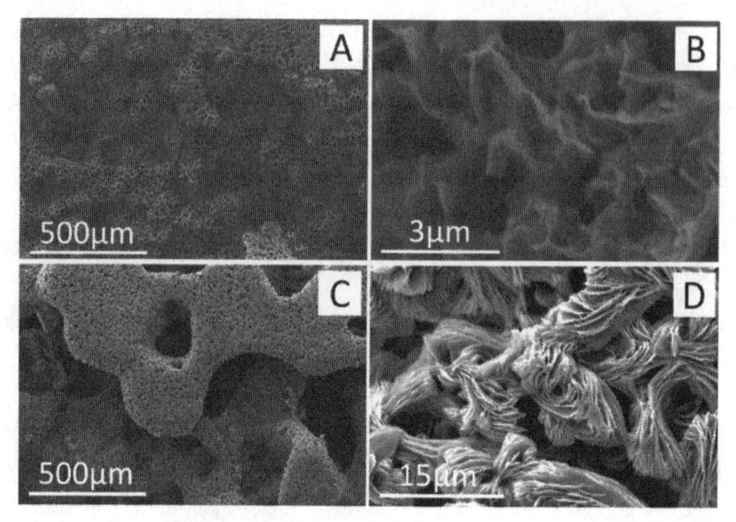

Figure 7.6 Porous PCL scaffolds constitute by aggregated microparticles prepared by thermally induced phase separation (A, B) without and (C, D) with NaCl particles as leachable porogen. Adapted from [77].

Advanced TE approaches rely upon the employment of biodegradable porous scaffolds able to trigger the on-demand release of molecular agents fulfilling the specific needs of transplanted cells and the integrating tissue. Polypeptide growth factors are powerful regulators of biological function. They modulate many cellular functions including migration, proliferation, differentiation, and survival. Therefore, additional direction over cell fate, beyond control of biomaterials chemistry and structure, can be achieved through the spatial and temporal controlled incorporation and release of morphogens and growth factors [78]. The design of biocompatible and biodegradable scaffolds that are able to sequester and the delivery of bioactive molecules in a controlled fashion is therefore a key issue in TE. Porous PCL scaffolds can be used as a carrier for growth factors or antibiotics to accelerate tissue growth or healing or to prevent infection. The utility of PCL scaffolds for *in vitro* and *in vivo* cell growth/tissue engineering and drug release is promising owing to its ability to maintain structural integrity and drug releasing over long periods. In the work by Wang *et al.*, porous PCL scaffolds, in form of tube and cylinder, comprising a continuous phase of PCL and a dispersed phase of lactose or gelatin particles with defined size range (45–90, 90–125, and 125–250 μm) were obtained by precipitation casting technique [79]. The scaffolds enabled the release of 80% of the lactose content after 3 days of soaking in phosphate buffer saline solution at 37°C. Conversely, samples containing gelatin particles of 90–125 and 125–250 μm size range displayed gradual and highly efficient release of around 90% of the protein phase over 21 days. In another work, scaffold-mediated gene delivery of transforming growth factor $\beta3$ was achieved by using a 3D woven PCL scaffold seeded with mesenchymal stem cells for *in vitro* cartilage formation [80]. Bioactive PCL scaffolds were also obtained through the thermal assembly of protein-activated and protein-free PCL microspheres fabricated by emulsion technique [81]. The main advantage of this approach is that the pore size and interconnectivity as well as the mechanical properties of the scaffold can be modulated by an appropriate choice of the size of the protein-free microparticles and process conditions. Furthermore, the scaffolds supported a sustained delivery of BSA over 28 days *in vitro* while the release profile can be controlled by the modulation of the internal structure of protein-loaded microparticles.

Novel PCL scaffolds have been not so far developed by integrating drug delivery devices inside the scaffold. For instance, by combining fused deposition modeling and electrospinning techniques porous scaffolds with localized delivery of rhodamine as model protein were obtained [82]. The process involved the use of a core-shell electrospun mat of hydrophobic PCL and hydrophilic PEO/rhodamine fibers embedded in a 3D PCL scaffold fabricated by a melt-plotting system. The release of rhodamine was modulated by increasing the thickness of the PCL layer, finally leading to the achievement of a controlled release profile without the initial burst. In another work, resveratrol-loaded albumin nanoparticles were synthesized and entrapped into a PCL scaffold prepared by a solvent casting and leaching method [83]. The as prepared scaffold provided a sustained release of resveratrol, up to a cumulative release of 64% at day 12, which enhanced mesenchymal stem cell differentiation and ECM mineralization for bone TE. Another interesting approach to improve drug encapsulation and release from PCL scaffolds is based on the formation of appropriate inorganic or organic bioactive coatings onto the pores of the scaffolds [49, 84–86]. Coated scaffolds able to release vascular endothelial growth factor and bone morphogenetic protein were prepared to stimulate bone regeneration [7, 52, 84–86]. By using fibrin glue loaded with transforming growth factor $\beta 1$ as drug releasing coating for porous PCL scaffolds it was possible to recruit mesenchymal cells and induce the process of cartilage formation when implanted in ectopic sites [7]. This approach offered the advantage to avoid *in vitro* cell culture and cell seeding into the scaffold before implantation, finally reducing the morbidity from the donor sites and improving costs efficiency.

7.4 Conclusions

This chapter provides an overview of the use of drug delivery PCL devices in TE and HC. The slow degradation rate of PCL coupled with its hydrophobic nature enabled the preparation of long term release devices. Furthermore, blending PCL with hydrophilic materials, such as proteins and calcium phosphate ceramics, allowed the encapsulation of hydrophilic compounds.

The excellent chemical stability and processability of PCL were demonstrated by the possibility to fabricate a large variety of nano- and microstructured biomedical devices in form of particles, fibers and porous scaffolds. These materials were used for the successful encapsulation and release of therapeutic drugs, such as antibiotics, anti-tumor drugs and growth factors. The main challenges for drug delivery for TE and HC applications are related to the successful translation of PCL-based drug delivery platforms to the clinic.

References

1. Stupp, S. I. (2005). Biomaterials for regenerative medicine, *MRS Bull,* **30**, 546–553.
2. Lee, K., Silva, E. A., and Mooney, D. J. (2011). Growth factor delivery-based tissue engineering: general approaches and a review of recent developments, *J R Soc Interface,* **8**, 153–170.
3. Biondi, M., Ungaro, F., Quaglia, F., and Netti, P. A. (2008). Controlled drug delivery in tissue engineering, *Biomaterials,* **60**, 229–242.
4. Lee, S. and Shin, H. (2007). Matrices and scaffolds for delivery of bioactive molecules in bone and cartilage tissue engineering, *Adv Drug Deliv Rev,* **59**, 339–359.

5. Oommen, S., Gupta, S. K., and Vlahakis, N. E. (2011). Vascular endothelial growth factor A (VEGF-A) induces endothelial and cancer cell migration through direct binding to integrin $\alpha 9\beta 1$, *J Biol Chem,* **286**, 1083–1092.
6. Leipzig, N. D. Wylie, R. G., Kim, H and Shoichet, M. S. (2011). Differentiation of neural stem cells in three-dimensional growth factor-immobilized chitosan hydrogel scaffolds, *Biomaterials*, **32**, 57–64.
7. Huang, Q., Goh, J. C. H., Hutmacher, D. W., and Lee, E. H. (2002). *In vivo* mesenchymal cell recruitment by a scaffold loaded with transforming growth factor b1 and the potential for *in situ* chondrogenesis, *Tissue Eng,* **8**, 469–482.
8. Kaigler, D., Wang, Z., Horger, K., Mooney, D. J., and Krebsbach, P. H. (2006). VEGF scaffolds enhance angiogenesis and bone regeneration in irradiated osseous defects, *J Bone Mineral Res,* **21**, 735–744.
9. Okada, M. (2002). Chemical syntheses of biodegradable polymers, *Progr Polym Sci,* **27**, 87–133.
10. Serrano, M. C., Pagani, R., Vallet-Regí, M., Peña, J., Rámila, A., Izquierdo, I., and Portolés, M. T. (2004). *In vitro* biocompatibility assessment of poly(ε-caprolactone) films using L929 mouse fibroblasts, *Biomaterials,* **25**, 5603–5611.
11. Choong, C. S. N., Hutmacher, D. W., and Triffitt, J. T. (2006). Co-culture of bone marrow fibroblasts and endothelial cells on modified polycaprolactone substrates for enhanced potentials in bone tissue engineering, *Tissue Eng,* **12**, 2521–2531.
12. Park, J., Pérez, R. A., Jin, G., Choi, S., Kim, H., and Wall, I. B. (2013). Microcarriers designed for cell culture and tissue engineering of bone, *Tissue Eng Part B,* **19**, 172–190.
13. Totaro, A., Salerno, A., Imparato, G., Domingo, C., Urciuolo, F., and Netti, P. A. (2015). Engineered PCL-HA microscaffolds for bone modular tissue engineering *in vitro*, Submitted.
14. Salerno, A., Guarnieri, D., Iannone, M., Zeppetelli, S., and Netti, P. A. (2010). Effect of micro- and macroporosity of bone tissue three-dimensional-poly(ε-caprolactone) scaffold on human mesenchymal stem cells invasion, proliferation, and differentiation *in vitro, Tissue Eng Part A,* **16**, 2661–2673.
15. Farrugia, B. L., Brown, T. D., Upton, Z., Hutmacher, D. W., Dalton, P. D., and Dargaville, T. R. (2013). Dermal fibroblast infiltration of poly(ε-caprolactone) scaffolds fabricated by melt electrospinning in a direct writing mode, *Biofabrication,* **5**, 025001.

16. Siemionow, M., Bozkurt, M., and Zor, F. (2010). Regeneration and repair of peripheral nerves with different biomaterials: review, *Microsurgery*, **30**, 574–588.
17. Luong-Van, E., Grøndahl, L., Chua, K. N., Leong, K. W., Nurcombe, V., and Cool, S. M. (2006). Controlled release of heparin from poly(ε-caprolactone) electrospun fibers, *Biomaterials*, **27**, 2042–2050.
18. Woodruff, M. A. and Hutmacher, D. W. (2010). The return of a forgotten polymer—Polycaprolactone in the 21st century, *Progr Polym Sci*, **35**, 1217–1256.
19. Schantz, J., Hutmacher, D. W., Lam, C. X. F., Brinkmann, M., Wong, K. M., Lim, T. C., Chou, N., Guldberg, R. E., and Teoh, S. H. (2003). Repair of calvarial defects with customised tissue-engineered bone grafts II. Evaluation of cellular efficiency and efficacy *in vivo*, *Tissue Eng*, **9**, 127–139.
20. Sinha, V. R., Bansal, K., Kaushik, R., Kumria, R., and Trehan, A. (2004). Polycaprolactone microspheres and nanospheres: an overview, *Int J Phar*, **278**, 1–23.
21. Cipitria, A., Skelton, A., Dargaville, T. R., Dalton, P. D., and Hutmacher, D. W. (2011). Design, fabrication and characterization of PCL electrospun scaffolds-a review, *J Mater Chem*, **21**, 9419–9453.
22. Jenkins, M. J. and Harrison, K. L. (2006). The effect of molecular weight on the crystallization kinetics of polycaprolactone, *Polym Adv Tech*, **17**, 474–478.
23. Salerno, A. Zeppetelli, S., Di Maio, E., Iannace, S., and Netti, P. A. (2011). Design of bimodal PCL and PCL-HA nanocomposite scaffolds by two step depressurization during solid-state supercritical CO_2 foaming, *Macr Rapid Comm*, **32**, 1150–1156.
24. Izquierdo, R., Garcia-Giralt, N., Rodriguez, M. T., Cáceres, E., García, S. J., Gómez Ribelles, J. L., Monleón, M., Monllau, J. C., and Suay, J. (2008). Biodegradable PCL scaffolds with an interconnected spherical pore network for tissue engineering, *J Biomed Mater Res*, **85A**, 25–35.
25. Salerno, A. and Domingo, C. (2014). A novel bio-safe phase separation process for preparing open-pore biodegradable polycaprolactone microparticles, *Mater Sci Eng C*, **42**, 102–110.
26. Bittiger, H. and Marchessault, R. H. (1970). Crystal structure of poly-ε-caprolactone, *Acta Cryst*, **B26**, 1923–1927.
27. Ang, K. C., Leong, K. F., Chua, C. K., and Chandrasekaran, M. (2007). Compressive properties and degradability of poly(ε-caprolatone)/hydroxyapatite composites under accelerated hydrolytic degradation, *J Biomed Mater Res*, **80A**, 655–660.

28. Eshraghi, S. and Das, S. (2010). Mechanical and microstructural properties of polycaprolactone scaffolds with one-dimensional, two-dimensional, and three-dimensional orthogonally oriented porous architectures produced by selective laser sintering, *Acta Biomater,* **6**, 2467–2476.

29. Engelberg, I. and Kohn, J. (1991). Physico-mechanical properties of degradable polymers used in medical applications: a comparative study, *Biomaterials,* **12**, 292–304.

30. Dash, T. K. and Konkimalla, V. B. (2012). Poly-ε-caprolactone based formulations for drug delivery and tissue engineering: a review, *J Control Release* **158**, 15–33.

31. Chen, D. R., Bei, J. Z., and Wang, S. G. (2000). Polycaprolactone microparticles and their biodegradation, *Polym Deg Stability,* **67**, 455–459.

32. Iwata, T. and Doi, Y. (2002). Morphology and enzymatic degradation of poly(ε-caprolactone) single crystals: does a polymer single crystal consist of micro-crystals, *Polym Int,* **51**, 852–858.

33. Williams, D. F. (2003). Revisiting the definition of biocompatibility, *Med Dev Technol,* **14**, 10–13.

34. Peng, R., Yao, X., and Ding, J. (2011). Effect of cell anisotropy on differentiation of stem cells on micropatterned surfaces through the controlled single cell adhesion, *Biomaterials,* **32**, 8048–8057.

35. Miller, E. D., Li, K., Kanade, T., Weiss, L. E., Walker, L. M., and Campbell, P. G. (2011). Spatially directed guidance of stem cell population migration by immobilized patterns of growth factors, *Biomaterials,* **32**, 2775–2785.

36. Costa, P., Gautrot, J. E., and Connelly, J. T. (2014). Directing cell migration using micropatterned and dynamically adhesive polymer brushes, *Acta Biomater,* **10**, 245–2422.

37. Porter, J. R., Henson, A., Ryan, S., and Popat, K. C. (2009). Biocompatibility and mesenchymal stem cell response to poly(ε-caprolactone) nanowire surfaces for orthopedic tissue engineering, *Tissue Eng Part A,* **15**, 2547–2559.

38. Salerno, A., Guarnieri, D., Iannone, M., Zeppetelli, S., Di Maio, E., Iannace, S., and Netti, P. A. (2009). Engineered L-bimodal poly(ε-caprolactone) porous scaffold for enhanced hMSC colonization and proliferation, *Acta Biomater,* **5**, 1082–1093.

39. Prabhakaran, M. P., Venugopal, J. R., Chyan, T. T., Hai, L. B., Chan, C. K., Lim, A. Y., and Ramakrishna, S. (2008). Electrospun biocomposite

nanofibrous scaffolds for neural tissue engineering, *Tissue Eng Part A*, **14**, 1787–1797.

40. Williams, J. M., Adewunmi, A., Schek, R. M., Flanagan, C. L., Krebsbach, P. H., Feinberg, S. E., Hollister, S. J., and Das, S. (2005). Bone tissue engineering using polycaprolactone scaffolds fabricated via selective laser sintering, *Biomaterials*, **26**, 4817–4827.

41. Fechek, C., Yao, D., Kaçorri, A., Vasquez, A., Iqbal, S., Sheikh, H., Svinarich, D. M., Perez-Cruet, M., and Chaudhry, G. R. (2008). Chondrogenic derivatives of embryonic stem cells seeded into 3D polycaprolactone scaffolds generated cartilage tissue *in vivo*, *Tissue Eng Part A*, **14**, 1403–1413.

42. Ng, K. W., Achuth, H. N., Moochhala, S., Lim, T. C., and Hutmacher, D. W. (2007). *In vivo* evaluation of an ultra-thin polycaprolactone film as a wound dressing, *J Biomater Sci Polym Ed*, **18**, 925–938.

43. De valence, S., Tille, J., Mugnai, D., Mrowczynski, W., Gurny, R., Möller, M., and Walpoth, B. H. (2012). Long term performance of polycaprolactone vascular grafts in a rat abdominal aorta replacement model, *Biomaterials*, **33**, 38–47.

44. Nisbet, D. R., Rodda, A. E., Horne, M. K., Forsythe, J. S., and Finkelstein, D. I. (2009). Neurite infiltration and cellular response to electrospun polycaprolactone scaffolds implanted into the brain, *Biomaterials*, **30**, 4573–4580.

45. Boccaccini, A. R., Erol, M., Stark, W. J., Mohn, D., Hong, Z., and Mano, J. F. (2010). Polymer/bioactive glass nanocomposites for biomedical applications: a review, *Comp Sci Tech*, **70**, 1764–1776.

46. Salerno, A., Oliviero, M., Di Maio, E., Netti, P. A., Rofani, C., Colosimo, A., Guida, V., Dallapiccola, B., Palma, P., Procaccini, E., Berardi, A. C., Velardi, F., Teti, A., and Iannace, S. (2010). Design of novel three-phase PCL/TZ–HA biomaterials for use in bone regeneration applications, *J Mater Sci Mater Med*, **21**, 2569–2581.

47. Salerno A., Zeppetelli S., Oliviero M., Battista E., Di Maio E., Iannace S., Netti P. A. (2012). Microstructure, degradation and *in vitro* MG63 cells interactions of a new poly(ε-caprolactone), zein, and hydroxyapatite composite for bone tissue engineering, *J Bioact Comp Polym*, **27**, 210–226.

48. Sionkowska, A. (2011). Current research on the blends of natural and synthetic polymers as new biomaterials: review, *Progr Polym Sci*, **36**, 1254–1276.

49. López-Rodríguez, N., López-Arraiza, A., and Sarasua, J. R. (2006). Crystallization, Morphology, and mechanical behavior of polylactide/poly(ε-caprolactone) blends, *Polym Eng Sci*, **46**, 1299–1308.

50. Tsuji, H. and Ikada, Y. (1998). Blends of aliphatic polyesters. II. Hydrolysis of solution-cast blends from poly(L-lactide) and poly (1-caprolactone) in phosphate-buffered solution, *J Appl Polym Sci,* **67**, 405–415.

51. Shim, J., Huh, J., Park, J. Y., Jeon, Y., Kang, S. S., Kim, J. Y., Rhie, J., and Cho, D. (2013). Fabrication of blended polycaprolactone/poly (lactic-*co*-glycolic acid)/b-tricalcium phosphate thin membrane using solid freeform fabrication technology for guided bone regeneration, *Tissue Eng Part A,* **19**, 317–327.

52. Suárez-González, D., Barnhart, K., Migneco, F., Flanagan, C., and Hollister, S. J. (2012). Controllable mineral coatings on PCL scaffolds as carriers for growth factor release, *Biomaterials,* **33**, 713–721.

53. Rai, B., Teoh, S. H., and Ho, K. H. (2005). An *in vitro* evaluation of PCL–TCP composites as delivery systems for platelet-rich plasma, *J Control Release,* **107**, 330–342.

54. Lyons, J. G., Blackie, P., and Higginbotham, C. L. (2008). The significance of variation in extrusion speeds and temperatures on a PEO/PCL blend based matrix for oral drug delivery, *Int J Pharm,* **351**, 201–208.

55. Chawla, J. S. and Amiji, M. M. (2002). Biodegradable poly(ε-caprolactone) nanoparticles for tumor-targeted delivery of tamoxifen, *Int J Pharm,* **249**, 127–138.

56. Yang, J., Park, S. B., Yoon, H., Huh, Y., and Haam, S. (2006). Preparation of poly ε-caprolactone nanoparticles containing magnetite for magnetic drug carrier, *Int J Pharm,* **324**, 185–190.

57. Benoit, M. A., Baras, B., and Gillard, J. (1999). Preparation and characterization of protein-loaded poly(ε-caprolactone) microparticles for oral vaccine delivery, *Int J Pharm,* **184**, 73–84.

58. Yesil-Celiktas, O. and Cetin-Uyanikgil, E. O. (2012). *In vitro* release kinetics of polycaprolactone encapsulated plant extract fabricated by supercritical antisolvent process and solvent evaporation method, *J Supercrit Fluids,* **62**, 219–225.

59. Jeong, J., Lee, J., and Cho K. (2003). Effects of crystalline microstructure on drug release behavior of poly(q-caprolactone) microspheres, *J Control Release* **92**, 249–258.

60. Poletto, F. S., Jäger, E., Ré, M. I., Guterres, S. S., and Pohlmann, A. R. (2007). Rate-modulating PHBHV/PCL microparticles containing weak acid model drugs, *Int J Pharm,* **345**, 70–80.

61. Hattori, K., Tomita, N., Tamai, S., and Ikada Y. (2000). Bioabsorbable thread for tight tying of bones, *J Orthop Sci,* **5**, 57–63.

62. Bölgen, N., Vargel, I., Korkusuz, P., Menceloğlu, Y. Z., and Pişkin, E. (2007). *In vivo* performance of antibiotic embedded electrospun PCL membranes for prevention of abdominal adhesions, *J Biomed Mater Res Part B: Appl Biomater,* **81B**, 530–543.

63. Chen, M., Patra, P. K., Warner, S. B., and Bhowmick, S. (2007). Role of fiber diameter in adhesion and proliferation of NIH 3T3 fibroblast on electrospun polycaprolactone scaffolds, *Tissue Eng,* **13**, 579–587.

64. Liang, D., Hsiao, B. S., and Chu, B. (2007). Functional electrospun nanofibrous scaffolds for biomedical applications. *Adv Drug Deliv Rev,* **59**, 1392–1412.

65. Kim, T. G., Lee, D. S., and Park, T. G. (2007). Controlled protein release from electrospun biodegradable fiber mesh composed of poly(ε-caprolactone) and poly(ethylene oxide), *Int J Pharm,* **338**, 276–283.

66. Huang, Z., He, C., Yang, A., Zhang, Y., Han, X., Yin, J., and Wu, Q. (2006). Encapsulating drugs in biodegradable ultrafine fibers through co-axial electrospinning, *J Biomed Mater Res Part A,* **77A**, 169–179.

67. Kenawy, E., Abdel-Hay, F. I., El-Newehy, M. H., and Wnek, G. E. (2009). Processing of polymer nanofibers through electrospinning as drug delivery systems, *Mater Chem Phys,* **113**, 296–302.

68. Del Valle, L. J., Camps, R., Díaz, A., Franco, L., Rodríguez-Galán, A., and Puiggalí, J. (2011). Electrospinning of polylactide and polycaprolactone mixtures for preparation of materials with tunable drug release properties, *J Polym Res,* **18**, 1903–1917.

69. Zhang, Y. Z., Wang, X., Feng, Y., Li, J., Lim, C. T., and Ramakrishna, S. (2006). Coaxial Electrospinning of (fluorescein isothiocyanate-conjugated bovine serum albumin)-encapsulated poly(ε-caprolactone) nanofibers for sustained release, *Biomacromolecules,* **7**, 1049–1057.

70. Salerno, A., Di Maio, E., Iannace, S., and Netti, P. A. (2009). Engineering of foamed structures for biomedical application, *J Cell Plastics,* **45**, 103–117.

71. Salerno, A., Di Maio, E., Iannace, S., and Netti, P. A. (2011). Solid-state supercritical CO_2 foaming of PCL and PCL-HA nano-composite: effect of composition, thermal history and foaming process on foam pore structure, *J Supercrit Fluids,* **58**, 158–167.

72. Salerno, A., Fanovich, A., and Domingo, C. (2014). The effect of ethyl-lactate and ethyl-acetate plasticizers on PCL andPCL–HA composites foamed with supercritical CO_2, *J Supercrit Fluids,* **95**, 394–406.

73. Salerno, A., Iannace, S., and Netti, P. A. (2008). Open-pore biodegradable foams prepared via gas foaming and microparticulate templating, *Macromol Biosci,* **8**, 655–664.

74. Salerno, A., Zeppetelli, S., Di Maio, E., Iannace, S., Netti, P. A. (2011). Processing/structure/property relationship of multi-scaled PCL and PCL–HA composite scaffolds prepared via gas foaming and NaCl reverse templating, *Biotech Bioeng,* **108**, 963–976.

75. Salerno, A. and Domingo, C. (2013). A clean and sustainable route towards the design and fabrication of biodegradable foams by means of supercritical CO_2/ethyl lactate solid-state foaming, *RSC Adv,* **3**, 17355–17363.

76. Salerno, A., Zeppetelli, S., DI Maio, E., Iannace, S., and Netti, P. A. (2010). Novel 3D porous multi-phase composite scaffolds based on PCL, thermoplastic zein and ha prepared via supercritical CO_2 foaming for bone regeneration, *Comp Sci Tech,* **70**, 1838–1846.

77. Salerno, A. and Domingo, C. (2015). Pore structure properties of scaffolds constituted by aggregated microparticles of PCL and PCL-HA processed by phase separation, *J Porous Mater,* **22**, 425–435.

78. Sands, R. W. and Mooney, D. J. (2007). Polymers to direct cell fate by controlling the microenvironment, *Curr Opin Biotechnol,* **18**, 448–453.

79. Wang, Y., Chang, H., Wertheim, D. F., Jones, A. S., Jackson and C., Coombes, A. G. A. (2007). Characterization of the macroporosity of polycaprolactone-based biocomposites and release kinetics for drug delivery, *Biomaterials,* **28**, 4619–4627.

80. Brunger, J. M., Huynh, N. P. T., Guenther, C. M., Perez-Pinera, P., Moutos, F. T., Sanchez-Adams, J., Gersbach, G. A., and Guilak, F. (2014). Scaffold-mediated lentiviral transduction for functional tissue engineering of cartilage, *Proc Natl Acad Sci U S A,* **18**, E798–E806.

81. Luciani, A., Coccoli, V., Orsi, S., Ambrosio, L., and Netti, P. A. (2008). PCL microspheres based functional scaffolds by bottom-up approach with predefined microstructural properties and release profiles, *Biomaterials,* **29**, 4800–4807.

82. Yoon, H. and Kim, G. (2011). A three-dimensional polycaprolactone scaffold combined with a drug delivery system consisting of electrospun nanofibers, *J Pharm Sci,* **100**, 424–430.

83. Kamath, M. S., Ahmed, S. S., Dhanasekaran, M., and Santosh, S. W. (2014). Polycaprolactone scaffold engineered for sustained release of resveratrol: therapeutic enhancement in bone tissue engineering, *Int J Nanomedicine,* 9, 183–195.

84. Singh, S., Wu, B. M., and Dunn, J. C. Y. (2012). Delivery of VEGF using collagen-coated polycaprolactone scaffolds stimulates angiogenesis, *J Biomed Mater Res Part A,* **100A**, 720–727.

85. Liu, X., Zhao, K., Gong, T., Song, J., Bao, C., Luo, E., Weng, J., and Zhou, S. (2014). Delivery of growth factors using a smart porous nanocomposite scaffold to repair a mandibular bone defect, *Biomacromolecules*, **15**, 1019–1030.

86. Rai, B., Teoh, S. H., Hutmacher, D. W., Cao, T., and Ho, K. H. (2005). Novel PCL-based honeycomb scaffolds as drug delivery systems for rhBMP-2, *Biomaterials*, 26, 3739–3748.

Chapter 8

Engineered PLGA Nanosystems for Enhanced Blood Compatibility

Y. M.Thasneem and Chandra P. Sharma
Division of Biosurface Technology, Biomedical Technology Wing,
Sree Chitra Tirunal Institute for Medical Sciences and Technology,
Trivandrum, Kerala, India
sharmacp@sctimst.ac.in

8.1 Introduction

Progress in material science and nanotechnology has propelled the evolution of biomaterials with improved quality and extended patent life [1]. However, even in the 21st century, the struggle to create surfaces that interact actively with the biological environment and evoke the same sequence of effects as the corporal tissues do, still continues. Tailoring the interface compatibility of biomaterials through surface modification thus becomes indispensable in the optimization of physiological interactions and clinical performance. Since the original discovery by Folkman and Long in 1964, polymeric materials, with their endless diversity in topology and architecture,

Handbook of Polyester Drug Delivery Systems
Edited by MNV Ravikumar

ISBN 978-981-4669-65-8 (Hardcover), 978-981-4669-66-5 (eBook)
www.panstanford.com

have occupied a major status in the fabrication of controlled drug delivery systems [2]. A variety of polymers, ranging from synthetic, natural, to hybrid, have been scrutinized for the synthesis of biodegradable nanoparticles. The optimal formulation for every particular drug delivery application is delineated by the specificities of the polymer used and the process of nanoparticle preparation. The minimal toxicity and biodegradability of aliphatic polyesters like poly(lactic-*co*-glycolide) (PLGA) has propelled its wide spread usage in controlled release technology [3].

8.2 Poly(Lactic-*co*-Glycolic Acid): The Need to Engineer the Surface

The biomaterial history of PLGA, which began as biodegradable sutures in the 1970s, has now extended to various clinical and basic science research fields, including drug delivery, diagnostics, cardiovascular diseases, vaccine, cancer, and tissue engineering. The biocompatibility, biodegradability, and tailor-made physicochemical characteristics are the basic factors that have led to the success story of PLGA family of thermoplastic polymers, including its Food and Drug Administration approval for clinical use in humans [3, 4]. Though PLGA-based technology opens new vistas in theranostics and tissue engineering applications, especially for cardiovascular diseases and cancer, it suffers from certain intrinsic drawbacks, including the following [5].

- Relative hydrophobicity compared to the natural extracellular matrix
- Poor *in vivo* blood circulation time
- Rapid recognition and removal of PLGA by body's defence mechanism
- Negatively charged PLGA surface limiting the transport across biological barriers such as plasma membrane or tight junctions
- Passive diffusion into tumor via enhanced permeability and retention (EPR) effect, which doesn't meet the required clinical efficacy

Taking into consideration that the mechanical and degradative characteristics of PLGA are apparently ideal, a promising approach seems to be the surface engineering of bare PLGA nanoparticles, post synthesis. This can help provide the required surface characteristics to overcome these delivery challenges and improve the *in vivo* performance, without altering the properties of the bulk polymer, PLGA [6].

Irrespective of the nanoparticle's material origin or surface characteristics, evaluating the biohematocompatibility parameters becomes mandatory before the widespread application of nanoparticles in humans. This can help to execute appropriate safety precautions concerning human health [7]. Hence, before proceeding to the surface engineering aspects, let us understand the terms of biocompatibility and various processes occurring on the PLGA nanoparticle surface, once it is in contact with the biological system.

8.3 Biocompatibility

The expression of the benign coexistence of materials within the biological environment along with adequate functionality has been the foundation of the subject biocompatibility [8]. The intrinsic physicochemical properties of nanoparticles while interfacing with the cells and extracellular environment can elicit a sequence of biological events and consequences that dictate the biocompatibility of the system as a whole. Initially, biocompatibility was equated in terms of inertia or no changes in the surrounding tissue. In this insight, Williams has reviewed the biocompatibility concept in 1986 as "the ability of a material to perform with an appropriate host response in a specific situation" [9]. Later, in 1993, Ratner redefined biocompatibility as the body's acceptance of a material, i.e., the ability of an implant surface to interact with the cells and fluids of the biological system and to kindle reactions homologous to the body tissues [10]. It can be generalized that a high level of biocompatibility is attained when material interacts within the host without generating unwanted toxic, immunogenic, carcinogenic, and thrombogenic consequences. The term *biocompatibility* incorporates material properties that include toxicity, tissue compatibility, and hemocompatibility and provides a measure to screen materials in terms of *in vitro* cytotoxicity and hemocompatibility parameters. Biocompatibility can be broadly divided into two elements, biosafety, i.e., appropriate host response at both systemic and localized levels, and biofunctionality, i.e., the capacity of the material to fulfill the particular purpose for which it is intended [11].

8.4 What Makes PLGA a Biocompatible Polymer?

Biocompatibility of a polymer comprises physiological, physicochemical, and molecular considerations. Unlike metallic nanoparticles, *in vitro* cytotoxicity studies have confirmed that the surface area of PLGA nanoparticles does not contribute to the toxic potential. Moreover, the chemical composition of PLGA nanoparticles does not affect the cellular viability. Being polyester, PLGA is subjected to hydrolysis and enzymatic degradation once administered inside the human body. The nonspecific hydrolytic scission of the ester bonds of PLGA yields lactic and glycolic acid as degradation by-products. These get incorporated into the citric acid cycle in a way similar to anaerobic metabolism and get eventually removed as carbon dioxide and water. Furthermore, as PLGA degradation products are formed at a very slow rate, they do not affect normal cell viability [12].

8.5 Hemocompatibility

The primary body fluid that comes in contact with any biomaterial, once inside the body, is the blood. In 1992, the International Organization for Standardization adopted and published a multipart international standard on the biological evaluation of medical devices, ISO 109939. Among the various tests stated in ISO 10993, hemocompatibility assessment forms an important yardstick for nano-enabled drug delivery programs [13].

In simple terms, hemocompatibility can be considered as a measure of the particle's capacity to interact safely with the different components of blood. Hemocompatibility and subsequent cytocompatibility and the resulting biological behavior are largely dictated by the process of nanoparticle interface to the blood, cells, and tissues after an intravenous injection, unlike the material's bulk properties. In order to understand pathophysiological processes occurring at the blood-material interface, careful correlation of materials' chemical composition, charge, flexibility, and wettability to the conditions of blood flow has to be performed [14]. Such an assessment of hemocompatibility factors can provide reliable *in vitro/in vivo* extrapolation and thereby help us foresee the complications involved under clinical situations.

8.6 Events Occurring at the Material–Blood Interface

A material in contact with blood for a specific time can evoke several reactions that collectively contribute to its general compatibility/incompatibility criteria. Parenteral injection of nanoparticles result in their high, localized concentration in the blood stream and can trigger a complex series of interrelated events, including protein adsorption, platelet and leukocyte activation/adhesion, or the activation of complement and coagulation cascade. All these events depend on the manner in which the particles interface with the blood proteins and the cellular components.

8.7 Bio–Nano Interface

Surface adsorption can be considered as a universal rule of materials in biology. All materials in contact with the biological environment are immediately adsorbed by a selected set of biomolecules, and nanoparticles are no exception. The higher surface area and surface free energy of nanoparticles can further promote this formation of molecular corona. The corona formation within the blood happens in less than 0.5 minutes and consists of 300 different plasma proteins whose identity and quantity, arrangement, orientation, conformation, and affinity define the hydrodynamic volume and the biological identity of a nanoparticle. Indeed, this nano–bio interface can affect the uptake, transportation, cytotoxicity, and in short the pathophysiology or fate of nanoparticles in the biological system [15]. Generally, the corona proteins that promote the recognition by reticuloendothelial system are called opsonins and include IgG, complement proteins, and fibrinogen. This can result in the accumulation of the nanoparticles in the liver following recognition by the kupffer cells. In contrast, dysopsonins or passivating agents such as albumin prevent this interaction with the mononuclear phagocytic system. Thus the qualitative and quantitative knowledge about protein corona is of considerable significance to biodistribution and *in vivo* performance. In fact, Nel has described the nano–bio interface as the dynamic physicochemical interactions, kinetics, and thermodynamic exchanges between the surfaces of nanoparticles and biological components [16]. The physicochemical characteristics of nanoparticles and exposure time specify this corona which can further influence the downstream hemocompatibility parameters such as hemolysis, thrombosis, and recognition by the immune system as depicted in Fig. 8.1. [17].

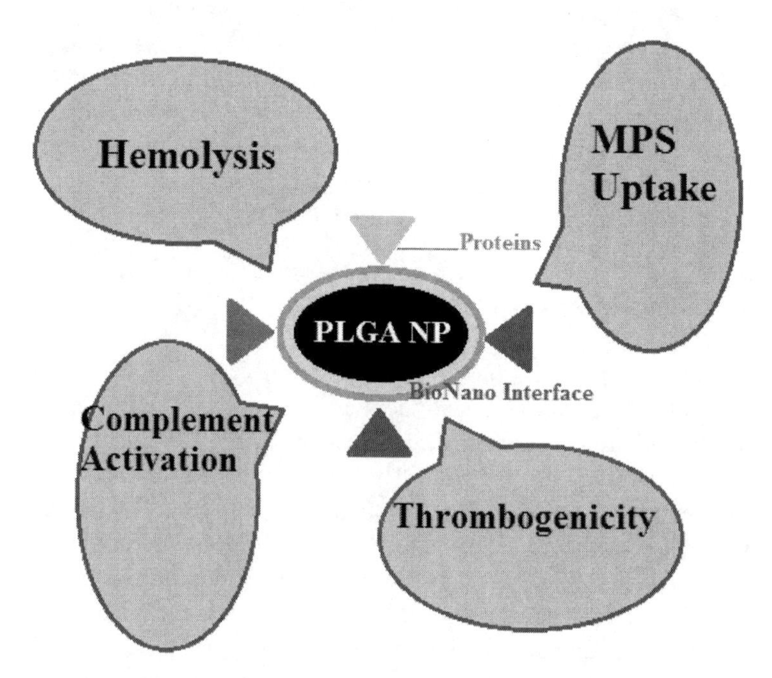

Figure 8.1 Hemocompatibility events at the bio–nano interface.

8.8 An Insight into the Protein Corona on PLGA Nanoparticles

According to the reports on protein adsorption, the generally hydrophobic PLGA nanoparticle surface carrying a negative charge must undergo opsonisation. A systematic study by Sempf *et al.* using matrix-assisted laser desorption/ionization time of flight (MALDI-TOF) has identified 15 different plasma proteins which are associated with the PLGA nanoparticles, as provided in Table 8.1; seven of them belong to the less abundant protein series. Apolipoprotein E, vitronectin, histidine-rich glycoprotein, and kininogen1 were the major proteins in term of high MASCOT score, a software package from Matrix Science [18]. The type and specificity of this proteome can vary according to the size, synthesis procedure, and surface modification. As the size determines the curvature and surface free energy of particles, the protein adsorption pattern on PLGA nanoparticles vs. microparticles can be different. Coming to the synthesis procedure, apolipoproteins dominated the protein corona of spray-dried PLGA nanoparticles to 25%, in contrast to less than 10% on particles prepared by solvent evaporation technique [19]. An interesting report by Sobczynski *et al.* has pointed out that the adsorption of certain high molecular weight plasma proteins such as immunoglobulins can inhibit the adhesion of vascular-targeted PLGA nanoparticles [20].

Table 8.1 Major proteins associated with PLGA nanoparticles

Histidine-rich glycoprotein	Plasminogen
Kininogen 1	Fibrinogen alpha chain
Apolipoprotein E	Complement C4 A
Vitronectin	IgG kappa chain C
Complement C3	Ig gamma 1 chain C
Beta 2 glycoprotein 1	Coagulation factor 5
Serum amyloid A4 protein	IgG gamma 3 chain C region
	Ig mu heavy chain disease protein

Source: Reprinted from Sempf *et al.*, Copyright 2013, with permission from Elsevier.

8.9 Hemolysis

Hemolysis is a life-threatening condition characterized by the rupture of red blood cells. As a consequence, the iron-containing protein hemoglobulin is released into the plasma. The small size-to-surface area ratio and unique physicochemical properties of nanoparticles mandate the *in vitro* evaluation of the biocompatibility with red blood cells, a prerequisite for early preclinical development [13]. Mostly, *in vitro* hemolysis studies pertain to the evaluation of the percent hemolysis through spectrometric detection of plasma-free hemoglobulin, released from akaryocyte following the incubation of particles with blood for a stipulated time period [21]. This is considered to be a degree of nanoparticle-induced exosomatic hemocytolysis. Experimental conditions such as incubation time, wavelength of hemoglobulin quantification, inclusion of anticoagulants, use of red blood cells instead of whole blood, difference in relative centrifugal force, can complicate the test result interpretation. Nevertheless, good correlation exists between *in vitro* hemolysis studies and *in vivo* toxicity profile. Furthermore, the debris released from damaged RBCs has the potential to activate the phagocytic cascade and results in their rapid clearance from the circulation, in addition to the risk of anaemia. A high degree of hemolytic activity is associated with nanoparticles with more hydrophobicity and cationic surface charge in comparison to neutral, anionic, and zwitterionic surface charges. The surface hydrophobicity of particles can further contribute cumulatively to the hemolytic potential. In addition, most nanoparticles demonstrate a dose-dependent increase of hemolysis. It is interesting to note that the protein corona, in general, can modulate and the dysopsonin albumin can indeed silence the hemolytic activity of nanoparticles [22]. The molecular basis of erythrocyte deformation and hemolysis shed light on the alteration of actin structure that forms the major scaffold by organizing the spectrin tetramers in a hexagonal structure.

8.10 Hemolytic Potential of PLGA Nanoparticles

In accordance to the American Society for Testing and Materials international protocol (ASTM) E2524-08, materials are classified as nonhemolytic, moderately hemolytic, and hemolytic when the hemolysis index is below 2%, between 2% and 5%, and above 5% respectively. Microscopic observation of erythrocytes have confirmed that the RBCs retain the typical biconcave or donut shape, with a diameter of around 7.5 μm following incubation with PLGA nanoparticles. This pertains to the nonhemolytic potential of PLGA nanoparticles [23]. A systematic study by Fornaguera *et al.* has confirmed the nearly zero hemolytic index of PLGA nanoparticles following 10 minutes and 24 hours of incubation with the erythrocytes [24].

8.11 Thrombogenecity

Under physiological conditions, activation of coagulation cascade or hemostasis is a mechanism that maintains the integrity of circulatory system following vascular damage. Hemostasis is initiated upon endothelial surface injury that exposes collagen or tissue factor to the flowing blood. However, when any pathological insult such as biomaterial exposure interface with the regulatory mechanism of coagulation cascade, excessive thrombin formation occur that leads to thrombosis. To be specific, thrombogenecity is defined as the capacity of a material to activate the coagulation cascade and result in the formation of a thrombus. A thrombus is a solid mass of aggregated platelets, fibrin, red blood cells, and other cellular components that can disintegrate to smaller clots called emboli [25]. As illustrated by Virchow's triad, the platelets or a nuclear disk-shaped cells that are derived from the megakaryocytes in the bone marrow form the major culprit behind nanoparticle-induced thromboembolism. The main role of platelets in blood clotting is to preserve the integrity of vascular walls through interconnecting fibrin fibers to form a platelet plug. In fact, platelets are also termed thrombocytes as their activation ultimately results in the formation of a thrombus. A sequence of events gets initiated upon platelet aggregation and degranulation that leads to the concomitant activation of leukocytes, endothelial cells, and other platelets. The integrin (glycoprotein) receptors on the platelet surface, especially GPIIb/IIIa, act as the focal point of attachment of other platelets and plasma proteins and it is the receptor upregulation following platelet activation that actually helps the clot mass to grow. The plasma protein fibrinogen present on the protein corona can promote crosslinking between the GPIIb/IIIa receptors on different platelets, leading to platelet–platelet and platelet–surface interaction, adhesion, and aggregation, respectively. In fact, carbon nanoparticles induce platelet aggregation through this mechanism [26]. In addition, the cross-talk between platelets and leukocytes is mediated by fibrinogen through the receptor, MAC-1. Under high shear rate, platelet activation proceeds through interaction with von Willibrand factor, while under low shear rates, fibrinogen mediates platelet activation [27]. Thromboembolic complications of prosthetic devices are initiated through surface-induced platelet activation. This process, which proceeds through antigen–antibody complexes is termed as intrinsic pathway, while tissue factor propelled extrinsic pathway forms the physiological inception of coagulation [28]. Both these pathways of coagulation are characterized by continuous changes in platelet morphology and degranulation and culminate in the formation of fibrin clot following activation of thrombin on fibrinogen surface. Concomitantly, thrombin activates factor XIII to crosslink and stabilize fibrin clot into an insoluble fibrin gel. As a result, the platelet compatibility limits the critical surface tension of biomaterials between 20 and 30 dyne/cm [29].

The compatibility of injected nanomaterials with the coagulation cascade is regulated by the physiochemical characteristics such as size, charge, surface hydrophobicity, and composition. All these characteristics are subjected to alterations depending on the conformation and types of proteins exposed on the nanoparticle corona. On an average, the interaction of nanoparticles could be initiated either through contact with the plasma coagulation factors or through the cellular components. The consequence can vary between the depletion of coagulation components, resulting in excessive bleeding or the contact activation leading to the formation of undesirable clots. The release of clotting factors from platelets or the tendency to cause platelet aggregation can serve as an *in vitro* measure of platelet activation by nanoparticles under *in vivo* conditions. Thrombogenicity can be estimated by various methods following incubation of the nanoparticles in platelet-rich plasma, obtained after separating erythrocytes from whole blood by centrifugation. Light transmittance aggregometry (LTA) following platelet stimulation by an exogenous agonist using aggregometer, quantification of the release of platelet factor 4 or thromboglobulin by enzyme-linked immunosorbent assay (ELISA), analyzing the morphological changes in platelets by lactate dehydrogenase assay or electron microscopy, estimating the upregulation of platelet receptors glycoprotein GPIIb/IIIa (CD 41) and p-selectin (CD 62p) using flow cytometry represent some of the common techniques used to determine nanoparticle-induced platelet activation. Nanoparticle-induced changes in the coagulation parameters such as prothrombin time (PT), activated partial thromboplastin time (APTT), and thrombin time (TT) can be performed by an automated analyzer. APTT and PT analysis are used to differentiate between intrinsic and extrinsic mode of coagulation by determining factor deficiencies such as factor VIII in the plasma and by analyzing the time required by plasma to clot after addition of factor III, respectively [30]. The anionic surface charge on a nanoparticle tend to self-activate coagulation factor XII, which in turn activates the intrinsic coagulation cascade in a size-dependent manner. Further studies have demonstrated that the platelet activation by cationic particles proceeds through membrane perturbations, while the anionic counterparts associate with the classical pathway is characterized by the upregulation of the adhesion receptors [31]. On a molecular level, the nanoparticles can stimulate phospholipase C and Rap1b, which leads to the integrin $\alpha_{IIb}\beta_3$ mediated platelet aggregation following the release of secondary messengers, adenosine diphosphate, and thromboxane A_2 [32].

8.12 PLGA Nanoparticles Interaction with the Coagulation Cascade

The systemically administered PLGA nanoparticles can come in contact with the circulating platelets, especially when the half-life of the particles in the blood is increased. Various reports confirm that the PLGA nanoparticles do not invoke any platelet aggregation at concentrations below 1 mg/ml, irrespective of surface charge [23, 33]. However, thrombin readily adsorbs onto negatively charged surfaces and degradable polyester nanoparticles (polylactic acid) have been shown to induce platelet aggregation [34]. A study by Cenni *et al.* has reported significant decrease in prothrombin activity without much change in partial thromboplastin time following the conjugation of PLGA nanoparticles with alendronate [35]. However, this has no clinical significance as prothrombin activity greater than 75% is considered within the normal range of test in humans. Nevertheless, this study highlights the extrinsic mode of coagulation of the mentioned nanoparticles via modification of factor VII.

8.13 Complement Activation

The complement activation system, which is a key effector of both the innate and cognate immune system, is responsible for the identification, opsonization, and elimination of foreign invasion through pattern recognition of danger signals. The system can be activated via three different initiation pathways termed classical, lectin, and alternate pathways, and all of them converge to generate the same set of effector molecules. Antigen–antibody complexes, bound C-reactive protein, and damaged self cells can trigger the classical pathway, while the specific carbohydrates on the surface of microorganisms (mannose-binding lectin, ficolins) activate the lectin pathway, independent of antibodies. On the basis of the increasing number of reports, it is the alternative pathway that gets activated on contact with the surfaces of artificial materials such as biomaterials and turns out to be a clinically relevant pathway for complement assessment [36]. The major event of activation cascade is the enzymatic cleavage of C3 into C3b and C3a. In fact, it is the two C3 convertases, C4b, 2a, C3b, Bb that mediates the activation. The activation proceeds through sequential cleavage of complement proteins that lead to the liberation of anaphylatoxins C3a and C5a, respectively. As these products are chemotactic in nature, they can recruit and stimulate other immune cells such as granulocytes to secrete pro-inflammatory mediators and thereby sustain the activation till the target is ingested or cleared through receptor-mediated endocytic or phagocytic processes. In addition, the terminal complement proteins polymerize to form C5b-9 or the lytic complex that can create pores in the pathogenic membrane. Thus, the effector functions culminate at either direct lysis or it mediates leukocyte function in inflammation. Being a complex system of over 35 soluble and cell surface proteins in blood, it forms the first line of defence against generally all intrusions including nanomedicine infusion. The physiological regulation of complement is mediated through two key complement proteins, factor H and C4b-binding protein, which constitute the primary inhibitors of complement activation.

On an average, nanoparticles do share several structural features with microorganisms such as the surface display of repeated patterns, and this can increase the probability of cascade activation. The priming of nanoparticle surface by the complement initiating molecules such as C1-Q, C-reactive protein, or mannose-binding lectin is regulated by several interfacial dynamic forces and physicochemical characteristics that include chemical composition, topology, reactivity, shape, hydrodynamic size, and surface features. Those nanoscale features and motifs that incite the complement pathway can only be elucidated through structure activity and high-throughput approaches. It should also be kept in mind that the plasma protein corona formed on nanoparticle surface can modulate complement activation. In fact, the opsonized C3, which is present in the protein corona, can mimic the configuration of bound C3b and is able to generate an initiating C3 convertase, which further activates the alternative pathway [37].

8.14 Physicochemical Characteristics of Nanoparticle Modulate the Complement Activation

8.14.1 Size and Curvature

The interacting proteins of the complement system have their characteristic size, which varies between 7 nm for C3b to 37 nm for the pentameric IgM. With this cross-sectional area, they are able to sense and interact with nanoparticles in the range of 1–1000 nm and, consequently, initiate distinct activation pathways of varying magnitude. Particles of the size range 50–100 nm are comparable to the larger molecules of complement system (C1Q, mannose-binding lectin, IgM) and 12–20 nm^2 is typically required for epitope recognition by antibodies. In case of dextran-coated iron oxide nanoparticles, potent complement activation was reported on 250 nm sized particles in comparison to larger (650 nm) particles in diameter. This is because the tendency of the IgG molecule to adopt the staple conformation which promotes C1Q binding and complement activation increases as size decreases. In contrast, using differently sized particles (200, 500, 1000 nm) prepared with PLGA, it was reported that opsonization, serum IgG response, phagocytosis, and clearance by mononuclear phagocytic system increased with increasing size [38]. The reason being as size increases, more surface area is available for the assembly of alternative pathway amplification loop. To be specific, 40 nm^2 is occupied by a surface-bound C3b molecule on any nanoparticle. Hence, as size decreases a major portion of activated C3 molecules tend to release in to the surrounding medium than getting deposited on the nanoparticle surface. Thus, it is noteworthy to understand that topological features, including surface curvature or epitope density, can impose geometrical strain on the sensing molecules and thereby affect the size-mediated complement response [39].

8.14.2 Surface Charge and Functional Groups

It is well known that the presence of functional groups such as amine and hydroxyl moieties mediate the complement activation via the alternative pathway. The significance of surface charge in complement activation is no less as the recognition unit of classical pathway, C1Q, opsonize surfaces through electrostatic and hydrophobic interactions. The influence of charge and functional groups in mediating complement activation were studied using hybrid nanoparticles of PLGA-lipid-PEG polymer with methoxyl, carboxyl, and amine functional groups. The results indicated that amine functional groups activated the complement to the maximum, while methoxyl groups elicited very minimal levels of activation. The comparatively low level of activation by carboxyl and methoxyl functionalisation could be mediated through factor H as it has lower binding affinity to negatively charged surfaces.

However, a combinatorial functionalization using amine and methoxyl groups produced the highest level of complement activation [40]. Interestingly, methylation of phosphate oxygen moiety on PEGylated liposomes prevents complement activation, possibly via blocking the adsorption of sensing molecules C1Q and P. Coming to hydroxyl groups, several reports claim that surface hydroxyl groups that reduce the contact angle can mitigate the complement activation. However, the nucleophilic nature of hydroxyl groups could promote its interaction with the thioester group of C3b and subsequent activation via the alternative pathway.

8.14.3 Surface Projected Polymers

A common strategy adopted to modulate complement activation is through surface modification, the basic mechanism being the steric hindrance provided by surface-adsorbed polymers. Here, the conformational state of the polymer chain, defined by grafting density, can act as the molecular switch for complement activation and amplification. A threshold grafting density is required for substantial complement activation, below which minimal activation is induced. One reason could be the increasing accumulation of factor B and C3 at higher polymer density. Significance of conformation is further illustrated by the fact that alteration from mushroom to brush type could switch the initiating pathway from classical to the lectin mode. In case of dextran-coated nanoparticles, the complement activation became more prominent with loop-like conformation in comparison to brush model. The contribution of sialic acid to the complement inhibitor effects of the sugar glycocalyx coat of vascular endothelial cells is mediated through the recruitment of factor H. Carbohydrate chemistry of the glycopolymer chain also determines the complement activation. Literature cites contradictory reports for glucose- and galactose-induced complement activation [41].

8.15 Complement Activation Mediated by PLGA Nanoparticles

PLGA nanoparticles have been used as a negative reference for complement activation as they consume minimal complement proteins at a concentration below 3 mg nanoparticles and with a surface area of 40 cm^2. However, the surface hydrophobicity of PLGA polymer is responsible for invoking a weak to high level of complement activation.

8.16 Interaction with the Lympho-Reticular/ Mononuclear Phagocyte System

The mononuclear phagocyte system is a part of the immune system that consists of the cells derived from the progenitor cells of the bone marrow and are characterized by their avid phagocytic or pinocytic activity. These cells differentiate to form the monocytes which enter the blood circulation and are termed as resident macrophages once it get localized in specific tissues such as lymph node, liver (kupffer cells), spleen, brain (microglia), bone (osteoclasts), and skin (langerhans). The pattern recognition receptors present on these cells help identify the unique molecular structures present on biomaterials, in a way similar to microbial recognition. Macrophages or the dominant cells type of the mononuclear phagocytic system (MPS) are the primary infiltrating cells against biomaterial implantation in soft and hard tissues. When more than one macrophage attached to the same endocytic material, fusion of individual macrophages can take place. This results in the formation of a morphological variant termed foreign-body giant cells or multinucleated giant cells. These cells are often associated with the degradation of polymeric devices.

Unlike fibroblasts, macrophages can adhere to material surface through pattern recognition and specific integrin receptors. Following attachment, macrophages proliferate and fuse to form multinucleated giant cells (MNGCs) or foreign-body giant cells (FBGCs) on PLGA surface and, as a consequence, the degradation process undergoes a shift from the standard bulk mode to surface activated erosion. The interface between PLGA and macrophages was analyzed by focussed ion beam or FIB method, and it revealed the formation of a sealing zone as the penetrating pseudopodia forms a closed compartment between the FBGCs and the underlying substrata. This fusion following cell adhesion initiates the biodegradation process as many enzymes, reactive oxygen intermediates, and other products are released into this sealing zone. Following differentiation, macrophages can also form the variant cell type, osteoclasts, which play an active role in bone mineral resorption by secreting protons [42].

8.17 Physicochemical Characteristics Influencing Macrophage-Mediated Phagocytosis

The capability of macrophages to ingest particulate matter by phagocytosis is central to both innate and adaptive immunity. This mechanism of host defence system is initiated by surface ligation of receptors such as Fcγ (IgG/immunoglobulin receptors) or CR1/CR2 (complement receptors) and is facilitated by the opsonization process. This ligand–receptor interaction leads to the activation of tyrosine phosphorylation-dependent signalling events that can evoke distinct morphological changes such as the coordinated actin polymerization and extension of pseudopods, all of which culminating in the engulfment of particles within a phagosome [43].

PLGA microparticles undergo a conventional kind of phagocytosis where the pseudopods move circumferentially around the microparticle in an attempt to fuse the distal ends. The process begins within 15 minutes of particle incubation and in a time-dependent manner gets completed by 120 minutes [44].

Macrophage-mediated particle uptake is dependent on the physicochemical characteristics such as size, surface charge, hydrophobicity, and also on the concentration of particles in the medium. The inflammatory response generated by the particles can be manipulated by modulating the size factor. To be specific, the sequential analysis of polymer-entrapped antigens has revealed that the microparticles promote humoral immune response, while nanoparticles lead to cellular mediated immunity. It has been reported that better uptake for PLGA nanoparticles were observed after 4 hours of incubation, while microparticles were still found to attach to cell membrane. In fact, the larger diameter of microparticles demands more time for proper cell engulfment. Also, the intracellular analysis of PLGA nanoparticles following phagocytosis by murine macrophage has confirmed their co-localization with lysosomes, which also suggest their timely entrance to the cell cytoplasm.

Coming to the shape of particles, spherical shape promotes faster and higher rate of endocytosis compared to rods and disks, while some reports also favor the preferential uptake of rod/ cylindrical shape over other structures. Other physical properties such as particle rigidity also influences macrophage uptake and self-signalling process as adhesion induced activation of myosin-II is maximized by contact with a rigid rather than flexible substrate, such as the stiff cancer cells or the rigid RBC discocytes [45].

8.18 Cytokines as a Biomarker of Nanoparticle Mediated Immunomodulation

Following particle phagocytosis, the activated mononuclear phagocytic system produces cytokines and reactive oxygen intermediates that induce an immune response as part of host defence. Determination of the cytokine production can provide a conclusive data regarding the acute inflammatory response following nanoparticle administration. In case of PLGA, the microparticles are more inflammatory compared to the nanoparticles, as the former induced tissue necrosis factor-α and interlekin-1-β production when incubated with murine macrophages under *in vitro* conditions. Further, the PLGA microparticles are capable of activating the intracellular stress sensing pathway known as the nucleotide-binding oligomerization domain like receptors or the NALP3 inflammasome, in dendritic cells. Polymeric nanoparticles like PLGA have a tendency to polarize the immune response towards the T-helper cell 1 type, and these cytokines have a tendency to induce the macrophages toward a pro-inflammatory M1 phenotype [46]. PLGA nanoparticles were found to elicit minimal inflammatory response during the process of macrophage maturation, as exemplified by the similarity in release profile of pro-inflammatory cytokines interleukin-2, interleukin-6, interleukin-12, and TNF-α to the negative control, saline. Moreover, the anti-inflammatory cytokine IL-10 and chemokines interferon-γ, interleukin-4, and interleukin-5 remained at normal levels following PLGA treatment in Balb/c mice [47].

8.19 Clearance of Nanoparticles in Circulation

Once injected, nanoparticles directly reach the heart through the venous network. Following pulmonary circulation, the lung capillaries form the primary constraints for the injected nanoparticles. Unlike microparticles, the lower radius of curvature enables the escape of nanoparticles from pulmonary retention and they enter the systemic circulation via the pulmonary vein of the organ, the heart. The pore size of normal intact endothelium is 5 nm, and as a result, nanoparticles less than 5 nm enter into rapid equilibrium with the extravascular extracellular space. While in systemic circulation, the nanoparticles interact with the various serum proteins and cells and depending on their surface features and size can undergo opsonization and related phagocytosis. Moreover, certain organs with fenestrated vasculature like spleen, kidney, and liver present further sieving constraints for those that escape from the phagocytosis in the blood. The spleen, being the primary organ responsible for filtering the blood from antigens, microorganisms, and stiff old red blood cells, presents a unique micro-anatomical challenge for circulating nanoparticles. The nanoparticles, depending on size (>200 nm), shape (elongated or irregular), and elasticity, can get trapped in the sinus slits (0.2–0.5 μm) during their circulation in the splenic sinusoids and eventually succumb to phagocytosis by the splenic macrophages positioned in the cord space. Next in line are the kidneys, which are responsible for eliminating particles with minimal catabolism via glomerular filtration, tubular secretion, and urinary excretion. The filtration size threshold in renal clearance as imposed by the glomerular filtration unit is in the size range of 6–8 nm. In addition, surface charge also decides the renal clearance as the filtration is least for anionic particles and greatest for cationic particles followed by neutral particles. Those that escape renal clearance confront the hepatobiliary unit. The ciliated kupffer cells with stellate branches and ruffled surface are specifically adapted for the mechanical sequestration of colloidal and foreign substrates. As the primary function of the liver is to eliminate viruses, particles of size range 10–20 nm can undergo rapid biliary excretion. Additionally, hepatocytes are also involved in endocytosis and enzymatic degradation of foreign particles [48].

8.20 Strategies to Enhance Circulation Half-Time of Nanoparticles

Limiting the particle clearance from blood circulation is a prerequisite for engineering long-circulating nanoparticles, and a number of approaches have been developed in that direction. The basic theory followed in developing protein repellent biomaterials is to create either a super hydrophilic surface with a water contact angle approaching zero or a super hydrophobic surface with a surface tension approaching to zero [49]. Table 8.2 briefs about the different strategies followed to enhance hemocompatibility.

8.20.1 PEGylation

Surface modification with the nontoxic, non-immunogenic polymer PEG is the gold standard among the various protein-repellent strategies. The characteristics of PEG, such as solubility in aqueous and organic solvents, flexibility owing to the absence of bulky moieties in the polymer backbone, high hydration capacity which increases the hydrodynamic volume, and availability of a range of molecular species with low polydispersity index, have all contributed to its success. Repeating units of methoxy PEG, CH_3O-$(CH_2CH_2O)_n$-H, in its linear or branched form, constitute the majority of PEGylated products. The mechanism behind is the brush-induced steric repulsion that directly prevents contact between proteins and underlying surface and the formation of hydration layer that energetically resists protein adsorption. Moreover, the polar gauche conformation of hydrated PEG strongly resembles that of the structure of a water molecule. PEGylation of PLGA nanoparticles has proved to enhance circulation half-life, solubility, kinetic stability, and drug payload. For example, Gref *et al.* have reported that 50% of PLGA nanoparticles remained in the circulation following PEGylation with molecular weight 20,000 in the first 5 minutes and hence supports the stealth properties provided by PEGylation [50].

Table 8.2

Strategies	Advantages	Disadvantages
PEGylation	Nontoxic, non-immunogenic	Accelerated blood clearance, compromised cellular uptake
Non-ionic hydrophilic polymers	Lower accelerated blood clearance	Poor efficiency in comparison to PEG
Recombinant PEG mimetics	Enable mass production & purification from *E. coli*	Potential immunogenicity
HESylation	Low toxicity & high efficiency compared to pluronics	Bulky HES can hinder the interactions with other proteins
Cell-based-biomimetic systems	Elicit compatibility reactions in common to innate biological cells, to an extent	Poor scalability
Pathogen-inspired-immune evasion	Attempt to induce the immune evasion strategies of pathogen through material design	Poor scalability

However, bio-accumulation- and auto-oxidation-mediated toxicity, accelerated blood clearance, and associated immunogenicity interference with cellular uptake, and endosomal escape of nanoparticles has propelled the search for better alternatives to PEGylation.

8.20.2 Non-ionic Hydrophilic Polymers

On the basis of the success of PEG, non-ionic hydrophilic polymer based alternatives, such as poly-*N*-vinyl-2-pyrollidine, poly-4-acryloylmorpholine, poly-*N*-*N*-dimethylacrylamide, polyvinylalcohol, polyoxazoline, polyamonoacids, and polyglycerol, have been attempted to suppress accelerated blood clearance. However, their eminence over PEG has been yet not verified [51].

8.20.3 Recombinant PEG Mimetics

Fusion of the therapeutic protein or polysaccharide to a polypeptide based on neutral amino acids such as proline, alanine, and serine is gaining interest. Such PASylation and XTEN technology provide the advantage of one-step modification and mass production in *E. coli* without further purification steps [52].

8.20.4 HESylation

The plasma volume expander hydroxyethyl starch is a biodegradable semisynthetic polymer synthesized by hydroxyethylation of starch fragments. Basically, HESylation is a half-life extension technology mediated by coupling of hydroxyethyl starch. The biodegradation and circulation half-life of HESylated products can be controlled by varying the molar mass and the molar substitution value. The safety profile of HESylated products is established from the fact that the associated immunity is comparatively lower than albumin and dextran [52]. Besheer *et al.* has esterified different molar masses of HES with lauric acid and had used to stabilize PLGA nanosphere preparation through nanoprecipitation. Subsequently, the results had demonstrated that the HES laureates with higher molar mass could effectively prevent the uptake of nanospheres, and the stealth effect was comparable or in fact better than commercial pluronics [53].

8.20.5 Cell-Based Biomimetic Systems

Inspired by the 120 day circulation time of RBCs, several attempts have been made to mimic the mechanobiology and membrane features of RBCs to extend the circulation half-life of nanoparticles. Some of the developed RBCs mimicking platforms include the coupling of CD-47 to polymeric particles as a self-marker to resist phagocytosis and developing highly biconcave microparticles with increased elasticity. Camouflaging the nanoparticles with erythrocyte ghosts and a hitchhiking strategy, where the nanoparticles are covalently or physically coupled to RBC membrane has also been attempted and displayed excellent stability *in vitro* and greater circulation time to PEG under *in vivo* conditions [54]. For example, Hu *et al.* camouflaged the PLGA nanoparticle surface with RBC membrane for achieving long circulation, while inspired from this natural long-circulating entities, Doshi *et al.* synthesized biomimetic nanoparticles from the polymer PLGA which mimic the key structural and functional features of RBCs [55, 56]. The blood compatibility potential of RBC-mimicking nanoparticles is fully understood from the fact that the RBC membrane enclosure provide an extended circulatory half-life of 39.6 hours in a mouse model, which even surpass the 15.8 hour circulation provided by the PEGylation [54]. The zwitterionic feature of most cell membrane has inspired the surface modification of nanoparticles with zwitterionic disulfide and siloxane molecules [57]. In addition, phosphorylcholine, the electrically neutral head group present in cell membranes, has motivated the development of protein-repellent surfaces by forming a highly thermodynamic hydration barrier [58]. Inspired from the compatibility of glucose-predominant cellular glycocalyx, Thasneem *et al.* functionalized PLGA nanoparticles with glucose amine for enhanced hemocompatibility of PLGA nanoparticles [59]. Features of endothelial cell surface such as nano-sized topography, optimized mechanical properties, and the presentation of bio-responsive motifs, which enable the perfect coexistence with the circulating blood, have inspired a great deal of research toward blood vessel surface mimicking modification. As an example, nitric oxide mediated thromboresistance of the endothelial cell surface has inspired Thasneem *et al.* to modify the PLGA nanoparticle surface with L-cysteine, in order to enhance the hemocompatibility of PLGA nanoparticles [60].

8.20.6 Pathogen-Inspired Immune Evasion

The strategy of pathogens to escape immune recognition and infect cells with maximum efficacy has inspired the design of long-circulating and immune-evading nanoparticles. Coating of nanoparticles with dextran, hyaluronan, heparin, and polysialic acid is all based on pathogen-mediated camouflaging of antigenic surface with capsular polysaccharides and glycoproteins. Self-assembled particles mimicking the architecture of virus envelopes like influenza or herpes virus have also been developed [61]. On a similar line, Thasneem *et al.* has modified the PLGA nanoparticle surface with the heterogeneous mucin molecule for imparting the immune dodging functionalities of a mucylated pathogen [62].

8.21 Conclusion

The interfacial phenomenon of an ideal biomaterial inside the body must simulate the corporal tissues. This mandates a thorough evaluation of the overlapping hemocompatibility parameters before any envisioned medical intervention, especially those involving long-term contact with the blood. The concept of enhancing the hemocompatibility of a known FDA-approved polymer, PLGA, is significant from the viewpoint of the material's hydrophobicity. Well-defined surface modification of PLGA nanosystems could control the individual parameters involved in the chain of hemocompatibility sequence, in order to achieve an overall tolerance with the blood cells and the plasma. Several approaches to modify the PLGA nanoparticle surface have been attempted and published. However, their success is limited by immunogenicity, compromised targetability, and accelerated blood clearance mechanisms. This is often associated with the PEGylated nanocarriers, which in turn compelled the search for better surface stabilizing alternatives. Among the different approaches attempted, biomimetic surface modification, including thiolation, glucosylation, and mucylation, is the most promising as the physiological conditions are closely mimicked.

Acknowledgements

The authors wish to thank Prof. Jagan Mohan Tharakan, director, and Mr Neelakantan Nair O. S., head, BMT Wing, Sree Chitra Tirunal Institute for Medical Sciences and Technology, Poojapura,

Thiruvananthapuram, for providing facilities. This work was supported by the Department of Science and Technology, Government of India, through the project "Facility for Nano/microparticle-Based Biomaterials: Advanced Drug Delivery Systems 8013," under the Drugs & Pharmaceuticals Research Programme, New Delhi. The authors would also like to thank UGC for Junior Research Fellowship.

References

1. Zhang, Y., Chan, H. F., and Leong, K. W. (2013). Advanced materials and processing for drug delivery: the past and the future. *Adv Drug Deliv Rev,* **65**, 104–120.
2. Qiu, L. Y., and Bae, Y. H. (2006). Polymer architecture and drug delivery. *Pharm Res,* **23**, 1–30.
3. Lü, J.-M., Wang, X., Marin-Muller, C., Wang, H., Lin, P. H., Yao, Q., and Chen, C. (2009). Current advances in research and clinical applications of PLGA-based nanotechnology. *Expert Rev Mol Diagn,* **9**, 325–341.
4. Makadia, H. K., and Siegel, S. J. (2011). Poly lactic-*co*-glycolic acid (PLGA) as biodegradable controlled drug delivery carrier. *Polymers (Basel),* **3**, 1377–1397.
5. Sah, H., Thoma, L. A., Desu, H. R., Sah, E., and Wood, G. C. (2013). Concepts and practices used to develop functional PLGA-based nanoparticulate systems. *Int J Nanomedicine,* **8**, 747.
6. Croll, T. I., O'connor, A. J., Stevens, G. W., and Cooper-White, J. J. (2004). Controllable surface modification of poly(lactic-*co*-glycolic acid) (PLGA) by hydrolysis or aminolysis I: physical, chemical, and theoretical aspects. *Biomacromolecules,* **5**, 463–473.
7. Gaspar, R., and Duncan, R. (2009). Polymeric carriers: preclinical safety and the regulatory implications for design and development of polymer therapeutics. *Adv Drug Deliv Rev,* **61**, 1220–1231.
8. Ratner, B. D., Hoffman, A. S., Schoen, F. J., and Lemons, J. E. (2013). Biomaterials science: an evolving, multidisciplinary endeavor (Introduction). In *Biomaterials Science (Third Edition).* Ratner, B. D., Hoffman, A. S., Schoen, F. J., and Lemons, J. E. (eds.), Academic Press.
9. Williams, D. F. (2008). On the mechanisms of biocompatibility. *Biomaterials,* **29**, 2941–2953.
10. Ratner, B. D. (1993). New ideas in biomaterials science—a path to engineered biomaterials. *J Biomed Mater Res,* **27**, 837–850.

11. Naahidi, S., Jafari, M., Edalat, F., Raymond, K., Khademhosseini, A., and Chen, P. (2013). Biocompatibility of engineered nanoparticles for drug delivery. *J Control Release,* **166**, 182–194.

12. Danhier, F., Ansorena, E., Silva, J. M., Coco, R., Le Breton, A., and Préat, V. (2012). PLGA-based nanoparticles: an overview of biomedical applications. *J Control Release,* **161**, 505–522.

13. Dobrovolskaia, M. A., Aggarwal, P., Hall, J. B., and Mcneil, S. E. (2008). Preclinical studies to understand nanoparticle interaction with the immune system and its potential effects on nanoparticle biodistribution. *Mol Pharm,* **5**, 487–495.

14. Wang, Y.-X., Robertson, J., Spillman, W., Jr., and Claus, R. (2004). Effects of the chemical structure and the surface properties of polymeric biomaterials on their biocompatibility. *Pharm Res,* **21**, 1362–1373.

15. Verderio, P., Avvakumova, S., Alessio, G., Bellini, M., Colombo, M., Galbiati, E., Mazzucchelli, S., Avila, J. P., Santini, B., and Prosperi, D. (2014). Delivering colloidal nanoparticles to mammalian cells: a nano–bio interface perspective. *Adv Healthc Mater,* **3**, 957–976.

16. Nel, A. E., Mädler, L., Velegol, D., Xia, T., Hoek, E. M., Somasundaran, P., Klaessig, F., Castranova, V., and Thompson, M. (2009). Understanding biophysicochemical interactions at the nano–bio interface. *Nat Mater,* **8**, 543–557.

17. Tenzer, S., Docter, D., Kuharev, J., Musyanovych, A., Fetz, V., Hecht, R., Schlenk, F., Fischer, D., Kiouptsi, K., and Reinhardt, C. (2013). Rapid formation of plasma protein corona critically affects nanoparticle pathophysiology. *Nat Nanotechnol,* **8**, 772–781.

18. Sempf, K., Arrey, T., Gelperina, S., Schorge, T., Meyer, B., Karas, M., and Kreuter, J. (2013). Adsorption of plasma proteins on uncoated PLGA nanoparticles. *Eur J Pharm Biopharm,* **85**, 53–60.

19. Lück, M., Pistel, K.-F., Li, Y.-X., Blunk, T., Müller, R. H., and Kissel, T. (1998). Plasma protein adsorption on biodegradable microspheres consisting of poly (d, l-lactide-*co*-glycolide), poly (l-lactide) or ABA triblock copolymers containing poly (oxyethylene): influence of production method and polymer composition. *J Control Release,* **55**, 107–120.

20. Sobczynski, D. J., Charoenphol, P., Heslinga, M. J., Onyskiw, P. J., Namdee, K., Thompson, A. J., and Eniola-Adefeso, O. (2014). Plasma protein corona modulates the vascular wall interaction of drug carriers in a material and donor specific manner. *PLoS One,* **9**, e107408.

21. Dobrovolskaia, M. A., Clogston, J. D., Neun, B. W., Hall, J. B., Patri, A. K., and Mcneil, S. E. (2008). Method for analysis of nanoparticle hemolytic properties *in vitro*. *Nano Lett,* **8**, 2180–2187.

22. Saha, K., Moyano, D. F., and Rotello, V. M. (2014). Protein coronas suppress the hemolytic activity of hydrophilic and hydrophobic nanoparticles. *Mater Horiz,* **1**, 102–105.

23. Luo, R., Neu, B., and Venkatraman, S. S. (2012). Surface functionalization of nanoparticles to control cell interactions and drug release. *Small,* **8**, 2585–2594.

24. Fornaguera, C., Caldero, G., Mitjans, M., Vinardell, M. P., Solans, C., and Vauthier, C. (2015). Interaction of PLGA nanoparticles with blood components: protein adsorption, coagulation, activation of the complement system and hemolysis studies. *Nanoscale,* **7**, 6045–6058.

25. Furie, B., and Furie, B. C. (2008). Mechanisms of thrombus formation. *N Engl J Med,* **359**, 938–949.

26. Radomski, A., Jurasz, P., Alonso-Escolano, D., Drews, M., Morandi, M., Malinski, T., and Radomski, M. W. (2005). Nanoparticle-induced platelet aggregation and vascular thrombosis. *Br J Pharmacol,* **146**, 882–893.

27. Ilinskaya, A. N., and Dobrovolskaia, M. A. (2013). Nanoparticles and the blood coagulation system. Part I: benefits of nanotechnology. *Nanomedicine,* **8**, 773–784.

28. Gorbet, M. B., and Sefton, M. V. (2004). Biomaterial-associated thrombosis: roles of coagulation factors, complement, platelets and leukocytes. *Biomaterials,* **25**, 5681–5703.

29. Rao, G., and Chandy, T. (1999). Role of platelets in blood-biomaterial interactions. *Bull Mater Sci,* **22**, 633–639.

30. Osoniyi, O., and Onajobi, F. (2003). Coagulant and anticoagulant activities in Jatropha curcas latex. *J Ethnopharmacol,* **89**, 101–105.

31. Ilinskaya, A. N., and Dobrovolskaia, M. A. (2013). Nanoparticles and the blood coagulation system. Part II: safety concerns. *Nanomedicine,* **8**, 969–981.

32. Guidetti, G. F., Consonni, A., Cipolla, L., Mustarelli, P., Balduini, C., and Torti, M. (2012). Nanoparticles induce platelet activation *in vitro* through stimulation of canonical signalling pathways. *Nanomedicine,* **8**, 1329– 1336.

33. Ramtoola, Z., Lyons, P., Keohane, K., Kerrigan, S. W., Kirby, B. P., and Kelly, J. G. (2011). Investigation of the interaction of biodegradable

microand nanoparticulate drug delivery systems with platelets. *J Pharm Pharmacol,* **63**, 26–32.

34. Sahli, H., Tapon-Bretaudière, J., Fischer, A.-M., Sternberg, C., Spenlehauer, G., Verrecchia, T., and Labarre, D. (1997). Interactions of poly (lactic acid) and poly (lactic acid-*co*-ethylene oxide) nanoparticles with the plasma factors of the coagulation system. *Biomaterials,* **18**, 281–288.
35. Cenni, E., Granchi, D., Avnet, S., Fotia, C., Salerno, M., Micieli, D., Sarpietro, M. G., Pignatello, R., Castelli, F., and Baldini, N. (2008). Biocompatibility of poly (d, l-lactide-*co*-glycolide) nanoparticles conjugated with alendronate. *Biomaterials,* **29**, 1400–1411.
36. Salvador-Morales, C., and Sim, R. B. (2013). Complement activation. In *Handbook of Immunological Properties of Engineered Nanomaterials,* Dobrovolskaia, M. A., and McNeil, S. E. (eds.), 357–384.
37. Nilsson, B., Ekdahl, K. N., Mollnes, T. E., and Lambris, J. D. (2007). The role of complement in biomaterial-induced inflammation. *Mol Immunol,* **44**, 82–94.
38. Oyewumi, M. O., Kumar, A., and Cui, Z. (2010). Nano-microparticles as immune adjuvants: correlating particle sizes and the resultant immune responses. *Expert Rev Vaccines,* **9**, 1095–1107
39. Pedersen, M. B., Zhou, X., Larsen, E. K. U., Sørensen, U. S., Kjems, J., Nygaard, J. V., Nyengaard, J. R., Meyer, R. L., Boesen, T., and Vorup-Jensen, T. (2010). Curvature of synthetic and natural surfaces is an important target feature in classical pathway complement activation. *J Immunol,* **184**, 1931–1945.
40. Salvador-Morales, C., Zhang, L., Langer, R., and Farokhzad, O. C. (2009). Immunocompatibility properties of lipid–polymer hybrid nanoparticles with heterogeneous surface functional groups. *Biomaterials,* **30**, 2231–2240.
41. Yu, K., Lai, B. F., Foley, J. H., Krisinger, M. J., Conway, E. M., and Kizhakkedathu, J. N. (2014). Modulation of complement activation and amplification on nanoparticle surfaces by glycopolymer conformation and chemistry. *ACS Nano,* **8**, 7687–7703.
42. Xia, Z., Huang, Y., Adamopoulos, I. E., Walpole, A., Triffitt, J. T., and Cui, Z. (2006). Macrophage-mediated biodegradation of poly (dl-lactide-coglycolide) *in vitro. J Biomed Mater Res A,* **79**, 582–590.
43. Flannagan, R. S., Harrison, R. E., Yip, C. M., Jaqaman, K., and Grinstein, S. (2010). Dynamic macrophage "probing" is required for the efficient capture of phagocytic targets. *J Cell Biol,* **191**, 1205–1218.

44. Gomes, A. D. J., Lunardi, C. N., Caetano, F. H., Lunardi, L. O., and Machado, A. E. D. H. (2006). Phagocytosis of PLGA microparticles in rat peritoneal exudate cells: a time-dependent study. *Microsc Microanal,* **12**, 399– 405.

45. Sosale, N. G., Rouhiparkouhi, T., Bradshaw, A. M., Dimova, R., Lipowsky, R., and Discher, D. E. (2015). Cell rigidity and shape override CD47's 'self'signaling in phagocytosis by hyperactivating myosin-II. *Blood,* **125**, 542–552.

46. Elsabahy, M., and Wooley, K. L. (2013). Cytokines as biomarkers of nanoparticle immunotoxicity. *Chem Soc Rev,* **42**, 5552–5576.

47. Semete, B., Booysen, L., Kalombo, L., Venter, J. D., Katata, L., Ramalapa, B., Verschoor, J. A., and Swai, H. (2010). *In vivo* uptake and acute immune response to orally administered chitosan and PEG coated PLGA nanoparticles. *Toxicol Appl Pharmacol,* **249**, 158–165.

48. Longmire, M., Choyke, P. L., and Kobayashi, H. (2008). Clearance properties of nano-sized particles and molecules as imaging agents: considerations and caveats. *Nanomedicine (Lond),* **3**, 703–717.

49. Vladkova, T. G. (2010). Surface engineered polymeric biomaterials with improved biocontact properties. *Int J Polym Sci,* **2010**, Article ID 296094, 22 pages.

50. Gref, R., Minamitake, Y., Peracchia, M., Trubetskoy, V., Torchilin, V., and Langer, R. (1994). Biodegradable long-circulating polymeric nanospheres. *Science,* **263**, 1600–1603.

51. Ishihara, T., Maeda, T., Sakamoto, H., Takasaki, N., Shigyo, M., Ishida, T., Kiwada, H., Mizushima, Y., and Mizushima, T. (2010). Evasion of the accelerated blood clearance phenomenon by coating of nanoparticles with various hydrophilic polymers. *Biomacromolecules,* **11**, 2700–2706.

52. Besheer, A., Liebner, R., Meyer, M., and Winter, G. (2013). Challenges for PEGylated proteins and alternative half-life extension technologies based on biodegradable polymers. In *Tailored Polymer Architectures for Pharmaceutical and Biomedical Applications,* Scholz, C., and Kressler, J., eds., 215–233.

53. Besheer, A., Vogel, J. R., Glanz, D., Kressler, J. R., Groth, T., and MäDer, K. (2009). Characterization of PLGA nanospheres stabilized with amphiphilic polymers: hydrophobically modified hydroxyethyl starch vs pluronics. *Mol Pharm,* **6**, 407–415.

54. Fang, R. H., Hu, C.-M. J., and Zhang, L. (2012). Nanoparticles disguised as red blood cells to evade the immune system. *Expert Opin Biol Ther,* **12**, 385–389.

55. Hu, C.-M. J., Zhang, L., Aryal, S., Cheung, C., Fang, R. H., and Zhang, L. (2011). Erythrocyte membrane-camouflaged polymeric nanoparticles as a biomimetic delivery platform. *Proc Natl Acad Sci U S A,* **108**, 10980–10985.

56. Doshi, N., Zahr, A. S., Bhaskar, S., Lahann, J., and Mitragotri, S. (2009). Red blood cell-mimicking synthetic biomaterial particles. *Proc Natl Acad Sci,* **106**, 21495–21499.

57. Estephan, Z. G., Schlenoff, P. S., and Schlenoff, J. B. (2011). Zwitteration as an alternative to PEGylation. *Langmuir,* **27**, 6794–6800.

58. Jordan, M., and Elbayoumi, T. (2013). Phosophorylcholine-based biomimetic coatings for nanomedical applications. *J Membra Sci Technol,* **3**, e113.

59. Thasneem, Y., Sajeesh, S., and Sharma, C. P. (2013). Glucosylated polymeric nanoparticles: a sweetened approach against blood compatibility paradox. *Colloids Surf B Biointerfaces,* **108**, 337–344.

60. Thasneem, Y., Sajeesh, S., and Sharma, C. P. (2011). Effect of thiol functionalization on the hemo-compatibility of PLGA nanoparticles. *J Biomed Mater Res A,* **99**, 607–617.

61. Sawdon, A., and Peng, C.-A. (2013). Engineering antiphagocytic biomimetic drug carriers. *Ther Deliv,* **4**, 825–839.

62. Thasneem, Y., Rekha, M., Sajeesh, S., and Sharma, C. P. (2013). Biomimetic mucin modified PLGA nanoparticles for enhanced blood compatibility. *J Colloid Interface Sci,* **409**, 237–244.

Chapter 9

Peroral Polyester Drug Delivery Systems

Matilde Durán-Lobato,[a] Maria Ángeles Holgado,[a] Josefa Álvarez-Fuentes,[a] José L. Arias,[b] Mercedes Fernández-Arévalo,[a] and Lucía Martín-Banderas[a]

[a]*Department of Pharmacy and Pharmaceutical Technology, Faculty of Pharmacy, University of Sevilla, C/Profesor García González 2, 41012 Sevilla, Spain*

[b]*Department of Pharmacy and Pharmaceutical Technology, Faculty of Pharmacy, University of Granada, Campus de Cartuja s n, 18071 Granada, Spain*

luciamartin@us.es

9.1 Introduction

Polyesters are a very important group of biodegradable polymers that are characterized by the existence of esters bonds in their main chain. They are extensively investigated given their great diversity and versatility of synthesis, and due to the fact that the ester groups are hydrolytically degradable. Polycondensation of difunctional monomers primarily leads to the formation of polymers with low molecular weight (MW). On the other hand, ring-opening polymerization results in polymers with high MW. In this line, biodegradable polyesters are synthesized via ring-opening

Handbook of Polyester Drug Delivery Systems
Edited by MNV Ravikumar

ISBN 978-981-4669-65-8 (Hardcover), 978-981-4669-66-5 (eBook)
www.panstanford.com

polymerization of medium-size lactones (six- and seven-membered) [1, 2].

Thanks to the generation of nontoxic degradation products and to an adjustable degradation rate, the aliphatic polyesters most commonly used in drug delivery are poly(D,L-lactide) (PLA), poly(glycolide) (PGA), poly(D,L-lactide-*co*-glycolide) (PLGA), poly(3-hydroxybutyrate) (PHB), and poly(ε-caprolactone) (PCL) (Fig. 9.1) [3]. In fact, they have been approved by the Food and Drug Administration (FDA) for the preparation of several medicines (including those administered systemically), and in tissue engineering (to synthesize scaffolds that can control and direct cell growth and differentiation). Regarding the formulation of targeted drug delivery systems, those biodegradable polyesters have been extensively studied for diverse therapeutic applications [4]. In the field of bone and tissue engineering, they have been also used in the preparation of resorbable medical implants in the shape of rods, plates, fibers, and beads [5].

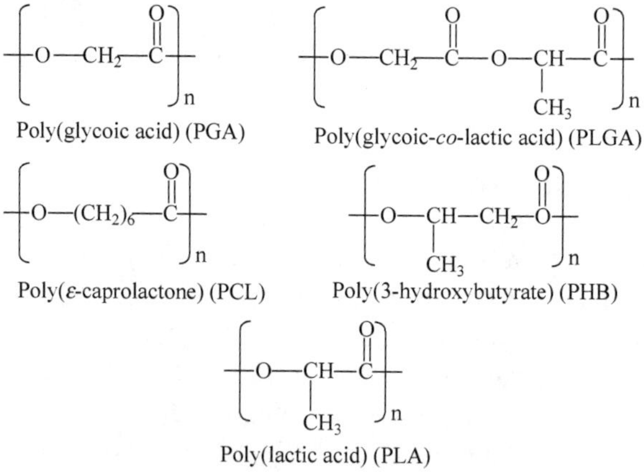

Poly(glycoic acid) (PGA)

Poly(glycoic-*co*-lactic acid) (PLGA)

Poly(ε-caprolactone) (PCL)

Poly(3-hydroxybutyrate) (PHB)

Poly(lactic acid) (PLA)

Figure 9.1 Chemical structure of aliphatic polyesters widely employed in drug delivery engineering.

PLA, PLGA, and PCL have found very promising applications in the field of drug delivery given the possibility of preparing a nanoparticulate system with controlled drug release kinetics by basically modifying the MW of these biodegradable polymers. In fact, their *in vitro* and *in vivo* degradation kinetics and *in vivo* fate have been characterized comprehensively [7]: the slow degradation rate of these polymers can be catalyzed by lipases and induces an almost negligible immune response. Regarding the effect of the MW on the kinetics of degradation, it has been described that the hydrolytic degradation of PLA is faster as the MW increases.

PGA is the simplest biodegradable poly(R-hydroxy acid) which was described in 1954 to be a low-cost tough fiber forming polymer. This linear aliphatic polyester is characterized by a highly crystalline structure and by a hydrophilic character. It has been described that PGA can undergo hydrolytic degradation both *in vitro* and *in vivo* [2, 8]. PGA is prepared by ring-opening polymerization of a cyclic lactone, glycolide. It is characterized by a high melting point (224–227°C) and by a very low solubility in organic solvents (except for hexafluoro-2-propanol). Because of the excellent biocompatibility and biodegradability, PGA and its copolymers have been utilized in dental, drug delivery, and orthopaedic applications. For instance, PGA was utilized in the 1960s in the development of the first synthetic absorbable suture (Dexon®, American Cyanamid Corp.) [9]. Unfortunately, biomedical use of PGA is limited by a low solubility in aqueous and organic media, and by a rapid degradation rate leading to acidic products that can induce an intense inflammatory response [10].

PLA is usually obtained by polycondensation of D- or L-lactic acid, or by ring-opening polymerization of lactide, a cyclic dimer of lactic acid [2]. First reports on the biomedical applications of PLA can be found in the 1960s. Since then, PLA has found a very promising use in the preparation of sutures, drug delivery devices, prosthetics, scaffolds, vascular grafts, bone screws, and in the design of pins and plates for temporary internal fracture fixation [6, 8]. The hydrophobic character of this aliphatic polyester comes from the existence of -CH3 side groups in its chemical structure. Compared to PGA, PLA is more resistant to hydrolysis thanks to the steric shielding effect of these -CH3 side groups. However, and due to this property, it presents a slow degradation rate leading to a slow drug release rate [2, 11].

PLGA has become one of the most studied copolymer biomaterials for drug encapsulation because of its excellent tissue compatibility, biodegradable nature, and safety profile for use in humans [12]. The popularity of PLGA can be further explained by its approval for a number of clinical applications by the FDA. This diblock copolymer can be obtained by copolymerization of glycolic acid and lactic acid, being less stiff than its components since the degree of crystallinity decreases as the content of either co-monomer is greater. The hydrophilic character of PLGA is greater as the percentage of glycolic acid in the chemical composition is increases. More specifically, PLGA composed of 25–70% in glycolic acid is amorphous, a desirable property for many drug delivery applications. This biodegradable copolymer has been used in the engineering of micro- and nanomatrices for drug delivery applications due to its ease preparation, commercial availability at a reasonable cost, versatility, biocompatibility, and hydrolytic degradation into resorbable and harmless products [6].

The degradation rate of PLGA is defined by its MW, distribution, lactic acid/glycolic acid ratio, polymer end-group, size, and pH of the (physiological) medium. Generally, low MW PLGAs degrade faster (rapid drug release and greater initial burst release) [13]. Probably, one of the significant disadvantages of PLGA is the generation of an acidic environment during its degradation that contributes to the bulk erosion of the matrix a circumstance can affect the stability of peptides and proteins [14]. With respect to the influence of the pH of the external medium on the degradation rate, contradictory results have been reported: low and high pH values may accelerate the degradation rate, while it has been further suggested a pH-independent hydrolysis *in vitro*. This is probably the consequence of the combined effects of the hydrophobic/hydrophilic balance and of the degree of crystallinity of the copolymer. The extent of water hydration can determine the degradation rate. In fact, amorphous PLGA has been reported to degrade more quickly than semicrystalline PLGA [8].

PCL is a hydrophobic and semicrystalline aliphatic polyester which degree of crystallinity decreases with increasing MW. It is synthesized by ring-opening polymerization of ε-caprolactone in presence of tin octoate catalyst [15]. PCL is biocompatible (being completely excreted upon bioresorbtion), highly permeable to small drug molecules, and its slow degradation rate makes it suitable for controlled drug delivery applications. Functional groups could be incorporated to the chemical structure to render the polymer more hydrophilic, adhesive, and biocompatible, thus enabling positive cell responses [11, 16].

Interestingly, PCL can be modified by copolymerization or blending with other polymers, this allowing the modification of its physical chemistry and mechanical properties, e.g., degree of crystallinity, solubility, and degradation rate [17]. Such copolymers, PCL with, e.g., cellulose propionate, cellulose acetate butyrate, PLA, or PLGA, will exhibit drug release kinetics [18]. Concretely, these drug release properties depend on the type of formulation, the method of preparation, the PCL content, and on the size and amount of drug loaded. The slow degradation rate, coming from a progressive and homogeneous erosin and hydrolysis (compared to PLA and PLGA), can be advantageously used to formulate long-term drug release/ delivery nanosystems.

PHB is a highly crystalline (degree of crystallinity >50%) has been proposed as an alternative to PLGA in the formulation of drug delivery systems (and implants) that can be administered through the parenteral route of administration [19]. This is a linear homopolymer biosynthesized by diverse strains of bacteria, or prepared by chemical synthesis.

The crystallinity (and resulting brittleness) of PHB can be reduced by copolymerization with 3-hydroxyvaleric acid. Copolymerisation offers may be advantageously used to gain control of the degradation properties. PHB and its copolymers with HV are soluble in a number of common organic solvents, and they can be processed into membranes, fibers, or microspheres. In addition, and thanks to their adequate biocompatibility and biodegradability, these polymers have been used to formulate matrices for drug delivery and tissue engineering [6]. The piezoelectric properties associated to these polymers make them attractive materials for orthopaedic applications, such as bone plates, as they can stimulate bone growth [20].

Finally, polyester dendrimers offer an attractive alternative in the preparation of drug delivery systems, given their high drug loading capacity [21, 22]. Encapsulation or conjugation of hydrophobic, labile, and small drugs within dendrimers may improve their solubility in aqueous media, while protecting them from harsh surroundings, and keeping to very minimum the characteristic (and negative) biodistribution into healthy tissues and rapid clearance [23, 24]. The name dendrimer derived from the Greek word *dendron*, meaning "tree," which refers to the unique tree-like branching architecture of these polymers. They are characterized by layers between each cascade point, popularly known as "generations." The polymer architecture is basically composed of an inner core moiety followed by radially attached "generations" that contains chemical functional groups at the exterior terminal surface [25]. Several dendrimers have been developed for drug delivery applications, e.g., polyamidoamine (PAMAM), poly(propyleneimine) (PPI), poly-L-lysine, melamine, poly(etherhydroxylamine) (PEHAM), poly(esteramine) (PEA), and polyglycerol [26].

The interest in the use of dendrimers in biomedicine has recently increased and numerous review articles on dendrimers and their biomedical applications have been published [27–29]. The most significant advantages of dendrimers are polyvalency, high degree of branching, high solubility, globular architecture, and, more importantly, their well-defined architecture (monodispersity) which translate into a more consistent well-defined polymer that provides a better reproducibility of results [30].

It has been described that the physical chemistry of dendrimers is better than analogous linear polyester, e.g., chemical reactivity, aqueous and nonpolar solubilities, ionic conductivity, etc. [31]. Dendrimers based on the monomer 2,2-bis(hydroxymethyl) propionic acid and hybrids with polyethylene oxide (PEO) has been used to prepare drug delivery systems against malignancies [32]. Diverse polyester dendrimers incorporating monomers such as glycerol, succinic acid, phenylalanine, and lactic acid have been prepared by Grinstaff *et al.* [33].

The oral use of dendrimers may be very promising especially in the delivery of anticancer and antihypertensive drugs [34, 35]. It has been described that upon oral administration, the transport of a dendrimer through the epithelial layer of the gastrointestinal tract depends on its characteristics. Absorption of the dendrimers through both Peyer's patches and enterocytes has been measured after administration of a dose of 28 mg/kg. Results demonstrated the preferential uptake of the dendrimer through Peyer's patches at the small intestine [36, 37].

Further consideration of the chemical structure and modifications of polyesters is taken when it comes to optimize the interaction with the intestinal barrier to improve the absorption of drugs, particularly in the case of nano-sized drug delivery systems.

Two main pathways are differentiated in this case, known as paracellular and transcellular transport [38]. In order to increase the paracelullar transport of nanocarriers, chemical modifications involve the attaching of molecules capable to open tight junctions between intestinal cells to the surface of nanocarriers, being chitosan and thiomers the most employed ones. On the other hand, the transcellular pathway has been more extensively explored; in this sense, both nonspecific and specific cell targeting strategies have been developed (Fig. 9.2) [39]. For a nonspecific targeting of both enterocytes and microfold (M) cells, the combination of adequate particle size and surface charge plays they key role for the interaction with these intestinal cells. In addition, the employment of excipients with mucoadhesive properties (e.g., chitosan, PVA, Vit E, TPGS) as coatings for nanoparticles to improve the interaction with hydrophilic and negatively charged mucus has been also explored so as to increase the residence time of carriers at the site of absorption.

Regarding the specific targeting of intestinal cells, a variety of molecules have been explored to facilitate the cell interaction. In the case of enterocytes specific targeting, lectins (WGA, Con-A), lecithins and vitamins have been the most studied ligands [40, 41], to mention a few. In the case of M cells targeting, the same strategy is followed, along with employing receptor ligands, usually molecules that are present in pathogens surface or produced by them (lectins, RGD peptide, adhesins, antibodies and toxins, to mention a few) [42, 43].

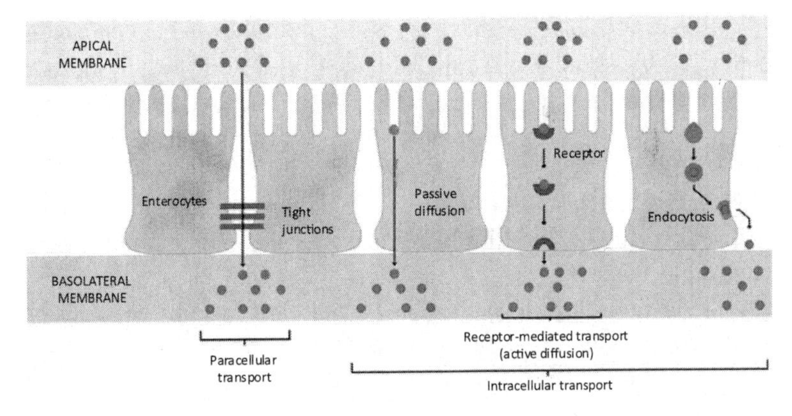

Figure 9.2 Different pathways for the transport of nanoparticles through the gastrointestinal barrier.

On the other hand, biodegradable polyester nanoparticles with poly(ethylene glycol) (PEG) chains at their surface have attracted growing interest as promising drug delivery systems [44, 45]. Some researchers have pointed out that this PEG coating could affect the interaction of nanoparticles with biological surfaces and facilitate their rapid penetration across the human mucosal barrier [46, 47].

In relation to this, Song *et al.* [48] characterized the cellular transport mechanism of the well-known pegylated polyester nanoparticles and determined the effect of polymer architecture (PEG chain length and core material) on its cellular interaction and transcellular transport. The pegylated polyester nanoparticles were found to undergo an energy-dependent, lipid raft-mediated, but caveolae-independent endocytosis. Variance in PEG chain length (from 2000 to 5000 Da) was found to hardly affect the cellular interaction and the intracellular itinerary of the nanoparticles. In fact, PEG molecular weight can strongly influence the efficacy of oral nanomedicines and, usually, the lower PEG molecular weight is, the higher oral bioavailability is attained [49].

Moreover, pegylated nanoparticles may also represent an effective strategy to protect the drug agiantst liver cytochrome P450 metabolic enzymes [49] and gastrointestinal harsh enzymes [50]. The mechanism of PEG stabilization effect is not clear, but it can be reasonably attributed to a reduced interaction between the nanoparticles and the enzymes of the digestive fluids due to the PEG cloud.

Other effect of the PEGylation process is related to its potential to cross the loosely adherent mucus layer, which is continuously removed by peristalsis and replaced, and to reach the firmly adherent layer deposed on the gastrointestinal epithelium [51]. Hanes and coworkers have recently demonstrated that mucoinert pegylated nanocarriers of adequate size (200–500 nm), the so-called mucus-penetrating particles (MPPs), may deeply penetrate human mucus, whereas uncoated particles of comparable size are immobilized by the mucus meshes [46].

In view of the above exposed scientific findings, it is clear that biomedicine is constantly evolving and new approaches are incorporated to the diagnosis and treatment of diseases. In this context, the design of efficient and safe nanomedicines to be administered orally is still needed. In this respect, aliphatic polyesters are contributing to beat that challenge. This chapter is devoted to the use of the most common, PLGA and PCL, in the development of oral drug delivery nanoformulations. Special attention is given to their application in antitumor and antibiotic therapies, as well as in oral vaccination.

9.2 Oral Chemotherapy

Cancer treatment is based on surgery, radiation, chemotherapy, hormone therapy, immune therapy, targeted therapy, and combinations of these strategies. Oral chemotherapy is the most appealing approach against cancer since it may provide a prolonged exposure of malignant cells to cytotoxic agents with a much better patient compliance and improvement of the quality of life. Unfortunately, upon oral administration the majority of antitumor molecules undergo an intensive first-pass metabolic effect leading to a poor oral bioavailability (paclitaxel ≈ 1%, docetaxel < 10%, doxorubicin < 5%). Dose-limiting toxicities are further contributing to the failure of oral chemotherapy, e.g., rashes, handfoot syndrome, even described in new anticancer agents. For instance, erlotinib (Tarceva®, against lung and pancreas cancer) can induce uncomfortable rash which may lead to non-adherence to the treatment. To beat the challenges, it has been proposed the development of nanoparticulate systems in oral chemotherapy.

In this line, PLGA-based drug nanocarriers are promising tools to (i) optimize the oral bioavailability of anticancer molecules, (ii) minimize the associated toxicity by increasing drug uptake by tumor cells and thus by reducing the drug dose needed to treat the malignancy, and (iii) provide a controlled drug release profile that can prolong the contact time between the cancer cells and the chemotherapy agent [52]. Numerous research reports have described the potential use of PLGA-based nanosystems in oral chemotherapy. Similarly, PCL-based drug nanocarriers have recently entered preclinical and clinical stages of development [53–55].

9.2.1 PLGA-Based Nanoformulations

In a recent investigation [56], an original oral polysorbate-free formulation of docetaxel nanocapsules (NCs) embedded in gastro-resistant microparticles has been reported. NCs 300 nm in diameter were synthesized by emulsion-solvent evaporation. After this, NCs were dispersed in a solution of Eudragit L and HPMC. Finally, PLGA suspension was spray-dried (2–10 microns in diameter).

The oral delivery in minipigs of this novel docetaxel formulation elicited a significantly enhanced relative bioavailability compared with oral Taxotere, resulting in a calculated area under the plasma concentration–time curve (AUC) and $C_{\max}$ values 10- and 8.4-fold higher.

This way, docetaxel, a P-glycoprotein (P-*gp*) drug substrate, showed an improved oral bioavailability without affecting the physiological activity of the P-gp and CYP3A gut metabolism. Figure 9.3 shows plasma level–time profiles of docetaxel in different formulations after oral and IV administration. The authors demonstrated a prolonged blood circulation of docetaxel NCs via lymphatic absorption.

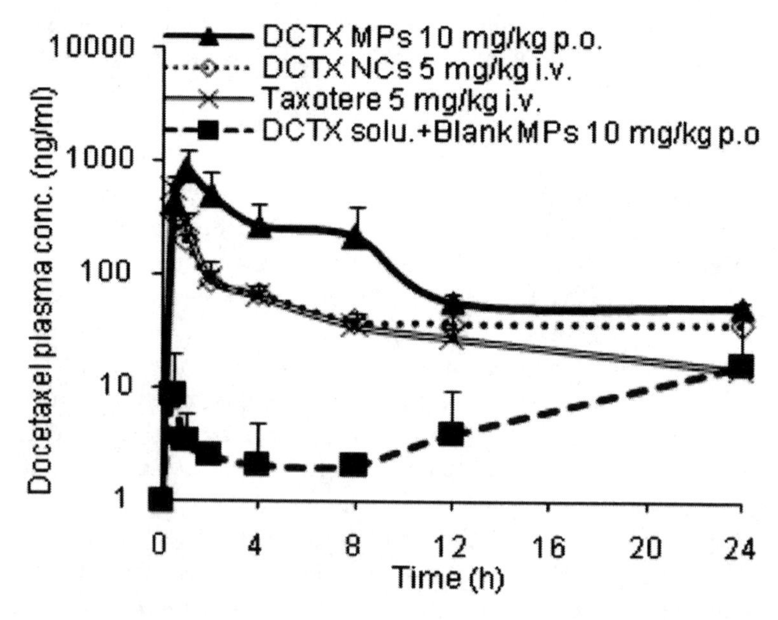

Figure 9.3 Mean plasma docetaxel level-time profiles following IV administration of docetaxel NCs at a dose of 5 mg/kg, oral administration of docetaxel solution with blank micorparticles, or docetaxel formulation at a dose of 10 mg/kg (n = 4). Reprinted with permission from [56].

It is well known that P-*gp* inhibitors can improve drug oral bioavailability. As an example, a PLGA nanoplatform was formulated for the oral delivery of exemestane against breast cancer [57]. In this case, the hydrophilic vitamin E derivate D-α-tocopheryl poly(ethylene glycol) 1000 succinate (TPGS) was incorporated. *In vitro* assays in MCF-7 cells revealed a greater cytotoxic activity compared to the free drug, which was reported to depend on the drug concentration and incubation time. TPGS has been further used in the formulation of paclitaxel-loaded PLGA NPs. This vitamin E derivate can be incorporated to the formulation during the synthesis procedure [58] or can be included within the polymer structure [59]. In the latter case, TPGS can improve drug permeability through cell membranes by inhibiting P-*gp*. Alternatively, TPGS has been described to contribute to the antitumor effect by inducing apoptosis. Compared to Taxol®, the nanoparticulate system reduced the viability of MCF-7 human breast cancer cell by ~1.3 times after 24 hours [58]. In this investigation it was also demonstrated that the nanoformulation improved the oral bioavailability of paclitaxel.

Oral bioavailability of doxorubicin has further been improved by its vehiculization into PLGA-based nanoparticulate systems. In a recently published research report, the *in vivo* antitumor effect of doxorubicin-loaded PLGA NPs has been investigated [60]. Interestingly, it was established a meaningful *in vitro–in vivo* correlation between the intestinal epithelial permeability and the anticancer efficacy.

PLGA NPs have also been postulated to overcome P-*gp*-mediated efflux of vincristine sulphate, a clear resistance mechanism developed by tumor cells against this drug [61]. In this investigation, dextran sulphate-PLGA nanohybrids (<150 nm in size) were prepared by combining a self-assembly technique with a nanoprecipitation method (Fig. 9.4). Negatively charged dextran formed a complex with the positively charged drug, leading to greater encapsulation efficiencies. This nanoformulation was evaluated *in vitro* in MCF-7 and P-*gp* overexpressing MCF-7/Adr cells. Compared to the free drug (in aqueous solution), it was found that the oral bioavailability of vincristine was markedly enhanced when loaded into the nanoparticulate system (equivalent drug dose: 4 mg/kg), probably thanks to an inhibition of the P-*gp* efflux.

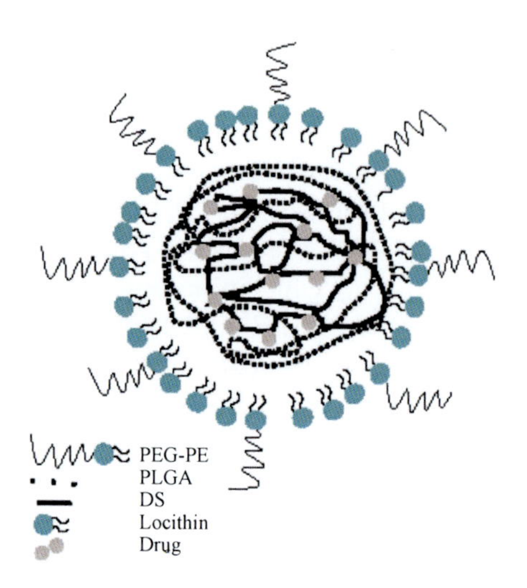

Figure 9.4 Schematic illustration of a novel self-assembled dextran sulphate-PLGA hybrid nanoparticle (DPN) to improve the encapsulation efficiency and oral bioavailability of vincristine sulphate. Adapted from [61].

Gemcitabine has also taken advantage of PLGA nanoformulations [62]. The preparation procedure was based on a multiple emulsification-solvent evaporation methodology. Oral absorption of the nanoparticulate system took place through M cells of Peyer's patches. *In vitro* studies analyzed the cytotoxic effect of the gemcitabine-loaded NPs in Caco-2 cells and K562 cells. Gemcitabine-loaded nanocarrier was found to be ~8.6-fold more cytotoxic to K562 cancer cells after 48 hours of incubation compared to the free drug. This could be attributed to a higher NP cell uptake via endocytosis which cannot take place in the case of the free drug given its hydrophilic character. In addition, drug-loaded NPs found to be noncytotoxic to Caco-2 cells probably because these cells have a molecular "fingerprint" different to malignant cells and quite similar to enterocytes.

Andrographolide isolated from *Andrographis paniculata* exhibits anti-inflammatory and anticancer activities, but its high hydrophobic character and poor oral bioavailability hamper its clinical use. PLGA-PEG-PLGA micelles have been developed to improve the oral bioavailability of this drug [63]. *In vitro* cytotoxic investigations demonstrated that andrographolide-loaded micelles inhibited more efficiently cell proliferation inhibition, arrested cell cycle at the G_2/M phase, and displayed pro-apoptosis effects in MAD-MB-231 cells. Furthermore, and compared with the drug aqueous suspension, the area under curve of plasma concentration (AUC) and the mean residence time of the drug-loaded micelles were increased by ~2.7- and ~2.5-fold, respectively.

Bromelain is a natural enzyme that may reduce the toxicity associated to the clinical use of anticancer drug molecules. Unhappily, its activity is significantly reduced upon oral administration. As an alternative, bromelain was loaded to PLGA NPs previously surface coated with Eudragit® L30D to increase their gastric stability. The oral administration of the nanoparticulate system reduced tumor burden of Ehrlich ascites carcinoma in Swiss albino mice, thus increasing the life-span (~160 ± 6%) when compared with free bromelain (~24 ± 3%) [64, 65].

9.2.2 PCL-Based Nanoformulations

PCL nanoparticulate systems have been also investigated as drug delivery devices against cancer. However, very little has been done up to now to implement their use in oral chemotherapy. Hopefully, some research reports have been published very recently (Table 9.1).

It has been investigated the development of PCL microparticles for the oromucosal delivery of Δ^9-tetrahydrocannabinol (THC) and cannabidiol (CBD) [66]. The microparticles were analyzed as an alternative delivery system for long-term cannabinoid administration in a murine xenograft model of glioma. *In vitro* studies demonstrated a sustained release of the two cannabinoids during several days. Local administration of THC-, CBD-, or a mixture (1:1 w:w) of THC- and CBD-loaded microparticles every 5 days to mice bearing.

More recently, a composite nanosystem (micelles incorporated into a pH-responsive hydrogel) was engineered for the oral delivery of docetaxel [67]. Micelles were characterized by a small particle size (~20 nm) and high drug loading values (~7.8%).

Cytotoxicity experiments of in 4T1 cells demonstrated their effective antitumor activity. It was reported that the composite nanosystem showed much quicker diffusion of micelles in simulated intestinal fluid in comparison with other physiological pHs. Docetaxel

Table 9.1 Illustrative examples of the potential use of PCL nanosystems in drug delivery through the oral route

Synthesis methodology	Anticancer drug	Nanostructure	*In vitro* evaluation	*In vivo* evaluation	References
Solvent displacement in an acetone-water system containing Pluronic® F- 68	Tamoxifen	PCL NP	Increased local drug concentration in estrogen receptor positive MCF-7 breast cancer cells, after 1 hour of incubation	–	[72]
Heat-initiated free radical polymerization	–	pH-sensitive hydrogel based on based on methoxyl PEG-PCL-acryloyl chloride, PEG methyl ether methacrylate, and methacrylic acid	–	–	[53]
Drug intercalation in the interlayer gallery montmorillonite, which is further compounded with PCL	Tamoxifen	Montmorillonite/PCL composites. Controlled drug release pattern	Enhanced antitumor activity against HeLa and A549 cancer cells	Optimized pharmacokinetic profile when the drug is loaded to the NPs	[54]
Heat-initiated free radical method	–	pH-sensitive hydrogel based on methoxyl PEG-PCL-acryloyl chloride		Maximum tolerance dose of the hydrogel was > 10,000 mg/kg body weight in BALB/c mice. The hydrogel was nontoxic after gavage	[55]

release was determined to be quite slowly. The pharmacokinetic study revealed that the composite nanosystem significantly enhanced the oral drug bioavailability (~75%), probably thanks to the small intestine targeting release of the pH-responsive hydrogel. The nanosystem was effective in inhibiting tumor growth in a subcutaneous 4T1 breast cancer model, and reduced the systemic toxicity of the drug compared with the intravenous treatment.

9.2.3 Oral Chemoprevention

The use of curcumin and quercetin against cancer is seriously limited by their low solubility and physicochemical stability, rapid plasma clearance, and low cellular uptake. In view of that, diverse nanoparticulate systems have been developed to overcome these problems. For instance, curcumin-loaded PLGA NPs have been investigated as oral chemopreventive agents against diethylnitrosamine (DEN)-induced hepatocellular carcinoma in rats [68].

The NPs (~15 nm in diameter) were prepared by a modified emulsion-diffusion-evaporation method. Drug equivalent doses were 20 mg/kg were administered once per week during 16 weeks. It was found that were more efficient against cancer at much lower doses than the free drug. Such promising results may be the consequence of an improved oral bioavailability.

With respect to quercetin, it has been recently proposed its co-encapculation into PLGA NPs along with tamoxifen [69] in order to obtain a synergistic antitumor effect while reducing the hepatotoxicity induced by tamoxifen. *In vivo* antitumor evaluations were conducted in DMBA-induced breast tumor bearing female SD rats. Oral co-administration of free tamoxifen (3 mg/kg) and free quercetin (6 mg/kg) resulted in an insignificant reduction in tumor growth as compared to free tamoxifen ($p > 0.05$). On the contrary, tamoxifen+quercetin-loaded NPs significantly suppressed tumor growth in comparison with the free tamoxifen ($p < 0.01$) and tamoxifen+quercetin (1:2 w/w) ($p < 0.05$), as a consequence of the greater oral bioavailability provided by the nanoparticulate system. In addition, the nanoformulation did not induce the hepatotoxic effects characteristic of tamoxifen.

Finally, PLGA NPs have been investigated as delivery systems for SR13668, an Akt inhibitor with antineoplastic and antiangiogenic activities and a promising cancer chemopreventive activity [70]. This molecule is characterized by a very low oral bioavailability (< 1% in rats). The nanosystem was prepared by flash nanoprecipitation integrated with spray drying [71]. It was reported an enhanced oral bioavailability compared to controls (Labrasol®, and the drug in 0.5% methylcellulose), when drug equivalent doses of 2.8 mg/kg were administered to beagle dogs ($n = 8$).

9.3 Oral Antibiotic Therapy

Antimicrobial resistances limit the therapeutic efficacy of conventional therapy against microorganisms. Although many investigations are devoted to the discovery of new antibacterial drugs, other studies are focused on improving the clinical outcomes of current antibacterial drugs by incorporating them into novel formulations. In this line, drug carriers are engineered to minimize the induction of resistances against antibiotics, and for the efficient delivery of antibiotics presenting low plasma half-lives or with difficulties in crossing across biological barriers [73, 74]. In addition, several infectious diseases are caused by microorganisms capable of surviving inside phagocytic cells. Such intracellular localization protects them from the immune system and from antibiotics with poor penetration efficiencies into phagocytic cells [75]. In fact, hydrophilic antibiotics do not exhibit high intracellular concentrations and, furthermore, they are concentrated inside lysosomes upon internalization, where the bioactivity of the drug is low. Thus, a limited intracellular activity against sensitive bacteria is often displayed [76].

Nanoparticulate-based drug carriers may significantly help in the selective delivery of antibiotics into phagocytic cells, where the pathogen can be found [74–77]. Such a targeted and controlled drug delivery may further reduce the toxicity associated to the mechanism of action of the drug [73, 74, 76, 78]. However, little is known about the toxic effects of antimicrobial NPs at the central nervous system [79], and there is a need for new characterization techniques that are not affected by NP size properties [80].

Ideally, formulation of these nanocarriers should be based on passive targeting strategies, involving the development of long-circulating NPs that can selectively extravasate into the site of infection where inflammation has led to the enhanced retention and permeability effect [76, 81]. Complementarily, NP design can be based on active targeting strategies primarily involving the surface functionalization of the drug nanocarrier with ligands that can bind antigens at site of infection [76].

9.3.1 PLGA-Based Nanoformulations

Probably, PLGA, PGA, and PLA are the biodegradable polymers most widely investigated in antibiotic therapy development [73, 74, 77, 78]. In fact, numerous research articles highlight that they can improve the pharmacokinetics and pharmacodynamics of antimicrobial drugs. Special insight has recently gained the optimization of the oral administration of antimicrobial drug. For example, clarithromycin-loaded PLGA NPs prepared by nanoprecipitation have been described to be more effective than the free drug against *Staphylococcus aureus*. They can display *in vitro* a similar antibacterial effect but at 1/8 concentration of the free drug [82]. These promising findings were not confirmed *in vivo*. Analogous results were obtained with azithromycin-loaded PLGA NPs against *Salmonella typhi* [83].

PLGA NPs can be formulated by employing multiple emulsion methodologies to encapsulate rifampicin, isoniazid, pyrazinamide, and ethambutol. These NPs have been administered orally for cerebral drug delivery in a murine model (*Mycobacterium tuberculosis* H_{37}Rv-infected mice). Results demonstrated that a single oral dose of the nanoformulation maintained drug plasma levels for 5–8 days, and for 9 days in brain. Furthermore, five oral doses of the nanoparticulate system administered every 10 days resulted in undetectable bacilli in the meninges. According to these results, such polymeric NPs have a promising potential for drug delivery to the brain [84, 85].

Oral administration amphotericin B can be improved by its loading to PLGA NPs. *In vivo* therapeutic efficacy of this formulation was explored in a neutropenic murine model of disseminated and invasive pulmonary aspergillosis [86]. It was shown that the nanosystem was very useful in an invasive pulmonary aspergillosis where intraperitoneal Fungizone™ and Ambisome® did not induce successful results.

Hydrophobic natural antimicrobial compounds (i.e., eugenol and trans-cinnamaldehyde) have been also incorporated to PLGA nanospheres by emulsion evaporation in presence of the surfactant polyvinyl alcohol [87]. The nanoparticulate system satisfactorily inhibited the growth of *Salmonella* spp. and *Listeria* spp. with drug equivalent concentrations ranging from 10 to 20 mg/mL [74].

Lectin can be employed to formulate drug nanocarriers on the basis of active drug targeting strategies. This protein can selectively bind to carbohydrates found in the external membranes of microorganisms [76, 77]. For instance, it has been demonstrated that NPs surface modified with lectin moieties can bind to the surface of *Helicobacter pylori*, thus releasing the antimicrobial agent into the bacteria [88]. Lectin-functionalized PLGA NPs (loaded with rifampicin, isoniazid, and pyrazinamide) have been also synthesized by a two-step carbodiimide procedure [89]. The oral administration of the NPs to guinea pigs significantly increased the bioavailability of the antitubercular drugs ($p < 0.001$).

Surface functionalization with PEG chains can further be advantageously applied to antimicrobial therapy. For instance, *in vitro* antibacterial assays to PLGA NPs loaded with roxithromycin and surface modified with PEG chains (PEGylated) determined that the minimal inhibitory concentration of such nanosystem compared to roxythromicin was: (i) 9-fold lower on *S. aureus*; (ii) 4.5-fold lower on *Bacillus subtilis*; and, (iii) 4.5-fold lower on *Staphylococcus epidermidis* [90].

9.3.2 PCL-Based Nanoformulations

Compared to PLGA, the introduction of PCL in the formulation of nanomedicines against infectious diseases could be considered unsatisfactory. The list of antimicrobial agents formulated in PCL-based nanoparticulate systems is clearly short (Table 9.2). It is clear that even if *in vitro* results suggest the efficacy and safety of the PCL nanoformulations, *in vivo* investigations are needed to define the potential clinical used of PCL-based nanosystems.

PCL-based nanocomposites have been also proposed for the local (buccal) delivery of metronidazole benzoate [91, 92] and chlorhexidine [93].

Table 9.2 Representative examples of antimicrobial agents incorporated to PCL-based platforms that can be administered through the oral route

Anti-infective drug	PCL platform	*In vitro* evaluation	*In vivo* evaluation	References
	PCL-*b*-PEG-*b*-PCL "flower-like" polymeric micelles	–	Statistically significant increase of drug oral bioavailability (up to 3.3 times) with respect to the free drug in the presence of isoniazid	[94]
Rifampicin	PCL-*b*-PEG-*b*-PCL flower-like polymeric micelles coated with chitosan or hydrolyzed galatomannan/chitosan	Galatomannan/chitosan surface modification leads to a significant increase of the intracellular concentration in RAW 264.7 murine macrophages	–	[95]
	PLA-*co*-PCL matrix (70/30)	Enhanced activity against activity against a common osteomyelitis causing bacteria *Pseudomonas aeruginosa*	–	[96]
Isoniazid	PCL microspheres and nanospheres	–	–	[97]

(*Continued*)

Table 9.2 (*Continued*)

Anti-infective drug	PCL platform	*In vitro* evaluation	*In vivo* evaluation	References
Amphotericin B	Diblock copolymers composed of PCL and poly(*N,N*-dimethylamino-2-ethyl methacrylate), or methoxy PEG	Drug-loaded NPs were 10 times less cytotoxic than free drug in *in vitro* hemolysis tests with human red blood cells	–	[98]
	Methoxy PEO-*b*-PCL nanomatrices containing stearyl substituents on PCL	Encapsulated drug showed a reduction in its hemolytic activity against rat red blood cells compared to Fungizone®	–	[99]
	PCL nanospheres coated with the non-ionic surfactant poloxamer 188	The nanospheres were more active than free drug against amphotericin B-susceptible strains of *Leishmania donovani* amastigotes in thioglycolate-elicited peritoneal macrophages	–	[100]
	TPGS-*b*-PCL-ran-PLGA diblock copolymer	The actual minimum inhibitory concentration of the drug-loaded NPs against *Candida albicans* was significantly lower than that of free drug, and the NPs were found to be less toxic on blood cells	Drug-loaded NPs achieved significantly better and prolonged antifungal effects when compared with free drug	[101]

9.4 Oral Vaccination

Oral delivery of vaccines differs from oral drug delivery in that a pharmacological effect is generally expected at the absorption site. Mucosal delivery the only route of vaccination inducing both mucosal and systemic immune responses [102], as well as allowing the dissemination of the immunization thanks to the antigen-sensitized precursor B and T lymphocytes [103]. However, most antigens are protein- or peptide-based molecular structures, thus facing problems of low mucosal permeability and lack of stability in the gastrointestinal tract [102, 104]. The adjuvants classically utilized to solve the problem often display toxicity effects [105, 106]. As an alternative, nanoparticulate systems can combine a protective effect on antigens at the gastrointestinal tract with the activation of the immune system at the mucose, without the need of adjuvants [102].

Activation of the immune system relies in a specific uptake at the mucose [107] by microfold (M) cells [108] or by enterocytes and dendritic cells (DCs) [109]. Then, the NPs are transported to the mucosa-associated lymphoid tissue (MALT), where antigen-presenting cells (APCs) (lymphocytes, macrophages, and DCs) are activated to generate the protective mucosal immune response [104]. Of course, surface chemistry and polymer composition play a key role in the activation and modulation of the immune system [106, 110], since the NP can be engineered to emulate pathogen associated molecular patterns toward APCs and modulate in turn the immune response (Fig. 9.5) [103, 111].

9.4.1 PLGA-Based Nanoformulations

PLGA and PLA NPs have been investigated as antigen carriers for oral vaccine delivery. They can be tailored adequately to encapsulate both lipophilic and hydrophilic compounds by modifications in the nanoprecipitation method and solvent-evaporation technique [107, 113–116]. A number of proteins, peptides, bacterial toxoids, inactivated bacteria, and deoxyribonucleic acid (DNA) plasmids have been loaded to the NPs [102, 117–133], frequently inducing both mucosal and systemic immune responses after oral or intragastric administration.

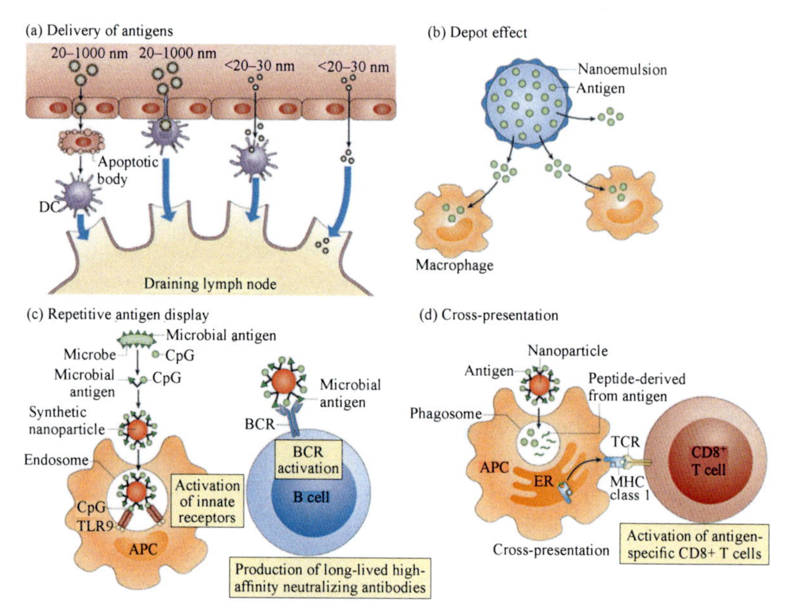

Figure 9.5 Mechanisms by which nanoparticles alter the induction of immune responses. Reprinted with permission from [112].

Particle engineering may influence the efficiency of oral vaccination. Concretely, factors favoring the immune response induced by the particle-based vaccine are the size (nano-sized systems may be uptaken more selectively by M cells) [102], and the particle architecture (PEGylation, functionalization with targeting molecules onto the surface, and/or with antigen stabilizers within the particle matrix [134–136].

Preclinical investigations on the development of PLGA particles for oral vaccine delivery have analyzed the benefits coming from the vehiculization of tetanus toxoid [129], cholera toxin B [137], *Helicobacter pylori* lysates [134], or rotavirus (strain SA11) [138]. In another investigation it was described how ovalbumin and the immunostimulant monophosphoryl lipid A can induce a stronger immunoglobulin (Ig) G immune response, compared to controls, when they are incorporated into PLGA NPs [139], thus obtaining higher IgA titers. PLGA NPs surface functionalized with *Ulex europaeus* 1 lectin moieties have been used for the improved delivery of the Hepatitis B antigen to M cells [135]. *In vivo* studies conducted in BALB/c mice suggested that the nanoformulation can lead to a fourfold increase in the degree of interaction with bovine submaxillary mucin, as well as mucosal secretion of interleukin (IL)-2 and interpheron (IFN)-γ in spleen homogenates. Finally, it has been identified the protection against *Salmonella typhimurium* upon oral administration of typhimurium phosphorylcholine antigen-loaded PLGA particles [130].

Clinical trials have revealed the limited contribution to oral immunization displayed by nanoparticulate systems. For example, only 30% of vaccine efficacy after oral administration of PLGA microspheres loaded with *Escherichia coli* colonization factor antigen П has been reported [140]. These results are questioning whether M cell uptake actually takes place in humans [102, 141]. Therefore, results so far achieved in oral vaccine delivery by using polyester-based carriers still do not satisfy the requirements of efficacy and safety for the development of marketable nanoformulations. Results from investigations devoted to the oral delivery of proteins (on particle stability, analysis of mucus penetration, and protein integrity in the NP matrix) may improve the knowledge on oral vaccine delivery by polyester particulate systems [142–144] (see Section 9.5). It is expected that the large number of publications in the field and the increased knowledge will help in understanding the mechanisms and limitations faced, hence adequately directing the development of valuable immunization therapies.

9.4.2 PCL-Based Nanoformulations

Preliminary studies on the use of PCL particles in oral drug delivery started with bovine serum albumin as a water-soluble model antigen for encapsulation [145]. In this research report, spherical, smooth, and homogeneously distributed PCL microparticles (5 to 10 μm in size) were formulated by a double emulsion-solvent evaporation technique. Protein loading and entrapment efficiency were reported to be 5% and 30%, respectively. Interestingly, polyacrylamide gel electrophoresis and isoelectric focusing analyses of the protein released from these particles confirmed that bovine serum albumin seemed to remain unaltered by the protein encapsulation process.

In another investigation, a hot saline antigenic extract from *Brucella ovis* was encapsulated in PCL microparticles [146]. The vaccine administered orally protected BALB/c mice against *B. ovis* infection similarly to the reference living attenuated *B. melitensis* Rev. 1 vaccine. In addition, it was determined that the use of the free hot saline antigenic extract or the empty PCL microparticles did not produce any protective effect.

Very few *in vitro* and *in vivo* studies can be found describing the use of PCL in oral vaccine delivery. Optimistically, the promising results coming from immune system targeting by PCL NPs after parenteral administration [147] may attract the attention toward the oral route of administration.

9.5 Oral Delivery of Proteins

Oral delivery of proteins and peptides has always been an enticing field of research, probably being insulin the most representative biomacromolecule investigated. The use of polyesters for the encapsulation and delivery of proteins in general, and specifically in the case of insulin, displays key challenges: (*i*) the hydrophobic character of the polymer generally determines very low (negligible) entrapment efficiencies for water-soluble proteins, (*ii*) the generation of a highly acidic microenvironment within the nanocarrier matrix by (bio)degradation can be seriously harmful for the encapsulated proteins [148, and (*iii*) the control of the *in vivo* fate of the nanoparticulate system in the gastrointestinal tract demands an optimized (surface) engineered. In this line, overcoming the mucus barrier must be also addressed [107, 143, 144, 149].

For instance, the stability of insulin during the erosion of PLA-PLGA microspheres was analyzed [150]. It was observed that deamidation (and thus degradation) of insulin was triggered by the acidic conditions within the particle matrix. Previous studies on PEGylated PLA NPs loaded with tetanus toxoid confirmed the negative effect of polymer degradation on protein stability [142]. Regarding the influence of the physical chemistry and particle engineering on the control of the *in vivo* behavior in the gastrointestinal tract, it is accepted that a wise surface engineering is needed. In fact, it has been recently reported that negatively charged surfaces hinder the interaction of the nanocarrier with the negatively charged intestinal mucosa [107]. Additionally, NP stability in gastric and intestinal fluids containing enzymes must be evaluated [151]. In this investigation the effect of digestive fluids on the stability of PLA and PEGylated PLA NPs was evaluated. It was determined that the NPs do interact with pepsin and pancreatin enzymes, leading to particle agglomeration and precipitation. Interestingly, it was suggested that PEGylation may display a protein repellent stabilizing effect against enzymatic degradation.

In order to maximize the interaction the nanoformulation with gastrointestinal epithelial cells, it has been proposed the surface functionalization of the particulate system (i.e., PEG-*b*-PCL) with a transferrin receptor specific 7 peptide [151]. It was demonstrated that such a surface decoration can increase the transport of the NPs across a human colon carcinoma cell line (Caco-2) monolayer.

9.5.1 PLGA-Based Nanoformulations

Polyesters offer the basis of a biocompatible structure with advantageous properties for oral insulin delivery. Recently published investigations have been conducted to beat all the challenges highlighted above.

Cui *et al.* [152], improved the loading efficiency of insulin in PLGA NPs by increasing its liposolubility, which was achieved by producing an insulin—soybean phosphatidylcholine complex via an anhydrous co-solvent lyophilization method. The complex was subsequently encapsulated in NPs following a reverse micelle-solvent evaporation method, attaining entrapment efficiencies up to 90%, corresponding to a theoretical loading capacity of 3.6%. After intragastric administration of insulin-loaded PLGA NPs (insulin equivalent dose 20 IU/kg) to diabetic rats, a decrease in fasting plasma glucose levels to 57.4% within the first 8 hours was obtained, continued for a 12-hour period. The relative oral bioavailability of insulin loaded into NPs and assayed in pharmacokinetic studies was 7.7% related to SC injection of its solution.

Alternative production methods also seemed to offer the opportunity to be used as means to improve insulin loading. Liu *et al.* [153], proposed the production of PLA/PLGA microcapsules by combining a Shirasu porous glass (SPG) membrane emulsification method with a double emulsion-evaporation technique, employing recombinant human insulin (rhI). Much higher encapsulation efficiency was claimed to be obtained by applying this production technique compared to conventional mechanical stirring methods. In this case, the higher encapsulation efficiency value obtained after optimization of the preparation process was claimed to be 91.82%. However, the calculation of the LC from the provided values seems to be, in the best case, of 1.97% (7.5 mg of insulin encapsulated in 0.25 g of polymer, with a 65.61% EE value).

The premature release of insulin after administration and consequent losses due to acidic degradation in the GI tract was addressed as well, aiming at improving ultimate bioavailability and pharmacological effect. In this sense, Cui *et al.* [154], employed the Hp55 cellulose derivative (hydroxypropyl methylcellulose phthalate, HPMCP) to prevent the burst release of insulin in the stomach. Hp55 is a pH-sensitive polymer employed as enteric coating, which would protect insulin-loaded PLGA nanospheres from acidic gastric media, yet allowing for release at the pH of the small intestine. Its inclusion in the formulation also increased drug recovery values up to 65.41%, which correspond to a theoretical 2.7% of LC, probably due to charge interactions or hydrogen bond formation between Hp55 phthalate groups and insulin amino groups. This strategy reduced the initial release of protein from the NPs in simulated gastric fluid (SFG) from 50% to 20%, still the relative bioavailability of insulin in diabetic rats was 6.2% compared to SC injection.

Later, Naha *et al.* [155], followed a similar approach employing Eudragit-L30 as enteric coating in PLGA microparticles, as well as including serum albumin as protein stabilizer and employing sucrose in the production to obtain a porous matrix. A 40% EE was obtained with this approach, resulting in a theoretical LC of 10% (10 mg mL^{-1} of initial insulin referred to polymer particle). A reduction in blood glucose levels of 62.7% was obtained with this delivery system, versus 37.1% of an Eudragit-L30-free system produced under the same conditions (25 IU insulin/kg dose). No relative bioavailability data is available.

Some years later, Wu *et al.* employed Hp55 once again as enteric coating to selectively release insulin in the intestinal tract [156]. This time Eudragit®RS (RS) was also included in the formulation, which would presumably enhance the penetration of the protein across the mucosal surface via opening of tight junctions between intestinal cells. Values of EE of 73.9% and LC of 6.7% were obtained.

Burst release at pH 1.2 was markedly reduced and, following oral administration in diabetic rat models, a pharmacological availability of 9.2% was attained.

Davaran *et al.* [157], explored a different strategy to prevent insulin burst release, based on linear PLGA-PEG and star-branched b-cyclodextrin-PLGA (b-CD-PLGA), and glucose-PLGA (Glu-PLGA) copolymer NPs, prepared by the double emulsion method. Insulin stability within the carriers was attributed to the b-CD segment of star-branched PLGA, and once again the highest blood glucose reduction was also achieved at 4.5 hours and maintained for 18.5 hours.

Issues related to particle stability in simulated fluids were addressed by Santander-Ortega *et al.* [158], who studied blend matrices comprised of PLGA and polyoxyethylene derivatives for this purpose. Two types of blend formulations, PLGA:poloxamer (Pluronic F68®) and PLGA:poloxamine (Tetronic T904®), were analyzed and compared to PLGA NPs. Blend formulations were capable of considerably reducing the interaction between NPs and digestive enzymes, while pure PLGA NPs were mostly affected by their presence. The inclusion of polyoxyethylene derivatives also influenced encapsulation efficiency. No bioavailability records from studies with these formulations are available yet.

With regard to carrier modifications aiming at improving the interaction with the intestinal mucosal, numerous strategies have been reported in the literature mainly related to PLGA NPs [116]. In the specific case of insulin-loaded PLGA NPs, one of the first approaches reported in this sense also consisted of the employment of PEO-block-PLA copolymers [159] PEO blocks present a high permeation capability due to their amphiphilic properties [160] and hence, insulin-loaded poly(lactic acid)-b-Pluronic-b-poly(lactic acid) (PLA-F127-PLA) NPs were produced. A loading capacity of 0.08% was obtained and the formulations were tested in diabetic mice; a decrease in blood glucose concentration from 18.5 to 5.3 mmol/L (70%) within 4.5 hours was obtained, and the minimum blood glucose concentration (4.5 mmol/L) was maintained for 18.5 hours. No relative bioavailability records were reported though.

Some years later, Reix *et al.* [161], provided *in vitro* data of the interaction between insulin-loaded PLGA NPs and Caco-2 culture as gastrointestinal cell model, as well as *in vivo* studies carried out with the same formulations. A 97% EE was obtained with formulations prepared with 400 μL of 3.5 mg/mL insulin solution and 200 mg of polymer (which would account for a 0.67% LC). In order to track the particles, FITC-insulin- and ^{125}I/insulin-loaded NPs were produced as well. The kinetic of particle absorption in cells was ascribed to clathrin-mediated endocytosis, and the intracellular traffic resulted in basolateral exocytosis of NPs reflected in a plateau in uptake kinetics, where absorption was counterbalanced with rapid exocytosis.

The same year, Zhang *et al.* [162], produced cationic chitosan-PLGA NPs to improve the interaction with negatively charged groups of the intestinal mucosa. The introduction of CS into the formulation led to a shift in the zeta potential values of NPs to positive values (from -1.72 mV ± 0.2 mV for PLGA-NPs to +43.1 ± 0.3 mV in the case of CS-PLGA-NPs). The encapsulation efficiency of CS-PLGA-NPs was similar ($p>0.05$) to PLGA NPs (52.76 ± 3.48% and LC 1.29 ± 0.29 in CS-PLGA-NPs versus 46.87 ± 2.23% LC 1.14 ± 0.36 in PLGA NPs). Thus, the improvement in pharmacological availability observed was attributed to the improved interaction with the mucosa due to the presence of CS: PLGA-NP and CS-PLGA-NP presented pharmacological availability values relative to SC injection of 7.6% and 10.5%, respectively, at an equivalent insulin dose of 15 IU/kg.

In a different approach Jain *et al.* [163], aimed at targeting folate (FA) receptors expressed in the GI tract [164]. To do so, insulin was encapsulated in FA-coupled PEGylated PLGA NPs. EE and LC of 87% and 6.5% respectively were obtained and the stability in SGF and SIF was tested, demonstrating a protective effect of insulin degradation by NPs, and the capability of the FA-coupled NPs to remain dispersed in suspension. Following, insulin-loaded NPs in doses equivalent to 50 IU/kg of insulin were administered to diabetic rats, with the FA-coupled NPs exhibiting the highest hypoglycemic effect and pharmacological availability of 19.62 ± 1.68 relative to SC injection, also the highest reported until the moment, to the best of our knowledge.

Later, Alonso-Sande *et al.* [165], developed PLGA-mannosamine NPs aiming at selectively targeting NPs to M cells, thus increasing their affinity with the intestinal mucosa. PLGA was chemically modified with mannosamine (MN) via gel permeation chromatography, and NPs were produced via the double emulsion technique employing either PLGA or MN-PLGA. The presence of MN molecules led to a 10%-enhancement in encapsulation efficiency (68% ± 2% in PLGA NPs versus 77% ± 15% in MN-PLGA NPs, corresponding to theoretical values of LC of 3.4% and 3.85% respectively), which was attributed to higher affinity of proteins for a polymer with increased hydrophilicity due to saccharide modification. In addition, MN molecules provided some additional improvement in particle stability in SIF, that was assimilated to that of the PEGylation strategy, based on a protein repellent effect of MN residues toward enzymes [166]. Finally, the systems were studied in terms of interaction with epithelial cells in a human intestine follicle associated epithelium cell culture model; MN modification proved to significantly enhance the transport of NPs across the cell monolayer in both Caco-2 cells monoculture and Caco-2 and Raji cells co-culture. This observation suggested the improvement in the affinity of PLGA NPs toward intestinal cells by the inclusion of MN residues, via nonspecific and specific interactions with both enterocytes and M cells. No pharmacological availability results from *in vivo* studies are yet available for this system prototype.

Although numerous strategies have been followed to overcome the challenges associated with oral delivery of insulin in polyester-based carriers, the results reported until the moment are still insufficient for the real development of an oral therapy, and the ultimate bioavailability of insulin attained with the systems tested still remains unsatisfactory. The door remains open to research for carrier delivery systems that offer reproducible absorption and demonstrate superiority over injectable insulin formulations and oral hypoglycemic agents, supported by efficient and sufficient manufacture of insulin for oral delivery in a cost-conscious pharmaceutical marketplace [166].

9.5.2 PCL-Based Nanoformulations

The use of PCL in the development of nanoparticulate systems for the oral delivery of proteins relies in its adequate permeability to proteins, and in its slow biodegradation rate which does not produce the acidic environment where peptides and proteins are disrupted. Therefore, numerous investigations have been devoted to the use of this biocompatible polymer in oral drug delivery of peptides and proteins, e.g., heparin [167]. Preparation procedures may be based on melt encapsulation, solvent evaporation, or double-emulsion solvent extraction/evaporation methodologies. A proper modification of the methodologies can generate high protein/peptide entrapment efficiencies. In fact, one of the difficulties associated to the incorporation of these biomolecules to the NP matrix comes from their hydrophilic character. As a consequence, proteins and peptides preferentially partition out into the aqueous phase, thus determining low entrapment efficiencies. An interesting approach to the problem is based on the investigation of the capability of copolymer micelles in organic solvents to sequester peptides/proteins [168]. In this investigation PCL-block-poly(2-vinyl pyridine) micelles were efficiently loaded with ovalbumin and bovine serum albumin. It was observed a sustained protein release profile that ended after 12–30 h. Data suggested that the proteins are sequestered in the poly(2-vinyl pyridine) corona block of the micelles.

9.6 Conclusions

The use of polyesters for the development of drug delivery systems for peroral administration has been widely explored by the scientific community, enticed by the beneficial properties of these polymers determining a tunable structure and controlled release, along with biocompatibility and biodegradability. Oral administration has always been a route of choice on account of the benefits that it provides for both patient (ease of administration and avoidance of pain and discomfort, leading to a higher compliance) and pharmaceutical industry (ease of distribution, reduction of production costs). In this field, polyester-based formulations have been extensively proved to be highly advantageous for the optimization of oral chemotherapy (minimization of toxicity and increase of bioavailability), the minimization of antibiotics resistance and the targeting against intracellular microorganisms, the engineering of nanoplatforms to elicit mucosal immunization responses and the design of oral protein carriers, to mention a few. Although the success of formulations intended for the delivery of lipophilic drugs has been remarkably higher than that of formulations designed for hydrophilic molecules, the choice of polyesters for formulation development remains a safe and promising choice in both cases. Furthermore, their use for nanocarriers development has clearly surpassed polyester-based traditional formulations in the actual scientific horizon.

In spite of the extensive knowledge and results generated, oral delivery is still a very much challenging field for nanotechnology; an exponential and highly fructuous development of novel formulations and therapies was expected since nanotechnology became a prominent field in the arena of pharmaceutical sciences and alike. However, nowadays the yields from nanotechnology are considered to be below those initial expectations and a much more critical perspective toward this field has arisen. This vision seems to be more pronounced when it comes to oral delivery by means of nanotechnology, specially regarding hydrophilic molecules and more specifically those with a critical structure sensitive to the harmful conditions of the gastrointestinal tract (peptides, proteins, DNA, RNA).

It is however worth considering that nanotechnology is a very recent research field that opened up a whole new array of possibilities; when new opportunities are proposed, new challenges, questions and barriers are discovered. Alternative toxicity issues and testing, mechanistic matters and biological barriers that long ago were unbeatable by traditional formulations, are now object of discussion. Hence, although the success so far achieved in the field still do not satisfy all the expectations, the results generated are bound to improve the knowledge and understanding of these new challenges and barriers, so the limitations can be overcome and valuable peroral-based therapies can be developed. In this enterprise, among the varied and large number of materials employed for nanocarriers formulation, polyesters continue to stand out as a safe and advantageous choice supported by extensive and well documented information. There is no doubt they will continue to be a reference for pharmaceutical development in peroral delivery.

References

1. Lofgren, A., Albertsson, A. C., Dubois, P., Herome, R. (1995). Recent advances in ring opening polymerization of lactones and related compounds. *J Macromol Sci Rev Macromol Chem Phys*, **35**, 379–418.

2. Vroman, I. Tighzert, L. (2009). Biodegradable polymers. *Materials,* **2**, 307–344.
3. Vilar, G., Tulla-Puche, J., Albericio, F. (2012). Polymers and drug delivery systems. *Curr Drug Deliv,* **9**, 367–394
4. Bikiaris, D., Karavelidis, V., Karavas, E. (2009). Novel biodegradable polyesters. synthesis and application as drug carriers for the preparation of raloxifene HCl loaded nanoparticles. *Molecules,* **14**, 2410–2430
5. Behera, B. K. (2013). Pharmaceutical applications of lactides and glycolides: a review. *J Med Pharm Innov,* **1**, 1–5
6. El-Fattah, A. A., Kenawy, E. R., Kandil, S. (2014). Biodegradable polyesters as biomaterials for biomedical applications. *Int J Chem Appl Biol Sci,* **1**, 2–11.
7. Sinha, V. R., Bansal, K., Kaushik, R., Kumria, R., Trehan, A. (2004). Poly-ε-caprolactone microspheres and nanospheres: an overview. *Int J Pharm,* **278**, 1–23
8. Ha, C. S., Gardella, J. A. (2005). Surface chemistry of biodegradable polymers for drug delivery systems. *Chem Rev,* **105**, 4205–4232
9. Singh, V., Tiwari, M. (2010). Structure-processing-property relationship of poly(glycolic acid) for drug delivery systems 1: synthesis and catalysis. *Int J Pol Sci,* Article ID 652719, 23 pages.
10. Nair, L. S., Laurencin, C. T. (2007). Biodegradable polymers as biomaterials. *Progr Polym Sci,* **32**, 762–798.
11. Mansour, H. M., Sohn, M., Al-Ghananeem, A., DeLuca, P. P. (2010). Materials for pharmaceutical dosage forms: molecular pharmaceutics and controlled release drug delivery aspects. *Int J Mol Sci,* **11**, 3298–3322.
12. Berkland, C., Kipper, M. J., Narasimhan, B., Kim, K. Y., Pack, D. W. (2004). Microsphere size, precipitation kinetics and drug distribution control drug release from biodegradable polyanhydride microspheres. *J Control Release,* **94**, 129–141
13. Jaraswekin, S., Prakongpan, S., Bodmeier, R. (2007). Effect of poly(lactide-*co*-glycolide) molecular weight on the release of dexamethasone sodium phosphate from microparticles. *J Microencapsul,* **24**, 117–128.
14. Wang, L., Chaw, C. S., Yang, Y. Y., Moochhala, S. M., Zhao, B., Ng, S., Heller, J. (2004). Preparation, characterization, and *in vitro* evaluation of physostigmine-loaded poly(ortho ester) and poly(ortho ester)/poly(D,L-lactide-*co*-glycolide) blend microspheres fabricated by spray drying. *Biomaterials,* **25**, 3275–3282.

15. Mochizuki, M., Hirami, M. (1997). Structural effects on biodegradation of aliphatic polyesters. *Polym Adv Technol*, **8**, 203–209.

16. Behera, A. K., Barik, B. B., Joshi S. (2012). Poly-ε-caprolactone based microspheres and nanospheres: a review. *J Pharm Res*, **1**, 38–45.

17. Chang, R. K., Price, J. C., Whitworth, C. W. (1986). Dissolution characteristics of poly (ε-caprolactone)-polylactide microspheres of chlorpromazine. *Drug Dev Ind Pharm*, **12**, 2355–2380.

18. Freiberg, S., Zhu, X. (2004). Polymer microspheres for controlled drug release. *Int J Pharm*, **282**, 1–18.

19. Koosha, F., Muller, R. H., Davis, S. S. (1989). Polyhydroxybutyrate as a drug carrier. *Crit Rev Ther Drug Carrier Syst*, **6**, 117–130.

20. Reddy, C. S., Ghai, R., Rashmi., Kalia, V. C. (2003). Polyhydroxyalkanoates: an overview. *Bioresour Technol*, **87**, 137–146.

21. Baker, J. R. (2013). Why I believe nanoparticles are crucial as a carrier for targeted drug delivery. *Rev Nanomed Nanobiotechnol*, **5**, 423–429.

22. Kolhe, P., Misra, E., Kannan, R. M., Kannan, S., Lieh-Lai, M. (2003). Drug complexation, *in vitro* release and cellular entry of dendrimers and hyperbranched polymers. *Int J Pharm*, **259**, 143–160.

23. Gu, L., Wu, Z. H., Qi, X. L., He, H., Ma, X. L., Chou, X. H., Wen, X. G., Zhang, M., Jiao, F. (2013). Polyamidomine dendrimers: an excellent drug carrier for improving the solubility and bioavailability of puerarin. *Pharm Dev Technol*, **18**, 1051–1057.

24. Zhou, Z. Y., D'Emanuele, A., Attwood, D. (2013). Solubility enhancement of paclitaxel using a linear-dendritic block copolymer. *Int J Pharm*, **452**, 173–179.

25. Madaan, K., Kumar, S., Poonia, N., Lather, V., Pandita, D. (2014). Dendrimers in drug delivery and targeting: drug-dendrimer interactions and toxicity issues. *J Pharm Bioallied Sci*, **6**, 139–150.

26. Wolinsky, J. B., Grinstaff, M. W. (2008). Therapeutic and diagnostic applications of dendrimers for cancer treatment. *Adv Drug Deliv Rev*, **60**, 1037–1055.

27. Svenson, S., Tomalia, D. A. (2005). Commentary-dendrimers in biomedical applications—reflections on the field. *Adv Drug Deliv Rev*, **57**, 2106–2129.

28. Nanjwade, B. K., Bechra, H. M., Derkar, G. K., Manvi, F. V., Nanjwade, V. K. (2009). Dendrimers: emerging polymers for drug-delivery systems. *Eur J Pharm Sci*, **38**, 185–196.

29. Svenson, S. (2009). Dendrimers as versatile platform in drug delivery applications. *Eur. J Pharm Biopharm*, **71**, 445–462.
30. Twibanire, J. K., Grindley, T. B. (2014). Polyester dendrimers: smart carriers for drug delivery. *Polymers*, **6**, 179–213.
31. Vedha-Hari, B. N., Kalaimagal, K., Porkodi, R., Kumar, P., Ajay, J. Y. (2012). Dendrimer: globular nanostructured materials for drug delivery. *Int J Pharm Tech Res*, **4**, 432–451.
32. Gillies, E. R., Fréchet, J. M. J. (2005). Dendrimers and dendritic polymers in drug delivery. *Drug Discov Today*, **10**, 35–43.
33. Grinstaff, M. W. (2002). Biodendrimers: new polymeric biomaterials for tissue engineering. *Chemistry*, **8**, 2839–2846.
34. Kulhari, H., Kulhari, D. P., Prajapati, S. K., Chauhan, A. S. (2013). Pharmacokinetic and pharmacodynamic studies of poly(amidoamine) dendrimer based simvastatin oral formulations for the treatment of hypercholesterolemia. *Mol Pharm*, **10**, 2528–2533.
35. Sadekar, S., Thiagarajan, G., Bartlett, K., Hubbard, D., Ray, A., McGill, L. D., Ghandehari, H. (2013). Poly(amidoamine) dendrimers as absorption enhancers for oral delivery of camptothecin. *Int J Pharm*, **456**, 175–185.
36. Florence, A. T., Hussain, N. (2001). Transcytosis of nanoparticle and dendrimer delivery systems: evolving vistas. *Adv Drug Deliv Rev*, **50**, S69–S89.
37. Cheng, Y., Xu, Z., Ma, M., Xu, T. (2008). Dendrimers as drug carriers: applications in different routes of drug administration. *J Pharm Sci*, **97**, 123–143.
38. Yun, Y., Cho, Y. W., Park, K. (2013). Nanoparticles for oral delivery: targeted nanoparticles with peptidic ligands for oral protein delivery. *Adv Drug Deliv Rev*, **65**, 822–832.
39. des Rieux, A., Pourcelle, V., Cani, P. D., Marchand-Brynaert J., Préat V. (2013). Targeted nanoparticles with novel non-peptidic ligands for oral delivery. *Adv Drug Deliv Rev*, **65**, 833–844.
40. Chan, J. M., Zhang, L., Yuet, K. P., Liao, G., Rhee, J., Langer, R., Farokhzad, O. C. (2009). PLGA-lecithin-PEG core-shell nanoparticles for controlled drug delivery. *Biomaterials*, **30**, 1627–1634.
41. Yin, Y., Chen, D., Qiao, M., Wei, X., Hu, H. (2007). Lectin-conjugated PLGA nanoparticles loaded with thymopentin: *ex vivo* bioadhesion and *in vivo* biodistribution. *J Control Release*, **123**, 27–38.
42. Fievez, V., Plapied, L., des Rieux, A, Pourcelle, V., Freichels, H., Wascotte, V., Vanderhaeghen, M. L., Jérôme, C., Vanderplasschen, A., Marchand-

Brynaert, J., Schneider, Y. J., Préat, V. (2009). Targeting nanoparticles to M cells with non-peptidic ligands for oral vaccination. *Eur J Pharm Biopharm,* **73**, 16–24.

43. Gupta, P. N., Khatri, K., Goyal, A. K., Mishra, N., Vyas, S. P. (2007). M-cell targeted biodegradable PLGA nanoparticles for oral immunization against hepatitis B. *J Drug Target,* **15**, 701–713.
44. Wei, Q., Li, T., Wang, G., Li, H., Qian, Z., Yang, M. (2010). Fe(3)O(4) nanoparticles loaded PEG-PLA polymeric vesicles as labels for ultrasensitive immunosensors. *Biomaterials,* **31**, 7332–7339.
45. Mei, H., Shi, W., Pang, Z., Wang, H., Lu, W., Jiang, X., Deng, J., Guo, T., Hu, Y., (2010). EGFP-EGF1 protein-conjugated PEG-PLA nanoparticles for tissue factor targeted drug delivery. *Biomaterials,* **31**, 5619–5626.
46. Tang, B. C., Dawson, M., Lai, S. K., Wang, Y. Y., Suk, J. S., Yang, M., Zeitlin, P., Boyle, M. P., Fu, J., Hanes, J. (2009). Biodegradable polymer nanoparticles that rapidly penetrate the human mucus barrier. *Proc Natl Acad Sci U S A,* **106**, 19268–19273.
47. Conte, C., d'Angelo, I., Miro, A., Ungaro, F., Quaglia, F. (2014). PEGylated polyester-based nanoncologicals. *Curr Top Med Chem,* **14**, 1097–1114.
48. Song, Q., Wang, X., Hu, Q., Huang, M., Yao, L., Qi, H., Qiu, Y., Jiang, X., Chen, J., Chen, H., Gao, X. (2013). Cellular internalization pathway and transcellular transport of pegylated polyester nanoparticles in Caco-2 cells. *Int J Pharm,* **445**, 58–68.
49. Zabaleta, V., Ponchel, G., Salman, H., Agueros, M., Vauthier, C., Irache, J. M. (2012). Oral administration of paclitaxel with pegylated poly(anhydride) nanoparticles: permeability and pharmacokinetic study. *Eur J Pharm Biopharm,* **81**, 514–523.
50. Tobio, M., Sanchez, A., Vila, A., Soriano, I., Evora, C., VilaJato, J. L., Alonso, M. J. (2000). The role of PEG on the stability in digestive fluids and *in vivo* fate of PEG-PLA nanoparticles following oral administration. *Colloids Surf B Biointerfaces,* **18**, 315–323.
51. Ensign, L. M., Cone, R., Hanes, J. (2012). Oral drug delivery with polymeric nanoparticles: the gastrointestinal mucus barriers. *Adv Drug Deliv Rev,* **64**, 557–570.
52. Jain, A. K., Das, M., Swarnakar, N. K., Jain, S. (2011). Engineered PLGA nanoparticles: an emerging delivery tool in cancer therapeutics critical reviews™. *Ther Drug Carrier Syst,* **28**, 1–45.
53. Wang, K., Xu, X., Wang, Y., Yan, X., Guo, G., Huang, M., Luo, F., Zhao, X., Wei, Y., Qian, Z. (2010). Synthesis and characterization of poly(methoxyl

ethylene glycol-caprolactone-*co*-methacrylic acid-*co*-poly(ethylene glycol) methyl ether methacrylate) pH-sensitive hydrogel for delivery of dexamethasone. *Int J Pharm,* **389**, 130–138.

54. Kevadiya, B. D., Thumbar, R. P., Rajput, M. M., Rajkumar, S., Brambhatt, H., Joshi, G. V., Dangi, G. P., Mody, H. M., Gadhia, P. K., Bajaj, H. C. (2012). Montmorillonite/poly-(ε-caprolactone) composites as versatile layered material: reservoirs for anticancer drug and controlled release property. *Eur J Pharm Sci,* **47**, 265–272.
55. Tan, L., Xu, X., Song, J., Luo, F., Qian, Z. (2013). Synthesis, characterization, and acute oral toxicity evaluation of pH-sensitive hydrogel based on MPEG, poly(ε-caprolactone), and itaconic acid. *Biomed Res Int,* Article ID 239838, 9 pages.
56. Attili-Qadri S., Karra N., Nemirovski A., Schwob O., Talmon Y., Nassar T., Benita S. (2013). Oral delivery system prolongs blood circulation of docetaxel nanocapsules via lymphatic absorption. *Proc Natl Acad Sci U S A,* **110**, 17498–17503.
57. Li, Z., Liu, K., Sun, P., Mei, L., Hao, T., Tian, Y., Tang, Z., Li, L., Chen, D. (2013). Poly(D, L-lactide-*co*-glycolide)/montmorillonite nanoparticles for improved oral delivery of exemestane. *J Microencapsul,* **30**, 432–440.
58. Zhao, L., Feng, S. S. (2010). Enhanced oral bioavailability of paclitaxel formulated in vitamin E-TPGS emulsified nanoparticles of biodegradable polymers: *in vitro* and *in vivo* studies. *J Pharm Sci,* **99**, 3552–3560.
59. Chen, H., Zheng, Y., Tian, G., Tian, Y., Zeng, X., Liu, G., Liu, K., Li, L., Li, Z., Mei, L., Huang, L. (2011). Oral delivery of DMAB-modified docetaxel-loaded PLGA-TPGS nanoparticles for cancer chemotherap. *Nanoscale Res Lett,* **6**, 4.
60. Jain, A. M., Swarnakar, N. K., Das, M., Godugu, C., Singh, R. P., Rao, P. R., Jain, S. (2011). Augmented anticancer efficacy of doxorubicin-loaded polymeric nanoparticles after oral administration in a breast cancer induced animal model. *Mol Pharm,* **8**, 1140–1151.
61. Ling, G., Zhang, P., Zhang, W., Sun, J., Meng, X., Qin, Y., Deng, Y., He, Z. (2010). Development of novel self-assembled DS-PLGA hybrid nanoparticles for improving oral bioavailability of vincristine sulfate by P-gp inhibition. *J Control Release,* **148**, 241–248.
62. Joshi, G., Kumar, A., Sawant, K. (2014). Enhanced bioavailability and intestinal uptake of Gemcitabine HCl loaded PLGA nanoparticles after oral delivery. *Eur J Pharm Sci,* **60**, 80–89.

63. Zhang, J., Li, Y., Gao, W., Repka, M. A., Wang, Y., Chen, M. (2014). Andrographolide-loaded PLGA-PEG-PLGA micelles to improve its bioavailability and anticancer efficacy. *Expert Opin Drug Deliv*, **11**, 1367–1380.

64. Bhatnagar, P., Gupta, K. C. (2013). Oral administration of eudragit coated bromelain encapsulated PLGA nanoparticles for effective delivery of bromelain for chemotherapy *in vivo*. *Proc. 29th South. Biomed. Eng. Conf.*, 47–48.

65. Bhatnagar, P., Patnaik, S., Srivastava, A. K., Mudiam, M. K. R., Shukla, Y., Panda, A. K., Pant, A. B., Kumar, P., Gupta, K. C. (2014). Anti-cancer activity of bromelain nanoparticles by oral administration. *J Biomed Nanotechnol*, **10**, 3558–3575.

66. Hernán Pérez de la Ossa, D., Lorente, M., Gil-Alegre, M. E., Torres, S., García-Taboada, E., Aberturas, M. R., Molpeceres, J., Velasco, G., Torres-Suárez, A. I. (2013). Local delivery of cannabinoid-loaded microparticles inhibits tumour growth in a murine xenograft model of glioblastoma multiforme. *PLoS One*, **8**, e54795.

67. Wang, Y., Chen, L., Tan, L., Zhao, Q., Luo, F., Wei, Y., Qian, Z. (2014). PEG-PCL based micelle hydrogels as oral docetaxel delivery systems for breast cancer therapy. *Biomaterials*, **35**, 6972–6985.

68. Ghosh, D., Choudhury, S. T., Ghosh, S., Mandal, A. K., Sarkar, S., Ghosh, A., Das Saha, K., Das, N. (2012). Nanocapsulated curcumin: oral chemopreventive formulation against diethylnitrosamine induced hepatocellular carcinoma in rat. *Chem Biol Interact*, **195**, 206–214.

69. Jain, A. K., Thanki, K., Jain, S. (2013). Co-encapsulation of tamoxifen and quercetin in polymeric nanoparticles: implications on oral bioavailability, antitumor efficacy, and drug-induced toxicity. *Mol Pharm*, **10**, 3459–3474.

70. Green, C. E., Swezey, R., Bakke, J., Shinn, W., Furimsky, A., Bejugam, N., Shankar, G. N., Jong, L., Kapetanovic, I. M. (2011). Improved oral bioavailability in rats of SR13668, a novel anti-cancer agent. *Cancer Chemother Pharmacol*, **67**, 995–1006.

71. Banerjee, A. A., Shen, H., Hautman, M., Anwer, J., Hong, S., Kapetanovic, I. M., Liu, Y., Lyubimov, A. V. (2013). Enhanced oral bioavailability of the hydrophobic chemopreventive agent (SR13668) in beagle dogs. *Curr Pharm Biotechnol*, **14**, 464–469.

72. Chawla, J. S., Amiji, M. M. (2003). Cellular uptake and concentrations of tamoxifen upon administration in poly(ε-caprolactone) nanoparticles. *AAPS Pharm Sci*, **5**, E3.

73. Gao, P., Nie, X., Zou, M., Shi, Y., Cheng, G. (2011). Recent advances in materials for extended-release antibiotic delivery system. *J Antibiot,* **64**, 625–634.

74. Abdollahi, S., Lotfipour, F. (2012). PLGA- and PLA-based polymeric nanoparticles for antimicrobial drug delivery. *Biomed Int,* **3**, 1–11.

75. Prior, S., Gander, B., Blarer, N., Merkle, H. P., Subira, M. L., Irache, J. M., Gamazo, Z. (2002). *In vitro* phagocytosis amd monocyte-macrophage activation with poly(lactide) and poly(lactide-*co*-glycolide) microspheres. *Eur J Pharm Sci,* **15**, 197–207.

76. Salouti, M., Ahangari, A. (2014). Nanoparticle based drug delivery systems for treatment of infectious diseases. In *Application of Nanotechnology in Drug Delivery,* Sezer, A. D. (ed.), InTech.

77. Zhang, L., Pornpattananangkul, D., Hu, M. J., Huang, M. (2010). Development of nanoparticles for antimicrobial drug delivery. *Curr Med Chem,* **17**, 585–594.

78. Kalluru, R., Fenaroli, F., Westmoreland, D., Ulanova, L., Maleki, A., Roos, N., Paulsen, M., Madsen, M. P., Koster, G., Jacobsen, W. E., Wilson, S., Roberg-Larsen, H., Khuller, G. K., Singh, A., Nyström, B., Griffiths, G. (2013). Poly(lactide-*co*-glycolide)-rifampicin nanoparticles efficiently clear Mycobacterium bovis BCG infection in macrophages and remain membrane-bound in phagolysosomes. *J Cell Sci,* **126**, 3043–3054.

79. Hu, Y. L., Gao, J. Q. (2010). Potential neurotoxicity of nanoparticles. *Int J Pharm,* **394**, 115–121.

80. De Jong, W. H., Borm, P. J. (2008). Drug delivery and nanoparticles: applications and hazards. *Int J Nanomedicine,* **3**, 133–149.

81. Moghimi, S. M., Hunter, A. C., Murray, J. C. (2001). Long-circulating and target-specific nanoparticles: theory to practice. *Pharm Rev,* **53**, 283–328.

82. Mohammadi, G., Nokhodchi, A., Barzegar-Jalali, M., Lotfipour, F., Adibkia, K., Ehyaei, N., Valizadeh, H. (2011). Physicochemical and anti-bacterial performance characterization of clarithromycin nanoparticles as colloidal drug delivery system. *Colloids Surf B,* **88**, 39–44.

83. Mohammadi, G., Valizadeh, H., Barzegar-Jalali, M., Lotfipour, F., Adibkia, K., Milani, M., Azhdarzadeh, M., Kiafar, F., Nokhodchi, A. (2010). Development of azithromycin-PLGA nanoparticles: physicochemical characterization and antibacterial effect against Salmonella typhi. *Colloids Surf B,* **80**, 34–39.

84. Sharma, A., Pandey, R., Sharma, S., Khuller, G. K. (2004). Chemotherapeutic efficacy of poly (DL-lactide-*co*-glycolide)

nanoparticle encapsulated antitubercular drugs at sub-therapeutic dose against experimental tuberculosis. *Int J Antimicrob Agents*, **24**, 599–604.

85. Pandey, R., Khuller, G. K. (2006). Oral nanoparticle-based antituberculosis drug delivery to the brain in an experimental model. *J Antimicrob Chemother*, **57**, 1146–1152.

86. Italia, J. L., Sharp, A., Carter, K., Warn, P., Kumar, M. (2011). Peroral amphotericin B polymer nanoparticles lead to comparable or superior *in vivo* antifungal activity to that of intravenous ambisome (R) or Fungizone (TM). *PLoS One*, **6**, 625744.

87. Gomes, C., Moreira, R. G., Castell-Perez, E. (2011). Poly (DL-lactide-*co*-glycolide) (PLGA) nanoparticles with entrapped trans-cinnamaldehyde and eugenol for antimicrobial delivery applications. *J Food Sci*, **76**, 16–24.

88. Umamaheshwari, R. B., Jain, N. K. (2003). Receptor mediated targeting of lecitin conjugated gliadin nanoparticles in the treatment of Helicobacter pylori. *J Drug Target*, **11**, 415–423.

89. Sharma, A., Sharma, S., Khuller, G. K. (2004). Lectin-functionalized poly (lactide-*co*-glycolide) nanoparticles as oral/aerosolized antitubercular drug carriers for treatment of tuberculosis. *J Antimicrob Chemoth*, **54**, 761–766.

90. Koopaei, M. N., Maghazei, M. S., Mostafaviz, S. H., Jamalifar, H., Samadi, N., Amini, N., Malek, S. J., Darvishi, B., Atyabi, F., Dinarvand, R. (2012). Enhanced antibacterial activity of roxithromycin loaded pegylated poly lactide-*co*-glycolide nanoparticles. *Daru J Pharm Sci*, **20**, 92.

91. El-Kamel, A. H., Ashri, L. Y., Alsarra, I. A. (2007). Micromatricial metronidazole benzoate film as a local mucoadhesive delivery system for treatment of periodontal diseases. *AAPS PharmSciTech*, **14**, E75.

92. Dabhi, M. R., Sheth, N. R. (2013). Formulation development of physiological environment responsive periodontal drug delivery system for local delivery of metronidazole benzoate. *Drug Dev Ind Pharm*, **39**, 425–436.

93. Medlicott, N. J., Holborow, D. W., Rathbone, M. J., Jones, D. S., Tucker, I. G. (1999). Local delivery of chlorhexidine using a tooth-bonded delivery system. *J Control Release*, **20**, 337–343.

94. Moretton, M. A., Hocht, C., Taira, C., Sosnik, A. (2014). Rifampicin-loaded "flower-like" polymeric micelles for enhanced oral bioavailability in an extemporaneous liquid fixed-dose combination with isoniazid. *Nanomedicine (Lond)*, **9**, 1635–1650.

95. Moretton, M. A., Chiappetta, D. A., Andrade, F., das Neves, J., Ferreira, D., Sarmento, B., Sosnik, A. (2013). Hydrolyzed galactomannan-modified nanoparticles and flower-like polymeric micelles for the active targeting of rifampicin to macrophages. *J Biomed. Nanotechnol,* **9**, 1076–1087.

96. Ahola, N., Veiranto, M., Männistö, N., Karp, M., Rich, J., Efimov, A., Seppälä, J., Kellomäki, M. (2012). Processing and sustained *in vitro* release of rifampicin containing composites to enhance the treatment of osteomyelitis. *Biomatter,* 2, 213–225.

97. Duran, N., De Oliveira, A. F., De Azevedo, M. M. (2006). *In vitro* studies on the release of isoniazid incorporated in poly(ε-caprolactone). *J Chemother,* **18**, 473–479.

98. Shim, Y. H., Kim, Y. C., Lee, H. J., Bougard, F., Dubois, P., Choi, K. C., Chung, C. W., Kang, D. H., Jeong, Y. I. (2011). Amphotericin B aggregation inhibition with novel nanoparticles prepared with poly(ε-caprolactone)/poly(n,n-dimethylamino-2-ethyl methacrylate) diblock copolymer. *J Microbiol Biotechnol,* **21**, 28–36.

99. Falamarzian, A., Lavasanifar, A. (2010). Chemical modification of hydrophobic block in poly(ethylene oxide) poly(caprolactone) based nanocarriers: effect on the solubilization and hemolytic activity of amphotericin B. *Macromol Biosci,* **11**, 648–656.

100. Espuelas, M. S., Legrand, P., Loiseau, P. M., Bories, C., Barratt, G., Irache, J. M. (2002). *In vitro* antileishmanial activity of amphotericin B loaded in poly(ε-caprolactone) nanospheres. *J Drug Target,* **10**, 593–599.

101. Tang, X., Zhu, H., Sun, L., Hou, W., Cai, S., Zhang, R., Liu, F. (2014). Enhanced antifungal effects of amphotericin B-TPGS-b-(PCL-ran-PGA) nanoparticles *in vitro* and *in vivo*. *Int J Nanomedicine,* **24**, 5403–5413.

102. des Rieux, A., Fievez, V., Garinot, M., Schneider, Y., Preat, V. (2006). Nanoparticles as potential oral delivery systems of proteins and vaccines: a mechanistic approach. *J Control Release,* **116**, 1–27.

103. Azizi, A., Kumar, A., Diaz-Mitoma, F., Mestecky, J. (2010). Enhancing oral vaccine potency by targeting intestinal M cells. *Plos Pathog,* **6**, e1001147.

104. Clark, M. A., Jepson, M. A., Hirst, B. H. (2001). Exploiting M cells for drug and vaccine delivery. *Adv Drug Deliv Rev,* **50**, 81–106.

105. Wilson-Welder, J. H., Torres, M. P., Kipper, M. J., Mallapragada, S. K., Wannemuehler, M. J., Narasimhan, B. (2009). Vaccine adjuvants:

current challenges and future approaches. *J Pharm Sci,* **98**, 1278–1316.

106. Woodrow, K. A., Bennett, K. M., Lo, D. D. (2012). Mucosal vaccine design and delivery. *Annu Rev Biomed Eng,* **14**, 17–46.

107. Durán-Lobato, M., Carrillo-Conde, B., Khairandish, Y., Peppas, N. A. (2014). Surface-modified P (HEMA-*co*-MAA) nanogel carriers for oral vaccine delivery: design, characterization, and *in vitro* targeting evaluation. *Biomacromolecules,* **15**, 2725–2734.

108. Neutra, M. R. (1998). Current concepts in mucosal immunity. V Role of M cells in transepithelial transport of antigens and pathogens to the mucosal immune system. *Am J Physiol,* **274**, G785–G791.

109. Kaiserlian, D., Etchart, N. (1999). Entry sites for oral vaccines and drugs: A role for M cells, enterocytes and dendritic cells?. *Semin Immunol,* **11**, 217–224.

110. Carrillo-Conde, B., Song, E., Chavez-Santoscoy, A., Phanse, Y., Ramer-Tait, A. E., Pohl, N. L. B., Wannemuehler, M. J., Bellaire, B. H., Narasimhan, B. (2011). Mannose-functionalized "pathogen-like" polyanhydride nanoparticles target C-type lectin receptors on dendritic cells. *Mol Pharm,* **8**, 1877–1886.

111. Pashine, A., Valiante, N. M., Ulmer, J. B. (2005). Targeting the innate immune response with improved vaccine adjuvants. *Nat Med,* **11**, S63–S68.

112. Smith, D. M., Simon, J. K., Baker, J. R. (2013). Applications of nanotechnology for immunology. *Nat Rev Immunol,* **13**, 592–605.

113. Bilati, U., Allémann, E., Doelker, E. (2005). Development of a nanoprecipitation method intended for the entrapment of hydrophilic drugs into nanoparticles. *Eur J Pharm Sci,* **24**, 67–75.

114. Durán-Lobato, M., Muñoz-Rubio, I., Holgado, M. A., Álvarez-Fuentes, J., Fernández-Arévalo, M., Martín-Banderas, L. (2014). Enhanced cellular uptake and biodistribution of a synthetic cannabinoid loaded in surface-modified poly(lactic-*co*-glycolic acid) nanoparticles. *J Biomed Nanotechnol,* **10**, 1068–1079.

115. Martín-Banderas, L., Sáez-Fernández, E., Holgado, M. A., Durán-Lobato, M., Prados, J. C., Melguizo, C., Arias, J. L. (2013). Biocompatible gemcitabine-based nanomedicine engineered by Flow Focusing (R) for efficient antitumor activity. *Int J Pharm,* **443**, 103–109.

116. Martin-Banderas, L., Duran-Lobato, M., Muñoz-Rubio, I., Alvarez-Fuentes, J., Fernández-Arévalo, M., Holgado, M. A. (2013). Functional

PLGA NPs for oral drug delivery: recent strategies and developments. *Mini Rev Med Chem*, **13**, 58–69.

117. Eldridge, J. H., Meulbroek, J. A., Staas, J. K., Tice, T. R., Gilley, R. M. (1989). Vaccine-containing biodegradable microspheres specifically enter the gut-associated lymphoid tissue following oral administration and induce a disseminated mucosal immune response. In *Immunobiology of Proteins and Peptides*, (Springer) pp. 191–202.
118. Esparza, I., Kissel, T. (1992). Parameters affecting the immunogenicity of microencapsulated tetanus toxoid. *Vaccine*, **10**, 714–720.
119. Maloy, K. J., Donachie, A. M., O'Hagan, D. T., Mowat, A. M. (1994). Induction of mucosal and systemic immune responses by immunization with ovalbumin entrapped in poly(lactide-*co*-glycolide) microparticles. *Immunology*, **81**, 661–667.
120. Jones, D. H., McBride, B. W., Thornton, C., O'Hagan, D. T., Robinson, A., Farrar, G. H. (1996). Orally administered microencapsulated Bordetella pertussis fimbriae protect mice from B. pertussis respiratory infection. *Infect Immun*, **64**, 489–494.
121. Allaoui-Attarki, K., Pecquet, S., Fattal, E., Trolle, S., Chachaty, E., Couvreur, P., Andremont, A. (1997). Protective immunity against Salmonella typhimurium elicited in mice by oral vaccination with phosphorylcholine encapsulated in poly(DL-lactide-*co*-glycolide) microspheres. *Infect Immun*, **65**, 853–857.
122. Challacombe, S., Rahman, D., O'hagan, D. (1997). Salivary, gut, vaginal and nasal antibody responses after oral immunization with biodegradable microparticles. *Vaccine*, **15**, 169–175.
123. Kofler, N., Ruedl, C., Rieser, C., Wick, G., Wolf, H. (1997). Oral immunization with poly-(D, L-lactide-*co*-glycolide) and poly-(L-lactic acid) microspheres containing pneumotropic bacterial antigens. *Int Arch Allergy Immunol*, **113**, 424–431.
124. Allaoui-Attarki, K., Fattal, E., Pecquet, S., Trollé, S., Chachaty, E., Couvreur, P., Andremont, A. (1998). Mucosal immunogenicity elicited in mice by oral vaccination with phosphorylcholine encapsulated in poly (D, L-lactide-*co*-glycolide) microspheres. *Vaccine*, **16**, 685–691.
125. Chen, S. C., Jones, D. H., Fynan, E. F., Farrar, G. H., Clegg, J. C., Greenberg, H. B., Herrmann, J. E. (1998). Protective immunity induced by oral immunization with a rotavirus DNA vaccine encapsulated in microparticles. *J Virol*, **72**, 5757–5761.

126. Herrmann, J. E., Chen, S. C., Jones, D. H., Tinsley-Bown, A., Fynan, E. F., Greenberg, H. B., Farrar, G. H. (1999). Immune responses and protection obtained by oral immunization with rotavirus VP4 and VP7 DNA vaccines encapsulated in microparticles. *Virology*, **259**, 148–153.

127. Kim, S., Doh, H., Jang, M., Ha, Y., Chung, S., Park, H. (1999). Oral immunization with Helicobacter pylori-loaded poly (D, L-lactide-co-glycolide) nanoparticles. *Helicobacter*, **4**, 33–39.

128. Conway, M. A., Madrigal-Estebas, L., McClean, S., Brayden, D. J., Mills, K. H. (2001). Protection against Bordetella pertussis infection following parenteral or oral immunization with antigens entrapped in biodegradable particles: effect of formulation and route of immunization on induction of Th1 and Th2 cells. *Vaccine*, **19**, 1940–1950.

129. Jung, T., Kamm, W., Breitenbach, A., Hungerer, K., Hundt, E., Kissel, T. (2001). Tetanus toxoid loaded nanoparticles from sulfobutylated poly (vinyl alcohol)-graft-poly (lactide-*co*-glycolide): evaluation of antibody response after oral and nasal application in mice. *Pharm Res*, **18**, 352–360.

130. Fattal, E., Pecquet, S., Couvreur, P., Andremont, A. (2002). Biodegradable microparticles for the mucosal delivery of antibacterial and dietary antigens. *Int J Pharm*, **242**, 15–24.

131. Gutierro, I., Hernandez, R., Igartua, M., Gascon, A., Pedraz, J. L. (2002). Influence of dose and immunization route on the serum IgG antibody response to BSA loaded PLGA microspheres. *Vaccine*, **20**, 2181–2190.

132. Gutierro, I., Hernandez, R., Igartua, M., Gascon, A., Pedraz, J. L. (2002). Size dependent immune response after subcutaneous, oral and intranasal administration of BSA loaded nanospheres. *Vaccine*, **21**, 67–77.

133. Foster, N., Hirst, B. H. (2005). Exploiting receptor biology for oral vaccination with biodegradable particulates. *Adv Drug Deliv Rev*, **57**, 431–450.

134. Garinot, M., Fievez, V., Pourcelle, V., Stoffelbach, F., des Rieux, A., Plapied, L., Theate, I., Freichels, H., Jerome, C., Marchand-Brynaert, J., Schneider, Y., Preat, V. (2007). PEGylated PLGA-based nanoparticles targeting M cells for oral vaccination. *J Control Release*, **120**, 195–204.

135. Gupta, P. N., Khatri, K., Goyal, A. K., Mishra, N., Vyas, S. P. (2007). M-cell targeted biodegradable PLGA nanoparticles for oral immunization against hepatitis B. *J Drug Target*, **15**, 701–713.

136. Mishra, N., Tiwari, S., Vaidya, B., Agrawal, G. P., Vyas, S. P. (2011). Lectin anchored PLGA nanoparticles for oral mucosal immunization against hepatitis B. *J Drug Target*, **19**, 67–78.

137. O'Hagan, D. T., McGee, J. P., Lindblad, M., Holmgren, J. (1995). Cholera toxin B subunit (CTB) entrapped in microparticles shows comparable immunogenicity to CTB mixed with whole cholera toxin following oral immunization. *Int J Pharm*, **119**, 251–255.

138. Nayak, B., Panda, A. K., Ray, P., Ray, A. R. (2009). Formulation, characterization and evaluation of rotavirus encapsulated PLA and PLGA particles for oral vaccination. *J Microencapsul*, **26**, 154–165.

139. Sarti, F., Perera, G., Hintzen, F., Kotti, K., Karageorgiou, V., Kammona, O., Kiparissides, C., Bernkop-Schnuerch, A. (2011). *In vivo* evidence of oral vaccination with PLGA nanoparticles containing the immunostimulant monophosphoryl lipid A. *Biomaterials*, **32**, 4052–4057.

140. Tacket, C. O., Reid, R. H., Boedeker, E. C., Losonsky, G., Nataro, J. P., Bhagat, H., Edelman, R. (1994). Enteral immunization and challenge of volunteers given enterotoxigenic E. coli CFA/II encapsulated in biodegradable microspheres. *Vaccine*, **12**, 1270–1274.

141. Katz, D. E., DeLorimier, A. J., Wolf, M. K., Hall, E. R., Cassels, F. J., van Hamont, J. E., Newcomer, R. L., Davachi, M. A., Taylor, D. N., McQueen, C. E. (2003). Oral immunization of adult volunteers with microencapsulated enterotoxigenic Escherichia coli (ETEC) CS6 antigen. *Vaccine*, **21**, 341–346.

142. Tobio, M., Gref, R., Sanchez, A., Langer, R., Alonso, M. J. (1998). Stealth PLA-PEG nanoparticles as protein carriers for nasal administration. *Pharm Res*, **15**, 270–275.

143. Lai, S. K., Wang, Y., Hanes, J. (2009). Mucus-penetrating nanoparticles for drug and gene delivery to mucosal tissues. *Adv Drug Deliv Rev*, **61**, 158–171.

144. Beloqui, A., Solinís, M. Á., Rieux, A., Préat, V., Rodríguez-Gascón, A. (2014). Dextran–protamine coated nanostructured lipid carriers as mucus-penetrating nanoparticles for lipophilic drugs. *Int J Pharm*, **468**, 105–111.

145. Benoit, M. A., Baras, B., Gillard, J. (1999). Preparation and characterization of protein-loaded poly(ε-caprolactone) microparticles for oral vaccine delivery. *Int J Pharm*, **184**, 73–84.

146. Murillo, M., Grilló, M. J., Reñé, J., Marín, C. M., Barberán, M., Goñi, M. M., Blasco, J. M., Irache, J. M., Gamazo, C. (2001). A Brucella ovis antigenic

complex bearing poly-ε-caprolactone microparticles confer protection against experimental brucellosis in mice. *Vaccine,* **19**, 4099–4106.

147. Prashant, C. K., Bhat, M., Srivastava, S. K., Saxena, A., Kumar, M., Singh, A., Samim, M., Ahmad, F. J., Dinda, A. K. (2014). Fabrication of nanoadjuvant with poly-ε-caprolactone (PCL) for developing a single-shot vaccine providing prolonged immunity. *Int J Nanomedicine,* **9**, 937–950.

148. Fu, K., Pack, D. W., Klibanov, A. M., Langer, R. (2000). Visual evidence of acidic environment within degrading poly (lactic-*co*-glycolic acid) (PLGA) microspheres. *Pharm Res,* **17**, 100–106.

149. Tang, B. C., Dawson, M., Lai, S. K., Wang, Y. Y., Suk, J. S., Yang, M., Zeitlin, P., Boyle, M. P., Fu, J., Hanes, J. (2009). Biodegradable polymer nanoparticles that rapidly penetrate the human mucus barrier. *Proc Natl Acad Sci U S A,* **106**, 19268–19273.

150. Ibrahim, M. A., Ismail, A., Fetouh, M. I., Göpferich, A. (2005). Stability of insulin during the erosion of poly(lactic acid) and poly(lactic-*co*-glycolic acid) microspheres. *J Control Release,* **106**, 241–252.

151. Tobio, M., Sanchez, A., Vila, A., Soriano, I., Evora, C., Vila-Jato, J., Alonso, M. (2000). The role of PEG on the stability in digestive fluids and *in vivo* fate of PEG-PLA nanoparticles following oral administration. *Colloids Surf B,* **18**, 315–323.

152. Cui, F., Shi, K., Zhang, L., Tao, A., Kawashima, Y. (2006). Biodegradable nanoparticles loaded with insulin–phospholipid complex for oral delivery: preparation, *in vitro* characterization and *in vivo* evaluation. *J Control Release,* **114**, 242–250.

153. Liu, R., Huang, S., Wan, Y., Ma, G., Su, Z. (2006). Preparation of insulin-loaded PLA/PLGA microcapsules by a novel membrane emulsification method and its release *in vitro. Colloids Surf B,* **51**, 30–38.

154. Cui, F., Tao, A., Cun, D., Zhang, L., Shi, K. (2007). Preparation of insulin loaded PLGA-Hp55 nanoparticles for oral delivery. *J Pharm Sci,* **96**, 421–427.

155. Naha, P. C., Kanchan, V., Manna, P. K., Panda, A. K. (2008). Improved bioavailability of orally delivered insulin using Eudragit-L30D coated PLGA microparticles. *J Microencapsul,* **25**, 248–256.

156. Wu, Z. M., Zhou, L., Guo, X. D., Jiang, W., Ling, L., Qian, Y., Luo, K. Q., Zhang, L. J. (2012). HP55-coated capsule containing PLGA/RS nanoparticles for oral delivery of insulin. *Int J Pharm,* **425**, 1–8.

157. Davaran, S., Omidi, Y., Rashidi, M. R., Anzabi, M., Shayanfar, A., Sohrab, G., Vesal, N., Davaran, F. (2008). Preparation and *in vitro* evaluation of linear and star-branched PLGA nanoparticles for insulin delivery. *J Bioact Compatible Polym*, **23**, 115–131.

158. Santander-Ortega, M., Bastos-Gonzalez, D., Ortega-Vinuesa, J., Alonso, M. (2009). Insulin-loaded PLGA nanoparticles for oral administration: an *in vitro* physico-chemical characterization. *J Biomed Nanotechnol*, **5**, 45–53.

159. Xiong, X. Y., Li, Y. P., Li, Z. L., Zhou, C. L., Tam, K. C., Liu, Z. Y., Xie, G. X. (2007). Vesicles from pluronic/poly (lactic acid) block copolymers as new carriers for oral insulin delivery. *J Control Release*, **120**, 11–17.

160. Iwanaga, K., Ono, S., Narioka, K., Kakemi, M., Morimoto, K., Yamashita, S., Namba, Y., Oku, N. (1999). Application of surface-coated liposomes for oral delivery of peptide: effects of coating the liposome's surface on the GI transit of insulin. *J Pharm Sci*, **88**, 248–252.

161. Reix, N., Parat, A., Seyfritz, E., van der Werf, R., Epure, V., Ebel, N., Danicher, L., Marchioni, E., Jeandidier, N., Pinget, M., Frère, Y., Sigrist, S. (2012). *In vitro* uptake evaluation in Caco-2 cells and *in vivo* results in diabetic rats of insulin-loaded PLGA nanoparticles. *Int J Pharm*, **437**, 213–220.

162. Zhang, X., Sun, M., Zheng, A., Cao, D., Bi, Y., Sun, J. (2012). Preparation and characterization of insulin-loaded bioadhesive PLGA nanoparticles for oral administration. *Eur J Pharm Sci*, **45**, 632–638.

163. Jain, S., Rathi, V. V., Jain, A. K., Das, M., Godugu, C. (2012). Folate-decorated PLGA nanoparticles as a rationally designed vehicle for the oral delivery of insulin. *Nanomedicine*, **7**, 1311–1337.

164. Ashokkumar, B., Mohammed, Z. M., Vaziri, N. D., Said, H. M. (2007). Effect of folate oversupplementation on folate uptake by human intestinal and renal epithelial cells. *Am J Clin Nutr*, **86**, 159–166.

165. Alonso-Sande, M., des Rieux, A., Fievez, V., Sarmento, B., Delgado, A., Evora, C., Remuñán-López, C., Préat, V., Alonso, M. J. (2013). Development of PLGA-mannosamine nanoparticles as oral protein carriers. *Biomacromolecules*, **14**, 4046–4052.

166. Iyer, H., Khedkar, A., Verma, M. (2010). Oral insulin–a review of current status. *Diabetes Obes Metab*, **12**, 179–185.

167. Jiao, Y., Ubrich, N., Marchand-Arvier, M., Vigneron, C., Hoffman, M., Lecompte, T., Maincent, P. (2002). *In vitro* and *in vivo* evaluation of oral

heparin-loaded polymeric nanoparticles in rabbits. *Circulation*, **105**, 230–235.

168. Miller, A. C., Bershteyn, A., Tan, W., Hammond, P. T., Cohen, R. E., Irvine, D. J. (2009). Block copolymer micelles as nanocontainers for controlled release of proteins from biocompatible oil phases. *Biomacromolecules*, **10**, 732–741.

Chapter 10

Overview of Polyester Nanosystems for Nasal Administration

Imran Vhora, Sushilkumar Patil, Hinal Patel, Jitendra Amrutiya, Rohan Lalani, and Ambikanandan Misra
Pharmacy Department, Faculty of Technology and Engineering, The Maharaja Sayajirao University of Baroda, Kalabhavan, Vadodara 390001, Gujarat, India
misraan@hotmail.com, a.n.r.misra-pharmacy@msubaroda.ac.in

Apart from the treatment of local diseases, nasal drug delivery has caught attention of formulation scientists since long for direct delivery to brain as well as systemic delivery not only of small drug molecules but also of proteins and peptides. The classical advantages of nasal drug administration, such as ease of local treatment, enhancing bioavailability of proteins and peptides, direct brain delivery, and evasion of unwanted adverse effects associated with other routes of administration, have been further augmented by nanosystems which also avail us with the possibility of providing controlled release of drugs, modifying the formulation for targeted delivery, and protecting the therapeutic agent. The choice of materials used to formulate nasal nanosystems is often influenced by toxicity, biodegradability, biocompatibility, and ease of manufacture

Handbook of Polyester Drug Delivery Systems
Edited by MNV Ravikumar

ISBN 978-981-4669-65-8 (Hardcover), 978-981-4669-66-5 (eBook)
www.panstanford.com

for ultimate application. Polyesters have shown their potential in nasal drug delivery to fit the criteria of biodegradability, nontoxicity, and ability to control drug release through co-polymerization with other polymers of same or different nature at various degrees to suit the biological needs. Several polyester polymers have evolved in drug delivery with copolymer of lactic acid and glycolic acid already approved by USFDA for drug delivery application. Therefore, further explorations of polyester nanosystems in nasal drug delivery will extend the clinical applicability and provide better treatment alternatives in future. In this chapter, the physicochemical diversity of polyesters and methods of preparation of polyester nanosystems have been detailed with further discussions on the drug delivery applications of these nanosystems, giving emphasis on the influence of their physicochemical properties.

10.1 Introduction

From its initial recognition as an alternative to oral and intravenous route due to its ease of accessibility, intranasal drug route is gaining a supreme focus for systemic, local as well as brain drug delivery. Its utilization has been further augmented by the unique features offered by nasal cavity, namely high absorptive capacity due to the presence of epithelium surface containing microvilli, highly vascularized subepithelium and direct access to venous blood which bypasses the first-pass metabolism resulting in rapid attainment of therapeutic concentration with lower doses [1]. Moreover, the olfactory region of nasal cavity lies in contiguity with CSF flow providing a direct way to deliver drugs to CSF circumventing the blood–brain barrier [2]. Along with olfactory nerve, the trigeminal nerves innervating the respiratory region can further augment the transport to brain [3, 4].

The advantage of close proximity of CSF flow tracts with olfactory lobe can only be utilized in application when the drug is transportable across the intervening nasal epithelium separating the nasal cavity from CSF compartment [2]. Nanoparticles are often the choice of delivery system to enhance drug transport across this barrier. Nanoparticles can achieve this by transporting drug in encapsulated form or by retaining the drug load in nasal cavity for extended period of time or by protecting the drug from enzymes which ultimately results in improved therapeutic efficacy [5, 6].

Still, when choosing the nanoparticles for nasal delivery, it would be essential that they are biodegradable and also do not induce any allergic or immune response due to the presence of nasal-associated lymphoid tissue (NALT) in nasopharynx which mediates nasal immune system [7]. Various naturally occurring polymers like polysaccharides gums, proteins are inherently degradable and biocompatible, but suffer from lack of consistent properties. In contrast, synthetic polymers follow a precisely controlled and consistent degradation profile and reproducible physicochemical properties such as glass transition temperature (T_g), elasticity, hydrophilicity, etc. Biocompatibility is not an inherent characteristic of any material, but depends on the biological milieu that the polymer is exposed to and the extent of the tolerability that exists with regards to the interaction of that polymer with the tissue [8]. In spite of extensive research in material science, copolymer poly(lactic-*co*-glycolic acid) (PLGA) is the only biodegradable polymer approved by USFDA and European Medicine Agency (EMA) in marketed products [9]. The reason behind biodegradability of PLGA is the presence of hydrolytically susceptible ester linkages. The hydrolysis produces lactic and glycolic acids which are usually present in body under physiological conditions and do not pose any sensitivity or toxicity. Similar to PLGA, there are a lot of other polyesters which are biodegradable and are being explored for drug delivery applications and degrade into components which are easily excreted from body. However, there is limited application of these newer polymers in nasal drug delivery.

Although use of polyester nanosystems for nasal delivery may provide advantages of biocompatible delivery system, there are several challenges associated with nasal drug delivery, such as mucous secretion, mucociliary clearance, epithelial layer, intervening mucosa associated lymphatic tissue, etc., which need to be overcome in order to achieve desired outcomes from these nanosystems, whether it be delivery to submucosal dendritic/antigen presenting cells, delivery to systemic circulation, or delivery to brain.

The present chapter focuses on polyester as a broad category of biodegradable and biocompatible polymers discussing their physicochemical properties. The physicochemical properties of polyester nanosystems also have a great influence on the ultimate biological outcomes on nasal administration. Hence, further section discusses their preparation techniques emphasizing the processing parameters which can be regulated in order to govern the physicochemical properties of the polyester nanosystems followed by their influence on their safety and efficacy *in vitro* and *in vivo*. Section 10.6 discusses the current advances giving directions for future opportunities in the field. Overall, this chapter will provide a hands-on guide for designing polyester nanoparticulate systems for the development of effective delivery systems for nasal administration.

10.2 Physicochemical Diversity of Polyesters

Polyesters belong to the group of biodegradable polymers, which are used for delivery of a vast variety of drugs. A good chemical and physical diversity exists in polyesters which can be used to develop drug delivery carriers. Hence, understanding their physicochemical properties is of prime importance in order to develop a suitable drug delivery system. Classification of these polyester polymers can be made into two major groups, namely polyesters from biological sources and polyesters from petrochemical sources. The structure of different polyesters, their commercial sources, and physical properties are given in Table 10.1.

10.2.1 Polyesters from Biological Sources

This group of polymer includes polyesters which are synthesized from either biotechnologically synthesized monomer, e.g., polylactic acid, or either directly extracted from the bacterial/fungal cultures, e.g., polyhydroxyalkanoates. Most widely available and used polymers of this class are poly-α-hydroxy acids (polylactic acid, polyglycolic acid, and polylactic acid-*co*-glycolic acid), and poly hydroxyalkanoates.

10.2.1.1 Polymers of α-hydroxy acids

Polylactic acid and polyglycolic acid are the key polymers of this class. However, both have been utilized now to produce copolymers or polymer mixes that are being used extensively in drug delivery. These polymers are considered as both biodegradable (degrades in body after administration) and biocompatible (raising no toxicity issues in contact with living cells), which makes them the most useful polymer in drug delivery. It is synthesized from the lactic acid monomers, which are either derived from fermentation process of carbohydrates involving microorganisms, usually of genus *Lactobacillus*, or fungi and further polymerization to high or low molecular weight polyesters. As the monomer synthesized exists in the L-lactic acid form, the polymerization of it leads exclusively to poly-L-lactic acid (PLLA). Polycondensation reaction has been reported, which produced low molecular PLLA; however, use of chain extenders/adjuvants can yield high molecular weight PLLA [10]. Azeotropic dehydration polycondensation has been used alternatively to obtain high molecular weight polylactic acid (PLA) [11, 12]. Apart from these processes, the most widely used method now is the ring-opening polymerization reaction (patented by Cargill, USA) of lactide, which leads to pure high molecular weight (molecular weights ≥ 100000) PLA [13].

In contrast to biosynthesized L-lactic acid, use of chemically synthesized lactic acid yields polymer with usually different percentage of L-lactic acid and D-lactic acid. The method of ring-opening polymerization, with an intermediate step of synthesis of lactide, also produces PLA with different proportions of D- and L-lactic acids, although allowing control of proportions and sequences of D- and L-lactic acids in polylactic acid [14, 15].

Another polymer of this category is polyglycolic acid, which is made of glycolic acid monomers by similar synthesis methods [16]. Its high crystallinity and insolubility in organic solvents has limited its use in drug delivery, but it has been used in combination with PLA as a copolymer, PLGA for drug delivery applications due to its additional advantage of controlling physicochemical properties by PLA and PLGA content and due to low cost commercial availability as compared to other polyesters [17]. PLGA is available in various ratios of lactic acid and glycolic acid content, which ranges from 50:50 to 85:15. Up to 50:50 ratio, PLGA polymers are soluble in common organic solvents; however, further increasing PGA content above 50% poses solubility problem [18]. Apart from PLGA, copolymers of PLA can also be prepared using ε-caprolactone, ε-valerolactone, etc., to get desired drug delivery systems.

Physical characteristics of poly-α-hydroxy acids (melting temperature, crystallinity, solubility) are dependent on the molecular weight, thermal treatments, stereoisomeric purity, and copolymerization with other monomers. Among all the physical characteristics, the crystallinity is the most important governing factor of drug release. A generalization can be made that the higher the crystallinity, the lower the drug release rate is. Pure PLA is semicrystalline in nature, while polyesters of meso compound and racemic mixture of D- and L-lactides (Fig. 10.1) are amorphous in nature. Moreover, blends containing equivalent amounts of L-lactic acid and D-lactic acid produce racemic crystallite (stereocomplexation) having higher mechanical properties that of PLLA and poly-D-lactic acid (PDLA). In general, semicrystalline poly-α-hydroxy acids degrade in two stages, initially with hydrolysis of amorphous areas followed by degradation of crystalline areas. Breaking of ester bonds through hydrolysis leads to decrease in molecular weight of the polymer. Apart from this, several other factors contribute to different degradation patter of these polymers. For example, branched PLA, in contrast to linear PLA, degrades by bulk erosion [19]. Also, it has been demonstrated that size-dependent degradation occurs with PLA, PGA, and PLGA systems [20]. Nanoparticles degrade slowly than microparticles of PLA due to diffusion mechanism playing role in degradation of microparticles [21, 22]. Similar to degradation mechanisms, drug release mechanisms can be correlated with two stages, i.e., initial burst release followed by gradual controlled release thereafter [23].

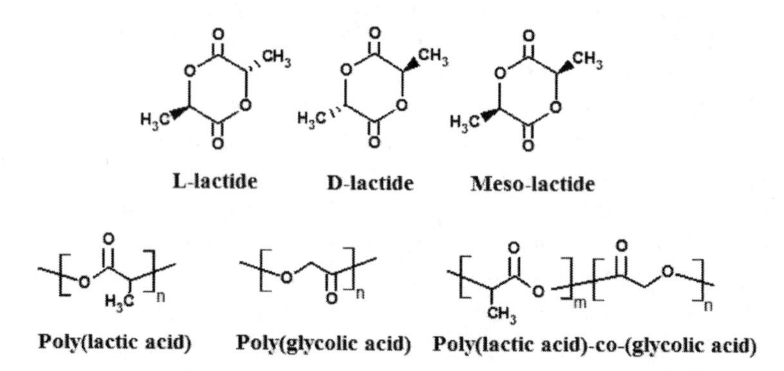

Figure 10.1 Structures of L-, D- and meso-lactide and homo- and co-polymers of α-hydroxy acids (subscripts m and n represent monomer content).

10.2.1.2 Polyhydroxyalkanoates

Polyhydroxyalkanoates can be synthesized chemically or can be extracted from bacterial cell cultures where they occur as intracellular energy reservoirs. Homopolymer polyhydroxybutyrate (PHB) and copolymers of PHB (polyhydroxybutyrate-*co*-hydroxyvalerate) (PHBV), polyhydroxybutyrate-*co*-hydroxyhexanoate (PHBH), polyhydroxybutyrate-*co*-hydroxyoctanoate (PHBO), and polyhydroxybutyrate-*co*-hydroxyoctadecanoate (PHBOD) belong to this group of polyesters. These polymers are generally nontoxic due to their biodegradation product being naturally occurring in humans [24, 25]. Among these polymers, only few have been evaluated in drug delivery, which includes PHB and PHBV (Fig. 10.2). PBHV polymers are available with varying content of valarate. These brittle polymers have a strong tendency to crystallize, high melting temperature [26] and slower rate of biodegradation as compared to other biodegradable polyesters [27]. However, copolymers are suitable for biomedical and drug delivery applications [26, 28, 29]. In PHBV, as the valerate content increases the melting temperature, T_g, as well as crystallinity decreases [30, 31]. Low molecular weight and high valerate content accelerates the degradation rate [32].

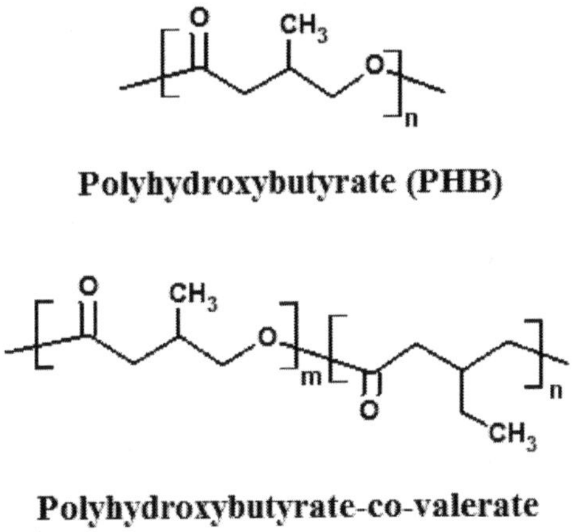

Polyhydroxybutyrate (PHB)

Polyhydroxybutyrate-co-valerate

Figure 10.2 Structures of most commonly employed polyhydroxyalkanoates (subscripts m and n represent monomer content).

10.2.2 Polyesters from Petrochemical Sources

Structures of polyesters derived from petrochemical sources are given in Fig. 10.3. Their physical properties are described in Table 10.1.

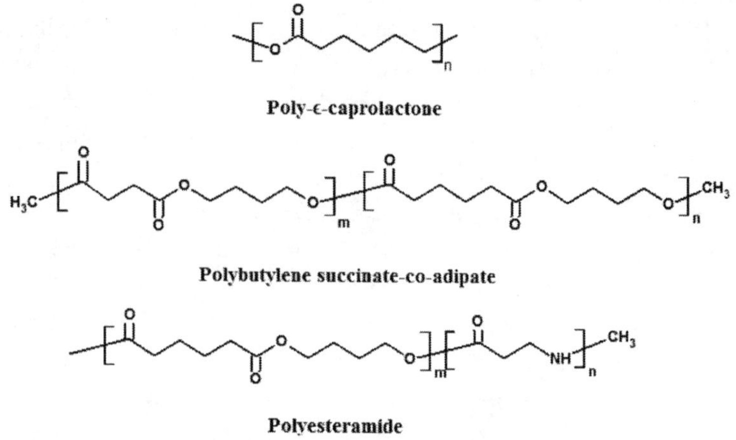

Poly-ε-caprolactone

Polybutylene succinate-co-adipate

Polyesteramide

Figure 10.3 Structures of polyesters derived from petrochemical sources (subscripts m and n represent monomer content).

10.2.2.1 Poly-ε-caprolactone

Successful application of poly-hydroxy acids in suture manufacturing, drug delivery systems and medical device industry boosted the development of other polyesters. Among them, polycaprolactone (PCL) gained a wide attention in drug delivery applications due to its desirable characteristics. It can be synthesized easily by ring-opening polymerization of ε-caprolactone.

Degradation of PCL follows the similar pattern as that of lactide polymers through bulk hydrolysis *in vitro* as well as *in vivo* [33, 34] by initial hydrolytic cleavage of ester groups followed by degradation of low molecular weight crystalline phases. Additionally, shape and size of the system as well as the presence of other components may have influence on the degradation rate [33]. Copolymers as well as graft polymers have been synthesized using PCL and PLA, which show physical properties different from homopolymers and random copolymers [35]. These can be used to get polymers of desired physicochemical properties. For example, high water permeability can be conferred to PLA by copolymerization of PCL with PLA to get controlled release of drug through combined balance of diffusion release from PCL matrix and erosion release from PLA matrix [36]. Copolymers of PCL with poly-δ-valerolactone and DL-ε-decalactone also show higher degradation rate when the copolymer contents is sufficient enough to bring the melting point of ε-caproate chains down to body temperature [33]. Also, addition of moieties like oleic acid and tertiary amines in the polymer can also be used to augment the rate of hydrolysis of PCL [33, 34].

10.2.2.2 Aliphatic copolyesters

Aliphatic copolyesters of diols and dicarboxylic acids have been synthesized. Various diols such as ethanediol, propanediol, butanediol can be reacted with diacids like succinic acid, cibacic acid, or adipic acid to obtain different copolymers. Among these, polybutylene succinate (PBS) synthesized from butanediol and succinic acid and polybutylene succinate adipate synthesized from butanediol and adipic acid are mainly utilized polymers in drug delivery [37, 38]. These polymers are marketed under the names of Bionolle®, EnPoll, Skygreen, and PBS. The degradation characteristics of these copolymers depend on their structures [39]. Presence of adipic acid in PBS makes the copolymer more crystalline [40] and also influences the degradation rate [37].

10.2.2.3 Other polyesters

Several other polyesters, such as poly(dihydropyrans), poly(propylene fumarate), and poly(p-dioxanone), have been synthesized for multiple applications [41, 18]. Polyesteramides are manufactured by copolycondensation of polyamide monomers and adipic acid [42, 43]. Figure 10.3 and Table 10.1 show the structure and physical properties of the polyesteramide, respectively. PEA exhibits highest polar component demonstrating compatibility with polar products. This polymer has not been extensively studied in drug delivery. Polydihydropyrans have been studied for *in vitro* and *in vivo* steroid and antimalarial drug delivery.

Table 10.1 Commonly used polyesters, their commercial sources and physicochemical properties

Polymer	Trade name (Company)	Physicochemical properties
Poly-L-lactic acid / Poly-D-lactic acid	Purasorb PD (PDLA) and Purasorb PL (PLLA) (Corbion-Purac, USA/ Thailand, Brazil, The Netherlands and Spain) Lactel L-PL (PLLA) and Lactel DL-PL (PDLA) (Lactel® Absorbable Polymers - Durect, USA), Medisorb 100-L (PLLA) (Alkemers, USA) PLA (PolySciTech, Akina Inc., USA)	Semicrystalline, T_m:173–178°C, T_g: 60–65°C Solubility: soluble in methylene chloride, chloroform, dioxane, dioxolane and furan Nonsolvent: water, ethanol, hexane and heptane MW: >100,000 Tensile strength: 55–82 MPa Breaking elongation %: 5–10% Modulus: 2.8–4.1 GPa
Poly-DL-lactic acid	Purasorb PDL (Corbion-Purac, USA/Thailand, Brazil, The Netherlands and Spain) Lactel DL-PL (Lactel® Absorbable Polymers - Durect, USA), Medisorb 100-DL (Alkemers, USA) PDLLA (PolySciTech, Akina Inc., USA)#	Amorphous T_m:173–178°C, T_g: 55–60°C Solubility: soluble in methylene chloride, chloroform, methyl ethyl ketone, pyridine, ethyl lactate, tetrahydrofuran, acetone xylene, ethyl acetate, dimethylformamide Nonsolvent: water, ethanol, hexane and heptane MW: 4000–6000 Tensile strength: 27–41 MPa Breaking elongation %: 3–10% Modulus: 1.4–2.8 GPa

Polymer	Trade name (Company)	Physicochemical properties
Polyglycolic acid	Purasorb PG (Corbion-Purac, USA/Thailand, Brazil, The Netherlands and Spain) Lactel PG (Lactel Absorbable Polymers - Durect, USA)[$] Medisorb 100-PG (Alkemers, USA) PGA (PolySciTech, Akina Inc., USA)	Semicrystalline T_m: 225–230°C, T_g: 35–40°C Solubility: soluble in hexafluoroisopropanol and hexafluoroacetone Insoluble in most commonly used organic solvents and water MW: >100,000 Tensile strength: >68 MPa Breaking elongation %: 15–20% Modulus: 6.9 GPa
Poly-lactic acid-*co*-glycolic acid (ratios available 50:50, 60:40, 85:15, 90:10)	Purasorb PDLG (Corbion-Purac, USA/Thailand, Brazil, The Netherlands and Spain) Lactel DL-PLGA (Lactel Absorbable Polymers - Durect, USA), Medisorb DL 2A (Alkermes, USA) PLLGA DL and PLGA LG (PolySciTech, Akina Inc., USA)[#]	Amorphous T_g: 45–55°C Solubility: soluble in dichloromethane, chloroform, tetrahydrofuran, acetone and ethylacetate Antisolvents: water MW: 40,000– 100,000 Tensile strength: 41–55 MPa Breaking elongation %: 3–10% Modulus: 1.4–2.8 GPa
Polyhydroxybutyrate	Tirel (Metabolix, USA), Enmat (Tianan, China), Biocycle (Copersucar, Brazil), Biomer L (Biomer, Germany)	Semicrystalline T_m: 175°C, T_g: 4°C Solvent: Chloroform, dichloromethane, tetrachloroethanol, acetic anhydride, dimethyl formamide

(*Continued*)

Table 10.1 (*Continued*)

Polymer	Trade name (Company)	Physicochemical properties
		Antisolvents: water, methanol, ethanol, isopropyl alcohole, alkanes, methyl ethyl ketone, tetrahydrofuran MW: >100,000 Tensile strength: 0 MPa Breaking elongation %: 5% Modulus: 3.5 MPa
Polyhydroxybutyrate-*co*-valerate (valerate content 7 mole%)	Biopol D400G, D411G, D300G, D311G, 600G, 611G (Metabolix, USA), Enmat (Tianan, China), Biocycle (Copersucar, Brazil), Biomer L (Biomer, Germany)	Semicrystalline T_m: 153°C, T_g: 5°C Solvent: Chloroform, dichloromethane, tetrachloroethanol, acetic anhydride, dimethyl formamide Antisolvents: water, methanol, ethanol, isopropyl alcohole, alkanes, methyl ethyl ketone, tetrahydrofuran MW: 300,000 Tensile strength: 0 MPa Breaking elongation %: 15% Modulus: 0.9 GPa
Polycaprolactone	Lactel PCL (Corbion-Purac, USA/Thailand, Brazil, The Netherlands and Spain) PCL (PolySciTech, Akina Inc., USA)# Celgreen (Daicel, Japan) CAPA (Solvay, UK)	Amorphous T_m: 58–63°C, T_g: –65 to –60°C Solubility: soluble in dichloromethane, chloroform, acetone, aromatic hydrocarbons, cyclohexanone Antisolvents: water MW: 80–150,000 Tensile strength: 21–34 MPa Breaking elongation %: 300–500% Modulus: 206–344 GPa

Polymer	Trade name (Company)	Physicochemical properties
Polybutylene succinate-*co*-adipate (PBSA)	Bionolle (Showa Denco, Japan) EnPol (Enpol Engineering Resins, USA) Skygreen (SK Chemicals, Korea) PBS (Anqing Hexing Chemical Co., China) Lunare SE (Nippon Shokubai, Japan)	Semicrystalline T_m: 114°C, T_g: -45°C Solubility: soluble in dichloromethane, chloroform, carbon tetrachloride, tetrahydrofuran Antisolvents: water, diethyl ether, hexane, methanol MW: 40,000 to >100,000 Tensile strength: 21–34 MPa Breaking elongation %: >500% Modulus: 250 MPa
Polyesteramide	BAK (Bayer, Germany)*	Semicrystalline T_m: 112°C, T_g: –29°C MW: >20000 Tensile strength: 21–34 MPa Breaking elongation %: >500% Modulus: 250 MPa

Physicochemical properties compiled from Refs. [37, 44–48].
*Discontinued.
#PolySciTech provides a wide range of PEGylated, copolymerized, and special-purpose polyesters with a variety of end groups.
$Durect provides ester terminated PLGA as well as acid-terminated PLGA.

10.3 Preparation of Polyester Nanosystems

As the particle size of polyester nanoparticles plays a major role in the nasal delivery (described later), an appropriate method of preparation should be chosen. Also, the method affects the entrapment efficiency of the therapeutic moieties and hence judicious choice of method is a prerequisite. Methods for the preparation of polyester nanoparticles consist of two main methods, namely the two-step method and the one-step method. For the two-step method, an emulsified system is prepared in the first step and the second step involves the process being employed for formation of nanoparticles. In the other method, formation of emulsion prior to nanoparticle formation is not required and is carried out in a single step. Such methods are based on precipitation of polymer in solution by the virtue of self-assembly mechanism induced by ionic gelation, polyelectrolyte complex formation or hydrophobic interaction in solution. Apart from these techniques, newer supercritical fluid technology based techniques have also emerged for preparation of nanoparticles. The most important methods which have been used for the synthesis of polyester nanosystems are described below, and the important process parameters are discussed for their effect on the properties of nanosystems. Table 10.2 describes the representative examples of each method used for the preparation of polyester nanoparticles.

10.3.1 Emulsification–Solvent Evaporation Method

It was the first method employed for the formation of polymeric nanoparticles [49]. Polymer along with the drug (which is hydrophobic in nature) is dissolved in volatile nonpolar solvents like dichloromethane/chloroform, which act as an oil phase. Oil phase is dispersed in solution of surfactants in a polar solvent, which in most cases is water. Organic volatile solvent is evaporated by applying temperature and/or reduced pressure or even at ambient temperature based on the stability of the entity being entrapped and the solvent used. Solvent evaporation leads to precipitation of solubilized polymer/drug forming nanoparticles stabilized by surfactant.

The concern with this method is that it involves use of organic solvents. This use is limited by regulatory bodies that have laid down the limits for the daily intake of such solvents. This needs that complete evaporation of organic solvent be ensured by method and subsequent evaluation of residual content in the final formulation. A method can also be modified by use of solvents which are less toxic, such as ethyl acetate [50].

As described earlier, the single emulsification technique yields good results with only hydrophobic drugs. However, the method can be modified for entrapment of hydrophilic drugs through double emulsification [51–54], in which the hydrophilic drugs are dissolved in water, which is formulated into a primary w/o emulsion using a lipophilic surfactant and subsequently into a double emulsion using hydrophilic surfactant [52, 55]. Organic solvent is evaporated, which yields nanoparticles. Controlling the emulsification energy at each emulsification step yields particles of desired size [53, 54]. In the mini-emulsion technique, o/w mini-emulsion is prepared at high shear rates using a hydrophilic emulsifier resulting in formation of monodisperse emulsion droplets. Subsequently, the organic solvent can be evaporated from droplets to get homogeneous nanoparticles [56, 57].

In both the emulsification methods, particle size can be set to desired values by controlling the stirring rate or sonication rate used for preparation of emulsion, by controlling the surfactant concentration employed and by concentration of polymer in the organic solution which governs the polymer amount in the droplets. The process, however, requires complete removal of solvent, thus making the nanoparticle preparation a time-consuming process. And also, coalescence of droplets during the evaporation stage may occur, affecting the morphology of particles.

10.3.2 Emulsification–Solvent Diffusion Method

Emulsification–solvent diffusion method is a modified version of solvent evaporation method [58]. A prerequisite for this method is the partial solubility in water of the polymer solvent used to prepare the emulsion [59]. In brief, polyester/drug solution prepared in partially water soluble organic solvent is emulsified using polyvinyl alcohol (PVA), cetyltrimethyl ammonium bromide (CTAB), poloxamer, or any other stabilizer in a sufficient quantity of water to get w/o emulsion [60–63]. Further addition of water shifts the solubility equilibrium, causing the partially soluble solvent to diffuse into the water, leading to precipitation of polymer-forming nanoparticulate system. Ideally, the initial process of preparation of emulsion involves the use of organic solvent and water, mutually saturated with each other to provide better control of diffusion [63, 64]. Partially water-soluble solvents which can be used for this method include propylene carbonate, benzyl alcohol, methyl acetate, ethyl acetate, isopropyl acetate, butyl lactate, methyl ethyl ketone, etc., which can be chosen on the basis of the solubility of polyester and drug. Here also, controlling the homogenization rate for the preparation of o/w emulsion, surfactant type, its concentration, and concentration of polymer in the diffusing solvent [63, 64] can be used to tune the particle size [64]. The process provides high encapsulation efficiencies (generally >70%), high batch-to-batch reproducibility, ease of scale-up, simplicity, and narrow particle size distribution. Nevertheless, the high volumes of water needs to be eliminated from the suspension after nanoparticle preparation and the leakage of water-soluble drug into the saturated aqueous phase during emulsification may occur, reducing encapsulation efficiency [65].

10.3.3 Salting Out

Chlorinated organic solvents which are harmful for environment as well as for humans used in the above two methods have been a major concern since the inception of these techniques. To overcome the issue of employing such solvents and surfactants, Bindschaedler *et al.* developed a modified version of emulsion process, termed the salting-out process [66]. Briefly, a polymer solution is prepared in a water-miscible organic solvent (acetone or THF) and is dispersed by sonication in aqueous phase, which is saturated with/contains a very high concentration of an electrolyte. The emulsion prepared doesn't allow diffusion of organic solvent in water due to the presence of high concentration of soluble components in water. Further dilution with water removes the electrolyte and allows diffusion of the organic solvent in water, forming nanoparticles [67]. Electrolytes having high saturation solubility can be employed for such method, which includes magnesium chloride, calcium chloride, and magnesium acetate. The advantage of this process is that no temperature is involved in the preparation hence the method can be used for temperature sensitive biomolecules like proteins [68, 69]. Concentration of salt, concentration of polymer in organic solvent, stirring rate, and surfactant concentration can be optimized to get desired particle size with uniform distribution. A shortcoming of the process is its applicability to lipophilic drugs only and requirement of extensive washing of nanoparticle [70].

Modification of the method has also been proposed which uses combination of emulsification-solvent evaporation and salting-out process, which includes preparation of emulsion with two organic solvents, one water miscible and other nonmiscible (i.e., acetone: dichloromethane) in saturated salt solution [71–73]. Organic solvent is then evaporated either at room temperature or at reduced pressure. Process parameters same as that of the salting-out process along with miscible/nonmiscible organic solvent ratio may need optimization for the control of nanoparticle-size and encapsulation efficiency. However, generally, the process allows precise particle size control with very low polydispersity.

10.3.4 Nanoprecipitation (Solvent Displacement) Method

The basic principle of this technique is interfacial precipitation of a polymer after displacement of a water-miscible organic solvent from a polymer solution [74, 75]. Organic solvent rapidly diffuses into aqueous phase, which causes fall of interfacial tension forming nanocapsules consisting of core of organic solvent globule and surrounding layer generated by precipitation of polymer at interface [76]. Afterwards, diffusion of organic solvent in aqueous medium through surrounding layer forms nanoparticles.

Use of surfactant prevents aggregation of nanoparticle suspension has shown to affect the particle size [77]. Surfactant concentration should be chosen appropriately in order to achieve balance between the low aggregation and low solubilization capacity [78]. However, nanoparticles have also been prepared without use of any surfactant [79]. Additionally, organic/aqueous phase volume ratio and drug/polymer ratio needs to be optimized to get desirable particle size with higher entrapment efficiency [78, 80, 81] and homogenization can be used to get smaller particle size [80]. The method is of limited use for entrapment of hydrophilic drugs.

Table 10.2 Method of preparation of polyester nanoparticles

Drug encapsulated	Polymer	Solvent	Surfactant/salt	Comment	Particle size (nm)	Reference
Emulsification-solvent evaporation method						
Amphotericin B	PLGA	Methylene chloride, dimethyl sulfoxide	Polyvinyl alcohol (PVA)	Low efficiency of loading due to nonmiscibility of drug and polymer and due to DMSO, which extracts the drug from dichloromethane phase into aqueous phase.	130	[82]
Human serum albumin	PLA	Methylene chloride	PVA	A low emulsification temperature and solvent removal under vacuum found suitable for heat sensitive materials; higher surfactant concentration led to small particle size with polydispersity index of <0.1.	200	[83]
Praziquantel	PLGA	Methylene chloride, ethyl acetate	PVA	Different solvents showed different particle sizes. Ethyl acetate led to smaller particle size and low entrapment efficiency due to loss of drug during diffusion step	200 and 280	[84]

Drug encapsulated	Polymer	Solvent	Surfactant/salt	Comment	Particle size (nm)	Reference
Tetanus toxoid	PLA-PEG	Ethyl acetate	PEG	The entrapment efficiency and size were independent of the stabilizer concentration and had an acceptable narrow range of size and showed ~35% entrapment and good coating efficiency.	130–150	[85]
Nimodipine	mPEG-PLA	Acetone	Sodium cholate	The system was devoid of initial burst effect.	~75	[86]
Emulsification-solvent diffusion method						
Doxorubicin	PLGA	Acetone	Pluronic 127	High loading efficiency (~96%) was achieved due to aqueous insolubility of PLGA-doxorubicin complex than free drug	250–350	[87]
Plasmid DNA	PLA-PEG	Methylene chloride	PVA/PVP	Nanoparticles obtained from w/o Emulsion diffusion showed rapid release, with the plasmid encapsulated in polymer at 80–90% efficiency due to complex between plasmid and PVA in nanoparticle leading to hydrophobicity or charge neutralization	~300	[88]

(*Continued*)

Table 10.2 (*Continued*)

Drug encapsulated	Polymer	Solvent	Surfactant/salt	Comment	Particle size (nm)	Reference
Plasmid DNA	PLGA-PEO	Methylene chloride	Ethanol	The NPs are absorbed via transcellular route, with PEO imparting properties similar to PEG	160–190	[89]
			Salting out			
Dexamethasone or rapamycin	PLGA-PEO	Acetone	Magnesium chloride hexahydrate	A high loading efficiency is explained by crystallization of drug in NPs. A decrease in volume: area ratio due to hydrophobicity of drug, which decreases the interfacial tension at the solvent interphase is responsible for lower particle size of NPs	120–250	[90]
Insulin	Polymalic acid and chitosan	Dil. HCL	Sodium chloride	Amorphous form was salted out as aggregates which showed pH dependent release after layer by layer deposition with weak polyelectrolytes in NPs. At low pH proteins like insulin get ionized, which enhances repulsion between insulin chains and aggregates, thus lowering particle size	100–230	[91]

Drug encapsulated	Polymer	Solvent	Surfactant/salt	Comment	Particle size (nm)	Reference
Nanoprecipitation (solvent displacement) method						
Nattokinase	PHB	Acetone	Tween 80	The enzyme was immobilized on PHB nanoparticles leading to 20% increase in enzymatic activity	100–125	[92]
Olanzapine	PLGA	Acetonitrile	Poloxamer 407	Smaller particle size obtained at optimum stabilizer concentration that coated the NPs uniformly Drug:polymer and non-aqueous: aqueous ratio were also found to affect the nanoparticle properites	121–200	[21]
Tetanus toxoid	Sulfobutylated-PVA-PLGA	Acetone, ethyl acetate and dichloromethane	Pluronic F -68	Organic solvent/solvent mixtures used showed different particle size with acetone showing lowest size and ethylacetate/ dichloromethane showing highest	100, 500 and 1000 nm	[93]

(Continued)

Table 10.2 (*Continued*)

Drug encapsulated	Polymer	Solvent	Surfactant/salt	Comment	Particle size (nm)	Reference
			Dialysis			
Epirubicin hydrochloride	PLA	Dimethylformamide (DMSO)	–	Higher concentration of PLA dissolved in solvent led to increase in particle size	~320	[94]
Carvacrol	PHB	Trifluoroethanol	Tween 80	Increasing surfactant concentration led to lower particle size		
Paclitaxel and doxorubicin	PCL-PVA	DMSO	–	The polymer has a high negative zeta potential. The tendency to encapsulate both hydrophobic and hydrophilic drugs	~360	[95]
Supercritical fluid technology						
Ketoprofen	PLA	Supercritical solvent: $scCO_2$ Antisolvent: air	–	Pulsed RESS process was used; particles formed were of core–shell configuration with drug core and PLA shell	700 nm	[96]
Naproxen	PLA	Supercritical solvent: $scCO_2$ Antisolvent: air	–	Pulsed RESS process was used; particles formed were of core-shell configuration with drug core and PLA shell		[97]

Drug encapsulated	Polymer	Solvent	Surfactant/salt	Comment	Particle size (nm)	Reference
Dexamethasone	PLGA	Solvent: Supercritical dichlorofluoromethane Antisolvent: toluene	–	RESS and RESOLV process was used; RESS at high pressure yielded particles ranging from 100 to 500 nm distribution while RESOLV yielded 100 nm particles with narrow distribution	100–500 nm and 100 nm	[98]
–	PCL	Solvent-Acetone $scCO_2$ mixture Antisolvent: Water	–	SAILA process was used; organic solvent mixed with SCF to achieve higher concentration of polyester Decrease in CO_2 mole fraction, increase in supercritical pressure and increase temperature led to smaller particle sizes	160–220 nm	[99]
Paclitaxel	PLL	Solvent: dichloromethane Supercritical Antisolvent: $scCO_2$	–	SASEM process was used; ultrasonic vibrations during expansion process to obtain nanoparticles Increasing supercritical pressure and solvent flow rate gives smaller particle sizes	400–800 nm	[100]

10.3.5 Dialysis

Dialysis is based on a solvent displacement mechanism but includes, in contrary to the conventional nanoprecipitation technique, additional tools such as dialysis tubes or semipermeable membranes with suitable molecular weight cutoff, which serves as a physical barrier for the polymer [101, 102]. Thus, dialysis is performed against an aqueous phase, in which the organic solvent (i.e., trifluoroethanol or acetone) is soluble. The displacement of the organic solvent through the membrane induces a progressive loss of polymer solubility and subsequently forming homogeneous suspensions of nanoparticles. According to the solvents used, the morphology and size of the particles can be affected [103]. Moreover, similar to nanoprecipitation method, the dialysis method can be used with or without surfactant. Nonetheless, it has been noted that the concentration of surfactant can be used as a parameter to control the particle size of the nanoparticles [104]. This is a simple and effective method for the preparation of small-sized nanoparticles with narrow distribution, but it yields limited entrapment of hydrophilic molecules such as proteins and peptides because of their decreased solubility in organic or hydrophobic phase.

10.3.6 Supercritical Fluid (SCF) Technology

SCF offers an effective technique for nanoparticle formation with potential for high purity and devoid of toxic organic solvents ensuring environmental friendliness as opposed to previous production processes employed. One can also employ SCF systems with carbon monoxide, *n*-pentane, water, ammonia, carbon dioxide, diflorochloromethane, etc. [105]. The SCF technology involves two principal techniques for nanoparticle production. A vast literature is available on manufacture of polyester microparticles through the use of SCF technology; however, only a few reports cite the use of SCF for preparation polyester nanoparticles.

Among different techniques employed, SAS (supercritical anti-solvent) and RESS (rapid expansion of supercritical solution) have been most widely used. However, they are used with modifications (described herein) to get processes like SASEM (supercritical anti-solvent with enhanced mass transfer), RESOLV (rapid expansion of supercritical solution into liquid solvent), or SAILA (supercritical-assisted injection in liquid antisolvent) and pulsed RESS, which are more suitable for preparation of nanoparticles. The SAS method requires that the drug and polymer should be soluble in a suitable solvent (dichloromethane, methanol, chloroform, etc.) which the supercritical fluid can extract, while the latter two methods require that polymers and drug have solubility in supercritical solvent. The SAS process is carried out by injecting, at controlled flow rate, the polymer/drug solution in the supercritical fluid which extracts the solvent leading to the formation of polymer/drug composites. Other RESS processes involve solubilization of polyester and drug in supercritical solvent ($scCO_2$, i.e., supercritical CO_2) or in mixture of organic solvent and supercritical solvent and then expansion in air (RESS) or in another liquid solvent (toluene or water) (RESOLV and SAILA) [98, 99]. Supercritical fluid loses its solvent power during expansion step, causing a high degree of supersaturation and subsequent homogeneous nucleation, inducing particle formation. In SAILA, the organic solvent chosen is water miscible and, hence, diffuses into water during expansion. SAILA can be applied to achieve higher concentrations of polymer in the expanding liquid [99]. The RESS-based method has been shown to produce core–shell nanoparticles with polyester coat on the drug core [97, 96, 106].

Although SAS and RESS methods are usually employed for polyester microparticle preparation [107, 108], there are reports which show preparation of nanoparticles with polyester polymers [98, 100]. Systems prepared thus are not absolutely free from presence of microparticles [98]. Furthermore, RESS yields particle size with wide distribution, while RESOLV provides better regulation of particle size within a narrow size range and it even makes it possible to achieve particle size of 100 nm and below [98]. This is attributed to the suppression of the particle growth at the jet in the expansion chamber. The SAS method has also been modified with additional provision for sonication to promote mass transfer during solvent extraction, which enables the manufacture of nanoparticles of polyesters (SASEM process) [100].

Although it is not possible to cover in detail the effect of all parameters involved in the preparation of polyester nanoparticles through SCF technology, some generalizations can be made which can be of use initially during product development. In case of

SAS, controlling the concentration of polymer/drug solution, the introduction rate of polymer/drug solution in supercritical fluid and flow rate of supercritical solvent may govern the particle size [100]. In case of SAILA and RESOLV, increasing the supercritical pressure and flow rate of organic solution in supercritical fluid yields smaller particle sizes [99, 100]. Conclusively, these methods can be suitably employed for small organic compounds as well as for the biological molecules like proteins which are sensitive to temperature.

Preparation of nanoparticles sometimes requires additional processing in order to attain desirable characteristics in specific needs of therapeutic molecules like DNA/RNA. In such situations, nanoparticles can be coated with cationic moieties like chitosan or PEI for making cationic surface which can be used to complex ionically with therapeutic genes like siRNA [109] or pDNA [63, 110]. Apart from this, several block polymers have been synthesized which can be used for preparation of nanosystems like micelles. Such polymeric micelles can be used for delivery of siRNA by preparing complex with cationically modified polymer with siRNA [111, 112]. Additionally, such polymer can be used to load the drugs like camptothein using appropriate technique like self-assembly [113].

10.4 Drug Release and Degradation of Polyester Nanosystems

Soon after the design of a drug delivery system, it is obligatory to ensure that release of loaded drug out of the carrier system occurs in a predictable manner. But, predictability of release requires a thorough understanding of realistic influencing variables such as physical, chemical, and biological properties. The mechanism of drug transport or release highly depends on the selected polymer matrix. Polyesters owe their biodegradability to presence of chemically labile ester backbone, which is variably sensitive to different environmental conditions such as presence of water, pH, temperature, enzymes, etc. [114]. This makes polyester matrix respond dynamically in response to *in vivo* conditions such that the matrix properties change with time with progressive water ingress and backbone cleavage. Therefore, knowledge of this dynamic behavior of polyester may be useful in anticipating the drug release from nanosystems composed of same.

Drug release from polyester-based carrier system occurs through a series of steps kindled with water absorption into the hydrophobic polymer, hydrolysis of the ester bonds, followed by erosion with liberation of broken fragments. The hydration or absorption of water inside a nanosystem initiates a pore-forming process through volume occupied by water. The number and size of pores containing water gradually increases, which initiates diffusion of drug through these pores. The contact of water prompts another important phenomena, i.e., hydrolysis of ester bonds, leading to generation of low molecular weight fragments of polymer [115]. The acidic products of ester hydrolysis show autocatalytic effect due to the specific acid catalysis nature of most of polyester polymers, leading to further increase in rate of hydrolysis [116]. Finally, the erosion process starts when the fragments are reduced to molecular weight small enough (1100 Da) to become soluble and diffuse into release medium [114]. Over the entire process, affinity of polymer changes from hydrophobic to hydrophilic as the molecular weight of polymer decreases with degradation.

Now each of the above process is sensitive to the physicochemical milieu. The rate of uptake of water has been correlated with the hydrophilicity of matrix. Therefore, any process or formulation variable which affects the hydrophilicity of matrix results in increased rate of water uptake such as higher amount of hydrophilic monomer or copolymer, low molecular weight, T_g, crystallinity, etc. Depending on rate of ingress of water the subsequent polyester degradation may proceed by either homogenous or bulk degradation. The bulk degradation is favored when rate of ingress is higher than degradation rate, which leads to uniform degradation throughout matrix [117]. The ester hydrolysis inevitably produces acidic by-product which tends to accumulate in the center and induce auto-catalysis leading to heterogeneous degradation of polymer matrix, i.e., more in the acid-rich center than surface, which is further accentuated by large size of matrix making the degradation products difficult to diffuse out of particle [118, 119]. Furthermore, the plasticity of polymer as reflected by T_g, a function of chemical makeup, water content, and additives also affects the degradation. The penetration of water usually occurs through amorphous regions [117]. The contact with water reduces the Tg and increases the plasticity, which may lead to chain rearrangement throughout the process of degradation, which affects the pore characteristics and drug release through them. The final process of erosion consisting of mass transport of polymer fragments is facilitated by reduced Tg. The erosion process leads to growth of water filler pores due to mass loss of polymer, which subsequently coalesce with neighboring pores to form larger pores [114]. The frequent consequence of this process is the second phase of burst of drug release observed from such matrices.

The dependence of degradation of ester backbone over physicochemical make up indicates that physicochemically different polyesters differ in their degradation kinetics. Highly crystalline (50%) PGA generally erodes by bulk erosion with a biphasic degradation kinetics with major mass loss in second phase from initially resistant crystalline regions [120]. In contrast, PLA is more amorphous and major mass loss occurs in first phase. The degradation also varies with relative composition of D and L enantiomers in polymer with racemic mixtures degrading at a much faster rate than pure forms [121]. PLGA copolymers, though they offer the advantage of controlled physicochemical properties, are generally amorphous and their degradation rates depend on the relative ratios of monomers. The degradation is faster for equimolar mixtures due to low crystallinity compared to co-monomer-rich polyesters. PCL, semicrystaline in nature, also undergoes random hydrolysis at a rate three times slower than poly-D,L-lactic acid due to five hydrophophobic methylene repeat units and an increase in crystallinity with a decrease in molecular weight during hydrolysis [122]. The degradation of poly(ortho ester) follows a preferential surface erosion pattern, and less acid-sensitive since acid hydrolysis products are diffused away from the surface, and the matrix retains its geometry, which could be advantageous to drug delivery applications.

Focusing on the way drug molecules make out of the polyester-based drug delivery system, it can be argued that it largely depends on the location of drug within the matrix and structure of the nanosystem employed. For hydrophilic drugs, diffusion through water-filled pores is the principle mechanism at the beginning, which subsequently becomes degradation and erosion dependent. For hydrophobic drugs, diffusion through the polymer and subsequent partitioning out into dissolution medium is the principle mechanism and later by dissolution of polymer carrier [118, 123, 124]. The transport through water-filled pores may occur by either the diffusion process driven by chemical potential gradient or the convection due to the force of osmotic pressure, which is generally observed due to influx of water in nonswelling polymers. PLGA may absorb water, leading to an increase in osmotic pressure, but swelling can compensate it by polymer chains rearrangement, making diffusion the primary mechanism.

Considering the anticipated shapes of drug release profile, it is rarely observed to be monophasic in nature due to the dynamic nature of polymeric matrix and is generally bi-phasic or tri-phasic [114]. Heterogeneous degradation of a large particulate delivery system has been attributed to tri-phasic release, while small particles result in a bi-phasic profile [125, 126]. Phase 1 is generally observed as an initial burst effect due to the release of free drug molecule associated surfaces and cracks [114, 127]. Phase 2 of drug release is governed by matrix degradation properties, during which drug diffuses slowly out of small pores in the slowly degrading dense polymer matrix. Depending on the rate of polymer degradation and the rate of drug diffusion, a near-zero-order release has been achieved [128, 129]. Phase 3 is usually observed as second burst release and it marks the onset of rapid erosion and sudden mass loss due to disintegration of particles [130]. Nonetheless, it should be considered that owing to the complexity of the process and the factors affecting drug release from biodegradable polyester, the release profiles may not follow the traditional pattern.

10.5 Therapeutic Outcomes of Nasal Administration of Polyester Nanoparticles

Polyester nanoparticles have been used in the delivery of several therapeutic agents ranging from drugs to vaccines, siRNA, as well as DNA. Table 10.3 details the *in vitro* and *in vivo* outcomes of the polyester nanosystems. However, the outcomes of these nanosystems are governed by the physicochemical properties of the nanosystems; hence this section discusses the correlation between the physicochemical properties of these nanosystems

Table 10.3 Therapeutic potential of polyester nanosystems in nasal delivery

Delivery system	Payload	Type of study	Study outcome	Ref.
PLA-PEG nanoparticles	Tetanus toxoid	*In vitro* and *in vivo*	Increasing PEG coating density increased absorption through nasal mucosa PEG stabilizes PLA nanoparticles in lysozyme Generation of both humoral as well as mucosal immunity *in vivo*.	[53, 54]
PLA-PEG nanoparticles	Tetanus toxoid	*In vitro* and *in vivo*	Biodistribution in lymph nodes, lung, liver, and spleen 3 fold higher than PLA nanoparticles 10 fold higher absorption in blood after 1 hr than PLA nanoparticles Uptake by M cells of NALT as well as by trancellular and paracellular pathways into blood and lymph Nanoparticles of toxoid cleared from blood by macrophage uptake only No effect of plasma on nanoparticle stability Immune response higher, increasing and long lasting	[85]
Lectin-conjugated PLA-PEG nanoparticles	–	*In vivo* brain uptake study in rats	Two fold higher uptake in brain as compared to nonconjugated nanoparticles Safe and effective vector with negligible ciliatoxicity and safety to brain cells No immunogenicity of nanoparticles	[131, 132]

Delivery system	Payload	Type of study	Study outcome	Ref.
PLGA nanoparticles	Olanzapine	Histopathological study on sheep nasal mucosa, *In vivo* brain uptake study in rats	No significant toxicity to nasal mucosa, Significantly increased brain uptake than intranasal and intravenous solution	[78]
Solanum tuberosum lectin Conjugated PLGA nanoparticles	–	*In vivo* brain uptake study	Mild cytotoxicity and negligible cilia irritation compared to unconjugated nanoparticles	[133]
Wheat germ agglutinin conjugated PLA nanoparticles	Coumarin	*In vivo*	2 fold increase in brain uptake of coumarin 4.5–5.6 fold and 5.6–7.7 fold higher brain levels of PLA nanoparticles and agglutinin modified PLA nanoparticles as compared to pain solution	[134]
PEG-PLA nanoparticles	Nimodipine	*In vivo*	Demonstrated olfactory pathway of absorption of nanoparticles 1.56 times higher AUC in brain as compared to solution	[135]
Cell-penetrating peptide modified PLGA nanoparticles	Insulin	*In vivo* brain uptake study	Enhanced brain uptake of insulin in Alzheimer's disease 6.5 times higher olfactory bulb disposion and 3.5 times higher cerebrum as compared to nonmodified PLGA nanoparticles Total accumulated dose of 6%	[136]

(*Continued*)

Table 10.3 (*Continued*)

Delivery system	Payload	Type of study	Study outcome	Ref.
Sulfobutylated poly(vinyl alcohol)-graft-PLGA nanoparticles	Tetanus toxoid	*In vivo* efficacy study	More antibody titer observed for intranasal immunization than per oral. 10 times less dose required for nasal route than oral route to induce similar immune response	[137]
Surface-modified PLGA microspheres	Recombinant hepatitis B antigen	*In vivo* vaccine efficacy study in Balb/c mice	Lowest nasal clearance rate No loss of antigen integrity Systemic and mucosal immunization	[138]
PLGA and N-trimethyl chitosan coated PLGA nanoparticles	Ovalbumin as model antigen	Nasal toxicity study, *In vivo* efficacy study	Slow releasing particles with undetectable immunization Cationical, mucoadhesive and fast releasing delivery system leads to high serum immunization	[139]
PLGA-poloxamer nanoparticles	Plasmid DNA	*In vitro* and *in vivo*	Cellular transport of DNA Controlled release of plasmid DNA	[89]
PCL nanoparticles	Carboplatin	*In vitro* and *in vivo*	Initial burst release followed by slow and continuous release indicating biphasic pattern, more cytotoxicity and better nasal absorption compared to solution	[140]

Delivery system	Payload	Type of study	Study outcome	Ref.
PCL-polyethylenimine complexes	pGL3 gene	*In vitro*	Controlled degradation of complexes, significant low cytotoxicity, and high transfection efficiency	[110]
Cell-penetrating peptide modified PEG–PCL nano-micelles	siRNA or dextran as a model siRNA	*In vitro* and *in vivo*	Improved brain delivery along the olfactory and trigeminal nerve pathway compared to intravenous delivery	[111]
Tat modified PCL-PEG nanomicelles	Camptothecin	*In vitro* and *in vivo*	Higher *in vitro* cytotoxicity to glioma cells Improved nose to brain uptake Better survival rates in intracranial glioma tumor bearing mice than nonmodified nanomicelles	[113]
Tat modified PCL-PEG nanomicelles	Marker	*In vitro* and *in vivo*	Improved brain uptake 100 nm micelles showed higher uptake than 600 nm micelles Concentration in tumor incubated side of brain than the other side Higher brain levels as compared to other organs	[112]

with their pharmacokinetic and pharmacodynamics outcomes. The section is divided in two parts, namely polyesters of poly-α-hydroxy acids and other polyesters, as the literature available is mostly on polyhydroxy acids. However, the generalizations made can be applied to other delivery systems as well, except for the effect of their controlled release on pharmacodynamics. Figures 10.4 and 10.5 depict the biological outcomes of these systems correlating the physicochemical features of the polyester nanosystems.

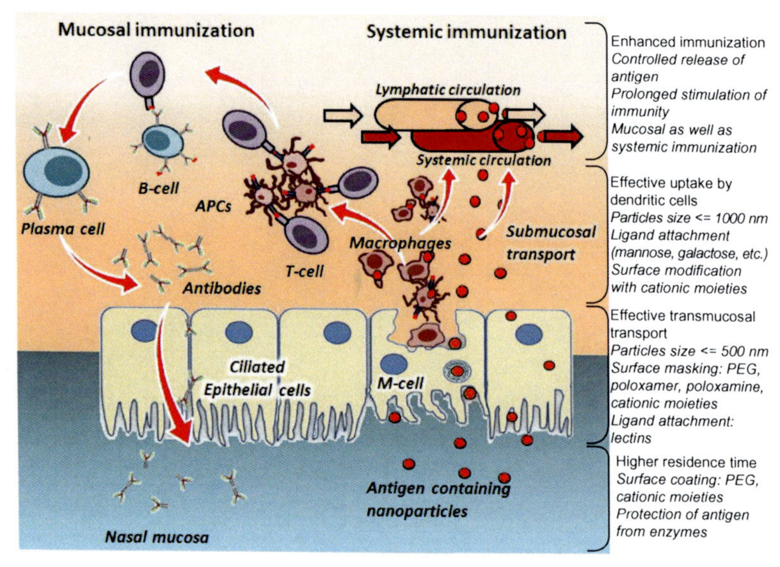

Figure 10.4 Transport and therapeutic outcomes of antigen-containing polyester nanoparticles and influence of physicochemical properties of nanoparticles. Uptake of nanoparticles takes place through M cells and through transcellular and paracellular transport via the epitelical cell layer of mucosa. The uptake of transported nanoparticles by macrophages and antigen-presenting cells (APCs) guides the ultimate antibody production by plasma cells through antigen presentation and identification by T cells and recognition of antigen by B cells and activation of B cells and production of antibody producing plasma cells. The antibodies produced (secretary IgAs) are transported through the epithelial cell layer into the mucosa for mucosal protection. The submucosal transport of APCs and macrophages and also of the nanoparticles to the systemic circulation and lymphatic circulation provides the systemic immunization following the same pathway as that of mucosal immunization.

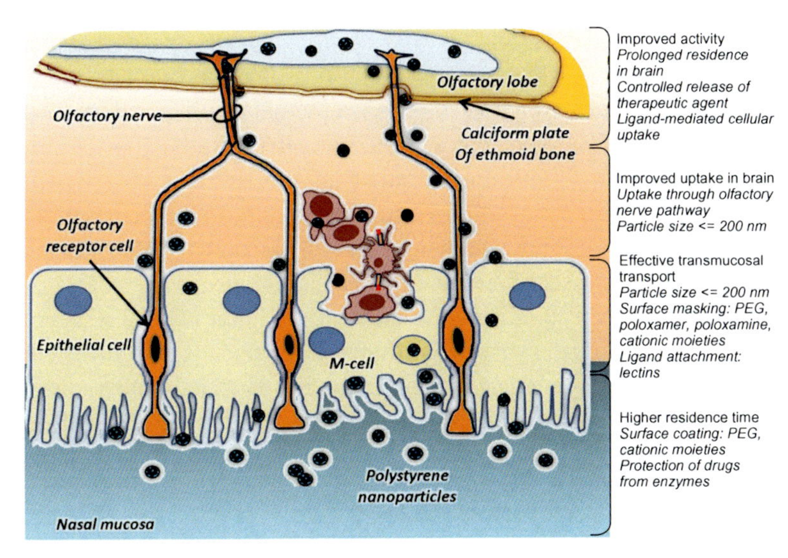

Figure 10.5 Nose-to-brain transport of therapeutic agent containing polyester nanoparticles in brain and influence of physicochemical properties of nanoparticles. The figure depicts an area of olfactory epithelium lining the calciform plate of ethmoid bone, which separates the olfactory lobe from nasal mucosa. Uptake of nanoparticles takes place through M cells and through transcellular and paracellular transport via the epitelical cell layer of mucosa. Transported cells are transported to the olfactory lobe via olfactory nerve pathway and then distributed throughout the brain. Nose-to-brain transport of nanoparticles also takes place through trigeminal nerve pathway through respiratory epithelium (similar to olfactory nerve pathway through olfactory epithelium).

10.5.1 Polyesters of Poly-α-Hydroxy Acids

Among all the polyesters, poly-α-hydroxy acids, in particular PLA and PLGA, are extensively used for nasal delivery of drugs, antigens, proteins as well as genes. Therapeutic outcomes of poly-α-hyxroxy acid nanoparticulate systems used for nasal delivery are given in the Table 10.2. The role of nanoparticulate systems in nasal delivery of drugs in context of their physicochemical properties is described here.

Use of PLA and PLGA is increasing due to their growing role in delivery of vaccines. PLA has been used for delivery of antigens like tetanus toxoid [53, 54, 85], *Schistosoma mansoni* antigen [141], and ricin toxin [142]. Formulations of these vaccines range from nanoparticles of few hundred nanometers to particles of a few micrometers. Also, there exist several contrasting results in vaccine delivery with some literature demonstrating enhancement of immunity after nasal administration of vaccine in nanoparticulate systems while some showing no enhancement after using nanosystems. Nasal delivery of therapeutic drugs in polyester nanosystems is also of growing interest. The fate of such delivery is either systemic delivery or brain delivery, which is also governed by the physicochemical characteristics of the nanosystems. Here, detailed evaluation of such systems with respect to their physicochemical characteristics and biological outcomes has been made.

Firstly, the particle size of the nanoparticulate systems is the most important regulator of their therapeutic effects. PLA microspheres of tetanus toxoid have been evaluated in rats, rabbits, and guinea pigs after intranasal administration in comparison with free antigen. The microspheres demonstrated improved immunogenicity as compared to free antigen [5]. PLA-PEG particulate systems of different sizes for delivery of tetanus toxoid demonstrated that particles with size of ~200 nm and ~1500 nm were similar [143]. This indicates that there exists a particle size threshold, which governs the transport through nasal mucosa and the size range evaluated was well below the threshold showing similar immunization with particles of both sizes [143]. This deduction can be supported by the similar results observed with nanoparticles of chitosan used for similar purpose showing similar transport for nanoparticles and 1 μm particles while significantly reduced transport of 5–10 μm particles [144] or even negative effect of 30 μm PLGA particles on transport [145]. Intranasal delivery of various vaccines has also been assessed with the antigen-loaded PLGA drug delivery systems. One of the earlier studies on nasal antigen delivery was performed on tetanus toxoid adsorbed on nanoparticles prepared from surface modified polyesters such as sulfobutylated poly(vinyl alcohol)-graft-PLGA [137]. Tetanus toxoid was adsorbed on nanoparticles through ionic interaction and effect of particle size (100, 500, and 1500 nm) on immune response was assessed. It was concluded that particles of intermediate size range are more effective in immunization using intranasal route

[137]. It has been demonstrated that particles having size of 500 nm and below are efficiently taken up by dendritic cells [146]. Similar results have been quoted for PLA nanoparticles (<1 μm), which were phagocytosed, while larger-sized particles were not [147].

Apart from particle size, surface charge of particles also affects the delivery using polyester nanoparticles. PLGA/PLA nanoparticles expose anionic surface due to the ionization of surface localized carboxylic acid groups. In order to alter the interaction of PLA with mucosal membranes of the nose, coating with PEG has been proposed [53]. After PEG coating, a drop in the negative zeta potential value of PLA was reported, which required higher density of surface-coated PEG. It was observed that small size with high density coating of PEG was more efficient in transmucosal transport than particles with lower density/no PEG coating. In another study, no effect of PEG coating on the zeta potential of nanoparticles was observed, but particles varied in their surface hydrophilicity, PLA nanoparticles being more hydrophobic [85]. Thus, PEG stabilizes particles in mucosal fluids apart from assisting in the passage of the antigen cargo into systemic and lymphatic circulation and thus providing colossal and long-lasting immunity [85]. PLA-PEG nanoparticles of tetanus toxoid have shown enhanced systemic and lymphatic absorption through nasal route in contrast to the PLA nanoparticles in other studies as well demonstrating the importance of PEG coating [53, 54]. *In vitro* study with PLA nanoparticles showed problem of aggregation of nanoparticles in presence of lysozyme, which was overcome by PEG-coated nanoparticles [53, 54]. The role of PEG-modified PLA nanoparticles on stability and improvement of mucosal membrane transport can be attributed to the PEG chains projected from the exterior of the nanoparticles in the surrounding milieu during the preparation of nanoparticles, which would reduce the nasal clearance of the nanoparticles [85, 138]. But it has been shown that PEG may also affect the release characteristics of the nanoparticles [85]. PEG chains, apart from being projected on the surface of the nanoparticles, may also be oriented inside of the nanoparticles interacting with encapsulated moieties. This might be the cause of higher protein release from PEG-PLA nanoparticles than from PLA nanoparticles, where apart from the degradation rate of PLA, PEG also plays a role [85].

Moreover, one can also modify the surface of the nanoparticles with cationic moieties. Cationic surface charge (obtained by coupling to polylysine or protamine sulphate) enhances the uptake of nanoparticles in dendritic cells [146]. However, surface modification with cationic moieties would also increase nonspecific uptake in other cells of the body. Thus, nanoparticles of smaller size regardless of its anionic charge could provide larger surfaces for interaction with dendritic cells/macrophages and will avoid the problem of nonspecific uptake by other cells. Hence, surface coating with a targeting ligand for dendritic cells/macrophages would be a better option to opt rather than cationic modification of surface.

The effect of particles size on absorption through nasal mucosa is already described. But, in order to generalize for the brain delivery, particle size is usually required to be as small as possible, which has been demonstrated though improved outcomes with drug delivery systems like microemulsions, liposomes, nanoparticles, and so on. Most of the nanoparticulate systems used for brain delivery are described below and all have reported particle size lower than 200 nm. Olanzapine PLGA nanoparticles delivered by the intranasal route have been found to transport drug directly to the brain after intranasal delivery without any significant toxicity to nasal mucosa. Also, *in vivo* pharmacokinetic studies showed significantly higher uptake of intranasally delivered PLGA nanoparticles compared to intravenous and intranasal solution [78]. PEG-PLA nanoparticles of vasoactive intestinal peptide, a neuroprotective peptide, have been shown to provide vasoactive intestinal peptide (VIP) in the brain, with 3.5–4.7 fold higher level than simple solution. Further, it has been established that nimodipine nanoparticles of PLA-PEG are transported in the brain through the olfactory nerve pathway into the olfactory lobe [135]. The study also demonstrated that the drug provided constant and prolonged blood levels of nimodipine. All studies described above also show the importance of PEG coating on nanoparticles, and it is now an indispensable ingredient in the development of nanoparticulate systems. This effect of enhanced uptake can be attributed to the phenomenon of its enhanced association with the mucosa of nasal tissue.

Further, lectin-conjugation to nanoparticles can be used, taking the advantage of the fact that *N*-acetyl-D-glucosamine and sialic acid are in copious amounts in nasal cavity, to which lectins can specifically bind and improve the absorption into the brain [148]. The nanoparticles of PEG-PLA conjugated with wheat germ agglutinin presented 2 fold higher brain uptake than nonconjugated nanoparticles [131]. Conjugation of wheat germ agglutinin with VIP-loaded PEG-PLA nanoparticles has shown to enhance the efficiency of nonconjugated PEG-PLA nanoparticles efficiency further to 5.6–7.7 fold through targeting [134]. Conjugation of PLGA nanoparticles with *Solanum tuberosum* lectin has shown brain targeting efficiency as well as enhanced accumulation in brain with mild cytotoxicity and negligible cilia irritation compared to unconjugated nanoparticles [133]. In another study, insulin delivered by cell-penetrating peptide-coated PLGA nanoparticles showed higher brain transport of insulin for Alzheimer's disease [136]. However, such transport was not shown to be due to cell-penetrating peptide. Such system can provide improved uptake in brain cells after intranasal administration.

PLGA:poloxamer- and PLGA:poloxamine-based nanoparticulate systems have also been developed with encapsulated plasmid DNA vaccine [89, 149]. The particles have demonstrated maintenance of biological activity of pDNA, and poloxamer and poloxamine have shown stability in nasal mucosa and transcellular absorption in nasal epithelial tissue by a similar mechanism as that of PEG [89]. Nanoparticles with poloxamer exhibited higher antibody titres, while those containing poloxamine showed response similar to naked DNA. Though the cause for the difference was not evaluated and further evaluation of these nanoparticles is needed, it was noteworthy that controlled release of pDNA without any burst effect was effective in providing higher systemic immunization [89]. Delivery of plasmid DNA can also be achieved through surface modification of PLA-PEG nanoparticles with cationic CTAB with substantial preservation of biological activity of pDNA and could be of potential use in future gene delivery through the nasal route [150]. The system has demonstrated lower cytotoxicity as compared to lipofectamine-2000 and also shown transfection in the presence of serum. Yet, only two polyester-based systems have been noted in the delivery of gene. This suggests that future development of similar strategies for gene delivery through nasal mucosa needs to be evaluated.

In general, PLGA nanoparticles offer several advantages, making them a good carrier/adjuvant for vaccine delivery. These include

(i) protection of antigen from proteolytic cleavage, (ii) selective delivery to phagocytic/antigen presenting cells, avoiding their entry to systemic circulation, (iii) tunable loading of antigen in nanoparticle, and (iv) prolonged stimulation of dendritic cells by sustained antigen presentation [151]. In contrast, in case of delivery of therapeutic agents to the brain, PLGA systems would provide (i) improved uptake of drug in the brain after nasal administration through suitable modification of particle size and surface, (ii) improved retention of drug in nasal mucosa, and (iii) controlled release of drug for prolonged activity.

10.5.2 Other Polyesters

We have discussed the effect of the physicochemical properties of poly-hydroxy acid nanoparticles (in particular PLA and PLGA nanoparticles). The generalizations made for these systems also apply to other polyesters. Other polyester nanosystems and their therapeutic outcomes are discussed here. PCL has high permeability and very low *in vivo* degradation rate, and it has been used for long-term delivery. However, very slow degradation rate of PCL, 2–3 years, makes pure PCL based delivery systems unsuitable for FDA approval [152]. Hence, PCL is mostly used in amalgamated forms with other biodegradable polymers like PLA, PLGA, and other polyesters to accelerate its degradation. Carboplatin entrapped in PCL nanoparticles showed typical burst and slow release phases during *in vitro* release and *ex vivo* permeation studies [140]. Carboplatin-loaded PCL nanoparticles show better nasal absorption and more cytotoxicity than carboplatin solution.

While only one publication is available for brain delivery using PCL nanoparticles, there are more publications available on nose-to-brain delivery done using PEG-modified PCL micelles. Studies with coumarin-loaded PEG-PCL micelles have shown effect of particle size of micelles on brain distribution and brain uptake of micelles [112]. Brain uptake was significantly higher for 100 nm micelles than that with 600 nm size, the reason being the improved transcellular absorption of micelles from nasal epithelium to brain. In contrast, particle size ranging from 100 to 600 nm did not have an effect on the tumor distribution in the brain. Further, it has been shown that PCL can be used to accelerate transnasal transport in collusion with targeting peptides. Cell-penetrating peptide, Tat, modified polyethylene glycol–polycaprolactone (PEG-PCL) copolymeric nanomicelles containing siRNA, or dextran showed improved brain delivery after intranasal delivery compared to intravenous delivery, and these nanomicelles were found to accelerate transport of siRNA in the brain by high transmucosal permeation and subsequent penetration to brain through olfactory and trigeminal nerve pathway [111]. Nanomicelle systems prepared with PEG-PCL modified with Tat peptide have also shown enhanced brain uptake of camptothecin and higher cytotoxicity to tumor cells *in vitro* as compared to nonmodified nanomicelles [113].

PCL has also been used for gene delivery as biodegradable nonviral vector for gene delivery with controlled release properties [110]. Biodegradable poly(ester amine) complexes made of PCL and polyethylenimine for pGL3 gene exhibited controlled degradation pattern with high transfection efficiency without any significant cytotoxicity.

Apart from PCL, some polyester like PHB, PHBV, polyesteramides, other aliphatic copolyesters based drug delivery systems have demonstrated their potential in controlled release of drugs ranging from small organic molecules to large biological molecules. Some literature available on these polymers is described below.

Release pattern of PHBV can be tuned by quenching at room temperature or by treatment at higher temperatures (120°C), which crystallizes the PHBV, leaving less amorphous sites in the nanostructure [29]. The controlled release of ciprofloxacin from PHBV nanoparticles and nanoparticles of modified PHB has also been demonstrated through degradation and generation of the tortuous pathway, which slows the diffusion of drug from nanoparticle core to the diffusion media. Additionally, construction of nanohybrids by using PHBV and clays (silicates) led to an increased degradation rate of polymer matrix. Such processing and formulation parameters can be tuned for modifying the drug release and ultimately the therapeutic outcome of PHB nanocarriers [29]. Poly(esteramide)s are biodegradable cationic polymers developed primarily for biomedical applications [153]. The most widely studied subgroup of these polymers are poly(β-amino esters) (PBAEs). The presence of ester as well as amide linkage imparts it a good degradability and thermomechanical stability. Further, the cationic group endows it advantage of improved cell interaction [154]. PBAEs, due to their positively charged amide bonds, also hold tremendous promise in DNA delivery [155].

Some polyesters, such as poly(dihydropyrans), poly(propylene fumarate), poly(p-dioxanone), polymaleic acid, and so on, are evolving. This necessitates that these new polymers and also polymers like PBH, PBHV and polyestaramides be evaluated for their safety and efficacy extensively in order to expedite their use in nasal drug delivery.

10.6 Future Perspective

In order to augment the applications of polyester nanoparticles, it is considered to be prudent to carry out certain modifications so as to get more control over the release rate, mechanical properties, drug loading, physiologic barrier permeation, cell interaction, etc. Although bulk-modified polymers which generally provide improvement over the mechanical properties such as toughness, Tg, crystallinity, ductility, etc., can be used as such, these modifications are generally performed for biomedical applications rather than for drug delivery. In case of drug delivery, surface modifications are often preferred which lead to an altered way of nanoparticle interaction with the biological environment with improved efficacy.

Polyester-based micelle-like nanoparticles which are also called amphiphilic block copolymers, core–shell nanoparticles, crew-cut micelles, polymersomes, etc., find wide utility in drug delivery systems. They are popular as emerging carriers for poorly water-soluble drugs in their inner core and at the same instance bear hydrophilic surface characteristics to evade scavenging systems. A typical A-B diblock system consists of a hydrophilic shell (A) and a hydrophobic core (B). Although several hydrophobic polymer such as polyester, polystyrene, and poly(methyl methacrylate) find use as hydrophobic core, polyesters are the most preferred due their biodegradability [156]. Zhang *et al.* prepared PCL triblock copolymers, PEG-PCL-PEGs, which showed core–shell structure and particle size below 100 nm owing to their micellar nature. The hydrophobic drug 4′-demethyl-epipodophyllotoxin was easily loaded into the structure. The reduction in crystallinity, as studied by X-ray diffractogram, was proportional to PEG block length. More interesting was the fact that the drug release was easily controlled through variation of copolymer composition due to change in crystallinity and hydrophilicity [157]. Tamboli *et al.* prepared similar pentablock copolymer nanoparticles systems (PLA-PCL-PEG-PCL-PLA) and obtained sustained delivery of steroids. Their hypothesis was based on the fact that use of high molecular weight PCL to sustain the release results in increased crystallinity. Therefore, another polyester (PLA) was incorporated into the core to reduce the crystallinity and hydrophobicity of the existing PCL-PEG-PCL triblock [158]. The pentablock copolymer showed significant reduction in initial burst release after 2 days, i.e., 64% and 34% release from triblock and pentablock copolymer, respectively.

Applications of biodegradable polyesters in triggered drug release systems are also fascinating. Thermoresponsive triggers often utilize non-biodegradable poly(*N*-isopropylacrylamide) [159] and therefore there is need of systems based on biodegradable polymers. Recently, Hong *et al.* reported a dual-responsive, pH and thermal, pentablock copolymer based on PLGA [160]. The polymeric micelles were prepared so as to have a core consisting of PLGA and deprotonated L-hystidine and outer shell of PEG. The micellar system responds to decreasing pH around the pK_a of histidine (~7.0), leading to protonation and subsequent disruption due to hydration and swelling. Similarly, to increasing temperature, the systems respond by progressive dehydration of the PEG shell, leading to disruption of sensitive balance between the PLGA-PEG block. The shell shrinks and allows water to penetrate inside to disrupt the micellar structure. Similar to PLGA, the high crystallinity of PCL has been utilized to prepare thermally triggered systems. The PEG-PCL diblock system with crystalline cores at room temperature having a melting point around 40–45°C has been reported [161]. The system was prepared by ring-opening polymerization of PCL with PEG_{2000}. The system can be triggered by heating above melting point of core. The system can be made to respond to a radio frequency AC magnetic field by incorporating magnetite inside the core, which heats upon magnetic induction.

For intranasal applications, it essential that the delivery system contain mucoadhesive property or *in situ* gelling property so that system can reside at place for longer after administration as spray and provides sustained drug release. Mucoadhesive properties have been reported for PLGA nanoparticles by surface coating with chitosan, trimethyl chitosan (TMC), carbopol [162, 163]. PLGA–TMC nanoparticles prepared had positive zeta potential compared to that of PLGA, which interacts with mucin and resulted in increased immunogenicity of hepatitis B surface antigen depending on size after nasal administration. PEG-coated PLA nanoparticles and chitosan-coated PLGA nanoparticles have been examined for their ability to load protein and transport across mucosa. Findings suggested that intranasal delivery of tetanus toxoid encapsulated nanoparticles resulted in long-lasting antibody response. Additionally, chitosan coating on the PLGA nanoparticles exhibited stability against the lysozyme and increased transport across nasal mucosa [164]. Although PEG coating helps cross nasal mucosa by preventing the aggregation at the nasal mucosa; it is believed that PEG might inhibit the cell surface interactions due to its presence on surface. So, modifying the surface with biorecognitive ligands seems to be a promising approach for targeting. *In situ* gelling systems of PEG-PLGA-PEG triblock have been reported for parenteral administration, which has advantages over poloxamer-based systems, which are non-biodegradable [165].

The abundance of *N*-acetyl-D-glucosamine- and sialic acid-containing glycoproteins in tumor, intestine, and nasal cavity is encouraging for development of targeted delivery of immunogens to NALT [166]. Different lectins have been attached to PLGA nanosystems to promote their transcellular transport [167]. Transcytosis inhibition studies have proved that PEG(3000)-PLA(40000)-based nanoparticles occur through clathrin-mediated mechanism. Several other ligands, including lectins, have been used with polyester-based biodegradable nanosystems like wheat germ agglutinin [131], lactoferrin [168], concanavalin A [169], odorranalectin [170], *Solanum tuberosum* lectin [133], *Ulex europaeus* agglutinin I [171], and cell-penetrating peptides like penetratin [172], low molecular weight protamine [173], etc. Wen *et al.* prepared odorranalectin-conjugated urocortin peptide loaded PEG-PLGA nanoparticles for brain-targeting purpose. Findings suggested that odorranalectin-conjugated nanoparticles increased the brain delivery and therapeutic efficacy of the nanoparticles [170]. It has been reported that wheat germ agglutinin (WGA) functionalized PEG-PLA nanoparticles exhibited negligible nasal ciliatoxicity and about two times higher brain uptake of nanoparticles in the brain than unmodified nanoparticles [131]. Similarly, Xi *et al.* conjugated PEG-PLA nanoparticles with low molecular weight protamine for the brain targeting by nasal delivery. Results revealed that significant accumulation of functionalized NPs into the cells than unmodified NPs by the cellular uptake mechanism of raft-mediated endocytosis as well as direct translocation. Distribution study also suggested significant brain uptake after intranasal administration [173].

Hybrid nanoparticles of lipid-PLGA have been suggested to combine the advantages of both lipid and PLGA. The short circulation half-life of PLGA can be overcome by such hybrid nanoparticles. PLGA-lecithin-PEG nanoparticles for the controlled drug delivery consisted of hydrophobic PLGA core due to biodegradability and high encapsulation ability to hydrophobic drugs, surrounding monolayer of lecithin with PEG shell providing the steric stability and long circulation *in vivo*. The system was found to be better than simple PLGA nanoparticles in terms of stability, loading, and release and cell uptake [174]. In a recent study it was pointed out that cell uptake of such nanoparticles can be modulated by adjusting lipid composition, making it more cationic for efficient uptake [175].

10.7 Conclusion

Polyester nanosystems, available in a range of polymeric forms with varying physicochemical properties, along with their biodegradability, biocompatibility, and controlled degradation behavior, offer a handy tool in the development of novel drug delivery systems. These systems are further being optimized with newer polymer compositions and newer manufacturing technologies to extend the applications beyond that of simple controlled release. Nasal drug delivery systems, in order to utilize the unique benefits of this route, such as local delivery, immunization, and brain delivery, as well as to meet the specific set of formulation requirements, such as mucoadhesion, controlled release, permeation, and cellular delivery, find nanoparticulate systems as the final recourse. Polyester nanosystems have made their position in nasal delivery of drugs, which range from small organic drugs to larger biomolecules like proteins, peptides, and therapeutic genes. Although voluminous progress has been made in the field of nasal drug delivery using nanoparticles, polyester nanosystems are yet evolving. Even though the availability of several polyesters and possibility of synthesizing their copolymers, only few polymers including PLA, PLGA, and PCL have been studied in nasal delivery at a sufficient extent. This infers that the future needs to be directed toward use of polyesters that are yet to be evaluated for their intranasal therapeutic application and finding novel applications such as triggered release, targeting, etc. Also, the application of such systems in nasal delivery of biomolecules like therapeutic genes and proteins is limited, and this needs to addressed in current research. However, one thing is apparent from the growing interest and evolving newer strategies for improving outcomes of polyester nanosystems of different physicochemical and biological properties: polyester nanosystems will take up a firm place in nasal drug delivery.

References

1. Turker, S., Onur, E., and Ozer, Y. (2004). Nasal route and drug delivery systems. *Pharm World Sci,* **26**, 137–142.
2. Pardridge, W. M. (2012). Drug transport across the blood–brain barrier. *J Cereb Blood Flow Metab,* **32**, 1959–1972.
3. Djupesland, P. G., Mahmoud, R. A., and Messina, J. C. (2013). Accessing the brain: the nose may know the way. *J Cereb Blood Flow Metab,* **33**, 793–794.
4. Lochhead, J. J., and Thorne, R. G. (2012). Intranasal delivery of biologics to the central nervous system. *Adv Drug Deliv Rev,* **64**, 614–628.
5. Almeida, A. J., Alpar, H. O., and Brown, M. R. (1993). Immune response to nasal delivery of antigenically intact tetanus toxoid associated with poly(L-lactic acid) microspheres in rats, rabbits and guinea-pigs. *J Pharm Pharmacol,* **45**, 198–203.
6. Illum, L. (2007). Nanoparticulate systems for nasal delivery of drugs: a real improvement over simple systems? *J Pharm Sci,* **96**, 473–483.

7. Debertin, A. S., Tschernig, T., Tonjes, H., Kleemann, W. J., Troger, H. D., and Pabst, R. (2003). Nasal-associated lymphoid tissue (NALT): frequency and localization in young children. *Clin Exp Immunol,* **134**, 503–507.

8. Shive, M. S., and Anderson, J. M. (1997). Biodegradation and biocompatibility of PLA and PLGA microspheres. *Adv Drug Deliv Rev,* **28**, 5–24.

9. Lu, J. M., Wang, X., Marin-Muller, C., Wang, H., Lin, P. H., Yao, Q., and Chen, C. (2009). Current advances in research and clinical applications of PLGA-based nanotechnology. *Expert Rev Mol Diagn,* **9**, 325–341.

10. Moon, S. I., Lee, C. W., Miyamoto, M., and Kimura, Y. (2000). Melt polycondensation of L-lactic acid with Sn(II) catalysts activated by various proton acids: a direct manufacturing route to high molecular weight poly(L-lactic acid). *J Polym Sci Polym Chem,* **38**, 1673–1679.

11. Abiko, A., and Iwahashi, H. (2014). *Method for producing poly-L-lactic acid*. EP2028209B1.

12. Hartmann, M. H. (1998). High molecular weight polylactic acid polymers. In *Biopolymers from Renewable Resources,* Kaplan, D., ed., (Springer Berlin Heidelberg, New York, USA) pp. 367–411.

13. Gruber, P. R., Hall, E. S., Kolstad, J. J., Iwen, M. L., Benson, R. D., and Borchardt, R. L. (1994). *Incorporates removal of water or solvent carrier to concentrate the lactic acid feed followed by polymerization to a low molecular weight prepolymer*. US5357035.

14. Albertsson, A.-C., and Varma, I. K. (2002). Aliphatic polyesters: synthesis, properties and applications. In *Degradable Aliphatic Polyesters,* Albertsson, A.-C., ed., (Springer Berlin Heidelberg, USA) pp. 1–40.

15. Okada, M. (2002). Chemical syntheses of biodegradable polymers. *Prog Polym Sci,* **27**, 87–133.

16. Singh, V., and Tiwari, M. (2010). Structure-processing-property relationship of poly(glycolic acid) for drug delivery systems 1: synthesis and catalysis. *Int J Polym Sci,* **2010**, Article ID 652719, 23 pages.

17. Lewis, D. H. (1990). Controlled release of bioactive agents from lactide/glycolide polymers. In *Biodegradable Polymers as Drug Delivery Systems*, Chasin, M., and Langer, R. eds., (Marcel Dekker, New York) pp. 1–41.

18. Domb, A. J., Kumar, N., Sheskin, T., Bentolila, A., Slager, J., and Teomim, D. (2001). Biodegradable polymers as drug carrier systems. In *Polymeric*

Biomaterials, Revised and Expanded, Dumitriu, S., ed., (CRC Press, USA) p. 94.

19. Kissel, T., Brich, Z., Bantle, S., Lancranjan, I., Nimmerfall, F., and Vit, P. (1991). Parenteral depot-systems on the basis of biodegradable polyesters. *J Control Release,* **16**, 27–41.
20. Vert, M., Mauduit, J., and Li, S. (1994). Biodegradation of PLA/GA polymers: increasing complexity. *Biomaterials,* **15**, 1209–1213.
21. Grizzi, I., Garreau, H., Li, S., and Vert, M. (1995). Hydrolytic degradation of devices based on poly(DL-lactic acid) size-dependence. *Biomaterials,* **16**, 305–311.
22. Park, T. G. (1995). Degradation of poly(lactic-*co*-glycolic acid) microspheres: effect of copolymer composition. *Biomaterials,* **16**, 1123–1130.
23. Hazrati, A. M., Akrawi, S., Hickey, A. J., Wedlund, P., Macdonald, J., and Deluca, P. P. (1989). Tissue distribution of indium-111 labeled poly(glycolic acid) matrices following jugular and hepatic portal vein administration. *J Control Release,* **9**, 205–214.
24. Reusch, R. N., Sparrow, A. W., and Gardiner, J. (1992). Transport of poly-*β*-hydroxybutyrate in human plasma. *Biochim Biophys Acta,* **1123**, 33–40.
25. Saito, T., Tomita, K., Juni, K., and Ooba, K. (1991). *In vivo* and *in vitro* degradation of poly(3-hydroxybutyrate) in pat. *Biomaterials,* **12**, 309–312.
26. Misra, S. K., Valappil, S. P., Roy, I., and Boccaccini, A. R. (2006). Polyhydroxyalkanoate (PHA)/inorganic phase composites for tissue engineering applications. *Biomacromolecules,* **7**, 2249–2258.
27. Bergstrand, A., Andersson, H., Cramby, J., Sott, K., and Larsson, A. (2012). Preparation of porous poly (3-Hydroxybutyrate) films by water-droplet templating. *J Biomater Nanobiotechnol,* **3**, 431.
28. Sabir, M. I., Xu, X., and Li, L. (2009). A review on biodegradable polymeric materials for bone tissue engineering applications. *J Mater Sci,* **44**, 5713–5724.
29. Singh, N. K., Das Purkayastha, B. P., Roy, J. K., Banik, R. M., Gonugunta, P., Misra, M., and Maiti, P. (2011). Tuned biodegradation using poly(hydroxybutyrate-*co*-valerate) nanobiohybrids: emerging biomaterials for tissue engineering and drug delivery. *J Mater Chem,* **21**, 15919–15927.
30. Holmes, P. A. (1988). Biologically produced PHA polymers and copolymers. In *Developments in Crystalline Polymers,* Basset, D. C., ed., (Elsevier, London) pp. 1–65.

31. Marchessault, R. H., and Yu, G.-E. (2005). Crystallization and material properties of polyhydroxyalkanoates PHAs. In *Biopolymers Online*, (Wiley-VCH Verlag GmbH & Co. KGaA) 3b.

32. Brandl, H., Gross, R. A., Lenz, R. W., and Fuller, R. C. (1988). Pseudomonas oleovorans as a source of poly (β-hydroxyalkanoates) for potential applications as biodegradable polyesters. *Appl Environ Microbiol,* **54**, 1977–1982.

33. Pitt, C. G., Gratzl, M. M., Kimmel, G. L., Surles, J., and Schindler, A. (1981). Aliphatic polyesters II. The degradation of poly (DL-lactide), poly (ε-caprolactone), and their copolymers *in vivo*. *Biomaterials,* **2**, 215–220.

34. Woodward, S. C., Brewer, P. S., Moatamed, F., Schindler, A., and Pitt, C. G. (1985). The intracellular degradation of poly(ε-caprolactone). *J Biomed Mater Res,* **19**, 437–444.

35. Reiss, G., Hurtrez, C., and Bahadur, P. (1985). Block copolymers. In *Encyclopedia of Polymer Science and Engineering,* Mark, H. F., Bikales, N. M., Overberger, C. G., Menges, G., and Kroschwitz, J. I., eds., (John Wiley and Sons, New York, USA) p. 398.

36. Song, C. X., Sun, H. F., and Feng, X. D. (1987). Microspheres of biodegradable block copolymer for long-acting controlled delivery of contraceptives. *Polym J,* **19**, 485–491.

37. Fujimaki, T. (1998). Processability and properties of aliphatic polyesters, 'BIONOLLE', synthesized by polycondensation reaction. *Polym Degrad Stabil,* **59**, 209–214.

38. Ishioka, R., Kitakuni, E., and Ichikawa, Y. (2005). Aliphatic polyesters: "Bionolle". In *Biopolymers Online,* Doi, Y., and Steinbüchel, A., eds., (Wiley-VCH Verlag GmbH & Co. KGaA, Weinheim) p. 4.

39. Muller, R. J. Witt, U., Rantze, E., and Deckwer, W. D. (1998). Architecture of biodegradable copolyestyers containing aromatic constituents, *Polym Degrad Stab*. **59**, 203–208.

40. Yokota, Y., and Marechal, H. (1999). Processability of biodegradable poly(butylene) succinate and its derivates. A case study. *Biopolym Conf,* Wurzburg, Germany.

41. DiBenedetto, L. J., and Huang, S. J. (1988). Biodegradable hydroxylated polymers as controlled release agents. *Polym Mat Sci Eng,* **59**, 812–816.

42. Grigat, E., Koch, R., and Timmermann, R. (1998). BAR 1095 and BAK 2195: completely biodegradable synthetic thermoplastics. *Polym Degrad Stabil,* **59**, 223–226.

43. Steinbuchel, A., and Doi, Y. (2002). *Biopolymers, Volume 4: Polyesters III- Applications and Commercial products*, (Wiley-VCH: Weinheim, Germany) p. 398.

44. Domb, A., Amselem, S., Shah, J., and Maniar, M. (1992). Degradable polymers for site-specific drug delivery. *Polym Adv Technol,* **3**, 279–292.

45. Kulve, H. (2014). Anticipating market introduction of nanotechnology-enabled drug delivery systems. In *Application of Nanotechnology in Drug Delivery*, Sezer, A.D., ed., (InTech) pp. 501–524.

46. Prabha, L. S., Nanthini, R., and Prabhu, K. (2012). Design, synthesis and characterisation of novel biodegradable aliphatic copolyesters-poly (ethylene sebacate-*co*-butylene succinate) and poly (ethylene sebacate-*co*-butylene adipate). *Oriental J Chem,* **28**, 1659–1671.

47. Södergård, A., and Stolt, M. (2002). Properties of lactic acid based polymers and their correlation with composition. *Prog Polym Sci,* **27**, 1123–1163.

48. Sudesh, K., Abe, H., and Doi, Y. (2000). Synthesis, structure and properties of polyhydroxyalkanoates: biological polyesters. *Prog Polym Sci,* **25**, 1503–1555.

49. El-Aasser, M. S., Ugelstad, J., and Vanderhoff, J. W. (1979). *Polymer emulsification process*. US4177177.

50. Rao, J. P., and Geckeler, K. E. (2011). Polymer nanoparticles: preparation techniques and size-control parameters. *Prog Polym Sci,* **36**, 887–913.

51. Gurny, R., Peppas, N., Harrington, D., and Banker, G. (1981). Development of biodegradable and injectable latices for controlled release of potent drugs. *Drug Develop Ind Pharm,* **7**, 1–25.

52. Hu, S., and Zhang, Y. (2010). Endostar-loaded PEG-PLGA nanoparticles: *in vitro* and *in vivo* evaluation. *Int J Nanomedicine,* **5**, 1039–1048.

53. Vila, A., Gill, H., Mccallion, O., and Alonso, M. J. (2004). Transport of PLA-PEG particles across the nasal mucosa: effect of particle size and PEG coating density. *J Control Release,* **98**, 231–244.

54. Vila, A., Sanchez, A., Evora, C., Soriano, I., Mccallion, O., and Alonso, M. J. (2005). PLA-PEG particles as nasal protein carriers: the influence of the particle size. *Int J Pharm,* **292**, 43–52.

55. McCall, R. L., and Sirianni, R. W. (2013). PLGA nanoparticles formed by single- or double-emulsion with vitamin E-TPGS. *J. Vis. Exp.,* e51015.

56. Musyanovych, A., and Landfester, K. (2014). Polymer micro- and nanocapsules as biological carriers with multifunctional properties. *Macromol Biosci*, **14**, 458–477.

57. Musyanovych, A., Schmitz-Wienke, J., Mailander, V., Walther, P., and Landfester, K. (2008). Preparation of biodegradable polymer nanoparticles by miniemulsion technique and their cell interactions. *Macromol Biosci,* **8**, 127–139.

58. Niwa, T., Takeuchi, H., Hino, T., Kunou, N., and Kawashima, Y. (1993). Preparations of biodegradable nanospheres of water-soluble and insoluble drugs with D,L-lactide/glycolide copolymer by a novel spontaneous emulsification solvent diffusion method, and the drug release behavior. *J Control Release,* **25**, 89–98.

59. Leroux, J.-C., Allemann, E., Doelker, E., and Gurny, R. (1995). New approach for the preparation of nanoparticles by an emulsification-diffusion method. *Eur J Pharm Biopharm,* **41**, 14–18.

60. Hariharan, S., Bhardwaj, V., Bala, I., Sitterberg, J., Bakowsky, U., and Ravikumar, M. N. V. (2006). Design of estradiol loaded PLGA nanoparticulate formulations: a potential oral delivery system for hormone therapy. *Pharm Res,* **23**, 184–195.

61. Hassou, M., Couenne, F., Le Gorrec, Y., and Tayakout, M. (2009). Modeling and simulation of polymeric nanocapsule formation by emulsion diffusion method. *AIChE J,* **55**, 2094–2105.

62. Khemani, M., Sharon, M., and Sharon, M. (2012). Encapsulation of berberine in nano-sized PLGA synthesized by emulsification method. *ISRN Nanotechnol,* **2012**, Article ID 187354, 9 pages.

63. Messai, I., and Delair, T. (2005). Adsorption of chitosan onto poly (D, L-lactic acid) particles: a physico-chemical investigation. *Macromol Chem Phys,* **206**, 1665–1674.

64. Trimaille, T., Pichot, C., Elaïssari, A., Fessi, H., Briançon, S., and Delair, T. (2003). Poly(D,L-lactic acid) nanoparticle preparation and colloidal characterization. *Colloid Polym Sci,* **281**, 1184–1190.

65. Pinto Reis, C., Neufeld, R. J., Ribeiro, A. J., and Veiga, F. (2006). Nanoencapsulation I. Methods for preparation of drug-loaded polymeric nanoparticles. *Nanomedicine,* **2**, 8–21.

66. Bindschaedler, C., Doelker, E., and Gurny, R. (1990). *Process for preparing a powder of water-insoluble polymer which can be redispersed in a liquid phase, the resulting powder and utilization thereof.* US4968350.

67. Konan, Y. N., Gurny, R., and Allémann, E. (2002). Preparation and characterization of sterile and freeze-dried sub-200 nm nanoparticles. *Int J Pharm,* **233**, 239–252.

68. Ibrahim, H., Bindschaedler, C., Doelker, E., Buri, P., and Gurny, R. (1992). Aqueous nanodispersions prepared by a salting-out process. *Int J Pharm,* **87**, 239–246.

69. Jung, T., Kamm, W., Breitenbach, A., Kaiserling, E., Xiao, J. X., and Kissel, T. (2000). Biodegradable nanoparticles for oral delivery of peptides: is there a role for polymers to affect mucosal uptake? *Eur J Pharm Biopharm,* **50**, 147–160.

70. Couvreur, P., Dubernet, C., and Puisieux, F. (1995). Controlled drug delivery with nanoparticles: current possibilities and future trends. *Eur J Pharm Biopharm,* **41**, 2–13.

71. McCarron, P. A., Donnelly, R. F., and Marouf, W. (2006). Celecoxib-loaded poly(D,L-lactide-*co*-glycolide) nanoparticles prepared using a novel and controllable combination of diffusion and emulsification steps as part of the salting-out procedure. *J Microencapsul,* **23**, 480–498.

72. Song, X., Zhao, Y., Wu, W., Bi, Y., Cai, Z., Chen, Q., Li, Y., and Hou, S. (2008). PLGA nanoparticles simultaneously loaded with vincristine sulfate and verapamil hydrochloride: Systematic study of particle size and drug entrapment efficiency. *Int J Pharm,* **350**, 320–329.

73. Song, X. R., Cai, Z., Zheng, Y., He, G., Cui, F. Y., Gong, D. Q., Hou, S. X., Xiong, S. J., Lei, X. J., and Wei, Y. Q. (2009). Reversion of multidrug resistance by co-encapsulation of vincristine and verapamil in PLGA nanoparticles. *Eur J Pharm Sci,* **37**, 300–305.

74. Barichello, J. M., Morishita, M., Takayama, K., and Nagai, T. (1999). Encapsulation of hydrophilic and lipophilic drugs in plga nanoparticles by the nanoprecipitation method. *Drug Dev Ind Pharm,* **25**, 471–476.

75. Galindo-Rodriguez, S., Allemann, E., Fessi, H., and Doelker, E. (2004). Physicochemical parameters associated with nanoparticle formation in the salting-out, emulsification-diffusion, and nanoprecipitation methods. *Pharm Res,* **21**, 1428–1439.

76. Fessi, H., Puisieux, F., Devissaguet, J. P., Ammoury, N., and Benita, S. (1989). Nanocapsule formation by interfacial polymer deposition following solvent displacement. *Int J Pharm,* **55**, R1–R4.

77. Moinard-Chécot, D., Chevalier, Y., Briançon, S., Beney, L., and Fessi, H. (2008). Mechanism of nanocapsules formation by the emulsion–diffusion process. *J Colloid Interface Sci,* **317**, 458–468.

78. Seju, U., Kumar, A., and Sawant, K. (2011). Development and evaluation of olanzapine-loaded PLGA nanoparticles for nose-to-brain delivery: *in vitro* and *in vivo* studies. *Acta Biomater,* **7**, 4169–4176.

79. Nanthakasri, W., Srisa-Ard, M., and Baimark, Y. (2011). Biodegradable blend nanoparticles of amphiphilic diblock copolymers prepared by nano-precipitation method. *J Biomater Nanobiotechnol,* **2**, 561.

80. Jayanta, K. (2012). Critical process parameters evaluation of modified nanoprecipitation method on lomustine nanoparticles and cytostatic activity study on L132 human cancer cell line. *J Nanomed Nanotechnol,* **3**, 149.

81. Kara, A., Ozturk, N., Sarisozen, C., and Vural, I. (2014). Investigation of Formulation Parameters of PLGA Nanoparticles Prepared By Nanoprecipitation Technique. *Proc 5th Int Conf Nanotechnol: Fundam Appl,* 941–943.

82. Venier-Julienne, M., and Benoit, J. (1996). Preparation, purification and morphology of polymeric nanoparticles as drug carriers. *Pharm Acta Helv,* **71**, 121–128.

83. Zambaux, M., Bonneaux, F., Gref, R., Maincent, P., Dellacherie, E., Alonso, M., Labrude, P., and Vigneron, C. (1998). Influence of experimental parameters on the characteristics of poly (lactic acid) nanoparticles prepared by a double emulsion method. *J Control Release,* **50**, 31–40.

84. Mainardes, R. M., and Evangelista, R. C. (2005). Praziquantel-loaded PLGA nanoparticles: preparation and characterization. *J Microencapsul,* **22**, 13–24.

85. Tobio, M., Gref, R., Sanchez, A., Langer, R., and Alonso, M. (1998). Stealth PLA-PEG nanoparticles as protein carriers for nasal administration. *Pharm Res,* **15**, 270–275.

86. Zhang, Q.-Z., Zha, L.-S., Zhang, Y., Jiang, W.-M., Lu, W., Shi, Z.-Q., Jiang, X.-G., and Fu, S.-K. (2006). The brain targeting efficiency following nasally applied MPEG-PLA nanoparticles in rats. *J Drug Target,* **14**, 281–290.

87. Yoo, H. S., Oh, J. E., Lee, K. H., and Park, T. G. (1999). Biodegradable nanoparticles containing doxorubicin-PLGA conjugate for sustained release. *Pharm Res,* **16**, 1114–1118.

88. Perez, C., Sanchez, A., Putnam, D., Ting, D., Langer, R., and Alonso, M. J. (2001). Poly(lactic acid)-poly(ethylene glycol) nanoparticles as new carriers for the delivery of plasmid DNA. *J Control Release,* **75**, 211–224.

89. Csaba, N., Sanchez, A., and Alonso, M. J. (2006). PLGA: poloxamer and PLGA: poloxamine blend nanostructures as carriers for nasal gene delivery. *J Control Release,* **113**, 164–172.

90. Zweers, M. L. T., Engbers, G. H. M., Grijpma, D. W., and Feijen, J. (2006). Release of anti-restenosis drugs from poly(ethylene oxide)-poly(D,L-lactic-*co*-glycolic acid) nanoparticles. *J Control Release,* **114**, 317–324.

91. Fan, Y. F., Wang, Y. N., Fan, Y. G., and Ma, J. B. (2006). Preparation of insulin nanoparticles and their encapsulation with biodegradable

polyelectrolytes via the layer-by-layer adsorption. *Int J Pharm,* **324**, 158–167.

92. Deepak, V., Pandian, S., Kalishwaralal, K., and Gurunathan, S. (2009). Purification, immobilization, and characterization of nattokinase on PHB nanoparticles. *Bioresour Technol,* **100**, 6644–6646.

93. Jung, T., Kamm, W., Breitenbach, A., Hungerer, K.-D., Hundt, E., and Kissel, T. (2001). Tetanus toxoid loaded nanoparticles from sulfobutylated poly (vinyl alcohol)-graft-poly (lactide-*co*-glycolide): evaluation of antibody response after oral and nasal application in mice. *Pharm Res,* **18**, 352–360.

94. Liu, M., Zhou, Z., Wang, X., Xu, J., Yang, K., Cui, Q., Chen, X., Cao, M., Weng, J., and Zhang, Q. (2007). Formation of poly(L,D-lactide) spheres with controlled size by direct dialysis. *Polymer,* **48**, 5767–5779.

95. Sheikh, F. A., Barakat, N. A., Kanjwal, M. A., Aryal, S., Khil, M. S., and Kim, H. Y. (2009). Novel self-assembled amphiphilic poly(ε-caprolactone)-grafted-poly(vinyl alcohol) nanoparticles: hydrophobic and hydrophilic drugs carrier nanoparticles. *J Mater Sci Mater Med,* **20**, 821–831.

96. Imran Ul-Haq, M., Chasovskikh, E., and Signorell, R. (2010). Phase behavior of ketoprofen–poly(lactic acid) drug particles formed by rapid expansion of supercritical solutions. *Langmuir,* **26**, 14951–14957.

97. Gadermann, M., Kular, S., Al-Marzouqi, A. H., and Signorell, R. (2009). Formation of naproxen-polylactic acid nanoparticles from supercritical solutions and their characterization in the aerosol phase. *Phys Chem Chem Phys,* **11**, 7861–7868.

98. Asandei, A. D., Erkey, C., Burgess, D. J., Saquing, C., Saha, G., and Zolnik, B. S. (2005). Preparation of drug delivery biodegradable plga nanocomposites and foams by supercritical cosub2 expanded ring opening polymerization and by rapid expansion from CHCIF2 supercritical solutions. *Mater Res Soc Symp Proc,* **845**, 243–248.

99. Campardelli, R., Adami, R., Della Porta, G., and Reverchon, E. (2012). Nanoparticle precipitation by supercritical assisted injection in a liquid antisolvent. *Chem Engg J,* **192**, 246–251.

100. Lee, L. Y., Smith, K. A., and Wang, C.-H. (2006). Fabrication of micro and nanoparticles of paclitaxel-loaded poly L lactide for controlled release using supercritical antisolvent method: effects of thermodynamics and hydrodynamics. http://hdl.handle.net/1721.1/30387

101. Jeon, H.-J., Jeong, Y.-I., Jang, M.-K., Park, Y.-H., and Nah, J.-W. (2000). Effect of solvent on the preparation of surfactant-free poly (DL-lactide-

co-glycolide) nanoparticles and norfloxacin release characteristics. *Int J Pharm,* **207**, 99–108.

102. Kostag, M., Köhler, S., Liebert, T., and Heinze, T. (2010). Pure cellulose nanoparticles from trimethylsilyl cellulose. *Macromol Symposia,* **294**, 96–106.

103. Akagi, T., Kaneko, T., Kida, T., and Akashi, M. (2005). Preparation and characterization of biodegradable nanoparticles based on poly(gamma-glutamic acid) with L-phenylalanine as a protein carrier. *J Control Release,* **108**, 226–236.

104. Shakeri, F., Shakeri, S., and Hojjatoleslami, M. (2014). Preparation and characterization of carvacrol loaded polyhydroxybutyrate nanoparticles by nanoprecipitation and dialysis methods. *J Food Sci,* **79**, N697–N705.

105. Hutchenson, K. W. (2002). Organic chemical reactions and catalysis in supercritical fluid media. In *Supercritical Fluid Technology in Materials Science and Engineering,* Sun, Y. P., ed., (Marcel Dekker, New York) pp. 87–188.

105. Moon, S. I., Lee, C. W., Taniguchi, I., Miyamoto, M., and Kimura, Y. (2001). Melt/solid polycondensation of L-lactic acid: an alternative route to poly(L-lactic acid) with high molecular weight. *Polymer,* **42**, 5059–5062.

106. Thakur, R., and Gupta, R. B. (2006). Formation of phenytoin nanoparticles using rapid expansion of supercritical solution with solid cosolvent (RESS-SC) process. *Int J Pharm,* **308**, 190–199.

107. Patomchaiviwat, V., Paeratakul, O., and Kulvanich, P. (2008). Formation of inhalable rifampicin–poly(L-lactide) microparticles by supercritical anti-solvent process. *AAPS PharmSciTech,* **9**, 1119–1129.

108. Zhan, S., Chen, C., Zhao, Q., Wang, W., and Liu, Z. (2013). Preparation of 5-Fu-loaded PLLA microparticles by supercritical fluid technology. *Ind Eng Chem Res,* **52**, 2852–2857.

109. Su, W. P., Cheng, F. Y., Shieh, D. B., Yeh, C. S., and Su, W. C. (2012). PLGA nanoparticles codeliver paclitaxel and Stat3 siRNA to overcome cellular resistance in lung cancer cells. *Int J Nanomedicine,* **7**, 4269–4283.

110. Choi, M.-K., Arote, R., Kim, S.-Y., Chung, S.-J., Shim, C.-K., Cho, C.-S., and Kim, D.-D. (2007). Transfection of primary human nasal epithelial cells using a biodegradable poly (ester amine) based on polycaprolactone and polyethylenimine as a gene carrier. *J Drug Target,* **15**, 684–690.

111. Kanazawa, T., Akiyama, F., Kakizaki, S., Takashima, Y., and Seta, Y. (2013). Delivery of siRNA to the brain using a combination of nose-to-brain delivery and cell-penetrating peptide-modified nano-micelles. *Biomaterials,* **34**, 9220–9226.

112. Kanazawa, T., Taki, H., Tanaka, K., Takashima, Y., and Okada, H. (2011). Cell-penetrating peptide-modified block copolymer micelles promote direct brain delivery via intranasal administration. *Pharm Res,* **28**, 2130–2139.

113. Taki, H., Kanazawa, T., Akiyama, F., Takashima, Y., and Okada, H. (2012). Intranasal delivery of camptothecin-loaded tat-modified nanomicells for treatment of intracranial brain tumors. *Pharmaceuticals,* **5**, 1092–1102.

114. Fredenberg, S., Wahlgren, M., Reslow, M., and Axelsson, A. (2011). The mechanisms of drug release in poly(lactic-*co*-glycolic acid)-based drug delivery systems—a review. *Int J Pharm,* **415**, 34–52.

115. Makadia, H. K., and Siegel, S. J. (2011). Poly lactic-*co*-glycolic acid (PLGA) as biodegradable controlled drug delivery carrier. *Polymers,* **3**, 1377–1397.

116. Kaur, M. A. S. (2008). *Poly(D,L-lactide-co-glycolide) 50:50-hydrophilic Polymer Blends: Hydrolysis, Bioadhesion and Drug Release Characterization,* (University of Iowa) publication number 3323434.

117. Alexis, F. (2005). Factors affecting the degradation and drug-release mechanism of poly(lactic acid) and poly[(lactic acid)-*co*-(glycolic acid)]. *Polym Int,* **54**, 36–46.

118. Ford Versypt, A. N., Pack, D. W., and Braatz, R. D. (2013). Mathematical modeling of drug delivery from autocatalytically degradable PLGA microspheres—a review. *J Control Release,* **165**, 29–37.

119. Ramalingam, M., Vallittu, P., Ripamonti, U., and Li, W. J., eds. (2012). *Tissue Engineering and Regenerative Medicine: A Nano Approach,* (Taylor & Francis).

120. Mark, J. E., ed. (2007). *Physical Properties of Polymers Handbook,* (Springer New York).

121. Miller, R. A., Brady, J. M., and Cutright, D. E. (1977). Degradation rates of oral resorbable implants (polylactates and polyglycolates): rate modification with changes in PLA/PGA copolymer ratios. *J Biomed Mater Res,* **11**, 711–719.

122. Buchanan, F., (ed.), Institute of Materials, Minerals, and Mining (2008). *Degradation Rate of Bioresorbable Materials: Prediction and Evaluation,* (Woodhead Publishing Limited, Cambridge, England).

123. Reis, R. L., and San Román, J., eds. (2004) *Biodegradable Systems in Tissue Engineering and Regenerative Medicine,* (CRC Press).

124. Steele, T. W. J., Huang, C. L., Kumar, S., Iskandar, A., Baoxin, A., Chiang Boey, F. Y., Loo, J. S. C., and Venkatraman, S. S. (2013). Tuning drug release in polyester thin films: terminal end-groups determine specific rates of additive-free controlled drug release. *NPG Asia Mater,* **5**, e46.

125. Berchane, N. S., Carson, K. H., Rice-Ficht, A. C., and Andrews, M. J. (2007). Effect of mean diameter and polydispersity of PLG microspheres on drug release: experiment and theory. *Int J Pharm,* **337**, 118–126.

126. Sansdrap, P., and Moës, A. J. (1997). *In vitro* evaluation of the hydrolytic degradation of dispersed and aggregated poly(D,L-lactide-*co*-glycolide) microspheres. *J Control Release,* **43**, 47–58.

127. Fu, Y., and Kao, W. J. (2010). Drug release kinetics and transport mechanisms of non-degradable and degradable polymeric delivery systems. *Expert Opin Drug Deliv,* **7**, 429–444.

128. Okada, H. (1997). One- and three-month release injectable microspheres of the LH-RH superagonist leuprorelin acetate. *Adv Drug Deliv Rev,* **28**, 43–70.

129. Su, Z. X., Shi, Y. N., Teng, L. S., Li, X., Wang, L. X., Meng, Q. F., Teng, L. R., and Li, Y. X. (2011). Biodegradable poly(D,L-lactide-*co*-glycolide) (PLGA) microspheres for sustained release of risperidone: Zero-order release formulation. *Pharm Dev Technol,* **16**, 377–384.

130. Matsumoto, A., Matsukawa, Y., Horikiri, Y., and Suzuki, T. (2006). Rupture and drug release characteristics of multi-reservoir type microspheres with poly(DL-lactide-*co*-glycolide) and poly(DL-lactide). *Int J Pharm,* **327**, 110–116.

131. Gao, X., Tao, W., Lu, W., Zhang, Q., Zhang, Y., Jiang, X., and Fu, S. (2006). Lectin-conjugated PEG-PLA nanoparticles: preparation and brain delivery after intranasal administration. *Biomaterials,* **27**, 3482–3490.

132. Liu, Q., Shao, X., Chen, J., Shen, Y., Feng, C., Gao, X., Zhao, Y., Li, J., Zhang, Q., and Jiang, X. (2011). *In vivo* toxicity and immunogenicity of wheat germ agglutinin conjugated poly(ethylene glycol)-poly(lactic acid) nanoparticles for intranasal delivery to the brain. *Toxicol Appl Pharmacol,* **251**, 79–84.

133. Chen, J., Zhang, C., Liu, Q., Shao, X., Feng, C., Shen, Y., Zhang, Q., and Jiang, X. (2012). Solanum tuberosum lectin-conjugated PLGA nanoparticles for nose-to-brain delivery: *in vivo* and *in vitro* evaluations. *J Drug Target,* **20**, 174–184.

134. Gao, X., Wu, B., Zhang, Q., Chen, J., Zhu, J., Zhang, W., Rong, Z., Chen, H., and Jiang, X. (2007). Brain delivery of vasoactive intestinal peptide enhanced with the nanoparticles conjugated with wheat germ agglutinin following intranasal administration. *J Control Release*, **121**, 156–167.

135. Zhang, Q. Z., Zha, L. S., Zhang, Y., Jiang, W. M., Lu, W., Shi, Z. Q., Jiang, X. G., and Fu, S. K. (2006). The brain targeting efficiency following nasally applied MPEG-PLA nanoparticles in rats. *J Drug Target*, **14**, 281–290.

136. Yan, L., Wang, H., Jiang, Y., Liu, J., Wang, Z., Yang, Y., Huang, S., and Huang, Y. (2013). Cell-penetrating peptide-modified PLGA nanoparticles for enhanced nose-to-brain macromolecular delivery. *Macromol Res*, **21**, 435–441.

137. Jung, T., Kamm, W., Breitenbach, A., Hungerer, K. D., Hundt, E., and Kissel, T. (2001). Tetanus toxoid loaded nanoparticles from sulfobutylated poly(vinyl alcohol)-graft-poly(lactide-*co*-glycolide): evaluation of antibody response after oral and nasal application in mice. *Pharm Res*, **18**, 352–360.

138. Jaganathan, K. S., and Vyas, S. P. (2006). Strong systemic and mucosal immune responses to surface-modified PLGA microspheres containing recombinant hepatitis B antigen administered intranasally. *Vaccine*, **24**, 4201–4211.

139. Slutter, B., Bal, S., Keijzer, C., Mallants, R., Hagenaars, N., Que, I., Kaijzel, E., Van Eden, W., Augustijns, P., Lowik, C., Bouwstra, J., Broere, F., and Jiskoot, W. (2010). Nasal vaccination with N-trimethyl chitosan and PLGA based nanoparticles: nanoparticle characteristics determine quality and strength of the antibody response in mice against the encapsulated antigen. *Vaccine*, **28**, 6282–6291.

140. Alex, A. T., Joseph, A., Shavi, G., Rao, J. V., and Udupa, N. (2014). Development and evaluation of carboplatin-loaded PCL nanoparticles for intranasal delivery. *Drug Deliv*, 1–10.

141. Baras, B., Benoit, M. A., Dupre, L., Poulain-Godefroy, O., Schacht, A. M., Capron, A., Gillard, J., and Riveau, G. (1999). Single-dose mucosal immunization with biodegradable microparticles containing a Schistosoma mansoni antigen. *Infect Immun*, **67**, 2643–2648.

142. Yan, C., Rill, W. L., Malli, R., Hewetson, J., Naseem, H., Tammariello, R., and Kende, M. (1996). Intranasal stimulation of long-lasting immunity against aerosol ricin challenge with ricin toxoid vaccine encapsulated in polymeric microspheres. *Vaccine*, **14**, 1031–1038.

143. Vila, A., Sanchez, A., Evora, C., Soriano, I., Vila Jato, J. L., and Alonso, M. J. (2004). PEG-PLA nanoparticles as carriers for nasal vaccine delivery. *J Aerosol Med*, **17**, 174–185.

144. Vila, A., Sanchez, A., Janes, K., Behrens, I., Kissel, T., Vila Jato, J. L., and Alonso, M. J. (2004). Low molecular weight chitosan nanoparticles as new carriers for nasal vaccine delivery in mice. *Eur J Pharm Biopharm,* **57**, 123–131.

145. Tobio, M., Schwendeman, S. P., Guo, Y., Mciver, J., Langer, R., and Alonso, M. J. (1999). Improved immunogenicity of a core-coated tetanus toxoid delivery system. *Vaccine,* **18**, 618–622.

146. Foged, C., Brodin, B., Frokjaer, S., and Sundblad, A. (2005). Particle size and surface charge affect particle uptake by human dendritic cells in an *in vitro* model. *Int J Pharm,* **298**, 315–322.

147. Kanchan, V., and Panda, A. K. (2007). Interactions of antigen-loaded polylactide particles with macrophages and their correlation with the immune response. *Biomaterials,* **28**, 5344–5357.

148. Berger, G., Kogan, T., Skutelsky, E., and Ophir, D. (2005). Glycoconjugate expression in normal human inferior turbinate mucosa: a lectin histochemical study. *Am J Rhinol,* **19**, 97–103.

149. Csaba, N., Caamaño, P., Sánchez, A., Domínguez, F., and Alonso, M. J. (2004). PLGA: poloxamer and PLGA: poloxamine blend nanoparticles: new carriers for gene delivery. *Biomacromolecules,* **6**, 271–278.

150. Zou, W., Liu, C., Chen, Z., and Zhang, N. (2009). Preparation and characterization of cationic PLA-PEG nanoparticles for delivery of plasmid DNA. *Nanoscale Res Lett,* **4**, 982–992.

151. Elamanchili, P., Diwan, M., Cao, M., and Samuel, J. (2004). Characterization of poly(D,L-lactic-*co*-glycolic acid) based nanoparticulate system for enhanced delivery of antigens to dendritic cells. *Vaccine,* **22**, 2406–2412.

152. Gunatillake, P., Mayadunne, R., and Adhikari, R. (2006). Recent developments in biodegradable synthetic polymers. *Biotechnol Annu Rev,* **12**, 301–347.

153. Barrera, D. A., Zylstra, E., Lansbury, P. T., and Langer, R. (1993). Synthesis and RGD peptide modification of a new biodegradable copolymer: poly(lactic acid-*co*-lysine). *J Am Chem Soc,* **115**, 11010–11011.

154. Fonseca, A. C., Gil, M. H., and Simões, P. N. (2014). Biodegradable poly(ester amide)s—a remarkable opportunity for the biomedical area: Review on the synthesis, characterization and applications. *Progress Polym Sci,* **39**, 1291–1311.

155. Jere, D., Yoo, M. K., Arote, R., Kim, T. H., Cho, M. H., Nah, J. W., Choi, Y. J., and Cho, C. S. (2008). Poly (amino ester) composed of poly (ethylene glycol) and aminosilane prepared by combinatorial chemistry as a gene carrier. *Pharm Res,* **25**, 875–885.

156. Jones, M.-C., and Leroux, J.-C. (1999). Polymeric micelles—a new generation of colloidal drug carriers. *Eur J Pharm Biopharm,* **48**, 101–111.

157. Zhang, Y., and Zhuo, R. X. (2005). Synthesis and *in vitro* drug release behavior of amphiphilic triblock copolymer nanoparticles based on poly (ethylene glycol) and polycaprolactone. *Biomaterials,* **26**, 6736–6742.

158. Tamboli, V., Mishra, G. P., and Mitra, A. K. (2013). Novel pentablock copolymer (PLA-PCL-PEG-PCL-PLA) based nanoparticles for controlled drug delivery: effect of copolymer compositions on the crystallinity of copolymers and *in vitro* drug release profile from nanoparticles. *Colloid Polym Sci,* **291**, 1235–1245.

159. Wei, H., Cheng, S.-X., Zhang, X.-Z., and Zhuo, R.-X. (2009). Thermo-sensitive polymeric micelles based on poly(N-isopropylacrylamide) as drug carriers. *Prog Polym Sci,* **34**, 893–910.

160. Hong, W., Chen, D., Jia, L., Gu, J., Hu, H., Zhao, X., and Qiao, M. (2014). Thermo- and pH-responsive copolymers based on PLGA-PEG-PLGA and poly(L-histidine): synthesis and *in vitro* characterization of copolymer micelles. *Acta Biomater,* **10**, 1259–1271.

161. Glover, A. L., Nikles, S. M., Nikles, J. A., Brazel, C. S., and Nikles, D. E. (2012). Polymer micelles with crystalline cores for thermally triggered release. *Langmuir,* **28**, 10653–10660.

162. Krishnakumar, D., Kalaiyarasi, D., Bose, J. C., and Jaganathan, K. S. (2012). Evaluation of mucoadhesive nanoparticle based nasal vaccine. *J Pharm Invest,* **42**, 315–326.

163. Zou, W., Liu, C., Chen, Z., and Zhang, N. (2009). Studies on bioadhesive PLGA nanoparticles: a promising gene delivery system for efficient gene therapy to lung cancer. *Int J Pharm,* **370**, 187–195.

164. Vila, A., Sánchez, A., Tobío, M., Calvo, P., and Alonso, M. J. (2002). Design of biodegradable particles for protein delivery. *J Control Release,* **78**, 15–24.

165. Jeong, B., Bae, Y. H., and Kim, S. W. (2000). *In situ* gelation of PEG-PLGA-PEG triblock copolymer aqueous solutions and degradation thereof. *J Biomed Mater Res,* **50**, 171–177.

166. Giannasca, P. J., Boden, J. A., and Monath, T. P. (1997). Targeted delivery of antigen to hamster nasal lymphoid tissue with M-cell-directed lectins. *Infect Immun,* **65**, 4288–4298.

167. Song, Q., Yao, L., Huang, M., Hu, Q., Lu, Q., Wu, B., Qi, H., Rong, Z., Jiang, X., Gao, X., Chen, J., and Chen, H. (2012). Mechanisms of transcellular

transport of wheat germ agglutinin-functionalized polymeric nanoparticles in Caco-2 cells. *Biomaterials,* **33**, 6769–6782.

168. Liu, Z., Jiang, M., Kang, T., Miao, D., Gu, G., Song, Q., Yao, L., Hu, Q., Tu, Y., Pang, Z., Chen, H., Jiang, X., Gao, X., and Chen, J. (2013). Lactoferrin-modified PEG-*co*-PCL nanoparticles for enhanced brain delivery of NAP peptide following intranasal administration. *Biomaterials,* **34**, 3870–3881.

169. Shao, X., Liu, Q., Zhang, C., Zheng, X., Chen, J., Zha, Y., Qian, Y., Zhang, X., Zhang, Q., and Jiang, X. (2013). Concanavalin A-conjugated poly(ethylene glycol)-poly(lactic acid) nanoparticles for intranasal drug delivery to the cervical lymph nodes. *J Microencapsul,* **30**, 780–786.

170. Wen, Z., Yan, Z., Hu, K., Pang, Z., Cheng, X., Guo, L., Zhang, Q., Jiang, X., Fang, L., and Lai, R. (2011). Odorranalectin-conjugated nanoparticles: preparation, brain delivery and pharmacodynamic study on Parkinson's disease following intranasal administration. *J Control Release,* **151**, 131–138.

171. Gao, X., Chen, J., Tao, W., Zhu, J., Zhang, Q., Chen, H., and Jiang, X. (2007). UEA I-bearing nanoparticles for brain delivery following intranasal administration. *Int J Pharm,* **340**, 207–215.

172. Xia, H., Gao, X., Gu, G., Liu, Z., Hu, Q., Tu, Y., Song, Q., Yao, L., Pang, Z., Jiang, X., Chen, J., and Chen, H. (2012). Penetratin-functionalized PEG-PLA nanoparticles for brain drug delivery. *Int J Pharm,* **436**, 840–850.

173. Xia, H., Gao, X., Gu, G., Liu, Z., Zeng, N., Hu, Q., Song, Q., Yao, L., Pang, Z., Jiang, X., Chen, J., and Chen, H. (2011). Low molecular weight protamine-functionalized nanoparticles for drug delivery to the brain after intranasal administration. *Biomaterials,* **32**, 9888–9898.

174. Chan, J. M., Zhang, L., Yuet, K. P., Liao, G., Rhee, J. W., Langer, R., and Farokhzad, O. C. (2009). PLGA-lecithin-PEG core-shell nanoparticles for controlled drug delivery. *Biomaterials,* **30**, 1627–1634.

175. Hu, Y., Ehrich, M., Fuhrman, K., and Zhang, C. (2014). *In vitro* performance of lipid-PLGA hybrid nanoparticles as an antigen delivery system: lipid composition matters. *Nanoscale Res Lett,* **9**, 434.

Chapter 11

Polyester Particles for Drug Delivery to the Skin: Local and Systemic Applications

Ana Luiza S. Aguillera Forte,[a,b] **Ana Melero,**[b] **Ruy C. Ruver Beck,**[c] **and Claus-Michael Lehr**[d,e]

[a]*Faculty of Pharmaceutical Sciences of Ribeirão Preto, University of São Paulo, Ribeirão Preto, SP, 14040-903, Brazil*

[b]*Faculty of Pharmacy, University of Valencia, Valencia, 46100, Spain*

[c]*Faculdade de Farmácia, Universidade Federal do Rio Grande do Sul, Porto Alegre, RS, 90610-000, Brazil*

[d]*Biopharmaceutics and Pharmaceutical Technology, Saarland University, Saarbrücken, Saarland, 66123, Germany*

[e]*Department of Drug Delivery, Helmholtz Institute for Pharmaceutical Research Saarland (HIPS), Helmholtz Center for Infection Research (HZI), Saarbrücken, 66123, Germany*

aanaluiza2@gmail.com, ana.melero@uv.es

The skin is the largest human organ and absorption tissue in the body. However, the outermost layer of the skin (*stratum corneum,* SC) is a very effective barrier to the entrance of xenobiotics, and not all kinds of drugs can passively diffuse through this organ. The possibility of diffusion is mainly limited to the physicochemical properties of the molecules, such as molecular weight, partition coefficient, and

Handbook of Polyester Drug Delivery Systems
Edited by MNV Ravikumar

ISBN 978-981-4669-65-8 (Hardcover), 978-981-4669-66-5 (eBook)
www.panstanford.com

ionization degree in the formulation, among others. Therefore, it is possible to modulate the access of drugs to the deeper skin layers, if the therapeutic target is local, or even achieve plasma concentrations to provide systemic effects. If the intention is a local therapy, drugs should not reach systemic concentrations, as the effects would be regarded as undesired side effects. In this case, matrix systems can be a good approach to control drug release and to reduce or retard drug diffusion. On the contrary, if the desired effect is systemic and the drug is not absorbed in a sufficient rate and amount, there is a need to enhance penetration. For these purposes, polyester particles have been deeply investigated. Particles can provide additional advantages, as for example, protecting the drug from chemical or light degradation, increasing drug solubility in aqueous solvents or the stability. Recently, studies have been started to focus on hair follicle targeting, as it has been observed that some kinds of nanoparticles can accumulate into the hair follicle and release the drug for a longer period with topical and systemic applications as well. Furthermore, some groups are also investigating the possibility to develop immune responses using nanoparticles that can be phagocytized by skin immune cells. This fact would represent a new very promising approach to develop non-invasive vaccines that would ease the procedures and increase patient's acceptability.

11.1 Introduction

The skin is not just the largest human organ, but also an excellent biological barrier. It has developed defensive mechanisms to inhibit attacks by microbes, toxic chemicals, UV radiation, and particulate matter [1]. Thus, although some authors demonstrate that systems like nanoemulsions, micelles, or even some kinds of liposomes can enhance the passage of drugs, the potential for solid microparticles and nanoparticles (>10 nm) skin penetration is in general faulty, unless the skin barrier is disrupted [1]. Many strategies have been used to deliver sufficient amount of drugs into deeper layers of the skin, including chemical penetration enhancers and different physical enhancement techniques, such as iontophoresis [2], laser ablation [3], electroporation [4], ultrasound [5], and use of microneedles [6]. Although skin-targeted drug delivery still presents limitations, the skin offers a relatively large and readily accessible surface area. Furthermore, the disadvantages of oral administration such as enzymatic degradation or first-pass metabolism can be avoided [7]. More advantages of the transdermal drug delivery systems have emerged in recent years, as the potential for sustained release and controlled input kinetics, applicable properties for drugs with short biological half-live and for drugs with narrow therapeutic indices, respectively [8].

Biodegradable polymer–drug complexes seem to be an option for controlled drug delivery to the skin layers. Drugs can be incorporated into the carrier either dissolved, entrapped, adsorbed, attached, or encapsulated in the polymeric material to be delivered in a controlled way [9]. Biodegradable polymers offer certain advantages over lipid carriers, such as their stability upon storage, to be able to incorporate both hydrophobic and hydrophilic drugs, to be functionalized, and their physicochemical properties can be adjusted to achieve a controlled release [10]. Therefore, natural polymers, such as polysaccharides and proteins, and synthetic biodegradable polymers, such as aliphatic polyesters and polyphosphoester, have attracted considerable attention for these purposes. Especially the synthetic ones, since the biologically derived polymers sometimes can initiate an immune response in the human body and it is difficult to modify them chemically without modifying their bulk proportions [11]. In addition, degradation byproducts of the synthetic biodegradable polymers are generally considered to be low skin irritant and to have limited adverse effects [12]. The most extensively investigated polyesters for skin application are poly-(lactic-*co*-glycolic acid) (PLGA), poly(ε-caprolactone) (PCL), polylactic acid (PLA), and polyglycolic acid (PGA) [13].

Biodegradable polymeric nanoparticles seem to improve topical delivery to the skin for highly lipophilic drugs. Alvarez-Román *et al.* [14] confirmed that octyl methoxycinnamate (OMC), a high lipophilic sunscreen, encapsulated in PCL nanoparticles increase the levels of OMC by 3.4 folds into the SC. In addition, polymeric nanoparticles might protect the drug from the chemical degradation, as observed by Ourique *et al.* [15], proving that tretinoin-loaded lipid–core polymeric nanocapsules are less sensitive to photodegradation than the non-encapsulated and decreased the skin permeability coefficient, retaining tretinoin for a longer time on the skin surface.

To understand how drug-loaded polymeric nanoparticles interact with the skin, it is necessary to briefly describe the particularities on skin anatomy and its drug absorption pathways. Therefore, in the chapter, the skin structure and the possible routes to deliver drugs in polymeric particles for topical and systemic effects will be discussed, and several examples of the researches focusing in polyester particles will be summarized.

11.2 Skin Structure and Transport of Substances

The skin consists of two main layers. The underlying one, the dermis, is composed of connective tissue elements (collagen, elastin, glycosaminoglycans) and contains a variety of cell types (fibroblasts, mast cells, infiltrating leucocytes), nerves, blood and lymphatic vessels, pilosebaceous units, and sweat glands. The outer skin part, the epidermis, is separated from the dermis by the basement membrane and is constituted by keratinocytes (about 95%), melanocytes (melanin producer), Langerhans cells (immunological cells), and Merkel cells (mechanoreceptors). The lower keratinocytes are anchored to the basement membrane via hemidesmosomes [16].

Keratinocytes go through a keratinisation process in which the cell differentiates and moves upward from the basal layer (*stratum basale*), through the *stratum spinosum* and *stratum granulosum*, to the outermost layer, the SC (or horny layer) [1]. The final step in keratinocytes differentiation is associated with a radical change in their structure, resulting in their transformation into corneocytes: flat dead cells filled with keratin filaments and water, which are surrounded by a densely crosslinked protein layers (cell enveloped) [17].

SC is a composite of coneocytes and the secreted contents of lamellar bodies (elaborated by the keratinocytes) with a brick-and-mortar organization aspect [16]. The corneocytes comprise the "bricks" embedded in a "mortar," composed of multiple lipid bilayers of ceramides, fatty acids, cholesterol, and cholesterol esters [18]. This arrangement creates a tortuous path through which substances have to traverse in order to cross the SC [16]. Figure 11.1 presents a schematic representation of skin struture and the possible routes to drug penetration.

Among the main factors that influence the transdermal drug delivery, biological conditions of skin and physicochemical properties of drug (diffusion coefficient, concentration, partition coefficient, and molecular size and shape) can be cited. The flux of the drug through the SC is substantially influenced by the SC-to-vehicle partition coefficient. This parameter is crucial to establish a high initial concentration of the active in the first layer of the membrane [19]. Molecules showing intermediate partition coefficients (log $P_{\text{octanol/water}}$ of 1–3) can adequately diffuse into the lipid domains of the SC while still having being able to further partition into the viable epidermis [20]. Moreover, the molecular weight seems to be inversely proportional to the absorption as generally molecules should weigh less than 600 Da to effectively diffuse through the SC. On the other hand, these factors that influence skin penetration cannot be considered individually. The ideal molecular properties of a substance to penetrate through the SC include a low molecular mass, an adequate solubility in oil and water, a balanced partition coefficient, and a low melting point [19].

The transport of substances across the SC occurs mainly by passive diffusion and via three possible routes: transcellular, intercellular, and appendageal (where the structure of SC is interrupted). The most SC penetrants use the intercellular route [1, 18].

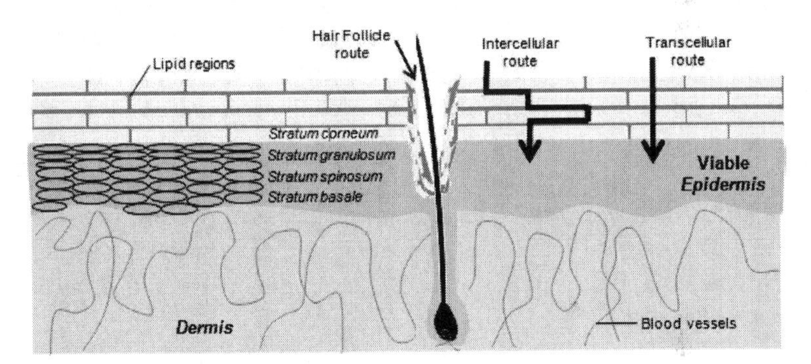

Figure 11.1 Schematic skin structure and routes of drug penetration. Adapted from [17].

Besides the barrier organization, the SC has an inherent mechanism for preventing foreign bodies from penetrating via the skin, as the outer layers of the SC undergo a process of desquamation renewing the SC completely in 14 days in humans, depending on the anatomical site or age [1]. Consequently, epidermis is a dynamic and constantly self-renewing tissue, in which a loss of the cells from the surface is balanced by cell growth in the lower epidermis [17].

Whereas the intercellular penetration of particles seems to be improbable, the hair follicle has been shown to be a relevant penetration pathway for particles as well as an important long-term reservoir [21]. In the case of particles, a few size-dependency studies have been carried out, but Patzelt *et al.* [22] investigated various types of particles at different sizes and concluded that the best range for follicular penetration is between 300 and 600 nm, depending on the material used [22]. Therefore, this route will be discussed in more detail in this chapter.

The hair follicle consists of a permanent, superficial structure, and a transient cycling component, which include the hair bulb. The permanent part of the hair follicle can be subdivided in two parts: the infundibulum and the isthmus. The infundibulum is the part between the skin surface to the point where the sebaceous gland duct opens to the hair canal. The isthmus is the part between the sebaceous gland duct opening and the bulge region, where the stem and the skin mast cell precursors are mainly located. The part beneath, the transient, extends from the bulge to the base of the hair follicle bulb and is located into the bulb and the suprabulbar region [23]. The hair follicle structure is represented in Fig. 11.2. Hair fibers are composed of dead cells filled with keratin. The outer layer is the cuticle followed by cortex and medulla, in the central region. Cuticle cells are flat and overlapping, like shingles on a roof, fixed at the root. The cortex contains cortical cells and the intercellular binding material, which is composed of cell membranes and adhesive material that binds the cuticle and cortical cells together. The medulla, if present, represents just a minor percentage of the mass of the hair [24].

Historically, follicular penetration was disregarded, mainly because it was assumed that hair follicles occupy less than 0.1% of the total skin area [25]. However, recently, hair follicles have attracted attention by representing an important pathway for intra- and transdermal drug delivery [26]. In addition, the hair follicle acts as a reservoir for topically applied substances, especially for particulate materials [27]. Micro- and nano-sized particles accumulated in the hair follicle are protected from SC turnover and persist for days before they are eliminated [28]. Polymeric particles can act as matrix systems controlling the delivery of the drug over hours or even days, the drug being bioavailable to its diffusion through the SC over prolonged periods of time [29]. The sebum flow is responsible for transporting particles stored in the hair follicle to the skin surface, but it is a slow process [30]. The development of optical analytical methods with high spatial resolution, such as laser scanning microscopy, made it possible to analyze follicular penetration [31]. Polyester nanoparticles were shown to penetrate deeply into the hair follicles, where they are stored up to 10 days. In addition, the penetration into the hair follicles is a fast process (1 h) in comparison to the release of nanoparticles out of the follicles, which continues for some days [32].

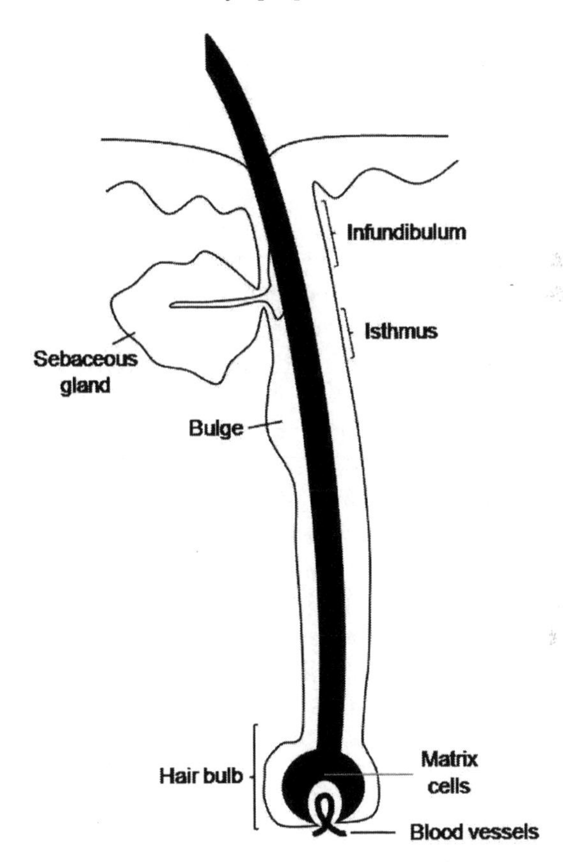

Figure 11.2 Morphology of the human hair follicle and its target structures of interest. Adapted from [21].

Nevertheless, penetration is not observed in all hair follicles. Lademann and colleagues [33] found that penetration is only possible if the follicles are active. Active follicles are characterized by sebum production and/or hair growth that seem to be relevant processes to facilitate the access of the particles into the follicle, whereas inactive follicles do not present growth or sebum productions.

It is necessary to select the suitable particle size to obtain the desired follicular penetration and to reach specific sites inside the hair follicle. Patzelt *et al.* [22] observed that the penetration in the hair follicle of PLGA particles was deeper with increasing particle size among the size range tested (122 nm < 230 nm < 300 nm < 470 nm < 643 nm). They also found that smaller and larger particles (122 and 860 nm) can be utilized for targeting the infundibular region, larger particles (230 nm and 300 nm) are qualified for penetrating into the region of the sebaceous gland, and the medium-size particles (470 and 643 nm) can penetrate down the bulge region. Lademann *et al.* [30] discussed that cuticle thickness can also have an effect on the follicular penetration of the nanoparticles, since the hair surface consists in a zigzag structure formed by the cuticle similar to the corneocytes structure that extends into the orifices of the hair follicle, and their thickness are similar. Therefore, in this case, the optimal particle size for follicular penetration in human skin should be in the region of 500 nm, once cuticle thickness in human hair is approximately 530 nm, in contrast to 320 nm in porcine hair, in which the optimal size should be in the region of 300 nm. However, Raber *et al.* [34] tested PLGA nanoparticles, chitosan-coated PLGA nanoparticles, and phospholipid-coated chitosan-PLGA nanoparticles of around 160 nm in both excised pig ear skin and human volunteers. They found an excellent *in vitro–in vivo* correlation (IVIVC, $r_2 = 0.987$) for both models [34].

Lademann *et al.* [30] also suggested that the hair movement acts as a geared pump, pushing the nanoparticles to the hair bulb, if their size is comparable to the surface structure of the hair and hair follicles. Accordingly, with this hypothesis, hair movement is essential for follicular penetration and experimental studies must be performed *in vivo* or on model tissues where the hairs should be moved artificially. Excised human skin is not applicable for follicular penetration studies, because human skin contracts instantly after excision, which conduces to a permanent occlusion of the hair follicles by the contracted elastic fibers. In turn, follicles in porcine ear skin remain open, since the skin remains fixed to the cartilage during the investigation [35].

Besides representing an important penetration pathway and reservoir, there are multiple target structures in the hair follicles for innovative therapeutic approaches. For example, hair follicles are connected with a network of blood capillaries and below the entrance of sebaceous gland duct to the hair canal, there is no mature SC. Around the infundibular epithelium, there are cells associated with the immune system, such as antigen-presenting cells and mast cells [36]. The sebaceous gland area is associated with the etiology of androgenic alopecia, acne, and other sebaceous gland dysfunctions; and the bulge region, where the epithelial stem cells are found, is therefore an area with a high proliferative capacity and multipotency [21].

11.3 Topical Application of Polyester Particles

11.3.1 Local Applications

The skin penetration of a drug to provide local effect in the skin is a complex process involving three major steps: first, the release of the drug from the vehicle; second, the penetration into the SC; and third, the partitioning from SC to target sites in the epidermis and dermis. The first step depends on the physicochemical properties of the drug and the vehicle and can be optimized by the correct design of the formulation. The second and third step also depend on the drug physicochemical properties and on the degree of drug saturation in the vehicle as well as on the partitioning of the drug between the vehicle and the skin. It is also of major importance to consider the condition of the SC, since it is the main biological barrier for the drug penetration into the skin and can be compromised by some skin diseases [37].

Anti-inflammatory drugs represent a broad range of molecules, many with the potential to provide a topical delivery. Vega *et al.* [38] developed flurbiprofen-loaded PLGA, PLGA with poly(ethylene) glycol (PLGA-PEG) nanospheres with and without hydroxylpropyl-β-cyclodextrin and observed that only (PLGA-PEG) nanospheres showed slight permeation improvement in *ex vivo* human skin. However, nanospheres were efficient in maintaining the drug on the viable epidermis for a long period, reducing the risk of systemic effects. After 24 hours, the application of nanospheres retained about 9 fold higher drug amounts in the skin compared with the control solution containing the non-encapsulated drug. This higher retention was attributed to the reservoir effect of the nanospheres that sustained the drug and limited its absorption. Additionally, it was observed that nanospheres with HPβCD were more efficient to decrease 12-O-tetradecanoylphorbol 13-acetate (TPA)-induced mouse ear edema than nanospheres without HPβCD. Luengo *et al.* [39] did not detect differences in drug transport into the SC using flufenamic acid loaded PLGA nanoparticles, compared to non-encapsulated drug, whereas in the deeper skin layers, drug penetration was slightly delayed for the nanoencapsulated drug in shorter incubation times (<12 h). In longer incubation times (>12 h), an increase of 50% in the drug amount transported into the deep skin layers was observed for the nanoencapsulated drug compared to the formulation containing the non-encapsulated drug. Özcan *et al.* [40] compared betamethasone-17-valerate (BMV)-loaded PLGA and lecithin/chitosan (LC) nanoparticles in topical application, and the results in *ex vivo* experiments showed that both polymeric nanoparticles enhanced the amount of BMV in the epidermis, which is the target site of topical steroidal treatment, compared with commercial formulation, which contained a higher amount (10 times) of drug. Moreover, the transepidermal water loss (TEWL) measurement exhibited no barrier function changes upon the application of nanoparticles on the skin. However, LC nanoparticles had higher drug accumulation when compared to PLGA nanoparticles in the epidermis, and no drug was found in the receptor compartment. Unlike for PLGA nanoparticles, the amount of drug accumulated in the dermis was significantly higher than commercial cream formulation, which was probably due to the transfollicular route. Although PLGA nanoparticles produced an inhibitory effect on paw edema formation, especially after the fourth hour with respect to the control group, there was no statistically significant difference with respect to commercial cream, such as in LC nanoparticles. Thus, LC nanoparticles could be suggested as more suitable carriers for dermal delivery of BMV.

A key advantage in the use of polyester nanoparticles is not only sustained release of the drug but also increasing its penetration while mitigating the damage to the skin barrier function. Cyclosporine A is a potent immunosuppressant that, although used clinically, is orally administered for the treatment of various dermatological diseases like psoriasis, alopecia areata, lichen planus, and contact hypersensitivity. Its long-term systemic administration produces detrimental effects like nephrotoxicity, hypertension, hyperlipidemia, hepatitis, etc. [41]. Considering the side effects of the drug, due to its systemic toxicity, it is, nowadays, only used for skin disorders in those cases where all other treatments fail. In this kind of cases, topical application of the drug avoiding access to systemic circulation is desirable [42]. However, the drug has a poor skin permeability. The rigid cyclic structure, the hydrophobicity, and the high molecular weight of cyclosporine A combined with the barrier properties of SC do not allow its topical application [43]. For this reason, Jain *et al.* [44] develop cyclosporine A–loaded PLGA nanoparticles for enhancing its topical delivery. The *in vitro* permeation study showed that polymeric nanoparticles increased the amount of drug in SC and dermis by 4.8 and 2.62 times respectively over control, and no drug was detected in the receptor compartment, decreasing the risk of drug diffusion to the blood vessels and undesirable systemic effects. It was also demonstrated that the nanoparticles mainly followed the pilosebaceous route to entry into the skin layers. Furthermore, the system did not cause any changes in the morphology structure and integrity of skin. Therefore, it may be effectively used as a topical delivery vehicle for the treatment of various dermatological diseases.

The encapsulation of antimicrobial drugs using polyester particles can also improve their efficacy in topical therapy, decreasing their side effects. Valizadeh *et al.* [45] studied clarithromycin, a broad-spectrum macrolide antibiotic widely used in skin infections. PLGA nanoparticles enhanced its antimicrobial activity *in vitro* in comparison with the non-encapsulated drug. The increase in the antibacterial efficacy can be attributed to the physicochemical properties of the nanoscale particles, such as fusion of nanoparticles with the microbial cell wall and release of the loaded drug or adsorption of nanoparticles to the cell wall. Thus serving as a reservoir for continuous release of drug molecules, which will diffuse into the microorganism. Furthermore, the drug stability was shown to be greater, since the drug is protected against rapid enzymatic/ hydrolytic degradation.

Santoyo *et al.* [46] showed that cidofovir, a new antiviral agent effective against herpes viruses, loaded in PLGA microparticles penetrated less in the porcine skin than the non-encapsulated drug solution. The profile of drug distribution into the skin, 24 hours after the topical application, demonstrated that microparticles concentrated the drug in the outer layers, decreasing in the deeper layers. This effect could be because the microparticles form a film on the skin surface, reducing the TEWL and favoring drug penetration due to occlusion. Therefore, the drug amount found in the basal epidermis, site of the virus lesion, was higher with the microparticles than with the cidofovir solution. Kumar *et al.* [47], using clotrimazole-loaded PLGA microparticles, reached a reservoir effect in the skin higher than commercial formulation. This is an interesting effect as patients frequently stop applying the medicine before the fungal infection completely disappears, leading to re-infection in the body. Thus, a system able to maintain therapeutic amounts of the drug for a long time can support the complete eradication of the infection.

In the last two decades, a progress in nanotechnology focused on the incorporating agents in nanoparticles for detecting, preventing, and treating oncological diseases has been observed. Nanoparticles might both protect the encapsulated drug and improve its bioavailability to promote a localized action to treat skin cancer [48]. Several studies have been focused on investigating novel chemopreventive agents, especially dietary antioxidants that inhibit the development and progression of the non-melanoma skin cancer (NMSC). Das *et al.* [49] utilized the combination of oral and topical administration of apigenin-loaded PLGA nanoparticles in ultraviolet B (UVB)- and benzo(a)pyrene (BaP)-induced skin tumor in mice. The results showed that the nanoparticles reduced tissue damage and frequency of chromosomal aberrations more efficiently that non-encapsulated apigenin, possibly due to their nanoparticle size and fast mobility, pointing the potential to use in therapeutic management of NMSC with reduced toxicity. Srivastava *et al.* [50] also demonstrated that polyphenolic constituents of black (theaflavin) and green (epigallocatechin-3-gallate) tea in PLGA nanoparticles topically applied together were able to protect the mouse skin against 7,12-dimethylbenz[a]anthracene (DMBA)-induced DNA damage, in a dose-dependent way, more efficiently than the not encapsulated compounds. In addition, the nanoparticles seems to induce DNA repair genes and to suppress DNA damage responsive genes tested in this study, which the authors suggest to be in consequence of the higher capacity of the nanoparticles on penetrating across the cell surface.

Another promising treatment for a diversity of malignant and premalignant skin disorders, including NMSC, is the photodynamic therapy (PDT), which consists in the application of a photosensitizing (PS) drug followed by photoirradiation that leads to cell apoptosis or necrosis [51]. Nanoparticles are able to encapsulate photosensitizers (PSs) or prodrugs to be used in PDT, such as protoporphyrin IX (PpIX), modulating their lipophilia and preventing aggregation in aqueous solution [52]. Silva *et al.* [53] observed that PLGA nanoparticles provide a slower release of PpIX, compared to its control solution (non-encapsulated PpIX). In addition, encapsulated PpIX in polymeric nanoparticles seem to be protected from aggregation in aqueous medium for 30 days and may be administrated in liquid form. *In vitro* studies of skin permeability have shown that PpIX permeation through the skin was minimal and *in vivo* skin application of nanoparticles on hairless mice showed higher retention of the PS both in SC and in epidermis + dermis (EP + D) compared to control solutions, suggesting a minimal permeation through the skin and a localized effect. The increase of the skin permeability and retention by the nanoparticles make possible the topical application of PpIX in cases of skin tumors and in off-label uses, such as psoriasis, acne, and photorejuvenation [53].

Apart from the pharmaceutical application, nanoparticles are increasingly being incorporated in cosmetics products [54]. However, polyester nanocarriers are susceptible to hydrolysis in an aqueous environment and therefore are difficult to formulate in a water-based topical formulation. Nevertheless, nanoparticles of PLGA appear to be promising nanocarriers to enhance topical delivery of steroidal agents, such as the endogenous steroidal hormone dehydroepiandrosterone (DHEA), which modulate the expression of several dermal genes, especially those involved in collagen synthesis, and its administration can improve skin characteristics, like hydration, epidermal thickness, sebum production, and skin pigmentation. Therefore, the topical administration of DHEA can promote anti-skin-aging activities [55]. Badihi *et al.* [55] recorded higher amounts of [3H]-DHEA in viable skin layers following different periods of incubation of DHEA-loaded nanocapsules on excised pig skin in comparison with non-encapsulated molecule in solution. Significantly higher (4 fold) skin flux values were observed for nanocapsules as compared to the control solution, suggesting that PLGA nanocapsules, in a water-free topical formulation or in a freshly reconstitute dispersion, have promising potential to be used topically for the treatment and prevention of skin-aging processes.

On the other hand, Azarbayjani *et al.* [56] reported that PLGA and PLA microparticles increase the epidermis retention of Levothyroxine (T4), a synthetic hormone administered in cosmetic creams to reduce deposits of adipose tissues on skin. PLA and PLGA microparticles have been used as a skin penetration retardant able to cause skin accumulation of the active and may be used with sunscreens, insect repellants, and other ingredients that are meant to be concentrated on the skin surface.

Unlike, Teixeira *et al.* [57] evaluating flexible polymeric nanocapsules with a retinyl palmitate (a stable form of vitamin A) core and PLA shell. The authors observed that the drug was able to reach deeper layers of human excised skin, since the drug was found in the receptor compartment of the Franz diffusion cells after 24 hours, but the highest amount was in the viable epidermis and dermis. The authors attributed these results to the nanocapsules affinity for the SC and to their flexible characteristics, indicating the system applicability for retinyl palmitate delivery and its association with compounds that require to delivery into deeper skin layers.

Poly(ε-caprolactone) has a high permeability to many drugs and an excellent biocompatibility; however, as its biodegradation is slower than the other polyesters, PCL has been mainly regarded as more adequate for long-term delivering. For this reason, studies using PLC have been mostly focused in sutures, subdermal contraceptive devices, and wound dressings. In addition, this polymer is very useful in tissue engineering, including human skin engineering [58]. However, PCL nanocapsules, which have a lipophilic core and a hydrophilic shell, can be used as topical carriers since the SC is constantly renovated and particles do not passively penetrate to the deeper skin layers. Lboutounne *et al.* [59] showed that chlorhexidine-loaded PCL nanocapsules maintained an antimicrobial activity against several bacteria, which can be an effect of the attachment of the polymer on the bacterial surface. Moreover, the nanocapsules sustained the release of the drug in SC and hair follicle, decreasing the permeation through the skin. Nanometric PCL delivery systems for mometasone [42] and tretinoin [15] were prepared by our group and then incorporated in hydrogels. Skin permeation experiments conducted with heat-separated human epidermis mounted in a Franz diffusion cell revealed that nanoencapsulation resulted in reduced transdermal permeability, independent of type of nanocarrier and/or the encapsulated drug. Furthermore, the nanoformulations preserved the drug integrity from light instability in the case of tretinoin. In both cases, increasing the viscosity of the medium resulted in a synergistic effect in reducing drug permeability.

11.3.2 Systemic Applications

Although polyester particles can be deposited in the hair follicle, generally they are not able to penetrate the SC. Therefore, an increasing number of studies are investigating the combination of nanoparticles and active skin penetration techniques, which aim to achieve the intradermal drug/vaccine delivery.

Microneedles are solid or hollow needles, with length between 70 and 1000 μm, which are linked to a patch-like support. When the microneedles are pressed into the skin, they create temporary microconduits able to facility the permeation of drug into the skin [60]. The length of microneedles allows them to pass through the SC and reach the viable epidermis without stimulating the pain receptors, which are situated in the underlying dermis [61]. Vučen *et al.* [62] demonstrated that ketoprofen-loaded in poly(D,L-lactic acid) enables a higher *in vitro* permeation of the drug in porcine skin when microneedles were applied. It accumulated in the dermis and gradually permeated through this skin layer over 24 hours after exposure, being a potential system to sustain the transdermal delivery to ketoprofen.

It is also possible to use needle-free jet injectors to delivery polyester particles into the skin. These systems employ high speed to puncture the skin and delivery liquids and powders without using needles, which is less painful that the conventional needle systems and increase the compliance of the patients [63]. Michinaka *et al.* [64] utilized a commercial liquid jet injector to deliver coumarin-6-load PLGA microparticles into full thickness excised human skin. The results demonstrated that the amount of fluorescent compound into the skin was higher with the particle suspension than with the solution free of particles, but its dispersion area was smaller for the particulate dispersion than for the solution. Environmental scanning electron microscopy (ESEM) showed that the particle concentration into the skin increases with the depth until nearly 3 mm, below which it starts decreasing. The alteration of several parameters allows controlling the particle dispersion and efficiently delivering the drug or vaccine into human skin by jet injectors. The increase in the particle size, for example, can decrease the dispersion area and penetration of particles beyond dermis due to the resistance offered by the skin tissue. Thus, the jet injectors may be an efficient device to transdermal administration of particulate drugs.

The iontophoresis implicates the application of small electrical current (usually <500 mA/cm^2) to a drug reservoir to facilitate the transfer of the drug across the skin utilizing the principle of the repulsion forces of the same charges [8]. Tomoda *et al.* [65] reported that the permeability of indomethacin-loaded PLGA nanoparticles through rat skin was significantly higher than the polyvinyl alcohol (PVA)-coated indomethacin-loaded PLGA nanoparticles when iontophoresis was applied *ex vivo*. In addition, indomethacin transition to circulation and accumulation in muscle by the transdermal delivery of nanoparticles were significantly enhanced by using the combination with iontophoresis *in vivo*. Thus, the authors suggest that PLGA nanoparticles without any polymeric stabilizer layer on the surface have more effective potential to deliver encapsulated drugs by applying iontophoresis than a stabilizer-coated PLGA nanoparticle. In another study, Tomoda *et al.* [66] also demonstrated *in vivo* that PLGA nanoparticles increase the skin permeability of estradiol comparing with the free molecules. In addition, the application of iontophoresis enhanced the permeability of the nanoparticle system by means of intensifying the accumulation of the particles in the hair follicles in respect of the simple diffusion.

11.3.3 Transcutaneous Immunization (TCI)

The skin has a large immune network of cells, including epidermal keratinocytes (DKs) and Langerhans cells (LCs), dermal fibroblasts (FBs), dendritic cells (DCs) and mast cells (MCs), even local draining lymph nodes with T and B lymphocytes (T cells, B cells), and afferent and efferent lymph channels. Therefore, transcutaneous immunization (TCI) has attracted researchers' attention because of its potential benefits, such as the strong humoral and cellular response both in the site of application and at distant mucosal sites [67].

However, to be effective, the vaccine antigens require to surpass the SC barrier and to achieve the LCs in the epidermis, which behave as antigen-presenting cells (APCs). The complex organization of SC limits the amount of antigens that reach the LCs, providing an insufficient immunogenicity. For this reason, particulate carriers are an interesting strategy to TCI, since they can deliver antigens and adjuvants, increasing their stability and facilitating their absorption. Furthermore, nanoparticles can also increase their antigenicity by mimicking the size of microorganisms. Biodegradable polyester nanoparticles seem to be an alternative for TCI once these particles are able to accumulate into the hair follicles, where the antigens are readily uptaken by a large variety of peri-follicular APCs due to the default SC barrier in the lower follicular orifice [68].

Mattheolabakis *et al.* [69] demonstrated that ovalbumin (OVA)-loaded PLGA nanoparticles were able to enter into the mouse skin *in vivo* by the ducts of the hair follicle and stimulated a strong proliferative response, especially when the cholera toxin was used as adjuvant, suggesting that nanoparticles were effective in delivering the antigen to the APCs. Furthermore, the immunogenic behavior observed in this study also suggests that the antigen was released from the nanoparticles only after their phagocytosis *in vivo*. However, rodent skin is not appropriate for testing transdermal delivery or transfollicular penetration because its lipids and corneocytes are not similar to the human ones and form a leakier arrangement [70]. In addition, mice have a higher follicular density than man. Therefore, Mittal *et al.* [71] investigated the potential of transfollicular delivery of OVA using polymeric nanoparticles (NPs) without any pretreatment in excised pig ears. It was observed that the amount of OVA nanoparticles into the hair follicles was 2 to 3 times higher than OVA solution, reaffirming the potential use of the nanoparticles to improve TCI.

Physical barrier disruption methods are also being explored to be used with particles in TCI, including jet injection and gene gun devices, microneedle patches, exfoliation, electroporation, and thermal ablation techniques. The use of these techniques can either facilitate the penetration of the antigen as to lead to a nonspecific immunostimulation that improves antigen-specific immune response [67]. Nevertheless, some points should be considered for the application of some of these methods for mass vaccination, such as the costs of the devices needed and, more important, the difficulty to use them, as they require professional training. In addition, the gravity of the barrier disruption and the time necessary for its recuperation should also be taken into account, since there might be a pathogen invasion [68]. DeMuth *et al.* [72] developed a system based on PLGA microparticles incorporated in water-soluble microneedles of poly(acrylic acid) (PAA), which are disintegrated after their insertion in the skin, leaving a depot of microparticles that promote a long-term controlled and sustained drug and vaccine delivery. The *in vivo* results showed that the encapsulation of a protein vaccine in PLGA microparticles generated a strong humoral and cellular immunity, surpassing that obtained with the traditional needle-based vaccine administration, confirming the potential applicability of this system to TCI.

11.4 Conclusions

Polyester particles are very promising vehicles for the delivery of drugs via the skin both for local and systemic purposes. Among all investigated polyester polymers for skin application, polylactic acid, polyglycolic acid, poly-(lactic-*co*-glycolic acid), and poly(ε-caprolactone) have been accepted for human use by the FDA, as they are biodegradable, biocompatible, and generally regarded as safe. Many studies have been successfully conducted to modify drug absorption through the SC using polyester particles, which can control drug absorption and decrease systemic side effects associated with therapeutic plasma peaks after topical application of drugs onto the skin. On the other hand, innovative particles have been designed to enhance drug permeability of poorly bioavailable drugs to achieve systemic effects. These attempts have been unsuccessful due to the size of the particles, unless the SC had been previously damaged. It should always be taken into account that disrupting the skin barrier can always provide undesired effects and should be avoided if possible. Another possibility is to target hair follicles, where polyester particles tend to accumulate under certain conditions. The most recent and innovative applications of polyester particles intend to develop needle-free or at least pain-free vaccination by means of microneedles, as the skin is a very inmunoactive tissue, which could be a very interesting approach, but still needs optimization to become a clinical reality.

References

1. Prow, T. W., Grice, J. E., Lin, L. L., Faye, R., Butler, M., Becker, W., Wurm, E. M. T., Yoong, C., Robertson, T. A., Soyer, H. P., and Roberts, M. S. (2014). Nanoparticles and microparticles for skin drug delivery, *Adv Drug Deliv Rev,* **63**, 470–491.
2. Kalia, Y. N., Naik, A., Garrison, J., and Guy, R. H. (2004). Iontophoretic drug delivery, *Adv Drug Deliv Rev,* **56**, 619–658.
3. Chen, X., Shah, D., Kositratna, G., Manstein, D., Anderson, R. R., and Wu, M. X. (2012). Facilitation of transcutaneous drug delivery and vaccine immunization by a safe laser technology, *J Control Release,* **159**, 43–51.
4. Denet, A., Vanbever, R., and Préat, V. (2004). Skin electroporation for transdermal and topical delivery, *Adv Drug Deliv Rev,* **56**, 659–674.
5. Lavon, I., and Kost, J. (2004). Ultrasound and transdermal drug delivery, *Drug Discov Today,* **9**, 670–676.
6. Prausnitz, M. R. (2004). Microneedles for transdermal drug delivery, *Adv Drug Deliv Rev,* **56**, 581–587.
7. Lam, P. L., and Gambari, R. (2014). Advanced progress of microencapsulation technologies: *in vivo* and *in vitro* models for studying oral and transdermal drug deliveries, *J Control Release,* **178**, 25–45.
8. Naik, A., Kalia, Y. N., and Guy, R. H. (2000). Transdermal drug delivery: overcoming the skin's barrier function, *Pharm Sci Technol Today,* **3**, 318–326.

9. Wei, X. W., Gong, C. Y., Gou, M. L., Fu, S. Z., Guo, Q. F., Shi, S., Luo, F., Guo, G., Qiu, L. Y., and Qian, Z. Y. (2009). Biodegradable poly(ε-caprolactone)-poly(ethylene glycol) copolymers as drug delivery system, *Int J Pharm,* **381**, 1–18.
10. Hans, M. L., and Lowman, A. M. (2002). Biodegradable nanoparticles for drug delivery and targeting, *Curr Opin Solid St M,* **6**, 319–327.
11. Tian, H., Tang, Z., Zhuang, X., Chen, X., and Jing, X. (2012). Biodegradable synthetic polymers: preparation, functionalization and biomedical application, *Prog Polym Sci,* **37**, 237–280.
12. Nair, L. S., and Laurencin, C. T. (2007). Biodegradable polymers as biomaterials, *Prog Polym Sci,* **32**, 762–798.
13. Yang, Y., Bugnot, J., and Hong, S. (2013). Nanoscale polymeric penetration enhancers in topical drug delivery, *Polym Chem,* **4**, 2651–2657.
14. Alvarez-Román, R., Naik, A., Kalia, Y. N., Guy, R. H., and Fessi, H. (2004). Enhancement of topical delivery from biodegradable nanoparticles, *Pharm Res,* **21**, 1818–1825.
15. Ourique, A. F., Melero, A., Silva, C. B., Schaefer, U. F., Pohlmann, A. F., Guterres, S. S., Lehr, C. M., Kostka, K. H., and Beck, R. C. R. (2011). Improved photostability and reduced skin permeation of tretinoin: development of a semisolid nanomedicine, *Eur J Pharm Biopharm,* **79**, 95–101.
16. Menon, G. K. (2002). New insights into skin structure: scratching the surface, *Adv Drug Deliv Rev,* **54**, Suppl. 1, S3–S17.
17. Bouwstra, J. A., and Ponec, M. (2006). The skin barrier in healthy and diseased state, *Biochim Biophys Acta,* **1758**, 2080–2095.
18. Barry, B. W. (2001). Novel mechanisms and devices to enable successful transdermal drug delivery, *Eur J Pharm Sci,* **14**, 101–114.
19. Aulton, M. E. (2007). *Aulton's Pharmaceutics – The design and manufactured of medicines* (Churchill Livingstone, London).
20. Benson, H. A. E. (2005). Transdermal drug delivery: penetration enhancement techniques, *Curr Drug Deliv,* **2**, 23–33.
21. Lademann, J., Richter, H., Schanzer, S., Knorr, F., Meinke, M., Sterry, W., and Patzelt, A. (2011). Penetration and storage of particles in human skin: perspectives and safety aspects, *Eur J Pharm Biopharm,* **77**, 465–468.
22. Patzelt, A., Richter, H., Knorr, F., Schäfer, U., Lehr, C. M., Dähne, L., Sterry, W., and Lademann, J. (2011). Selective follicular targeting by modification of the particle sizes, *J Control Release,* **150**, 45–48.

23. Patzelt, A., Knorr, F., Blume-Peytavi, U., Sterry, W., and Lademann, J. (2008). Hair follicles, their disorders and their opportunities, *Drug Discov Today Dis Mech,* **5**, e173–e181.

24. Wei, G., Bhushan, B., and Torgerson, P. M. (2005). Nanomechanical characterization of human hair using nanoindentation and SEM, *Ultramicroscopy,* **105**, 248–266.

25. Otberg, N., Richter, H., Schaefer, H., Blume-Peytravi, U., Steerry, W., and Lademann, J. (2004). Variations of hair follicle size and distribution in different body sites, *J Invest Dermatol,* **122**, 14–19.

26. Lademann, J., Knorr, F., Richter, H., Blume-Peytavi, U., Vogt, A., Antoniou, C., Sterry, W., and Patzelt, A. (2008). Hair follicles—an efficient storage and penetration pathway for topically applied substances, *Skin Pharmacol Physiol,* **21**, 150–155.

27. Wosicka, H., and Cal, K. (2010). Targeting to the hair follicles: current status and potential, *J Dermatol Sci,* **57**, 83–89.

28. Rancan, F., Blume-Peytavi, U., and Vogt, A. (2014). Utilization of biodegradable polymeric materials as delivery agents in dermatology, *Clin Cosmet Investig Dermatol,* **7**, 23–34.

29. Marin, E., Briceño, M. I., and Caballero-George, C. (2013). Critical evaluation of biodegradable polymers used in nanodrugs, *Int J Nanomedicine,* **8**, 3071–3091.

30. Lademann, J., Patzelt, A., Richter, H., Antoniou, C., Sterry, W., and Knorr, F. (2009). Determination of the cuticula thickness of human and porcine hairs and their potential influence on the penetration of nano particles into the hair follicles, *J Biomed Opt,* **14**, 1–4.

31. Alvarez-Román, R., Naik, A., Kalia, Y. N., Guy, R.H., and Fessi, H. (2004). Skin penetration and distribution of polymeric nanoparticles, *J Control Release,* **99**, 53–62.

32. Lademann, J., Richter, H., Teichmann, A., Otberg, N., Blume-Peytavi, U., Luengo, J., Weiß, B., Schaefer, U. F., Lehr, C. M., Wepf, R., and Sterry, W. (2007). Nanoparticles – an efficient carrier for drug delivery into the hair follicles, *Eur J Pharm Biopharm,* **66**, 159–164.

33. Lademann, J., Otberg, N., Richter, H., Weigmann, H. J., Lindemann, U., Schaefer, H., and Sterry,W. (2001). Investigation of follicular penetrationof topically applied substances, *Skin Pharmacol Appl Skin Physiol,* **14**, Suppl. 1, 17–22.

34. Raber, A. S., Mittal, A., Schäfer, J., Bakowsky, U., Reichrath, J., Vogt, T., Schaefer, U. F., Hansen, S., and Lehr, C. M. (2014). Quantification of nanoparticles uptake into hair follicles in pig ear and human forearm, *J Control Release,* **179**, 25–32.

35. Mak, W. C., Patzelt, A., Richter, H., Renneberg, R., Lai, K. K., Rühl, E., Sterry, W., and Lademann, J. (2012). Triggering of drug release of particles in hair follicles, *J Control Release,* **160**, 509–514.
36. Knorr, F., Lademann, J., Patzelt, A., Sterry, W., Blume-Peytavi, U., and Vogt, A. (2009). Follicular transport route – research progress and future perspectives, *Eur J Pharm Biopharm,* **71**, 173–180.
37. Jensen, L. B., Petersson, K., and Nielsen, H. M. (2011). *In vitro* penetration properties of solid lipid nanoparticles in intact and barrier-impaired skin, *Eur J Pharm Biopharm,* **79**, 68–75.
38. Vega, E., Egea, M. A., Garduño-Ramírez, M. L., García, M. L., Sánchez, E., Espina, M., and Calpena, A. C. (2013). Flurbiprofen PLGA-PEG nanospheres: role of hydroxyl-β-cyclodextrin on *ex vivo* human skin permeation and *in vivo* topical anti-inflammatory efficacy, *Colloids Surf B Biointerfaces,* **110**, 339–346.
39. Luengo, J., Weiss, B., Schneider, M., Ehlers, A., Stracke, F., König, K., Kostka, K. H., Lehr, C. M., and Schaefer, U. F. (2006). Influence of nanoencapsulation on human skin transport of flufenamic acid, *Skin Pharmacol Physiol,* **19**, 190–197.
40. Özcan, İ., Azizoğlu, E., Şenyiğit, T., Özyazıcı, M., and Özer, Ö. (2013). Comparison of PLGA and lecithin/chitosan nanoparticles for dermal targeting of betamethasone valerate, *J Drug Target,* **21**, 542–550.
41. Fradin, M. S., Ellis, C. N., and Voorhees, J. J. (1990). Management of patients and side effects during cyclosporine therapy for cutaneous disorders, *J Am Acad Dermatol,* **23**, 1265–1275.
42. Melero, A., Ourique, A. F., Guterres, S. S., Pohlmann, A. R., Lehr, C. M., Beck, R. C. R., and Schaefer, U. (2014). Nanoencapsulation in lipid-core nanocapsules controls mometasone furoate skin permeability and its penetration to deeper skin layers, *Skin Pharmacol Physiol,* **27**, 217–228.
43. Choi, H. K., Flynn, G. L., and Amidon, G. L. (1995). Percutaneous absorption and dermal delivery of cyclosporin A, *J Pharm Sci,* **84**, 581–583.
44. Jain, S., Mittal, A., and Jain, A. K. (2011). Enhanced topical delivery of cyclosporin A using PLGA nanoparticles as carrier, *Curr Nanosci,* **7**, 524–530.
45. Valizadeh, H., Mohammadi, G., Ehyaei, R., Milani, M., Azhdarzadeh, M., Zakeri-Milani, P., and Lotfipour, F. (2012). Antibacterial activity of clarithromycin loaded PLGA nanoparticles, *Pharmazie,* **67**, 63–68.

46. Santoyo, S., Jalón, E. G., Ygartua, P., Renedo, M. J., and Blanco-Príeto, M. J. (2002). Optimization of topical cidofovir penetration using microparticles, *Int J Pharm,* **242**, 107–113.

47. Kumar, L., Verma, S., Jamwal, S., Vaidya, S., and Vaidya, B. (2014). Polymeric microparticles-based formulation for the eradication of cutaneous candidiasis: development and characterization, *Pharm Dev Technol,* **19**, 318–325.

48. Dianzani, C., Zara, G. P., Maina, G., Pettazzoni, P., Pizzimenti, S., Rossi, F., Gigliotti, C. L., Ciamporcero, E. S., Daga, M., and Barrera, G. (2014). Drug delivery nanoparticles in skin cancers, *Biomed Res Int,* **2014**, 1–13.

49. Das, S., Das, J., Samadder, A., Paul, A., and Khuda-Bukhsh, A. R. (2013). Efficacy of PLGA-loaded apigenin nanoparticles in Benzo[a]pyrene and ultraviolet-B induced skin cancer of mice: mitochondria mediated apoptotic signalling cascades, *Food Chem Toxicol,* **62**, 670–680.

50. Srivastava, A. K., Bhatnagar, P., Singh, M., Mishra, S., Kumar, P., Shukla, Y., and Gupta, K. C. (2013). Synthesis of PLGA nanoparticles of tea polyphenols and their strong *in vivo* protective effect against chemically induced DNA damage, *Int J Nanomedicine,* **8**, 1451–1462.

51. Paszko, E., Ehrhardt, C., Senge, M. O., Kelleher, D. P., and Reynolds, J. V. (2011). Nanodrug applications in photodynamic therapy, *Photodiagn Photodyn Ther,* **8**, 14–29.

52. Li, B., Moriyama, E. H., Li, F., Jarvi, M. T., Allen, C., and Wilson, B. C. (2007). Diblock copolymer micelles deliver hydrophobic Protoporphyrin IX for photodynamic therapy, *Photochem Photobiol,* **83**, 1505–1512.

53. Silva, C. L., Ciampo, J. O., Rossetti, F. C., Bentley, M. V. L. B., andPierre, M. B. R. (2013). Improved *in vitro* and *in vivo* cutaneous delivery of Protoporphyrin IX from PLGA-based nanoparticles, *Photochem Photobiol,* **89**, 1176–1184.

54. Beck, R., Guterres, S., and Pohlmann, A. (2011). *Nanocosmetics and Nanomedicines - New Approaches for Skin Care* (Springer, Berlin).

55. Badihi, A., Debotton, N., Frušić-Zlotkin, M., Soroka, Y., Neuman, R., and Benita, S. (2014). Enhanced cutaneous bioavailability of dehydroepiandrosterone mediated by nano-encapsulation, *J Control Release,* **189**, 65–71.

56. Azarbayjani, A. F., Khu, J. V., Chan, Y. W., and Chan, S. Y. (2011). Development and characterization of skin permeation retardants and enhancers: a comparative study of levothyroxine-loaded PNIPAM, PLA, PLGA and EC microparticles, *Biopharm Drug Dispos,* **32**, 380–388.

57. Teixeira, Z., Zanchetta, B., Melo, B. A. G., Oliveira, L. L., Santana, M. H. A., Paredes-Gamero, E. J., Justo, G. Z., Nader, H. B., Guterres, S. S., and Durán, N. (2010). Retinyl palmitate flexible polymeric nanocapsules: characterization and permeation studies, *Colloids Surf B Biointerfaces*, **81**, 374–380.

58. Woodruff, M. A., and Hutmacher, D. W. (2010). The return of a forgotten polymer – Polycaprolactone in the 21st century, *Prog Polym Sci*, **35**, 1217–1256.

59. Lboutounne, H., Chaulet, J. F., Ploton, C., Falson, F., and Pirot, F. (2002). Sustained *ex vivo* skin antiseptic activity of chlorhexidine in poly(ε-caprolactone) nanocapsule encapsulated form and as a digluconate, *J Control Release*, **82**, 319–334.

60. Gomaa, Y. A., El-Khordagui, L. K., Garland, M. J., Donnelly, R. F., McInnes, F., and Meidan, V. M. (2012). Effect of microneedle treatment on the skin permeation of a nanoencapsulated dye, *J Pharm Pharmacol*, **64**, 1592–1602.

61. Zhang, W., Ding, B., Tang, R., Ding, X., Hou, X., Wang, X., Gu, S., Lu, L., Zhang, Y., Gao, S., and Gao, J. (2011). Combination of microneedles with PLGA nanoparticles as a potential strategy for topical drug delivery, *Curr Nanosci*, **7**, 545–551.

62. Vučen, S. R., Vuleta, G., Crean, A. M., Moore, A. C., Ignjatović, N., and Uskoković, D. (2013). Improved percutaneous delivery of ketoprofen using combined application of nanocarriers and silicon microneedles, *J Pharm Pharmacol*, **65**, 1451–1462.

63. Alexander, A., Dwivedi, S., Ajazuddin, Giri, T. K., Saraf, S., and Tripathi, D. K. (2012). Approaches for breaking the barriers of drug permeation through transdermal drug delivery, *J Control Release*, **164**, 26–40.

64. Michinaka, Y., and Mitragotri, S, (2011). Delivery of polymeric particles into skin using needle-free liquid jet injectors, *J Control Release*, **153**, 249–254.

65. Tomoda, K., Yabuki, N., Terada, H., and Makino, K. (2014). Application of polymeric nanoparticles prepared by an antisolvent diffusion with preferential solvation for iontophoretic transdermal drug delivery, *Colloid Polym Sci*, **292**, 3195–3203.

66. Tomoda, K., Watanabe, A., Suzuki, K., Inagi, T., Terada, H., and Makini, K. (2012). Enhanced transdermal permeability of estradiol using combination of PLGA nanoparticles system and iontophoresis, *Colloids Surf B Biointerfaces*, **97**, 84–89.

67. Hansen, S., and Lehr, C. M. (2012). Nanoparticles for transcutaneous vaccination, *Microb Biotechnol*, **5**, 156–167.

68. Mittal, A., Raber, A. S., Lehr, C. M., and Hansen, S. (2013). Particle based vaccine formulations for transcutaneous immunization, *Hum. Vaccin Immunother*, **9**, 1950–1955.

69. Mattheolabakis, G., Lagoumintzis, G., Panagi, Z., Papadimitriou, E., Partidos, C. D., and Avgoustakis, K. (2010). Transcutaneous delivery of a nanoencapsulated antigen: induction of immune responses, *Int J Pharm*, **385**, 187–193.

70. Fartasch, M. (1996). The nature of the epidermal barrier: structural aspects, *Adv Drug Deliv Rev*, **18**, 273–282.

71. Mittal, A., Raber, A. S., Schaefer, U. F., Weissmann, S., Ebensen, T., Schulze, K., Guzmán, C. A., Lehr, C. M., and Hansen, S. (2013). Non-invasive delivery of nanoparticles to hair follicles: a perspective for transcutaneous immunization, *Vaccine*, **31**, 3442–3451.

72. DeMuth, P. C., Garcia-Beltran, W. F., Ai-Ling, M. L., Hammond, P. T., and Irvine, D. J. (2013). Composite dissolving microneedles for coordinated control of antigen and adjuvant delivery kinetics in transcutaneous vaccination, *Adv Funct Mater*, **23**, 161–172.

Chapter 12

Aliphatic Polyester Protein Drug Delivery Systems

Brian Amsden
Department of Chemical Engineering, Queen's University Kingston, Ontario K7L 3N6, Canada
amsden@queensu.ca

12.1 Introduction

Proteins are an important drug class because of their potential to treat a wide range of conditions. Compared to small drug entities, protein therapeutics are highly specific in their action, and are expected to be less toxic than synthetically derived molecules as well as to behave more predictably *in vivo* [1]. They thus represent a significant potential market. The global therapeutic proteins market has been estimated to have been worth $93 billion in 2010 and has been forecast to grow to $141.5 billion by 2017 [2].

Currently, parenteral administration is the route of choice of protein drugs. This is because of the stability requirements of proteins and the barriers to their absorption by other conventional

Handbook of Polyester Drug Delivery Systems
Edited by MNV Ravikumar

ISBN 978-981-4669-65-8 (Hardcover), 978-981-4669-66-5 (eBook)
www.panstanford.com

routes such as oral, nasal, buccal, and transdermal [3, 4]. However, parenteral delivery may not be suitable for a number of protein drugs. Most therapeutic proteins have a short *in vivo* half-life (Table 12.1) and, upon injection, are unevenly distributed in the interstitial fluid and unable to reach the desired physiologic sites. They may also bind unselectively to cellular receptors and thus cause undesirable side effects. Furthermore, many therapeutic proteins are produced locally to act on cells in the immediate environment. These proteins are often active at very low concentrations, of from 10^{-9} to 10^{-11} M [5], and are required at the local tissue site for a prolonged period of time [6, 7]. Thus, administration regimens typically consist of multiple injections, often at supraphysiologic concentrations, which pose problems with patient compliance and possible complications when administered in a non-clinical setting. A long-term continuous and localized protein drug delivery depot could therefore provide numerous and distinct advantages, both therapeutic and financial, for many therapeutic proteins, and in many cases, is required for their clinical application.

Table 12.1 Serum half-lives of selected proteins [8, 9]

Protein	Function	Half-life (min)
Insulin	Controls glucose metabolism	<25
Interferon-γ	Immunomodulator	25–35
Growth hormone	Stimulates skeletal growth	18
Erythropoietin	Stimulates erythrocyte production	360
Granulocyte colony stimulating factor	Immunomodulator	300
Interleukin-2	Immunonodulator	14–198
Vascular endothelial growth factor A	Stimulates blood vessel growth	<3

In this chapter, the use of aliphatic polyesters in the development of protein-releasing depot formulations will be reviewed. First, the unique nature of protein molecules and the resulting stability issues that arise from their properties are presented. Following that, an overview of the physical properties of the most commonly used aliphatic polyesters is given. The discussion of the formulations that follows is organized as to the nature of the polymer form, starting with solid micro- and nanoparticles, then injectable semisolids, followed by viscous hydrophobic liquids and elastomers, and concluding with hydrogels.

12.2 The Nature of Proteins

When designing a depot system for protein drugs, it is important to consider the nature of protein molecules. Proteins are linear polymers consisting of amino acids linked via a peptide bond. Each amino acid unit in the polymer backbone is called a residue. The linear polymers formed by the linking of amino acids together are called polypeptides. Proteins are polypeptides consisting of greater than 50 amino acid residues, or polypeptides with a molecular weight of greater than about 5500 g/mol. Polypeptides consisting of fewer than 50 amino acid residues are commonly referred to as peptides [10]. There are 20 amino acids found in proteins, in different molar ratios as well as sequence. These amino acids may be hydrophobic or hydrophilic, or may be ionizable in aqueous solution. In solution, proteins assume a three-dimensional or folded shape that arises from the interactions of the amino acids along the polypeptide chains as well as with their environment. The conformation that is achieved is a result of a minimization of the Gibbs free energy at the existing environmental conditions.

The specific function of a protein therapeutic is governed by its folded, three-dimensional conformation. A protein's conformation is dependent on disulfide linkages and noncovalent forces, which include hydrogen bonding, electrostatics, hydrophobic interactions and van der Waals interactions. These forces are relatively weak, under the most favorable conditions the folded state is stabilized by only 5–15 kcal/mol, and are easily disrupted by changes in the environment of the protein such as pH, ionic strength, and temperature [11]. When these interactions are disrupted, the protein is likely to undergo a structural change, potentially rendering it inactive or immunogenic [12]. Proteins are also susceptible to other inactivating influences, including aggregation and adsorption at interfaces, as well as chemical reactions such as deamidation, acylation catalyzed by the presence of acids, isomerization, cleavage, oxidation, thiol disulfide exchange, and β-elimination in aqueous solutions [3, 13, 14]. There are, therefore, a number of processing and polymer factors to be considered in the formulation of therapeutic proteins, such as temperature, shear, the presence of surfactants, the use of appropriate buffers, the presence of organic solvents, control of ionic strength, and the presence of oxidizers such as ions, radicals and peroxide, and light.

12.3 Aliphatic Polyesters

Since the first report of sustained drug delivery from poly(lactide) in 1970 [15], there has been considerable effort devoted to the investigation of drug delivery from biodegradable aliphatic polyesters, and in particular poly(glycolide-*co*-lactide). The popularity of these polymers stems from their long history of *in vivo* safety as part of approved biomedical devices. For example, many commercial biodegradable sutures are composed of these copolymers [16].

Polyesters used in drug formulation are generally prepared through ring-opening polymerization. The reaction is typically carried out in the melt, using a catalyst along with a low molecular weight nucleophile, usually an alcohol, to initiate the polymerization and control the molecular weight of the resulting polymer [17]. A variety of monomers can be incorporated into the polymer in this way, including ε-caprolactone, lactide, and glycolide, (Table 12.2), which provides for facile manipulation of polymer properties.

The homopolymers poly(ε-caprolactone), poly(L-lactide), and poly(glycolide) are all semi-crystalline, while poly(D,L-lactide) is amorphous due to its atactic nature. Copolymerization is often used to manipulate thermal properties, and in particular the degree of crystallinity and glass transition temperature of the polymer. These thermal properties have strong influences on the degradation rate of the copolymer, as discussed below.

Aliphatic polyesters degrade *in vivo* principally through hydrolysis of the labile ester linkage, ultimately leading to loss of mass, or erosion, of the sample [19]. The erosion process is dependent on the relative rates of water diffusion into the bulk of the polymer and the kinetics of the hydrolysis reaction [20]. When the rate of hydrolysis is faster than the rate of diffusion of water into the polymer, polymer backbone degradation is restricted to a thin region near the surface. Thus, the polymer exhibits loss of mass primarily from the surface in a process termed surface erosion (Fig. 12.1A). Conversely, when the rate of hydrolysis is slower than the rate at which water diffuses into the polymer, the polymer sample is relatively rapidly hydrated, and bulk erosion occurs (Fig. 12.1B). For many of the polyesters that have been examined for protein drug delivery, water penetration into the polymer occurs much more rapidly than does hydrolysis and these polymers therefore undergo bulk erosion.

Table 12.2 Structure of monomers commonly used in the preparation of polyesters examined for protein drug delivery and the glass transition temperatures (T_g) and melting points (T_m) of their homopolymers

Monomer	Structure	T_g (°C)	T_m (°C)
ε-caprolactone	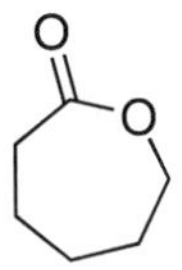	−60 to −65 [18]	60
D,L-lactide	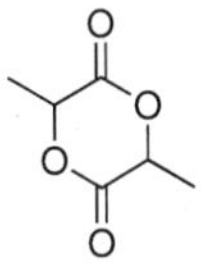	57 [16]	–
L-lactide	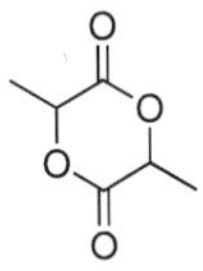	65 [16]	175
glycolide	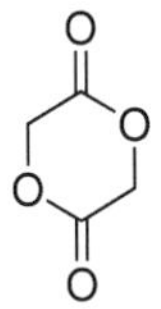	35 [16]	225

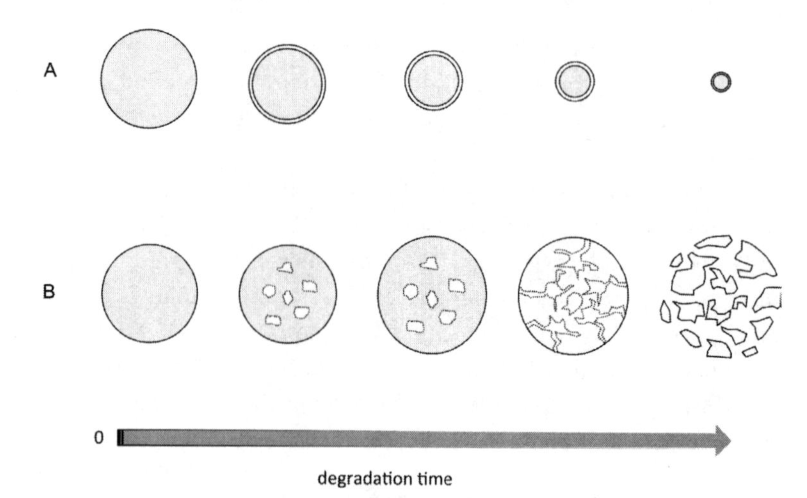

Figure 12.1 Schematic representation of the changes in polymer sample appearance during (A) surface erosion and (B) bulk erosion. During surface erosion, the erosion zone is restricted to the surface and relatively thin and the sample becomes smaller with time while retaining its three-dimensional shape. During bulk erosion, polymer chain cleavage occurs throughout the bulk following rapid hydration of the sample. As a result of the accumulation of acidic degradation products in the interior, accelerated degradation may occur in isolated regions within the sample (represented as the lighter regions outlined with dashed lines). The acidic products also induce an osmotic pressure driven water transport into the sample, causing it to swell and cracks or pores to form. Ultimately, the sample breaks apart and the smaller particles continue to undergo bulk erosion.

In the case of bulk erosion, the polymer undergoes hydrolysis throughout the bulk of the sample. However, appreciable mass loss is not observed for a prolonged period of time, despite the fact that the polymer chains are undergoing cleavage and thus the average sample molecular weight is decreasing. The hydrolysis reaction occurs randomly along the polymer backbone, and mass is not lost until a fragment of low enough molecular weight is formed that is water-soluble. Furthermore, this low molecular weight fragment must also be able to be transported out of the bulk of the sample. As the diffusion of these low molecular weight fragments through the polymer itself is very slow, appreciable mass loss only results if a pathway is available for transport to the surface, such as pores, are present in the polymer bulk. Additionally, as a result of the cleavage of the ester group, the lower molecular weight polymer or oligomer fragments that are formed bear either a carboxylic acid or alcohol end group. The acids catalyze the hydrolysis reaction and hence, aliphatic polyesters are susceptible to auto-catalyzed degradation. Thus, degradation rates in the interior of the device can be greater than at the surface, due to the accumulation of acidic degradation products in the interior with the accompanying acceleration of the hydrolysis reaction, and the presence of buffering agents in the external aqueous environment [19]. As a result of the presence of the acidic degradation products within the interior of the sample, water may be drawn into the polymer through osmotic action, if there are no pores within the sample. This increase in water content causes the polymer sample to swell, and then break into smaller fragments, at which point significant mass loss is observed.

A consideration of mass transport principles indicates that the rate of water penetration into the polymer sample is governed by the diffusivity of water within the polymer and the solubility of water within the polymer. Water diffusivity within the polymer is dependent on polymer chain flexibility, increasing as chain flexibility increases and hence as the glass transition temperature of the polymer decreases. Water diffusion is also dependent on the degree of crystallinity of the polymer. The water molecules can only move within the amorphous regions of the polymer [19, 21]. Water solubility within the polymer is determined by the nature of the comonomers used and the nature of the incorporated drug and/or excipients. For example, glycolide is more hydrophilic than ε-caprolactone, as a result of the five membered alkane segment of ε-caprolactone, and so poly(glycolide) will absorb more water than does poly(ε-caprolactone). Hydrophilic drugs or excipients will either dissolve to form pores that increase the surface area over which water diffusion can occur, or act in an osmotic fashion to draw water into the polymer bulk. Hydrophobic drugs or excipients can either act as plasticizers, reducing the polymer glass transition temperature and increase water penetration into the polymer, or can retard water penetration into the bulk [22].

The hydrolysis rate is also determined by the ability of water to access the ester group. For example, the steric hindrance posed by the methyl group of lactide is one reason why poly(lactide) degrades more slowly than does poly(glycolide). For these reasons, the rate of cleavage of monomer–monomer ester linkages along the polyester backbone of the monomers listed in Table 12.2 is, from fastest to slowest, glycolide > lactide > ε-caprolactone.

12.4 Formulations

An effective formulation for protein delivery must possess a number of features. The protein structure and bioactivity must be retained during fabrication, storage, and following release, as patient safety and drug efficacy can be compromised if even a small fraction of the protein molecules is degraded [12, 23]. An appropriate release rate must be generated, in terms of local concentration and duration, to achieve the desired therapeutic response. Although often desirable, the release rate does not need to be constant. For example, it has been demonstrated that vascular endothelial growth factor-A (VEGF) needs a higher initial release to begin angiogenesis, followed by a steady but lower release rate that is still within the therapeutic window to continue the angiogenesis process [24]. Conversely, epidermal growth factor (EGF) requires prolonged continuous release to be effective [25]. The protein must also be completely released from the device. Incomplete protein release is commonly reported, and is often a result of protein-polymer adsorption or complexation, protein aggregation, or degradation of the protein [12, 23, 26]. Finally, the formulation should be capable of being manufactured in relatively large-scale so as to yield a reproducible, sterile product. It is also often desirable for the formulation to be capable of administration via simple injection through standard gauge needles for minimally invasive localization to the desired site of action.

12.4.1 Micro- and Nanoparticles

One of the most studied formulation approaches is the encapsulation of the protein as dispersed solid particles within aliphatic polyester microspheres [27, 28]. These formulations are typically made using poly(lactide-*co*-glycolide) (PLG). PLG is used most often because it is available commercially in varying molecular weight and composition, can be amorphous, depending on monomer composition, and because of the hydrolytic susceptibility of the glycolide and lactide units. Hence its degradation duration can be manipulated from hours to several months, increasing with decreased glycolide content [16].

Protein release from microspheres generally occurs in three phases: an initial burst, followed by diffusion controlled release, and then erosion controlled release (Fig. 12.2) [29]. The initial burst is due to surface resident protein particles being dissolved and transported quickly away from the depot surface. The burst release phase can be problematic, as it may result in the protein concentration exceeding the minimum toxic concentration of the protein drug or inducing downregulation of the protein drug's cell receptors.

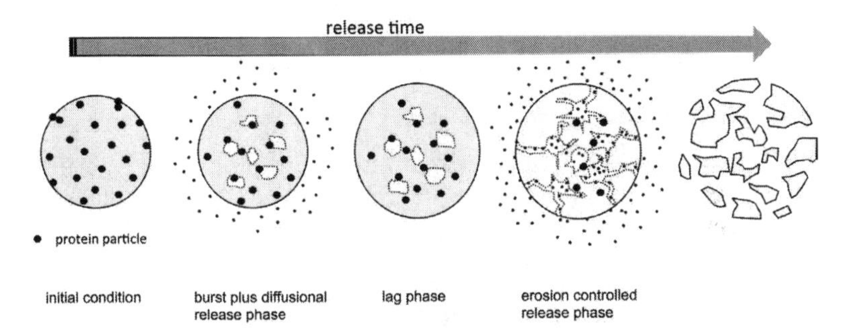

Figure 12.2 Schematic of the release phases from PLG microspheres. Note that a lag phase may not be present, depending on the total particle loading and/or amount of a pore-forming agent initially present, or the initial porosity of the microsphere.

The diffusion-controlled release phase is a result of dissolved protein diffusing through the water-filled pores and channels within the microspheres. These pores can be present as a result of the manufacturing process, formed by the dissolution of dispersed protein-containing particles connected to each other such that they form a cluster that reaches the surface, or formed through the incorporation of hydrophilic excipients such as water-soluble, low molecular weight compounds or polymers, e.g., poly(ethylene glycol). The diffusional release phase dominates until all the interconnected particles are dissolved. If the total particle loading (protein-containing plus excipient particles) is above a critical volumetric fraction, called the percolation threshold, then diffusion dominates throughout the majority of the release phase, and the release duration is typically short. The percolation threshold is defined as the volume fraction of dispersed particles at which enough particles are touching so as to form a path spanning the thickness of the device (Fig. 12.3) [30]. For a spherical geometry, this percolation threshold value is at a volume loading of approximately 30–40%, depending on the size of the incorporated particles relative to the radius of the microsphere [31]. For a given microsphere radius, larger particles will exhibit a lower percolation threshold.

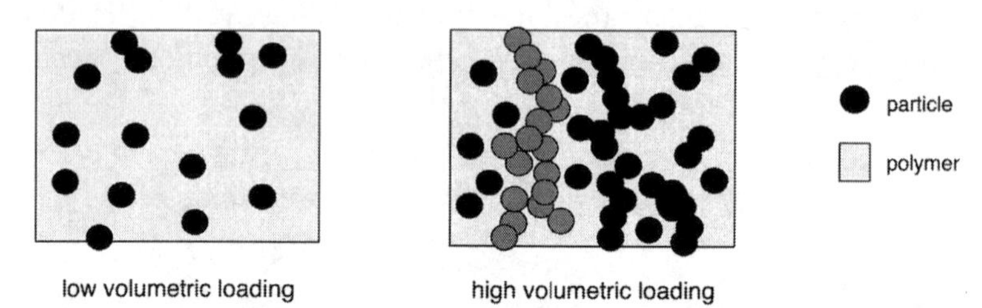

Figure 12.3 Schematic representation of the influence of drug particle loading on the fraction of particles released during the diffusion phase. In the image on the left, wherein the particle loading is low, very few particles are connected to the surface directly or by connection to a particle in contact with the surface. For this volumetric loading, the diffusional phase will be short and the majority of the drug will be released through polymer erosion. In the image on the right, the particle volumetric loading is above the percolation threshold, which is illustrated by the cluster of particles that are interconnected to each other in such a way as to span the thickness of the device (colored gray). At this volumetric loading, the majority of the drug particles in the device will be released by first dissolving to form a pore followed by diffusion through the pore to the surface.

If particle loading is low such that very few particles are connected to the surface (less than around 5–10%), the diffusional phase will be short in duration and a lag phase will be exhibited following the diffusional phase (Fig. 12.4). This lag phase may also be a result of closure of the pores initially present and/or formed by the dissolution of the protein particles [32], although the influence of this effect would depend on the size of the initial pores.

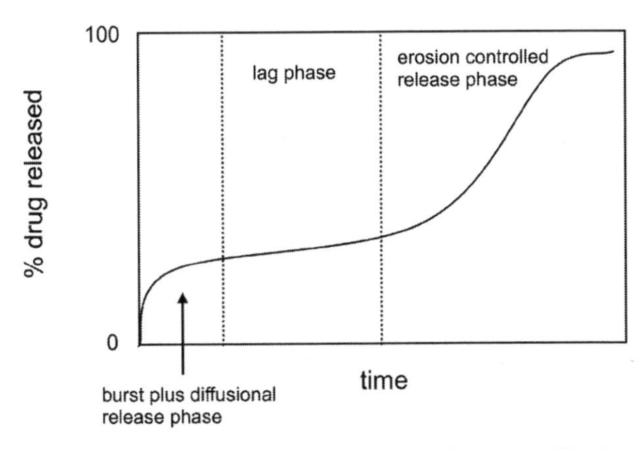

Figure 12.4 Representative triphasic protein release profile from PLG microspheres.

At some point, polymer mass loss is obtained, pores and cracks are formed in the microsphere as a result of the polymer bulk erosion process, and the release rate increases, a process referred to as the secondary burst. The release rate will also increase as a result of the fragmentation of the microsphere. Although this triphasic release pattern is often observed, a continuous, and sometimes nearly constant, release rate from PLG microspheres (Fig. 12.5) has been demonstrated to be achievable [33–36]. To obtain a continuous release rate, the diffusion phase must overlap with the erosion release phase.

PLG microspheres have the advantages of not only providing a continuous release, but also of being easily injected to the target site, providing a long term release duration, and of having a reasonable shelf-life. Nonetheless, they also have disadvantages, including the need for reconstitution before injection and the possibility of migration from the site of injection. However, the most significant problem with PLG microspheres as a delivery system is maintenance of protein stability, both during microsphere preparation and release [28, 37, 38].

Conventionally, incorporation of the protein drug into the polymer was achieved through either an emulsion approach, where the protein was dissolved in the aqueous phase while the polymer was dissolved in an organic solvent, or a suspension approach wherein the protein was suspended as solid particles in a polymer–

organic solvent solution [27]. The latter approach consistently yielded higher protein encapsulation efficiencies and bioactivity retention as protein interactions with the polymer, solvent, or shear effects at the solvent–water interface are reduced [38]. New processing techniques have recently been developed that use either mild solvents such as dimethyl sulfoxide or are solvent free and which produce uniform particles in a large scale that are suitable for protein encapsulation with high encapsulation efficiency and retained bioactivity. For an overview of these manufacturing processes, the interested reader is referred to the excellent review of Mao *et al.* [39].

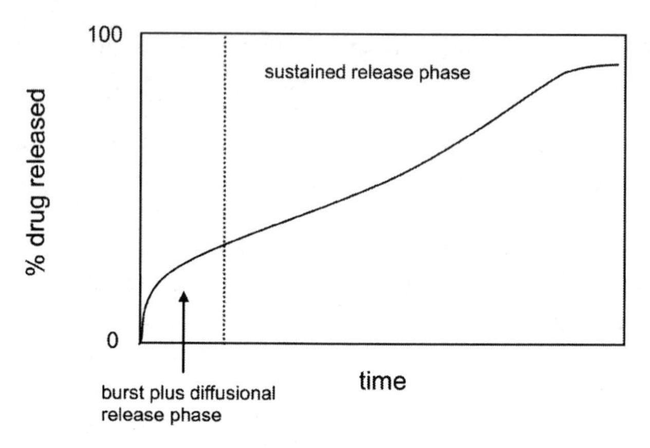

Figure 12.5 Representative sustained protein release profile achieved through balance of erosion phase and the creation of pores throughout the microsphere, resulting in an increase in the rate of diffusional release with time.

A more significant issue however, is maintenance of protein activity during the release period. When polymers such as PLG degrade, they liberate oligomers and monomers with carboxylic acid end groups. The presence of these oligomers and monomers has been shown to decrease the local pH at the surface of the polymer and in the pores and channels of the device. In fact, the pH at the centre of PLG microspheres has been determined to be as low as from 1.5 [40] to 1.8 [41]. This reduction in the pH of the inner environment of the microspheres has been linked to inactivation and denaturation of many proteins within PLG microspheres prior to being released. For example, when fast-degrading PLG microspheres were used to deliver interleukin-1α, the cytokine lost its activity during incubation, and the extent of cytokine inactivation was consistent with the microsphere degradation rate [42]. Other protein drugs for which this effect has been noted are carbonic anhydrase [43], atriopeptin III [44], and VEGF [45]. Besides being potentially acid-labile, most protein therapeutics are prepared through recombinant techniques, and many recombinant proteins are non-glycosylated. This non-glycosylated state makes them susceptible to aggregation with pH changes, and aggregation can lead to precipitation from solution and/or a loss of biological activity [3, 46]. For example, interleukin-2 remains unaggregated at pH 7, but undergoes aggregation at pH < 5 [47]. This aggregation can be irreversible and lead to immunogenic complications [12].

Strategies that have been employed to reduce the impact of the lowered microenvironmental pH have included blending poly(ethylene glycol) (PEG) [48] into PLG, utilizing PEG-PLG block copolymers [49], and the co-incorporation of particles of basic salts [50–52]. The inclusion of a basic excipient, which upon dissolution would neutralize the acidic products, has been explored using salts such as $Mg(OH)_2$, $Ca(OH)_2$, and $ZnCO_3$ [50]. Nevertheless, inclusion of basic salt particles to control the acidic microenvironment has been largely unsuccessful, as distributing the basic salt particles homogeneously throughout the microsphere is difficult, and so pH varied throughout the interior of the polymer [53]. Furthermore, the presence of the salt particles in the formulation would make controlling the release kinetics of the protein, and thus manufacture of reproducible microspheres, more complicated. With respect to blending PEG into PLG, PEG has only a limited solubility in PLG, and forms a two-phase system. Blending PEG with PLG in a microsphere formulation maintained the internal pH at between 5 and 5.8 over four weeks in PBS likely due to enhanced diffusion of the acidic degradation products out of the microsphere [54]. However, when this formulation strategy was utilized for the release of VEGF, a significant burst of nearly 60% of the loaded growth factor was observed by the first measured time point of 4 days [51]. The use of PEG-PLG block copolymers for forming microspheres for protein delivery was considered to be able to provide both prevention of protein adsorption to the PLG as PEG is highly protein adsorption resistant, as well as providing a gel-like environment through which the acidic degradation products could readily be transported out of the interior. Within the microsphere bulk, the PEG chains would orient themselves towards the aqueous regions wherein the dissolved protein resides [49]. Nevertheless, incomplete release of proteins such as erythropoietin and tetanus toxoid were achieved, likely a result of aggregation induced by the presence of free PEG molecules in the aqueous regions [55]. Further, in microspheres prepared from diblock copolymers of PEG and poly(D,L-lactide), acid catalyzed acylation of atrial natriuretic peptide still occurred [56].

Recognizing that many therapeutic proteins of interest are recombinant and thus non-glycosylated, covalent attachment of PEG to the protein (termed PEGylation) is another approach that has been explored to overcome issues with respect to aggregation and adsorption [57]. Conjugating PEG to proteins is known to enhance protein stability [58], and PEGylation has been demonstrated to limit protein adsorption to blank PLG microspheres [59] and increase the total fraction of protein released [59–62]. Moreover, in most of the reports of the release of PEGylated proteins, a decreased burst release *in vitro* was observed [59, 61, 62]. Thus, protein PEGylation is a promising approach for overcoming aggregation and adsorption issues in developing PLG and other aliphatic polyester microsphere formulations. That said, it is essential to control the site and extent of PEG conjugation to the protein to prevent significant modifications to its physicochemical properties and its activity relative to the native protein.

Another approach that has been explored to improve the release of stable and bioactive proteins from PLG microspheres is to first prepare porous microspheres, load the protein drug into the pores in a subsequent step, then close the pores at the surface. Kim *et al.* achieved a sustained release of recombinant human growth hormone for over 1 month by generating porous PLG microspheres in this manner [63]. The pores were formed by incorporating Poloxamer 407 as an extractable pore-forming agent. Poloxamers are triblock copolymers composed of a central poly(propylene oxide) block flanked at each end with poly(ethylene glycol) blocks. Following removal of the Poloxamer 407, the microspheres were loaded with protein by dipping into a solution of the growth hormone. The pores were then closed using a water miscible solvent that partially dissolves PLG. In a similar strategy Chung *et al.* achieved a continuous release of basic fibroblast growth factor (FGF-2) by first functionalizing the surface of the pores with primary amine groups and using the amines to conjugate heparin to the surface of the pores. FGF-2 release was controlled by taking advantage of the reversible binding between heparin and the growth factor [64]. In a variation on this approach, the Schwendeman group employed a pore self-healing strategy. In this technique, following immersion of the porous PLG microspheres in the protein drug solution and allowing sufficient time for the solution to fill the pores, the pores at the surface are closed by raising the temperature to above the glass transition temperature of the hydrated PLG. This temperature ranges from 37 to 43°C, depending on the molecular weight and composition of the polymer [65, 66]. This is, however, a complicated manufacturing strategy, that may be difficult to scale up.

More recently, PLG nanoparticles have been examined for therapeutic protein delivery, as they are more readily injectable, and thus less prone to clogging the needle, than microspheres [67, 68]. Protein incorporation into these nanoparticles is accomplished through first generating a water/oil/water double emulsion and employing high shear homogenization. These high shear conditions are responsible for protein denaturation during PLG microsphere preparation, and so steps must be taken to ensure that the protein remains non-denatured during nanoparticle formation. Other common potential issues with nanoparticle delivery systems are that release begins with a large initial burst effect and release durations are typically short. For example, d'Angelo *et al.* entrapped FGF-2 and platelet-derived growth factor-BB (PDGF-BB) in PLG:Poloxamer blends [68]. Under optimized conditions, the encapsulation efficiency of PDGF-BB was 87%, that of FGF-2 was 63%, while the production yield was 73%. Moreover, the incorporation of Poloxamer resulted in high growth factor bioactivity following release. Nevertheless, a large initial burst of 50–60% was observed within the first 24 hours, and release was essentially complete at only 78 ± 7% by 7 days.

As the degradation rate of poly(ε-caprolactone) (PCL) is much slower than that of PLG [69, 70], PCL has been examined for protein delivery with the reasoning that acidic degradation products would be less problematic. There are, however, few reports of PCL being used to prepare protein loaded microspheres [71–75], and in a recent report, it was determined that protein aggregation was problematic and the reason for incomplete release of encapsulated recombinant human epidermal growth factor [74].

12.4.2 Modified Monomers

As one of the principal issues with the use of PLG is the accumulation of acidic degradation products within the device, recently glycolide [76] and ε-caprolactone [77] have been modified so as to be more hydrophilic and used to form polymers examined for protein delivery (Fig. 12.6). The rationale for this strategy was that water absorption would be enhanced, thereby swelling the device and resulting in more rapid transport of the acidic degradation products out of the device interior.

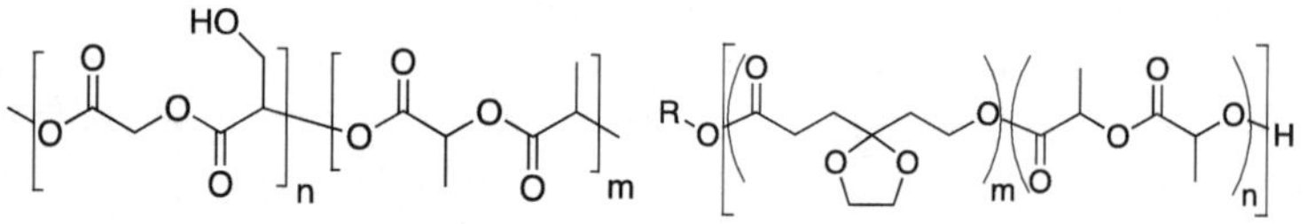

Figure 12.6 Chemical structures of poly(hydroxymethyl glycolide-*co*-D,L-lactide) (left) and poly(5-ethylene ketal ε-caprolactone-*co*-D,L-lactide) (right).

Ghassemi *et al.* prepared copolymers of hydroxymethyl glycolide and lactide and used these copolymers to prepare microspheres containing the model protein drugs lysozyme and bovine serum albumin (BSA). These copolymers were amorphous with glass transition temperatures ranging between 34 and 38°C. They degraded in a nearly linear fashion with time with the degradation rate increasing with increased hydroxymethyl lactide composition [78]. Both the release of lysozyme and BSA were governed principally by the degradation of the polymers, and both model proteins appeared to maintain structural integrity [78, 79]. Moreover, the microenvironmental pH within the microspheres was far less acidic than in PLG microspheres. In particular, the pH within poly(hydroxymethyl glycolide-*co*-D,L-lactide) of 35 molar % hydroxymethyl glycolide remained consistently above 5 throughout the time period studied [80]. Nevertheless, acid catalyzed acylation of an encapsulated peptide, octreotide, was observed, albeit to a lower extent than the same peptide encapsulated in PLG [13].

Babasola and Amsden have recently examined low molecular weight (~2400 g/mol) copolymers of 5-ethylene ketal ε-caprolactone and D,L-lactide as a delivery vehicle for both VEGF and hepatocyte growth factor (HGF) [81]. These copolymers were amorphous with glass transition temperatures of from –24 to –30°C, degraded in a nearly linear fashion with time both *in vitro* and *in vivo* [82, 83], and were well tolerated following subcutaneous implantation in rats [83]. These copolymers were designed to be liquid at body temperature, a delivery strategy that is discussed in Section 12.4.4.

12.4.3 Injectable *in situ* Forming Semisolid Implants

In contrast to micro- or nanoparticle formulations, for which protein incorporation is complicated, liquid based polymer implants can be injected through standard gauge needles and allow for the straightforward incorporation of thermally sensitive drugs such as proteins by simple mixing of protein solid particles into the liquid at room temperature. With these advantages in mind, injectable liquid polyester formulations based on *in situ* solidification approaches first introduced by Dunn *et al.* [84] have been investigated for protein delivery. In these formulations, the polymer, typically PLG although poly(lactide) and poly(ε-caprolactone-*co*-lactide) have been used [85], is dissolved in a water-soluble organic solvent throughout which solid protein particles have been suspended. Upon injection into tissue, the solvent diffuses into the surrounding tissue and the polymer precipitates, entrapping the protein particles within a solid or semisolid depot. Protein release would then occur under similar driving forces as have been described for microspheres, i.e., a mixture of diffusion through porous channels and depot erosion. This formulation approach has been commercialized for the delivery of the peptide leuprolide acetate (Eligard©), which uses *N*-methyl-2-pyrrolidone (NMP) as the solvent, and has been examined for the delivery of insulin [86], tumor necrosis factor [87], and bone morphogenetic proteins [88].

One limitation of this formulation approach is the high initial burst that can occur as a consequence of the time required for polymer precipitation following injection. Various parameters have been studied for controlling the burst effect, including polymer molecular weight, polymer concentration, and the hydrophobicity of the solvent system used. Each of these parameters will influence the rate at which polymer precipitation occurs. In general, a slower polymer precipitation rate results in a reduced burst effect with the added benefit of a more sustained release profile [89]. The reduced burst effect is a result of the fact that the depot remains a viscous solution for a longer period following injection, and a thicker solidified polymer skin surrounding the depot is formed that has a lower porosity. Thus, the use of a higher molecular weight PLG and/or a higher polymer concentration will result in a reduction in the burst amount. However, these approaches would result in a polymer solution with a higher viscosity, which may make it more difficult to inject. Another strategy is to use a more hydrophobic solvent or solvent combination. Several solvents have been explored, with the most common being NMP and dimethyl sulfoxide (DMSO) due to pharmaceutical precedence in approved parenteral products [85]. However, they are both freely miscible with water and cause a high initial burst effect. For that reason, more hydrophobic solvents such as triacetin, benzyl alcohol, and benzyl benzoate have been investigated [89]. Selection of the organic solvent used also requires consideration of protein–solvent interactions as well as potential toxicity and induction of local inflammation upon injection into the tissue. NMP, DMSO, benzyl alcohol, and benzyl benzoate are all components of injectable formulations for human use, while triacetin is approved in veterinary products [89], although there has been little evaluation of their toxicological profile in parenteral administration. The preliminary preclinical data of their toxicity and host tolerance in *in situ* precipitating PLG formulations indicate that they are well tolerated [89], although further studies are required.

12.4.4 Injectable Viscous Liquid Implants

With respect to the local tissue implantation of protein drug depots, viscous, amorphous, hydrophobic, liquid polymers based on polyesters possess potential advantages over micro- and nanoparticulate formulations, as well as liquid *in situ* forming implants. These advantages include avoiding the use of organic solvents, and the viscous, amorphous nature of the polymer may limit irritation when implanted in soft tissue.

To form an aliphatic polyester such that it is a liquid at body temperature but can also be readily injected through a needle requires that the molecular weight of the polymer is kept fairly low (usually less than a few thousand g/mol), and also that it be amorphous or have a melting point less than 37°C and that its glass transition temperature should be less than about 0–10°C [90]. For these reasons, the polymers used often contain ε-caprolactone as its homopolymer has a very low glass transition temperature (Table 12.2). For example, Ethicon has patented low molecular weight copolymers of ε-caprolactone and lactide for local drug delivery that are viscous liquids [91].

Polymer viscosity can also be adjusted by the choice of initiator used in the polymerization, as for low molecular weight polymers, the initiator can represent a major portion of the overall macromolecule formed. For example, it has been demonstrated that the longer the chain length of an alkanol initiator used, the lower the melt viscosity of the polymer, until a chain length of 8 carbons, after which there was no noticeable effect. Additionally, the use of secondary alcohols resulted in higher viscosities while the use of an unsaturated alcohol also reduced the melt viscosity of the polymer [92, 93].

Despite their potential advantages, to date, none of these polymers have been used in the delivery of protein therapeutics. This is likely because, as their molecular weights are low and hydrolytic degradation increases as polymer molecular weight decreases, they will rapidly degrade to generate an acidic microenvironment, detrimental to the stability of the entrapped protein.

Reasoning that problems associated with polymer microenvironmental acidity could be avoided if the protein was released prior to extensive degradation of the polymer, osmotic pressure driven release has been exploited to deliver highly bioactive protein therapeutics from liquid injectable polymers [81, 94, 95]. The use of osmotic pressure as the driving force results in a sustained release profile that does not rely on polymer erosion. In these formulations, either hepatocyte growth factor (HGF) or VEGF was lyophilized with serum albumin and trehalose. Both albumin and trehalose act in multiple roles: to protect the protein therapeutic during lyophilization, to prevent aggregation and adsorption of the protein to the polymer, and to draw water into the polymer via an osmotic activity gradient. The lyophilized particles were reduced to diameters of 2–5 μm, and then mixed via stirring into poly(D,L-lactide-*co*-5-ethylene ketal ε-caprolactone). Following injection into an aqueous medium, protein release was determined to occur through regions of excess hydration within the polymer, formed around the dispersed particles as a result of enhanced water absorption via osmosis, as shown in Fig. 12.7 [94]. With this mechanism, because of the low particle loadings used (10%), *in vitro* release from a spherical geometry occurred with no burst effect and was sustained for 40 days for VEGF and 90 days for HGF. Furthermore, the microenvironmental pH within the polymer remained at greater than 5.8 over the 56 days measured. As a result, high growth factor bioactivity was retained throughout the release duration.

12.4.5 Elastomers

Elastomers are amorphous, crosslinked polymers that have a glass transition temperature far below their operating temperature and that are capable of recovering their original dimensions following stretching. Biodegradable elastomers based on ε-caprolactone and D,L-lactide have been synthesized and explored, mostly as scaffolds for tissue engineering [96], but also as a means for growth factor delivery [97–101]. The use of elastomers as a delivery vehicle for proteins allows for the utilization of osmotic pressure for driving release and generating a constant release rate.

In elastomers, the osmotic release mechanism proceeds as follows (Fig. 12.8) [102–104]. Water partitions into and diffuses through the polymer until it encounters a polymer-surrounded drug particle (hereinafter referred to as a capsule). At the particle/polymer interface, the water phase separates and dissolves a portion of the solid particle to form a saturated solution. The water activity in the saturated solution is much less than that in the surrounding aqueous medium, setting up an appreciable activity gradient. As a result of this activity gradient, water is drawn into the capsule and the capsule swells, generating a pressure equal to the osmotic pressure of the dissolved solid. This pressure is resisted by the elastic nature of the polymer. As the polymer is strained, energy is stored by polymer chain extension, bond bending, or bond stretching. This energy is dissipated if bond breakage or viscoelastic flow occurs.

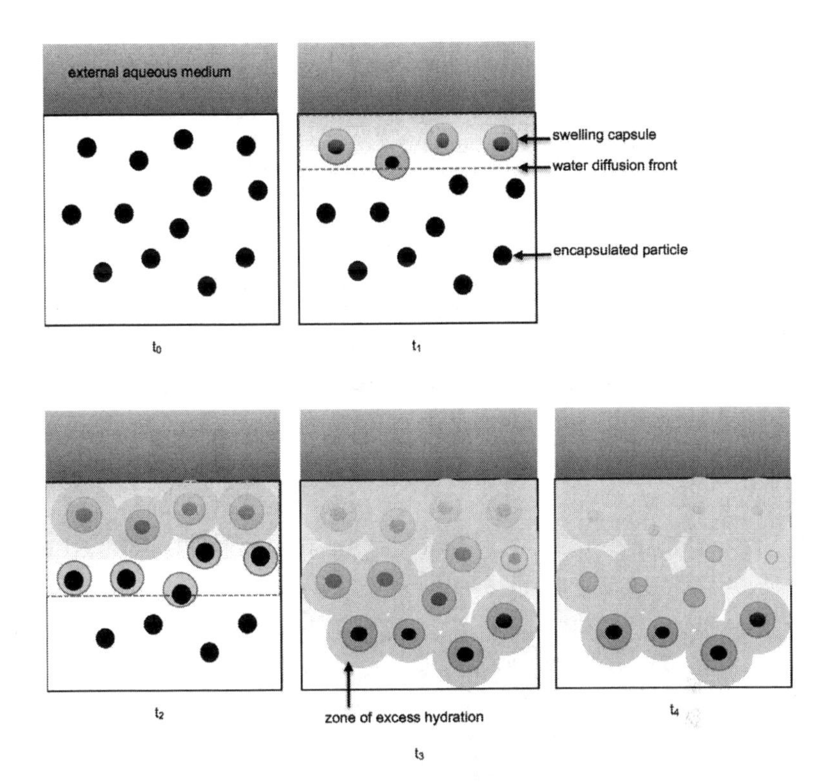

Figure 12.7 Graphical representation of the osmotic release mechanism from liquid hydrophobic polymers. At t_0, the polymer/particle suspension is first introduced into the aqueous medium. At t_1, water from the surrounding medium dissolves into and diffuses through the polymer matrix until it encounters a polymer-enclosed particle. At the particle/polymer interface, the water phase dissolves a portion of the particle to form a saturated solution. The activity gradient between the solution at the particle surface and the surrounding medium draws water into the polymer to generate a pressure equal to the osmotic pressure of the saturated solution. The osmotic pressure forms zones of excess hydration in the surrounding polymer (t_2). The zones of excess hydration eventually overlap as the water diffusion front proceeds through the polymer (t_3). Solute is transported through the superhydrated regions that reach the polymer/aqueous medium interface. Once solid particles no longer remain, the osmotic driving force decreases. Reprinted from Ref. 94, Copyright 2013, with permission from Elsevier.

Bond breakage initiates crack formation in the polymer bulk. The crack or rupture formed connects the contents in the capsule to a pore network that ultimately extends to the surface of the device. The capsule contents are forced through the pore network under the pressure differential between the capsule and the external medium. This process occurs in a particle layer-by-particle layer manner throughout the device. If the intracapsule osmotic pressure is insufficient to initiate crack formation, thermodynamic equilibrium is reached and the capsule contents are not released until sufficient polymer degradation occurs. The release rate achieved from the osmotic mechanism is nearly zero order for much of the release period, provided the volume fraction of dispersed solid is below a critical level. Osmotic pressure driven release only dominates if the total volumetric loading of the particles in the polymer matrix is less than the percolation threshold [105].

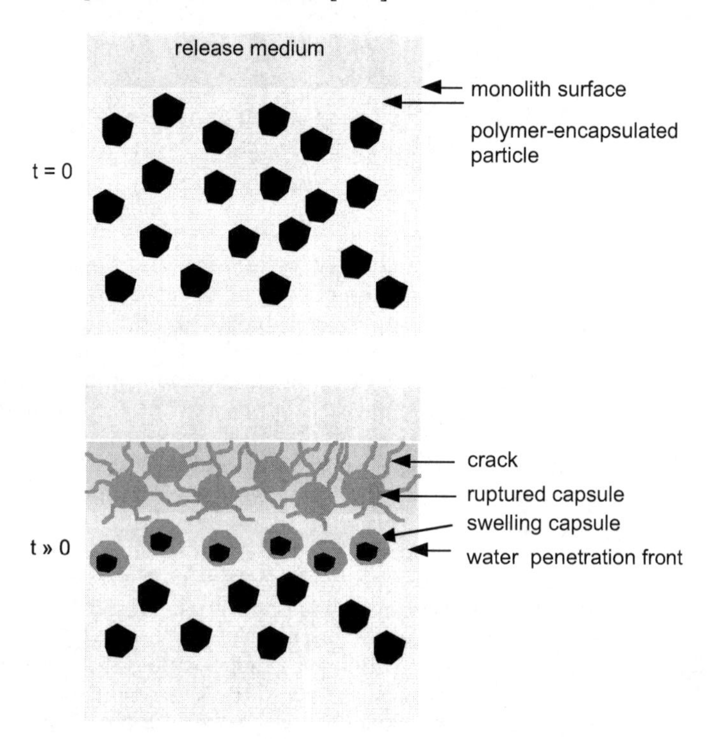

Figure 12.8 Schematic of the osmotic release mechanism for elastomer devices.

The important factors controlling this osmotically driven release are solute osmotic activity, the solute loading, particle size, device geometry, and polymer properties such as hydophobicity, Young's modulus and tear resistance. Increasing polymer hydrophobicity, modulus, or tear resistance results in a decrease in release rate. Polymer hydrophobicity is determined by the monomers chosen to synthesize the polymer, while modulus and tear resistance are controlled by manipulating both crosslink density of the elastomer as well as its glass transition temperature. Increasing the osmotic activity of the solute results in an increase in release rate, for a given polymer, solute size and volumetric loading. An increase in volume fraction of the particles produces a faster release, and as the particle decreases, the release rate decreases. For slab geometries, the release rate is zero order for much of the release duration. However, for cylinders a zero order release rate only can be approximated for up to a mass fraction released of 60% of the initially loaded particles [104], and for spheres, only for approximately a mass fraction released of 25% [31].

Using this release strategy, Chapanian *et al.* photoencapsulated VEGF and HGF within cylinders composed of 3-arm *star*-poly(ε-caprolactone-*co*-trimethylene carbonate-*co*-D,L-lactide) terminally functionalized with acrylate groups. Photocuring was chosen as it occurs rapidly, at room temperature and with minimal heat generation, and so minimal denaturation of the solid protein occurred during device fabrication. The growth factors were co-lyophilized with trehalose and BSA in a pH succinate buffer that also contained 5 % w/v NaCl, then ground and sieved so as to be less than 25 μm in diameter. These particles were then loaded into the 3 mm diameter × 12 mm long cylinders at a volume fraction of 10%. When assessed in an *in vitro* release medium (phosphate buffered saline, pH 7.4) the growth factors were released in a sustained fashion for 10 days, the release rate of the growth factors was independent of the polymer degradation rate, and the released growth factors were highly bioactive, retaining on average 80% of the bioactivity of as-received growth factor [97].

Elastomers are also highly suitable for combined tissue engineering and growth factor delivery. In these applications, the elastomer provides both a scaffold of appropriate mechanical properties to support cell growth and differentiation while also combining protein drug delivery to drive tissue formation. This combined approach was demonstrated by Guan *et al.* [101]. In that study, a thermoplastic poly(ester-urethane)urea elastomer was prepared by reacting poly(ε-caprolactone) with butyldiisocyanate, which was then chain extended with 1,4-diaminobutane. Composite particles of basic fibroblast growth factor (FGF-2) plus bovine serum albumin with or without heparin were distributed throughout the polymer. In this case, release was primarily diffusionally controlled. FGF-2 release began with a high initial burst of roughly 37% with subsequent release being much slower and reaching approximately 70% after 28 days. The FGF-2 bioactivity was retained for only up to 21 days, after which no bioactivity remained. Nevertheless, this work demonstrates the potential of combining controlled protein release and a cell scaffold of appropriate mechanical properties to generate *de novo* tissue.

12.4.6 Hydrogels

ABA triblock copolymers of hydrophilic PEG (block A) and hydrophobic aliphatic polyesters such as PLG and PCL (block B) have also been explored for protein delivery in the form of hydrogels. In aqueous solution, these triblock copolymers are thermosensitive, such that they can exhibit a lower critical solution temperature and thus are solutions at room temperature but quickly phase separate at higher temperatures to form a weak hydrogel [106]. Phase separation temperature is concentration, block configuration (ABA or BAB), and PEG and PLG or PCL block length dependent [106]. For example, Jeong *et al.* have shown that PEG-PLG-PEG (ABA) copolymers with PEG block molecular weight of 550 g/mol and a PLG block molecular weight of 2810 g/mol undergoes a phase transition under the appropriate concentrations near body temperature [107] (Fig. 12.9). In contrast, PLG-PEG-PLG (BAB) triblock copolymers (PLG molecular weight 1500 g/mol, PEG molecular weight 1000 g/mol) exhibit lower gelation concentrations and a lower solution-to-gel transition temperature (Fig. 12.10) [108]. Following subcutaneous injection in rats the PLG-PEG-PLG gels persisted for over 4 weeks [109], and were well tolerated [110]. These polymers have been trade-named ReGel®, and have been examined in clinical trials for formulations containing paclitaxel [111]. The delivery of a number of proteins from ReGel® has been examined, which have included interleukin-2 (IL-2) [112], insulin [113], and granulocyte colony stimulating factor (G-CSF) [110].

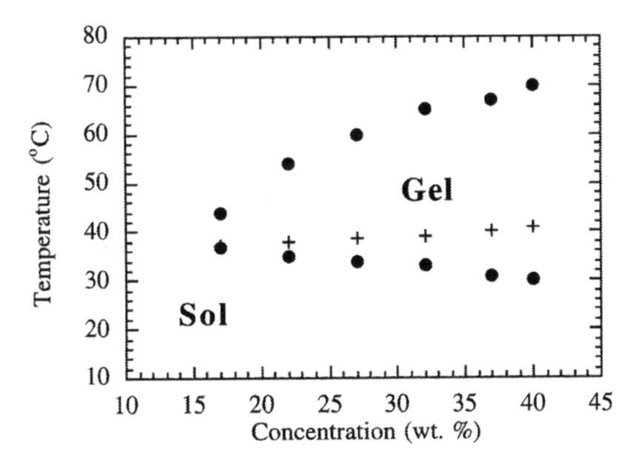

Figure 12.9 Phase diagram of PEG-PLG-PEG in aqueous solution. The "+" indicates the temperature at which the gel phase becomes turbid. Reprinted from Ref. 107, Copyright 2000, with permission from Elsevier.

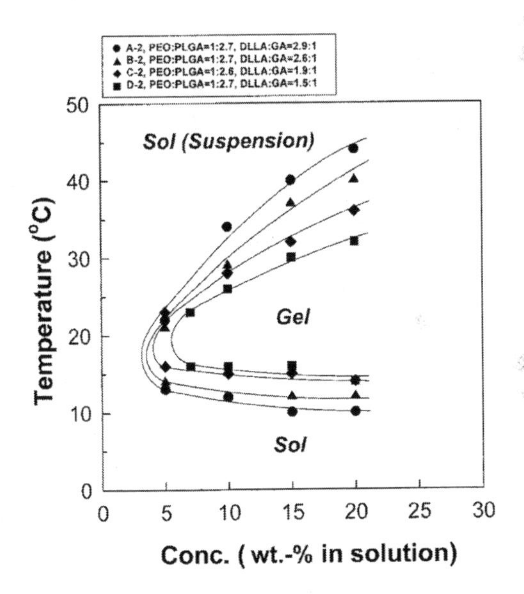

Figure 12.10 Phase diagrams of PLG-PEG-PLG (1500-1000-1500) triblock copolymers of varying glycolide:lactide monomer ratios in aqueous solution. Reprinted from Ref. 108, Copyright 2001, with permission from Elsevier.

The success of the ReGel® formulation led to the development of other polymers utilizing different hydrophobic polyester blocks [106]. As an example, Hyun *et al.* synthesized PCL-PEG (750 g/mol PEG, 2490 g/mol PCL) diblocks that undergo thermogelation and demonstrated that these hydrogels could release FITC-labeled bovine serum albumin *in vitro* in a sustained manner over 20 days, and *in vivo* for up to 30 days [114].

In these thermosensitive hydrogel formulations, protein loading is straightforward and little loss of protein bioactivity following release has been reported. However, release is diffusionally controlled. Protein diffusion within hydrogels occurs within the aqueous regions and therefore is strongly dependent on the water content of the hydrogel and the molecular weight of the protein, increasing as the water content increases and as the size of the protein decreases [115]. Given the high water contents of the gels, relatively short release durations of proteins are obtained. For example, IL-2 formulated in ReGel® was released over 3–4 days, but apparently without loss of bioactivity [112]. The release rate can be slowed by distributing the protein as a particle of the lower solubility salt form, as demonstrated by Kim *et al.* for insulin by employing a Zn-insulin salt. *In vitro*, the release of Zn-insulin from ReGel® was zero-order and began without an initial burst. Moreover, plasma insulin levels were maintained over 15 days after a subcutaneous injection of ReGel® containing 0.2 wt% Zn-insulin in rats [113]. However, this strategy of utilizing poorly soluble protein salts is not readily transposable to other protein therapeutics.

In a different approach, hydrogels were prepared by first dissolving PEG-poly(lactide)-PEG triblock copolymers in tetraglycol, then gradually introducing water to phase separate the triblock copolymer, forming a hydrogel [116]. In this hydrogel system, named MedinGel, the poly(lactide) formed crosslink points through assembly into hydrophobic microdomains. In this formulation approach, the protein drug is added to the polymer solution prior to the water addition. The formulation was assessed using bovine serum albumin and fibrinogen as model protein drugs. The release of the lower molecular weight (66.7 kg/mol) bovine serum albumin (BSA) began with a significant initial burst of between 30 and 40%, while the burst effect was nearly eliminated when the much larger fibrinogen (400 kg/mol) was entrapped and released. Following the burst effect, protein release was sustained, and in the case of fibrinogen, nearly linear. Furthermore, complete protein release was reported, indicating little adsorption and/or aggregation of these model proteins within these systems. Drawbacks of this approach are the poor encapsulation efficiencies achieved, which for BSA ranged from 37 to 52.5%, issues with protein stability and hydrogel stability on storage, the large burst effect, and the protein molecular weight dependent release rates.

12.4.7 Formulation Manufacture and Quality by Design

When designing a formulation, the principles of Quality by Design (QbD) should be implemented. In the QbD approach, the formulation is designed in such a manner that, from conceptualization of the formulation to its use in the clinic, attributes that are critical to the performance of the formulation, or critical quality attributes, are identified and means of controlling variability in these attributes are considered to ensure consistent and reliable clinical performance. Critical quality attributes are physical, chemical, biological, or microbiological properties or characteristics that need to be maintained within an appropriate limit, range, or distribution to meet the desired product quality [117].

Common critical quality attributes for formulations based on aliphatic polyesters would include the molecular weight and molecular weight distribution of the polymer, the polymer composition (i.e., monomer molar ratios), concentration levels of residual solvents, protein encapsulation amount and efficiency, maintenance of protein bioactivity during storage and following release, a reproducible release profile, moisture content, degradation time, and sterility of the product. It should be noted that manufacture of the formulation may have to be done under aseptic conditions as standard sterilization protocols denature proteins. Critical quality attributes for microspheres would also include microsphere size and size distribution, porosity and average pore size, the concentration and distribution of excipients such as pore-forming agents or compounds to modulate microenvironmental pH, and the composition and viscosity of the diluent used to suspend the microspheres before injection. Nanoparticles would include the same critical quality attributes, but have in addition a target value for zeta potential. Formulations based on *in situ* precipitation and viscous liquid polyesters and elastomers would have as additional critical quality attributes the viscosity of the polymer, the composition of lyophilized protein particles, and the protein particle size and size distribution. Elastomer formulations would also include mechanical properties such as modulus, and crosslinking efficiency. Finally, hydrogel formulations would have as additional critical quality attributes dissolution time at a given temperature, gelation temperature, and gelation time.

12.5 Conclusions

Although tremendous efforts have been expended and progress has been made, the successful development of clinically translated aliphatic polyester formulations for protein drug delivery still faces challenges. The discussion above has highlighted some strategies that have been explored to overcome the obstacles towards successful translation of the different formulation approaches. The primary obstacles include eliminating the impact of acidic degradation products on protein stability, reducing protein aggregation and denaturation as a result of adsorption to the hydrophobic polymer phase, and reduction of the burst effect.

PLG microspheres have received the greatest attention of all formulations to date, with one product making it to market. Nutropin Depot™ was a PLG microsphere formulation designed to delivery recombinant human growth hormone for an extended time. This formulation was approved in 1999 to treat growth deficiency in children, but withdrawn from the market in 2004, principally as a result of the cost of reproducibly manufacturing the microspheres. Furthermore, a frequent adverse reaction with the Nutropin Depot® were injection-site reactions, which occurred in nearly all patients. Once injected, the microspheres caused an acute tissue reaction, manifested as nodule formation. These nodules gradually disappeared as the microspheres degraded. The development of more hydrophilic monomers for the preparation of micro- and nanoparticulate delivery systems, as well as for liquid depots or thermogelling approaches appears promising and should be examined in greater detail. However, the development cost of this strategy may be high as the newly developed monomers will have to be assessed for safety *in vivo*, delaying the emergence of these formulations into the clinic and increasing the cost of clinical trials.

Reproducibility in drug release characteristics using *in situ* precipitating, viscous liquid hydrophobic, and thermogelling formulations will be an issue, as protein release rate will be influenced by the shape, size, and structure of the depot that forms. Control over these parameters is difficult to achieve with administration being performed by varying personnel and depot formation in different tissue locations. Moving forward, this challenge may be overcome through the design of appropriate administration devices, such as autoinjectors, along with tissue imaging during administration.

New formulation approaches for effective protein therapeutics are needed, but must be developed by keeping in mind the scale-up and manufacturing costs of the formulation, ensuring that it is competitive with less costly alternatives, such as, for example, injectable and long-circulating PEGylated proteins.

References

1. Frokjaer, S., Otzen, D. (2005). Protein drug stability: a formulation challenge. *Nat Rev Drug Discov*, **4**, 298–306.
2. GBI Research (2011). Therapeutic Proteins Market to 2017 - High Demand for Monoclonal Antibodies will Drive the Market. Report code: GBIHC080MR.
3. Cleland, J., Powell, M., Shire, S. (1993). The development of stable protein formulations: a close look at protein aggregation, deamidation, and oxidation. *Critical Rev Therap Drug Carrier Sys*, **10**, 307–377.
4. Wu, F., Jin, T. (2008). Polymer-based sustained-release dosage forms for protein drugs, challenges, and recent advances. *AAPS PharmSciTech*, **9**, 1218–1229.
5. Gurdon, J. B., Bourillot, P. Y. (2001). Morphogen gradient interpretation. *Nature*, **413**, 797–803.
6. Amsden, B. G. (2011). Delivery approaches for angiogenic growth factors in the treatment of ischemic conditions. *Expert Opin Drug Del*, **8**, 873–890.
7. Tayalia, P., Mooney, D. J. (2009). Controlled growth factor delivery for tissue engineering. *Adv Mater Weinheim*, **21**, 3269–3285.

8. Kompella, U., Lee, V. (1999). Pharmacokinetics of peptide and protein drugs. In: Lee V (ed.) *Peptide and Protein Drug Delivery*, pp 391–484.
9. Lee, V. (1999). Changing needs in drug delivery in the era of peptide and protein drugs. In: *Peptide and Protein Drug Delivery*. CRC Press, pp 1–56.
10. Berg, J. M., Tymoczko, J. L., Stryer, L. (2002). *Biochemistry*, 5th ed., W H Freeman, New York.
11. Fasman, G. D. (1989). *Prediction of Protein Structure and the Principles of Protein Conformation*, Springer Science & Business Media.
12. Jiskoot, W., Randolph, T. W., Volkin, D. B., *et al.* (2011). Protein instability and immunogenicity: roadblocks to clinical application of injectable protein delivery systems for sustained release. *J Pharm Sci*, **101**, 946–954.
13. Ghassemi, A. H., van Steenbergen, M. J., Barendregt, A., *et al.* (2012). Controlled release of octreotide and assessment of peptide acylation from poly(D,L-lactide-*co*-hydroxymethyl glycolide) compared to PLGA microspheres. *Pharm Res*, **29**, 110–120.
14. Lucke, A., Kiermaier, J., Gopferich, A. (2002). Peptide acylation by poly(alpha-hydroxy esters). *Pharm Res*, **19**, 175–181.
15. Yolles, S., Eldridge, J., Woodland, J. (1970). Sustained delivery of drugs from polymer/drug mixtures. *Polym News*, **1**, 9–15.
16. Perrin, D. E., English, J. P. (1997). Polyglycolide and polylactide. In: Domb AJ, Kost J, Wiseman DM (eds) *Handbook of Biodegradable Polymers*, Hardwood Academic Publishers, New York, pp 3–27.
17. Albertsson, A.-C., Varma, I. K. (2002). Aliphatic polyesters: synthesis, properties and applications. In: *Degradable Aliphatic Polyesters*, Springer Berlin Heidelberg, Berlin, Heidelberg, pp 1–40.
18. Brode, G., Koleske, J. (1972). Lactone polymerization and polymer properties. *J Macromol Sci Chem A*, **6**, 1109–1144.
19. Anderson, J. M., Shive, M. S. (2012). Biodegradation and biocompatibility of PLA and PLGA microspheres. *Adv Drug Deliv Rev*, **64**, 72–82.
20. Burkersroda, F., Schedl, L., Gopferich, A. (2002). Why degradable polymers undergo surface erosion or bulk erosion. *Biomaterials*, **23**, 4221–4231.
21. Crank, J., Park, G. (1968). *Diffusion in Polymers*, Academic Press, New York.
22. Timmins, M., Liebmann-Vinson, A. (2003). Biodegradable polymers: degradation mechanisms, part 1. In: *Biodegradable Polymers*, Citus Books, London, pp 285–328.

23. Carpenter, J. F., Randolph, T. W., Jiskoot. W., *et al.* (2009). Overlooking subvisible particles in therapeutic protein products: gaps that may compromise product quality. *J Pharm Sci*, **98**, 1201–1205.

24. Silva, E. A., Mooney, D. J. (2010). Effects of VEGF temporal and spatial presentation on angiogenesis. *Biomaterials*, **31**, 1235–1241.

25. Sheardown, H., Wedge, C., Chou, L., *et al.* (1993). Continuous epidermal growth factor delivery in corneal epithelial wound healing. *Invest Ophthalmol Vis Sci*, **34**, 3593–3600.

26. Jiang, G., Woo, B., Kang, F., *et al.* (2002). Assessment of protein release kinetics, stability and protein polymer interaction of lysozyme encapsulated poly (D,L-lactide-*co*-glycolide) microspheres. *J Control Release*, **79**, 137–145.

27. Sinha, V., Trehan, A. (2003). Biodegradable microspheres for protein delivery. *J Control Release*, **90**, 261–280.

28. Schwendeman, S. P., Shah, R. B., Bailey, B. A., Schwendeman, A. S. (2014). Injectable controlled release depots for large molecules. *J Control Release*, **190**, 240–253.

29. Arifin, D. Y., Lee, L. Y., Wang, C.-H. (2006). Mathematical modeling and simulation of drug release from microspheres: implications to drug delivery systems. *Adv Drug Deliv Rev*, **58**, 1274–1325.

30. Siegel, R. A. (2011). Porous systems. In: Siepmann J, Siegel R A, Rathbone M J (eds) *Fundamentals and Applications of Controlled Release Drug Delivery*, Springer US, Boston, MA, pp 229–251.

31. Amsden, B. (1996). Osmotically activated protein release from electrostatically generated polymer microbeads. *AICHE J*, **42**, 3253–3266.

32. Wang, J., Wang, B. A., Schwendeman, S. P. (2002). Characterization of the initial burst release of a model peptide from poly(D,L-lactide-*co*-glycolide) microspheres. *J Control Release*, **82**, 289–307.

33. Takada, S., Uda, Y., Toguchi, H., Ogawa, Y. (1994). Preparation and characterization of copoly(D,L-lactic/glycolic acid) microparticles for sustained release of thyrotropin releasing hormone by double nozzle spray drying method. *J Control Release*, **32**, 79–85.

34. Mehta, R., Thanoo, B., DeLuca, P. (1996). Peptide containing microspheres from low molecular weight and hydrophilic poly(D,L-lactide-*co*-glycolide). *J Control Release*, **41**, 249–257.

35. Sah, H., Toddywala, R., Chien, Y. W. (1995). Continuous release of proteins from biodegradable microcapsules and *in vivo* evaluation of their potential as a vaccine adjuvant. *J Control Release*, **35**, 137–144.

36. Yeh, M., Jenkins, P., Davis, S., Coombes, A. (1995). Improving the delivery capacity of microparticle systems using blends of poly(DL-lactide-*co*-glycolide) and poly(ethylene glycol). *J Control Release,* **37**, 1–9.

37. van de Weert, M., Hennink, W. E., Jiskoot, W. (2000). Protein instability in poly(lactic-*co*-glycolic acid) microparticles. *Pharm Res,* **17**, 1159–1167.

38. van der Walle, C. F., Sharma, G., Ravikumar, M. (2009). Current approaches to stabilising and analysing proteins during microencapsulation in PLGA. *Expert Opin Drug Del*, **6**, 177–186.

39. Mao, S., Guo, C., Shi, Y., Li, L. C. (2012). Recent advances in polymeric microspheres for parenteral drug delivery—part 2. *Expert Opin Drug Del*, **9**, 1209–1223.

40. Fu, K., Pack, D., Klibanov, A., Langer, R. (2000). Visual evidence of acidic environment within degrading poly(lactic-*co*-glycolic acid) (PLGA) microspheres. *Pharm Res*, **17**, 100–106.

41. Shenderova, A., Burke, T., Schwendeman, S. (1998). Evidence for an acidic microclimate in PLGA microspheres. *Proceed Intl Symp Control Rel Bioact Mater*, **25**, 265–266.

42. Chen, L., Apte, R., Cohen, S. (1997). Characterization of PLGA microspheres for the controlled delivery of IL-1alpha for tumor immunotherapy. *J Control Release,* **43**, 261–272.

43. Crotts, G., Park, T. (1998). Protein delivery from poly(lactic-*co*-glycolic acid) microspheres: release kinetics and stability issues. *J Microencaps,* **15**, 699–713.

44. Johnson, R., Lanaski, L., Gupta, V., *et al.* (1991). Stability of atriopeptin-III in poly(D,L-Lactide-*co*-glycolide) microspheres. *J Control Release,* **17**, 61–67.

45. Kim, T., Burgess, D. (2002). Pharmacokinetic characterization of 14C-vascular endothelial growth factor controlled release microspheres using a rat model. *J Pharm Pharmacol*, **54**, 897–905.

46. Cumming, D. (1992). Improper glycosylation and the cellular editing of nascent proteins. In: Ahern, T J, Manning, M C (eds), *Stability of Protein Pharmaceuticals, Part B: In vivo Pathways of Degradation and Strategies for Protein Stabilization,* Plenum Press, New York, pp 1–42.

47. Watson, E., Kenney, W. (1988). High-performance size-exclusion chromatography of recombinant derived proteins and aggregated species. *J Chromatog*, **436**, 289–298.

48. Jiang, W., Schwendeman, S. (2001). Stabilization and controlled release of bovine serum albumin encapsulated in poly(D,L-lactide) and poly(ethylene glycol) microsphere beads. *Pharm Res*, **18**, 878–885.

49. Kissel, T., Li, Y. X., Unger, F. (2002). ABA-triblock copolymers from biodegradable polyester A-blocks and hydrophilic poly(ethylene oxide) B-blocks as a candidate for *in situ* forming hydrogel delivery systems for proteins. *Adv Drug Deliv Rev*, **54**, 99–134.

50. Zhu, G., Mallery, S., Schwendeman, S. (2000). Stabilization of proteins encapsulated in injectable poly (lactide-*co*-glycolide). *Nat Biotechnol*, **18**, 52–57.

51. King, T., Patrick, C. (2000). Development and *in vitro* characterization of vascular endothelial growth factor (VEGF)-loaded poly(DL-lactic-*co*-glycolic acid)/poly(ethylene glycol) microspheres using a solid encapsulation/single emulsion/solvent extraction technique. *J Biomed Mater Res*, **51**, 383–390.

52. Lavelle, E. C., Yeh, M. K., Coombes, A. G., Davis, S. S. (1999). The stability and immunogenicity of a protein antigen encapsulated in biodegradable microparticles based on blends of lactide polymers and polyethylene glycol. *Vaccine*, **17**, 512–529.

53. Li, L., Schwendeman, S. (2005). Mapping neutral microclimate pH in PLGA microspheres. *J Control Release*, **101**, 163–173.

54. Ding, A. G., Schwendeman, S. P. (2008). Acidic microclimate pH distribution in PLGA microspheres monitored by confocal laser scanning microscopy. *Pharm Res*, **25**, 2041–2052.

55. Giteau, A., Venier-Julienne, M. C., Aubert-Pouessel, A., Benoit, J. P. (2008). How to achieve sustained and complete protein release from PLGA-based microparticles? *Int J Pharm*, **350**, 14–26.

56. Lucke, A., Fustella, E., Tessmar, J., *et al.* (2002). The effect of poly(ethylene glycol)-poly(D,L-lactic acid) diblock copolymers on peptide acylation. *J Control Release*, **80**, 157–168.

57. Pai, S. S., Tilton, R. D., Przybycien, T. M. (2009). Poly(ethylene glycol)-modified proteins: implications for poly(lactide-*co*-glycolide)-based microsphere delivery. *AAPS J*, **11**, 88–98.

58. Jevsevar, S., Kunstelj, M., Porekar, V. G. (2010). PEGylation of therapeutic proteins. *Biotechnol J*, **5**, 113–128.

59. Diwan, M., Park, T. G. (2001). Pegylation enhances protein stability during encapsulation in PLGA microspheres. *J Control Release*, **73**, 233–244.

60. Diwan, M., Park, T. G. (2003). Stabilization of recombinant interferon-alpha by pegylation for encapsulation in PLGA microspheres. *Int J Pharm*, **252**, 111–122.

61. Castellanos, I. J., Al-Azzam, W., Griebenow, K. (2005). Effect of the covalent modification with poly(ethylene glycol) on alpha-chymotrypsin stability upon encapsulation in poly(lactic-*co*-glycolic) microspheres. *J Pharm Sci*, **94**, 327–340.

62. Byeon, H. J., Kim, I., Choi, J. S., *et al.* (2015). PEGylated apoptotic protein-loaded PLGA microspheres for cancer therapy. *Int J Nanomedicine*, **10**, 739–748.

63. Kim, H., Chung, H., Park, T. (2006). Biodegradable polymeric microspheres with "open/closed" pores for sustained release of human growth hormone. *J Control Release*, **112**, 167–174.

64. Chung, H. J., Kim, H. K., Yoon, J. J., Park, T. G. (2006). Heparin immobilized porous PLGA microspheres for angiogenic growth factor delivery. *Pharm Res*, **23**, 1835–1841.

65. Shah, R. B., Schwendeman, S. P. (2014). A biomimetic approach to active self-microencapsulation of proteins in PLGA. *J Control Release*, **196**, 60–70.

66. Reinhold, S. E., Desai, K.-G. H., Zhang, L., *et al.* (2012). Self-healing microencapsulation of biomacromolecules without organic solvents. *Angew Chem Int Ed Engl*, **51**, 10800–10803.

67. Golub, J. S., Kim, Y.-T., Duvall, C. L., *et al.* (2010). Sustained VEGF delivery via PLGA nanoparticles promotes vascular growth. *Am J Physiol Heart Circ Physiol*, **298**, H1959–H1965.

68. d'Angelo, I., Garcia-Fuentes, M., Parajó, Y., *et al.* (2010). Nanoparticles based on PLGA:Poloxamer blends for the delivery of proangiogenic growth factors. *Mol Pharmaceut*, **7**, 1724–1733.

69. Perrin, D., English, J. (2010). Polycaprolactone. In: Domb A, Kost J, Wiseman D (eds), *Handbook of Biodegradable Polymers*. CRC Press, Amsterdam, pp 63–77.

70. Lemoine, D., Francois, C., Kedzierewicz, F., *et al.* (1996). Stability study of nanoparticles of poly(ε-caprolactone), poly(D,L-lactide) and poly(D,L-lactide-*co*-glycolide). *Biomaterials*, **17**, 2191–2197.

71. Youan, B., Benoit, M. A., Baras, B., Gillard, J. (1999). Protein-loaded poly(ε-caprolactone) microparticles. I. Optimization of the preparation by (water-in-oil)-in water emulsion solvent evaporation. *J Microencapsulation*, **16**, 587–599.

72. Shenoy, D. B., D'Souza, R. J., Tiwari, S. B., Udupa, N. (2003). Potential applications of polymeric microsphere suspension as subcutaneous depot for insulin. *Drug Dev Ind Pharm*, **29**, 555–563.

73. Jameela, S. R., Suma, N., Jayakrishnan, A. (1997). Protein release from poly(ε-caprolactone) microspheres prepared by melt encapsulation and solvent evaporation techniques: a comparative study. *J Biomat Sci-Polym E*, **8**, 457–466.

74. Haushey, L. A., Bolzinger, M. A., Fessi, H., Briançon, S. (2010). rhEGF microsphere formulation and *in vitro* skin evaluation. *J Microencapsulation*, **27**, 14–24.

75. Coccoli, V., Luciani, A., Orsi, S., *et al.* (2008). Engineering of poly(ε-caprolactone) microcarriers to modulate protein encapsulation capability and release kinetic. *J Mater Sci - Mater Med*, **19**, 1703–1711.

76. Leemhuis, M., van Nostrum, C., Kruijtzer, J., *et al.* (2006). Functionalized poly(alpha-hydroxy acid)s via ring-opening polymerization: toward hydrophilic polyesters with pendant hydroxyl groups. *Macromolecules*, **39**, 3500–3508.

77. Tian, D., Dubois, P., Grandfils, C., Jerome, R (1997). Ring-opening polymerization of 1,4,8-trioxapirol[4,5]-9-undecanone: a new route to aliphatic polyesters bearing functional pendent groups. *Macromolecules*, **30**, 406–409.

78. Ghassemi, A. H., van Steenbergen, M. J., Talsma, H., *et al.* (2010). Hydrophilic polyester microspheres: effect of molecular weight and copolymer composition on release of BSA. *Pharm Res*, **27**, 2008–2017.

79. Ghassemi, A. H., van Steenbergen, M. J., Talsma, H., *et al.* (2009). Preparation and characterization of protein loaded microspheres based on a hydroxylated aliphatic polyester, poly(lactic-*co*-hydroxymethyl glycolic acid). *J Control Release*, **138**, 57–63.

80. Liu, Y., Ghassemi, A. H., Hennink, W. E., Schwendeman, S. P. (2012). The microclimate pH in poly(D,L-lactide-*co*-hydroxymethyl glycolide) microspheres during biodegradation. *Biomaterials*, **33**, 7584–7593.

81. Babasola, I. O., Rooney, M., Amsden, B. G. (2013). Corelease of bioactive VEGF and HGF from viscous liquid poly(5-ethylene ketal ε-caprolactone-*co*-D,L-lactide). *Mol Pharmaceut*, **10**, 4552–4559.

82. Babasola, O. I., Amsden, B. G. (2011). Surface eroding, liquid injectable polymers based on 5-ethylene ketal ε-caprolactone. *Biomacromolecules*, **12**, 3423–3431.

83. Babasola, I. O., Bianco, J., Amsden, B. G. (2012). *In vivo* degradation and tissue response to poly(5-ethylene ketal ε-caprolactone-*co*-D,L-lactide). *Biomacromolecules*, **13**, 2211–2217.
84. Dunn, R. L., English, J. P., Cowsar, D. R., Vanderbilt, D. P. (1990). Biodegradable *in situ* forming implants and methods of producing the same. US Patent 4,938,763.
85. Hatefi, A., Amsden, B. (2002). Biodegradable injectable *in situ* forming drug delivery systems. *J Control Release*, **80**, 9–28.
86. Dhawan, S., Kapil, R., Kapoor, D. N. (2011). Development and evaluation of *in situ* gel-forming system for sustained delivery of insulin. *J Biomater Appl*, **25**, 699–720.
87. Eliaz, R., Kost, J. (2000). Characterization of a polymeric PLGA-injectable implant delivery system for the controlled release of proteins. *J Biomed Mater Res*, **50**, 388–396.
88. Andriano, K., Chandrasekhar, B., McEnery, K., *et al.* (2000). Preliminary *in vivo* studies on the osteogenic potential of bone morphogenetic proteins delivered from an absorbable puttylike polymer matrix. *J Biomed Mater Res*, **53**, 36–43.
89. Parent, M., Nouvel, C., Koerber, M., *et al.* (2013). PLGA *in situ* implants formed by phase inversion: critical physicochemical parameters to modulate drug release. *J Control Release*, **172**, 292–304.
90. Amsden, B. G. (2010). Liquid, injectable, hydrophobic and biodegradable polymers as drug delivery vehicles. *Macromol Biosci*, **10**, 825–835.
91. Scopelianos, A., Bezwada, R., Arnold, S. (1998). Injectable liquid copolymers for soft tissue repair and augmentation. Patent number 5,824,333.
92. Amsden, B., Hatefi, A., Knight, D., Bravo-Grimaldo, E. (2004). Development of biodegradable injectable thermoplastic oligomers. *Biomacromolecules*, **5**, 637–642.
93. Mikhail, A., Sharifpoor, S., Amsden, B. (2006). Initiator structure influence on thermal and rheological properties of oligo(ε-caprolactone). *J Biomat Sci-Polym E*, **17**, 291–301.
94. Babasola, I. O., Zhang, W., Amsden, B. G. (2013). Osmotic pressure driven protein release from viscous liquid, hydrophobic polymers based on 5-ethylene ketal ε-caprolactone: potential and mechanism. *Eur J Pharm Biopharm*, **85**, 765–772.

95. Amsden, B. G., Timbart, L., Marecak, D., *et al.* (2010). VEGF-induced angiogenesis following localized delivery via injectable, low viscosity poly(trimethylene carbonate). *J Control Release,* **145**, 109–115.

96. Amsden, B. (2007). Curable, biodegradable elastomers: emerging biomaterials for drug delivery and tissue engineering. *Soft Matter,* **3**, 1335–1348.

97. Chapanian, R., Amsden, B. G. (2010). Combined and sequential delivery of bioactive VEGF165 and HGF from poly(trimethylene carbonate) based photo-cross-linked elastomers. *J Control Release,* **143**, 53–63.

98. Chapanian, R., Amsden, B. G. (2010). Osmotically driven protein release from photo-cross-linked elastomers of poly(trimethylene carbonate) and poly(trimethylene carbonate-*co*-D,L-lactide). *Eur J Pharm Biopharm,* **74**, 172–183.

99. Gu, F., Younes, H. M., El-Kadi, A., *et al.* (2005). Sustained interferon-gamma delivery from a photocrosslinked biodegradable elastomer. *J Control Release,* **102**, 607–617.

100. Gu, F., Neufeld, R., Amsden, B. (2006). Osmotic driven release kinetics of bioactive therapeutic proteins from a biodegradable elastomer are linear, constant, similar and adjustable. *Pharm Res,* **23**, 782–789.

101. Guan, J., Stankus, J., Wagner, W. (2007). Biodegradable elastomeric scaffolds with basic fibroblast growth factor release. *J Control Release,* **120**, 70–78.

102. Schirrer, R., Thepin, P., Torres, G. (1992). Water absorption, swelling, rupture and salt release in salt-silicone rubber compounds. *J Mater Sci,* **27**, 3424–3434.

103. Soulas, D. N., Sanopoulou, M., Papadokostaki, K. G. (2009). Comparative study of the release kinetics of osmotically active solutes from hydrophobic elastomeric matrices combined with the characterization of the depleted matrices. *J Appl Polym Sci,* **113**, 936–949.

104. Amsden, B. (2003). A model for osmotic pressure driven release from cylindrical rubbery polymer matrices. *J Control Release,* **93**, 249–258.

105. Amsden, B., Cheng, Y., Goosen, M. (1994). A mechanistic study of the release of osmotic agents from polymeric monoliths. *J Control Release,* **30**, 45–56.

106. He, C., Kim, S., Lee, D. (2008). *In situ* gelling stimuli-sensitive block copolymer hydrogels for drug delivery. *J Control Release,* **127**, 189–207.

107. Jeong, B., Bae, Y. H., Kim, S. W. (2000). Drug release from biodegradable injectable thermosensitive hydrogel of PEG-PLGA-PEG triblock copolymers. *J Control Release,* **63**, 155–163.

108. Lee, D. S., Shim, M. S., Kim, S. W., *et al.* (2001). Novel thermoreversible gelation of biodegradable PLGA-block-PEO-block-PLGA triblock copolymers in aqueous solution. *Macromol Rapid Comm,* **22**, 587–592.

109. Jeong, B., Bae, Y., Kim, S. (2000). *In situ* gelation of PEG-PLGA-PEG triblock copolymer aqueous solutions and degradation thereof. *J Biomed Mater Res,* **50**, 171–177.

110. Zentner, G. M., Rathi, R., Shih, C., *et al.* (2001). Biodegradable block copolymers for delivery of proteins and water-insoluble drugs. *J Control Release,* **72**, 203–215.

111. DuVall, G. A., Tarabar, D, Seidel, R. H., *et al.* (2009). Phase 2: a dose-escalation study of OncoGel (ReGel/paclitaxel), a controlled-release formulation of paclitaxel, as adjunctive local therapy to external-beam radiation in patients with inoperable esophageal cancer. *Anti-Cancer Drugs,* **20**, 89–95.

112. Samlowski, W. E., McGregor, J. R., Jurek, M., *et al.* (2006). ReGel polymer-based delivery of interleukin-2 as a cancer treatment. *J Immunother,* **29**, 524–535.

113. Kim, Y. J., Choi, S., Koh, J. J., *et al.* (2001). Controlled release of insulin from injectable biodegradable triblock copolymer. *Pharm Res,* **18**, 548–550.

114. Hyun, H., Kim, Y. H., Song, I. B., *et al.* (2007). *In vitro* and *in vivo* release of albumin using a biodegradable MPEG-PCL diblock copolymer as an *in situ* gel-forming carrier. *Biomacromolecules,* **8**, 1093–1100.

115. Amsden, B. (1998). Solute diffusion within hydrogels. Mechanisms and models. *Macromolecules,* **31**, 8382–8395.

116. Molina, I., Li, S., Martinez, M. B., Vert, M. (2001). Protein release from physically crosslinked hydrogels of the PLA/PEO/PLA triblock copolymer-type. *Biomaterials,* **22**, 363–369.

117. IEWG (2009). ICH Harmonised Tripartite Guideline. Pharmaceutical Development Q8 (R2).

Chapter 13

Biodegradable Polyester-Based Multi-Compartmental Delivery Systems for Oral Nucleic Acid Therapy

Husain Attarwala and Mansoor Amiji
Department of Pharmaceutical Sciences, School of Pharmacy, Northeastern University, Boston, MA 02115, USA
m.amiji@neu.edu

Gene and small interference RNA (siRNA) therapy has tremendous potential for the treatment of numerous diseases, including cancer as well as genetic, metabolic, and inflammatory diseases. Realization of this potential has been primarily halted by challenges associated with the delivery of these materials at desired biological target sites. Oral administration of nucleic acids is even more challenging since additional physiological barriers of the GI tract which rapidly degrade administered naked nucleic acids have to be bypassed. In order to protect nucleic acids from rapid degradation in the GI tract, a delivery vector is needed which can adequately protect the encapsulated payload, along with capability to facilitate uptake and *in vivo* expression of delivered material by desired target cells. In

Handbook of Polyester Drug Delivery Systems
Edited by MNV Ravikumar

ISBN 978-981-4669-65-8 (Hardcover), 978-981-4669-66-5 (eBook)
www.panstanford.com

this chapter, we will discuss polyester-based multicompartmental systems for oral gene and siRNA delivery developed in our laboratories. A solid-in-solid multicompartmental system referred to as **n**anoparticle-**i**n-**m**icrosphere **o**ral **s**ystem (NiMOS) consisting of type B gelatin nanoparticles encapsulated within poly(ε-caprolactone) (PCL)-based microsphere was developed for intestinal nucleic acid delivery via the oral route. This system was evaluated for local anti-inflammatory nucleic acid therapy for the treatment of inflammatory bowel disease (IBD) in colitis bearing mice. Results obtained from these studies demonstrate utility of multicompartmental systems for overcoming physiological barriers and enabling site-specific release of encapsulated material for alleviating disease symptoms.

13.1 Oral Nucleic Acid Therapies

13.1.1 Oral Gene Therapy

The concept of gene therapy refers to prevention or treatment of diseases by means of transferring genetic material to specific cells, resulting in changes in protein expression for disease treatment or prevention. Gene therapy is a very attractive field under active research both in industry and academia since all of the fundamental physiological functions are regulated by genetic expression [1]. The major barrier hindering breakthroughs in gene therapy is the delivery of genetic material to the desired cells for therapeutic effect. Nucleic acid delivery vehicles capable of transferring exogeneous genetic material *in vivo* can be divided in to two categories: (i) viral vectors and (ii) nonviral vectors. Viral vectors such as retrovirus, herpes simplex virus, adenovirus and adeno-associated virus are in general capable of producing efficient gene transfer and expression; however, they have potential toxicity concerns such as oncogenicity, immune stimulation, and mutagenesis caused by integration of exogenous genes in the host genome, resulting in limited clinical usefulness of these vectors [2–4].

Safe and effective gene delivery vectors must be developed to enable clinical translation of nucleic acid therapies. Toward this goal, nonviral vectors have been under active research with increasing investigations on polymeric nanoparticles as one way of delivering nucleic acids [5]. Polymeric nanoparticles have advantages such as ease of fabrication in to different shapes as per delivery requirement, ease of synthesis and possibility of surface modification for active targeting. Additionally, particles made up of polyester based polymers such as poly(lactic-*co*-glycolic acid) (PLGA) and poly(ε-caprolactone) (PCL) can protect nucleic acids against degradation due to extreme pH and physiological enzymes, hence allowing for adequate extended exposure of genetic payload at relevant physiological sites of drug action [6–9]. An effective gene delivery vehicle would require rationally designed and optimized formulation capable of enabling efficient uptake by desired cell types with high and extended payload exposures, along with efficient intracellular trafficking, resulting in required pharmacodynamic effects for disease treatment or prevention [10].

13.1.2 Opportunities in Oral Nucleic Acid Therapy

The route of drug administration is an important aspect to be considered during material selection and formulation fabrication, and depends upon location and physiological environment at the target site, along with considering transit barriers (extracellular and intracellular) and formulation stability characteristics required for efficient transit and delivery of encapsulated nucleic acid material. Majority of the diseases have target sites in the body which are accessible only by the systemic route of drug delivery through the bloodstream [11, 12]. For systemic delivery of nucleic acids, delivery system should be capable of remaining stable in the bloodstream, avoid filtration by kidney and clearance by reticulo-endothelial system, including liver, kidney, lungs, spleen, and capture by nontarget cells such as phagocytes [13]. However, undesirable passive uptake and accumulation of particles in tissues such as liver, kidney, and tumors can be intentionally utilized for diseases requiring targeted nucleic acid delivery to those same sites. In contrast to systemic delivery, local delivery can result in reduced off-target effects with increased on-target bioavailability, as the locally delivered drug is in close proximity with the target site of action. In addition, local delivery can result in site-limited biodistribution with minimal systemic exposure, hence helping with reduction in drug dosage and/or dosing frequency, with better safety profile and patient acceptance. Tissues such as skin [14, 15], eye [16, 17], mucous membranes [18], and local tumors [19] can benefit from localized drug delivery. In all cases, choice of drug delivery route, whether localized or systemic, and material used for engineering delivery vectors, will be dictated by the type and site of target tissue, where drug delivery is desired [10].

Local nucleic acid delivery through the oral route can be very useful for treating diseases of the gastrointestinal (GI) tract, such as inflammatory bowel disease (IBD), duodenal ulcers, GI infections, celiac disease, along with colon and gastric cancers, since the delivered material can come in direct contact with diseased sites at high concentrations before getting absorbed in to the systemic circulation [6, 8, 10]. In addition, oral route can also be used for delivering plasmid-DNA and RNA replicon-based vaccines, where uptake by the small intestinal M cells in the payer's patches can help elicit local as well as systemic immune responses against expressed antigens [4, 20–22]. However, oral delivery of nucleic acids poses potential challenges such as low transfection efficiencies attributed to aggressive enzymatic degradation in the physiological environment of the GI tract, along with tight junction between the epithelial cells, which limit the entry of foreign particles. Amiji's group [7–9, 23, 24] have developed multicompartmental systems for oral delivery of nucleic acids, which are discussed in detail in further sections of this chapter.

13.1.3 Extracellular and Intracellular Barriers to Oral Nucleic Acid Therapy

Oral route of drug administration has several advantages over systemic routes such as ease of administration, patient compliance, noninvasiveness, and overall reduced healthcare costs, along with increased bioavailability at local sites of GI tract. This is especially beneficial for GI diseases, where target sites are located in the GI tract [12]. However, there are only a few studies utilizing oral route for nucleic acid delivery via nonviral vectors since it is not as straight forward because nucleic acids face several anatomical and physiological barriers in the GI tract limiting their stability and uptake in the GI tract. For example, low gastric pH can cause rapid depurination of nucleotides, while intestinal endo- and exonucleases can rapidly degrade administered nucleic acids. Additionally, nucleic acids have to overcome physical barriers of mucous layer and epithelial cells and avoid uptake by phagocytes. Hence, only a very small percent of delivered dose is available at the target site, ultimately resulting in reduced or sub therapeutic gene transfection efficiencies. Furthermore, even after successful cellular uptake by target cell types, nucleic acids have to bypass endolysosomal degradation and either get delivered in the cellular cytosol (for siRNA therapy) or penetrate the cell nucleus (for plasmid DNA delivery and protein expression) to exert its therapeutic action [25]. Figure 13.1 illustrates some of the challenges and opportunities encountered in oral nucleic acid therapy.

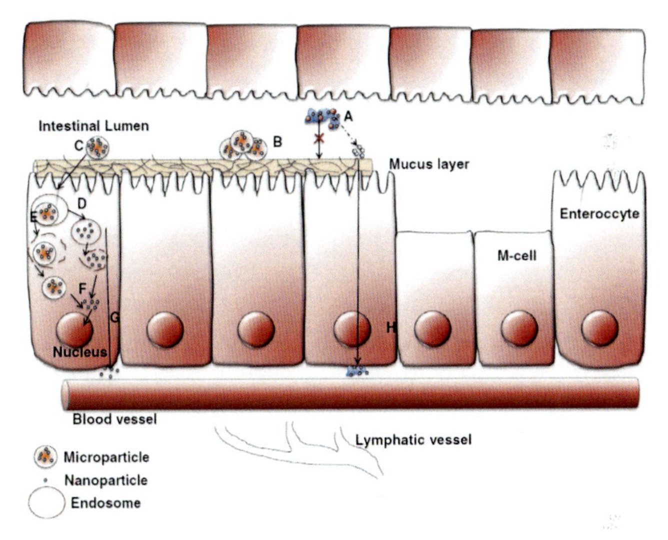

Figure 13.1 Opportunities and challenges in oral delivery of nanoparticles using microscale carriers. (A) Aggregated particles (i.e., with blood components or carrier matrix molecules) leading to restricted release; (B) physical docking and/or accumulation on cell surface; (C) microparticle internalization into the cell followed by endosomal release of (D) nanoparticles within the microscale device or (E) the microparticle itself; (F) particle uptake into the nucleus; (G) crossing of the particles into the bloodstream; (H) particles that are contaminated with the microscale carrier matrix, interfering with further delivery. Reprinted from [44] with permission from Springer.

13.2 Polyester-Based Multicompartmental Formulations

13.2.1 Biodegradable Polyesters: Chemistry and Properties

Polyester-based advanced drug delivery systems primarily utilize PCL-, PLGA-, and PLA-based matrices. These matrices allow for modification of physicochemical and mechanical properties, such as polymer molecular weight, functionalization, and copolymerization, for addressing specific biological and physiological necessities, required for delivering drugs to different physiological sites. In recent years, polymeric nano- and microparticles have been engineered to achieve delivery objectives with greater control over sustained release and target specificity [26]. The major advantage with these materials is their biodegradability, since they can get degraded either enzymatically or non-enzymatically, and produce toxicologically safe metabolites that can be further cleared by regular metabolic pathways.

PCL polymer can be synthesized with a wide molecular weight range, along with possibilities of modifying its physicochemical and mechanical properties by copolymerization or blending with a number of other polymers. For example, PCL hydrogels, dendrimers, and micelles are formulated after copolymerization, while formulations for tissue engineering such as scaffolds, films, and fibers are fabricated essentially after blending with other polymers for modifying its physical and biodegradation properties. PCL is reported to be compatible with a number of natural polymers such as hydroxyl apatite, starch, and chitosan, as well as with synthetic polymers such as polyethylene glycol (PEG), oxazolines, polyvinyl alcohol, polyethylene oxide, and oxazolines. These PCL modifications can satisfy wide range of biophysical requirements, needed for drug delivery, and hence have been widely used for development of different drug delivery systems [27–30]. PCL microspheres are largely used for controlled drug release, but can also be used for targeted drug delivery by surface modification of microspheres with targeting agents [27, 31]. In our laboratories, we have developed PCL-based systems for a wide range of therapeutic applications, including tumor-targeted drug delivery and anti-inflammatory gene therapy [7, 32–36].

PLGA is a Food and Drug Administration approved biodegradable polyester material for use in humans and has been successfully used for the development of safe drug delivery nanosystems as it undergoes hydrolysis in the body to form lactic acid and glycolic acid. These degradation products get further metabolized through regular biochemical metabolic pathways. Since the body actively metabolizes these degradation products, PLGA maintains a very clean toxicity profile, making it a material of choice for delivering therapeutic agents. PLGA has been used for delivering various drug substances ranging from small molecules and biologics, such as proteins, nucleic acids, and vaccine antigens, by modifying its physicochemical properties through blending with a wide of materials such as PEG or PEO, polyvinyl alcohol, chitosan, alginate, and pectin. In our laboratories, we have developed PLGA-based systems for tumor-targeted drug delivery and cardiometabolic gene therapy [37–40].

13.2.2 Rationale of Designing Polyester Multicompartmental Systems for Gene Delivery

An ideal oral nucleic acid delivery system should have the following three properties: it should (i) provide protection against the physiological environment and degrading enzymes of the GI tract, (ii) enable efficient targeted cellular uptake by desired target cell types, and (iii) facilitate release of nucleic acids from the endolysosomal compartment into the cellular cytosol or nucleus. In order to have all of the above needed characteristics, a formulation containing different compartments can be engineered, such that all of the above requirements are fulfilled. Multicompartmental systems can be designed such that nucleic acids are encapsulated within the inner hydrophilic phase protected by surrounding hydrophobic phase(s). The inner phase can be utilized for harboring nucleic acid cargo, protected within the surrounding outer protective phase. These systems can be targeted toward specific cell types by attaching targeting ligands either to outer hydrophobic casing or inner hydrophilic compartment [41, 42]. For instance, hydrophilic nanoparticles which are surface modified with tufsin can be used for active targeting towards macrophages [42, 43]. Another advantage of multicompartmental systems is their ability to encapsulate both hydrophilic and hydrophobic payloads in respective compartments for synergistic benefits in disease treatment [44]. Different types of multicompartmental systems can be divided into three categories: (i) liquid-in-liquid multicompartmental systems, (ii) solid-in-liquid multicompartmental systems, and (iii) solid-in-solid multicompartmental systems (Fig. 13.2). Liquid-in-liquid multicompartmental systems are primarily water-in-oil-in-water multiple emulsions comprising of an aqueous liquid internal phase, surrounded by an oily outer liquid external phase, further surrounded by an outermost aqueous phase [45]. Solid-in-liquid multicompartmental systems consist of solid nanoparticles encapsulated within liquid internal phase, protected by one or more outer phase(s). An example of this system is nanoparticles-in-emulsion (NiE) system, consisting of gelatin nanoparticles encapsulated in the innermost phase of a water-in-oil-in water multiple emulsion [41]. Solid-in-solid multicompartmental systems consist of solid nanoparticles encapsulated within solid outer microsphere casing. An example of this system is the NiMOS consisting of gelatin nanoparticles encapsulated within a polyester (PCL)-based microsphere for oral plasmid DNA and siRNA delivery [8]. The NiMOS formulation will be discussed in detail in further sections of this chapter.

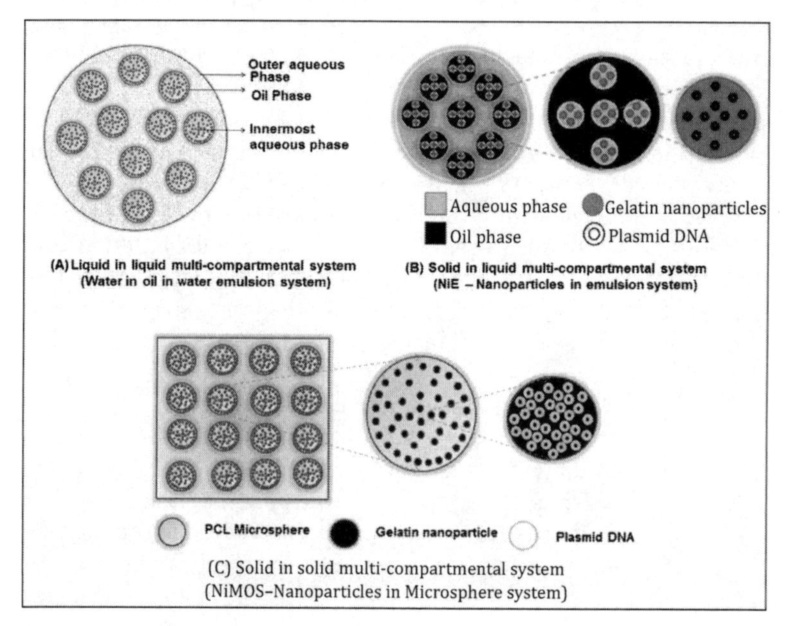

Figure 13.2 Schematic representation of multicompartmental gene delivery systems. Reprinted from [44] with permission from Springer.

13.2.3 Examples of Polyester-Based Systems for Gene Therapy

A multicompartmental system consisting of type B gelatin nanoparticles encapsulated within PCL microspheres was developed for oral delivery of plasmid DNA and siRNA for the treatment of IBD. Gelatin nanoparticles have been developed in our laboratories for plasmid DNA and siRNA delivery over the past decade and we have shown efficient gene transfection efficiencies *in vitro* and *in vivo* [8, 24, 32, 44, 46–49]. Gelatin nanoparticles administered as such by oral route can rapidly get digested in the GI tract by proteases such as pepsin and trypsin, thereby exposing the encapsulated siRNA to low gastric pH and nucleases, resulting in their rapid degradation. In order to protect gelatin nanoparticles against these physiological barriers, they were encapsulated within PCL microspheres. PCL microspheres, composed of polyester-based lipid polymer, remains stable in the presence of gastric enzymes and low gastric pH, and also protects the encapsulated material. Upon their transit to the small intestine, M cells located in the intestinal payer's patches can act as entry portals for access to immune cells and lymphatics, thus delivering the material at target sites important for inflammatory diseases. In addition, the PCL matrix can also get degraded by the intestinal lipases, resulting in the release of gelatin nanoparticles, which can be endocytosed by intestinal cells such as epithelial cells, intraepithelial lymphocytes, and macrophages. Furthermore, in the case of intestinal inflammation, PCL microspheres can also pass through the leaky epithelial junctions through passive diffusion, where they can come in direct contact and get internalized by phagocytes such as macrophages [50].

13.3 Illustrative Example of Gene Therapy with Nanoparticles-in-Microsphere Oral System

13.3.1 NiMOS Formulation

Bhavsar *et al.* [7, 8, 23, 24] developed the NiMOS formulation for the oral delivery of nucleic acids for the treatment of intestinal inflammatory diseases. The formulation consisted of two solid compartments, (i) gelatin nanoparticles encapsulating nucleic acids, and (ii) PCL microspheres encapsulating gelatin nanoparticles.

Gelatin nanoparticles with encapsulated nucleic acids were formulated by an ethanol–water solvent displacement method. Different formulation variables such as pH, rate of ethanol addition, speed of stirring, and final ethanol:water ratio were optimized to obtain less than 300 nm sized particles with greater than 95% nucleic acid loading efficiency. NiMOS were formulated using a double emulsion technique. In the first step, aqueous suspension of gelatin nanoparticles was homogenized with 0.5% (w/v) PCL in dichloromethane at 9,000 rpm for 10 minutes. In the second step, the above system was further homogenized with 0.1% (w/v) polyvinyl alcohol in water at 9000 rpm for 5 minutes. After this step, organic phase was evaporated, and the hardened microspheres were obtained after freeze-drying. The above formulation parameters were optimized using a 3^3 factorial experimental design using particles size of NiMOS as the response [24]. The major variables affecting the particle size of NiMOS were identified as follows: (i) concentration of PCL polymer in the organic phase, (ii) homogenization speed used for formation of double emulsion, and (iii) concentration of gelatin particles to be encapsulated in the internal phase of NiMOS. Experimental data obtained by modifying above three parameters were used to derive a model equation, which was then used to obtain NiMOS within the size range of 2–5 μm, suitable for oral nucleic acid delivery (Fig. 13.3) [8, 10, 24].

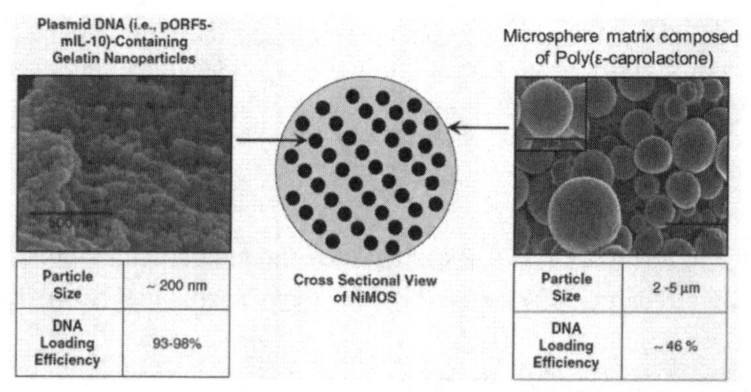

Figure 13.3 Schematic illustration showing the cross-sectional view of NiMOS. On the left is the scanning electron microscopy (SEM) image of gelatin nanoparticles, which are less than 200 nm in diameter, and can physically encapsulate plasmid DNA at a loading efficiency of >93%. On the right is the SEM image of 2–5 mm NiMOS with the overall DNA encapsulation efficiency of ~46%. Reprinted by permission from Macmillan Publishers Ltd: Ref. 8, Copyright 2008.

13.3.2 Oral Biodistribution and Gene Transfection Studies

After successful formulation development and characterization, NiMOS containing plasmid DNA were tested for their capability of effectively transfecting desired regions of the GI tract. Wistar rats [23] and Balb/C mice [7] were dosed with NiMOS, gelatin nanoparticles, and naked plasmid DNA by oral route, and the reporter gene expressions were measured. The plasmid DNA used for these studies were enhanced green fluorescent protein (EGFP-N1) or β-galactosidase (CMV-βgal), both of these which are capable of generating reporter gene products upon successful transfection [7, 23]. The results obtained from these studies showed that gelatin nanoparticles and naked plasmid DNA as such were not able to produce any measurable levels of reporter gene products, possibly due to rapid degradation by proteases and nucleases in the GI tract. Hence these formulations needed additional modifications or layer of protection for enhancing their stability in the GI tract. On the other hand, NiMOS-treated groups showed high levels of reporter gene (EGFP-N1 or CMV-βgal) expression in the small and large intestines even after 5 days post oral administration in both mouse and rat studies. These results clearly suggested that NiMOS were capable of protecting the encapsulated payload against different extracellular and intracellular barriers of small and large intestinal gene delivery, along with facilitating efficient integration of the transgene in the intracellular gene expression machinery for efficient protein expression.

Biodistribution studies conducted using ^{111}In-labeled gelatin in Wistar rats and Balb/c mice [7, 23] showed that naked gelatin nanoparticles were rapidly cleared from the GI tract in both mice and rats with predominant exposure to the large intestine at 1 hour post oral dose mice [7, 23]. For example, approximately greater 85% of delivered dose was recovered from the large intestine at 1 hour post oral dose, in the rat study [23], indicating rapid proteolytic degradation of gelatin nanoparticles in the GI tract, resulting in reduced GI exposure and a short residence time. On the other hand, majority of the ^{111}In-labeled gelatin nanoparticles administered dose resided in the small and large intestines at 1 and 4 hours post dose, in the case of NiMOS, administered orally. For example, after 1 hour, 55% and 10% of the administered dose was recovered from the small and large intestines, respectively. Even after 4 hours post dose approximately 25% of administered dose was recovered from the large intestine (Fig. 13.4). Moreover, negligible systemic exposure of gelatin nanoparticles was observed, as evidenced by negligible radioactive recoveries from plasma and liver. These results show that NiMOS are capable of producing stable and sustained GI nucleic acid delivery, with localized exposures to small and large intestines [7, 23].

13.3.3 Efficacy of Oral IL-10 Gene Therapy in Inflammatory Bowel Disease Model

IL-10 plays an important role in regulating various mucosal immune responses by modulating both innate and cell-mediated inflammatory pathways. IL-10 knockout mice and mice lacking IL-10 2 receptor show development of enterocolitis by 2 to 3 months of age, along with development of inflammatory lesions in different regions of the GI tract [51–54]. IL-10 plays an important role in down regulating different pro-inflammatory cytokines and chemokines such as tumor necrosis factor alpha (TNF-α), IL-1, IL-8, and IL-6 along with augmenting activation of regulatory T cells [54]. However, no improvement in clinical symptoms and disease severity were observed when systemic IL-10 protein therapy was tested in clinical trials for Crohn's disease [55, 56]. These unsatisfactory results may be attributed to the low mucosal exposures of recombinant IL-10 protein in the GI tract upon systemic administration, where drug activity is desired. If this was the case, a delivery systems capable of locally producing high IL-10 exposures in the affected sites of the GI tract may prove to be efficacious in alleviating inflammatory disease symptoms [54].

To achieve this goal, NiMOS were evaluated for delivering murine IL-10 plasmid DNA (mIL-10) aimed at treating 2,4,6-trinitrobenzenesulfonic acid (TNBS)-induced acute colitis in Balb/C mice [8]. Colitis was induced by rectally injecting 8–10-week-old female Balb/C mice at the start of the study and after 4 days. Development of colitis was confirmed by observing stool consistency, body weight, rectal bleeding, and tissue histology. After the colitis model was established, the mice were orally administered

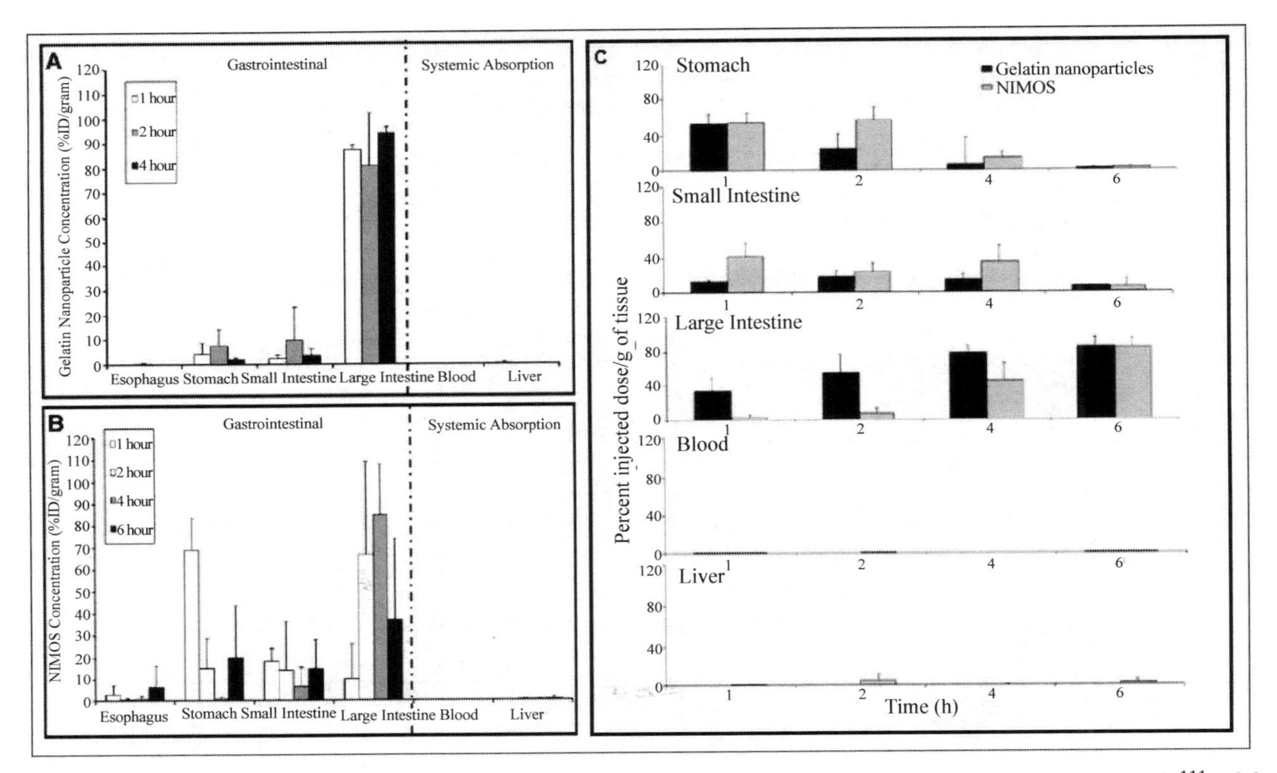

Figure 13.4 Gastrointestinal distribution following oral administration of ^{111}In-labeled gelatin nanoparticles and ^{111}In-labeled gelatin nanoparticles encapsulated in the NiMOS in 24 h fasted female Balb/C mice (A & B) or Wistar rats (C). Reprinted from [44] with permission from Springer.

with naked mIL-10, mIL-10 containing gelatin nanoparticles, or mIL-10 containing NiMOS. Animals administered with formulations not containing mIL-10 served as vehicle controls. All animals were euthanized 4 days post treatment followed by collection of blood and tissue samples.

IL-10 gene expression was evaluated by analyzing IL-10 mRNA and protein expression levels in all animals. After oral administration, mice treated with mIL-10 containing NiMOS showed maximum IL-10 mRNA expression in the large intestine compared with gelatin nanoparticles and naked mIL-10 plasmid DNA-treated groups. These results correlated with ELISA results for murine IL-10 protein, where NiMOS-treated groups showed maximum IL-10 protein expression in the large intestine, compared to all of the other groups (Fig. 13.5). For example, mean IL-10 protein levels for the vehicle control group was only 25 pg/mg of total protein, which was significantly lower than the mIL-10 containing NiMOS treated group, where the mean IL-10 protein expression level was 180 pg/mg of total protein. Further, results obtained from cytokine profiling showed an overall reduction in the expression of pro-inflammatory cytokines such as TNF-α, interferon gamma (IFN-γ), IL-12, IL-1α, IL-1β, and chemokines such as monocyte chemotactic protein (MCP)-1 and monocyte inflammatory protein (MIP)-1 for the NiMOS-treated group when compared with control groups. In addition, the NiMOS-treated group showed reduced weight loss and increased colon length compared to the control groups. Further, the intestinal tissue histology for the NiMOS-treated group showed a healthy cellular architecture similar to that of naïve mice (non-colitis bearing mice), whereas control groups receiving either gelatin nanoparticles or naked mIL-10 or vehicle controls showed villus atrophy, thickening of luminal wall, loss of mucosal architecture, infiltration of immune cells, and loss of protective epithelial layer (Fig. 13.5). These results suggest that mIL-10 containing NiMOS were capable of delivering encapsulated payload at the target site, facilitate robust and sustained transgene expression, and finally help in putting forth a strong anti-inflammatory response for alleviating symptoms of colitis. These results also suggest that NiMOS can prove to be a very useful strategy for delivering other forms of nucleic acid therapeutics such as siRNAs, mRNAs, micro-RNAs (miRNAs), and DNA/RNA vaccines, locally to the small and large intestines, where therapeutic effect may be desired, especially for intestinal chronic inflammatory conditions [8].

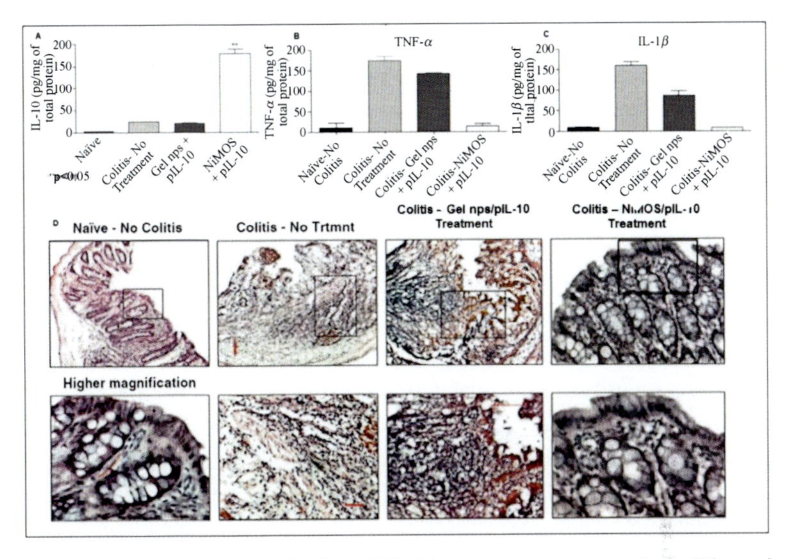

Figure 13.5 Murine interleukin (IL)-10 transgene expression (A) and reduction in the levels of proinflammatory cytokines TNFα (B) and IL-1β (C) upon delivery of murine interleukin (IL)-10-expressing plasmid DNA administered in NiMOS. (D) The therapeutic benefits of oral interleukin (IL)-10 gene therapy as determined by tissue histology upon oral administration of murine IL-10-expressing plasmid DNA in NiMOS. Reprinted by permission from Macmillan Publishers Ltd: Ref. 8, Copyright 2008.

13.4 Illustrative Example of Gene Silencing with Nanoparticles-in-Microsphere Oral System

13.4.1 NiMOS Formulation for Delivery of siRNA Duplexes

Inflammatory bowel disease consists of two main subgroups: (i) Crohn's disease and (ii) ulcerative colitis. Both of these are chronic inflammatory disorders resulting from dysregulation of the mucosal immune system in genetically susceptible individuals [57, 58]. Biological therapy with monoclonal antibodies is needed for patients who fail to respond conventional therapy. TNF-α plays a crucial role in the pathogenesis of various inflammatory disorders, and its inhibition by monoclonal antibodies has already been successfully utilized in mitigating disease symptoms for many conditions including IBD [59–61]. Although monoclonal antibodies are highly effective in disease treatment and maintaining clinical remission, many patients fail to respond or lose response over the duration of therapy, which is attributed to low mucosal bioavailability and anti-drug antibodies produced after systemic delivery. Additionally, cytokine therapy has severe side effects such as risk of infections and cancers such as skin cancer and lymphoma [57]. Ribonucleic acid interference (RNAi) therapy is a new and exciting approach for treating a wide range of diseases, including gastrointestinal diseases, infections, and cancers [62]. Similar to other nucleic acid therapies, progress of RNAi therapy is halted by the challenge of safe and efficacious delivery to the target cells. Although significant progress has been made for systemic liver specific siRNA therapy [63–65], not much progress has been made with oral route of siRNA administration. This is mainly because the unprotected siRNA rapidly degrades in physiological environment of the GI tract due to low gastric pH and degrading GI enzymes, thus necessitating the development of formulation systems capable of overcoming these barriers. In order to address these unmet needs, NiMOS were developed and evaluated for oral TNF-α and cyclin D1 siRNA delivery for the treatment of IBD.

13.4.2 Oral Administration of TNF-α and Cyclin D1 siRNA and Gene Silencing Studies

Kriegel *et al.* [32, 50] developed NiMOS formulation containing TNF-α and cyclin D1 siRNAs for the treatment of IBD. The major study hypothesis was that, by down regulating one or more proinflammatory disease markers such as TNF-α and cyclin D1 using specific siRNAs, balance between pro- and anti-inflammatory markers could be restored, resulting in an improvement in the diseased state. TNF-α is an important cytokine exerting various proinflammatory triggers for development of IBD. Several previous studies have shown significant improvement in IBD symptoms by inducing TNF- α inhibition in different animal models and humans [61, 66], including our oral TNF-α and cyclin D1 silencing studies with NiMOS, where encouraging results were obtained in DSS-induced colitis bearing Balb/C mice for the treatment of IBD [32, 50]. In our studies, we had also included cyclin D1 inhibition by siRNA, as it is believed to have a role in inflammation with overexpression in many cancers and inflammatory diseases, along with having a major role in progressing cell cycle from G1 to S phase [67–71]. As such, cyclin D1 is an important inhibition target for oral anti-inflammatory therapy in the treatment of IBD.

The major aim of this study was to evaluate the oral siRNA delivery potential of NiMOS, along with evaluating the efficacy of TNF-α and cyclin D1 silencing, administered either individually or in combination, for the treatment of IBD in DSS-induced colitis bearing mice. siRNAs targeting murine TNF-α and cyclin D1 were encapsulated within NiMOS either individually or in combination. Mice were fasted overnight (to avoid any food interference with oral delivery) [72], and orally administered with either therapeutic or control siRNA(s) containing NiMOS, once every other day for a total of three doses. Additional control groups included mice receiving NiMOS with no siRNA, mice receiving naked scramble siRNA, and mice receiving normal tap water during the duration of the study, for accounting for any potential off-target effects. Three and five days post last dose, the mice were sacrificed and blood and tissues were harvested for analyzing different biological markers of IBD, including analysis for target gene knockdown [32, 50].

13.4.3 Efficacy of Oral Gene Silencing in Inflammatory Bowel Disease Model

RT-qPCR analysis was performed for determination of TNF-α and cyclin D1 mRNA expressions in all of the study animals. The results from these analysis showed siRNA specific knockdown of target genes in all of the treated DSS-colitis bearing mice, indicating strong efficiency of NiMOS in promoting a robust siRNA mediated oral gene silencing effect. The results obtained for the early time point of the study showed that TNF-α/cyclin D1 siRNA combination therapy resulted in a greater TNF-α knockdown, when compared to individual siRNA containing NiMOS treatment. At the later time point of the study, either combination or individual TNF-α/cyclin D1 NiMOS treatments resulted in similar levels of TNF-α knockdown. These results may suggest that the dual therapy has a greater capability of promoting a potent TNF-α inhibition-based anti-inflammatory response, with an early onset. Cyclin D1 silencing was most pronounced in the group receiving cyclin D1 siRNA containing NiMOS, compared to all other treatment groups. Further, analysis of TNF-α protein expression for the individual TNF-α and combination TNF-α/cyclin D1 NiMOS therapy showed reductions in protein expression for both groups, where TNF-α protein reduction was greater for the TNF-α/cyclin D1 dual treatment group. Interestingly, TNF-α protein downregulation for the individual cyclin D1 siRNA containing NiMOS treatment group was greatest, when compared to all of the other treatment groups in the study (Fig. 13.6). Further, evaluations for proinflammatory cytokine expression showed a reduction in levels of interferon gamma, IL-1α, IL-2, IL-15, IL-17, granulocyte macrophage colony stimulating factor, and monocyte chemotactic protein 1. Finally, intestinal tissue histology showed reductions in inflammatory damage and tissue regeneration, for

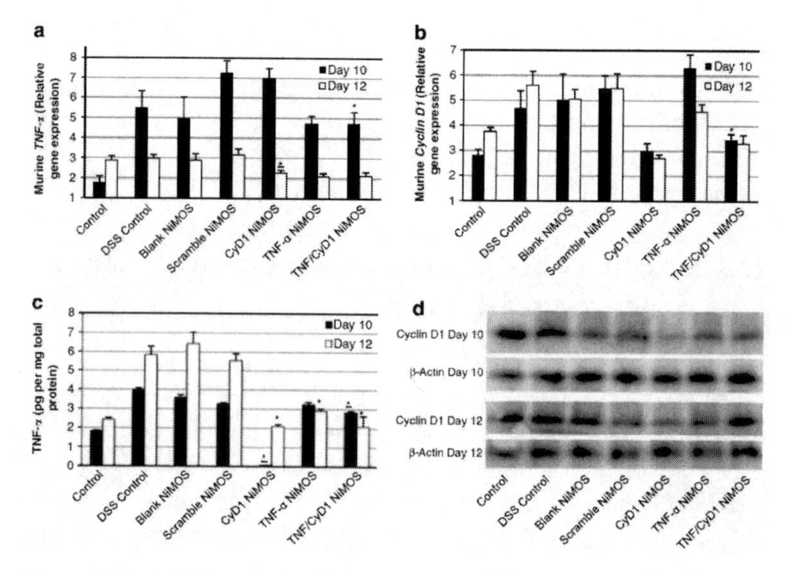

Figure 13.6 mRNA and protein expression profiles in control and siRNA-treated mice. Reprinted by permission from Macmillan Publishers Ltd: Ref. 50, Copyright 2011.

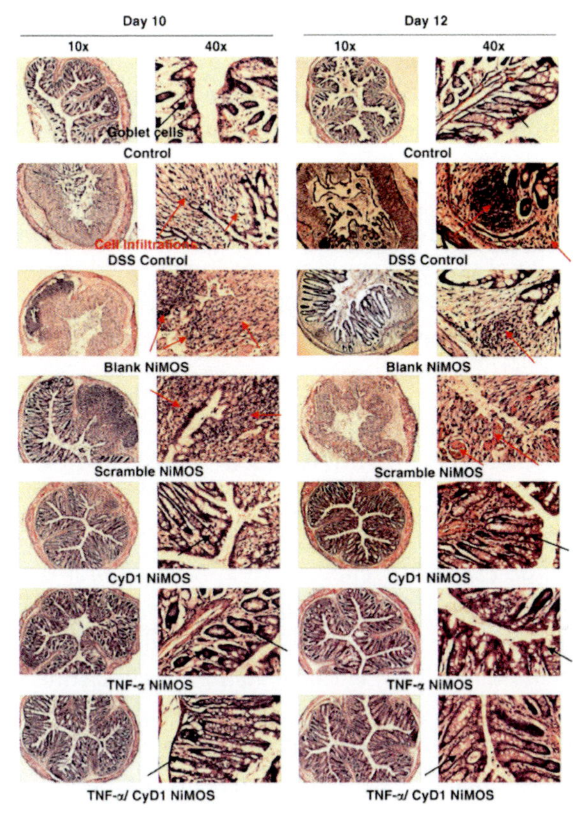

Figure 13.7 Microscopic evaluation of colonic tissue histopathology. Bright-field images of hematoxylin and eosin stained sections of the colon harvested from each control and test group. Images are shown at magnifications of ×10 and ×40 from tissue cryosections obtained on day 10 and day 12 of the study. Sections from the first control group show normal and healthy colon tissue. Intestinal tissues from the dextran sulfate sodium (DSS) control group, the group treated with blank and scrambled siRNA NiMOS showed a severe infiltration of white blood cells, abnormal mucosal structure, and a certain degree of goblet cell depletion. Tissue from the group receiving tumor necrosis factor-α (TNF-α), cyclin D1 (CyD1), or combined TNF/CyD1 silencing NiMOS showed signs of regeneration and exhibited a tissue architecture more closely resembling that of healthy tissue in the normal control group. Occurrence of goblet cells is indicated by red arrows; cell infiltration and abnormal tissue histology is indicated by black arrows. Reprinted by permission from Macmillan Publishers Ltd: Ref. 50, Copyright 2011.

groups treated with therapeutic siRNAs (TNF-α, cyclin D1, and cyclin D1/TNF-α NiMOS), regardless of individual or dual therapy, when compared to groups treated with control formulations (Fig. 13.7). These results have clearly showed that NiMOS were capable of efficiently delivering siRNAs by oral route, and capable of inhibiting expression of target genes, along with promoting downstream beneficial therapeutic effects for disease treatment. For further advancement of this strategy, targeting moieties can be attached to particle surface for active targeting to specific cells types, along with evaluations for determining knockdown potency using different siRNA doses [10].

13.5 Illustrative Examples of Polyester-Based Oral DNA Vaccines

Kenneth *et al.* [73] have reported a PLGA-based oral DNA vaccine delivery system for induction of mucosal immune responses. This system consisted of plasmid DNA complexed with PEG-modified poly(ethylene imine) (PEI), encapsulated within PLGA microparticles (PEI-PEG-PLGA), for oral DNA vaccine delivery to intestinal lymphoid tissues via the payer's patches. The microencapsulated polyplex system was formulated by a three-step process: (i) formation of plasmid DNA and PEI complex through electrostatic interactions, (ii) PEG modification of plasmid DNA/PEI complex by reaction with *N*-hydroxysuccininimidyl ester of PEG, and (iii) microencapsulation within PLGA polymer using a solvent evaporation double emulsion technique [73]. The mean diameter of PEG modification DNA/PEI complex and the PEI-PEG-PLGA microparticles was 84.3 nm and 2.18 μm. The DNA integrity was not damaged after the formulation process, which was shown by intact plasmid DNA bands on agarose gel. This system was evaluated for oral gene transfection in Wistar rats after a single oral dose of 24 μg plasmid DNA, formulated as PEI-PEG-PLGA complex or solubilized plasmid DNA in PBS. The rats were euthanized three days post single dose, and tissues were collected for analysis of transgene expression. The plasmid DNA used for this study had a transgene encoding for β-galactosidase protein. Results for gene transfection analysis, measured as expression of β-galactosidase, showed higher target protein expression in the spleen tissue obtained from rats treated with PEI-PEG-PLGA complex when compared with solubilized plasmid DNA. These results shows that the PEI-PEG-PLGA microparticulate system was capable of protecting the DNA payload in the GI tract, facilitate uptake by the intestinal lymphoid tissue, followed by migration to and transgene expression in the spleen. These results may encourage the use of PEI-PEG-PLGA as an oral DNA vaccine delivery system, where a plasmid DNA encoding vaccine antigen can be orally delivered to the spleen, possibly through lymphoid payer's patches, resulting in eliciting antigen specific mucosal as well as systemic immune responses [44, 73]. Future studies should include evaluations of antigen expression and generation of antigen specific immune responses in a disease model.

13.6 Summary

In summary, nucleic acid therapeutics have tremendous potential for treating various diseases, including inflammatory diseases of the GI tract. Oral delivery of nucleic acids is very challenging, since nucleic acids rapidly degrade in the GI tract when delivered as such. PCL-based systems such as NiMOS can be used for overcoming oral barriers for intestinal nucleic acid delivery. NiMOS demonstrated safe and efficacious oral *in vivo* gene transfection and therapeutic efficiencies for the treatment of IBD using both plasmid DNA (IL-10 plasmid DNA) and siRNAs (TNF-α/cyclin D1 siRNAs). These results warrant further development of this system for oral nucleic acid therapy and can be extended to other chronic inflammatory diseases of the GI tract, such as celiac disease.

Acknowledgments

Oral gene and RNAi therapies using NiMOS delivery system discussed in this chapter were supported by a grant R01-DK080477 from the National Institute of Diabetes, Digestive Diseases, and Kidney Diseases of the National Institutes of Health. Drs. Mayank Bhavsar, Christina Kriegel, and Shardool Jain are acknowledged for their contributions to these studies.

References

1. Friedmann, T. (1989). Progress toward human-gene therapy. *Science,* **244**(4910), 1275–1281.
2. Romano, G., Pacilio, C., and Giordano, A. (1999). Gene transfer technology in therapy: current applications and future goals. *Stem Cells,* **17**(4), 191–202.
3. Rothman, S., Tseng, H., and Goldfine, I. (2005). Oral gene therapy: a novel method for the manufacture and delivery of protein drugs. *Diabetes Technol Ther,* **7**(3), 549–557.
4. Loretz, B., *et al.* (2006). Oral gene delivery: strategies to improve stability of pDNA towards intestinal digestion. *J Drug Target,* **14**(5), 311–319.
5. Amiji, M. M., ed. (2005). *Polymeric Gene Delivery: Principles and Applications.* CRC Press: Boca Raton, FL.
6. Jabr-Milane, L., *et al.* (2008). Multi-functional nanocarriers for targeted delivery of drugs and genes. *J Control Release,* **130**(2), 121–128.
7. Bhavsar, M. D., and Amiji, M. M. (2008). Development of novel biodegradable polymeric nanoparticles-in-microsphere formulation for local plasmid DNA delivery in the gastrointestinal tract. *AAPS PharmSciTech,* **9**(1), 288–294.
8. Bhavsar, M. D., and Amiji, M. M. (2008). Oral IL-10 gene delivery in a microsphere-based formulation for local transfection and therapeutic efficacy in inflammatory bowel disease. *Gene Ther,* **15**(17), 1200–1209.
9. Bhavsar, M. D., and Amiji, M. M. (2007). Polymeric nano- and microparticle technologies for oral gene delivery. *Expert Opin Drug Deliv,* **4**(3), 197–213.
10. Kriegel, C., Attarwala, H., and Amiji, M. M. (2013). Multi-compartmental oral delivery systems for nucleic acid therapy in the gastrointestinal tract. *Adv Drug Deliv Rev,* **65**(6), 891–901.
11. Li, S. D., and Huang, L. (2006). Gene therapy progress and prospects: non-viral gene therapy by systemic delivery. *Gene Ther,* **13**(18), 1313–1319.
12. Whitehead, K. A., Langer, R., and Anderson, D. G. (2009). Knocking down barriers: advances in siRNA delivery. *Nat Rev Drug Discov,* **8**(2), 129–138.
13. Akhtar, S., and Agrawal, S. (1997). *In vivo* studies with antisense oligonucleotides. *Trends Pharmacol Sci,* **18**(1), 12–18.

14. Roos, A. K., *et al.* (2009). Optimization of skin electroporation in mice to increase tolerability of DNA vaccine delivery to patients. *Mol Ther*, **17**(9), 1637–1642.
15. Gothelf, A., *et al.* (2010). Duration and level of transgene expression after gene electrotransfer to skin in mice. *Gene Ther*, **17**(7), 839–845.
16. Cai, X., Conley, S., and Naash, M. (2008). Nanoparticle applications in ocular gene therapy. *Vision Res*, **48**(3), 319–324.
17. de la Fuente, M., *et al.* (2010). Chitosan-based nanostructures: a delivery platform for ocular therapeutics. *Adv Drug Deliv Rev*, **62**(1), 100–117.
18. Lai, S. K., Wang, Y. Y., and Hanes, J. (2009). Mucus-penetrating nanoparticles for drug and gene delivery to mucosal tissues. *Adv Drug Deliv Rev*, **61**(2), 158–171.
19. Janat-Amsbury, M. M., *et al.* (2004). Combination of local, nonviral IL12 gene therapy and systemic paclitaxel treatment in a metastatic breast cancer model. *Mol Ther*, **9**(6), 829–836.
20. Dubensky, T. W., Liu, M. A., and Ulmer, J. B. (2000). Delivery systems for gene-based vaccines. *Mol Med*, **6**(9), 723–732.
21. Jones, D. H., *et al.* (1997). Poly(DL-lactide-*co*-glycolide)-encapsulated plasmid DNA elicits systemic and mucosal antibody responses to encoded protein after oral administration. *Vaccine*, **15**(8), 814–817.
22. Howard, K. A., *et al.* (2004). Formulation of a microparticle carrier for oral polyplex-based DNA vaccines. *Biochim Biophys Acta*, **1674**(2), 149–157.
23. Bhavsar, M. D., and Amiji, M. M. (2007). Gastrointestinal distribution and *in vivo* gene transfection studies with nanoparticles-in-microsphere oral system (NiMOS). *J Control Release*, **119**(3), 339–348.
24. Bhavsar, M. D., Tiwari, S. B., and Amiji, M. M. (2006). Formulation optimization for the nanoparticles-in-microsphere hybrid oral delivery system using factorial design. *J Control Release*, **110**(2), 422–430.
25. Pouton, C. W., and Seymour, L. W. (2001). Key issues in non-viral gene delivery. *Adv Drug Deliv Rev*, **46**(1–3), 187–203.
26. Mohamed, F., and van der Walle, C. F. (2008). Engineering biodegradable polyester particles with specific drug targeting and drug release properties. *J Pharm Sci*, **97**(1), 71–87.
27. Dash, T. K., and Konkimalla, V. B. (2012). Poly-ε-caprolactone based formulations for drug delivery and tissue engineering: a review. *J Control Release*, **158**(1), 15–33.

28. Liu, C. B., *et al.* (2008). Thermoreversible gel-sol behavior of biodegradable PCL-PEG-PCL triblock copolymer in aqueous solutions. *J Biomed Mater Res B Appl Biomater*, **84**(1), 165–175.

29. Ma, Z., *et al.* (2008). Micelles of poly(ethylene oxide)-b-poly(ε-caprolactone) as vehicles for the solubilization, stabilization, and controlled delivery of curcumin. *J Biomed Mater Res A*, **86**(2), 300–310.

30. Sheikh, F. A., *et al.* (2009). Novel self-assembled amphiphilic poly(ε-caprolactone)-grafted-poly(vinyl alcohol) nanoparticles: hydrophobic and hydrophilic drugs carrier nanoparticles. *J Mater Sci Mater Med*, **20**(3), 821–831.

31. Varde, N. K., and Pack, D. W. (2004). Microspheres for controlled release drug delivery. *Expert Opin Biol Ther*, **4**(1), 35–51.

32. Kriegel, C., and Amiji, M. (2011). Oral TNF-alpha gene silencing using a polymeric microsphere-based delivery system for the treatment of inflammatory bowel disease. *J Control Release*, **150**(1), 77–86.

33. Devalapally, H., *et al.* (2007). Paclitaxel and ceramide co-administration in biodegradable polymeric nanoparticulate delivery system to overcome drug resistance in ovarian cancer. *Int J Cancer*, **121**(8), 1830–1838.

34. van Vlerken, L. E., *et al.* (2007). Modulation of intracellular ceramide using polymeric nanoparticles to overcome multidrug resistance in cancer. *Cancer Res*, **67**(10), 4843–4850.

35. Shah, L. K., and Amiji, M. M. (2006). Intracellular delivery of saquinavir in biodegradable polymeric nanoparticles for HIV/AIDS. *Pharm Res*, **23**(11), 2638–2645.

36. Shenoy, D. B., and Amiji, M. M. (2005). Poly(ethylene oxide)-modified poly(ε-caprolactone) nanoparticles for targeted delivery of tamoxifen in breast cancer. *Int J Pharm*, **293**(1–2), 261–270.

37. Milane, L., Duan, Z. F., and Amiji, M. M. (2011). Pharmacokinetics and biodistribution of lonidamine/paclitaxel loaded, EGFR-targeted nanoparticles in an orthotopic animal model of multi-drug resistant breast cancer. *Nanomedicine*, **7**(4), 435–444.

38. Milane, L., Duan, Z. F., and Amiji, M. M. (2011). Development of EGFR-targeted polymer blend nanocarriers for combination paclitaxel/lonidamine delivery to treat multi-drug resistance in human breast and ovarian tumor cells. *Mol Pharm*, **8**(1), 185–203.

39. Brito, L. A., *et al.* (2010). Non-viral eNOS gene delivery and transfection with stents for the treatment of restenosis. *Biomed Eng Online*, **9**, 56.

40. van Vlerken, L. E., *et al.* (2008). Biodistribution and pharmacokinetic analysis of paclitaxel and ceramide administered in multifunctional polymer-blend nanoparticles in drug resistant breast cancer model. *Mol Pharm*, **5**(4), 516–526.

41. Attarwala, H., and Amiji, M. (2012). Multi-compartmental nanoparticles-in-emulsion formulation for macrophage-specific anti-inflammatory gene delivery. *Pharm Res*, **29**(6), 1637–1649.

42. Amiji M., Kalariya, M., Jain S., and Attarwala H. (2011). Multi-compartmental macrophage delivery. WO Patent WO/2011/119,881.

43. Jain, S., and Amiji, M. (2012). Tuftsin-modified alginate nanoparticles as a noncondensing macrophage-targeted DNA delivery system. *Biomacromolecules*, **13**(4), 1074–1085.

44. Attarwala, H., and Amiji, M. (2012). Multi-compartmental nanoparticles-in-emulsion formulation for macrophage-specific anti-inflammatory gene delivery. *Pharm Res*, **29**(6), 1637–1649.

45. Shahiwala, A., and Amiji, M. M. (2008). Enhanced mucosal and systemic immune response with squalane oil-containing multiple emulsions upon intranasal and oral administration in mice. *J Drug Target*, **16**(4), 302–310.

46. Xu, J., and Amiji, M. M. (2012). Therapeutic gene delivery and transfection in human pancreatic cancer cells using epidermal growth factor receptor-targeted gelatin nanoparticles. *J Vis Exp*, **4**(59), e3612.

47. Magadala, P., and Amiji, M. M. (2008). Epidermal growth factor receptor-targeted gelatin-based engineered nanocarriers for DNA delivery and transfection in human pancreatic cancer cells. *AAPS J*, **10**(4), 565–576.

48. Kommareddy, S., and Amiji, M. M. (2007). Biodistribution and pharmacokinetic analysis of long-circulating thiolated gelatin nanoparticles following systemic administration in breast cancer-bearing mice. *J Pharm Sci*, **96**(2), 397–407.

49. Kaul, G., and Amiji, M. M. (2002). Long-circulating poly(ethylene glycol)-modified gelatin nanoparticles for intracellular delivery. *Pharm Res*, **19**(7), 1061–1067.

50. Kriegel, C., and Amiji, M. M. (2011). Dual TNF-alpha/cyclin D1 gene silencing with an oral polymeric microparticle system as a novel strategy for the treatment of inflammatory bowel disease. *Clin Transl Gastroenterol*, **2**(3), e2.

51. Moore, K. W., *et al.* (2001). Interleukin-10 and the interleukin-10 receptor. *Annu Rev Immunol*, **19**, 683–765.

52. Kuhn, R., *et al.* (1993). Interleukin-10-deficient mice develop chronic enterocolitis. *Cell*, **75**(2), 263–274.

53. Glocker, E. O., *et al.* (2009). Inflammatory bowel disease and mutations affecting the interleukin-10 receptor. *N Engl J Med*, **361**(21), 2033–2045.

54. Matsumoto, H., *et al.* (2014). Mucosal gene therapy using a pseudotyped lentivirus vector encoding murine interleukin-10 (mIL-10) suppresses the development and relapse of experimental murine colitis. *BMC Gastroenterol*, **14**, 68.

55. Fedorak, R. N., *et al.* (2000). Recombinant human interleukin 10 in the treatment of patients with mild to moderately active Crohn's disease. The Interleukin 10 Inflammatory Bowel Disease Cooperative Study Group. *Gastroenterology*, **119**(6), 1473–1482.

56. Schreiber, S., *et al.* (2000). Safety and efficacy of recombinant human interleukin 10 in chronic active Crohn's disease. Crohn's Disease IL-10 Cooperative Study Group. *Gastroenterology*, **119**(6), 1461–1472.

57. Xavier, R. J., and Podolsky, D. K. (2007). Unravelling the pathogenesis of inflammatory bowel disease. *Nature*, **448**(7152), 427–434.

58. Ordas, I., *et al.* (2012). Anti-TNF monoclonal antibodies in inflammatory bowel disease: pharmacokinetics-based dosing paradigms. *Clin Pharmacol Ther*, **91**(4), 635–646.

59. Strober, W., Fuss, I., and Mannon, P. (2007). The fundamental basis of inflammatory bowel disease. *J Clin Invest*, **117**(3), 514–521.

60. Papa, A., *et al.* (2009). Biological therapies for inflammatory bowel disease: controversies and future options. *Expert Rev Clin Pharmacol*, **2**(4), 391–403.

61. Hoentjen, F., and Van Bodegraven, A. A. (2009). Safety of anti-tumor necrosis factor therapy in inflammatory bowel disease. *World J Gastroenterol*, **15**(17), 2067–2073.

62. Bhavsar, M. D., Tiwari, S. B., and Amiji, M. M. (2006). Formulation optimization for the nanoparticles-in-microsphere hybrid oral delivery system using factorial design. *J Control Release*, **110**(2), 422–430.

63. Kanasty, R., *et al.* (2013). Delivery materials for siRNA therapeutics. *Nat Mater*, **12**(11), 967–977.

64. Fitzgerald, K., *et al.* (2014). A subcutaneous, potent and durable RNAi platform targeting metabolic diseases, genes PCSK9, ApoC3 and ANGPLT3. *Arterioscler Thromb Vasc Biol*, **34**(Suppl 1), A7.

65. Conde, J., and Artzi, N. (2015). Are RNAi and miRNA therapeutics truly dead? *Trends Biotechnol*, **33**, 141–144.

66. Armuzzi, A., *et al.* (2004). Infliximab in the treatment of steroid-dependent ulcerative colitis. *Eur Rev Med Pharmacol Sci*, **8**(5), 231–233.

67. Fu, M., *et al.* (2004). Minireview: cyclin D1: normal and abnormal functions. *Endocrinology*, **145**(12), 5439–5447.

68. Alao, J. P., *et al.* (2006). The cyclin D1 proto-oncogene is sequestered in the cytoplasm of mammalian cancer cell lines. *Mol Cancer*, **5**, 7.

69. Alao, J. P. (2007). The regulation of cyclin D1 degradation: roles in cancer development and the potential for therapeutic invention. *Mol Cancer*, **6**, 24.

70. Peer, D., *et al.* (2008). Systemic leukocyte-directed siRNA delivery revealing cyclin D1 as an anti-inflammatory target. *Science*, **319**(5863), 627–630.

71. Kornmann, M., *et al.* (1998). Increased cyclin D1 expression in chronic pancreatitis. *Pancreas*, **17**(2), 158–162.

72. Bhavsar, M. D., and Amiji, M. M. (2008). Oral IL-10 gene delivery in a microsphere-based formulation for local transfection and therapeutic efficacy in inflammatory bowel disease. *Gene Ther*, **15**(17), 1200–1209.

73. Howard, K. A., *et al.* (2004). Formulation of a microparticle carrier for oral polyplex-based DNA vaccines. *Biochim Biophys Acta*, **1674**(2), 149–157.

Chapter 14

PLGA Nano- and Microparticles for VEGF Delivery

Teresa Simon-Yarza,[a] Paula Diaz-Herraez,[b] Simon Pascual-Gil,[b,d] Elisa Garbayo,[b] Felipe Prosper,[c,d] and Maria J. Blanco-Prieto[b,d]
[a]*Institut Lavoisier, Universite de Versailles Saint Quentin en Yvelines, Versailles, France*
[b]*Department of Pharmacy and Pharmaceutical Technology, School of Pharmacy, University of Navarra, Pamplona, Spain*
[c]*Hematology, Cardiology and Cell Therapy, Clínica Universidad de Navarra, Foundation for Applied Medical Research, University of Navarra, Pamplona, Spain*
[d]*Instituto de Investigación Sanitaria de Navarra (IDISNA), Pamplona, Spain*
mjblanco@unav.es

14.1 Introduction

14.1.1 Discovery and Biological Aspects of Vascular Endothelial Growth Factor

The vascular endothelial growth factor (VEGF) was first described by Senger *et al.* in 1983, who purified the protein from tumor cells [1]. Due to its ability to enhance vascular permeability, this protein was named vascular permeability factor [2]. In 1989, this new protein was described completely by Dvorak *et al.*, who included

Handbook of Polyester Drug Delivery Systems
Edited by MNV Ravikumar

ISBN 978-981-4669-65-8 (Hardcover), 978-981-4669-66-5 (eBook)
www.panstanford.com

it in the growth factor (GF) family with its current name, VEGF [3, 4]. Since then, several VEGF family members have been identified in humans, including VEGF-B [5, 6], VEGF-C (also called VEGF-2) [7], VEGF-D [8], and placenta growth factor (PIGF) [9]. Among the VEGF family members, different subtypes have been reported. For instance, VEGF-A contains three subtypes containing a different number of amino acids, known as VEGF-121, VEGF-165, and VEGF-189, which are the major human VEGF isoforms and the ones that offer the strongest vascular permeability.

The VEGF family members are homodimeric glycoproteins with two subunits of about 120 to 200 amino acids in length [10] that are involved in angiogenesis and lymphangiogenesis [11]. They can also have prominent neurotropic effects, particularly VEGF-A and VEGF-B [11] (Table 14.1). However, not all of them act in the same way. For instance, VEGF-B and PIGF have a relatively minor role in the regulation of angiogenesis, whereas they play an important role in cardiac muscle function [12]. On the other hand, VEGF-C and VEGF-D are precursor forms that are involved in the regulation of lymphangiogenesis [13].

Table 14.1 Subtypes of VEGF and their reported biological effects (indicated by + symbol)

Processes	VEGF-A	VEGF-B	VEGF-C	VEGF-D	PIGF
Vasculogenesis	+	–	–	–	–
Angiogenesis	+	+	+	+	+
Lymphangiogenesis	–	–	–	+	–
Neuroprotective	+	+	+	–	–

Differences in biological activity can also be observed within the subtypes of one VEGF member. Among the VEGF-A subtypes, VEGF-121 can support the initial stage of vascular development but cannot replace the functions of the 165 isoform in the fine patterning of the vasculature. Both of them promote rapid growth of vessels, but are highly unstable and leaky. The 189 isoform, however, favors the slow growth of vessels which are relatively normal in appearance [14].

14.1.2 Therapeutic Possibilities of VEGF

The therapeutic use of VEGF is being explored for different types of tissue conditions outlined below and summarized in Fig. 14.1.

Updated information about the clinical trials (CT) performed to date is included in Table 14.2.

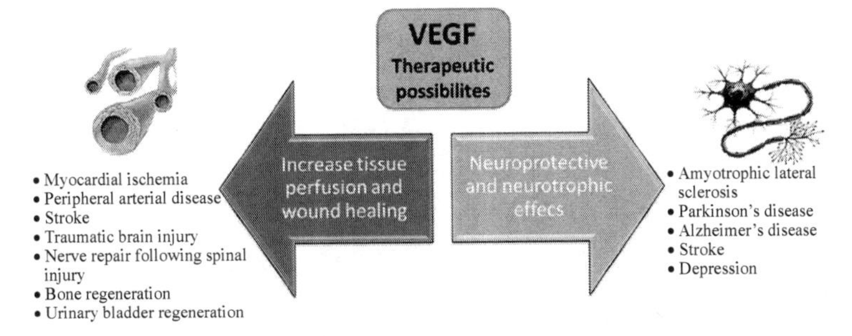

Figure 14.1 Therapeutic possibilities of VEGF. The diagram shows the tissue conditions for which therapeutic uses of VEGF are currently being explored.

14.1.2.1 Therapeutic uses of VEGF linked to its ability to increase tissue perfusion and wound healing

VEGF can mediate tissue restoration in cases of injury, ischemia, and wound healing. VEGF therapy could therefore be beneficial for diseases associated with tissue inflammation and necrosis. VEGF has been investigated for the clinical treatment of myocardial infarction (MI), chronic limb ischemia, tissue regeneration, ischemic stroke, or traumatic brain injury (see Table 14.2). Among the therapeutic possibilities of VEGF, the most widely investigated area has been MI treatment (reviewed in [37]). Studies in animal models demonstrated that VEGF administration to the infarct border limited myocardial damage, increased coronary vasculature, and reduced adverse cardiac remodeling [38]. Some CTs have shown certain benefits, such as of the study described by Losordo *et al.*, who found reduction of symptoms and improved tissue perfusion [16]. However, longer studies with larger numbers of patients were not so positive. In the CT reported by Stewart and coworkers, a lack of improvement in myocardial perfusion and absence of pericardial effusion on serial echocardiography were observed indicating that, at least at the dose used in the study, the vascular permeability-enhancing effect of VEGF was not clinically relevant [31]. Similar results were obtained by Gyöngyösi *et al.* [18]. Additionally, in another study by Stewart

Table 14.2 CT using VEGF for the treatment of different pathologies (*N* represents the number of patients included in the CT)

Ref./Clinical Trial Gov. Number	1.1 N	VEGF	Administration Route	Pathology	Phase
[15]	54	VEGF*	percutaneous transluminal angioplasty	lower limb ischemia	II
NCT00135850	48		intramyocardial	severe angina	I&II
NCT00351767	160		topical	diabetic foot ulcer	II
NCT00279539	12		intramyocardial	heart failure	I
NCT00143585	120			severe angina	II&III
[16]	5			coronary disease	I
NCT00134433	20				I&II
NCT00620217	52	VEGF-A 165			II
NCT00744315	20				II
[17]	93				II
[18]	40			chronic MI	
[19]	6			severe angina pectoris	I&II
[20]	80				II
[21]	178		intracoronary/ intravenous	vascular angiogenesis	II

Ref./Clinical Trial Gov. Number	1.1 N	VEGF	Administration Route	Pathology	Phase
[22]	14		intracoronary	severe atherosclerotic heart disease	I
[23]	15			severe MI	I
[24, 25]	103			coronary heart disease	II
[26]	21		intramuscular	chronic critical leg ischemia	I
[27]	9			critical limb ischemia	I
[28]	54	VEGF-A 165		diabetes mellitus and critical limb ischemia	
[29]	6			thromboangiitis obliterans	I
NCT01999803	18		intracerebroventricular	amyotrophic lateral sclerosis	I
NCT00800501	18				I&II
NCT01384162	18				I&II
NCT01174095	31		intramyocardial	diffuse coronary artery disease	I
[30]	21			severe coronary artery disease	I
[31]	67	VEGF-A 121		ischemic heart disease	II
[32]	15		intramuscular	critical limb ischemia	I
[33]	105			peripheral arterial disease	II

(*Continued*)

Table 14.2 (*Continued*)

Ref./Clinical Trial Gov. Number	1.1 N	VEGF	Administration Route	Pathology	Phase
NCT01757223	41	Ad VEGF-A 116	intramyocardial	diffuse coronary artery disease	I&II
NCT00056290	60		intramuscular	diabetic neuropathy	I&II
NCT00304837	–		–	critical limb ischemia	I
[34]	6	VEGF-C	intramyocardial	chronic MI	I
[35]	30			angina	I
[36]	19			chronic MI	I&II
NCT01002430	–	VEGF-D	endocardial	severe coronary heart disease	I
NCT00895479	250			avoid graft failure	III
NCT00069446	50	rhuM Ab VEGF	topical	diabetic foot ulcer	I

Note: VEGF isoform is not specified in these CT.

and coworkers, no increase in VEGF circulating levels or evidence of improvement in perfusion was observed, although higher doses of VEGF plasmid DNA were used [17]. The results of myocardial CT using VEGF delivery have thus prove to be disappointing in general.

VEGF has also been investigated to achieve therapeutic angiogenesis and vasculogenesis after peripheral arterial disease. This pathology is a circulatory problem in which narrowed arteries reduce blood flow to the limbs. In advanced stages, limb ischemia may result in amputation of the affected limb. In this area, Long *et al.* explored the potential of endothelial progenitor cells expressing VEGF-A on a rat hind limb ischemia model, and obtained promising results [39]. An increase in microvessel density and limb salvage was observed in animals treated with VEGF-A, which demonstrates the potential of this strategy for promoting angiogenesis. However, in the CT performed by Rajagopalan *et al.*, no differences between placebo and VEGF-121 transgene expression using adenoviral approaches were found, indicating that VEGF therapy may be ineffective for hind limb ischemia [33].

VEGF-A is essential for blood vessel formation in the developing and adult nervous system. Closely related to this, angiogenesis is a key feature of stroke recovery. Restoration of local blood flow to the brain can reverse the ischemic environment originated by hypoxic stroke. In addition, angiogenesis plays a vital role for striatal neurogenesis after stroke. Therapeutic angiogenesis and vasculogenesis therefore offer novel approaches that may help in the cure of both acute ischemic stroke and traumatic brain injury. The role of VEGF and other GF in post-stroke recovery has recently been revised by Talwar *et al.* [40]. It has been shown that VEGF administration can decrease brain infarct size, by promoting angiogenesis and neurogenesis near the penumbral area leading to improved functional recovery [41, 42]. However, because of the multiple VEGF effects on the central nervous system (CNS), the blood vessels formed through angiogenesis tend to be immature and permeable [43] and blood–brain barrier disruption has also been noticed [44]. It is therefore important to administer VEGF in combination with other neurotrophic factors and agents effective in reducing vascular permeability. At present, there is no ongoing CT for stroke involving this factor. VEGF has also been investigated for its ability to promote nerve repair following spinal injury leading to recovery of function associated with increased vessel density and reduced apoptosis [45]. Nevertheless, care must be taken since it was also found that VEGF-A therapy exacerbated lesion volume following spinal cord injury, probably through its effect on vascular permeability [46].

Vascularization, chondrogenesis, and osteogenesis underlie the success of bone regeneration [47]. Besides angiogenic properties, VEGF-A has chemotactic and mitogenic effects on osteoblast and osteogenic cells, presenting direct and indirect effects on bone regeneration [48]. Therefore, a number of studies have investigated VEGF application in combination with bone morphogenetic proteins (BMPs) in small and large animal models of bone regeneration (reviewed in [49]). Similarly, proper graft vascularization is a critical determinant of urinary bladder regeneration and has shown accelerated bladder regeneration, increasing angiogenesis and muscular repair [50].

14.1.2.2 Therapeutic uses of VEGF related to its neurotrophic and neuroprotective effects

VEGF plays multiple roles in the CNS. In addition to its angiogenic capacities in the brain, VEGF has nonvascular functions (reviewed in [51]). In support of this idea, several studies have shown that VEGF promotes neurogenesis, has trophic effects on neurons and glia, and is involved in axonal guidance [52]. Consequently, VEGF may be of therapeutic value in the treatment of neurological disorders. Amyotrophic lateral sclerosis (ALS) is a fatal neurodegenerative disease that results from the selective death of motor neurons. Current treatment has limited efficacy and other potential therapies based on VEGF are being investigated. The majority of these studies have focused on the use of VEGF-A and VEGF-B isoforms. Treatment of ALS animal models with VEGF yielded positive therapeutic outcomes such as slowed progression of the disease, motor function improvement, and extended survival (reviewed in [11]). Given the success in animal models, there are several ongoing CT using VEGF-A (see Table 14.2). Regarding Parkinson's disease (PD), current therapies are not able to slow or halt the disease progression. Several neurotrophic factors, particularly glial cell line derived neurotrophic factor (GDNF) (reviewed in [53]) and VEGF, have shown promise in this regard. There is evidence showing that both VEGF-B and VEGF-C isoforms promote dopaminergic cell survival in several *in vitro* and *in vivo* PD models [54–56]. However, microglia activation and blood–brain barrier disruption were reported after intracerebral delivery of VEGF-C, so these findings must be interpreted cautiously [54]. VEGF may also have therapeutic potential for the treatment of Alzheimer's disease (AD), the leading cause of dementia worldwide. The extracellular amyloid-β (Aβ) plaque is one of the neurotoxic proteins that are involved in the pathology. The accumulation of this protein in the brain is responsible for reduced vascular permeability that leads to the loss of blood flow. Recently, Garcia *et al.* demonstrated that VEGF overexpressing bone marrow stem cell administration increased neovascularization in the hippocampus and Aβ clearance leading to recovery of memory and repair of learning deficits in AD transgenic animals [57]. Neuroprotection by VEGF-A has also been demonstrated in animal models of stroke (revised in [58]). For instance, Hayashi *et al.* found that topical application of VEGF-A in the cortical surface reduced infarct volume, brain edema, and neural damage in rats [42]. Similar results were obtained by Sun *et al.* and Wang *et al.* when VEGF-A was administered intraventricularly [41, 59]. Finally, regarding CNS pathologies, VEGF has been found to be involved in adult hippocampal neurogenesis and offers potential as a treatment for depression [60].

In conclusion, numerous therapeutic possibilities of VEGF have been described but only a few of them have reached CT. Among the different VEGF family members, VEGF-121 and 165 isoforms are the ones which are being most widely tested in clinical studies (see Table 14.2). At present, VEGF-B has only been investigated *in vitro* [61] or in animal models [62]. In general, clinical studies show few positive results owing to the short half-life and high instability of the protein after intravenous/intracoronary administration [63]. In fact, this was the major cause of the unexpectedly low therapeutic effect reported in most of the CT carried out so far, and one of the main reasons why VEGF therapy is still not a viable treatment. Other critical aspects are the uncertainty about the dose that should be administered to produce an effective response and the limitations due to the route of administration. Initially, higher protein load, repeated doses, or alternative routes of administration were proposed as solutions. However, important side effects [64, 65], issues deriving from multiple dose administration [66] and difficulties due to the large molecular size [63] have made it necessary to find new approaches to enhance the bioavailability of injected proteins.

14.2 How to Improve VEGF Therapy

Most of the diseases for which protein therapy has been investigated require therapeutics to be protected from degradation and delivered in a prolonged and local manner, in order for them to act in a defined area. A great deal of effort has therefore been devoted to develop drug delivery systems (DDSs). DDSs were originally conceived as matrices with the function of vehicles able to deliver drugs at the desired site (specific organs, tissues or even cellular structures) and to modulate drug distribution in the body in order to achieve therapeutic levels [67]. However, with the evolution of physics, chemistry, materials science, and other biotechnologies, DDSs are now known to play a more multifunctional role and are considered as the cornerstone of protein therapy [68, 69].

Currently, DDSs can be broadly divided into four types: hydrogels, nanofibers, liposomes, and nano- and microparticles (NPs and MPs). In particular, NPs and MPs have attracted great interest because of their versatility and therapeutic potential [70–72]. NP and MP can be defined as spherical constructs on the nano- or micrometer scale in which the therapeutic agents are dissolved, encapsulated, or adsorbed [73]. One of the most interesting characteristics of particulate delivery systems, which constitutes an important advantage over other DDSs, is their relatively small size. It allows them to be administered using noninvasive techniques and to be taken up by various cell types [74, 75]. Cell internalization could result in higher bioavailability of proteins, since their sites of action are normally located in cell compartments, thus simultaneously reducing both the total injected dose required and the associated side effects related to free protein administration [74]. Besides, NPs and MPs also offer other important benefits for protein therapy. In fact, their capacity to enhance the therapeutic potential of protein therapy is mainly due to the fact that such devices allow

- Protein protection from degradation: MPs and NPs act as systems that entrap therapeutic agents, protecting them from environmental stresses and helping with protein stabilization

[76]. This is essential for their correct therapeutic effect, as injected peptides and proteins need to retain their structural integrity until they reach their delivery site and must not be degraded as a result of enzymatic interactions [77].

- Controlled release over time: MPs and NPs are able to release proteins in a controlled manner [78]. Encapsulated proteins are released by diffusion through the matrix during the initial stage [79] and later, depending on the degradation of the device [80]. Therefore, protein release can be modified by controlling drug attachment to the matrix or DDSs degradation profile. This makes it possible to maintain therapeutic agents in their therapeutic range concentration over prolonged periods of time, avoiding multiple dosages or elevated protein quantity administration.
- Specific administration of proteins to the desired body compartment: Many efforts have been made to achieve effective delivery of proteins through various routes of administration for successful therapeutic effects [76]. The NP and MP surface can be functionalized with several molecules in order to target particles towards specific organs, tissue, or cells. This strategy is known as active targeting. Interestingly, injected particles may have been detected in the body over lengthy periods of time after administration [81]. The potential to deliver therapeutic drugs to the desired body compartment, together with protein protection and controlled release, confirms the suitability of this kind of DDSs for maintaining therapeutic levels of drugs at their site of action.
- Minimization of toxic issues related to direct administration of proteins: The network and body responses activated by proteins are not well established yet, and it is not unusual to find undesired side effects [82]. DDSs play a key role in preventing treatment side effects, since they allow localized, controlled dose administration of VEGF that makes it possible to achieve the expected effect without inducing unwanted adverse effects.
- Multiple protein delivery: Nowadays it is known that several molecules are involved in triggering one single biological response [83]. Although VEGF is a powerful inductor of new vessel formation, it may lead to immature, leaky vasculature

when it is delivered alone [84]. Thus, the combination of more than one therapeutic protein is recommended in order to achieve the proper response. Nowadays it is possible to encapsulate more than one drug in the same NP [75] or MP [85], obtaining different release profiles if necessary [86]. The combination of two DDSs, one inside the other and each one containing one different drug, has also been studied, and has shown promising results (reviewed in [81]) which confirm that multiple drug delivery might be required for obtaining optimal patient outcomes.

All these advantages, represented in Fig. 14.2, might enhance drug bioavailability. Furthermore, one of the key steps towards the ideal DDSs is the use of a proper material for synthesizing the matrix that forms the DDSs. On the one hand, DDSs are thought to be present in the biological tissue for prolonged periods of time after administration, allowing constant drug therapeutic levels. On the other hand, DDSs must be degraded and eliminated from the organism to prevent tissue rejection or chronic inflammatory reactions. Thus, DDSs have to be biodegradable and biocompatible. Furthermore, a balance between degradation and permanence time must be achieved to satisfy all the therapeutic requirements. In this sense, poly(lactic-*co*-glycolic acid) (PLGA) polymer and its derivatives today constitute one of the most promising synthetic materials. These polymers are approved by the Food and Drug Administration for use as drug delivery platforms owing to their well established biodegradability and biocompatibility [87, 88]. Furthermore, PLGAs offer tunable release rates, ranging from a few days to several months. In addition, this biomaterial is useful to encapsulate both hydrophobic and hydrophilic drugs. Finally, DDSs made of PLGA can be administered through different routes of administration [80, 88].

As a result, PLGA has been extensively investigated for synthesizing polymeric NP and MP [87, 89]. Moreover, PLGA particles have proved to encapsulate drugs with high efficiency, to maintain protein bioactivity after encapsulation and also to be degraded and absorbed in the human body within several months [63]. These DDSs are therefore one of the most powerful strategies for delivering therapeutic proteins. A long list of proteins has been encapsulated in PLGA particles, showing great potential to improve the treatment of many diseases, including cancer and cardiovascular, inflammatory, and cerebral diseases (reviewed in [89]). From a clinical perspective, the application of particles delivering therapeutic agents specifically to the sites of injury for a long enough time is, in theory, a feasible treatment for diseases that nowadays still constitute challenging hurdles.

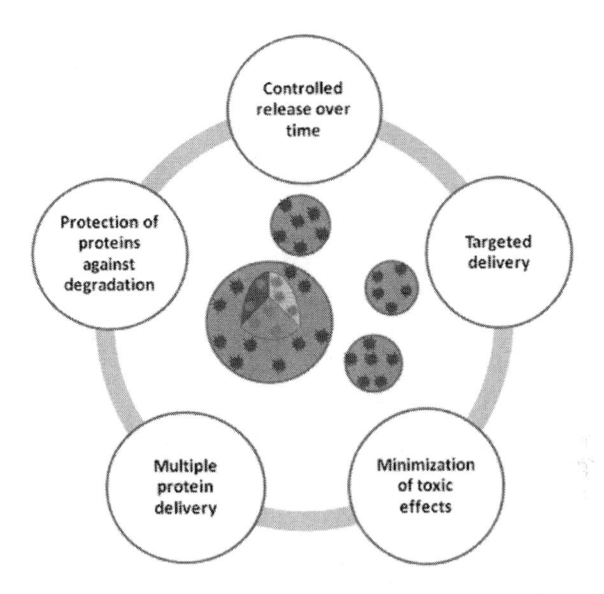

Figure 14.2 Advantages of using PLGA nano- and microparticles for protein delivery.

As previously explained in Section 14.1, VEGF is a cytokine with an elevated therapeutic potential for inducing angiogenesis and neuroprotection. However, VEGF is also a labile molecule which is rapidly degraded [90]. Nevertheless, since VEGF is considered a prime mediator of angiogenesis, it is therefore implicated in most carcinogenesis and metastasis processes [91]. Other toxic issues related to elevated VEGF dose injection have also been reported elsewhere, such as hypotension and a decrease in mean arterial pressure [92]. Therefore, VEGF, due to its inherent properties, is a perfect candidate to be incorporated into PLGA NPs and MPs. In addition, an important goal in VEGF therapy is to concentrate the protein specifically in the organ or tissue where the molecule is needed for its therapeutic action, for instance in the infarcted heart area after MI. Again, PLGA NPs and MPs play an essential role in meeting this requirement.

14.3 PLGA Particulates for VEGF Delivery: State of the Art

To facilitate a general overview of the state of the art regarding PLGA particles employed for VEGF delivery, work published in the last four years has been reviewed in this section (Table 14.3) (data from previous work were reviewed elsewhere [90]). Relevant information about methodological procedures in relation to particle preparation is reported in Table 14.4, whereas Table 14.5 aims to summarize the therapeutic applications.

Table 14.3 Number of studies published in the last 4 years including VEGF in PLGA particles classified considering the type of study, the combination of VEGF with other GF, and the combination of the particles with other DDSs

Type of study	Number of studies	Studies with several GF	Studies with combined DDSs	References
In vitro	10	3	3	[93–102]
In vivo	18	8	13	[103–120]

Table 14.4 Studies published in the last 4 years including VEGF in PLGA particles classified based on the type of particle, the type of DDS (particles alone-simple- or combined with other DDS), the method used to prepare the particles and the monomer ratio of the PLGA employed

Particles	DDS	Preparation Method	PLGA	References
MP (23)	Simple (14) Combined (9)	W/O/W (21) S/O/W (2)	50:50 (16) 37.5:25 (2) 85:15 (1) No data (4)	[93, 95–111, 113, 116, 118–120]
NP (5)	Simple (2) Combined (3)	W/O/W (3) No data (2)	50:50 (3) No data (2)	[94, 112, 114, 115, 117]

W/O/W corresponds to double emulsion and S/O/W corresponds to simple emulsion including solid VEGF in the inner organic phase. (*N*) Indicates number of studies.

14.3.1 Particle Preparation and Composition

In general, the particle preparation method and components of the formulation are the same in most studies. Particles are mostly prepared by the formation of a double emulsion followed by solvent evaporation. When encapsulating proteins, the labile nature of these molecules must be considered. This implies that during preparation, storage and release of the protein, the folding and 3D–4D structure must be maintained, since protein activity depends in all cases on the structure, its preservation being the big challenge of protein encapsulation [63]. Protein degradation can be the consequence of physical or chemical stress. Regarding physical stress, mechanical forces, and high temperature must be avoided, and so the systems used to form the double emulsion are designed to minimize them. An example is the Total Recirculation One Machine System®, in which emulsification takes place based on a turbulent circulation of the phases at room temperature, leading to high encapsulation while the bioactivity of the GF is preserved [119]. Another way to avoid high shear forces and temperature is the mild membrane emulsification technique, described by Ma G. [63]. Hydrophobic interactions of the protein with the organic phase are a frequent cause of protein unfolding and aggregation. Additives in the inner aqueous phase are included to reduce the interfacial tension and to attach to the surface of the protein, protecting the hydrophobic sites from interaction. Serum albumin is most usually employed to protect VEGF from this kind of degradation. In some cases PEG is also incorporated. Apart from protecting VEGF, PEG also permits us to obtain porous particles, affecting the release profile. A different way to avoid interaction with the W_1/O interphase is to include the protein in solid state in a S/O/W system. The Montero-Menei group is the only one which, in the last four years, has employed this approach, in which VEGF is first precipitated from a previously formed suspension [95, 98]. This protein in solid state is then dispersed in the organic solution. The method has been demonstrated to be effective in allowing very high encapsulation efficiency (97.4 ± 18.7%) and the preservation of the activity of the protein, demonstrated by a proliferation assay of human umbilical vein endothelial cells (HUVECs). This bioassay is an easy test in which HUVECs are treated with the native protein and with the equivalent dose of VEGF released from the particles, and proliferation is quantified and correlated. Compared to the bioactivity assays when encapsulating other proteins, this is a very easy, fast, and well-reproducible assay. Moreover, the dose needed to promote *in vitro* proliferation of HUVEC is very low (10 ng/ml). All these advantages make this assay the preferred one, and it is used in the majority of the studies reviewed.

Regarding the polymer in use, PLGA with 50:50 monomer ratio is the one that is most widely used in the publications reviewed here. The lactic-glycolic acid ratio is another parameter that affects the polymer degradation rate, and it should be taken into account to monitor the release profile of VEGF. Remarkably, none of the studies examined provides a proper discussion about the reasons why this polymer is chosen and about its impact on the properties of the DDSs. In our opinion this could be an interesting parameter to consider in order to optimize the formulations. Finally, once particles are prepared, lyophilization is almost the only method employed to dry and store the particles.

After several years of research in this field, control over GF release is nowadays a priority if we aim to succeed in simulating the physiological needs to efficiently promote tissue regeneration. In this sense, as has already been explained, protein release can be efficiently modified by encapsulation of proteins into PLGA particles, and for better release control particles can be incorporated into other DDSs. Recently published data have shown how hydrogels are the DDSs that are mostly employed for this aim, since they obtain a prolonged release over time and help to reduce the high burst effect generally observed with PLGA particles. This has been demonstrated, for instance, in the work by Wang *et al.* in which, after 6 days, the *in vitro* release of VEGF from the MP corresponded to 20–30%, whereas for the MP incorporated in the hydrogel it was only 12–13% [93]. Geng *et al.* also demonstrated 40% *in vitro* release from the particles vs. 25–30% from the hybrid system after 6 days [94]. A different way to delay the release of VEGF is the use of high molecular weight PLGA. This fact is discussed and explored in the work by Ju *et al.* to promote local angiogenesis in a rodent model [113]. In this work angiopoietin-1 (Ang-1) and VEGF were encapsulated in MP that were embedded in a hyaluronic acid (HA) hydrogel. Since it has been demonstrated that to efficiently promote angiogenesis a longer release of Ang-1 is needed, PLGA chosen to prepare Ang-1 MP had a higher molecular weight (38–54 KDa) than that used to encapsulate VEGF (5–15 KDa), both of which presented a monomer ratio of 50:50. Consequently, *in vitro* release of VEGF was greater after 13 days (26.9% vs. 9.71% for Ang-1). Finally, also with the aim of obtaining different release profiles of VEGF, preparation of PLGA MPs presenting different pore size has also been used [109]. Particles without pores or with small pore size (0.35 μm) presented a 35% VEGF release *in vitro* after 10 days, whereas particles with a pore size of 0.59 μm showed a 60% protein release.

Table 14.5 Studies published in the last 4 years including *in vivo* studies with VEGF encapsulated in PLGA particles, classified based on the disease under study, the combination with other GF and the combination with other DDS. (*N*) indicates number of studies

Organ	Combined GF	N studies combined devices (%)	References
Heart (2)	–	–	[119, 120]
Bone (4)	BMP-2 (2) PDGF (1)	4 (100%)	[103, 105, 106, 111]
Peripheral Ischemia (4)	Ang-1 (1)	2 (50%)	[109, 110, 114, 116]
CNS (5)	GDNF (2) Ang-1 (1)	2 (40%)	[107, 112, 113, 117, 118]
Nerve (2)	–	2 (100%)	[104, 108]
Urethral (1)	–	1 (100%)	[115]

14.3.2 Therapeutic Benefits

Studies with PLGA particles to deliver VEGF involve the same tissue defects in which VEGF has been administered directly (Fig. 14.1 and Table 14.2). Nevertheless, it is remarkable that in the last 4 years very few studies to treat **heart ischemia** using PLGA particles containing VEGF have been performed. Our group published a paper in 2010 in which VEGF was incorporated into PLGA MP and intramyocardially administered in a rat model of MI [119]. Despite the fact that no significant improvement on the cardiac function could be demonstrated, histology revealed an increased number of caveolin-1 and α-smooth muscle Actine positive vessels, representative of boosted angiogenesis and arteriogenesis processes respectively. In view of the results mentioned and data in the bibliography, VEGF itself seemed to be appropriate to promote vessel formation but insufficient to develop a more integrative response, in which other heart repair mechanisms are involved. Therefore in a second study we proposed two innovations [120]. On the one hand, the formulation was improved by incorporating a PEG-PLGA diblock polymer to give the particles a stealth effect. On the other hand, we tried combining VEGF's pharmacological effect with the antioxidant effect of Coenzyme-Q (CoQ), described as a cardioprotective agent. This CoQ was included in PLGA NPs as well, to improve the poor oral bioavailability which is a consequence of its lypophilic nature. In this case, VEGF PEG-PLGA MP succeeded in significantly improving heart function, probably due to a prolonged presence in the damaged tissue and a more effective prolonged release of the GF. Nevertheless, combination with CoQ NPs reduced this effect, maybe due to an antiangiogenic effect exerted by this molecule.

The small number of studies employing VEGF particles in heart repair contrasts with the large number of CT performed in this area with this GF (Table 14.2) and can be explained as the consequence of the emergence of recently discovered GF that appear to be more potent actors in the cardiovascular tissue regeneration processes.

On the other hand, **bone repair** has appeared as an interesting field of application for VEGF particles made of PLGA. In this area, there is nowadays great controversy concerning the beneficial effect of the combination of VEGF with other GF, preferentially with the well-known osteogenic BMP-2. In this context, Geuze *et al.* reported an interesting study in which a model of critical size ulnar defect in Beagle dogs was employed [103]. Animals were treated with BMP-2 and/or VEGF in PLGA MP. These particles were incorporated into two different scaffolds made of gelatin or of biphasic calcium phosphate, leading to a fast or slow release of the proteins respectively. The aim here was to evaluate the impact of the early/late release of both factors alone and in combination. BMP-2 release promoted bone formation in large rates and, even if authors observed a beneficial effect of VEGF when administered alone, the combination did not result in any significant improvement. The authors explained this result by the fact that the osteoconductive ceramic BCP is known to promote differentiation of MSC towards osteoblasts, which are able to secrete VEGF. Thus, administration of exogenous VEGF in this case should become ineffective [103]. This is a nice example of how the effect of VEGF PLGA particles can be drastically altered by the combination with other systems or materials. In contrast, in the work published by Reyes *et al.*, the conclusion was the opposite: in this case VEGF seemed to be the essential factor for successfully regenerating the damaged bone [105]. In this case, particles were also included in a brushite scaffold that allowed a slower release of VEGF. Different results, however, may possibly be attributed to the combination of VEGF with PlDGF, which seems to be a much more potent osteoregenerative factor. When comparing these two studies it should also be considered that the animal models employed were different, so comparisons between them must be considered cautiously.

The controversy regarding whether VEGF has a beneficial effect in combination with BMP-2 or not is addressed in the work by Hernandez *et al.* [106]. Here, the main conclusion was that VEGF effect in combination with BMP-2 is time dependent. After 4 weeks, multitherapy led to significantly increased bone formation compared to non-treated groups and to the BMP-2 group. However, after 12 weeks no differences could be detected between the BMP-2 group and the VEGF-BMP-2 treatments. This could be related to the results in the first-mentioned study, in which animals were sacrificed after 9 weeks [103]. In conclusion, considering the updated bibliography, it seems that VEGF release during the first weeks has an important role in bone regeneration but this effect tends to disappear after the first month.

Peripheral ischemia continues to attract the attention of researchers working with VEGF encapsulated in PLGA particles and, to our knowledge, 4 different studies have been published in the last 4 years to treat this disease. Two of them were performed by Shin and collaborators [109, 116]. In both cases, a mouse model was used and VEGF polymeric particles were embedded in a scaffold made of alginate and containing another factor: the antiapoptotic agent TAT-HSP27 or Ang-1. Treatment and analysis were performed in the same way: animals were treated one day after vessel occlusion, and the results were evaluated 4 weeks later. The use of the same model and conditions allows comparison between studies. In both cases better results were obtained when treating the animals with the hydrogel-particle system including both factors. The differences between combining VEGF with TAT-HSP27 or with Ang-1 are not clear. Arteriolar density appears to be higher with Ang-1 but the functional relevance of this result has not been demonstrated. In relation to the other two studies mentioned above, it is difficult to make a comparison of their effectiveness. In the work by Strauss *et al.* VEGF in MP was administered to a rabbit model of coronary total occlusion 12 weeks after the occlusion and the results were analyzed 3 weeks after treatment [110]. On the other hand, Geng *et al.* focused on demonstrating the reduction of the contracture when VEGF NP were included in a bladder acellular matrix allograft in a swine model [114]. This summary of the studies performed to treat peripheral ischemia shows how there are broad differences regarding the models in use and the different approaches to tackling this disease with VEGF particles.

CNS is the other main target for PLGA particles with VEGF. In the last few years two studies have been performed to treat a model of middle cerebral artery occlusion [107, 113]. Both studies reported an increase in vasculogenesis. Recently accepted for publication, the study by Ju *et al.* reports a sophisticated DDS in which particles are included in a HA hydrogel activated by the covalent attachment of Nogo receptor, known to bind several inhibitory myelin proteins promoting neural regeneration [113]. Both behavioral tests and histology performed to evaluate angiogenesis demonstrated a significant improvement with the hydrogel loaded with the MP. Nevertheless, this study did not evaluate the benefits of direct administration of the MP or the effect of monotherapy with VEGF particles. Groups of animals receiving (1) the PLGA particles directly, and (2) the DDS with each of the GFs separately, could clarify these points.

In the area of PD, the group of Herran *et al.* has published two studies in recent years [117, 118]. In both of them animals were treated with a combination of VEGF and GDNF particles. Combined therapy failed in the first study conducted in a severe model of PD lesion [118], but the second study succeeded in demonstrating the synergic effect of GF in a partial lesion rat model [117]. This confirmed the authors' hypothesis attributing the lack of significance after the first negative results to the severity of the model in use. The first study was performed in a severe lesion rat model, whereas for the second one, subregions in the caudoputamen complex were selected as a target for the lesion, causing more selective damage to the nigrostriatal dopaminergic pathway. These consecutive studies point to the importance of the choice of animal model and, at the same time, the limitations of animal models in general, and the need for critical analysis of the results obtained in these *in vivo* studies becomes patent once again.

As indicated in Table 14.5, two studies have been reviewed considering **nerve regeneration**, both of which included VEGF in PLGA MP in nerve prefabricated nerve conduits or grafts [104, 108]. The animal model was different in each case: Rui *et al.* used a rat sciatic nerve lesion (1 cm) [104] whereas Karagoz *et al.* bridged two healthy nerves to evaluate the lateral axonal sprouting into the vein graft [108]. In the work of Rui and collaborators, although no significant increase in the number of vessels in the group treated with VEGF particles was observed, an increase in the tibial nerve function was demonstrated. On the other hand, Karagoz and colleagues concluded that partial incision is not a useful method in clinical practice, since 4 weeks after the lesion was made, signs of tibial and peroneal nerve paralysis had healed naturally.

Numerous studies have been performed to explore the beneficial effect of VEGF incorporated in PLGA particles, involving various different injuries and diseases. In view of the results to date, we presume that these studies will continue to increase, in order to refine the nature and the mode of employment of these DDSs. This will ultimately lead to robust results that will make it possible to launch the first CT in humans in the near future.

14.4 Conclusions

The first clinical attempts to exploit GF therapy to regenerate damaged tissues taught us that direct administration of VEGF yields no benefits, for several reasons discussed in this chapter. DDSs were therefore proposed to overcome those limitations. Currently, we are witnessing a new era in which the efficacy of DDSs in allowing prolonged release while protecting VEGF from degradation *in vivo* has already been demonstrated. We are now moving towards the design of more complex DDSs that are able to exert a significant regenerative effect. Therefore, after this overview of the most recent studies published on the subject, we can conclude by emphasizing two main ideas: (i) VEGF particles are now generally proposed as a therapy in combination with other regenerative therapies, and (ii) complex building of DDSs, based on the properties of different materials and forms of dosage, is bringing about major progress in fulfilling the different requirements concerning the timing and dosage of each factor in combination with VEGF.

References

1. Senger, D. R., Galli, S. J., Dvorak, A. M., Perruzzi, C. A., Harvey, V. S., and Dvorak, H. F. (1983). Tumor cells secrete a vascular permeability factor that promotes accumulation of ascites fluid. *Science,* **219**, 983–985.
2. Dvorak, H. F. (2002). Vascular permeability factor/vascular endothelial growth factor: a critical cytokine in tumor angiogenesis and a potential target for diagnosis and therapy. *J Clin Oncol,* **20**, 4368–4380.
3. Ferrara, N., and Henzel, W. J. (1989). Pituitary follicular cells secrete a novel heparin-binding growth factor specific for vascular endothelial cells. *Biochem Biophys Res Commun,* **161**, 851–858.
4. Leung, D. W., Cachianes, G., Kuang, W. J., Goeddel, D. V, and Ferrara, N. (1989). Vascular endothelial growth factor is a secreted angiogenic mitogen. *Science,* **246**, 1306–1309.
5. Grimmond, S., Lagercrantz, J., Drinkwater, C., Silins, G., Townson, S., Pollock, P., Gotley, D., Carson, E., Rakar, S., Nordenskjöld, M., *et al.* (1996). Cloning and characterization of a novel human gene related to vascular endothelial growth factor. *Genome Res,* **6**, 124–131.
6. Olofsson, B., Pajusola, K., Kaipainen, A., von Euler, G., Joukov, V., Saksela, O., Orpana, A., Pettersson, R. F., Alitalo, K., and Eriksson, U. (1996). Vascular endothelial growth factor B, a novel growth factor for endothelial cells. *Proc Natl Acad Sci U S A,* **93**, 2576–2581.
7. Joukov, V., Pajusola, K., Kaipainen, A., Chilov, D., Lahtinen, I., Kukk, E., Saksela, O., Kalkkinen, N., and Alitalo, K. (1996). A novel vascular endothelial growth factor, VEGF-C, is a ligand for the Flt4 (VEGFR-3) and KDR (VEGFR-2) receptor tyrosine kinases. *EMBO J,* **15**, 1751.

8. Orlandini, M., Marconcini, L., Ferruzzi, R., and Oliviero, S. (1996). Identification of a c-fos-induced gene that is related to the platelet-derived growth factor/vascular endothelial growth factor family. *Proc Natl Acad Sci U S A,* **93**, 11675–11680.

9. Maglione, D., Guerriero, V., Viglietto, G., Delli-Bovi, P., and Persico, M. G. (1991). Isolation of a human placenta cDNA coding for a protein related to the vascular permeability factor. *Proc Natl Acad Sci U S A,* **88**, 9267–9271.

10. Shibuya, M. (2014). VEGF-VEGFR signals in health and disease. *Biomol Ther (Seoul),* **22**, 1–9.

11. Keifer, O. P., O'Connor, D. M., and Boulis, N. M. (2014). Gene and protein therapies utilizing VEGF for ALS. *Pharmacol Ther,* **141**, 261–271.

12. Bellomo, D., Headrick, J. P., Silins, G. U., Paterson, C. A., Thomas, P. S., Gartside, M., Mould, A., Cahill, M. M., Tonks, I. D., Grimmond, S. M., *et al.* (2000). Mice lacking the vascular endothelial growth factor-B gene (Vegfb) have smaller hearts, dysfunctional coronary vasculature, and impaired recovery from cardiac ischemia. *Circ Res,* **86**, E29–E35.

13. Alitalo, K., and Carmeliet, P. (2002). Molecular mechanisms of lymphangiogenesis in health and disease. *Cancer Cell,* **1**, 219–227.

14. Cheng, S. Y., Nagane, M., Huang, H. S., and Cavenee, W. K. (1997). Intracerebral tumor-associated hemorrhage caused by overexpression of the vascular endothelial growth factor isoforms VEGF121 and VEGF165 but not VEGF189. *Proc Natl Acad Sci U S A,* **94**, 12081–12087.

15. Mäkinen, K., Manninen, H., Hedman, M., Matsi, P., Mussalo, H., Alhava, E., and Ylä-Herttuala, S. (2002). Increased vascularity detected by digital subtraction angiography after VEGF gene transfer to human lower limb artery: a randomized, placebo-controlled, double-blinded phase II study. *Mol Ther,* **6**, 127–133.

16. Losordo, D. W., Vale, P. R., Symes, J. F., Dunnington, C. H., Esakof, D. D., Maysky, M., Ashare, A. B., Lathi, K., and Isner, J. M. Gene therapy for myocardial angiogenesis: initial clinical results with direct myocardial injection of phVEGF165 as sole therapy for myocardial ischemia. *Circulation,* **98**, 2800–2804.

17. Stewart, D. J., Kutryk, M. J. B., Fitchett, D., Freeman, M., Camack, N., Su, Y., Della Siega, A., Bilodeau, L., Burton, J. R., Proulx, G., *et al.* (2009). VEGF gene therapy fails to improve perfusion of ischemic myocardium in patients with advanced coronary disease: results of the NORTHERN trial. *Mol Ther,* **17**, 1109–1115.

18. Gyöngyösi, M., Khorsand, A., Zamini, S., Sperker, W., Strehblow, C., Kastrup, J., Jorgensen, E., Hesse, B., Tägil, K., Bøtker, H. E., *et al.* (2005).

NOGA-guided analysis of regional myocardial perfusion abnormalities treated with intramyocardial injections of plasmid encoding vascular endothelial growth factor A-165 in patients with chronic myocardial ischemia: subanalysis of the EUROINJECT-ONE multicenter double-blind randomized study. *Circulation*, **112**, I157–I165.

19. Sylvén, C., Sarkar, N., Rück, A., Drvota, V., Hassan, S. Y., Lind, B., Nygren, A., Källner, Q., Blomberg, P., van der Linden, J., *et al.* (2001). Myocardial Doppler tissue velocity improves following myocardial gene therapy with VEGF-A165 plasmid in patients with inoperable angina pectoris. *Coron Artery Dis.*, **12**, 239–243.
20. Kastrup, J., Jørgensen, E., Rück, A., Tägil, K., Glogar, D., Ruzyllo, W., Bøtker, H. E., Dudek, D., Drvota, V., Hesse, B., *et al.* (2005). Direct intramyocardial plasmid vascular endothelial growth factor-A165 gene therapy in patients with stable severe angina pectoris A randomized double-blind placebo-controlled study: the Euroinject One trial. *J Am Coll Cardiol*, **45**, 982–988.
21. Henry, T. D., Annex, B. H., McKendall, G. R., Azrin, M. A., Lopez, J. J., Giordano, F. J., Shah, P. K., Willerson, J. T., Benza, R. L., Berman, D. S., *et al.* (2003). The VIVA trial: Vascular endothelial growth factor in Ischemia for Vascular Angiogenesis. *Circulation*, **107**, 1359–1365.
22. Hendel, R. C., Henry, T. D., Rocha-Singh, K., Isner, J. M., Kereiakes, D. J., Giordano, F. J., Simons, M., and Bonow, R. O. (2000). Effect of intracoronary recombinant human vascular endothelial growth factor on myocardial perfusion: evidence for a dose-dependent effect. *Circulation*, 101, 118–121.
23. Henry, T. D., Rocha-Singh, K., Isner, J. M., Kereiakes, D. J., Giordano, F. J., Simons, M., Losordo, D. W., Hendel, R. C., Bonow, R. O., Eppler, S. M., *et al.* (2001). Intracoronary administration of recombinant human vascular endothelial growth factor to patients with coronary artery disease. *Am Hear J*, **142**, 872–880.
24. Hedman, M., Hartikainen, J., Syvänne, M., Stjernvall, J., Hedman, A., Kivelä, A., Vanninen, E., Mussalo, H., Kauppila, E., Simula, S., *et al.* (2003). Safety and feasibility of catheter-based local intracoronary vascular endothelial growth factor gene transfer in the prevention of postangioplasty and in-stent restenosis and in the treatment of chronic myocardial ischemia: phase II results of the Kuopio. *Circulation*, **107**, 2677–2683.
25. Hedman, M., Muona, K., Hedman, A., Kivelä, A., Syvänne, M., Eränen, J., Rantala, A., Stjernvall, J., Nieminen, M. S., Hartikainen, J., *et al.* (2009). Eight-year safety follow-up of coronary artery disease patients after local intracoronary VEGF gene transfer. *Gene Ther*, **16**, 629–634.

26. Shyu, K.-G., Chang, H., Wang, B.-W., and Kuan, P. (2003). Intramuscular vascular endothelial growth factor gene therapy in patients with chronic critical leg ischemia. *Am J Med,* **114**, 85–92.

27. Baumgartner, I., Pieczek, A., Manor, O., Blair, R., Kearney, M., Walsh, K., and Isner, J. M. (1998). Constitutive expression of phVEGF165 after intramuscular gene transfer promotes collateral vessel development in patients with critical limb ischemia. *Circulation*, **97**, 1114–1123.

28. Kusumanto, Y. H., van Weel, V., Mulder, N. H., Smit, A. J., van den Dungen, J. J. A. M., Hooymans, J. M. M., Sluiter, W. J., Tio, R. A., Quax, P. H. A., Gans, R. O. B., *et al.* (2006). Treatment with intramuscular vascular endothelial growth factor gene compared with placebo for patients with diabetes mellitus and critical limb ischemia: a double-blind randomized trial. *Hum Gene Ther,* **17**, 683–691.

29. Isner, J. M., Baumgartner, I., Rauh, G., Schainfeld, R., Blair, R., Manor, O., Razvi, S., and Symes, J. F. (1998). Treatment of thromboangiitis obliterans (Buerger's disease) by intramuscular gene transfer of vascular endothelial growth factor: preliminary clinical results. *J Vasc Surg,* **28**, 964–973; discussion 73–75.

30. Rosengart, T. K., Lee, L. Y., Patel, S. R., Sanborn, T. A., Parikh, M., Bergman, G. W., Hachamovitch, R., Szulc, M., Kligfield, P. D., Okin, P. M., *et al.* (1999). Angiogenesis gene therapy: phase I assessment of direct intramyocardial administration of an adenovirus vector expressing VEGF121 cDNA to individuals with clinically significant severe coronary artery disease. *Circulation*, **100**, 468–474.

31. Stewart, D. J., Hilton, J. D., Arnold, J. M. O., Gregoire, J., Rivard, A., Archer, S. L., Charbonneau, F., Cohen, E., Curtis, M., Buller, C. E., *et al.* (2006). Angiogenic gene therapy in patients with nonrevascularizable ischemic heart disease: a phase 2 randomized, controlled trial of AdVEGF(121) (AdVEGF121) versus maximum medical treatment. *Gene Ther*, **13**, 1503–1511.

32. Mohler, E. R., Rajagopalan, S., Olin, J. W., Trachtenberg, J. D., Rasmussen, H., Pak, R., and Crystal, R. G. (2003). Adenoviral-mediated gene transfer of vascular endothelial growth factor in critical limb ischemia: safety results from a phase I trial. *Vasc Med*, **8**, 9–13.

33. Rajagopalan, S., Mohler, E. R., Lederman, R. J., Mendelsohn, F. O., Saucedo, J. F., Goldman, C. K., Blebea, J., Macko, J., Kessler, P. D., Rasmussen, H. S., *et al.* (2003). Regional angiogenesis with vascular endothelial growth factor in peripheral arterial disease: a phase II randomized, double-blind, controlled study of adenoviral delivery of vascular endothelial growth factor 121 in patients with disabling intermittent claudication. *Circulation*, **108**, 1933–1938.

34. Vale, P. R., Losordo, D. W., Milliken, C. E., McDonald, M. C., Gravelin, L. M., Curry, C. M., Esakof, D. D., Maysky, M., Symes, J. F., and Isner, J. M. (2001). Randomized, single-blind, placebo-controlled pilot study of catheter-based myocardial gene transfer for therapeutic angiogenesis using left ventricular electromechanical mapping in patients with chronic myocardial ischemia. *Circulation*, **103**, 2138–2143.

35. Reilly, J. P., Grise, M. A., Fortuin, F. D., Vale, P. R., Schaer, G. L., Lopez, J., VAN Camp, J. R., Henry, T., Richenbacher, W. E., Losordo, D. W., *et al.* (2005). Long-term (2-year) clinical events following transthoracic intramyocardial gene transfer of VEGF-2 in no-option patients. *J Interv Cardiol*, **18**, 27–31.

36. Losordo, D. W., Vale, P. R., Hendel, R. C., Milliken, C. E., Fortuin, F. D., Cummings, N., Schatz, R. A., Asahara, T., Isner, J. M., and Kuntz, R. E. (2002). Phase 1/2 placebo-controlled, double-blind, dose-escalating trial of myocardial vascular endothelial growth factor 2 gene transfer by catheter delivery in patients with chronic myocardial ischemia. *Circulation*, **105**, 2012–2018.

37. Formiga, F. R., Tamayo, E., Simón-Yarza, T., Pelacho, B., Prósper, F., and Blanco-Prieto, M. J. (2012). Angiogenic therapy for cardiac repair based on protein delivery systems. *Heart Fail Rev*, **17**, 449–473.

38. Luo, Z., Diaco, M., Murohara, T., Ferrara, N., Isner, J. M., and Symes, J. F. (1997). Vascular endothelial growth factor attenuates myocardial ischemia-reperfusion injury. *Ann Thorac Surg*, **64**, 993–998.

39. Long, J., Wang, S., Zhang, Y., Liu, X., Zhang, H., and Wang, S. (2013). The therapeutic effect of vascular endothelial growth factor gene- or heme oxygenase-1 gene-modified endothelial progenitor cells on neovascularization of rat hindlimb ischemia model. *J Vasc Surg*, **58**, 756–765.e2.

40. Talwar, T., and Srivastava, M. V. P. (2014). Role of vascular endothelial growth factor and other growth factors in post-stroke recovery. *Ann Indian Acad Neurol*, **17**, 1–6.

41. Sun, Y., Jin, K., Xie, L., Childs, J., Mao, X. O., Logvinova, A., and Greenberg, D. A. (2003). VEGF-induced neuroprotection, neurogenesis, and angiogenesis after focal cerebral ischemia. *J Clin Invest*, **111**, 1843–1851.

42. Hayashi, T., Abe, K., and Itoyama, Y. (1998). Reduction of ischemic damage by application of vascular endothelial growth factor in rat brain after transient ischemia. *J Cereb Blood Flow Metab*, **18**, 887–895.

43. Proescholdt, M. A., Heiss, J. D., Walbridge, S., Mühlhauser, J., Capogrossi, M. C., Oldfield, E. H., and Merrill, M. J. (1999). Vascular endothelial

growth factor (VEGF) modulates vascular permeability and inflammation in rat brain. *J Neuropathol Exp Neurol*, **58**, 613–627.

44. Zhang, Z. G., Zhang, L., Jiang, Q., Zhang, R., Davies, K., Powers, C., Bruggen, N. v, and Chopp, M. (2000). VEGF enhances angiogenesis and promotes blood–brain barrier leakage in the ischemic brain. *J Clin Invest*, **106**, 829–838.
45. Widenfalk, J., Lipson, A., Jubran, M., Hofstetter, C., Ebendal, T., Cao, Y., and Olson, L. (2003). Vascular endothelial growth factor improves functional outcome and decreases secondary degeneration in experimental spinal cord contusion injury. *Neuroscience*, **120**, 951–960.
46. Benton, R. L., and Whittemore, S. R. (2003). VEGF165 therapy exacerbates secondary damage following spinal cord injury. *Neurochem Res*, **28**, 1693–1703.
47. Wang, H.-L., and Boyapati, L. (2006). "PASS" principles for predictable bone regeneration. *Implant Dent*, **15**, 8–17.
48. Mayr-Wohlfart, U., Waltenberger, J., Hausser, H., Kessler, S., Günther, K.-P., Dehio, C., Puhl, W., and Brenner, R. E. (2002). Vascular endothelial growth factor stimulates chemotactic migration of primary human osteoblasts. *Bone*, **30**, 472–477.
49. Gothard, D., Smith, E. L., Kanczler, J. M., Rashidi, H., Qutachi, O., Henstock, J., Rotherham, M., El Haj, A., Shakesheff, K. M., and Oreffo, R. O. C. (2014). Tissue engineered bone using select growth factors: A comprehensive review of animal studies and clinical translation studies in man. *Eur Cell Mater*, **28**, 166–207; discussion 207–208.
50. Youssif, M., Shiina, H., Urakami, S., Gleason, C., Nunes, L., Igawa, M., Enokida, H., Tanagho, E. A., and Dahiya, R. (2005). Effect of vascular endothelial growth factor on regeneration of bladder acellular matrix graft: histologic and functional evaluation. *Urology*, **66**, 201–207.
51. Zachary, I. (2005). Neuroprotective role of vascular endothelial growth factor: signalling mechanisms, biological function, and therapeutic potential. *Neurosignals*, **14**, 207–221.
52. Rosenstein, J. M., Krum, J. M., and Ruhrberg, C. (2010). VEGF in the nervous system. *Organogenesis*, **6**, 107–114.
53. Garbayo, E., Ansorena, E., and Blanco-Prieto, M. J. (2013). Drug development in Parkinson's disease: From emerging molecules to innovative drug delivery systems. *Maturitas*, **76**, 272–278.
54. Piltonen, M., Planken, A., Leskelä, O., Myöhänen, T. T., Hänninen, A.-L., Auvinen, P., Alitalo, K., Andressoo, J.-O., Saarma, M., and Männistö, P. T.

(2011). Vascular endothelial growth factor C acts as a neurotrophic factor for dopamine neurons *in vitro* and *in vivo*. *Neuroscience,* **192**, 550–563.

55. Falk, T., Yue, X., Zhang, S., McCourt, A. D., Yee, B. J., Gonzalez, R. T., and Sherman, S. J. (2011). Vascular endothelial growth factor-B is neuroprotective in an *in vivo* rat model of Parkinson's disease. *Neurosci Lett,* **496**, 43–47.
56. Yasuhara, T., Shingo, T., Muraoka, K., Kameda, M., Agari, T., Wen Ji, Y., Hayase, H., Hamada, H., Borlongan, C. V, and Date, I. (2005). Neurorescue effects of VEGF on a rat model of Parkinson's disease. *Brain Res,* **1053**, 10–18.
57. Garcia, K. O., Ornellas, F. L. M., Martin, P. K. M., Patti, C. L., Mello, L. E., Frussa-Filho, R., Han, S. W., and Longo, B. M. (2014). Therapeutic effects of the transplantation of VEGF overexpressing bone marrow mesenchymal stem cells in the hippocampus of murine model of Alzheimer's disease. *Front Aging Neurosci,* **6**, 30.
58. Greenberg, D. A., and Jin, K. (2013). Vascular endothelial growth factors (VEGFs) and stroke. *Cell Mol Life Sci,* **70**, 1753–1761.
59. Wang, Y., Galvan, V., Gorostiza, O., Ataie, M., Jin, K., and Greenberg, D. A. (2006). Vascular endothelial growth factor improves recovery of sensorimotor and cognitive deficits after focal cerebral ischemia in the rat. *Brain Res,* **1115**, 186–193.
60. Fournier, N. M., and Duman, R. S. (2012). Role of vascular endothelial growth factor in adult hippocampal neurogenesis: implications for the pathophysiology and treatment of depression. *Behav Brain Res,* **227**, 440–449.
61. Poesen, K., Lambrechts, D., Van Damme, P., Dhondt, J., Bender, F., Frank, N., Bogaert, E., Claes, B., Heylen, L., Verheyen, A., *et al.* (2008). Novel role for vascular endothelial growth factor (VEGF) receptor-1 and its ligand VEGF-B in motor neuron degeneration. *J Neurosci,* **28**, 10451–10459.
62. Sun, Y., Jin, K., Childs, J. T., Xie, L., Mao, X. O., and Greenberg, D. A. (2006). Vascular endothelial growth factor-B (VEGFB) stimulates neurogenesis: evidence from knockout mice and growth factor administration. *Dev Biol,* **289**, 329–335.
63. Ma, G. (2014). Microencapsulation of protein drugs for drug delivery: Strategy, preparation, and applications. *J Control Release,* **193**, 324–340.
64. Iwata, H., Mizutani, S., Tabei, Y., Kotera, M., Goto, S., and Yamanishi, Y. (2013). Inferring protein domains associated with drug side effects

based on drug-target interaction network. *BMC Syst Biol*, **7** Suppl 6, S18.

65. Kuhn, M., Al Banchaabouchi, M., Campillos, M., Jensen, L. J., Gross, C., Gavin, A.-C., and Bork, P. (2013). Systematic identification of proteins that elicit drug side effects. *Mol Syst Biol*, **9**, 663.

66. Ho, P. M., Bryson, C. L., and Rumsfeld, J. S. (2009). Medication adherence: its importance in cardiovascular outcomes. *Circulation*, **119**, 3028–3035.

67. Caldorera-Moore, M., Guimard, N., Shi, L., and Roy, K. (2010). Designer nanoparticles: incorporating size, shape and triggered release into nanoscale drug carriers. *Expert Opin Drug Deliv*, **7**, 479–495.

68. Alvarez-Lorenzo, C., and Concheiro, A. (2013). Bioinspired drug delivery systems. *Curr Opin Biotechnol*, **24**, 1167–1173.

69. Timko, B. P., and Kohane, D. S. (2012). Materials to clinical devices: technologies for remotely triggered drug delivery. *Clin Ther*, **34**, S25–S35.

70. Tan, M. L., Choong, P. F. M., and Dass, C. R. (2010). Recent developments in liposomes, microparticles and nanoparticles for protein and peptide drug delivery. *Peptides*, **31**, 184–193.

71. Parveen, S., Misra, R., and Sahoo, S. K. (2012). Nanoparticles: a boon to drug delivery, therapeutics, diagnostics and imaging. *Nanomedicine*, **8**, 147–166.

72. Kamaly, N., Xiao, Z., Valencia, P. M., Radovic-Moreno, A. F., and Farokhzad, O. C. (2012). Targeted polymeric therapeutic nanoparticles: design, development and clinical translation. *Chem Soc Rev*, **41**, 2971–3010.

73. Ravikumar, M. N. (2000). Nano and microparticles as controlled drug delivery devices. *J Pharm Pharm Sci*, **3**, 234–258.

74. Panyam, J., and Labhasetwar, V. (2003). Biodegradable nanoparticles for drug and gene delivery to cells and tissue. *Adv Drug Deliv Rev*, **55**, 329–347.

75. Sun, T., Zhang, Y. S., Pang, B., Hyun, D. C., Yang, M., and Xia, Y. (2014). Engineered nanoparticles for drug delivery in cancer therapy. *Angew Chemie Int Ed*, **53**, 12320–12364.

76. Jain, A., Jain, A., Gulbake, A., Shilpi, S., Hurkat, P., and Jain, S. K. (2013). Peptide and protein delivery using new drug delivery systems. *Crit Rev Ther Drug Carrier Syst*, **30**, 293–329.

77. Langer, R., and Peppas, N. A. (2003). Advances in biomaterials, drug delivery, and bionanotechnology. *AIChE J*, **49**, 2990–3006.

78. Wischke, C., and Schwendeman, S. P. (2008). Principles of encapsulating hydrophobic drugs in PLA/PLGA microparticles. *Int J Pharm,* **364**, 298–327.

79. Yeo, Y., and Park, K. (2004). Control of encapsulation efficiency and initial burst in polymeric microparticle systems. *Arch Pharm Res,* **27**, 1–12.

80. Casalini, T., Rossi, F., Lazzari, S., Perale, G., and Masi, M. (2014). Mathematical modeling of PLGA microparticles: from polymer degradation to drug release. *Mol Pharm,* **11**, 4036–4048.

81. Diaz-Herraez P., Pascual-Gil S., Garbayo E., Simon-Yarza T., Prosper F., Blanco-Prieto M. (2015). Cardiac drug delivery. In *Drug Delivery: An Integrated Clinical and Engineering Approach,* Y. Rosen, ed. (CRC Press Taylor & Francis Group).

82. Tatonetti, N. P., Liu, T., and Altman, R. B. (2009). Predicting drug side-effects by chemical systems biology. *Genome Biol,* **10**, 238.

83. Yancopoulos, G. D., Davis, S., Gale, N. W., Rudge, J. S., Wiegand, S. J., and Holash, J. (2000). Vascular-specific growth factors and blood vessel formation. *Nature,* **407**, 242–248.

84. Awada, H. K., Johnson, N. R., and Wang, Y. (2014). Dual delivery of vascular endothelial growth factor and hepatocyte growth factor coacervate displays strong angiogenic effects. *Macromol Biosci,* **14**, 679–686.

85. Formiga, F. R., Pelacho, B., Garbayo, E., Imbuluzqueta, I., Díaz-Herráez, P., Abizanda, G., Gavira, J. J., Simón-Yarza, T., Albiasu, E., Tamayo, E., *et al.* (2014). Controlled delivery of fibroblast growth factor-1 and neuregulin-1 from biodegradable microparticles promotes cardiac repair in a rat myocardial infarction model through activation of endogenous regeneration. *J Control Release,* **173**, 132–139.

86. Ruvinov, E., Leor, J., and Cohen, S. (2011). The promotion of myocardial repair by the sequential delivery of IGF-1 and HGF from an injectable alginate biomaterial in a model of acute myocardial infarction. *Biomaterials,* **32**, 565–578.

87. Pandita, D., Kumar, S., and Lather, V. (2015). Hybrid poly(lactic-*co*-glycolic acid) nanoparticles: design and delivery prospectives. *Drug Discov Today,* **20**, 95–104.

88. Mundargi, R. C., Babu, V. R., Rangaswamy, V., Patel, P., and Aminabhavi, T. M. (2008). Nano/micro technologies for delivering macromolecular therapeutics using poly(D,L-lactide-*co*-glycolide) and its derivatives. *J Control Release,* **125**, 193–209.

89. Danhier, F., Ansorena, E., Silva, J. M., Coco, R., Le Breton, A., and Préat, V. (2012). PLGA-based nanoparticles: an overview of biomedical applications. *J Control Release,* **161**, 505–522.

90. Simón-Yarza, T., Formiga, F. R., Tamayo, E., Pelacho, B., Prosper, F., and Blanco-Prieto, M. J. (2012). Vascular endothelial growth factor-delivery systems for cardiac repair: an overview. *Theranostics,* **2**, 541–552.

91. Claesson-Welsh, L., and Welsh, M. (2013). VEGFA and tumour angiogenesis. *J Intern Med,* **273**, 114–127.

92. Eppler, S. M., Combs, D. L., Henry, T. D., Lopez, J. J., Ellis, S. G., Yi, J.-H., Annex, B. H., McCluskey, E. R., and Zioncheck, T. F. (2002). A target-mediated model to describe the pharmacokinetics and hemodynamic effects of recombinant human vascular endothelial growth factor in humans. *Clin Pharmacol Ther,* **72**, 20–32.

93. Wang, Y., Wei, Y. T., Zu, Z. H., Ju, R. K., Guo, M. Y., Wang, X. M., Xu, Q. Y., and Cui, F. Z. (2011). Combination of hyaluronic acid hydrogel scaffold and PLGA microspheres for supporting survival of neural stem cells. *Pharm Res,* **28**, 1406–1414.

94. Geng, H., Song, H., Qi, J., and Cui, D. (2011). Sustained release of VEGF from PLGA nanoparticles embedded thermo-sensitive hydrogel in full-thickness porcine bladder acellular matrix. *Nanoscale Res Lett,* **6**, 312.

95. Musilli, C., Karam, J.-P., Paccosi, S., Muscari, C., Mugelli, A., Montero-Menei, C. N., and Parenti, A. (2012). Pharmacologically active microcarriers for endothelial progenitor cell support and survival. *Eur J Pharm Biopharm,* **81**, 609–616.

96. Qutachi, O., Shakesheff, K. M., and Buttery, L. D. K. (2013). Delivery of definable number of drug or growth factor loaded poly(DL-lactic acid-*co*-glycolic acid) microparticles within human embryonic stem cell derived aggregates. *J Control Release,* **168**, 18–27.

97. d'Angelo, I., Oliviero, O., Ungaro, F., Quaglia, F., and Netti, P. A. (2013). Engineering strategies to control vascular endothelial growth factor stability and levels in a collagen matrix for angiogenesis: the role of heparin sodium salt and the PLGA-based microsphere approach. *Acta Biomater,* **9**, 7389–7398.

98. Penna, C., Perrelli, M.-G., Karam, J.-P., Angotti, C., Muscari, C., Montero-Menei, C. N., and Pagliaro, P. (2013). Pharmacologically active microcarriers influence VEGF-A effects on mesenchymal stem cell survival. *J Cell Mol Med,* **17**, 192–204.

99. Simón-Yarza, T., Formiga, F. R., Tamayo, E., Pelacho, B., Prosper, F., and Blanco-Prieto, M. J. (2013). PEGylated-PLGA microparticles containing VEGF for long term drug delivery. *Int J Pharm,* **440**, 13–18.

100. Shah, R. B., and Schwendeman, S. P. (2014). A biomimetic approach to active self-microencapsulation of proteins in PLGA. *J Control Release,* **196C**, 60–70.

101. Devolder, R., Antoniadou, E., and Kong, H. (2013). Enzymatically cross-linked injectable alginate-g-pyrrole hydrogels for neovascularization. *J Control Release,* **172**, 30–37.

102. Choi, D. H., Subbiah, R., Kim, I. H., Han, D. K., and Park, K. (2013). Dual growth factor delivery using biocompatible core-shell microcapsules for angiogenesis. *Small,* **9**, 3468–3476.

103. Geuze, R. E., Theyse, L. F. H., Kempen, D. H. R., Hazewinkel, H. a W., Kraak, H. Y. a, Oner, F. C., Dhert, W. J. a, and Alblas, J. (2012). A differential effect of bone morphogenetic protein-2 and vascular endothelial growth factor release timing on osteogenesis at ectopic and orthotopic sites in a large-animal model. *Tissue Eng Part A,* **18**, 2052–2062.

104. Rui, J., Dadsetan, M., Runge, M. B., Spinner, R. J., Yaszemski, M. J., Windebank, A. J., and Wang, H. (2012). Controlled release of vascular endothelial growth factor using poly-lactic-*co*-glycolic acid microspheres: *In vitro* characterization and application in polycaprolactone fumarate nerve conduits. *Acta Biomater,* **8**, 511–518.

105. Reyes, R., De la Riva, B., Delgado, A., Hernández, A., Sánchez, E., and Évora, C. (2012). Effect of triple growth factor controlled delivery by a brushite-PLGA system on a bone defect. *Injury,* **43**, 334–342.

106. Hernández, A., Reyes, R., Sánchez, E., Rodríguez-Évora, M., Delgado, A., and Evora, C. (2012). *In vivo* osteogenic response to different ratios of BMP-2 and VEGF released from a biodegradable porous system. *J Biomed Mater Res A,* **100**, 2382–2391.

107. Bible, E., Qutachi, O., Chau, D. Y. S., Alexander, M. R., Shakesheff, K. M., and Modo, M. (2012). Neo-vascularization of the stroke cavity by implantation of human neural stem cells on VEGF-releasing PLGA microparticles. *Biomaterials,* **33**, 7435–7446.

108. Karagoz, H., Ulkur, E., Kerimoglu, O., Alarcin, E., Sahin, C., Akakin, D., and Dortunc, B. (2012). Vascular endothelial growth factor-loaded poly(lactic-*co*-glycolic acid) microspheres-induced lateral axonal sprouting into the vein graft bridging two healthy nerves: nerve graft prefabrication using controlled release system. *Microsurgery,* **32**, 635–641.

109. Shin, S.-H., Lee, J., Lim, K. S., Rhim, T., Lee, S. K., Kim, Y.-H., and Lee, K. Y. (2013). Sequential delivery of TAT-HSP27 and VEGF using microsphere/hydrogel hybrid systems for therapeutic angiogenesis. *J Control Release,* **166**, 38–45.

110. Teitelbaum, A. A., Qi, X., Osherov, A. B., Fraser, A. R., Ladouceur-Wodzak, M., Munce, N., Qiang, B., Weisbrod, M., Bierstone, D., Erlich, I., Sparkes, J., Wright, G., and Strauss, B. (2013). Therapeutic angiogenesis with VEGF164 for facilitation of guidewire crossing in experimental arterial chronic total occlusions. *EuroIntervention*, **8**, 1081–1089.

111. Zhang, L., Zhang, L., Lan, X., Xu, M., Mao, Z., Lv, H., Yao, Q., and Tang, P. (2014). Improvement in angiogenesis and osteogenesis with modified cannulated screws combined with VEGF/PLGA/fibrin glue in femoral neck fractures. *J Mater Sci Mater Med*, **25**, 1165–1172.

112. Herrán, E., Requejo, C., Ruiz-Ortega, J. A., Aristieta, A., Igartua, M., Bengoetxea, H., Ugedo, L., Pedraz, J. L., Lafuente, J. V., and Hernández, R. M. (2014). Increased antiparkinson efficacy of the combined administration of VEGF- and GDNF-loaded nanospheres in a partial lesion model of Parkinson's disease. *Int J Nanomedicine*, **9**, 2677–2687.

113. Ju, R., Wen, Y., Gou, R., Wang, Y., and Xu, Q. (2014). The experimental therapy on brain ischemia by improvement of local angiogenesis with tissue engineering in the mouse. *Cell Transplant*, **23** Suppl 1, S83–S95.

114. Xiong, Q., Lin, H., Hua, X., Liu, L., Sun, P., Zhao, Z., Shen, X., Cui, D., Xu, M., Chen, F., *et al.* (2015). A nanomedicine approach to effectively inhibit contracture during bladder acellular matrix allograft-induced bladder regeneration by sustained delivery of vascular endothelial growth factor. *Tissue Eng Part A*, **21**, 45–52.

115. Wang, J.-H., Xu, Y.-M., Fu, Q., Song, L.-J., Li, C., Zhang, Q., and Xie, M.-K. (2013). Continued sustained release of VEGF by PLGA nanospheres modified BAMG stent for the anterior urethral reconstruction of rabbit. *Asian Pac J Trop Med*, **6**, 481–484.

116. Shin, S.-H., Lee, J., Ahn, D.-G., and Lee, K. Y. (2013). Co-delivery of vascular endothelial growth factor and angiopoietin-1 using injectable microsphere/hydrogel hybrid systems for therapeutic angiogenesis. *Pharm Res*, **30**, 2157–2165.

117. Herrán, E., Pérez-González, R., Igartua, M., Pedraz, J. L., Carro, E., and Hernández, R. M. (2013). VEGF-releasing biodegradable nanospheres administered by craniotomy: a novel therapeutic approach in the APP/Ps1 mouse model of Alzheimer's disease. *J Control Release*, **170**, 111–119.

118. Herrán, E., Ruiz-Ortega, J. Á., Aristieta, A., Igartua, M., Requejo, C., Lafuente, J. V., Ugedo, L., Pedraz, J. L., and Hernández, R. M. (2013). *In vivo* administration of VEGF- and GDNF-releasing biodegradable polymeric microspheres in a severe lesion model of Parkinson's disease. *Eur J Pharm Biopharm*, **85**, 1183–1190.

119. Formiga, F. R., Pelacho, B., Garbayo, E., Abizanda, G., Gavira, J. J., Simon-Yarza, T., Mazo, M., Tamayo, E., Jauquicoa, C., Ortiz-de-Solorzano, C., *et al.* (2010). Sustained release of VEGF through PLGA microparticles improves vasculogenesis and tissue remodeling in an acute myocardial ischemia-reperfusion model. *J Control Release,* **147**, 30–37.

120. Simoń-Yarza, T., Tamayo, E., Benavides, C., Lana, H., Formiga, F. R., Grama, C. N., Ortiz-de-Solorzano, C., Ravikumar, M. N. V, Prosper, F., and Blanco-Prieto, M. J. (2013). Functional benefits of PLGA particulates carrying VEGF and CoQ10 in an animal of myocardial ischemia. *Int J Pharm,* **454**, 784–790.

Chapter 15

Polyester Nano- and Microsystems for Vaccine Delivery

Rajeev Sharma, Nishi Mody, Surbhi Dubey, Udita Agrawal, and Suresh P. Vyas

Drug Delivery Research Laboratory, Department of Pharmaceutical Sciences, Dr. H. S. Gour Vishwavidyalaya, Sagar, M.P., 470003, India

spvyas54@gmail.com

15.1 Introduction

A number of infectious diseases have been successfully cured in the last few decades by vaccination and thus they are gaining much interest as a tool for preventive medicine. Diptheria, neonatal tetanus, smallpox, and polio myelitis are a few diseases, where vaccination programs have successfully reduced their incidence worldwide. Vaccination or immunization can be defined as intended stimulation of resistance in the body against a specific pathogen. Vaccines can be killed or attenuated depending upon the virulence of the pathogen. A highly virulent one is given as killed vaccine, while the lesser ones are given in attenuated form. These treated microorganisms are unable to cause disease but have the capacity to induce the

Handbook of Polyester Drug Delivery Systems
Edited by MNV Ravikumar

ISBN 978-981-4669-65-8 (Hardcover), 978-981-4669-66-5 (eBook)
www.panstanford.com

immune system to develop a defense mechanism that continuously protects against the disease. Furthermore, the vaccination strategies are improved with the integration of genetics and nanotechnology with it, and as a result, specific antigens or gene sequences can be safely and effectively administered to the body in order to generate immune responses [1]. Conventional vaccines contain either treated microbes (killed or live attenuated) or components of microbes. Even though many of these vaccines play a vital role in controlling the infectious diseases, some do not come up with good protection against disease. Besides, it is not safe and wise to use live vaccines among the growing population of immunocompromised individuals in society. Also there are still a number of diseases for which licensed vaccines are not available yet. To surmount these challenges, vaccines based on naked DNA encoding a protective antigen or isolated polysaccharides and proteins are being developed. Despite the fact that these vaccines are more defined, less reactogenic, and safe, they require adjuvants to improve their efficacy as these vaccines are generally less immunogenic. To enhance the immunogenicity, aluminium-based adjuvants are commonly used but these vaccines may not produce strong cell-mediated immunity and can cause local reactions too [2, 3]. Thus novel adjuvants and delivery systems are needed to be developed for the formulation of effective vaccines. Nanovaccines have emerged out as a novel approach for vaccination in recent years.

15.2 Nanotechnology in Vaccine Delivery

Owing to their small size and large surface area, nanodevices perform better for early detection of infectious diseases and they can be easily tailored and introduced into the human body for therapeutic benefits at molecular levels [4]. Being small in size, they can be used for both *in vitro* and *in vivo* applications such as bioactive(s)/antigen delivery vehicles, analytical tools, diagnostic devices, and contrast agents. These can be of natural origin like spores, phage, etc., or can be developed in laboratory (Table 15.1). One such example of vaccine delivery vehicles are polymeric nanocarriers like nanoparticles, microparticles, micelles, and hydrogels. Antigens can be either attached to the surface of delivery system or can be entrapped within it. Encapsulation protects the antigen degradation from the hostile environment of gastrointestinal tract and can enhance the circulatory half-life. Immunogens conjugated to nanoparticle surface provokes the immune system in a manner similar to that of pathogens and thus these nanoparticles can serve as effective cargos for safe antigen delivery. With nanoparticles, sustained delivery of antigens can also be designed, which further maximizes the exposure of antigens to the immune system, thereby producing a long lasting effect. Antigens are required in low dose when they are administered with nanoparticles, which add to the efficiency as well as storage stability. It is expected that these nanotechnology-based particulate vaccines provide strong cellular and humoral immune response and a wide range of infectious diseases can be cured with these next-generation vaccines [37].

Table 15.1 Characteristic property of nanocarriers as vaccine delivery cargos [5–36]

Nanocarrier(s)	Source	Size (nm)	Immune response	Ref.
Spore	Bacterial	Various	Humoral, cellular	[5, 6]
Proteosome	Membrane protein-based	20–800	Humoral, cellular	[7–9]
Exosome	Cellular	50–100	Cellular	[10, 11]
Liposome	Lipid-based	Various	Humoral, cellular	[12–16]
Virosome	Liposome + viral env proteins	Various	Humoral, cellular	[17–20]
SuperFluid	Biodegradable polymer	25–250	Humoral, cellular	[21–23]
Nanobead	Inert nanomaterial	40	Humoral, cellular	[24–26]
VLPs	Viral	Various	Humoral	[27, 28]
Phage	Bacterial	Various	Humoral, cellular	[29–32]
Nanoparticles	Polymers	10–200	Humoral, cellular	[33, 34]
Hydrogel	Polymer/ lipids	Various	Humoral, cellular	[35, 36]

15.2.1 Routes of Administration of Nanovaccines

These carriers can be given by conventional as well as nontraditional routes, which are briefly discussed.

15.2.1.1 Oral route

Efficient oral gene/DNA therapy is possible for diseases such as inflammatory bowel disease [38]. Dilution occurs during the transport of the vaccine across GIT, and thus higher concentration of vaccines are required to achieve the desired effect. This need for high dose can act as a prominent drawback for oral administration of vaccines.

15.2.1.2 Nasal route

The nasal route offers the advantage of needle-free immunization. A safe and effective hepatitis B vaccine was developed in the form of nanoemulsion, which contains hepatitis B antigen. This nanoemulsion vaccine was nontoxic (as it was free from alum) and more stable than conventional counterpart with lesser frequency of administration [39]. Pipettes and spray pumps are used for manual delivery of bioactive(s) via the nasal route, but they suffer from many disadvantages like insufficient distribution of droplet/particles or local irritation. Developers are trying to overcome these drawbacks using single- or dual-dose delivery devices for vaccine delivery. An example of such devices is Flumist for influenza vaccine [40]. Nanoparticles present one more advantage over conventional nasal vaccines. With conventional vaccines, free antigens are rapidly cleared from the nasal cavity and are poorly absorbed from nasal epithelium cells, which results in poor immune response. With nanoparticles encapsulating antigen, these drawbacks were surmounted [41]. Although the nasal route serves as a good alternate for needle-free immunization, its immediate access to the brain warrants for further studies to ensure the safety of this route for immunization applications.

15.2.1.3 Intradermal route

The intradermal route delivers the antigens to the immunologically susceptible epidermal cells lying beneath the tough layer of skin. The loaded antigen upon administration via needle is emulsified with an adjuvant, resulting in depot formation, which allows the sustained delivery of the antigen followed by prolonged immune response. Generally, three types of anduvants are used: particulate, nonparticulate, and combination complexes. Oil emulsions, saponins, and immunostimulating complexes are the respective examples of the three classes. Adjuvants help not only in forming the reservoir but also in targeting antigens to pertinent antigen-presenting cells of the immune systems [42].

15.2.1.4 Intramuscular route

Administration of vaccine into the muscular mass is termed as intramuscular (i.m.) vaccination. The dorsogluteal and ventrogluteal muscles of the buttocks, the vastus lateralis muscle of the leg, and the deltoid muscle of the arm are frequently used sites for i.m. administration. It is used for vaccines that are administered in small volumes and depending on the site of injection; an administration is limited to between 2 and 5 mL of fluid. Vaccines containing adjuvants should be injected intramuscularly to reduce adverse local effects. The absorption with the i.m. route is more consistent compared with the oral or subcutaneous route, but once administered, the dose cannot be recovered. Gardasil, hepatitis A vaccine, rabies vaccine, and influenza vaccines based on inactivated viruses are commonly administered intramuscularly. Le and coworkers conducted a study on 20 volunteers who were given DNA vaccine encoded with malaria antigen, and they found excellent induction of CTL responses, but the administration via needle injection failed to induce detectable antigen-specific antibodies in any of the volunteers [43].

15.2.1.5 Subcutaneous route

When the vaccine is administered into the subcutaneous layer (above the muscles and below the skin) it is called subcutaneous (s.c.) administration. A subcutaneous injection is administered as a bolus into the subcutis, the layer of skin directly below the dermis and epidermis, collectively referred to as the cutis. Subcutaneous injections are highly effective in administering vaccines and medications such as insulin, morphine, diacetylmorphine and goserelin. They provide the advantage of self-administration and complete absorption. Small dose unit limits the frequent use of this route. Wang *et al.* subcutaneously vaccinated mice with formalin-killed *P. gingivalis* and then orally challenged with *P. gingivalis*. They found that vaccination protected the mice from alveolar bone resorption and inflammation, and thus they concluded that in this manner, an effective vaccine for the management of periodontitis can be developed [44].

15.3 Polymer-Based Vaccination

The efficiency of vaccines depends on their antigenicity. In general, proteins and peptides, either synthesized chemically or purified from microorganisms or tissues, or produced by recombinant DNA technology, are weakly antigenic. In order to increase their effectiveness, they require immune-stimulating compounds (ISCOMs) or adjuvants which enhance the immune stimulating potential of these antigens. These adjuvants need to be safe and effective, and in modern practice, numerous approaches have been designed to develop effective vaccines. One such approach is the development of nanoparticles formulated from biocompatible and biodegradable polymers as a vaccine delivery system to induce both cellular and humoral immune responses. Cell-mediated immunity along with cytotoxic T lymphocytes (CTL) can be induced by these particulate adjuvants to key pathogens, including viral infections. Nanobeads, gold-coated nanoparticles, microspheres, and liposomes have been successfully employed to deliver DNA. Polylactide-*co*-glycolide (PLGA) and polycaprolactone (PCL) are commonly used biodegradable and biocompatible materials used for preparing nanoparticles as vaccine delivery cargos [45]. The efficacy of these polymer-based vaccine delivery systems can be regulated by using appropriate polymer alone or in combination depending upon the desired release kinetics and other properties like morphology, entrapment efficiency, and particle size distribution. Reproducibility of formulation, purity, and safety of the formulation and stability during the manufacturing as well as storage are some important issues that are to be addressed in order to develop safe and effective nano- and microparticle-based vaccines. Toxicity can be developed as a result of accumulation of large particles into the vital organs, and the clearance of smaller particles is also a slow process. Thus, assessment of the safety of these polymer-based nanovaccines is equally important as the study of their efficacy.

15.4 Poly(D-L-Lactide-*co*-Glycolide) (PLGA)/PLA-Based Nano- and Microparticulate Systems as a Versatile Platform in Vaccine Delivery

PLGA is biocompatible and biodegradable synthetic polyester polymer investigated as one of the most striking polymeric materials for the synthesis of nano-/microparticulate carrier(s). In the body, upon hydrolysis, PLGA is converted into its successive biodegradable monomer units such as glycolic acid and lactic acid, since glycolic acid and lactic acid are normally present in the body, which takes part in the numerous biochemical and physiological pathways and are easily metabolized via the Krebs cycle [46]. Figure 15.1 illustrates the structure and metabolic degradation of PLGA. PLGA is commercially available with various copolymer compositions and molecular weights. For example, PLGA 75:25 depicts copolymer which consists of 75% lactic acid and 25% glycolic acid. However, PLGA 50:50 is the most commonly investigated polymer and is used to a higher extent than other copolymers. The degradation time can vary from several days to several months, depending on the copolymer ratio and molecular weight [47, 48].

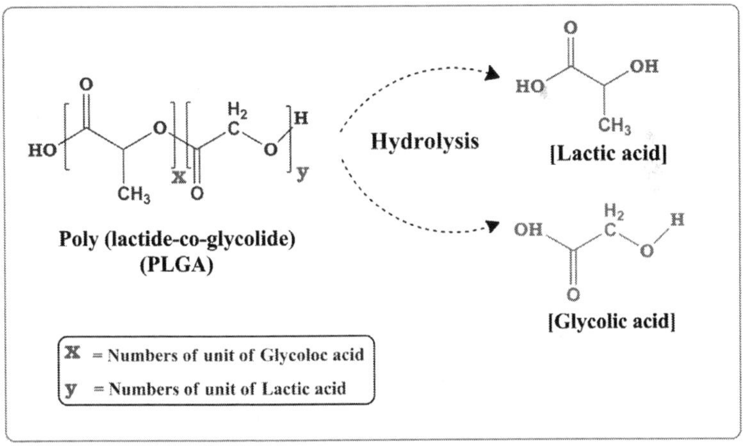

Figure 15.1 Structure and metabolic degradation of PLGA.

In the last two decades, PLGA/PLA-based particulate carrier-mediated vaccine delivery has gained remarkable attention owing to their exciting properties among the various investigated synthetic polymeric materials [49] (Table 15.2). PLGA/PLA particulate carrier systems have been used to deliver various encapsulated antigen(s) such as *hepatitis B virus (HBV), Bacillus anthracis, influenza, Plasmodium vivax*, and model antigens such as *bovine serum albumin, tetanus toxoid*, and *ovalbumin* [50, 51]. PLGA-based nano-/microparticulate are able to release encapsulated antigens over an expanded period of time in a controlled and sustained manner directly to phagocytic antigen-presenting cells (APCs) via nonspecific or receptor-mediated endocytosis, respectively. APCs such as dendritic cells preferentially internalize/uptake particles in the size range of 20–200 nm, while macrophages preferentially internalize larger particles of about 0.5–5 μm [52]. After immunization, expanded release of loaded antigens can provide significant immune responses, reduce the frequency of boosting administrations, and circumvent the risk of tolerance typically required to elicit defensive immunity [53].

Table 15.2 Properties of PLGA nano- and microparticles that make it a versatile vaccine delivery system

PLGA-based Nano-/ microparticles	✓ Biocompatible, biodegradable, and nontoxic ✓ Can protect structural and functional integrity of antigens ✓ Can act as adjuvant ✓ Can deliver encapsulated antigens to APCs ✓ Exhibits zero-order drug release kinetics ✓ US FDA and European Medicine Agency (EMA)-approved polymeric material for human use ✓ Well established for various types of vaccine delivery systems ✓ Can reduce the number of booster doses ✓ Can provide antigen release over weeks and months

In particular, the PLGA/PLA polymer-based particulate delivery systems are also well recognized as an adjuvant for the stimulation of significant humoral and cellular immune responses against encapsulated antigen(s) [54–57]. The selection of the method for fabricating vaccine antigen-loaded PLGA/PLA particulate system mainly depends on the nature (hydrophilic/hydrophobic) of immunogen, its physicochemical characteristics, and the desired association with the particles. Currently, the most commonly used method for the fabrication of antigen-loaded PLGA/PLA nanoparticles is the double emulsification solvent evaporation [49, 58]. Although the following fabrication techniques have been discussed in previous literature, a brief overview of these techniques is presented here:

- **Double emulsification**
  A double emulsion (or multiple emulsion) of water in oil in water type has been used. Following evaporation of the organic solvent, nanoparticles are formed which are then recovered by ultracentrifugation, washed repeatedly with buffer, and lyophilized. PLGA nanoparticles were prepared loaded with bovine serum albumin using the double emulsion solvent evaporation method. Owing to the high solubility of the protein in water, the double emulsion technique has been chosen as one of the appropriate methods [59].
- **Salting out**
  The salting out process has been used to prepare nanoparticles by various workers. It is based on the incorporation of a saturated aqueous solution of polyvinyl alcohol (PVA) into an acetone solution of the polymer under magnetic stirring to form an O/W emulsion. However, the process differs from the nanoprecipitation technique, as in the latter the polymeric solution (in acetone) was completely miscible with the external aqueous medium. But in the salting out technique, the miscibility of both the phases is prevented by the saturation of the external aqueous phase with PVA. The precipitation of the polymer occurs when sufficient amount of water is added to external phase to allow complete diffusion of the acetone from the internal phase into the aqueous phase. This technique is suitable for bioactive(s) and polymers which are soluble in polar solvents, such as acetone or ethanol [60].

- **Solvent displacement or nanoprecipitation**
  This method is based on the interfacial deposition of a polymer following displacement of a semipolar solvent miscible with water from a lipophilic solution. The solvent displacement method involves the use of an organic phase, which is completely soluble in the external aqueous phase. The organic solvent diffuses instantaneously into the external aqueous phase, inducing immediate polymer precipitation because of the complete miscibility of both the phases. Consequently, neither separation nor extraction of the solvent is required for the polymer precipitation. After nanoparticle preparation, the solvent is eliminated and free-flowing nanoparticles can be obtained under reduced pressure [61, 62].
- **Spray drying**
  Spray drying is a well-established method commonly used in the pharmaceutical industry for producing a dry powder from the liquid phase. In recent years, it has been identified as a suitable method for the preparation of proteins intended for pulmonary, nasal, and controlled oral delivery. It offers the advantage of drying and particle formation in a single-step continuous and scalable process with particle engineering possibilities. Furthermore, various particle properties such as particle size, bulk density, and flow properties can easily be tuned via simple manipulation of the process parameters or spray dryer configuration. Therefore, spray drying is potentially a versatile and commercially viable technique for formulating protein and peptide drugs [63].

15.4.1 PLGA/ PLA Nano- and Microparticles as Vaccine Adjuvants and Challenges

Adjuvants are immunopotentiators that have been added in the vaccine formulations to make them more immunogenic or to improve the vaccine efficiency or to improve the immune responses. Insoluble aluminum compounds (referred to as "alum") are the most commonly used adjuvants which have been extensively utilized in licensed human vaccines. However, aluminum salts-based adjuvants possess some restrictions: for example, they elicit IgE-mediated immune response, induce local reactions, and predominantly induce CTL immune responses. Therefore, they are not suitable for all types of antigens [64, 65]. In addition, while administration of aluminum-adjuvanted vaccine formulations effectively induce humoral immune responses (antibody titers), but they are unable to induce strong cellular immune responses (T-cell-mediated) that may be required for future vaccines, such as for malaria, tuberculosis, cancer, and HIV [66–68]. PLGA/PLA-based particulate delivery systems are instead used as an adjuvant similar to alum. Owing to ease surface engineering, surface stabilization and ease of loading of single or multiple antigen(s) explain the renewed interest in vaccine approaches using these particulate systems (Fig. 15.2) [69]. Despite promoting immunological responses, some problems encountered with PLGA/ PLA nano- and microparticles are unfavorable environments/conditions such as organic solvent, high shearing or cavitation forces, drying, freezing and excessive heat involved in the formulation development, antigen encapsulation, and storage process often causing significant degradation and aggregation of loaded antigen(s) [70].

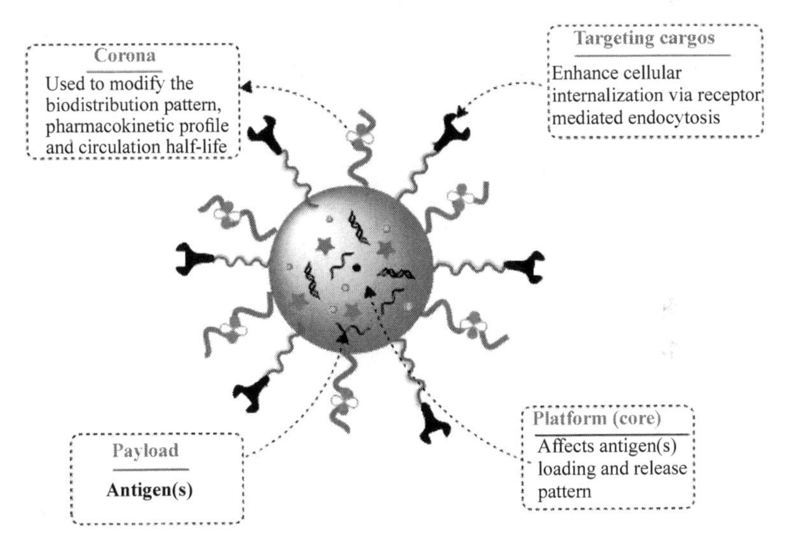

Figure 15.2 Schematic illustration of engineered PLGA nano-/microparticle.

In addition, after hydrolysis PLGA converts into successive monomer acidic units (lactic acid and glycolic acid); hence this inner acidic microenvironment can lose structural and functional integrity of encapsulated antigen(s) or denaturation of proteins, which is a challenging step in the development of these vaccine delivery systems [69, 71, 72]. Over the last years, various strategies were developed to resolve these hurdles by optimized preparation techniques or adding together stabilizing agents such as $Mg(OH)_2$, antacid excipients, using complexes of proteins with zinc, sugars, or surfactants [73–75]. Therefore, various efforts have been made to develop improved adjuvants that can address the unmet vaccination needs until holding an appropriate safety profile. Over the last three decades, numerous PLGA/PLA nano- and microparticle-based vaccine applications have been investigated but have not yet been developed as a commercial vaccine.

15.4.2 Recent Developments and Strategies to Enhance Immune Responses

Antigen-loaded PLGA nano- and microparticles can boost the immune response of administered antigens and more efficiently generate the immune response compared with soluble antigens. Antigen internalization takes place through the paracellular route and translocation mechanism. As uptake of antigen-loaded nano-/microparticles enhances, the immune response is augmented proportionally. Antigen-loaded NPs can provoke both humoral and cellular responses, as shown in Fig. 15.3 [76].

Several *in vitro* and *in vivo* experimental studies proved that after immunization, PLGA/PLA nano- and microparticles are efficiently phagocytosed by APCs and deliver antigens to macrophages and dendritic cells, directing the stimulation of considerable cytotoxic T lymphocyte (CTL)-mediated immune responses through MHC class I cross-presentation [77–80]. The exact mechanism of cross-presentation is not clear; however, it has been exclusively presented that PLGA/ PLA particles are capable of promoting cross-presentation of antigens in APCs [81]. In the recent years, various PLGA/PLA-based nano- and microparticulate vaccine delivery have been gaining significant attention, and numerous efforts are being made to develop novel engineered delivery systems and more efficient vaccine adjuvants for the development of modern vaccines against viral infections, intracellular pathogens, cancers, HIV, etc. Tables 15.3 and 15.4 summarize some studies of PLGA/PLA nanoparticle- and microparticle-mediated vaccine delivery.

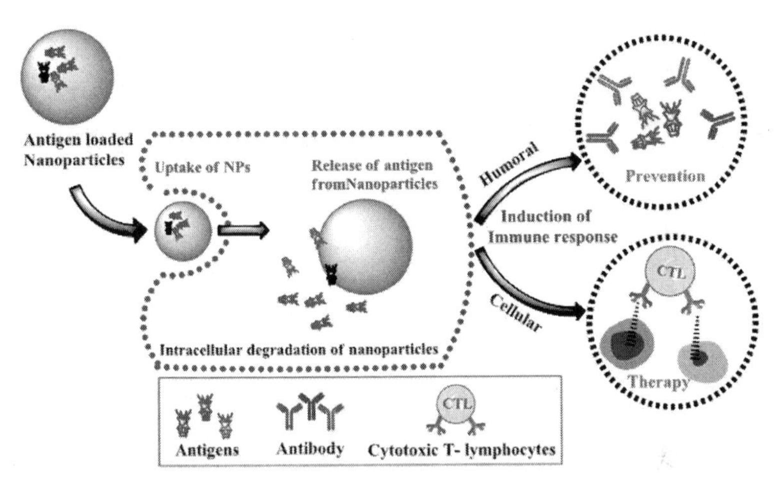

Figure 15.3 Diagrammatic representation of induction of immune responses by nanoparticle-based vaccine. Reprinted from Ref. 76, Copyright 2015, with permission from Elsevier.

Currently, a variety of PLGA/PLA nano- and microparticle-mediated vaccines are in early clinical trial stage, and some formulations have been clinically approved for vaccination. SEL-068 (Selecta Bioscience, Inc., USA) PLGA-based synthetic vaccine is the first nanovaccine which is under phase 1 clinical trial and is used for aversion of nicotine addiction and reversion and also cessation of the smoking habit. A universal peptide antigen which elicits helper T-cell-mediated immune response and TLR agonistic (immunopotentiating agent) are incorporated into the polymeric matrix of PLGA. B-cell antigen (nicotine) is covalently linked to the surface of nanoparticles. Good laboratory practice (GLP) safety and efficacy studies were performed in cynomolgus monkeys. The results revealed no dose-limiting toxicity even after repeated dosing [116].

In the last two decades, nano- and microparticle-mediated vaccine development has been widely explored for parenteral

Table 15.3 Some studies of PLGA nanoparticle-based vaccine delivery

Composition	Antigen(s)	Immunization route	Immune response	Cell line/ animal model	Ref.
PLA-PEG	Tetanus toxoid	Intranasal	Elicit high and long lasting immune response	Wistar rats	[82]
PVA grafted PLGA	Tetanus toxoid	Oral and nasal	NPs increased serum titers up to 3 × 103 (IgG) and 2 × 103 (IgA)	Female BALB/c mice	[83]
PLGA and PLA	Tetanus toxoid	i.m.	Long lasting immune response	Wistar rats	[84]
PLGA	Helicobacter pylori	Oral	Induce *H. pylori*-specific mucosal and systemic responses	Mice	[85]
PLGA (50:50)	Synthetic bovine para-influenza virus type-3 (BPI3V)	Intranasal	Induce stronger IgG antibody	BALB/c mice	[86]
PLGA	Inactivated PRRS virus	Intranasal	Stimulate IgG1 and IgG2 antibody, T-helper (Th)-1 and Th2 cytokines	Pigs	[87]
PLA and PLGA aerosolized NPs	Hepatitis B virus	Pulmonary	Enhance humoral, mucosal and cytokine responses	Rat	[88]
Lectin anchored PLGA NPs	Hepatitis B virus	Oral	Stimulate mucosal, isotyping response IgG1/ IgG2a and cell-mediated (IFN-γ and IL-2 level) immune responses	BALB/c mice	[89]

Composition	Antigen(s)	Immunization route	Immune response	Cell line/ animal model	Ref.
PEGylated PLGA NPs	Ovalbumin (OVA) as model antigen	Oral	Elicit IgG immune response	Mice	[90]
Stabilized lectinized PLGA NPs	Hepatitis B virus	Oral	Elicit sIgA in the mucosal secretion and IL-2 and IFN-g in the spleen homogenates.	Female BALB/c mice	[91]
PLGA NPs	OVA	Nasal	High serum antibody titers and sIgA levels	BALB/c mice	[92]
PLGA NPs	Tumor associated Ag	s.c.	Enhanced tumor control and prolonged survival of tumor-bearing mice with improved specific T cell responses	Mice	[93]
Mucoadhesive PLGA NPs	Newcastle disease virus DNA	Mucosal	Induce stronger cellular, humoral, and mucosal immune responses	Chickens	[50]
Multifunctional PLGA-based NPs	Bovine serum albumin (BSA)	Intranasal	To induce high systemic and humoral immune response	Mice	[94]
Surface modified PLGA NPs	Hepatitis B surface Antigen	Intranasal	To induce a potent immune response at mucosal surface(s) and systemic circulation.	BALB/c mice	[95]

Table 15.4 Some studies of PLGA-based microparticles (MPs) vaccine delivery

Composition	Antigen(s)	Immunization route	Immune response	Cell line/ animal model	Ref.
PLGA microspheres	Tetanus toxoid	s.c.	To elicit long-lasting immune responses	Wistar rats	[96]
PLA/PLGA microparticles	Tetanus toxoid	s.c.	Efficient antibody response	Mice	[97]
PLA/PLGA microspheres	Tetanus toxoid and synthetic malaria antigen	s.c.	Potential to elicit long-lasting immune responses	Mice	[98]
PLGA microspheres	Bovine serum albumin (BSA) as a model antigen	s.c. and oral	Significant serum IgG antibody responses to BSA	Mice	[99]
PLGA microspheres	Tetanus toxoid	Intranasal	-	Rats, rabbits and guinea-pigs	[100]
Mucoadhesive PLGA microparticles	Hepatitis B virus	Intranasal	Induce strong systemic and mucosal immune response	Mice	[101]
PLGA microspheres	Tumor antigens	Intradermal	Improve the efficacy of cancer immuno therapy	Mice	[102]

Composition	Antigen(s)	Immunization route	Immune response	Cell line/ animal model	Ref.
1,2-dioleoyl-3-trimethylammonium propane (DOTAP) blended PLGA MPs	Ovalbumin	–	Improve OVA-specific antibodies and cytokine production	BALB/c mice	[103]
PLGA and PLA microspheres	Hepatitis B virus	s.c.	Elicit adequate Th1 and CTL responses	BALB/c mice	[104]
PLGA microparticles	Ovalbumin	s.c.	High IgG1 and IFN-γ titre level	Mice	[105]
Novel PBAE/PLGA polymer blend microparticles	pDNA	–	Serve as an effective DNA vaccine delivery system	Transfect EL4 cells	[106]
PLGA microspheres	Ovalbumin peptide and vesicular stomatitis virus	Intradermal	Elicit cytotoxic T-lymphocyte response	C57BL/6 mouse model	[107]
Lipid-enveloped PLGA MPs	Thiolated ovalbumin	s. c.	Balanced Th1/Th2 immune responses	BALB/c mice	[108]
Porous PLGA MPs	Brachyspira hyodysenteriae (BmpB) as a model antigen against swine dysentery	Oral	Induce both Th1- and Th2-type responses based on elevated IgG1 and IgG2a titers	Mouse model	[109]

(*Continued*)

Table 15.4 (*Continued*)

Composition	Antigen(s)	Immunization route	Immune response	Cell line/ animal model	Ref.
Cationic PLGA MPs	Infectious bursal disease virus	Oral	To protect chickens against infectious bursal disease through lower neutralizing antibody titers and reduced IL-4 and IFN-α mRNA expression	Chickens	[110]
PLGA99 MPs	Chimeric protein rSAG1/2 (Toxoplasma gondii)	i.d.	Elicited significant long-term humoral and cell-mediated immune responses,	BALB/c mice	[111]
Cationic PLGA MPs	Plasmid (pVAC-1D) containing 1D gene FMD virus serotype Asia 1	i.m.	Longer humoral immune response	Guinea pigs	[112]
CpG- and protamine-containing PLGA MPs	Allergen of bee venom, phospholipase A2	s.c.	Strong induction of PLA2-specific antibody responses.	Mice	[113]
Polyethylenimine (PEI) functionalized PLGA MPs	Tumor associated, self Antigens encoded by plasmid DNA (pDNA)	i.m.	Improve the potency of self antigen-based cancer DNA vaccines.	BALB/c mice	[114]
Cationic poly(L-lactide) microspheres	Hepatitis B surface antigen	s.c.	Induce higher cellular immune responses	Mouse	[115]

immunization (intramuscular, intradermal, subcutaneous, etc). However, since the last five years, major research efforts are being made for the development of new mucosal vaccine candidates by designing targeted/multifunctional (essentially PRR ligands) delivery systems, using strong immune stimulating agents and selecting new mucosal immunization routes. Development of such types of new vaccine approaches improves the protection against a large number of diseases, including sexually transmitted diseases, respiratory viruses, influenza, and many others [117, 118]. Mucosal vaccinations possess certain advantages over the parenteral vaccination and can provoke both humoral and cellular immune responses at serosal and mucosal sites. Owing to their versatility, PLGA/PLA nano- and microparticles have emerged and been investigated as potential vaccine delivery systems for mucosal vaccination. Indeed, numerous experiential studies have shown that PLGA/PLA particulate efficiently delivered loaded antigens to M cells at the surface of Peyer's patches into the GIT mucosa, improving internalization mechanism and efficient delivery of encapsulated antigens to mucosal APCs, making them efficient vaccine delivery carriers for mucosal vaccination [88, 119–121]. In the recent years, various PLGA/PLA-based engineered nano- and microparticulate delivery systems have been investigated for efficient antigen delivery, which provides stronger immune response, immunomodulatory properties, cell-selective targeting, and greater stability (Fig. 15.4) [122]. Immunization studies in animal models like rodents and nonhuman primates have revealed that these approaches are capable of stimulating both cellular and humoral immune responses. For these reasons, PLGA/PLA-based engineered/multifunctional nano- and microparticulate delivery-based approaches provide a platform for the development of modern vaccines, thus attracting attention for future research [123].

15.5 Poly-ε-Caprolactone (PCL) for Vaccine Delivery

The major shortcomings with current vaccine therapy are repeated administrations, low immune response, and poor patient compliance which create an opportunity to budding researchers and scientists

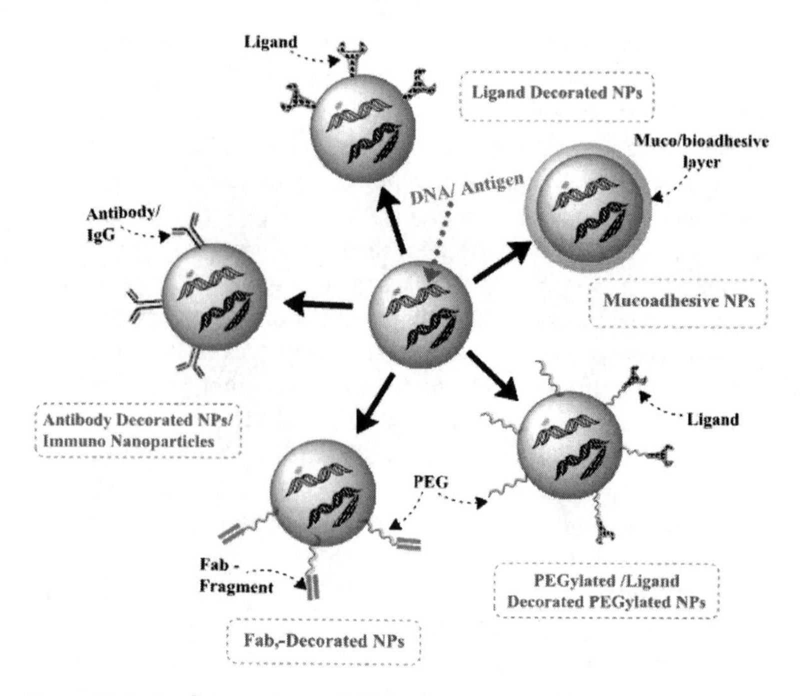

Figure 15.4 Surface-engineered PLGA/PLA nano- and microparticles as vaccine delivery candidate.

for developing a delivery package to overcome these faults. PCL is grasping the attention of investigators as one of the novel and nontoxic polymeric vaccine carrier, capable of offering its characteristic properties in vaccine delivery system fabrication. Compared with PLGA, PLA, and γ-PGA, PCL degrades at slower rate and generates least acidic environment upon degradation. Therefore, PCL is the polymer of choice for fabricating vaccine delivery systems which fulfill the above mentioned shortcomings to some extent. PCL and its PEGylated copolymer are the material of choice for fabricating nano- and microparticulate-based carrier-mediated efficient vaccines delivery avoiding frequent administration. Other advantages of PCL includes formation of compatible blends with other polymers, easily tailorable antigen release profile and mechanical properties using polymer blends, simplicity of shaping and molding with desired pore size, controlled and sustained antigen delivery, retainment of activity of vaccine delivery system for more than one year, and ease of incorporating surface functionality for targeting and rendering hydrophilicity [124]. Excellent biocompatibility, slow release profile, and ability of PCL to be fully excreted from the body once bioresorbed makes it suitable for controlled antigen delivery of biomacromolecules. Vaccines encapsulating PCL microspheres, hydrogels, and nanoparticles with or without adjuvants are being synthesized for oral, mucosal, and parentral vaccine delivery. Various methods employed for the preparation of PCL microsphere includes o/w emulsion solvent extraction/evaporation method, w/o/w emulsion solvent evaporation technique, spray drying technique, solution enhances dispersion technique, and hot melt technique. The integrity and immunogenicity of protein antigens is lost upon its release from PLGA and PLA polymeric systems because of acidic surrounding generated by their degradation. Unlike PLGA and PLA, biodegradation of PCL yield less acidic environment since upon hydrolysis in the body it yields 6-hydroxycaproic acid, which enters the citric acid cycle for metabolization. Therefore, the structural integrity and specific immunogenic response of the released antigen persists upon release from PCL system. Researchers have prepared PCL microspheres encapsulating recombinant hepatitis B surface antigen for intramuscular administration with the aim to induce humoral and cellular immunity [125]. High loading efficiency into the microspheres without loss of activity has been difficult to achieve and so adsorption or surface decoration of antigen offers an alternative approach. Various experimental investigations suggested that incorporation of antigens on the surface of PCL microspheres induces both humoral and adaptive immune responses. Several PCL-based formulations are being synthesized to evoke the various arms of immunity; some studies are summarized in Table 15.5.

For the same reason, PCL microspheres are utilized for protecting fish from marine infections like *Edwardsiellosis* [126]. PCL-based microparticulate system intended for antigenic protein delivery was designed for oral immunization. Microparticles were prepared by solvent evaporation method using bovine serum albumin (BSA) as a model antigen in which the optimal PCL particle size for efficient absorption along gastrointestine was observed to be between 5 and 10 μm. Another important property of PCL is its hydrophobic nature, responsible for rapid and efficient uptake by the antigen-presenting cells and arousal of intense immune response. But

Table 15.5 Some poly–ε- caprolactone and γ-PGA-based carrier systems for vaccine delivery

Composition	Antigen (s)	Immunization route	Immune response	Cell line/ animal model	Ref.
CL–alginate microspheres PCL–chitosan microspheres	OMVs of E. tarda	i.p.	Persistence and long lasting immune response, protects fish against Edwardsiellosis infection	Fish	[126]
PCL–PEG–PCL NPs	bFGF	s.c.	IgG, IgG1 mice, and IgG2a immune response	C57BL/6	[127]
γ-PGA NPs-Eriss	–	s.c.	Amplified and activated CTLs and interferon-c-secreting cells specific for the antigen, T type1 (Th1) biased cytokine production	C3H mice	[128]
γ-PGA NPs	OVA	Intranasal	Mice eliciting antigen-specific CTLs response	C57BL/6	[129]
γ-PGA NPs + hemagglutinin (HA)	–	Intranasal & s.c.	Enhanced the cross-protection against influenza virus infection upon intranasal immunization	–	[130]
γ-PGA NPs	*Japanese encephalitis*	i.p.	Enhanced the neutralizing antibody titer, survival at normally lethal JEV infection	BALB/c mice	[131]
PCL MPs	BSA	Oral	Protection of BSA integrity upon release	–	[132]

Composition	Antigen (s)	Immunization route	Immune response	Cell line/ animal model	Ref.
PCL, PLGA, PLGA-PCL MPs	DT	Intranasal	IFN-γ, IL-6, IgG IN/IM	BALB/c mice	[133]
PCL and PLGA bend	*Brucella ovis*	–	PCL caused higher uptake by J744 macrophages and cell respiratory burst	J744-macrophage	[134]
Freeze-dried chitosan/ PCL NPs	Antigenic protein and DNA	–	Higher uptake by A549 cells, great ability to form stable complexes, which protect DNA from nucleases.	A549 cells	[135]
PCL nanosphere	*S. equi* and cholera toxin	Intranasal	Increased immun ogenicity and mucosal immune response. Th 1 and Th 2 immune response generated	Mice	[136]
γ-PGA NPs	OVA and p24	s.c.	Potent and specific humoral and cellular immune responses. Increased INF γ production	BALB/c mice	[137]
γ-PGA NPs	OVA		Inhibited the growth of OVA transfected tumors with enhanced CTL response.	EL4, E.G7, YAC1, C57BL/6 mice and BALB/c-mice	[138]

intranasal clearance of the delivery vehicle reduces their efficacy, and so PCL microspheres are coated or blended with mucoadhesive polymers like alginate, chitosan, etc., rendering longer duration in intranasal passage. PCL microspheres showed prolonged residence time in upper respiratory tract explaining the potentials for efficient intranasal delivery of antigen [139]. An innovative delivery system comprising of nanoparticles-in-microsphere hybrid (NiMOS) has been investigated for effective gene delivery in gastrointestinal tract using PCL. Particles of size less than 10 μm were synthesized for oral vaccine delivery system in which gelatin nanoparticles were enclosed in PCL microspheres [140]. This system could effectively transfect the intestinal tract and, therefore, can be exploited for site-specific delivery of vaccine to intestinal infections.

Surface modification of PCL with PEG forms different block copolymers (diblock, triblock) imparting hydrophilicity, which makes PCL more compatible. These copolymers find application in vaccine delivery as micro-/nanoparticles or thermosensitive gel. A research group formulated mannan modified copolymer-based PCL–PEG–PCL nanoparticles by modified emulsion solvent evaporation method for human basic fibroblast growth factor (bFGF) antigen delivery to improve humoral immunity via dendritic cell targeting [127]. In the same fashion, Mannan loaded *in situ* gel forming thermosensitive poly(ε-caprolactone)-poly(ethylene glycol)-poly(ε-caprolactone) (PCL-PEG-PCL, PCEC) hydrogel (M-hydrogel) was prepared. A free flowing gel was obtained which converted to non flowing gel on incorporation of basic fibroblast growth factor (bFGF) at body temperature and serves as *in situ* depot delivery system. *In vivo* immunogenicity studies revealed that bFGF encapsulation in M-hydrogel evoked enhanced immune response as compared to normal saline (NS), blank hydrogel, pure bFGF, bFGF loaded hydrogel (bFGF-hydrogel), bFGF and mannan mixture (bFGF-M), and bFGF loaded Complete Freund's Adjuvant (bFGF-CFA) groups. Hence, M-hydrogel presents a potential carrier for *in situ* gel forming vaccine and could be used to deliver other antigens as well [141]. Prashant *et al.* prepared a single-shot vaccine for facilitating prolonged immunity by encapsulating tetanus toxoid (TT) as a model antigen, in poly-ε-caprolactone nanoparticles (PCL NPs). A single injection in mice generated a strong and continuous cell-mediated as well as humoral response two months later in absence of booster dose. The capability of enhanced cell-mediated immune (CMI) response may have high translational potential for immunization against intracellular infection by PCL nanoparticles [142].

15.6 γ–PGA for Vaccine Delivery

Biodegradable γ-PGA is powerful adjuvant and an excellent vaccine carrier capable of delivering antigenic proteins to APCs eliciting potent antigen-specific humoral and cellular immune responses. The amphiphilic nature of the polymer directs the delivery of antigens to the dendritic cell followed by their activation. Progress in nanoscopic systems for vaccine delivery has revealed self-assembling micelles and self-assembling nanoparticles as one of the potential carriers. Recent research is focused on self-assembled amphiphilic block or graft copolymers made from amphiphilic poly(amino acid) like several amphiphilic block and graft copolymers based on poly(amino acid)s have been employed. An outstanding characteristic of γ-PGA is insusceptibility to enzymatic degradation by several proteases. It is therefore the most widely studied polymer for preparing particulate vaccine delivery system for vaccination against diseases such as diverse tumors, *Japanese encephalitis* virus, and HIV [143]. Recently, γ-PGA nanoparticles have gained much attention in the field of cancer immunotherapy. In this way, Yoshikawa *et al.* prepared amphiphilic γ-PGA-nanoparticle-based tumor vaccine for cytosolic delivery of tumor antigen and studied the possible pathway of translocation in the cytoplasm, which can be exploited for preparing optimized vaccine carriers [144]. In another study, researchers investigated the efficacy of combining two vaccine strategies to promote MHC I presentation and CTL response (major effector cell in tumor immunity) against cancer. The approach involved γ-PGA nanoparticles for antigen delivery and an endoplasmic reticulum (ER) transport system containing an ER-insertion signal sequence (Eriss). Eriss was conjugated to the antigen, which aided the transport of cytosolic peptides to the ER. γ-PGA nanoparticles facilitated the presentation of antigen to MHC class I and Eriss efficiently induced CTL activity specific for the antigenic peptide [128]. In mice, subcutaneous immunization with OVA/γ-PGA NPs more effectively inhibits both the growth and metastasis of OVA-transfected tumors than immunization with OVA emulsified using Freund's complete adjuvant, which is the most effective among the existing adjuvants [145]. γ-PGA nanoparticles (NPs) are also investigated as a proficient carriers for nasal immunization eliciting potent and prolonged mucosal and systemic immune responses. According to one finding, nasal delivery of γ-PGA NPs encapsulating ovalbumin stimulated T lymphocytes (CTLs), predominant Th1 response and interferon-γ-secreting cells specific for OVA in the spleen and lymph nodes as compared to mucosal immunization. γ-PGA NPs are a promising antigen delivery carrier for the development of non-invasive cancer vaccines [129]. γ-PGA is also being utilized as adjuvant to exaggerate the antigenic response. Researchers observed the usefulness of γ-PGA NPs as a mucosal adjuvant for influenza virus hemagglutinin (HA) vaccine. Intranasal immunization with the mixture of PGA-NPs and HA vaccine from an influenza virus strain A/PR/8/34 (H1N1) or A/New Caledonia/20/99 (H1N1) enhanced protection of mice from A/PR/8/34 infection. Intranasal immunization with A/New Caledonia/20/99 HA vaccine and γ-PGA NPs induced cell-mediated immune responses and neutralizing antibody production for both A/New Caledonia/20/99 and A/PR/8/34. Thus, intranasal immunization with influenza virus hemagglutinin (HA) vaccines and γ-PGA NPs as adjuvant induces cross-protection against variants within a subtype and against different subtypes [130]. The polymer was also used as adjuvant to *Japanese encephalitis* (JE) vaccine against JE virus. Currently available vaccine provides short-term protection and requires booster doses rendering it inconvenient and uncomfortable. Number of adjuvants has been used in vaccine and most of them cause toxic effects. Therefore, in surge of biocompatible, safe and biodegradable adjuvant researchers examined natural γ-PGA NPs as adjuvant for vaccine delivery. Investigators administered a single dose of JE vaccine with γ-PGA-NPs as adjuvant in mice and observed 10 times increased level of the neutralizing antibody titer as compared to JE vaccine alone. Immunized mice displayed survival against the JE infection lethal dose while only 50% of the mice receiving JE vaccine alone survived. The formulation provides prolonged immunity as reveled by 100% immortality in mice 20 days after single dose [131]. γ-PGA-based nanoparticles were established as AIDS vaccine delivery and adjuvant system. *In vivo* immunization studies in mice showed that, as compared to p24 alone, sc injection of human immunodeficiency virus type 1 (HIV-1) p24-encapsulating PGA nanoparticles triggered antigen-specific IFN-γ-producing T cells in spleen cells and induced p24-specific serum antibodies. Adjuvant potential of γ-PGA was found to be comparable to complete Freund's adjuvant (CFA) in inducing p24-specific serum antibody. Complete Freund's adjuvant (CFA) is one of the most potent immunoadjuvant [129]. DNA vaccination with PEI/γ-PGA NPs loaded with a plasmid encoding *Plasmodium yoelii* merozoite surface protein 1:C terminus, administered i.v. in mice, has been shown to generate an antigen specific IgG response dominated by IgG1 and IgG2b and to induce weak Th1 (IFN-γ and IL-12 p40) and strong Th2 (IL-4) cytokines responses. In another study, the same complex when administered via intravenous and intraperitoneal route provided complete protection against lethal challenge with a significant increase in levels of immunoglobulins, Th1 and Th2 cytokines [130]. Table 15.5 summarizes some of the studies of γ-PGA delivery systems as vaccine carriers. In conclusion, γ-PGA carries an immense potential for formulation of nano- and microparticulate carriers for efficient antigen delivery. In addition to vaccine delivery, it also serves as a strong, biodegradable and nontoxic adjuvant capable of inducing both humoral and cellular immunity against number of infections. Distinctive properties of PGA such as universal use, preservation, controlled release, and safety will permit PGA carrier systems to emerge as essential machinery for protein and DNA-based vaccines.

15.7 Conclusion

The chapter focuses on the foremost concerns about polyester polymers, their types, uses and degradability as systems for delivery of vaccine. Antigen-based vaccines solely are deficient in activating the antigen-presenting cells (APCs) sufficiently, including dendritic cells (DCs) to produce considerably superior antigen-specific T-cell responses. To overcome this restriction, adjuvants have been used which improve the vaccines immunogenicity. However, limitations for the use of adjuvants have been reported for the diseases where CD8+ T-cell responses are crucial such as viral infections, cancer, or other intracellular infections.

Polyester particles have been studied for more than 25 years and have demonstrated enormous potential for the development of improved vaccines. The development of safe and effective nanosystems for vaccine delivery remains a most important objective in public health. Polyester polymers used to synthesize nanometer- or micrometer-sized vaccine delivery vehicles were useful in the controlled release of vaccine antigens and immunostimulant molecules and efficient antigen presenting cells targeting with elimination of booster doses. These particles made up of eco-compatible polymers are capable to induce both humoral and cellular immune response in animal models, which is a prerequisite for future vaccination strategies. However, complex regulatory requirements for the design of vaccine formulation and account in clinical studies pose limitations for the development of polyester particles as vaccine delivery systems for human use. It is therefore critical that further research on such delivery systems should be conducted for the generation of effortless design that may not impair cost, reproducibility and safety issues. Moreover, higher research must be performed on basic science for better understanding of the mechanisms of uptake by APCs, transport to the lymphoid system, biodistribution and degradation of nanoparticles in the body. Careful preclinical evaluation will be obligatory to screen polymeric nanocarriers before finalising the results in a real clinic. In conclusion, the next decade will be the time to test and optimize these polyester nanocarriers for vaccine delivery to maximize its efficacy.

References

1. Nandedkar, T. D. (2009). Nanovaccines: recent developments in vaccination. *J Biosci*, **34**, 1–9.
2. Guy, B. (2007). The perfect mix: recent progress in adjuvant research. *Nat Rev Microbio*, **5**, 505–517.
3. Harandi, A. M., Medaglini, D., and Shattock, R. J. (2010). Vaccine adjuvants: a priority for vaccine research. *Vaccine*, **28**, 2363–2366.
4. Seetharam, R. N. (2006). Nanomedicine - emerging area of nanobiotechnology research. *Curr Sci*, **91**, 260.
5. Duc, L. H., Hong, H. A., and Uyen, N. Q. (2004). Cutting immunogenicity and intracellular fate of B. subtilis spores. *Vaccine*, **22**, 1873–1885.

6. Ricca, E., and Cutting, S. M. (2003). Emerging applications of bacterial spores in nanobiotechnology. *J Nanobiotechnol,* **1**, 6.

7. Wu, X. F., Xie, Y. G., Yuan, Y., Wang, D., Yu, S. K., and Wang, X. L. (2007). Proteosome adjuvant and its application in anti-plague immunity induced by recombinant F1-V protein. *Chin J Microbiol Immunol,* **27**, 247–250.

8. Cyr, S. L., Jones, T., Stoica-Popescu, I., Brewer, A., Chabot, S., Lussier, M., Burt, D., and Ward B. J. (2007). Intranasal proteosome-based respiratory syncytial virus (RSV) vaccines protect BALB/c mice against challenge without eosinophilia or enhanced pathology. *Vaccine,* **25**, 5378–5389.

9. Kalantari, P., Harandi, O. F., Hankey, P. A., and Henderson A. J. (2008). HIV-1 Tat mediates degradation of RON receptor tyrosine kinase, a regulator of inflammation. *J Immunol,* **181**, 1548–1555.

10. Hwang, I., Shen, X., and Sprent, J. (2003). Direct stimulation of naive T cells by membrane vesicles from antigen-presenting cells: distinct roles for CD54 and B7 molecules. *Proc Natl Acad Sci U S A,* **100**, 6670–6675.

11. Vincent-Schneider, H., Stumptner-Cuvelette, P., Lankar, D., Pain, S., Raposo, G., Benaroch, P., and Bonnerot, C. (2002). Exosomes bearing HLA-DR1 molecules need dendritic cells to efficiently stimulate specific T cells. *Int Immunol,* **14**, 713–722.

12. Takagi, A., Matsui, M., Ohno, S., Duan, H., Moriya, O., Kobayashi, N., Oda, H., Mori, M., Kobayashi, A., Taneichi, M., Uchida, T., and Akatsuka, T. (2009). Highly efficient antiviral CD8+ T-cell induction by peptides coupled to the surfaces of liposomes. *Clin Vaccine Immunol,* **16**, 1383–1392.

13. Myc, L. A., Gamian, A., and Myc, A. (2011). Cancer vaccines. Any future? *Arch Immunol Ther Exp,* **59**, 249–259.

14. Henriksen-Lacey, M., Korsholm, K. S., Andersen, P., Perrie, Y., and Christensen, D. (2011). Liposomal vaccine delivery systems. *Exp Opin Drug Deliv,* **8**, 505–519.

15. Inoue, J., Ideue, R., Takahashi, D., Kubota, M., and Kumazawa Y. (2009). Liposomal glycosphingolipids activate natural killer T cell-mediated immune responses through the endosomal pathway. *J Control Release,* **133**, 18–23.

16. Schwendener, R. A., Ludewig, B., Cerny, A., and Engler, O. (2010). Liposome-based vaccines. *Methods Mol Biol,* **605**, 163–175.

17. Bungener, L., Idema, J., Ter Veer, W., Huckriede, A., Daemen, T., and Wilschut, J. (2002). Virosomes in vaccine development: induction of

cytotoxic T lymphocyte activity with virosome encapsulated protein antigens. *J Lipos Res,* **12**, 155–163.

18. Felnerova, D., Viret, J. F., Glück, R., and Moser C. (2004). Liposomes and virosomes as delivery systems for antigens, nucleic acids and drugs. *Curr Opin Biotechnol,* **15**, 518–529.
19. Christopher, M. E., and Wong, J. P. (2006). Recent developments in delivery of nucleic acid-based antiviral agents. *Curr Pharmaceut Design,* **12**, 1995–1906.
20. Kaneda, Y. (2012). Virosome: a novel vector to enable multi-modal strategies for cancer therapy. *Adv Drug Deliv Rev,* **64**, 730–738.
21. Gupta, U., and Jain, N. K. (2010). Non-polymeric nano-carriers in HIV/AIDS drug delivery and targeting. *Adv Drug Deliv Rev,* **62**, 478–490.
22. Castor, T. P. (2007). Polymer nanospheres for improved drug delivery of protein therapeutics and viral antigens. *Nanotech,* **2**, 362–365.
23. Martin Del Valle, E. M., and Galan, M. A. (2005). Supercritical fluid technique for particle engineering: Drug delivery applications. *Rev Chem Eng,* **21**, 33–69.
24. Kalkanidis, M., Pietersz, G. A., Xiang, S. D., Mottram, P. L., Crimeen-Irwin, B., Ardipradja, K., and Plebanski, M. (2006). Methods for nano-particle based vaccine formulation and evaluation of their immunogenicity. *Methods,* **40**, 20–29.
25. Scheerlinck, J. P. Y., and Greenwood, D. L. V. (2008). Virus-sized vaccine delivery systems. *Drug Discov Today,* **13**, 882–887.
26. Scheerlinck, J. P. Y., and Greenwood, D. L. V. (2006). Particulate delivery systems for animal vaccines. *Methods,* **40**, 118–124.
27. Tissot, A. C., Renhofa, R., Schmitz, N., Cielens, I., Meijerink, E., Ose, V., Jennings, J. E., Saudan, P., Pumpens, P., and Bachmann M. F. (2010). Versatile virus-like particle carrier for epitope based vaccines. *PLoS One,* **5**, e9809.
28. Buonaguro, F. M., Tornesello, M. L., and Buonaguro, L. (2011). New adjuvants in evolving vaccine strategies. *Expert Opin Biol Ther,* **11**, 827–832.
29. Chernyavskaya, A. S., Morozova, I. V., Lebedeva, S. A., and Zarenkov, M. I. (2005). Construction of variants of vaccinal strain Yersinia pestis EV76 (RIEG line) differing in antibiotic resistance spectra with stage-by-stage transduction of R-transposons. *Antibiot Khimioter,* **50**, 13–17.
30. March, J. B., Clark, J. R., and Jepson, C. D. (2004). Genetic immunisation against hepatitis B using whole bacteriophage particles. *Vaccine,* **22**, 1666–1671.

31. Clark, J. R., and March, J. B. (2006). Bacteriophages and biotechnology: vaccines, gene therapy and antibacterials. *Trends Biotechnol*, **24**, 212–218.

32. Zanghi, C. N., Lankes, H. A., Bradel-Tretheway, B., Wegman, J., and Dewhurst, S. (2005). A simple method for displaying recalcitrant proteins on the surface of bacteriophage lambda. *Nucleic Acid Res*, **33**, 1–7.

33. Teodora, M. (2014). Nanoparticles-based cancer vaccines. *Biotechnol Mol Biol Nanomed* **2**, 13–14.

34. Parlanea, N. A., Bernd, H. A., Rehmb, D., Wedlocka, N., and Buddle, B. M. (2014). Novel particulate vaccines utilizing polyester nanoparticles (bio-beads) for protection against Mycobacterium bovis infection—a review. *Vet Immunol Immunopatho*, **158**, 8–13.

35. Staats, H. F., and Leong, K. W. (2010). Polymer hydrogels: chaperoning vaccines. *Nat Mater*, **9**, 537–538.

36. Gubeli, R. J., Schöneweis, K., Huzly, D., Ehrbar, M., Charpin-El Hamri, G., El-Baba, M. D., Urban, S., and Weber, W. (2013). Pharmacologically triggered hydrogel for scheduling hepatitis B vaccine administration. *Sci Rep*, **3**, 2610.

37. Gregory, A. E., Titball, R., and Williamson, D. (2013). Vaccine delivery using nanoparticles. *Front Cell Infect Microbiol*, **3**, 13.

38. Bhavsar, M. D., and Aniji, M. M., (2007). Polymeric nano- and microparticle technologies for oral gene delivery. *Expert Opin Drug Deliv*, **4**, 197–213.

39. Makidon, P. E., Bielinska, A. V., Nigarekar, S. S., Janezak, K. W., Knowlton, J., Scott, A. J., Mank, N., Cao, Z., Rathinavelu, S., Beer, M. R., Wilkinson, J. E., Blanco, L. P., Landers, J. J., and Baker, J. R. Jr. (2008). Pre-clinical evaluation of a novel nanoemulsion-based hepatitis B mucosal vaccine. *PLoS One*, **3**, 2954.

40. ADIS, R&D profile. (2003). Influenza virus vaccine live intranasal-Med Immune vaccines: CAIV-T, influenza vaccine live intranasal. *Drugs R D*, **4**, 312–319.

41. Slutter, B., Hagenaars, N., and Jiskoot, W. (2008). Rational design of nasal vaccines. *J Drug Target*, **16**, 1–17.

42. Sinyakov, M. S., Dror, M., Lublin-Tennenbaum, T., Salzberg, S., Margel, S., and Avtation, R. R. (2006). Nano and microparticles as adjuvants in vaccine design: success and failure is related to host material antibodies. *Vaccine*, **24**, 6534–6541.

43. Le, T. P., Coonan, K. M., Hedstrom, R. C., Charoenvit, Y., Sedegah, M., Epstein, J. E., Kumar, S., Wang, R., Doolan, D. L., Maguire, J. D., Parker, S. E., Hobart, P., Norman, J., and Hoffman, S. L. (2000). Safety, tolerability and humoral immune responses after intramuscular administration of a malaria DNA vaccine to healthy adult volunteers. *Vaccine*, **18**, 1893–1901.

44. Wang, L., Guan, N., Jin, Y., Lin, X., and Gao, H. (2015). Subcutaneous vaccination with Porphyromonas gingivalis ameliorates periodontitis by modulating Th17/Treg imbalance in a murine model. *Int Immunopharmacol*, **25**(1), 65–73.

45. Jiang, W., Gupta, R. K., Deshpande, M. C., and Schwendeman, S. P. (2005). Biodegradable poly(lactic-*co*-glycolic acid) microparticles for injectable delivery of vaccine antigens. *Adv Drug Del Rev*, **57**, 391–410.

46. Kumari, A., Yadav, S. K., and Yadav, S. C. (2010). Biodegradable polymeric nanoparticles based drug delivery systems. *Colloids Surf B Biointerfaces*, **75**, 1–18.

47. Prokop, A., and Davidson, J. M. (2008). Nanovehicular intracellular delivery systems. *J Pharm Sci*, **97**, 3518–3590.

48. Vert, M., Mauduit, J., and Li, S. (1994). Biodegradation of PLA/GA polymers: increasing complexity. *Biomaterials*, **15**, 1209–1213.

49. Danhier, F., Ansorena, E., Silva, J. M., Coco, R., Le Breton, A., and Preat, V. (2012). PLGA-based nanoparticles: an overview of biomedical applications. *J Control Release*, **161**(2), 505–522.

50. Zhao, K., Zhang, Y., Zhang, X., Shi, C., Wang, X., Wang, X., Jin, Z., and Cui, S. (2014). Chitosan-coated poly(lactic-*co*-glycolic) acid nanoparticles as an efficient delivery system for Newcastle disease virus DNA vaccine. *Int J Nanomedicine*, **9**, 4609–4619.

51. Sivakumar, S. M., and Sukumaran, N. (2009). Induction of immune response of hepatitis B vaccine using polyester polymer as an adjuvant. *Procedia Vaccinol*, **1**, 164–173.

52. Xiang, S. D., Scholzen, A., Minigo, G., David, C., Apostolopoulos, V., Mottram, P. L., *et al.* (2006). Pathogen recognition and development of particulate vaccines: does size matter. *Methods*, **40**, 1–9.

53. Demento, S. L., Cui, W., Criscione, J. M., *et al.*, (2012). Role of sustained antigen release from nanoparticle vaccines in shaping the T cell memory phenotype. *Biomaterials*, **33**(19), 4957–4964.

54. Gupta, R. K., Singh, M., and O'Hagan, D. T. (1998). Poly(lactide-*co*-glycolide) microparticles for the development of single-dose controlled-release vaccines. *Adv Drug Deliv Rev*, **32**, 225–246.

55. Jaganathan, K. S., and Vyas, S. P. (2006). Strong systemic and mucosal immune responses to surface-modified PLGA microspheres containing recombinant hepatitis B antigen administered intranasally. *Vaccine,* **24**, 4201–4211.

56. Zhu, Q., Talton, J., Zhang, G., *et al.* (2012). Large intestine-targeted, nanoparticle-releasing oral vaccine to control genitorectal viral infection. *Nat Med,* **18**(8), 1291–1296.

57. Silva, J. M., Videira, M., Gaspar, R., Preat, V., and Florindo, H. F. (2013). Immune system targeting by biodegradable nanoparticles for cancer vaccines. *J Control Release,* **168**(2), 179–199.

58. Song, C., Labhasetwar, V., and Levy, R. J. (1997) Nanoparticle drug delivery system for restenosis. *Adv Drug Deliv Rev,* **24**, 63–85.

59. Nagavarma, B. V. N., Yadav, Hemant K. S., Ayaz, A., Vasudha, L. S., and Shivakumar, H. G. (2012). Different techniques for preparation of polymeric nanoparticles- a review. *Asian J Pharm Clin Res,* **5** Suppl 3, 35–49.

60. Masson, V., Maurin, F., Fessi, H., and Devissaguet, J. P. (1997). Influence of sterilization processes on poly(ε-caprolactone) nanospheres. *Biomaterials,* **18**, 327–335.

61. Molpeceres, J., Guzman, M., Aberturas, M. R., Chacon, M., and Berges, L. (1996) Application of central composite designs to the preparation of polycaprolactone nanoparticles by solvent displacement. *J Pharm Sci,* **82**(2), 206–213.

62. Lee, S. H., Heng, D., Ng, W. K., Chan, H. K., and Tan, R. B. (2011) Nano spray drying: A novel method for preparing protein nanoparticles for protein therapy. *Int J Pharm,* **403**, 192–200.

63. Johansen, P., Men, Y., Merkle, H. P., and Gander, B. (2000). Revisiting PLA/PLGA microspheres: an analysis of their potential in parenteral vaccination. *Eur J Pharm Biopharm,* **50**(1), 129–146.

64. Gupta, R. K., Rost, B. E., Relyveld, E., and Siber, G. R. (1995). Adjuvant properties of aluminum and calcium compounds, in *Vaccine Design: The Subunit and Adjuvant Approach,* (M. F. Powell, and M. J. Newman, eds.), Plenum Press, New York, 229–248.

65. Jain, S., O'Hagan, D. T., and Singh, M. (2011). The long-term potential of biodegradable poly(lactideco-glycolide) microparticles as the next-generation vaccine adjuvant. *Expert Rev Vaccines,* **10**(12), 1731–1742.

66. Cleland, J. L., Lim, A., Barron, L., Duenas, E. T., and Powell, M. F. (1997). Development of a single-shot subunit vaccine for HIV-1: part 4. Optimising microencapsulation and pulsatile release of MN rgp120 from biodegradable microspheres. *J Control Release,* **47**, 135–150.

67. Thomasin, C., Corradin, G., Men, Y., Merkle, H. P., and Gander, B. (1996). Tetanus toxoid and synthetic malaria antigen containing poly(lactide)/poly(lactide-*co*-glycolide) microspheres: importance of polymer degradation and antigen release for immune response. *J Control Release,* **41**, 131–145.

68. Hogenesch, H. (2002). Mechanisms of stimulation of the immune response by aluminum adjuvants. *Vaccine,* **20**, S34–S39.

69. Jain, S., O'Hagan, D. T., and Singh, M. (2011). The long-term potential of biodegradable poly(lactide-*co*-glycolide) microparticles as the next-generation vaccine adjuvant. *Expert Rev Vaccines,* **10**(12), 1731–1742.

70. Jiang, W., Gupta, R. K., Deshpande, M. C., and Schwendeman, S. P. (2005). Biodegradable poly(lactic-*co*-glycolic acid) microparticles for injectable delivery of vaccine antigens. *Adv Drug Deliv Rev,* **57**(3), 391–410.

71. Fu, K., Pack, D.W., Klibanov, A. M., and Langer, R. (2000). Visual evidence of acidic environment within degrading poly(lactic-*co*-glycolic acid) (PLGA) microspheres. *Pharm Res,* **17**(1), 100–106.

72. Schwendeman, S. P. (2002). Recent advances in the stabilization of proteins encapsulated in injectable PLGA delivery systems. *Crit Rev Ther Drug Carrier Syst,* **19**(1), 73–98.

73. Dailey, L. A., and Kissel, T. (2005). New poly(lactic-*co*-glycolic acid) derivatives: modular polymers with tailored properties. *Drug Discov Today Technol,* **2**, 7–13.

74. Zhu, G., Mallery, S. R., and Schwendeman, S. P. (2001). Stabilization of proteins encapsulated in injectable poly (lactide- coglycolide). *Nat Biotechnol,* **18**(1), 52–57.

75. Jiang, W., and Schwendeman, S. P. (2001). Stabilization and controlled release of bovine serum albumin encapsulated in poly(D,L-lactide) and poly(ethylene glycol) microsphere blends. *Pharm Res,* **18**(6), 878–885.

76. Sharma, R., Agrawal, U., Mody, N., and Vyas, S. P. (2015). Polymer nanotechnology based approaches in mucosal vaccine delivery: challenges and opportunities. *Biotechnol Adv,* **33**, 64–79.

77. Krishnamachari, Y., and Salem, A. K. (2009). Innovative strategies for co-delivering antigens and CpG oligonucleotides. *Adv Drug Deliv Rev,* **61**, 205–217.

78. Pavot, V., Rochereau, N., Primard, C., *et al.* (2013). Encapsulation of Nod1 and Nod2 receptor ligands into poly(lactic acid) nanoparticles potentiates their immune properties. *J Control Release,* 167(1), 60–67.

79. Newman, K. D., Elamanchili, P., Kwon, G. S., and Samuel, J. (2002). Uptake of poly(D,L-lactic-*co*-glycolic acid) microspheres by antigen presenting cells *in vivo*. *J Biomed Mater Res,* **60**(3), 480–486.

80. Primard, C., Poecheim, J., Heuking, S., Sublet, E., Esmaeili, F., and Borchard, G. (2013). Multifunctional PLGA-based nanoparticles encapsulating simultaneously hydrophilic antigen and hydrophobic immunomodulator for mucosal immunization. *Mol Pharm,* **10**(8), 2996–3004.

81. Men, Y., Audran, R., Thomasin, C., Eberl, G., Demotz, S., Merkle, H. P., Gander, B., and Corradin, G. (1999). MHC class I- and class II-restricted processing and presentation of microencapsulated antigens. *Vaccine,* **17**, 1047–1056.

82. Tobío, M., Gref, R., Sánchez, A., Langer, R., and Alonso, M. J. (1998). Stealth PLA-PEG nanoparticles as protein carriers for nasal administration. *Pharm Res,* **15**(2), 270–275.

83. Jung, T., Kamm, W., Breitenbach, A., Hungerer, K. D., Hundt, E., and Kissel, T. (2001). Tetanus toxoid loaded nanoparticles from sulfobutylated poly(vinyl alcohol)-graft-poly(lactide-*co*-glycolide): evaluation of antibody response after oral and nasal application in mice. *Pharm Res,* **18**, 352–360

84. Raghuvanshi, R. S., Katare, Y. K., Lalwani, K., Ali, M. M., Singh, O., and Panda, A. K. (2002). Improved immune response from biodegradable polymer particles entrapping tetanus toxoid by use of different immunization protocol and adjuvants. *Int J Pharm,* **245**, 109–121.

85. Kim, S. Y., Doh, H. J., Jang, M. H., Ha, Y. J., Chung, S. I., and Park, H. J. (1999) Oral immunization with Helicobacter pylori-loaded poly(D, L-lactide-*co*-glycolide) nanoparticles. *Helicobacter,* **4**(1), 33–39.

86. Mansoor, F., Earley, B., Joseph, P., Cassidy, Markey, B., *et al.* (2014). Intranasal delivery of nanoparticles encapsulating BPI3V proteins induces an early humoral immune response in mice. *Res Vet Sci,* **96**(3), 551–557.

87. Binjawadagi, B., Dwivedi, V., Manickam, C., *et al.* (2014). Adjuvanted poly (lactic-*co*-glycolic) acid nanoparticle-entrapped inactivated porcine reproductive and respiratory syndrome virus vaccine elicits cross-protective immune response in pigs. *Inter J Nanomed,* **9**(1), 679–694.

88. Thomas, C., Rawat, A., Hope-Weeks, L., and Ahsan, F. (2011). Aerosolized PLGA nanoparticles enhance humoral, mucosal and cytokine responses to hepatitis B vaccine. *Mol Pharm,* **8**, 405–415.

89. Mishra, N., Tiwari, S., Vaidya, B., Govind, P., Agrawal, and Vyas, S. P. (2011). Lectin anchored PLGA nanoparticles for oral mucosal immunization against hepatitis B. *J Drug Target,* **19**(1), 67–78.

90. Garinota, M., Fiéveza, V. B., Pourcellec, V., Stoffelbachd, F., Rieuxa, A. B., Plapieda, L. B., Theatee, I., Freichelsd, H., Jérômed, C., Marchand-Rynaertc, J., Yves-Jacques Schneiderc, Y. J., and Préa, V. (2007). PEGylated PLGA-based nanoparticles targeting M cells for oral vaccination. *J Control Release,* **120**, 195–204.

91. Gupta, P. N., Khatri, K., Amit, K., Goyal, Mishra, N., and Vyas, S. P. (2007). M-cell targeted biodegradable PLGA nanoparticles for oral immunization against hepatitis B. *J Drug Target,* **15**(10), 701–713.

92. Slütter, B., Bal, S., Keijzer, C., Mallants, R., Hagenaars, N., Que, I., Kaijzel, E., van Eden, W., Augustijns, P., Löwik, C., Bouwstra, J., Broere, F., and Jiskoot, W. (2010). Nasal vaccination with N-trimethyl chitosan and PLGA based nanoparticles: nanoparticle characteristics determine quality and strength of the antibody response in mice against the encapsulated antigen. *Vaccine,* **28**(38), 6282–6291.

93. Rosalia, R. A., Cruz, L. J., Duikeren, S. V., Tromp, A. T., Silva, A. L., Jiskoot, W., de Gruijl, T., Löwik, C., Oostendorp, J., van der Burg, S. H., and Ossendorp, F. (2015). CD40-targeted dendritic cell delivery of PLGA-nanoparticle vaccines induce potent anti-tumor responses. *Biomaterials,* **40**, 88–97.

94. Primard, C., Rochereau, N., Luciani, E., *et al.* (2010). Traffic of poly(lactic acid) nanoparticulate vaccine vehicle from intestinal mucus to sub-epithelial immune competent cells. *Biomaterials,* **31**(23), 6060–6068.

95. Pawar, D., Mangal, S., Goswami, R., and Jaganathan, K. S. (2013). Development and characterization of surface modified PLGA nanoparticles for nasal vaccine delivery: effect of mucoadhesive coating on antigen uptake and immune adjuvant activity. *Eur J Pharm Biopharm,* **85**(3 Pt A), 550–559.

96. Raghuvanshi, R. S., Singh, M., and Talwar, G. P. (1993). Biodegradable delivery system for single with tetanus toxoid. *Int J Pharm,* **93**, Rl–R5.

97. Esparza, I., and Kissel, T. (1992). Parameters affecting the immunogenicity of microencapsulated tetanus toxoid. *Vaccine,* **10**, 714–720.

98. Thomasin, C., Corradin, G., Men, Y., Merkle, H. P., and Gander, B. (1996). Tetanus toxoid and synthetic malaria antigen containing poly(lactide)/poly(lactide-*co*-glycolide) microspheres: importance of polymer degradation and antigen release for immune response. *J Control Release,* **41**(1–2), 131–145.

99. Igartua, M., Hernandez, R. M., Esquisabel, A., Gascon, A. R., Calvo, M. B., and Pedraz, J. L. (1998). Enhanced immune response after subcutaneous and oral immunization with biodegradable PLGA microspheres. *J Control Release*, **56**, 63–73.

100. Almeida, A. J., Alpar, H. O., and Brown, M. R. (1993). Immune response to nasal delivery of antigenically intact tetanus toxoid associated with poly(L-lactic acid) microspheres in rats, rabbits and guinea-pigs. *J Pharm Pharmacol*, **45**, 198–203.

101. Pawar, D., Goyal, A. K., Mangal, S., Mishra, N., Vaidya, B., Tiwari, S., Jain, A. K., and Vyas, S. P. (2010). Evaluation of mucoadhesive PLGA microparticles for nasal immunization. *AAPS J*, **12**(2), 130–137.

102. Waeckerle-Men, Y., and Groettrup, M. (2005). PLGA microspheres for improved antigen delivery to dendritic cells as cellular vaccines. *Adv Drug Deliv Rev*, **57**, 475–482.

103. Román, S. B., Gómez, S., Irache, J. M., and Espuelas, S. (2014). Co-encapsulated CpG oligodeoxynucleotides and ovalbumin in PLGA microparticles; an *in vitro* and *in vivo* study. *J Pharm Pharm Sci*, **17**(4), 541–553.

104. Qiu, S., Wei, Q., Liang, Z., Ma, G., Wang, L., An, W., Ma, X., Fang, X., He, P., Li, H., and Hu, Z. (2014). Biodegradable polylactide microspheres enhance specific immune response induced by Hepatitis B surface antigen. *Hum Vaccin Immunother*, **10**(8), 2350–2356.

105. Wang, Q., Tan, M. T., Keegan, B. P., Barry, M. A., and Heffernan, M. J. (2014). Time course study of the antigen-specific immune response to a PLGA microparticle vaccine formulation. *Biomaterials*, **35**(29), 8385–8393.

106. Balashanmugam, M. V., Nagarethinam, S., Jagani, H., Josyula, V. R., Alrohaimi, A., and Udupa, N. (2014). Preparation and characterization of novel PBAE/PLGA polymer blend microparticles for DNA vaccine delivery, *Sci World J*, **2014**, Article ID 385135, 9 pages.

107. Rubsamen, R. M., Herst, C. V., Lloyd, P. M., and Heckerman, D. E. (2014). Eliciting cytotoxic T-lymphocyte responses from synthetic vectors containing one or two epitopes in a C57BL/6 mouse model using peptide-containing biodegradable microspheres and adjuvants. *Vaccine*, **32**(33), 4111–4116.

108. Hanson, M. C., Bershteyn, A., Crespo, M. P., and Irvine, D. J. (2014). Antigen delivery by lipid-enveloped PLGA microparticle vaccines mediated by *in situ* vesicle shedding. *Biomacromolecules*, **15**(7), 2475–2481.

109. Jiang, T., Singh, B., Li, H. S., Kim, Y. K., Kang, S. K., Nah, J. W., Choi, Y. J., and Cho, C. S. (2014). Targeted oral delivery of BmpB vaccine using porous PLGA microparticles coated with M cell homing peptide-coupled chitosan. *Biomaterials*, **35**(7), 2365–2373.

110. Negash, T., Liman, M., and Rautenschlein, S. (2013). Mucosal application of cationic poly(D,L-lactide-*co*-glycolide) microparticles as carriers of DNA vaccine and adjuvants to protect chickens against infectious bursal disease. *Vaccine*, **31**(36), 3656–3662.

111. Chuang, S. C., Ko, J. C., Chen, C. P., Du, J. T., and Yang, C. D. (2013). Encapsulation of chimeric protein rSAG1/2 into poly(lactide-*co*-glycolide) microparticles induces long-term protective immunity against Toxoplasma gondii in mice. *Exp Parasitol*, **134**(4), 430–437.

112. Reddy, K. S., Rashmi, B. R., Dechamma, H. J., Gopalakrishna, S., Banumathi, N., Suryanarayana, V. V., and Reddy, G. R. (2012). Cationic microparticle [poly(D,L-lactide-*co*-glycolide)]-coated DNA vaccination induces a long-term immune response against foot and mouth disease in guinea pigs. *J Gene Med*, **14**(5), 348–352.

113. Gómez, M. J. M, Fischer, S., Csaba, N., Kündig, T. M., Merkle, H. P., Gander, B., and Johansen, P. (2007). A protective allergy vaccine based on CpG- and protamine-containing PLGA microparticles. *Pharm Res*, **24**(10), 1927–1935.

114. Kasturi S. P., Qin, H., Thomson, K. S., El-Bereir, S., Cha, S. C., Neelapu, S., Kwak, L, W., and Roy, K. (2006). Prophylactic anti-tumor effects in a B cell lymphoma model with DNA vaccines delivered on polyethylenimine (PEI) functionalized PLGA microparticles. *J Control Release*, **113**(3), 261–270.

115. Saini, V. Jain, V. Sudheesh, M. S., *et al.* (2010). Humoral and cell-mediated immune-responses after administration of a single-shot recombinant hepatitis B surface antigen vaccine formulated with cationic poly(L-lactide) microspheres. *J Drug Target*, **18**, 212–222.

116. Kim, M.-G., Park, J. Y., Shon, Y., Kim, G., Shim, G., and Oh, Y.-K. (2014). Nanotechnology and vaccine development. *Asian J Pharm Sci*, **9**(5), 227–235.

117. Li, A. V., Moon, J. J., Abraham, A., *et al.* (2013). Generation of effector memory T cell-based mucosal and systemic immunity with pulmonary nanoparticle vaccination. *Sci Transl Med*, **5**, 204ra130.

118. Nils, L. (2012). Recent progress in mucosal vaccine development: potential and limitations. *Nat Rev Immunol*, **12**, 592–605.

119. Kammona, O., and Kiparissides, C. (2012). Recent advances in nanocarrier-based mucosal delivery of biomolecules. *J Control Release,* **161**(3), 781–794.

120. Patrizia, P., Prego, C., Sanchez, A., and Alonso, M. J. (2010). Surface-modified PLGA-based nanoparticles that can efficiently associate and deliver virus-like particles. *Nanomedicine,* **5**(6), 843–853.

121. Mundargi, R. C., Babu, V. R., Rangaswamy, V., *et al.* (2008). Nano/micro technologies for delivering macromolecular therapeutics using poly(D, L-lactide-*co*-glycolide) and its derivatives. *J Control Release,* **125**, 193–209.

122. Levine, M. M. (2010). Immunogenicity and efficacy of oral vaccines in developing countries: lessons from a live cholera vaccine. *BMC Biol,* 129.

123. Zhao, L., Seth, A., Wibowoa, N., Zhao, C. X., Mitter, N., Yu, C., and Middelberg, A. P. (2014). Nanoparticle vaccines. *Vaccine,* **32**, 327–337.

124. Woodruff, M. A., and Hutmacher, D. W. (2010). The return of a forgotten polymer—Polycaprolactone in the 21st century. *Prog Polym Sci,* **30**, 1–40.

125. Tomar, P., Karwasara, V. S., and Dixit, V. K. (2011). Development characterizations and evaluation of poly(ε-caprolactone)-based microspheres for hepatitis B surface antigen delivery. *Pharm Dev Technol,* **16**(5), 489–496.

126. Bhavsar, M. D., Tiwari, S. B., and Amiji, M. M. (2006). Formulation optimization for the nanoparticles-in-microsphere hybrid oral delivery system using factorial design. *J Control Release,* **110**(2), 422–430.

127. Gou, M. L., Dai, M., Li, X. Y., Yang, L., Huang, M. J., Wang, Y. S., Kan, B., Lu, You., Wei, Y. Q., and Qian, Z. Y. (2008). Preparation of mannan modified anionic PCL–PEG–PCL nanoparticles at one-step for bFGF antigen delivery to improve humoral immunity. *Colloids Surf B Biointerfaces,* **64**, 135–139.

128. Matsuo, K., Yoshikawa, T., Oda, A., Akagi, T., Akashi, M., Mukai, Y., Yoshioka, Y., Okada, N., and Nakagawa, S. (2007). Efficient generation of antigen-specific cellular immunity by vaccination with poly(γ-glutamic acid) nanoparticles entrapping endoplasmic reticulum-targeted peptides. *Biochem Biophys Res Commun,* **362**, 1069–1072.

129. Matsuo, K, Koizumi, H., Akashi, M., Nakagawa, S., Fujita, T., Yamamoto, A., and Okada, N. (2011). Intranasal immunization with poly(γ-glutamic acid) nanoparticles entrapping antigenic proteins can induce potent tumor immunity. *J Control Release,* **152**, 310–316.

130. Okamoto, S., Matsuura, M., Akagi, T., Akashi, Mitsuru., Tanimoto, T., Ishikaw, T., Takahashi, M., Yamanishi, K., and Mori, Y. (2009). Poly(γ-glutamic acid) nano-particles combined with mucosal influenza virus hemagglutinin vaccine protects against influenza virus infection in mice. *Vaccine*, **27**, 5896–5905.

131. Okamotoa, S., Yoshii, H., Ishikawab, T., Akagi, T., Moria, Y., Akashi, M., Takahashie, M., and Yamanishia, K. (2008). Single dose of inactivated Japanese encephalitis vaccine with poly(γ-glutamic acid) nanoparticles provides effective protection from Japanese encephalitis virus. *Vaccine*, **26**, 589–594.

132. Benoit, M. A., and Baras, B. J. (1999). Preparation and characterization of protein-loaded poly(ε-caprolactone) microparticles for oral vaccine delivery. *Int J Pharm*, **184**, 73–84.

133. Singh, J., Pandit, S., Bramwell, V. W., and Alpar, H. O. (2006). Diphtheria toxoid loaded poly-(ε-caprolactone) nanoparticles as mucosal vaccine delivery systems. *Methods*, **38**, 96–105.

134. Gamazo, C., Goñi, M. M., Irache, J. M., Blanco-Príeto, M. J., and Murillo, M. (2002). Development of microparticles prepared by spray-drying as a vaccine delivery system against brucellosis. *Int J Pharm*, **242**(1), 341–344.

135. Jesus, S., Borchard, G., and Borges, O. (2013). Freeze dried chitosan/poly-ε-caprolactone and poly-ε-caprolactone nanoparticles: evaluation of their potential as DNA and antigen delivery systems. *J Genet Syndr Gene Ther*, **4**(7), 1–11.

136. Florindo, H. F., Pandit, S., Lacerda, L., Gonçalves, L. M., Alpar, H. O., and Almeida, A. J. (2009). The enhancement of the immune response against S. equi antigens through the intranasal administration of poly-ε-caprolactone-based nanoparticles. *Biomaterials*, **30**(5), 879–891.

137. Wang, X., Uto, T., Akagi, T., Akashi, M., and Baba M. (2008). Poly(γ-glutamic acid) nanoparticles as an efficient antigen delivery and adjuvant system: potential for an AIDS vaccine. *J Med Virol*, **80**, 11–19.

138. Yoshikawa, T., Okada, N., Oda, A., Matsuo, K., Matsuo, K., Kayamuro, H., Ishii, Y., Yoshinaga, T., Akagi, T., Akashi, M., and Nakagawa, S. (2008). Nanoparticles built by self-assembly of amphiphilic PGA can deliver antigens to antigen-presenting cells with high efficiency: a new tumor-vaccine carrier for eliciting effector T cells. *Vaccine*, **26**, 1303–1313.

139. Gorantla, Y., Palaniappan, R., Yeboah, M. K., Paulos, S., and Rajam, G. (2014). Mucoadhesive-polycaprolactone microspheres for delivery of protein subunit vaccines intranasally. *AAPS*, abstracts.aaps.org.

140. Bhavsar, M. D., Tiwari, S. B., and Amiji, M. M. (2006). Formulation optimization for the nanoparticles-in-microsphere hybrid oral delivery system using factorial design. *J Control Release,* **110**(2), 422–430.

141. Wu, Q. J., Gong, C. Y., Shi, S., Wang, Y. J., Huang, M. J., Yang, Li., Zhao, X., Wei Y. Q., and Qian, Z. Y. 2012. Mannan loaded biodegradable and injectable thermosensitive PCL-PEG-PCL hydrogel for vaccine delivery. *Soft Mater,* **10**(4), 472–486.

142. Prashant, C. K., Bhat, M., and Dinda A. K. (2014). Fabrication of nanoadjuvant with poly-ε-caprolactone (PCL) for developing a single-shot vaccine providing prolonged immunity. *Int J Nanomedicine,* **9**, 1–35.

143. Rice-Ficht A. C., Arenas-Gamboa A. M., Kahl-McDonagh M. M., and Ficht, T. A. (2010). Polymeric particles in vaccine delivery. *Curr Opin Microbiol,* **13**, 106–112.

144. Yoshikawa, T., Okada, N., Oda, A., Matsuo, K., Matsuo, K., Mukai, Y., Yoshioka, Y., Akagi, T., Akashi, M., and Nakagawa, S. (2008). Development of amphiphilic γ-PGA-nanoparticle based tumor vaccine: potential of the nanoparticulate cytosolic protein delivery carrier. *Biochem Biophys Res Commun,* **366**, 408–413.

145. Brown, D. M., Fisher, T. L., Wei, C., Frelinger, J. G., and Lord, E. M. (2001). Tumours can act as adjuvants for humoral immunity. *Immunology,* **102**(4), 486–497.

Chapter 16

Mucosal Immunization Using Polyester-Based Particulate Systems

Diana Gaspar,* Carina Peres,* Helena Florindo, and António J. Almeida

Research Institute for Medicines (iMed.ULisboa), Faculty of Pharmacy, Universidade de Lisboa, Av. Prof. Gama Pinto, 1649-003 Lisbon, Portugal

aalmeida@ff.ulisboa.pt

16.1 Introduction

In human adults, mucosal membranes represent an enormous surface area (up to 400 m^2 of surface), composed of single layers of epithelium covered by mucus, that line the respiratory, gastrointestinal, and urogenital tracts, as well as the eye conjunctiva, the inner ear, and the ducts of all exocrine glands [1, 2]. These surfaces constitute the most common portal of entry for pathogens; indeed, it is estimated that 70% of the potentially infectious agents invade and infect the host by mucosal routes [3]. Globally, mucosa-associated infections, such as acquired immunodeficiency syndrome

*Both authors contributed equally to this work.

Handbook of Polyester Drug Delivery Systems

Edited by MNV Ravikumar

ISBN 978-981-4669-65-8 (Hardcover), 978-981-4669-66-5 (eBook)

www.panstanford.com

(AIDS), tuberculosis, influenza, pneumonia, and sexually transmitted diseases are major causes of illness and serve big socioeconomic burdens in both developed and low- and middle-income countries [4].

An effective way of targeting pathogens before infection occurs is through vaccination. This concept was introduced with the cowpox virus in 1796 by Edward Jenner, and since then, vaccines have had a great impact on global health with the control of many infectious diseases, and even with the eradication of smallpox [5]. Nowadays, vaccines are routinely used for the prevention of several diseases, including tuberculosis, diphtheria, tetanus, hepatitis B, poliomyelitis, and meningitis, and vaccination is considered as the most appropriate method for the control of disease-causing pathogens from the economic, environmental, and ethical point of view. Apart from their current applications, vaccines hold great promise in the prevention of a myriad of diseases, including cancer [6–8], cystic fibrosis [9, 10], Alzheimer disease [11–13], human immunodeficiency virus (HIV) infection/AIDS [14–16], and, more recently, Ebola [17, 18].

Vaccines usually contain molecules that mimic disease-causing pathogens and are often composed by attenuated or killed forms of the target microorganism, as toxins or surface proteins [19]. However, their use raises several safety issues since pathogenic bioactive material can remain in vaccine formulation as contaminants [20–23]. In addition, to maintain the efficacy of these vaccines, strict storage conditions must be followed in order to avoid instability and degradation, and even a reversion to a virulent form. Moreover, subjects who have damaged or weakened immune systems cannot safely receive the vaccines [24, 25].

In order to overcome these issues, vaccines based on particulate formulations, such as whole-cell vaccines, virosomes, virus-like particles, or antigens formulated in particulate adjuvants, such as liposomes, microparticles (MPs), and nanoparticles (NPs) are being developed. These particulate vaccines can be made up of several materials like polymers (natural or synthetic), lipids, and inorganic compounds. Nowadays, the most common developed nanotechnology platforms can range from liposomes and micelles to carbon nanotubes and polymeric NPs, which are schematically represented in Fig. 16.1.

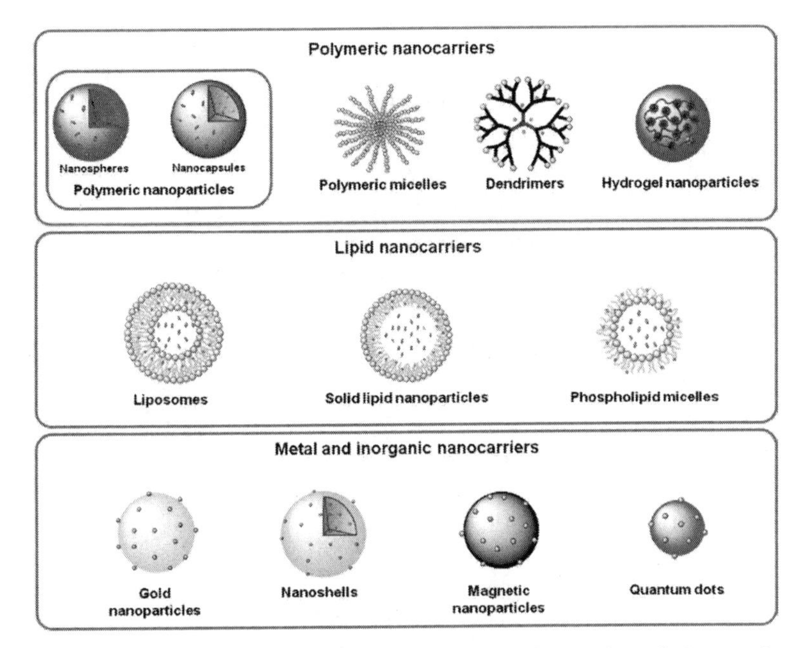

Figure 16.1 Examples of polymeric, lipid, and metal and inorganic nanocarriers. Reprinted from Ref. 8, Copyright 2007–2015 Frontiers Media S.A. All Rights Reserved.

In particular, biodegradable polymeric NPs are being widely explored as carriers for controlled and targeted delivery of a multitude of molecules, including proteins, peptides, nucleic acids, and oligonucleotides [26, 27], as they have been shown to modulate cellular and humoral immune responses [28]. Indeed, formulating protein antigens in polymeric NPs has emerged as one of the most promising strategies to trigger an immune response to vaccine antigens [26, 29]. Such NPs have the ability to stabilize vaccine antigens, but also to promote antigen sustained and controlled release to antigen-presenting cells (APCs), maximizing their capture, processing, and presentation, and consequently eliminating booster immunizations [30]. Their subcellular size, mimicking that of natural pathogens, and their large surfaces, which may present electrostatic or receptor-interacting properties, permit internalization and presentation of antigens by cells and may provoke better recognition by the immune system compared with soluble antigens [30]. In addition, their surface can be functionalized through the inclusion of anchoring devices that have adhesive properties, such as lectins and specific antibodies, establishing a close contact with the mucosal epithelium, or with appropriate ligands that allow to target specific cells and body locals and, consequently, to reduce toxicity and to increase biodistribution [8]. These nanoscale compounds are described as biocompatible with tissues and cells, stable in blood, nontoxic, non-immunogenic, non-inflammatory, not activating neutrophils, and avoiding the reticuloendothelial system, preventing premature phagocytosis by macrophages [31, 32].

Although a number of polymers have been developed and investigated for purposes of particle formulation, thermoplastic aliphatic polyesters are the most widely used polymers for NP and MP production [26, 33]. Aliphatic polyesters represent a diverse family of synthetic biodegradable polymers, which include the hydrophobic poly(lactic acid) (PLA), the hydrophilic poly(glycolic acid) (PGA), their copolymer poly(lactic-*co*-glycolic acid) (PLGA), and poly(ε-caprolactone) (PCL) [34–36] (Fig. 16.2). In the past three decades, these polymers have been extensively explored for different biomedical and pharmaceutical applications, such as surgical sutures, bone screws, tissue engineering scaffolds, and controlled release delivery systems [26, 33]. Contrary to natural polymers that present several drawbacks as high costs, low reproducibility, and questionable purity [37], these synthetic polymers can be easily produced at low costs (generally by polycondensation or ring-opening polymerization [ROP] routes), with tailored mechanical and degradative properties, and high polymer uniformity (in terms of molecular weight distribution, monomer orientation, sequence, stereo-regularity, polymer shape and morphology, and chemical functionality) [38]. These features, which are highly desired by the regulatory entities, allow their application as integral components in final dosage forms, delivery systems, and implantable devices [38].

Their successful application as delivery systems for hydrophobic and hydrophilic drugs, nucleic acids, peptide, and proteins is essentially due to two properties: biocompatibility and biodegradability [33, 36]. Biocompatibility is defined as the ability of a material to perform with an appropriate host response in a specific application [39], but may also be referred as the relationship between a material and the organism so that neither produces undesirable effects [40]. This concept includes different aspects of the material, namely its physical, chemical, and mechanical properties, but also its potential cytotoxic, mutagenic, and allergenic effects [40, 41]. It is important to highlight that biocompatibility is not the result of a single event, but rather a collection of processes involving different reaction mechanisms between a material and its host tissue [42]. Biodegradability refers to the ability of a material to degrade *in vitro* and *in vivo* either into products that are normal metabolites of the body or into products that can be completely eliminated from the body with or without further metabolic transformations [43, 44]. In the case of PLA, PGA, PLGA, and PLC, these polymers are hydrolyzed to form the corresponding hydroxy acids, which are byproducts of various metabolic pathways in the body under normal physiological conditions and are easily metabolized in the body via the Krebs cycle and physiologically eliminated [45]. Because of their excellent biocompatibility and biodegradability, these polymers have been approved by the US Food and Drug Administration (FDA) and European Medicine Agency (EMA) for a number of clinical applications, including delivery systems [36, 46, 47].

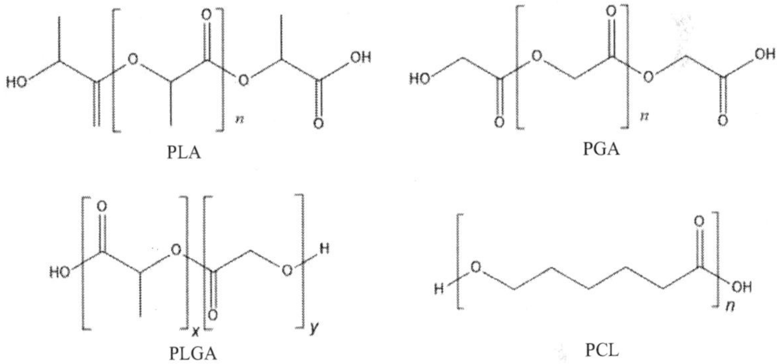

PLA

PGA

PLGA

PCL

Figure 16.2 Chemical structures of the most used aliphatic polyesters for nano- and microparticle formulation.

Although the selection of an appropriate polymer is important for a successful result, the route of vaccination is also crucial for an effective mucosal immunization using nanoparticulate-based vaccines. In contrast to traditional parenteral vaccination that are generally poor inducers of mucosal immunity, mucosal vaccination can elicit both systemic and mucosal immune responses, providing additional secretory antibody-mediated protection against pathogens at the mucosal site of entry [48, 49]. In addition, compared with parenteral vaccines, mucosal vaccines would be easier to administer, carry less risk of transmitting infections, and could simplify manufacturing, thereby increasing the potential for local vaccine production in developing countries [50].

Throughout this chapter, many examples of the use of MPs and NPs, in particular those made of biodegradable aliphatic polyesters, such as PLA, PCL, PLGA and PCL, will be presented to support their potential as mucosal delivery system.

16.2 Overcoming Mucosal Barriers: Mechanistic Issues

As mentioned before, mucosal surfaces are prone to viral and bacterial attacks and serve as entry ports for pathogens. Scientists are continuously working in order to ascertain protection through these surfaces as they constitute the first line of defense against infection. Mucosal vaccines are intended to utilize the elements of the immune responses associated with this route of administration [51]. The mucosal immune system can be divided into two general compartments known as inductive and effector sites. Inductive sites are areas where antigen sampling leads to initial activation of immune cells, while at effector sites, antibodies and cells of the immune system can perform their specific function upon activation (Table 16.1) [52].

The principal inductive sites for mucosal immune responses consist of region-specific lymphoid tissues, known as mucosa-associated lymphoid tissue (MALT), which is located along the surfaces of all mucosal tissues, as well as their surrounding regional lymph nodes. Nasopharynx-associated lymphoid tissue (NALT), gut-associated lymphoid tissue (GALT) and bronchus-associated lymphoid tissue (BALT) are the most common representatives of MALT, which comprises Peyer's patches (PPs) and isolated lymphoid follicles [48, 51–53].

Different routes of vaccination with associated features and the lymphoid tissue involved are shown in Fig. 16.3. Microfold cells (M cells) are the specialized cells present in the follicle-associated epithelium, which performs the function of transporting large and soluble particles and microorganisms to sub-epithelial dome region from the intestinal lumen [51].

Table 16.1 Components of the mucosal-associated lymphoid tissue (adapted from [52])

Induction site	Effector site	Lymphoid structures
Nasopharynx-associated lymphoid tissue (NALT)	Nasopharyngeal mucosa	Waldeyer's pharyngeal ring: Adenoids (pharyngeal tonsils) Palatine tonsils Lingual tonsils Tubal tonsils
Bronchus-associated lymphoid tissue (BALT)	Bronchial mucosa	Peyer's patches
	Lower Respiratory Tract	Isolated lymphoid follicles
Gut-associated lymphoid tissue (GALT)	Gastrointestinal mucosa	Peyer's patches Lymphoglandular complexes Isolated lymphoid follicles Cryptopatches Appendix
Conjunctiva-associated lymphoid tissue (CALT)	Conjunctiva	Lymphoepithelium Lymphoid follicles with B- and T-cell zones Adjacent blood vessels that have thickened endothelia Lymphoid vessels
Vulvovaginal-associated lymphoid tissue (VALT) or genital-associated lymphoid tissue (GENALT)	Urogenital tract	Lymphoid follicles
Rectal lymphoepithelial tissue	Gastrointestinal mucosa	Lymphoid follicles

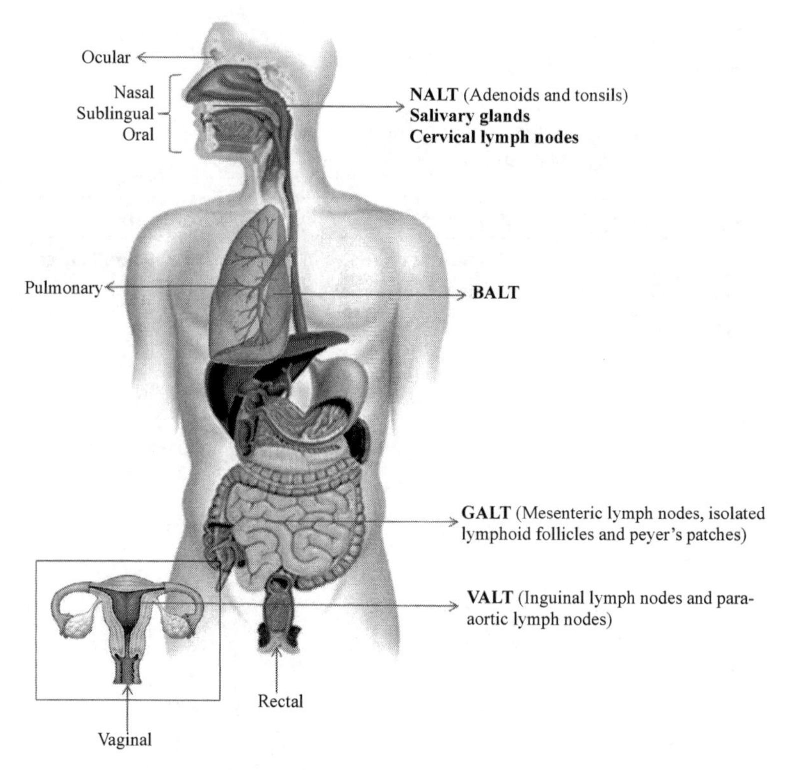

Figure 16.3 Mucosal immunization routes and compartmentalization of effector functions. Within the MALT, subcompartments can be identified, such as the NALT, BALT, GALT, and VALT (or GENALT). Intranasal vaccination is preferred for targeting the respiratory, gastric, and genital tracts; oral vaccination is effective for immunity in the gut and for the induction of mammary gland antibodies (which are secreted in milk); rectal immunization is best for the induction of colon and rectal immunity and to some extent genital tract immunity and intravaginal vaccination is the most effective for antibody and T-cell immunity in the genital tract. Adapted by permission from Macmillan Publishers Ltd: Ref. 48, Copyright 2012; and from Ref. 51, Copyright 2015, with permission from Elsevier.

The delivery by mucosal barriers has already become a reality for a number of drugs and antigens [54]. However, there is still much controversy surrounding the mechanism of particle transport across the epithelial barrier. For the transport of macromolecules and particulates through the mucosal barriers, different pathways can be distinguished (Fig. 16.4). Since the paracellular spaces, sealed by tight junctions, contribute less than 1% of the mucosal surface area and the pore diameter of these junctions was reported to be less than 10 Å, significant paracellular transport of macromolecules and particles is an unlikely event. Paracellular permeability for peptides can be enhanced, however, by polymers, such as chitosan, poly(acrylate), and/or starch. On the other hand, antigens associated or not with particles can also be taken up by transepithelial transport, where they are processed by APCs and presented to T lymphocytes, which can provide T helper (Th) cell to primed B cells in the same region [55, 56]. In addition, larger peptides and proteins have been shown to be able to pass the nasal membrane using an endocytotic transport process but only in low amounts, although this mechanism remains controversial [56–60].

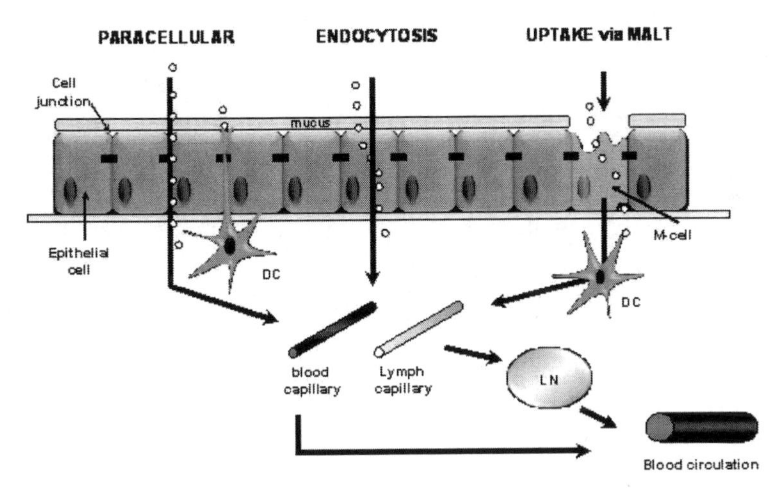

Figure 16.4 Routes and mechanisms of particle uptake and transport across the mucosal surfaces. Reproduced from [56] with permission of the Royal Society of Chemistry, Copyright 2012.

Generally, foreign antigens and pathogens are encountered through normal physiological functions, such as ingestion and inhalation, and the host thus has evolved organized lymphoid tissue in the regions that facilitate the initiation of antigen-specific immune responses following exposure to these mucosal antigens and pathogens [61]. After these antigens, associated or not with carriers, get in contact with mucosal surfaces, they are taken up by M cells at inductive site and are carried to APCs, including macrophages (Mφ), B cells, and dendritic cells (DCs). The M cells serve as specific ports of entry for antigens and have the ability to transcytose the antigenic material into the M-cell pocket located below, facilitating the delivery to APCs, including DCs and Mφ. Once taken up by these cells, they are directed toward conventional $CD4^+$ and $CD8^+$ T cells (effector cells) at the inducer site. Sometimes epithelial T cells can also process the antigen and present them to intraepithelial T cells including natural killer (NK) T cells.

There are certain factors like nature of the antigen, surrounding microenvironment and type of APC involved which can affect the immune response in mucosal tissue. After activation, both B and T cells reach the circulation via lymph and finally to the mucosal sites mainly mucosa of origin where the differentiation into memory or effector cells occurs. Following mucosal vaccination, the mucosal barriers to infection are reinforced, mainly through the induction of antigen-specific secretory-immunoglobulin A (sIgA) antibody production, which prevents pathogens and toxins from adhering to or infecting the epithelial cells and breaching the mucosal barrier [2, 48, 55]. Thus, sIgA plays an important role at the effector arm of the mucosal system, while IgG and IgM are actively produced at MALT and transported across mucosal cells. Detection of IgG at mucosal site can result from the passive leakage across the mucosal surface. The IgA is considered as the primary mucosal antibody and along with IgM it can be transported across the epithelial barrier, which provides potent immunity against large number of viral infections [2, 51, 62].

The type of immune response generated also depends upon the size of particles where the antigen is loaded [51, 63]. NPs account for induction of cellular responses, while MPs endorse humoral responses. Immunization with the 200–600 nm particles favored Th1 immune responses, whereas immunization with the 2–8 μm particles favored Th2 responses [51, 64]. This can be attributed to the fact that the small particles are taken up by APCs which induce cellular responses, while the MPs are only attached to the surface of Mφ and release the loaded antigens [51]. Regarding this, Jepson *et al.* showed that polystyrene MPs preferentially bound and were transcytosed by M cells, while PLGA MPs of similar size (0.5–0.6 μm) were less likely to be endocytosed by M cells. It was observed that particles bound to the surface were efficiently transcytosed [65, 66]. In another study, polystyrene particles (50–1000 nm) administered by oral gavage to rats were found only in PPs regions of the small intestine. This is supported by Hillery *et al.* using 0.059 μm PS particles in rats. Examination of lymphoid and non-lymphoid tissues after the administration of these particles by oral gavage revealed a total uptake of ~10% with 60% of the particles taken up into lymphoid tissue [67]. Using PLGA MPs (1–10 μm), Damgé *et al.* also reported a 12.7% total uptake of MPs predominantly by the PPs of rats [68, 69]. In another study, small PLA MPs (e.g., 4 μm in diameter) enhanced only plasma IgG responses without IgA responses in the intestine. In contrast, 8–10 μm PLA MPs enhanced IgA responses in the intestine [70], which suggest that the former size of particles is effectively transporting antigen to the systemic immune system (or peripheral lymph nodes) via epithelial cells for the initiation of IgG responses, while the latter sizes are successfully taken up by M cells for the initiation of mucosal IgA antibody responses.

16.2.1 Advances in the Search for New Antigen–Polyester Delivery Systems

Mucosal vaccination was initially dominated by oral vaccines. The development of intranasal vaccines followed, and today it is exploring many different routes for the delivery of mucosal vaccines, including aerosol inhalation, as well as intravaginal, rectal, and sublingual routes (Fig. 16.3), which are potentially able to invoke tolerance, resulting in a nonreactive immune response. This has been a hurdle for mucosal vaccine development and yet the desire to induce protective local and systemic responses, with pain-free and more convenient products. Nevertheless, few mucosal vaccines have reached the marketplace and products are still treated with caution, particularly where live organisms are utilized [71]. Vaccine delivery via most of these routes works well in experimental animal models, but only the oral and intranasal routes have so far been used for licensed human vaccines [48, 50]. Although the route of vaccination is important for a successful result, the selection of an appropriate formulation and adjuvant is also crucial for an effective mucosal response [48, 72]. Antigen protection can be afforded by a diverse range of methods, including encapsulation in lipid vesicles, use of polymeric materials and enteric coatings [71]. Owing to their heterogeneity and sometimes multifunctional activities, there is no ideal way of classifying adjuvants, but it is possible to separate them by their physical characteristics. O'Hagan and his collaborators have suggested that both traditional and novel adjuvants can be separated into two main categories (immune potentiators and delivery systems) on the basis of their dominant mechanisms of action and have stated the importance of this classification for the development and design of optimal vaccine formulations for systemic and mucosal administration [73]. In this chapter, we examine the use of polyester-delivery systems with adjuvant properties as key components in a vaccine strategy that does not require the use of live vectors to overcome tolerance and have exemplified their success in mucosal vaccines by different routes of administration.

16.2.2 Oral Immunization

The immunological characteristics of the gastrointestinal tract (GIT) have focused attention on the development of effective oral vaccines. Despite physical and biological barriers, the GIT is a major route of entry for numerous pathogens. Barriers include epithelial cells joined firmly by tight junction proteins, brush-border microvilli, and a dense layer of mucin. Antimicrobial peptides, such as defensins produced by epithelial cells and Paneth cells, are additional barrier to provide further protection. In addition to these barriers, the GIT tract includes immunological defense system, in particular sIgA, which is predominantly produced at intestinal mucosa by the harmonious interaction between epithelial cells and mucosal lymphocytes and blocks microbial infections by inhibiting adherence of mucosal pathogens at the intestinal lumen to host epithelial cells. sIgA can also neutralize toxins produced by gut pathogens by binding to biologically active sites of toxins [66].

Oral vaccine research represents the major challenge not only due to harsh gut milieu, which weakens the antigenicity of antigens that are delivered in soluble form, but also due to mucosal tolerance, which defends against unsought immune responses to digested antigens [48, 50, 51, 72]. Orally delivered vaccines have several advantages over other routes of antigen delivery, including convenience, cost-effectiveness, needle-free delivery, easy and comfortable administration, and the possibility of self-delivery. Currently, oral immunization has been shown to provide protection against a variety of bacterial pathogens, including *Vibrio cholerae*, *Salmonella enterica* serovar Typhi, and *Borrelia burgdorferi* [74, 75]. Protection of mice against *B. Burgdorferi* infection by oral immunization has been achieved using either *Lactobacillus plantarum* [74] or *Escherichia coli* [75] as delivery vehicles for the immunogen OspA.

However, the hostile environment of the GIT (low pH, presence of digestive enzymes, and the detergent activity of bile salts) often makes it difficult to induce protective immune responses by oral vaccination with antigen alone and antigen degradation can occur and their conformational epitopes destroyed. Additionally, effective oral delivery of antigen to the induction site of the mucosal immune system (e.g., GALT) is made difficult by the significant dilution, which creates a need of higher quantity of antigen/adjuvant formulation [51, 52, 66, 71]. Further, physical barriers, such as mucus and the tight junctions between the epithelial cells prevent the effective delivery of vaccine antigen. To overcome these obstacles, to provide antigen protection, and to potentiate a high enough response, effort has focused on development of effective antigen delivery systems for efficient oral vaccination [66]. Most importantly, oral vaccination through carriers can induce both mucosal and systemic immunity, making particulate antigens more effective than soluble ones due to an effective protection of the antigen until delivered to the immune cells from the harsh conditions of the GIT environment [48, 52, 66, 76]. In addition, as described before, particulate antigens are preferentially taken up in the GALT, especially by M cells serving as a gateway of the mucosal immune system, thus enhancing their antigenic activity. Several systems have been developed for targeting vaccine antigen selectively to the M cells in the follicle-associated epithelium of GALT [51, 66, 77]. A variety of biodegradable antigen delivery systems have been developed for oral vaccines. These include incorporation of antigens into polyester-based particles (e.g., PLGA, polyglycolic acid (PGA), PLA, PCL, polyethylene terephthalate [PET]), since these materials are stable in the GIT, interact with the intestinal epithelium and are made of safe materials and can be formulated for controlled drug release [78].

A pioneering work in the field of nanosystems for oral peptide immunization is the one reported by Damgé et al., in which the authors showed that poly(alkylcyanoacrylate) nanocapsules were able to increase the oral absorption of insulin. The authors attributed the success of these nanocapsules to the protection of the peptide against degradation [79, 80]. Lavelle *et al.* microencapsulated a model protein antigen, human gamma globulin (HGG), in PLGA MPs and administered them orally to rainbow trout. They proved that the association of HGG with PLGA MPs increased the retention time of the antigen in the stomach and delayed its entry into the intestinal region. After oral immunization, specific antibody was detected in the intestinal mucus of fish [81]. In another study performed by Ma *et al.*, PLGA-lipid NPs were prepared and conjugated to *Ulex europaeus* agglutinin-1 containing a Toll-like receptor (TLR)-agonist monophosphoryl lipid A (MPLA) as an oral vaccine delivery system. The prepared NPs could be effectively captured by mucosal DC after transportation by M cells. Mucosal IgA and serum IgG antibodies were stimulated after *in vivo* vaccination of the oral formulations [82]. Jung and his coworkers associated tetanus toxoid (TT) in sulfobutylated poly(vinyl alcohol)-graft-poly(lactide-*co*-glycolide) (SB(43)-PVAL-g-PLGA) NPs through an adsorption procedure, which led to an antigen delivery system with potential for oral vaccination. Results showed that TT loaded SB(43)-PVAL-g-PLGA NPs induced serum IgG and IgA immune responses in mice in a reproducible manner. They also concluded that NPs' surface properties, hydrophobicity, and charge also may have an important role in NPs absorption and induction of immune responses, since hydrophobic poly(styrene) NPs, showing the highest uptake, are selectively targeted to the M cell surface of the Payers patches, while more hydrophilic NPs are not only taken up via the M cells, but also via the normal intestinal enterocytes [83]. The same conclusions were obtained by Nayak and his collaborators, since they developed PLA and PLGA particles with rotavirus (strain SA11) entrapped and, after a single dose oral immunization with 20 μg of antigen entrapped in particles, they observed that the carriers improved and long-lasting IgA and IgG antibody titer in comparison to the soluble antigen [84]. However, there are some authors that refer that PLA/PLGA NPs have got limitation in oral vaccine delivery because of their sensitivity to harsh GIT. The aim of the study performed by Jain *et al.* was to improve the stability of PLA NPs in such environment by copolymerizing PLA with PEG.

For that, NPs were formulated using different block copolymers AB, ABA, and BAB (where "A" is PLA and "B" is PEG) encapsulating hepatitis B surface antigen (HBsAg) to evaluate their adjuvancity in generating immune response after oral administration. PLA NPs could not generate an effective immune response due to stability issues. On the other hand, oral administration of copolymeric NPs exhibited effective levels of humoral immunity along with the mucosal (sIgA) and cellular immune response (Th1). The results of *in vitro* and *in vivo* studies demonstrate that BAB NPs depict enhanced mucosal uptake leading to effective immune response as compared to other copolymeric NPs [85]. PLA/PLGA NPs have also been modified at their surface with specific ligands, such as RGD molecules [86] or lectins [87] in order to increase their uptake by the M cells associated to the intestinal mucosa. For example, in the work of Garinot and coworkers, ovalbumin (OVA) antigen was encapsulated in PLGA NPs surface-modified with RGD. Although an increased uptake of the modified particles *in vitro* and an effective targeting *in vivo* of the M cells were observed, the improvement of immune response was found to be minimal. A possible explanation for these results is the degradation of the ligand by the harsh conditions of the GIT. To avoid the ligand degradation in the stomach or in the gut, they proposed a novel strategy where non-peptidic β_1 integrin ligand could be used and grafted on the NPs in the place of RGD peptides [86]. On the other hand, Gupta and coworkers modified PLGA NPs with the UEA-1 lectin for the oral delivery of the recombinant hepatitis B surface antigen (rHBsAg). In this case, the surface modification resulted in a significantly enhanced immune response. Nevertheless, carbohydrates present in ingested food can complex with lectins and prevent its targeting activity which can be a source of interference in a clinical application of such systems [87]. Overall, it has been demonstrated that oral immunization with antigen-loaded NPs and MPs induces mucosal IgA and systemic IgG antibodies responses, providing a complete immune response. Besides protecting the antigen against the harsh environment of the GI tract, carriers are efficiently taken up by M cells, key players of the mucosal immunity induction. In addition, biodegradable particles allow a sustained release of the antigen, increasing the duration of the contact between antigen and immune cells, thus favoring an effective immune response [86].

16.2.3 Nasal Immunization

Many diseases, such as influenza, respiratory syncytial virus infection, measles, and meningitis, are associated with the entry of microorganisms across the respiratory mucosal surfaces [88]. Therefore, immunization across the nasal mucosa is gaining increasing attention as an alternative to the oral route due to important benefits of this modality of administration [51, 54, 78]. This is due to the greater permeability, high epithelium vascularization and the presence of numerous microvilli covering the epithelium, which generates a large absorption surface. Moreover, through this route, it is possible to avoid the enzymatic or acid degradation and the first-pass hepatic mechanism associated to the oral administration route, as the blood goes directly from the nose to the systemic circulation [56, 78, 89–91], assuming a greater importance as an alternative non-injectable route for local and systemic immunization. Moreover, in recent years it has emerged as a promising approach for the central nervous system (CNC) delivery of soluble and particulate drugs with low molecular weight, via the olfactory neuroepithelium, involving paracellular, transcellular and/or neuronal transport [56, 60, 91]. Furthermore, intranasal immunization efficiently stimulates a protective immune response in the lungs and upper respiratory tract and at distant sites, such as the gastric and genital tract mucosa, but it is rather poor at stimulating intestinal immune responses [48].

A range of different nasal vaccine systems has been described in the literature, mostly of them aimed for exploiting the advantage of a rapid onset of action when administered via this route [60, 92–94]. Nevertheless, the first evidence that the nasal administration of antigens can be a useful route of immunization has been provided by Peters and Allison (in 1929), who investigated the possibility of inducing immunity to scarlet fever by repeated applications of erythrogenic toxin through the nasal mucosa [95]. Additional examples of human efficacy testing of intranasal vaccines includes those targeted against adenovirus-vectored influenza [96], influenza A and B [97, 98] and a combination respiratory syncytial virus (RSV). More recently, Weiner *et al.* have suggested a novel mucosal immunological approach to Alzheimer's disease. With this study, it was possible to state that after nasal administration of amyloid-beta peptide cerebral amyloid burden in a mouse model of the disease decreased [11].

As mentioned before, the nasal cavity is an attractive route for immunization, but there are important issues to be taken into account for preparation of nasal formulations, namely the tightly impermeable epithelial cell layers, as well as the short residence time of formulations in the nasal cavity due to mucociliary. Thus, in order to overcome these obstacles, mucoadhesive MPs and NPs have been applied to enhance antigen absorption/uptake [59, 90, 99, 100]. These particulate carrier systems prolong the residence time of the proteins in the nasal cavity and significantly improve the uptake of antigen-loaded particles by epithelial cells. Subsequently, they activate the immune cells in the NALT and drain lymph nodes [101]. Consequently, these delivery systems should facilitate passage of free peptides/proteins through paracellular pathways, while uptake of the proteins by APC should be avoided [102].

Among the various approaches investigated until now, extensive research has been conducted on the development of biodegradable polyester particles for intranasal administration of antigens that are supposed to cross the nasal mucosa, mostly of them based on PLG [103–107] and PLGA [108–110]. According to these previous reports, this particulate antigenic material can be sampled by specialized cells that are similar in appearance to M cells, which overlay the NALT and then be transported to the posterior cervical lymph nodes, where the immune response is initiated (Fig. 16.5) [54, 111, 112]. Additionally, associated to the nasal mucosa there are underlying blood vessels, lymph vessels and lymphoid cells to which the particulate antigen may have direct access if it can cross the nasal epithelium [54, 113]. Briefly, nasal administration of particulate antigens results in uptake mainly by the M cells, while soluble antigens are mainly absorbed at the nasal epithelium. Antigens of the first type will be processed at the NALT and preferentially drained to the posterior cervical lymph nodes (PCLN). After uptake at the nasal epithelium, soluble antigens will be carried by antigen presenting cells to the superficial cervical lymph nodes (SCLN), which in turn drain to the PCLN. At the SCLN, soluble antigens may induce a systemic immune response or a status of specific tolerance. On the other hand, as PCLN are involved in the enhancement of a secretory immune response, these antigens can also induce this type of immunity. The final result of

NALT stimulation will depend on the balance between the activation in the posterior or superficial cervical lymph nodes. Therefore, given the sites of preferential uptake, the nature of the antigen plays an important role in the ultimate response [56, 90, 102, 114, 115].

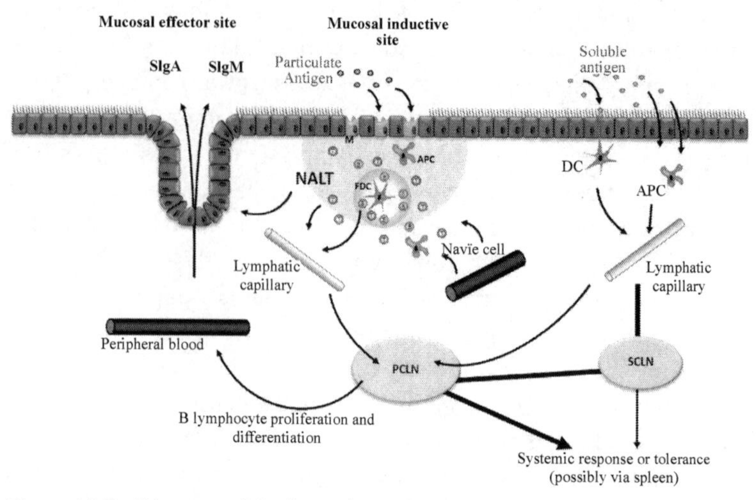

Figure 16.5 Diagram of the hypothetical mechanism of antigen processing and eliciting of the immune response after antigenic stimulation of the NALT. Adapted from Ref. 112, Copyright 1992, with permission from Elsevier. Legend: APC, antigen-presenting cells; FDC, follicular dendritic cells; M, M cells; PCLN, posterior cervical lymph node; SCLN, superficial cervical lymph node.

Almeida *et al.* for the first time reported the potential application of nanocarrier mediated nasal vaccine delivery. In this study, TT antigen adsorbed onto PLA microspheres were administered for nasal immunization in guinea pigs. The obtained data suggested the induction of potent immune response against TT when compared with the free antigens. The high level of systemic IgG titer was found in the group treated with the microsphere adsorbed TT, whereas the free antigen produced an immune response similar to that found in non-treated animals [103]. A few years later, Florindo *et al.* synthesized *Streptococcus equi*-loaded glycol chitosan surface-coated mucoadhesive PLA nanospheres. After nasal immunization, nanospheres evoked significant humoral, cellular, mucosal as well as a balanced Th1/Th2 immune response without coadministration of other adjuvants [116–118]. In another study, a strong evidence for dissemination of antigen-specific antibody-secreting cells from NALT to the cervical lymph nodes and spleen following intranasal immunizations has been provided by Heritage *et al.* [101]. These local and systemic humoral responses were generated by entrapment of human serum albumin (HSA) in polymer-grafted MPs [3-(triethoxysilyl)propyl-terminated polydimethylsiloxane (TS-PDMS)]. McDermott *et al.* reported that polymer-grafted starch MPs have been used as an alternative to PLG particles and were shown to effectively deliver antigens following intranasal immunization and elicit local and systemic humoral responses [119]. In another study, a single intranasal immunization with a *Schistosoma mansoni* antigen entrapped in PLG or PCL MPs resulted in sustained serum IgG responses as well as in both serum and BAL fluid IgA responses. However, this vaccine strategy failed to induce IgA responses in serum or BAL fluids following oral immunization. Interestingly, only PLG-entrapped, and not PCL-entrapped vaccine resulted in strong neutralizing antibody responses, following either intranasal or oral immunization. Moreover, the humoral responses were detectable earlier following PLG vs. PCL immunizations, presumably due to the physicochemical differences between the two polymers and different rates of antigen release [120].

More recently, Vajdy and O'Hagan explored the potential of cationic PLG MPs to induce local and systemic cell-mediated immunity following intranasal immunization with DNA encoding HIV-1 gag. In addition, they found that cationic PLG MPs with adsorbed DNA induced enhanced local and systemic cell-mediated, as well as humoral, immunity against HIV-1 gag. Thus, intranasal immunizations with DNA adsorbed onto cationic PLG appeared to be a novel approach for induction of enhanced local and systemic cell mediated as well as humoral immune responses [121]. Moreover, Csaba *et al.* produced new particle-mediated delivery systems based on polyoxyethylene derivatives, known as poloxamers and poloxamines, blended with PLGA, which create NPs with improved particle stability, payload protection, and controlled release after mucosal administration. In these particles, they encapsulated DNA, which were able to induce a strong and long-lasting serum IgG response after intranasal administration as compared to both naked DNA or DNA encapsulated in PLGA-poloxamine NPs. The different hydrophobic characteristics of the two polymer blend particles may be responsible for varying release rates and potential differences in immune responses elicited [122]. Table 16.2 lists some of the more important studies on the nasal administration of vaccines based on PLGA particles in recent years.

An alternative system that could hold promise for enhancing the uptake of particles when administered for nasal immunization is the one consisting of a polyester core and a chitosan-corona. TT was also introduced in this new kind of system in order to investigate its efficacy for the nasal transport of vaccines. The results made clear that, as expected, the chitosan coating helped the transport of the nanoencapsulated antigen across the nasal mucosa [78, 128]. The positive effect of chitosan in improving the nasal transport of macromolecules associated to nanosystems may be understood on the basis of the facilitated interaction and internalization of these nanosystems in the nasal epithelium, as it was shown by confocal fluorescence microscopy [129]. The possible effect of chitosan in opening the tight junction between epithelial cells is currently under investigation. However, the results obtained until now suggest that this effect might be dependent on the physical presentation of chitosan in the nanosystem (as a solid nanomatrix or as a soluble coating) [78]. Another option is taking into account the positive effect of PEG in preserving the stability of nanosystems in contact with mucosal components. For that, Tobío et al. produced MPs and NPs consisting of a hydrophobic core composed of PLA, in which the active molecule is well entrapped, and surrounded by a PEG corona for nasal protein administration. The results showed that, while nude NPs (without PEG corona) were totally inefficient at transporting the protein across the nasal mucosa, those with a PEG-corona led to a significant absorption of the protein [130]. With this interesting observation, it is possible to conclude that the PEG coating around the PLA particles might have a role in facilitating the interaction with the mucus layer and further transport of the encapsulated antigen across the nasal mucosal surfaces [54, 130]. The idea of using PEGylated PLA came from the observation that PLA nanoparticles aggregated upon contact with mucosal fluids and, therefore, PEGylation was conceived as a way to preserve the stability of NPs [130]. The results of the absorption and biodistribution of TT associated to these NPs clearly evidenced the positive effect of the PEGylation. In fact, it was

Table 16.2 Studies using particles made of polyesters as vaccine carriers for nasal administration

Particle composition	Antigen	Cell line or animal model	Results	Ref.
PLGA	Synthetic bovine parainfluenza virus type-3 (BPI3V) peptide motifs	BALB/c mice	Induction of stronge IgG antibody	[123]
PLGA	Inactivated PRRS virus	Pigs	IgG1 and IgG2 antibody, Th1 and Th2 cytokines	[124]
PLGA-coated gelatin	TT antigen	BALB/c mice	Humoral, cellular and mucosal immunity	[125]
Phosphoryl-choline linked to protein PLGA microspheres	*S. pneumoniae*	Mice	Protection against lethal challenge	[126]
Sulphobutylated poly(vinyl alcohol)-graf-PLGA NPs	TT	Mice	The smallest particles induced the most significant antibody responses	[83]
PLA–PEG NPs	TT	Mice	PEG-coated NPs induced significantly greater immune responses than PLA NPs	[54, 127]
PLA–PEG NPs	pDNA encoding β-galactosidase	Mice	PEG-coated NPs induced significantly greater immune responses against the encoded protein than the control plasmid solution	[54]

observed that the antigen could penetrate through the nasal mucosa much more efficiently when encapsulated into PLA-PEG particles. These results were corroborated by a work of Vila *et al.* in which it was found that both particle size and PEGylation degree influenced the uptake by the nasal mucosa and the subsequent biodistribution of the associated antigen [127]. The same authors found that this improved transport of the TT antigen was the explanation for the important and long-lasting response [54]. These results emphasize the importance of an adequate design of the delivery system to achieve an effective vaccine formulation. The same results were obtained by Lai *et al.* [131]. Another alternative strategy to the use of PLA-PEG–based NPs is to attach PLGA to a hydrophilic central chain of polyvinyl alcohol (PVA). The properties of this polymer can be modulated by substituting part of the PVA chain with negatively (sulfobutyl-PVA) or positively (diethylaminoethyl-PVA) charged derivatives. These studies revealed that branched copolymers based on sulfobutylated-PVA were promising for the nasal delivery of the model antigen TT, resulting in enhanced IgG and IgA antibody production as compared to a control TT solution in mice [78]. In fact, those absorption enhancers can act through a single or a combination of several mechanisms, such as the increase in membrane fluidity, decrease in mucous layer viscosity, disruption of tight junctions, inhibition of proteolytic enzymes, increase in blood flow, dissociation of protein aggregates, and increase in paracellular or transcellular transport [56, 91]. In addition, the presence of specific targeting ligands is also an important feature of nasally administered particulate drug or vaccine carriers. Monoclonal antibodies with specificity for M cells, or lectins, which bind to specific carbohydrate residues found on M cells, can increase the uptake of micro/nanoparticulate delivery systems [56, 132]. Furthermore, the efficiency of nanoparticulate nasal vaccine formulations based on PLGA or PCL, either with the antigens entrapped or adsorbed, is significantly increased by adding permeation enhancers to the formulations, such as chitosan, oleic acid and spermine, that ideally reversibly open the tight junctions are added to the formulations to facilitate the transport of macromolecules across the epithelium to the sub-mucosa and subsequently to the systemic and/or the lymphatic circulation [56, 102].

16.2.4 Ocular Immunization

Beyond oral and nasal routes, the ocular surface epithelium is also a possible route for immunization, since it is an important entry point for infectious agents [133, 134]. This surface, also known as conjunctiva, plays a central role in the protection of the eye from environmental factors and pathogens: first, acting as a physical barrier by defending the eye from injury and invasion of pathogens and allergens, and second by contributing to the homeostasis of the ocular surface [135]. The ocular mucosal surface shares many common immunologic features with other mucosal tissues, as it contains APCs, including DCs, Langerhans cells (LCs), Mφ, and B cells, but also CD4$^+$ and CD8$^+$ T cells and mast cells [136, 137] (Fig. 16.6). Moreover, the conjunctival epithelium is also able to produce pro-inflammatory cytokines, chemokines and antimicrobial peptides, which promotes the elimination of pathogens, and the protection of the eye from uncontrolled inflammatory responses [138–140]. In addition to conjunctival epithelium, many other tissues contribute to the ocular mucosal immune system, including the corneal epithelium, corneal stroma, the CALT and the lacrimal glands [135, 141]. Moreover, the mucosal immune system of this organ involves a complex set of interactions between local and systemic immunocompetent and parenchymal cells that communicate through specialized cell surface receptors and soluble mediators to protect the surface of the eye [135, 141].

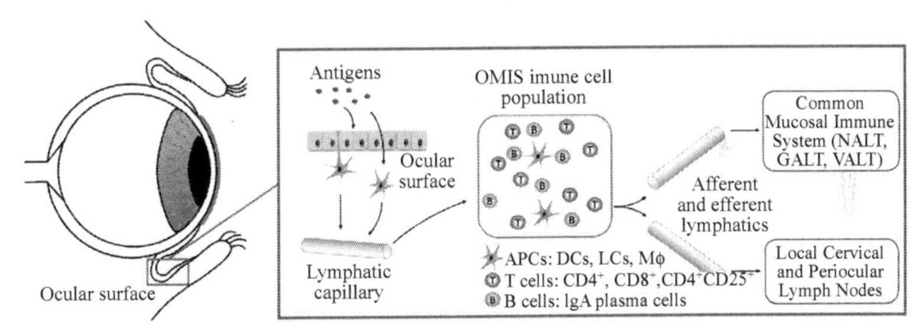

Figure 16.6 Once detected at the ocular surface, the antigen is taken up by local APCs, and home to regional lymph nodes where they stimulate effector T cells. The stimulation of the ocular mucosal immune system may also result in the induction of remote systemic and mucosal (e.g., NALT and GALT) immune responses.

In theory, the conjunctiva is an attractive choice for mucosal immunization, particularly against ocular infections, as eye drops are easily administered. In addition, and as described above, the conjunctiva is interconnected with the nasal mucosa via the tear ducts, whereby the administration of antigens to the conjunctival route would additionally drain to the NALT [141, 142]. However, this route has only been exploited in veterinary applications in the use of live attenuated vaccines, which have proven to be efficient against various infectious diseases, and some of them have effectively been licensed [143–145].

As described in the introduction, live attenuated vaccines have certain limitations for human uses; thus, particulate-based vaccines might be a safe alternative. However, very little is known about the conjunctival immunization and its immune response profile concerning this type of vaccines. Therefore, only a few studies using particulate-based vaccines in the ocular mucosal route have been published so far. One of these studies was performed by Rafferty *et al.* This research group formulated PLGA MPs containing dinitrophenylated bovine serum albumin and cytokines IL-5 and IL-6. Data obtained from this study demonstrated that ocular delivery of antigen/cytokine MPs can potentiate long-term mucosal antibody responses (serum IgG and tear IgA responses, up to 140 days post tertiary immunization) at ocular target and distal effector sites, as well as elicit circulating antibodies [146]. Hu and coauthors produced an ocular nanoparticulate-based vaccine against herpes simplex virus type 1 (HSV-1), which is responsible for the most common cause of corneal blindness in the world. In this study, NPs of Fe_3O_4 coated with glutamic acid, containing DNA plasmid pRSC-gD-IL-21 and polyethylenimine (PEI) were prepared and immunized in mice by ocular mucosal administration. As previously shown by Rafferty *et al.*, the results here obtained also demonstrated that NPs were able to induce strong specific immune responses (with high levels of specific neutralizing antibody, sIgA in tears, and IFN-γ, IL-4 in serum, and an increase in the cytotoxicity of natural killer cells) and effective inhibition of the disease in a HSV-1 infected murine model [147]. More recently, Nagesh and coauthors developed a PLGA MP-based vaccine containing DNA in order to immunize chickens against the infectious bursal disease virus (IBDV) [148]. This virus causes severe lymphoid cell depletion in the bursa of Fabricius (BF), having as main target the immature intrabursal B cells [149]. After administration by eye drop route, this vaccine loaded with CpG ODN or chicken IL-2 improved protection against IBDV in chickens. In addition, chickens treated with this vaccine showed less pathological and histopathological bursal lesions, a reduced IBDV antigen load as well as T-cell influx into the BF compared to the non-treated animals.

These studies reinforce the idea that ocular mucosal route and particulate-based vaccines may be a good alternative to conventional vaccination methods.

16.2.5 Vaginal Immunization

The vagina has been used as a delivery route since ancient times. In the last two decades, the vaginal route has been rediscovered as a potential route for systemic and mucosal delivery of peptides and other therapeutically important macromolecules, as reviewed in [150]. Traditionally, solutions, suppositories, gels, foams, and tablets have been used as vaginal formulations. Compared with conventional delivery routes, such as oral and nasal, the most advantageous features of the vaginal route for drug and peptide delivery are the ability to overcome first-pass metabolism and to avoid exposing vaccines to hostile environment, respectively, but also the ease of administration [151].

As other mucosal surfaces, the vagina is an important entry point for potential infectious agents. In this regard, several vaginal vaccine formulations are being developed against a variety of pathogens, including the HIV, HSV, and the human papilloma virus (HPV). MP- and NP-based vaccine composed by polyesters have already been tested for the immunization and treatment of these diseases. Indeed, Zhu and coauthors developed a targeted NP-releasing MP system to induce immunization against HIV in the female reproductive tract. These complex particles were made of PLGA containing a $CD4^+$ T-cell helper epitope fused with an HIV Env $CD8^+$ cytotoxic T-lymphocyte epitope and TLR ligands (MALP2, poly(I:C) and CpG ODN). After oral administration, this particulate system allowed the controlled release of the entrapped antigens over extended time periods, inducing T-cell and antibody immunities in the vaginal tract [152]. Kuo-Haller *et al.* also developed a vaccine delivery system composed of PLGA. These researchers observed that PLGA particles containing ovalbumin as model antigen administered by oral and nasal routes were able to induce sustained mucosal and systemic immunity in the mouse reproductive tract, presenting high vaginal antibody titers [153]. This polymer was also used by Thomas and coauthors to formulate microspheres of hepatitis B surface antigen (HBsAg) by a double-emulsion solvent-evaporation method with PEI in the external aqueous phase. In this study, the immunization was performed after pulmonary administration of the formulations to female rats and the immune response was monitored by measuring IgG levels in serum and sIgA levels in vaginal lavage fluids. These PLGA-PEI particles showed a continuous release of antigen over a period of 28–42 days and increased levels of IgG in serum and sIgA in vaginal lavage, which demonstrates its efficacy in eliciting systemic and mucosal immune responses. Despite a few more MP- and NP-based vaccine composed by polyester were tested for the immunization of the female reproductive tract [154–157], to our knowledge there is no polyester-based vaccine directly administered in the vaginal route. Indeed, several issues, which include cultural sensitivity, personal hygiene, gender specificity, local irritation, and influence of sexual intercourse, have to be taken in consideration during the design of a vaginal formulation and may be the cause of this poor clinical interest [150].

16.2.6 Rectal Immunization

The rectal mucosa of mice and humans are rich in antigen-transporting M cells and organized mucosal lymphoid follicles with characteristics of an immune-inductive site [158]. Initially, rectal mucosa was essentially used as a possible immunization route to establish protective immune responses in the female genital tract against sexually transmitted diseases [159–161]. However, it has been shown that rectal immunization of mice and macaques generate specific antibodies secretion at distant sites via the common mucosal immune system, but also locally [162–164]. In addition, there are evidences that local exposure to antigen in this route can result in much higher levels of sIgA in the region of exposure than at distant sites [165]. Therefore, these tissues start to be seen as a promising site for vaccine delivery, which may be convenient to exploit, particularly in young children. Moreover, and comparing to oral route, immunization via the rectal mucosa would avoid exposing vaccines to hostile environment, such as the low pH and abundant proteolytic enzymes present in the upper gastrointestinal tract [166].

This route might be successfully exploited for vaccine delivery using particulate delivery systems. However, to our knowledge, no studies using polyester-based delivery systems have been administered directly in the rectum. At the same time, there are serious limitations associated with the potential exploitation of the rectal route of immunization. Indeed, the vaccine may be expelled before it has time to be effective and this may be difficult to control. Additionally, there may be considerable cultural resistance to the acceptance of formulations delivered by rectal route [167]. Owing to these limitations, rectal immunization has not been studied as intensely as other mucosal sites. However, a few number of MPs and NPs administered in other routes and able to induce effective immune protection in the rectum have been performed so far [152, 156, 168]. For instance, Singh *et al.* demonstrated that DNA encoding HIV-1 gag adsorbed onto the surface of PLGA MPs were able to induce gag-specific T cell- and antibody-mediated responses in rectal mucosa after nasal immunization [156].

16.3 Conclusions

The mucosal immune system can induce humoral and cell-mediated immune responses in both systemic compartment and mucosal surfaces, being an ideal target for vaccination. However, mucosal route seems to be problematic because of the presence of enzymes, acidic conditions, insufficient absorption and degradation of antigens. Thus, in order to overcome these obstacles, particle-mediated carrier systems, such as nano- and microparticles have been developed. These carriers can provide protection of the immunogenic material and adjuvants during delivery, improve tissue targeting, and allow intracellular delivery. Various materials including polymers, lipids and metals can be used for the manufacture of particulate carrier systems. By careful consideration of the carrier, influenced by the type of antigen payload, the targeted tissue, as well as the desired effect (e.g., controlled release), it is possible to design and produce a successful and efficient delivery system. Polyester polymers are the primary candidates to develop carrier systems for immunization because of their biocompatible and biodegradable characteristics. These materials are degraded *in vivo*, either enzymatically or non-enzymatically or both to produce biocompatible, toxicologically safe by-products, which are further eliminated by the normal metabolic pathways. The nasal and oral routes are the most studied with highly successful results for several particulate-based vaccines. Nasal immunization is attractive due to its easy administration promoting a faster and stronger immunity and stimulation of Th1, Th2, and cytotoxic T lymphocytes. Moreover, the nasal mucosa allows not only the induction of immune responses within upper and lower respiratory tract, but also at the GIT and genital surfaces. In addition to the high patient compliance, oral immunization induces protective immune responses in GIT, salivary and mammary glands. Although ocular, vaginal, and rectal routes have been less explored for immunization purposes, these routes have been assessed as potential alternatives to overcome the enzymatic activity present at nasal and oral mucosae. Ocular mucosal surface is apparently an excellent via for immunization due to the easiness of the eye-drop administration, having shown to potentiate long-term mucosal antibody responses at ocular target, but also at distal effector sites. However, this route has been less well studied and few approaches have been performed using particulate-based vaccines. The vaginal and rectal mucosae are highly relevant for HIV and causative agents of other sexually transmitted diseases, having already demonstrated ability to induce both local and distal immune responses. However, there is poor patient compliance for general vaginal and rectal applications. It has been shown that these two immunization routes are often less immunogenic, and therefore require the addition of strong adjuvants in the vaccine formulations. For these reason, very few immunization studies using particle-based vaccines have been performed in these administration routes.

References

1. McGhee, J. R., and K. Fujihashi (2012). Inside the mucosal immune system. *PLoS Biol*, **10**(9), e1001397.

2. Holmgren, J., and C. Czerkinsky (2005). Mucosal immunity and vaccines. *Nat Med*, **11**, S45–S53.
3. Neutra, M. R., E. Pringault, and J.-P. Kraehenbuhl (1996). Antigen sampling across epithelial barriers and induction of mucosal immune responses. *Annu Rev Immunol*, **14**(1), 275–300.
4. Rhee, J. H., S. E. Lee, and S. Y. Kim (2012). Mucosal vaccine adjuvants update. *Clin Exp Vaccine Res*, **1**(1), 50–63.
5. Plotkin, S. A., and S. L. Plotkin (2011). The development of vaccines: how the past led to the future. *Nat Rev Microbiol*, **9**(12), 889–893.
6. Silva, J. M., *et al.* (2015). *In vivo* delivery of peptides and toll-like receptor ligands by mannose-functionalized polymeric nanoparticles induces prophylactic and therapeutic anti-tumor immune responses in a melanoma model. *J Control Release*, **198**, 91–103.
7. Cheever, M. A., and C. S. Higano (2011). PROVENGE (Sipuleucel-T) in prostate cancer: the first FDA-approved therapeutic cancer vaccine. *Clin Cancer Res*, **17**(11), 3520–3526.
8. Conniot, J., *et al.* (2014). Cancer immunotherapy: nanodelivery approaches for immune cell targeting and tracking. *Front Chem*, **2**, 105.
9. Le Moigne, V., *et al.* (2015). Bacterial phospholipases C as vaccine candidate antigens against cystic fibrosis respiratory pathogens: the Mycobacterium abscessus model. *Vaccine*, **33**(18), 2118–2124.
10. Grimwood, K., *et al.* (2015). Vaccination against respiratory Pseudomonas aeruginosa infection. *Hum Vaccin Immunother*, **11**(1), 14–20.
11. Lemere, C. A., *et al.* (2001). Nasal vaccination with β-amyloid peptide for the treatment of Alzheimer's disease. *DNA Cell Biol*, **20**(11), 705–711.
12. Alves, R. P., *et al.* (2014). Alzheimer's disease: is a vaccine possible? *Braz J Med Biol Res*, **47**(6), 438–444.
13. Guo, W., *et al.* (2013). A new DNA vaccine fused with the C3d-p28 induces a Th2 immune response against amyloid-beta. *Neural Regen Res*, **8**(27), 2581–2590.
14. Burton, D. R., *et al.* (2004). HIV vaccine design and the neutralizing antibody problem. *Nat Immunol*, **5**(3), 233–236.
15. Rose, N. F., *et al.* (2001). An effective AIDS vaccine based on live attenuated vesicular stomatitis virus recombinants. *Cell*, **106**(5), 539–549.

16. Locci, M., *et al.* (2013). Human circulating PD-1+CXCR3−CXCR5+ memory Tfh cells are highly functional and correlate with broadly neutralizing HIV antibody responses. *Immunity*, **39**(4), 758–769.
17. Rezza, G. (2015). A vaccine against Ebola: problems and opportunities. *Hum Vaccin Immunother*, **11**(5), 1258–1260.
18. Chowell, G., and C. Viboud (2015). Ebola vaccine trials: a race against the clock. *Lancet Infect Dis*, **15**(6), 624–626.
19. Dimier-Poisson, I., *et al.* (2015). Porous nanoparticles as delivery system of complex antigens for an effective vaccine against acute and chronic Toxoplasma gondii infection. *Biomaterials*, **50**, 164–175.
20. Victoria, J. G., *et al.* (2010). Viral nucleic acids in live-attenuated vaccines: detection of minority variants and an adventitious virus. *J Virol*, **84**(12), 6033–6040.
21. McCormick, M. C., *et al.* (2002). *Immunization Safety Review: SV40 Contamination of Polio Vaccine and Cancer*. National Academies Press.
22. Harasawa, R., and T. Tomiyama (1994). Evidence of pestivirus RNA in human virus vaccines. *J Clin Microbiol*, **32**(6), 1604–1605.
23. Cutrone, R., *et al.* (2005). Some oral poliovirus vaccines were contaminated with infectious SV40 after 1961. *Cancer Res*, **65**(22), 10273–10279.
24. Pliaka, V., Z. Kyriakopoulou, and P. Markoulatos (2012). Risks associated with the use of live-attenuated vaccine poliovirus strains and the strategies for control and eradication of paralytic poliomyelitis. *Expert Rev Vaccines*, **11**(5), 609–628.
25. Lee, S.-W., *et al.* (2012). Attenuated vaccines can recombine to form virulent field viruses. *Science*, **337**(6091), 188–188.
26. Akagi, T., M. Baba, and M. Akashi (2012). Biodegradable nanoparticles as vaccine adjuvants and delivery systems: regulation of immune responses by nanoparticle-based vaccine, in *Polymers in Nanomedicine*, S. Kunugi and T. Yamaoka, eds. (Springer, Berlin, Heidelberg), 31–64.
27. Silva, J. M., *et al.* (2013). Immune system targeting by biodegradable nanoparticles for cancer vaccines. *J Control Release*, **168**(2), 179–199.
28. Klippstein, R., and D. Pozo (2010). Nanotechnology-based manipulation of dendritic cells for enhanced immunotherapy strategies. *Nanomedicine*, **6**(4), 523–529.
29. Silva, J. M., *et al.* (2014). Development of functionalized nanoparticles for vaccine delivery to dendritic cells: a mechanistic approach. *Nanomedicine*, **9**(17), 2639–2656.

30. Gregory, A. E., R. Titball, and D. Williamson (2013). Vaccine delivery using nanoparticles. *Front Cell Infect Microbiol*, **3**, 13.
31. Kumari, A., S. K. Yadav, and S. C. Yadav (2010). Biodegradable polymeric nanoparticles based drug delivery systems. *Colloids Surf B Biointerfaces*, **75**(1), 1–18.
32. Elsabahy, M., and K. L. Wooley (2012). Design of polymeric nanoparticles for biomedical delivery applications. *Chem Soc Rev*, **41**(7), 2545–2561.
33. Albertsson, A.-C., and I. Varma (2002). Aliphatic polyesters: synthesis, properties and applications, in *Degradable Aliphatic Polyesters*. (Springer, Berlin, Heidelberg), 1–40.
34. Zhao, L., *et al.* (2014). A review of polypeptide-based polymersomes. *Biomaterials*, **35**(4), 1284–1301.
35. Lasprilla, A. J. R., *et al.* (2012). Poly-lactic acid synthesis for application in biomedical devices — a review. *Biotech Adv*, **30**(1), 321–328.
36. Danhier, F., *et al.* (2012). PLGA-based nanoparticles: an overview of biomedical applications. *J Control Release*, **161**(2), 505–522.
37. Hudson, D., and A. Margaritis (2014). Biopolymer nanoparticle production for controlled release of biopharmaceuticals. *Crit Rev Biotechnol*, **34**(2), 161–179.
38. Ali, M., and S. Brocchini (2006). Synthetic approaches to uniform polymers. *Adv Drug Deliv Rev*, **58**(15), 1671–1687.
39. Williams, D. F. (1989). A model for biocompatibility and its evaluation. *J Biomed Eng*, **11**(3), 185–191.
40. de Moraes Porto, I. (2012). Polymer biocompatibility, in *Polymerization*, A. d. S. Gomes, ed. (InTech).
41. Schmalz, G. (2002). Materials science: biological aspects. *J Dent Res*, **81**(10), 660–663.
42. Doherty, P. (2009). Inflammation, carcinogenicity and hypersensitivity, in *Biomedical Materials*, R. Narayan, ed. (Springer, US).
43. Edlund, U., and A. C. Albertsson (2002). Degradable polymer microspheres for controlled drug delivery, in *Degradable Aliphatic Polyesters*. (Springer, Berlin, Heidelberg), 67–112.
44. Arshady, R. (2003). Biodegradable polymers: concepts, criteria, and definitions, in *Biodegradable Polymers, the PBM Series*, R. Arshady, ed. (Citus Ltd, London). pp. 2–34.
45. El-Fattah, A. A., E. R. Kenawy, and S. Kandil (2014). Biodegradable polyesters as biomaterials for biomedical applications. *Int J Chem Appl Biol Sci*, **1**(Suppl S1), 2–11.

46. Seyednejad, H., *et al.* (2011). Functional aliphatic polyesters for biomedical and pharmaceutical applications. *J Control Release,* **152**(1), 168–176.

47. Athanasiou, K. A., G. G. Niederauer, and C. M. Agrawal (1996). Sterilization, toxicity, biocompatibility and clinical applications of polylactic acid/polyglycolic acid copolymers. *Biomaterials,* **17**(2), 93–102.

48. Lycke, N. (2012). Recent progress in mucosal vaccine development: potential and limitations. *Nat Rev Immunol,* **12**(8), 592–605.

49. Woodrow, K. A., K. M. Bennett, and D. D. Lo (2012). Mucosal vaccine design and delivery. *Annu Rev Biomed Eng,* **14**, 17–46.

50. Czerkinsky, C., and J. Holmgren (2012). Mucosal delivery routes for optimal immunization: targeting immunity to the right tissues, in *Mucosal Vaccines,* P. A. Kozlowski, ed. (Springer, Berlin, Heidelberg), 1–18.

51. Sharma, R., *et al.* (2015). Polymer nanotechnology based approaches in mucosal vaccine delivery: Challenges and opportunities. *Biotech Adv,* **33**(1), 64–79.

52. Chadwick, S., C. Kriegel, and M. Amiji (2010). Nanotechnology solutions for mucosal immunization. *Adv Drug Deliv Rev,* **62**(4), 394–407.

53. Chen, K., and A. Cerutti (2010). Vaccination strategies to promote mucosal antibody responses. *Immunity,* **33**(4), 479–491.

54. Vila, A., *et al.* (2004). PEG-PLA nanoparticles as carriers for nasal vaccine delivery. *J Aerosol Med,* **17**(2), 174–185.

55. Jung, T., *et al.* (2000). Biodegradable nanoparticles for oral delivery of peptides: is there a role for polymers to affect mucosal uptake? *Eur J Pharm Biopharm,* **50**(1), 147–160.

56. Almeida, A. J., and H. F. Florindo (2012). Nanocarriers overcoming the nasal barriers: physiological considerations and mechanistic issues, in *Nanostructured Biomaterials for Overcoming Biological Barriers,* Maria Jose Alonso and Noemi S. Csaba, eds. (RSC Publishing), 117–132.

57. Grass, G. M., and J. R. Robinson (1988). Mechanisms of corneal drug penetration II: Ultrastructural analysis of potential pathways for drug movement. *J Pharm Sci,* **77**(1), 15–23.

58. Inagaki, M., *et al.* (1985). Macromolecular permeability of the tight junction of the human nasal mucosa. *Rhinology,* **23**(3), 213–221.

59. Illum, L., *et al.* (2001)., Chitosan as a novel nasal delivery system for vaccines. *Adv Drug Deliv Rev,* **51**(1), 81–96.

60. Illum, L. (2003). Nasal drug delivery—possibilities, problems and solutions. *J Control Release,* 2003. **87**(1), 187–198.

61. Yuki, Y., and H. Kiyono (2003). New generation of mucosal adjuvants for the induction of protective immunity. *Rev Med Virol,* **13**(5), 293–310.

62. Wright, P. F. (2011). Inductive/effector mechanisms for humoral immunity at mucosal sites. *Am J Reprod Immunol,* **65**(3), 248–252.

63. Oyewumi, M. O., A. Kumar, and Z. Cui (2010). Nano-microparticles as immune adjuvants: correlating particle sizes and the resultant immune responses. *Expert Rev Vaccines,* **9**(9), 1095–1107.

64. Kanchan, V., and A. K. Panda (2007). Interactions of antigen-loaded polylactide particles with macrophages and their correlation with the immune response. *Biomaterials,* **28**(35), 5344–5357.

65. Jepson, M. A., *et al.* (2003). Comparison of poly (DL-lactide-*co*-glycolide) and polystyrene microsphere targeting to intestinal M cells. *J Drug Target,* **11**(5), 269–272.

66. Kunisawa, J., Y. Kurashima, and H. Kiyono (2012). Gut-associated lymphoid tissues for the development of oral vaccines. *Adv Drug Deliv Rev,* **64**(6), 523–530.

67. Hillery, A., P. Jani, and A. Florence (1994). Comparative, quantitative study of lymphoid and non-lymphoid uptake of 60 nm polystyrene particles. *J Drug Target,* **2**(2), 151–156.

68. Norris, D. A., N. Puri, and P. J. Sinko (1998). The effect of physical barriers and properties on the oral absorption of particulates. *Adv Drug Deliv Rev,* **34**(2), 135–154.

69. Damgé, C., *et al.* (1996). Intestinal absorption of PLAGA microspheres in the rat. *J Anat,* **189**(Pt 3), 491–501.

70. Tabata, Y., Y. Inoue, and Y. Ikada (1996). Size effect on systemic and mucosal immune responses induced by oral administration of biodegradable microspheres. *Vaccine,* **14**(17), 1677–1685.

71. Mann, J. F., *et al.* (2009). Delivery systems: a vaccine strategy for overcoming mucosal tolerance? *Expert Rev Vaccines,* **8**(1), 103–112.

72. Mowat, A. M., O. R. Millington, and F. G. Chirdo (2004). Anatomical and cellular basis of immunity and tolerance in the intestine. *J Pediatr Gastroenterol Nutr,* **39**, S723–S724.

73. O'Hagan, D. T., and N. M. Valiante (2003). Recent advances in the discovery and delivery of vaccine adjuvants. *Nat Rev Drug Discov,* **2**(9), 727–735.

74. del Rio, B., *et al.* (2008). Oral immunization with recombinant Lactobacillus plantarum induces a protective immune response in mice with Lyme disease. *Clin Vaccine Immunol*, **15**(9), 1429–1435.
75. Gomes-Solecki, M. J., D. R. Brisson, and R. J. Dattwyler (2006). Oral vaccine that breaks the transmission cycle of the Lyme disease spirochete can be delivered via bait. *Vaccine*, **24**(20), 4440–4449.
76. Lourdault, K., *et al.* (2014). Oral immunization with Escherichia coli expressing a lipidated form of LigA protects hamsters against challenge with Leptospira interrogans serovar copenhageni. *Infect Immun*, **82**(2), 893–902.
77. Slütter, B., *et al.* (2010). Conjugation of ovalbumin to trimethyl chitosan improves immunogenicity of the antigen. *J Control Release*, **143**(2), 207–214.
78. Alonso, M. J. (2004). Nanomedicines for overcoming biological barriers. *Biomed Pharmacother*, **58**(3), 168–172.
79. Damgé, C., *et al.* (1988). New approach for oral administration of insulin with polyalkylcyanoacrylate nanocapsules as drug carrier. *Diabetes*, **37**(2), 246–251.
80. Damgé, C., *et al.* (1990). Nanocapsules as carriers for oral peptide delivery. *J Control Release*, **13**(2), 233–239.
81. Lavelle, E., P. Jenkins, and J. Harris (1997). Oral immunization of rainbow trout with antigen microencapsulated in poly (DL-lactide-*co*-glycolide) microparticles. *Vaccine*, **15**(10), 1070–1078.
82. Ma, T., *et al.* (2014). M-cell targeted polymeric lipid nanoparticles containing a toll-like receptor agonist to boost oral immunity. *Int J Pharm*, **473**(1), 296–303.
83. Jung, T., *et al.* (2001). Tetanus toxoid loaded nanoparticles from sulfobutylated poly (vinyl alcohol)-graft-poly (lactide-*co*-glycolide): evaluation of antibody response after oral and nasal application in mice. *Pharm Res*, **18**(3), 352–360.
84. Nayak, B., *et al.* (2009). *Formulation, characterization and evaluation of rotavirus encapsulated PLA and PLGA particles for oral vaccination. J Microencapsulation*, **26**(2), 154–165.
85. Jain, A. K., *et al.* (2010). PEG–PLA–PEG block copolymeric nanoparticles for oral immunization against hepatitis B. *Int J Pharm*, **387**(1), 253–262.
86. Garinot, M., *et al.* (2007). PEGylated PLGA-based nanoparticles targeting M cells for oral vaccination. *J Control Release*, **120**(3), 195–204.

87. Gupta, P. N., *et al.* (2007). M-cell targeted biodegradable PLGA nanoparticles for oral immunization against hepatitis B. *J Drug Target,* **15**(10), 701–713.

88. Köping-Höggård, M., A. Sánchez, and M. J. Alonso (2005). Nanoparticles as carriers for nasal vaccine delivery. *Expert Rev Vaccines,* **4**(2), 185–196.

89. Almeida, A., and H. Alpar (1996). Nasal delivery of vaccines. *J Drug Target,* **3**(6), 455–467.

90. Davis, S. (2001). *Nasal vaccines. Adv Drug Deliv Rev,* **51**(1), 21–42.

91. Costantino, H. R., *et al.* (2007). Intranasal delivery: physicochemical and therapeutic aspects. *Int J Pharm,* **337**(1), 1–24.

92. FitzGerald, D., and R. J. Mrsny (2000). New approaches to antigen delivery. *Crit Rev Ther Drug Carrier Syst,* **17**(3), 165–248.

93. Cano, F., *et al.* (2000). Partial protection to respiratory syncytial virus (RSV) elicited in mice by intranasal immunization using live staphylococci with surface-displayed RSV-peptides. *Vaccine,* **18**(24), 2743–2752.

94. Matsuo, K., *et al.* (2000). Induction of innate immunity by nasal influenza vaccine administered in combination with an adjuvant (cholera toxin). *Vaccine,* **18**(24), 2713–2722.

95. Peters, B., and S. Allison (1929). Intranasal immunisation against scarlet fever. *Lancet,* **213**(5516), 1035.

96. Van Kampen, K. R., *et al.* (2005). Safety and immunogenicity of adenovirus-vectored nasal and epicutaneous influenza vaccines in humans. *Vaccine,* **23**(8), 1029–1036.

97. Treanor, J. J., *et al.* (1992). Protective efficacy of combined live intranasal and inactivated influenza A virus vaccines in the elderly. *Ann Intern Med,* **117**(8), 625–633.

98. Obrosova-Serova, N., *et al.* (1990). Evaluation in children of cold-adapted influenza B live attenuated intranasal vaccine prepared by reassortment between wild-type B/Ann Arbor/1/86 and cold-adapted B/Leningrad/14/55 viruses. *Vaccine,* **8**(1), 57–60.

99. Alpar, H. O., *et al.* (2005). Biodegradable mucoadhesive particulates for nasal and pulmonary antigen and DNA delivery. *Adv Drug Deliv Rev,* **57**(3), 411–430.

100. Illum, L. (2007). Nanoparticulate systems for nasal delivery of drugs: a real improvement over simple systems? *J Pharm Sci,* **96**(3), 473–483.

101. Heritage, P., *et al.* (1998). Intranasal immunization with polymer-grafted microparticles activates the nasal-associated lymphoid tissue and draining lymph nodes. *Immunology*, **93**(2), 249–256.

102. Amidi, M., *et al.* (2010). Chitosan-based delivery systems for protein therapeutics and antigens. *Adv Drug Deliv Rev*, **62**(1), 59–82.

103. Almeida, A., H. Alpar, and M. Brown (1993). Immune response to nasal delivery of antigenically intact tetanus toxoid associated with poly (L-lactic acid) microspheres in rats, rabbits and guinea-pigs. *J Pharm Pharmacol*, **45**(3), 198–203.

104. Alpar, H., *et al.* (1994). Immune responses to mucosally administered tetanus toxoid in biodegradable PLA microspheres. *Proc Int Symp Control Rel Bioact Mater*, **21**, 867–868.

105. Eyles, J. E., *et al.* (1998). Intra nasal administration of poly-lactic acid microsphere co-encapsulated Yersinia pestis subunits confers protection from pneumonic plague in the mouse. *Vaccine*, **16**(7), 698–707.

106. Eyles, J. E., E. D. Williamson, and H. O. Alpar (1999). Immunological responses to nasal delivery of free and encapsulated tetanus toxoid: studies on the effect of vehicle volume. *Int J Pharm*, **189**(1), 75–79.

107. Eyles, J., *et al.* (2000). Generation of protective immune responses to plague by mucosal administration of microsphere coencapsulated recombinant subunits. *J Control Release*, **63**(1), 191–200.

108. Shahin, R., *et al.* (1995). Adjuvanticity and protective immunity elicited by Bordetella pertussis antigens encapsulated in poly (DL-lactide-*co*-glycolide) microspheres. *Infect Immun*, **63**(4), 1195–1200.

109. Cahill, E., *et al.* (1995). Immune responses and protection against Bordetella pertussis infection after intranasal immunization of mice with filamentous haemagglutinin in solution or incorporated in biodegradable microparticles. *Vaccine*, **13**(5), 455–462.

110. Yan, C., *et al.* (1996). Intranasal stimulation of long-lasting immunity against aerosol ricin challenge with ricin toxoid vaccine encapsulated in polymeric microspheres. *Vaccine*, **14**(11), 1031–1038.

111. WU, H. Y., H. Nguyen, and M. Russell (1997). Nasal lymphoid tissue (NALT) as a mucosal immune inductive site. *Scand J Immunol*, **46**(5), 506–513.

112. Kuper, C. F., *et al.* (1992). The role of nasopharyngeal lymphoid tissue. *Immunol Today*, **13**(6), 219–224.

113. Florence, A. T. (1997). The oral absorption of micro-and nanoparticulates: neither exceptional nor unusual. *Pharm Res*, **14**(3), 259–266.

114. Sminia, T., and G. Kraal (1999). Nasal-associated lymphoid tissue. *Mucosal Immunol*, **2**, 357–364.

115. Tilney, N. L. (1971). Patterns of lymphatic drainage in the adult laboratory rat. *J Anat*, **109**(Pt 3), 369.

116. Florindo, H., *et al.* (2010). Surface modified polymeric nanoparticles for immunisation against equine strangles. *Int J Pharm*, **390**(1), 25–31.

117. Florindo, H., *et al.* (2009). The enhancement of the immune response against S. equi antigens through the intranasal administration of poly-ε-caprolactone-based nanoparticles. *Biomaterials*, **30**(5), 879–891.

118. Florindo, H., *et al.* (2009). New approach on the development of a mucosal vaccine against strangles: systemic and mucosal immune responses in a mouse model. *Vaccine*, **27**(8), 1230–1241.

119. McDermott, M. R., *et al.* (1998). Polymer-grafted starch microparticles for oral and nasal immunization. *Immunol Cell Biol*, **76**(3), 256–262.

120. Baras, B., *et al.* (1999). Single-dose mucosal immunization with biodegradable microparticles containing a Schistosoma mansoni antigen. *Infect Immun*, **67**(5), 2643–2648.

121. Vajdy, M., and D. T. O'Hagan (2001). Microparticles for intranasal immunization. *Adv Drug Deliv Rev*, **51**(1), 127–141.

122. Csaba, N., A. Sanchez, and M. J. Alonso (2006). PLGA: poloxamer and PLGA: poloxamine blend nanostructures as carriers for nasal gene delivery. *J Control Release*, **113**(2), 164–172.

123. Mansoor, F., *et al.* (2014). Intranasal delivery of nanoparticles encapsulating BPI3V proteins induces an early humoral immune response in mice. *Res Vet Sci*, **96**(3), 551–557.

124. Binjawadagi, B., *et al.* (2014). Adjuvanted poly (lactic-*co*-glycolic) acid nanoparticle-entrapped inactivated porcine reproductive and respiratory syndrome virus vaccine elicits cross-protective immune response in pigs. *Int J Nanomedicine*, **9**, 679–694.

125. Verma, A., A. Mittal, and A. Gupta (2013). Development and characterization of bipolymer based nanoparticulate carrier system as vaccine adjuvant for effecive immunization. *Int J Pharm Pharm Sci*, **5**(2), 188–195.

126. Trolle, S., *et al.* (2000). Intranasal immunization with protein-linked phosphorylcholine protects mice against a lethal intranasal challenge with Streptococcus pneumoniae. *Vaccine*, **18**(26), 2991–2998.

127. Vila, A., *et al.* (2004). Transport of PLA-PEG particles across the nasal mucosa: effect of particle size and PEG coating density. *J Control Release*, **98**(2), 231–244.

128. Vila, A., *et al.* (2002). Design of biodegradable particles for protein delivery. *J Control Release*, **78**(1), 15–24.

129. Vila, A., *et al.* (2004). Low molecular weight chitosan nanoparticles as new carriers for nasal vaccine delivery in mice. *Eur J Pharm Biopharm*, **57**(1), 123–131.

130. Tobio, M., *et al.* (1998). Stealth PLA-PEG nanoparticles as protein carriers for nasal administration. *Pharm Res*, **15**(2), 270–275.

131. Lai, S. K., *et al.* (2007). Rapid transport of large polymeric nanoparticles in fresh undiluted human mucus. *Proc Natl Acad Sci*, **104**(5), 1482–1487.

132. Giannasca, P. J., J. A. Boden, and T. P. Monath (1997). Targeted delivery of antigen to hamster nasal lymphoid tissue with M-cell-directed lectins. *Infect Immun*, **65**(10), 4288–4298.

133. Seo, K. Y., *et al.* (2010). Eye mucosa: an efficient vaccine delivery route for inducing protective immunity. *J Immunol*, **185**(6), 3610–3619.

134. Nelson, J., and J. Cameron (2005). The conjunctiva: anatomy and physiology. *Cornea*, **1**, 39–54.

135. Knop, E., and N. Knop (2007). Anatomy and immunology of the ocular surface. *Chem Immunol Allergy*, **92**, 36–49.

136. Akpek, E., and J. Gottsch (2003). Immune defense at the ocular surface. *Eye*, **17**(8), 949–956.

137. Sacks, E. H., *et al.* (1986). Lymphocytic subpopulations in the normal human conjunctiva: a monoclonal antibody study. *Ophthalmology*, **93**(10), 1276–1283.

138. Camelo, S., *et al.* (2006). Antigen from the anterior chamber of the eye travels in a soluble form to secondary lymphoid organs via lymphatic and vascular routes. *Invest Ophthalmol Vis Sci*, **47**(3), 1039–1046.

139. Ueta, M. (2008). Innate immunity of the ocular surface and ocular surface inflammatory disorders. *Cornea*, **27**, S31–S40.

140. Ueta, M., and S. Kinoshita (2010). Innate immunity of the ocular surface. *Brain Res Bull*, **81**(2), 219–228.

141. Nesburn, A. B., *et al.* (2006). Topical/mucosal delivery of sub-unit vaccines that stimulate the ocular mucosal immune system. *Ocul Surf*, **4**(4), 178–187.

142. Chentoufi, A. A., *et al.* (2010). Nasolacrimal duct closure modulates ocular mucosal and systemic CD4+ T-cell responses induced following topical ocular or intranasal immunization. *Clin Vaccine Immunol,* **17**(3), 342–353.

143. Gore, T. C., *et al.* (2005). Three-year duration of immunity in cats following vaccination against feline rhinotracheitis virus, feline calicivirus, and feline panleukopenia virus. *Vet Ther,* **7**(3), 213–222.

144. Müller, H., *et al.* (2012). Current status of vaccines against infectious bursal disease. *Avian Pathol,* **41**(2), 133–139.

145. Fensterbank, R., P. Pardon, and J. Marly (1985). Vaccination of ewes by a single conjunctival administration of Brucella melitensis Rev. 1 vaccine. *Ann Vet Res,* **16**(4), 351–356.

146. Rafferty, D., M. Elfaki, and P. Montgomery (1996). Preparation and characterization of a biodegradable microparticle antigen/cytokine delivery system. *Vaccine,* **14**(6), 532–538.

147. Hu, K., *et al.* (2011). An ocular mucosal administration of nanoparticles containing DNA vaccine pRSC-gD-IL-21 confers protection against mucosal challenge with herpes simplex virus type 1 in mice. *Vaccine,* **29**(7), 1455–1462.

148. Negash, T., M. Liman, and S. Rautenschlein (2013). Mucosal application of cationic poly(D,L-lactide-*co*-glycolide) microparticles as carriers of DNA vaccine and adjuvants to protect chickens against infectious bursal disease. *Vaccine,* **31**(36), 3656–3662.

149. Eterradossi, N., and Y. M. Saif (2008). *Infectious bursal disease,* in *Diseases of Poultry,* Y. M. Saif, A. M. Fadly, J. R. Glisson, L. R. McDougald, L. K. Nolan, and D. E. Swayne, eds. (Wiley-Blackwell, Ames, Iowa), 185–208.

150. Hussain, A., and F. Ahsan (2005). The vagina as a route for systemic drug delivery. *J Control Release,* **103**(2), 301–313.

151. das Neves, J., *et al.* (2015). Polymer-based nanocarriers for vaginal drug delivery. *Adv Drug Deliv Rev,* **92**, 53–70.

152. Zhu, Q., *et al.* (2012). Large intestine-targeted, nanoparticle-releasing oral vaccine to control genitorectal viral infection. *Nat Med,* **18**(8), 1291–1296.

153. Kuo-Haller, P., *et al.* (2010). Vaccine delivery by polymeric vehicles in the mouse reproductive tract induces sustained local and systemic immunity. *Mol Pharm,* **7**(5), 1585–1595.

154. Goodsell, A., *et al.* (2008). Beta7-integrin-independent enhancement of mucosal and systemic anti-HIV antibody responses following

combined mucosal and systemic gene delivery. *Immunology*, **123**(3), 378–389.

155. Hunter, S. K., M. E., Andracki, and A. M. Krieg (2001). Biodegradable microspheres containing group B Streptococcus vaccine: immune response in mice. *Am J Obstet Gynecol*, **185**(5), 1174–1179.

156. Singh, M., *et al.* (2001). Mucosal immunization with HIV-1 gag DNA on cationic microparticles prolongs gene expression and enhances local and systemic immunity. *Vaccine*, **20**(3–4), 594–602.

157. Allaoui-Attarki, K., *et al.* (1998). Mucosal immunogenicity elicited in mice by oral vaccination with phosphorylcholine encapsulated in poly (D,L-lactide-*co*-glycolide) microspheres. *Vaccine*, **16**(7), 685–691.

158. Mahajan, A., *et al.* (2005). Phenotypic and functional characterisation of follicle-associated epithelium of rectal lymphoid tissue. *Cell Tissue Res*, **321**(3), 365–374.

159. Crowley-Nowick, P., *et al.* (1997). Rectal immunization for induction of specific antibody in the genital tract of women. *J Clin Immunol*, **17**(5), 370–379.

160. Lehner, T., *et al.* (1993). T-and B-cell functions and epitope expression in nonhuman primates immunized with simian immunodeficiency virus antigen by the rectal route. *Proc Natl Acad Sci*, **90**(18), 8638–8642.

161. Tengvall, S., D. O'Hagan, and A. M. Harandi (2008). Rectal immunization generates protective immunity in the female genital tract against herpes simplex virus type 2 infection: relative importance of myeloid differentiation factor 88. *Antiviral Res*, **78**(3), 202–214.

162. Haneberg, B., *et al.* (1994). Induction of specific immunoglobulin A in the small intestine, colon-rectum, and vagina measured by a new method for collection of secretions from local mucosal surfaces. *Infect Immun*, **62**(1), 15–23.

163. Hopkins, S., *et al.* (1995). A recombinant Salmonella typhimurium vaccine induces local immunity by four different routes of immunization. *Infect Immun*, **63**(9), 3279–3286.

164. Zhou, F., J.-P. Kraehenbuhl, and M. R. Neutra (1995). Mucosal IgA response to rectally administered antigen formulated in IgA-coated liposomes. *Vaccine*, **13**(7), 637–644.

165. Kozlowski, P. A., *et al.* (1997). Comparison of the oral, rectal, and vaginal immunization routes for induction of antibodies in rectal and genital tract secretions of women. *Infect Immun*, **65**(4), 1387–1394.

166. Kantele, A., *et al.* (1998). Differences in immune responses induced by oral and rectal immunizations with Salmonella typhi Ty21a: evidence for compartmentalization within the common mucosal immune system in humans. *Infect Immun*, **66**(12), 5630–5635.

167. Sebastian, M., N. Ninan, and A. K. Haghi (2012). *Nanomedicine and Drug Delivery*. Apple Academic Press.

168. Manocha, M., *et al.* (2005). Enhanced mucosal and systemic immune response with intranasal immunization of mice with HIV peptides entrapped in PLG microparticles in combination with Ulex Europaeus-I lectin as M cell target. *Vaccine*, **23**(48–49), 5599–5617.

Chapter 17

Aliphatic Polyester Micro- and Nanosystems for Treating HIV, Tuberculosis, and Malaria

Alejandro Sosnik
Laboratory of Pharmaceutical Nanomaterials Science,
Department of Materials Science and Engineering,
Technion-Israel Institute of Technology, De-Jur Building,
Office 607, Technion City, Haifa, 3200003, Israel
sosnik@tx.technion.ac.il, alesosnik@gmail.com

17.1 Introduction

17.1.1 Poverty-Related Diseases

The development of novel, more potent, and broad-spectrum antibiotics has opened promising horizons in the treatment of infection. However, over the years this issue has become increasingly challenging owing to the emergence of resistant strains that withstand the current therapy and to a sharp decrease in the pace of new drug research and development. The situation is more concerning in the

Handbook of Polyester Drug Delivery Systems
Edited by MNV Ravikumar

ISBN 978-981-4669-65-8 (Hardcover), 978-981-4669-66-5 (eBook)
www.panstanford.com

so-called poverty-related diseases (PRDs), defined as infectious maladies of bacterial, viral, and parasitic origin that affect more than 1 billion people, mainly in low-income countries [1]. These diseases have a tremendous socioeconomic impact in the development of nations, especially those in the sub-Saharan region, that face a crisis of unprecedented consequences. Table 17.1 summarizes the most relevant PRDs classified by pathogen and describing the state-of-the-art therapy and the morbidity and mortality rates according to statistics of the World Health Organization (WHO).

The therapeutic arsenal available and the efficacy of the drugs used to treat PRDs is individual, ranging from the broad spectrum of antiretrovirals (ARVs) available to treat the chronic human immunodeficiency virus (HIV)/acquired immunodeficiency syndrome (AIDS) and the effective first-line combined therapy of standard (nonresistant) tuberculosis (TB) to hardly tractable ones such as extensively drug-resistant TB (XDR-TB) and Ebola.

Another issue pertains to intrinsic biopharmaceutical drawbacks of drugs such as low aqueous solubility and physicochemical instability in the biological environment that decrease bioavailability and complex administration regimens and severe adverse effects that jeopardize patient compliance and favor treatment interruption. Moreover, the cost of the medication is another relevant point as it represents a crucial hurdle in poor countries, especially for new drugs and innovative drug delivery systems (DDS).

In this complex scenario, different technological approaches are under investigation to improve the performance of drugs and by doing so to overcome, in a comprehensive manner, the above mentioned limitations. Among them, the encapsulation of drugs and vaccines within polyester microparticles (PMPs) and nanoparticles (PNPs) emerged as one of the most popular strategies [4, 5].

The source and therapy of PRDs is very eclectic. In fact, many of these infections has no chemotherapeutic treatment available. Among the PRDs, HIV/AIDS, TB, and malaria claim the largest death toll worldwide due to infection, a rate standing nowadays at approximately 4–5 million lives and they urgently demand the development of innovative DDS [6–8]. In this scenario, after a very brief description of the different diseases summarized in Table 17.1, the present chapter will overview the application of

Table 17.1 Most recurrent PRDs as defined by the WHO

Pathogen	Disease	Therapy	Infected population	Annual infections	Annual death toll
Bacterial	TB	**First line**: Rifampicin, isoniazid, pyrazinamide, ethambutol, streptomycin **Second line**: Rifabutin, clofazimine, ethionamide, clarithromycin, cycloserine, amikacin, kanamycin A, capreomycin, levofloxacin, moxifloxacin, gatifloxacin, linezolid, *p*-aminosalicylic acid, bedaquiline	9 billion (33% of the world population)[a]	9 million[a]	1.5 million[a] (360,000 co-infected with HIV)
	Leprosy	Rifampicin, clofazimine and dapsone for multibacilliary (MB) form and rifampicin, and dapsone for paucibacilliary (PB) form	12 million[b]	230,000[b]	14,000[b]
	Trachoma	Antibiotics to treat chlamydia trachomatis (e.g., azithromycin)	229 million[c]		2.2 million with visual impairment and 1.2 million blind[c]
Viral	HIV/AIDS	Over 25 approved ARVs. Combinations vary by region and individual	35 million[d] (70% in the sub-Saharan region)	2.4 million[d]	1.7 million[d]
	Dengue	No specific treatment. Early detection reduces fatality to less than 1%		2.5 billion at high risk 50–100 million[e]	12,500 (2.5% of severe dengue hospitalizations)[e]

(*Continued*)

Table 17.1 (*Continued*)

Pathogen	Disease	Therapy	Infected population	Annual infections	Annual death toll
	Ebola virus disease or hemorrhagic fever	No treatment or vaccine available. Some antivirals are being currently tested due to the 2014 outbreak	Different outbreaks registered since the discovery of the disease by Peter Piot in 1976, usually in the tropical region of the sub-Saharan region	Over 20,000 cases in 2014 outbreak[f]	8,000 in 2014[f]
Parasitic	Malaria	Treatment depends on stage of disease. Most common drugs are chloroquine, atovaquone-proguanil, artemisinin, dihydroartemisinin, artemether-lumefantrine, mefloquine, amodiaquine, quinine, quinidine, doxycycline (combined with quinine), clindamycin (combined with quinine), artesunate,	198 million[g]	3.2 billion at risk of infection and 1 billion at high risk of infection[g]	580,000[g] (90% of deaths in Africa; 78% of all deaths children below 5 years of age)
	Chagas disease (American trypanosomiasis)	Two phases, acute and chronic. Benznidazole, nifurtimox	8 million[h]	40,000 [2]	10,000 [3]
	Leishmaniasis	Treatment varies according to form of disease, region and Leishmania type. Usually injectable antimonials (meglumine antimoniate sodium stibogluconate), amphotericin B, ketoconazole, miltefosine, paromomycin, pentamidine	Three forms of the disease. The most common cutaneous (CL), the most lethal visceral (VL) and the mucocutaneous caused by over 20 types of Leishmania species[i]	1 million (CL)[i] 300,000 (VL)[i]	25,000[i]

Pathogen	Disease	Therapy	Infected population	Annual infections	Annual death toll
	Schistosomiasis	Praziquantel	Different species and regions provoke the intestinal and urogenital disease. The most severe form affect the nervous system. 42 million cases were treated in 2012[j]		200,000[j]
	Sleeping sickness (African human trypanosomiasis)	Treatment depends on the stage and is complicated and demands trained staff. First stage: pentamidine, suramin Second stage: melarsorpol, eflornithine, nifurtimox	30,000 Only found in sub-Saharan region[k]	10,000[k]	10,000[k]
	Lymphatic filariasis (or elephantiasis)	Albendazole with ivermectin or diethylcarbamazine citrate	120 million[l]	1.4 billion at risk[l]	Provokes chronic disability
	Onchocerciasis (river blindness)	Ivermectin. No prophylaxis is available	18 million[m]		Provokes blindness (2nd cause worldwide) in 270,00 and severe itching and dermatitis in 6.5 million of infected people[m]

(*Continued*)

Table 17.1 (*Continued*)

Pathogen	Disease	Therapy	Infected population	Annual infections	Annual death toll
	Echinococcosis (tapeworm)	Cystic and alveolar forms are treated with complex treatments (1) Percutaneous treatment of the hydatid cysts with the PAIR (Puncture, Aspiration, Injection, Re-aspiration) technique; surgery; (2) anti-infective drug treatment; (3) "watch and wait"	1 million. Four forms: cystic echinococcosis (hydatidosis); alveolar echinococcosis; polycystic echinococcosis; unicystic echinococcosis[n]		An average 2.2% postoperative death rate for surgical patients and about 6.5% of cases relapsing after intervention[n]

[a] Source: WHO (http://apps.who.int/iris/bitstream/10665/137094/1/9789241564809_eng.pdf)
[b] Source: WHO (http://www.who.int/mediacentre/factsheets/fs101/en/)
[c] Source: WHO (http://www.who.int/mediacentre/factsheets/fs382/en/)
[d] Source: WHO (http://www.who.int/mediacentre/factsheets/fs360/en/)
[e] Source: WHO (http://www.who.int/mediacentre/factsheets/fs117/en/)
[f] Source: WHO (http://www.who.int/mediacentre/factsheets/fs103/en/)
[g] Source: WHO (http://apps.who.int/iris/bitstream/10665/144852/2/9789241564830_eng.pdf)
[h] Source: WHO (http://www.who.int/mediacentre/factsheets/fs340/en/)
[i] Source: WHO (http://www.who.int/mediacentre/factsheets/fs375/en/)
[j] Source: WHO (http://www.who.int/mediacentre/factsheets/fs115/en/)
[k] Source: WHO (http://www.who.int/mediacentre/factsheets/fs259/en/)
[l] Source: WHO (http://www.who.int/mediacentre/factsheets/fs102/en/)
[m] Source: WHO (http://www.who.int/mediacentre/factsheets/fs374/en/)
[n] Source: WHO (http://www.who.int/mediacentre/factsheets/fs377/en/)

aliphatic polyester PMPs and PNPs to improve the therapy of these specific infectious diseases that will be addressed following a decreasing order of mortality rates.

17.1.1.1 Bacterial PRDs

As previously described for HIV, TB is the bacterial PRD with the largest incidence and mortality far ahead of other maladies such as leprosy and trachoma.

TB

TB is the second-most deadly infection after HIV, with approximately 1.5 million deaths every year [15]. The gold-standard therapy of nonresistant TB comprises two phases, phase I (2 months) with a combination of rifampicin, isoniazid, pyrazinamide, and ethambutol, and phase II (4 months) with rifampicin/isoniazid. Treatment interruption leads to therapeutic failure and development of multidrug resistant TB (MDR-TB). The disease is usually associated with the airways. However, resistant forms of the pathogen can colonize organs that are more distant [16]. The extrapulmonary form of the disease is more difficult to treat. Regardless of the efficacy of the first-line therapy to cure standard TB, it remains 25% and 2.4% of the preventable and all deaths, respectively [17, 18]. Moreover, most of the cases are concentrated in developing nations, TB being the leading cause of death in several African and Asiatic countries. Different biopharmaceutical drawbacks have been identified, though the fast hydrolysis of rifampicin in the gastric environment [19], a pathway that is catalyzed by isoniazid [20], another first-line anti-TB drug that is coadministered with the former according to all the clinical guidelines, is probably the most concerning one due to the significant decrease of the rifampicin oral bioavailability. The main reason of therapeutic failure and resistance development is that patients tend to abandon the long-term and high pill burden treatment after an initial and temporary improvement. In nonresistant TB, the implementation of DDS that sustain the release of anti-TB drug combinations over time and reduce the administration frequency would revolutionize the therapy of TB. Conversely, in resistant forms, a reduction in the required dose of novel drugs (e.g., bedaquiline) would represent a step forward in terms of treatment cost and patient affordability.

Leprosy

As TB, this infection is also caused by a mycobacterium. It is endemic in many regions of the developing world [21]. There exist two forms, the multibacilliary and the paucibacilliary that are distinguished by skin smears. Transmission takes place by contact between susceptible and genetically predisposed individuals and untreated multibacilliary patients. The main way of entry is the nasal mucosa. The treatment comprises antibiotics such as rifampicin, clofazimine, and dapsone.

Trachoma

Trachoma is a non-life-threatening infection provoked by the intracellular pathogen *Chlamydia trachomatis*, and it leads to visual impairment and blindness [22]. The transmission of the disease is favored by the presence of children, overcrowding, and the lack of running water. The most common treatment is based on azithromycin.

17.1.1.2 Viral PRDs

HIV/AIDS, dengue, and Ebola represent the most relevant viral infections hitting poor countries, though only the former has currently a chemotherapeutic treatment.

HIV/AIDS

Thirty-five million people are currently infected with HIV/AIDS, the most lethal infectious disease of our times with approximately 2.5 million annual deaths, 15% of them being children [9]. Until a cure is found, the chronic administration of at least three ARVs is the only way to maintain viral concentrations in plasma below detectable levels and to prevent the progress to the active phase of the disease and to reduce transmission rates among high-risk individuals [10]. In addition, high adherence levels (>95%) are critical to achieve therapeutic success [11]. This demand is challenged by complicated and frequent administration regimens and adverse side effects. It is worth stressing that more than 33% of the infected adults currently access the pharmacotherapy, these rates decreasing significantly to less than 25% in pediatric patients [9]. The pediatric population is being left behind in all the PRDs, becoming not only a serious health issue but also an ethical one [12]. New pharmaceutical products would play a fundamental role to (i) increase the oral bioavailability of the ARVs and reduce the required dose to attain therapeutic concentrations; a lower dose will reduce the cost of the medication and increase patient affordability in the poorest countries; (ii) prolong the half-life of the ARVs and enable a reduction of the frequency of administration; a more patient-compliant treatment due to less frequent administrations will redound in better adherence to the administration regimens and less treatment cessation (and development of resistance); (iii) better control the maximum plasma concentrations and reduce associated toxic effects; and (iv) significantly increase the oral pharmacokinetics and eliminate boosting agents used to improve the performance of protease inhibitors from the therapeutic cocktail.

Dengue

With 500 million annual infections and half a million hospitalizations (mainly children) due to haemorrhagic fever in Southeast Asia, the Pacific, and the Americas, dengue is the most recurrent arthropod-borne viral infection [13]. In some areas, the mortality rates could reach 5% of the hospitalized patients. Even though antivirals and vaccines are under development, there is no treatment available. The current management is by monitoring vital signs and an intensive intravenous rehydration therapy.

Ebola

Ebola is a zoonotic usually fatal infection that is transmitted by contact with fluids of infected individuals [14]. The incubation period is 1–21 days, and patients are infectious when they develop the symptomatology. Symptoms are not specific, and thus, diagnosis is difficult at the early stages. Treatment is only supportive and mortality rates can reach 30–90% of the cases.

17.1.1.3 Parasitic PRDs

Parasitosis probably represent the most heterogeneous group of PRDs, with malaria as leading representative, and a broad spectrum of regional diseases that together claim a large number of lives.

Malaria

Parasitic infections are endemic in many regions of the developing world, malaria claiming the highest global morbidity and mortality, mainly in sub-Saharan Africa, which accounts for 80% of the cases and 90% of the deaths [23]. While malaria can be prevented and cured, 580,000 people died of malaria in 2014, most of them children [24]. The current pharmacotherapy depends on the plasmodium strain and comprises the combination of two or more blood schizontocidal drugs with different biochemical targets in the parasite. The cocktails are classified into non-artemisinin and artemisinin [25]. The former includes sulfadoxine-pyrimethamine plus chloroquine or amiodaquine, while the latter artesunate, artemether, and dihydroartemisinin. Since the artemisinin therapy is effective, most innovation in malaria has to be mainly oriented to address the treatment of the vulnerable pediatric population [24, 26].

Chagas disease

The human American trypanosomiasis is an endemic tropical disease affecting populations in extensive regions of Latin America, from Argentina and Chile in the south to Mexico in the north. In recent years, migrants brought the disease to the United States [27]. Transmission can be by vectors, transfusion, oral or vertical (congenital). The incubation period could be up to several months. The acute phase leads to relatively low mortality rates in the 5–10% range, usually children, due to myocarditis and/or myeloencephalitis [28]. Then, the disease reaches a chronic phase with digestive, cardiac, or mixed forms. The pharmacological therapy in the acute phase (effective mainly in young patients and the closest to the infection time) is based on nifurtimox or benznidazole and leads to cure. Conversely, the treatment of the chronic disease relies on maintenance. A new drug, posaconazole, has shown good results in both acute and chronic Chagas, leading to elimination of amastigotes in cardiac cells [28].

Leishmaniasis

This disease is intracellular and affect macrophages. There are different species transmitted by the female sand fly, and the disease is found in two main forms: the visceral, which affects internal organs such as spleen, liver, and bone marrow, and the cutaneous, which provokes skin ulcerations, scarring, and deformities [29]. The former is treated with injectable antimonials, while the latter with liposomal amphotericin B. Untreated visceral leishmanisis is fatal. No prevention or vaccination exist. In addition, susceptibility to the therapy depends on the strain and regional variability in the therapeutic approach could be found.

Schistosomiasis

Schistosomiasis is a blood parasitic disease prevalent in Africa, the Middle East, South America, and Asia. It appears in an acute form known as Katayama syndrome and an advanced chronic one that affects the urinary system. The most severe manifestation is neuroschitosomiasis [30]. The treatment is based on oral praziquantel, which is also used in preventive chemotherapy.

Sleeping sickness

The human African trypanosomiasis refers to two different forms, one acute, caused by the *Trypanosoma brucei rhodesiense* and found in eastern and southern Africa, and a chronic one due to the infection with the *Trypanosoma brucei gambiense* in western and central Africa (98% of the cases), transmitted by the tsetse fly [31]. The former is a zoonosis, though humans are the main reservoir of the latter and play a key role in the transmission cycle. Treatments are difficult and classified into first (pentamidine and suramin) and second stage (melarsprol, eflornithine, and nifurtimox).

Lymphatic filarisis

This disease is endemic in 80 tropical and subtropical countries and currently affects approximately 120 million people, with 1 billion at risk [32]. Elimination programs aim to block the transmission cycle with chemotherapeutic campaigns with diethylcarbamazine and albendazole, or albendazole and ivermectin for periods of 4–6 years.

Onchocerciasis

This disease, also known as river blindness, is transmitted by infected black flies. The disease shows skin manifestations and acute and chronic lesions of the anterior and posterior eye, leading to blindness [33]. There is no prevention, and the treatment is based on ivermectin.

Echinococcosis

The disease in humans is provoked by the larval infections of taeniid cestodes of six *Echinococcus* species [34]. The treatment is very complex, including surgery, percutaneous aspiration, and injection of albendazole and mebendazole.

The development of micro- and nano-DDS for PRDs is a very small research niche that mainly addresses the research in HIV/AIDS, TB, and malaria. Thus, in the following section, the most relevant works at the interface of polyester particulate delivery systems, and these diseases will be discussed.

17.2 Polyesters in Drug Delivery

Aliphatic polyesters are probably the most extensively used biodegradable biomaterials, a phenomenon driven by the proven biocompatibility, the relatively easy synthetic pathways employed, and the ability to process them into constructs with a broad spectrum of different properties and fine-tune the degradation and release rate [35, 36]. The most significant contributions have been made for poly(lactic acid)s (PLA)s (in all the isomeric presentations) and their copolymers with poly(glycolic acid) (PGA) [37]. Depending on the steric isomerism of the precursor used, PLA can be semicrystalline or completely amorphous. Thus, L-lactide and D-lactide render the homopolymers crystallizable. Conversely, *meso*-D,L-lactide and the racemic L-lactide/D-lactide result in amorphous products [38]. At the same time, the more hydrophobic and hydrolysis-resistant poly(ε-caprolactone) (PCL) has undoubtedly gained a prominent position, especially in drug delivery [39]. The most outstanding features of PCL are relatively high permeability for the diffusion of both lipophilic and hydrophilic drugs, and thermal properties that enable the design of injectable systems [40] and the processing under mild heating conditions that do not comprise the integrity of sensitive cargos [41, 42]. PLA, PLGA, and PCL have been intensively investigated for the production of drug- and protein-loaded PMPs and PNPs. Moreover, short polyester blocks have been incorporated into intrinsically nondegradable polymers such as poly(ether-urethane)s to confer them hydrolytic sensitivity [43] or combined with hydrophilic blocks, e.g., poly(ethylene glycol), to produce copolymers with intermediate degradation rates [44–46]. From a different conceptual perspective, more hydrophilic polyester-poly(ethylene oxide) block copolymers that generate self-assembly nanocarriers such as polymeric micelles and polymeric vesicles (polymersomes) have been also designed [47, 48].

17.2.1 Polyester-Based Particles in the Release of ARVs

As mentioned above, the development of innovative DDSs in PRDs has been scarce. On the other hand, since the clinical translation is always in mind, a significant piece of the conducted work explored polyesters as possible ARV carriers. In this scenario, the production of particles has emerged as one of the most appealing approaches.

Most research was focused on the encapsulation of hydrophobic ARVs. Protease inhibitors (PIs) are ARVs that block the activity of the viral protease, and consequently the assembly, maturation, and infectivity of new virions [49]. PIs are currently recommended in initial anti-HIV regimens in combination with ARVs of other families and are always coadministered with ritonavir, another PI used in low (sub-therapeutic) concentrations as a boosting agent [50, 51]. Coadministration with ritonavir is critical to deplete the metabolism of the main PI by the hepatic cytochrome P-450 CYP3A4 [50]. Moreover, ritonavir competitively reduces the P-glycoprotein (P-gp)-mediated efflux of the main PI; P-gp is a pump of the ATP-binding cassette superfamily present in the intestinal epithelium that removes substrates (e.g., drugs) in the basolateral-to-apical direction against a concentration gradient and reduces their effective concentration in plasma. In a recent clinical trial, Cahn *et al.* showed that a boosted PI would enable the reduction of the number of ARVs in the therapeutic cocktail [52]. This study pointed out the key role of PIs [53]. Moreover, this could lead to a simpler regimen, less adverse effects, and a reduction of the treatment cost and an increase in patient compliance and affordability to gold-standard ARV combinations. Shah and Amiji encapsulated saquinavir into PEO-PCL NPs (200 nm) using the solvent displacement method [54].

In addition, the intracellular concentration of the drug increased significantly with respect to the free form in a THP-1 monocyte/macrophage cell line. A release model to predict the release behavior was also developed [55].

Another family of first-line poorly-water soluble ARVs is non-nucleoside reverse transcriptase inhibitors (NNRTIs). Efavirenz (EVF) is a first-line NNRTI recommended by WHO for children older than 3 years and adults [56]. Owing to the poor aqueous solubility and variable pharmacogenetic profiles, EFV displays low oral bioavailability of 40–45% [57, 58] and high inter- and intrasubject variability [53, 59–61]. Aiming to improve its oral pharmacokinetics, EFV was encapsulated into PCL NPs and MPs employing different technologies, from simple nanoprecipitation and single emulsion/solvent evaporation [62] to spray-drying [63]. The oral bioavailability was significantly increased with respect to the free drug and EFV-loaded micelles [56, 64, 65]. However, the particles could not sustain the release over time *in vivo*, a feature that would enable the change of the current once-a-day administration regimen. The incorporation of mucoadhesive poly(methacrylate) copolymers into the particles slightly prolonged the release in the gut though further investigations would be needed to support the usefulness of this strategy in clinics [66, 67].

Topical HIV microbiocides to prevent transmission following sexual intercourse are also being researched [68]. The group of Sarmento has extensively investigated the encapsulation of the dapivirine, a potent NNRTI prophylactic candidate, within PCL NPs for both vaginal and rectal administration in transmission prophylaxis [69, 70]. Particles were retained in epithelial monolayers and mucosa what could increase the microbiocide efficacy. Using another approach, Woodbrow *et al.* employed intravaginal gene silencing by means of silencing RNA (siRNA) loaded in PLGA NPs [71]. The efficacy in HIV was not tested though this platform appears as very versatile and promising due to the ability of the particles to penetrate the vaginal mucosa and release the cargo over time.

More recently, triplex-forming peptide nucleic acids and donor DNAs for recombination-mediated editing of the *CCR5* gene were synthesized for delivery into human peripheral blood mononuclear cells and conferring HIV-1 resistance [72].

The encapsulation of hydrophobic drugs is a relatively easy task and encapsulation efficiency is usually high. Conversely, water-soluble drugs are more challenging because common techniques employing single water-in-oil or double water-in-oil-in water emulsions cannot prevent the migration of the water-soluble cargo to the external aqueous phase and the decrease of the encapsulation extent. This is the case of ARVs of the nucleoside reverse transcriptase inhibitors (NRTIs). In this context, more advanced encapsulation technologies need to be implemented. Tshweu *et al.* investigated the encapsulation of lamivudine, a drug with a short half-life of 5–7 hours, within PCL NPs (215–450 nm) employing the spray-drying of a double emulsion [73]. The fast drying process prevented the migration of the drug and resulted in good encapsulation efficiency when compared to standard techniques. *In vivo* studies were not conducted and thus the benefit of the encapsulation could not be established. More recently, Seremeta *et al.* employed both spray-drying [74] and concentric electrohydrodynamic atomization (CEHDA) [75] and the same polyester to encapsulate the NRTI didanosine (ddI) and improve its chemical stability in the stomach; ddI undergoes very fast acid degradation, a phenomenon that decreases the oral bioavailability to 20–40%. Spray-drying from a suspension and a single emulsion resulted in the formation of MPs that contained the drug in the form of solid crystalline agglomerates dispersed in the polymeric matrix (Fig. 17.1). The yield was approximately 38–65% and the encapsulation efficiency 60–100% [66, 74, 75].

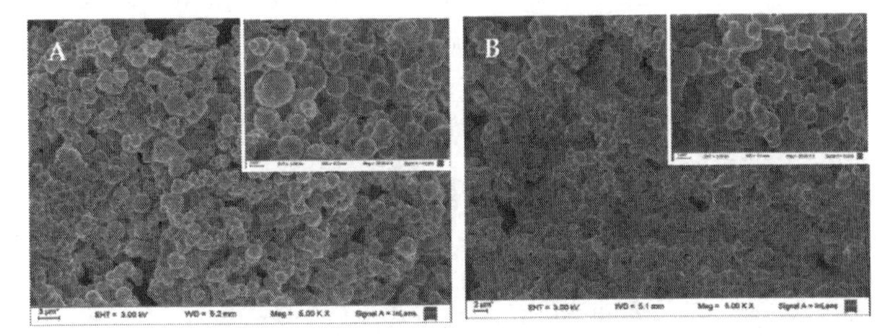

Figure 17.1 SEM micrographs of ddI-loaded PCL particles obtained by spray-drying of (A) emulsion and (B) suspension. The drug loading was 200 mg per gram of polymer.

The oral bioavailability in rats after the administration of 20 mg/kg (5 mg/mL) was increased 2.5 times with respect to a solution of the free drug. When the same drug was encapsulated employing a CEHDA setup that results in a drug core surrounded by a polymeric shell, the encapsulation was always above 95% and the more effective isolation of the drug from the medium increased its oral bioavailability by almost 4 times with respect to the free counterpart [75]. However, further investigations are required to increase the yield of this production method that was approximately 30–40%. These works represent a valuable starting point for the encapsulation of gold-standard hydrophilic ARVs, especially those displaying chemical instability in different portions the gastrointestinal tract.

Other works explored the encapsulation of ARVs into polyester particles for parenteral administration and more sustained release. For example, Srivastava and Sinha produced PLGA (PLGA 85:15 and PLGA 50:50) MPs for a depot parenteral delivery of the NRTI stavudine over 6–8 months [76].

Even more challenging is the strategy of co-encapsulating several ARVs in the same drug delivery system to ensure the timeous release of therapeutic dose. This concept has been introduced by Destache *et al.* who developed an injectable nano-DDS containing EFV, lopinavir/ritonavir, all poorly water-soluble ARVs, employing PLGA [77–79]. This system addressed one of the most critical drawbacks of the therapy, frequent administration, and sustained the release over one month by parenteral route (Fig. 17.2) [78]. NPs also increased the cellular uptake by monocyte-derived macrophages infected with HIV-1 *in vitro* and reduced the inhibitory concentration to the nM range. ARV co-encapsulation in the same nanocarrier aims to ensure the timeous release of all the drugs. On the other hand, it is still a controversy if this approach would be translatable, mainly because of the difficult control of the ARV relative ratios in the final product that would depend on the encapsulation capacity of the particles for each singular drug and the production method. On the other hand, the blending of NPs containing each one of the ARVs individually encapsulated that is more controllable and enables the combination of drugs in the final product according to the clinical guidelines cannot ensure the timeous intracellular release of the ARV in the right ratio. On the basis of this, additional studies where the efficacy in valid preclinical models of both types of DDS needs to be addressed. Finally, the co-encapsulation of both water-soluble and insoluble ARVs in the same carrier might be even more challenging, as supported by the work of Chaowanachan *et al.* that combined EFV- or saquinavir-loaded PLGA NPs with free tenofovir, a nucleotide reverse transcriptase inhibitor, under investigation in different pharmaceutical forms for HIV prophylaxis in sexual intercourse [80].

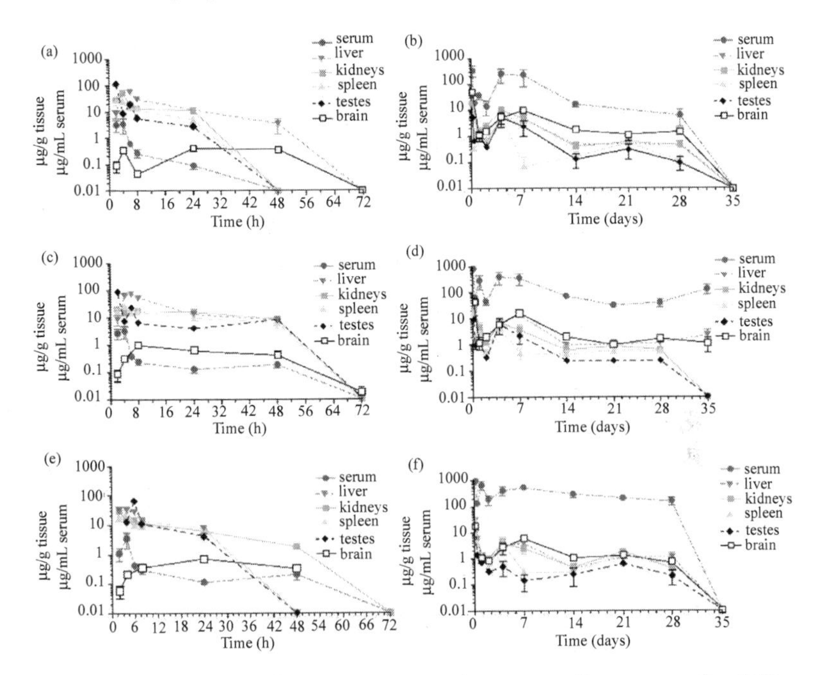

Figure 17.2 Serum and organ concentration versus time curves for ARVs after 20 mg/kg intraperitoneal injection to mice. (a, c, and e) Free drug injections. (b, d, and f) ARV NP injections. (a) Free ritonavir concentrations. (b) Ritonavir concentrations from NP formulation. (c) Free lopinavir concentrations. (d) Lopinavir concentrations from NP formulation. (e) Free efavirenz concentrations. (f) Efavirenz concentrations from NP formulation. Error bars represent SEM of three mice measurements by HPLC. Reproduced with permission from Ref. 78, Copyright 2010, Oxford University Press.

HIV reservoirs are a great challenge and increasing attention is being paid to the targeting of ARVs to the one in the central nervous system. In this context, Kuo *et al.* investigated the targeting of the

NNRTI nevirapine to the brain employing PLGA NPs surface-modified with transferrin, a protein that has receptors expressed in the blood–brain barrier [81]. On the other hand, this study was conducted employing the human brain microvascular endothelial cell monolayer model. Further studies *in vivo* would be needed to support the approach. Moreover, the clinical translation appears as very difficult, especially considering the contribution of a complex production process to the cost of the medication.

Other works used polyesters not as particulate systems but as monolithic release matrices. For example, PLA blends with polyethylene vinyl acetate have been used to develop vaginal rings for the sustained release of tenofovir [82]. This drug delivery system is expected to improve the performance of this water-soluble drug with respect to other topical formulations such as gels, where the drug is in free form.

17.2.2 Polyester Particles in the Release of Antituberculosis Drugs

The investigation of drug delivery systems in TB is mainly motivated by the urgent need to reduce the pill burden and the administration frequency and eventually shorten the treatment. These features are crucial to increase patient adherence, reduce treatment cessation, and develop resistance. For this, greater bioavailability and prolonged release need to be achieved. In this framework, the most outstanding contributions have been by the group of Khuller that explored different natural and synthetic biomaterials and administration routes to encapsulate first-line anti-TB drugs such as rifampicin, isoniazid, and pyrazinamide alone or co-encapsulate them in combination [83–85]. An outstanding characteristic of this research was the systematic assessment of the anti-TB activity in infected mice. Moreover, among the various polymers used, PLGA was the most prominent. Findings showed that free drugs were cleared from the systemic circulation after 12–24 hours. Conversely, nanoencapsulated drugs were found in plasma for up to 9 days and therapeutic concentrations in organs maintained for 9–11 days. Intriguingly, these results were obtained following oral administration, suggesting the absorption of the drugs in encapsulated form. Otherwise, the half-life should have been similar to the free counterpart. A more recent work by Semete *et al.* revealed that indeed PLGA NPs surpass the intestinal barrier upon oral administration and undergo biodistribution to different organs [86]. Remarkably, the preclinical evaluation of the NPs in a model of murine tuberculosis showed that five oral doses of the NPs administered every 10 days for 50 days sterilized the different organs [80]. The efficacy with the free forms was the same though after administration of 46 daily oral doses. Similar findings were observed PLGA NPs loaded with rifampicin, isoniazid, pyrazinamide, and ethambutol [87]. These results would open new opportunities to treat the disease, though the clinical studies are mandatory. This technology was also explored by subcutaneous injection [88], where one single dose of the NPs maintained the concentration of rifampicin, isoniazid, and pyrazinamide constant in plasma, lungs, and spleen for more than one month and resulted in undetectable bacterial counts in the different organs. Aiming to strengthen the adhesion to intestinal and lung mucosa and the drug absorption and bioavailability by both the oral and inhalation routes, PLGA NPs were also surface-decorated with lectins that recognize glycosylated clusters [89]. The plasma half-life of the different drugs was prolonged significantly from 4–9 days (uncoated NPs) to 6–14 days (coated NPs) upon oral/inhalation routes in mice, respectively. Moreover, the administration of three doses (one every 14 days) resulted in complete clearance of the mycobacterium. In a more recent work, rifampicin-loaded PLGA NPs efficiently cleared *M. bovis* BCG infection in macrophages *in vitro* [90]. Moreover, the authors revealed that they remained bounded to the membrane of phagolysosomes.

O'Hara and Hickey developed respirable rifampicin-loaded PLGA MPs [91]; alveolar macrophages are the main mycobacterium reservoir in nonresistant TB. Other researchers also explored the inhalation route [92, 93]. For example, Yoshida *et al.* evaluated the phagocytic uptake of rifampicin-loaded PLGA MPs by alveolar macrophages and the killing of *M. bovis* Calmette-Guérin [92].

More recently, Son and McConville coated rifampicin micro-crystals with PLGA or PLA for inhalation [94]. However, the performance of the delivery systems *in vivo* was not assessed. PLGA NPs were also used to adjust injectable antibiotics such as streptomycin to the oral route and prolong their release when compared to an intramuscular injection [95].

It is worth stressing that even though PLA is available in different forms, the fully amorphous P(D,L)LA was the most extensively used probably due to the faster release profile and biodegradation when compared to semi-crystalline counterparts.

PCL has been also the matter of research for the encapsulation of anti-TB drugs, though to a more limited extent. For example, Parikh and Dalwari encapsulated isoniazid in PCL MPs for inhalation therapy [96]. The production technique was spray-drying and the mass median aerodynamic diameter that governs the deposition in the airways ranged between 1.9 and 4 μm. However, the release was sustained only for a 1–2 hours, this phenomenon stemming from the high aqueous solubility of the drug. Other drug nanocarriers investigated to encapsulate antituberculosis drugs have been PCL-made polymeric micelles. In this framework, our research group reduced the gastric degradation of rifampicin in presence of soluble isoniazid and increased its oral bioavailability by up to 3.3 times [97, 98]. This constitutes the first attempt to develop a platform for the coadministration of rifampicin/isoniazid fixed dose combination in liquid form to children. The same nano-DDS was surface-decorated with chitosan and hydrolyzed galactomannan to confer mucoadhesiveness and active targeting features to deliver rifampicin to alveolar macrophages by inhalation [99].

Polyester particles were also investigated to encapsulate second-line antibiotics trialed for the treatment of resistant mycobacterium strains [100–102]. For example, ethionamide was detected in the lungs, liver, and spleen for 12 hours when administered orally in free form and 5–7 days when in PLGA NPs [100]. Similar results were obtained with the fluoroquinolone levofloxacin [102]. Other researchers developed other particles for inhalation containing antibiotics for MDR-TB such as ofloxacin-palladium complexes [103].

17.2.3 Polyester Particles in the Release of Antimalarials

Different nanotechnologies has been investigated to develop advanced antimalarial drug delivery systems [8]. Surprisingly, only a few works reported on the use of polyesters to encapsulate these drugs. Mosqueira *et al.* described PDLLA nanocapsules loaded with halofantrine, a poorly water-soluble drug related to quinine and lumefantrine, for intravenous release [104]. The efficacy of this formulation was evaluated in *P. berghei*-infected mice with similar or better activity, higher area-under-the-curve, and faster control of the parasite development during the first 48 hours post-infection than a solution [105]. Similarly, monensin, an antibiotic used in veterinary medicine, was encapsulated within PLGA NPs and the activity assessed *in vitro* against *Plasmodium falciparum* [106]. The particles showed a 10-fold increase of the efficacy with respect to the free drug. Others used PLGA particles to strengthen the immune response to antimalarial vaccines by the oral, intradermal, and intranasal routes, including the comprehensive attempts of Patarroyo and colleagues with the synthetic peptide SPf1066 [107–114].

17.3 Conclusions and Future Perspectives

High drug dose, poor oral drug bioavailability, frequent administration schedules, and prolonged treatments are a common characteristic of many curable and chronic infectious diseases affecting poor countries. This situation is aggravated by the fact that most of them demand combined therapies. Thus, the investigation of innovative pharmaceutical products has become crucial to improve the pharmacokinetic performance of drugs and enable a reduction of the administration regimens and a shortening of the treatment period. In this scenario, polyesters, the most extensively investigated and clinically trialed biodegradable polymers, will play a fundamental role. Other drawbacks of the therapy have to be urgently faced to improve efficacy. For example, the targeting of pathogen reservoirs and the design of novel prophylactic methods and vaccination. Polyesters will undoubtedly be preponderant also here. At the same time, it should be stressed that since only few research groups address these unique niches worldwide, in some cases these investigations may remain isolated efforts and the potential of these polymers undercapitalized. The awareness raised by the WHO and a broad spectrum of nongovernmental organizations in recent years on the necessity to strengthen these investigations has placed polyesters within easy reach because their biocompatibility by different administration routes has been extensively proven, as opposed to that of new experimental polymers. However, the impact that the incorporation of these biomaterials have on medication cost has to be pondered, especially in the case of PLA and PLGA derivatives that are more expensive than PCL. Only then, a balanced and realistic evaluation of their possible role in overcoming the main drawbacks of drugs in the therapy of PRDs could be assessed and by doing so to envision the most efficient pathways to reach the bench-to-bedside translation.

References

1. Murray, H. W., Pépin, J., Nutman, T. B., Hoffman, S. L., and Mahmoud, A. A. F. (2000). Tropical medicine, *BMJ*, **320**, 490–494.
2. Rassi Jr., A., Rassi, A., and Marin-Neto, J. A. (2010). Chagas disease, *Lancet*, **375**, 1388–1402.
3. Lozano, R. (2012). Global and regional mortality from 235 causes of death for 20 age groups in 1990 and 2010: a systematic analysis for the Global Burden of Disease Study 2010, *Lancet*, **380**, 2095–128.
4. Ravikumar, M. N. (2000). Nano and microparticles as controlled drug delivery devices, *J Pharm Pharm Sci*, **3**, 234–258.
5. Sosnik, A., Carcaboso, A., and Chiappetta, D. A. (2008). Polymeric nanocarriers: new endeavors for the optimization of the technological aspects of drugs, *Recent Pat Biomed Eng*, **1**, 43–59.
6. Sosnik, A., Chiappetta, D. A., and Carcaboso, A. (2009). Drug delivery systems in HIV pharmacotherapy: what has been done and the challenges standing ahead, *J Control Release*, **138**, 2–15.
7. Sosnik, A., Chiappetta, D. A., Moretton, M. A., Glisoni, R. J., and Carcaboso, A. (2010). New old challenges in tuberculosis: potentially effective nanotechnologies in drug delivery, *Adv Drug Deliv Rev*, **62**, 547–559.
8. Santos-Magalhães, N. S., and Mosqueira, V. C. F. (2010). Nanotechnology applied to the treatment of malaria, *Adv Drug Deliv Rev*, **62**, 560–575.
9. HIV/AIDS, WHO Fact Sheet. http://www.who.int/mediacentre/factsheets/fs360/en/ (accessed January 2015).
10. Thompson, M. A., Aberg, J. A., Hoy, J. F., Telenti, A., Benson, C., Cahn P., Eron, J. J., Günthard, H. F., Hammer, S. M., Reiss, P., Richman, D. D., Rizzardini, G., Thomas, D. L., Jacobsen, D. M., and Volberding, P. A. (2012). Antiretroviral treatment of adult HIV infection: 2012

recommendations of the International Antiviral Society-USA panel, *J Am Med Assoc*, **308**, 387–402.

11. Andrews, L., and Friedland, G. (2000). Progress in the HIV therapeutics and the challenges of adherence to antiretroviral therapy, *Inf Dis Clin N Am*, **14**, 1–26.
12. Sosnik, A. (2010). Nanotechnology contributions to the pharmacotherapy of pediatric HIV: a dual scientific and ethical challenge and a still pending agenda, *Nanomedicine (Lond)*, **5**, 833–837.
13. Guzman, M. G., Halstead, S. B., Artsob, H., Buchy, P., Farrar, J., Gubler, D. J., Hunsperger, E., Kroeger, A., Margolis, H. S., Martinez, E., Nathan, M. B., Pelegrino, J. L., Simmons, C., Yoksan, S., and Peeling, R. W. (2012). Dengue: a continuing global threat, *Nat Rev Microbiol*, **8**, S7–S16.
14. Beeching, N., Fenech, M., and Houlihan, C. F. (2014). Ebola virus disease, *BMJ*, **349**, g7348.
15. Global Tuberculosis Report, Geneva, Switzerland: World Health Organization, 2011, in, http://www.who.int/tb/publications/global_report/2011/gtbr11_full.pdf, (last accessed June 2013).
16. Norbis, L., Alagna, R., Tortoli, E., Codecasa, L. R., Migliori, G. B., and Cirillo, D. M. (2014). Challenges and perspectives in the diagnosis of extrapulmonary tuberculosis, *Expert Opin Anti-infective Ther*, **12**, 633–647.
17. Cole, S. T., Brosch, R., Parkhill, J., Garnier, T., Churcher, C., Harris, D., Gordon, S. V., Eiglmeier, K., Gas, S., Barry, 3rd, C. E., Tekaia, F., Badcock, K., Basham, D., Brown, D., Chillingworth, T., Connor, R., Davies, R., Devlin, K., Feltwell, T., Gentles, S., Hamlin, N., Holroyd, S., Hornsby, T., Jagels, K., Krogh, A., McLean, J., Moule, S., Murphy, L., Oliver, K., Osborne, J., Quail, M. A., Rajandream, M. A., Rogers, J., Rutter, S., Seeger, K., Skelton, J., Squares, R., Squares, S., Sulston, J. E., Taylor, K., Whitehead, S., and Barrell, B. G. (1998). Deciphering the biology of Mycobacterium tuberculosis from the complete genome sequence, *Nature*, **393**, 537–544.
18. WHO, The top ten causes of death. World Health Organization: Geneva, Switzerland, 2011, in, http://www.who.int/mediacentre/factsheets/fs310/en/index.html, (last accessed January 2015).
19. Singh, S., Mariappan, T. T., Sharda N., Kumar, S., and Chakraborti, A. K. (2000). The reason for an increase in decomposition of rifampicin in the presence of isoniazid under acid conditions, *Pharm Pharmacol Commun*, **6**, 405–410.

20. Sankar, R., Sharda, N., and Singh, S. (2003). Behavior of decomposition of rifampicin in the presence of isoniazid in the pH range 1–3, *Drug Dev Ind Pharm*, **29**, 733–738.

21. Lastória, C. J., and de Abreu, M. A. M. M. (2014). Leprosy: review of the epidemiological, clinical, and etiopathogenic aspects - Part 1, *An Bras Dermatol*, **89**, 205–218.

22. Tabbara, K. F. (2001). Trachoma: a review, *J Chemother*, **13** (Suppl. 1), 18–22.

23. World Malaria Report. World Health Organization, in, http://www.who.int/malaria/publications/world_malaria_report_2012/wmr2012_no_profiles.pdf, (last accessed January 2015).

24. WHO, World Malaria Report 2014, http://apps.who.int/iris/bitstream/10665/144852/2/9789241564830_eng.pdf, (last accessed January 2015).

25. WHO, Guidelines for the treatment of malaria, 2nd edition, http://whqlibdoc.who.int/publications/2010/9789241547925_eng.pdf?ua=1, (last accessed January 2015).

26. Kurth, F., Belard, S., Adegnika, A. A., Gaye, O., Kremsner, P. G., and Ramharter, M. (2010) Do paediatric drug formulations of artemisinin combination therapies improve the treatment of children with malaria? A systematic review and meta-analysis, *Lancet Infect. Dis.*, **10**, 125–132.

27. Rodrigues Coura, J., and Albajar Vinas, P. (2010). Chagas disease: a new worldwide challenge, *Nature,* **465**, S6–S7.

28. Pereira, P. C. M., and Navarro, E. C. (2013). Challenges and perspectives of Chagas disease: a review, *J Venom Anim Toxins Incl Trop Dis*, **19**, Art. 34.

29. Clem, A. (2010). A current perspective on leishmaniasis, *J. Glob. Infect. Dis.*, **2**, 124–126.

30. Gray, D. J., Ross, A. G., Li, Y.-S., and McManus, D. P. (2011). Diagnosis and management of schistosomiasis, *BMJ*, **342**, d2651.

31. Franco, J. R., Simarro, P. P., Diarra, A., and Jannin, J. G. (2014). Epidemiology of human African trypanosomiasis, *Clin Epidemiol*, **6**, 257–275.

32. Wynd, S., Melrose, W. D., Durrheim, D. N., Carron, J., and Gyapong, M. (2007). Understanding the community impact of lymphatic filariasis: a review of the sociocultural literature, *Bull World Health Organ*, **85**, 493–498.

33. Lazdins-Helds, J. K., Remme, J. H. F., and Boakye, B. (2003). Onchocerciasis, *Nat Rev Microbiol*, **1**, 178–179.

34. Moro, P., and Schantz, P. M. (2009). Echinococcosis: a review, *Int J Infect Dis*, **13**, 125–133.

35. Holland, S. J., Tighe, B. J., and Gould, P. L. (1986). Polymers for biodegradable medical devices. 1. The potential of polyesters as controlled macromolecular release systems, *J Control Release*, **4**, 155–180.

36. Kumar, N., Ravikumar, M. N., and Domb, A. J. (2001). Biodegradable block copolymers, *Adv Drug Deliv Rev*, **53**, 23–44.

37. Bala, I., Hariharan, S., and Kumar, M. N. (2004). PLGA nanoparticles in drug delivery: the state of the art, *Crit Rev Ther Drug Carrier Syst*, **21**, 387–422.

38. Garlotta, D. (2001). A literature review of poly(lactic acid), *J Polym Environ*, **9**, 63–84.

39. Woodruff, M. A., and Hutmacher, D. W. (2010). The return of a forgotten polymer: polycaprolactone in the 21st century, *Prog Polym Sci*, **35**, 1217–1256.

40. Sosnik, A., and Cohn, D. (2003). Poly(ethylene glycol)-poly(ε-caprolactone) block oligomers as injectable materials, *Polymer*, **44**, 7033–7042.

41. Carcaboso, A., Chiappetta, D. A., Höcht, C., Blake, M. M., Boccia, M. M., Baratti, C. M., and Sosnik, A. (2008). Melt-molding/compression manufacturing and *in vitro–in vivo* characterization of gabapentin-loaded poly(ε-caprolactone) implants for sustained release in animal studies, *Eur J Pharm Biopharm*, **70**, 666–673.

42. Carcaboso, A., Chiappetta, D. A., Opezzo, A. W., Höcht, C., Fandiño, A. C., Croxatto, J. O., Rubio, M. C., Sosnik, A., Bramuglia, G. F., Abramson, D. H., and Chantada, G. L. (2010). Episcleral implants for topotecan delivery to the posterior segment of the eye, *Invest Ophthamol Vis Sci*, **51**, 2126–2134.

43. Cohn, D., Lando, G., Sosnik, A., Garty, S., and Levi, A. (2006). PEO–PPO–PEO-based poly(ether ester urethane)s as degradable reverse thermo-responsive multiblock copolymers, *Biomaterials*, **27**, 1718–1727.

44. Cohn, D., and Younes, H. (1998). Biodegradable PEO/PLA block copolymers, *J Biomed Mater Res A*, **22**, 993–1009.

45. Cohn, D., and Hotovely-Salomon, A. (2005). Biodegradable multiblock PEO/PLA thermoplastic elastomers: molecular design and properties, *Polymer*, **46**, 2068–2075.

46. Cohn, D., Stern, T., González, M. F., and Epstein, J. (2002). Biodegradable poly(ethylene oxide)/poly(ε-caprolactone) multiblock copolymers, *J Biomed Mater Res A,* **59**, 273–281.

47. Liggins, R. T., and Burt, H. M. (2002). Polyether-polyester diblock copolymers for the preparation of paclitaxel loaded polymeric micelle formulations, *Adv Drug Deliv Rev,* **54**, 191–202.

48. Ahmed, F., and Discher, D. E. (2004). Self-porating polymersomes of PEG–PLA and PEG–PCL: hydrolysis-triggered controlled release vesicles, *J Control Release,* **96**, 37–53.

49. Ford, J., Khoo, S. H., and Back, D. J. (2004). The intracellular pharmacology of antiretroviral protease inhibitors, *J Antimicrob Chemother,* **54**, 982–990.

50. Boffito, M., Jackson, A., Amara, A., Back, D., Khoo, S., Higgs, C., Seymour, N., Gazzard, B., and Moyle, G. (2011). Pharmacokinetics of once-daily darunavir-ritonavir and atazanavir-ritonavir over 72 hours following drug cessation, *Antimicrob Agents Chemother,* **55**, 4218–4223.

51. Cahn, P., Andrade-Villanueva, J., Arribas, J. R., Gatell, M., Lama, J. R., Norton, M., Patterson, P., Sierra Madero, J., Sued, O., Figueroa, M. I., Rolon, M. J., and GARDEL Study Group (2014). Dual therapy with lopinavir and ritonavir plus lamivudine versus triple therapy with lopinavir and ritonavir plus two nucleoside reverse transcriptase inhibitors in antiretroviral-therapy-naive adults with HIV-1 infection: 48 week results of the randomised, open label, non-inferiority GARDEL trial, *Lancet Infect Dis,* **14**, 572–580.

52. Sax, P. A 2-drug approach: a shift in the HIV treatment paradigm. http://medscape.com/viewarticle/813931.

53. ter Heine, R., Scherpbier, H. J., Crommentuyn, K. M., Bekker, V., Beijnen, J. H., Kuijpers, T. W., Huitema, A. D. (2008). A pharmacokinetic and pharmacogenetic study of efavirenz in children: dosing guidelines can result in subtherapeutic concentrations, *Antiv Ther,* **13**, 779–787.

54. Shah, L. K., and Amiji M. M. (2006). Intracellular delivery of saquinavir in biodegradable polymeric nanoparticles for HIV/AIDS, *Pharm Res,* **23**, 2638–2645.

55. Ece Gamsiz, D., Shah, L. K., Devalapally, H., Amiji, M. M., and Carrier, R. L. (2008). A model predicting delivery of saquinavir in nanoparticles to human monocyte/macrophage (Mo/Mac) cells, *Biotechnol Bioeng,* **101**, 1072–1082.

56. Brichard, B., and Van der Linden, D. (2009). Clinical practice treatment of HIV infection in children, *Eur J Pediatr,* **168**, 387–392.

57. Rabel, S. R., Patel, M., and Sun, S. (2001). Electronic and resonance effects on the ionization of structural analogues of efavirenz, *AAPS Pharm Sci*, **3**, E28.

58. Gao, J. Zh., Hussain, M. A., Motheram, R., Gray D. A., Benedek, I. H., Fiske, W. D., Doll, W. J., Sandefer, E., Page, R. C., and Digenis, G. A. (2007). Investigation of human pharmacoscintigraphic behavior of two tablets and a capsule formulation of a high dose, poorly water soluble/highly permeable drug (Efavirenz), *J Pharm Sci*, **96**, 2970–2977.

59. Friedland, G., Khoo, S., Jack, C., and Lalloo, U. (2006). Administration of efavirenz (600 mg/day) with rifampicin results in highly variable levels but excellent clinical outcomes in patients treated for tuberculosis and HIV, *J Antimicrob Chemother*, **58**, 1299–1302.

60. Darwich, L., Esteve, A., Ruiz, L., Bellido, R., Clotet, B., and Martinez-Picado J. (2008). Variability in the plasma concentration of efavirenz and nevirapine is associated with genotypic resistance after treatment interruption, *Antivir Ther*, **13**, 945–951.

61. Fabbiani, M., Di Giambenedetto, S., Bracciale, L., Bacarelli, A., Ragazzoni E., Cauda, R., Navarra, P., and De Luca, A. (2009). Pharmacokinetic variability of antiretroviral drugs and correlation with virological outcome: 2 years of experience in routine clinical practice, *J Antimicrob Chemother*, **64**, 109–17.

62. Seremeta, K. P., Chiappetta, D. A., and Sosnik, A. (2013). Poly(ε-caprolactone), Eudragit® RS 100 and poly(ε-caprolactone)/Eudragit® RS 100 blend submicron particles for the sustained release of the antiretroviral efavirenz, *Colloids Surf B: Biointerfaces*, **102**, 441–449.

63. Tshweu, L., Katata, L., Kalombo, L., Chiappetta, D. A., Hocht, C., Sosnik, A., and Swai, H.(2014). Enhanced oral bioavailability of the antiretroviral efavirenz nano-encapsulated in poly(ε-caprolactone) nanoparticles by a spray-drying method, *Nanomedicine (Lond)*, **9**, 1821–1833.

64. Chiappetta, D. A., Hocht, C., Taira, C., and Sosnik, A.(2010). Efavirenz-loaded polymeric micelles for pediatric anti-HIV pharmacotherapy with significantly higher oral bioavailaibility, *Nanomedicine (Lond)*, **5**, 11–23.

65. Chiappetta, D. A., Hocht, C., Taira, C., and Sosnik, A. (2011). Oral pharmacokinetics of efavirenz-loaded polymeric micelles, *Biomaterials*, **32**, 2379–2387.

66. Seremeta, K. P. (2013). Encapsulation of antiretrovirals inpolymeric nano/microparticles for the optimization of the pharmacotherapy in the infection by the human inmunodeficiency virus (HIV). PhD Thesis.

Buenos Aires: Faculty of Pharmacy and Biochemistry, University of Buenos Aires; pp. 188.

67. Sosnik, A., das Neves, J., and Sarmento, B. (2014). Mucoadhesive polymers in the design of nano-drug delivery systems for administration by non-parenteral routes: a review, *Prog Polym Sci,* **39**, 2030–2075.

68. Hendrix, C. W., Cao, Y. J., and Fuchs, E. J. (2009). Topical microbicides to prevent HIV: Clinical drug development challenges, *Annu Rev Pharmacol Toxicol*, **49**, 349–375.

69. das Neves, J., Michiels, J., Ariën, K. K., Vanham, G., Amiji, M. M., Bahia, M. F., and Sarmento, B. (2012). Polymeric nanoparticles affect the intracellular delivery, antiretroviral activity and cytotoxicity of the microbicide drug candidate dapivirine, *Pharm Res,* **29**, 1468–1484.

70. das Neves, J., Araújo, F., Andrade, F., Michiels, J., Ariën, K. K., Vanham, G., Amiji, M., Bahia, M. F., and Sarmento, B. (2013). *In vitro* and *ex vivo* evaluation of polymeric nanoparticles for vaginal and rectal delivery of the anti-HIV drug dapivirine, *Mol Pharmaceutics,* **10**, 2793–2807.

71. Woodrow, K. A., Cu, Y., Booth, C. J., Saucier-Sawyer, J. K., Wood, M. J., and Saltzman, W. M. (2009). Intravaginal gene silencing using biodegradable polymer nanoparticles densely loaded with small-interfering RNA, *Nat Mater,* **8**, 526–533.

72. Schleifman, E. B., McNeer, N. A., Jackson, A., Yamtich, J., Brehm, M. A., Shultz, L. D., Greiner, D. L., Kumar, P., Saltzman, W. M., and Glazer P. M. (2013). Site-specific genome editing in PBMCs with PLGA nanoparticle-delivered PNAs confers HIV-1 resistance in humanized mice, *Mol Ther Nucleic Acids,* **2**, e135.

73. Tshweu, L., Katata, L., Kalombo, L., and Swai, H. (2013). Nanoencapsulation of water-soluble drug, lamivudine, using a double emulsion spray-drying technique for improving HIV treatment, *J Nanoparticle Res,* **15**, 1–11.

74. Seremeta, K. P., Martínez Pérez, S., López Hernández, O. D., Höcht, C., Taira, C., Reyes Tur, M. I., and Sosnik, A. (2014). Spray-dried didanosine-loaded polymeric particles for enhanced oral bioavailability, *Colloids Surf B: Biointerfaces,* **123**, 515–523.

75. Seremeta, K. P., Höcht, C., Taira, C., Cortez Tornello, P. R., Abraham, G. A., and Sosnik, A. (2015). Didanosine-loaded poly(ε-caprolactone) microparticles by a coaxial electrohydrodynamic atomization (CEHDA) technique, *J Mater Chem B,* **3**, 102–111.

76. Srivastava, S., and Sinha, V. R. (2011). Development and evaluation of stavudine loaded injectable polymeric particulate systems, *Curr Drug Del*, **8**, 436–447.

77. Destache, C. J., Belgum, T., Christensen, K., Shibata, A., Sharma, A., and Dash, A. (2009). Combination antiretroviral drugs in PLGA nanoparticle for HIV-1, *BMC Infect Dis*, **9**, Art. 198.

78. Destache, C. J., Belgum, T., Goede, M., Shibata, A., and Belshan, M. A. (2010). Antiretroviral release from poly(DL-lactide-*co*-glycolide) nanoparticles in mice, *J Antimicrob Chemother*, **65**, 2183–2187.

79. Shibata, A., McMullen, E., Pham, A., Belshan, M., Sanford, B., Zhou, Y., Goede, M., Date, A. A., and Destache, C. J. (2013). Polymeric nanoparticles containing combination antiretroviral drugs for HIV type 1 treatment, *AIDS Res Hum Retroviruses*, **29**, 746–754.

80. Chaowanachan, T., Krogstad, E., Ball, C., and Woodrow, K. A. (2013). Drug synergy of tenofovir and nanoparticle-based antiretrovirals for HIV prophylaxis, *PLoS One*, **8**, Art. e61416.

81. Kuo, Y. C., Lin, P. I., and Wang, C. C. (2011). Targeting nevirapine delivery across human brain microvascular endothelial cells using transferrin-grafted poly(lactide-*co*-glycolide) nanoparticles, *Nanomedicine (Lond)*, **6**, 1011–1026.

82. McConville, C., Major, I., Friend, D. R., Clark, M. R., Woolfson, D., and Malcolm, R. K. (2012). Development of polylactide and polyethylene vinyl acetate blends for the manufacture of vaginal rings, *J Biomed Mater Res Part B: Appl Biomater*, **100B**, 891–895.

83. Dutt, M., and Khuller, G. K. (2001). Sustained release of isoniazid from a single injectable dose of poly(DL-lactide-*co*-glycolide) microparticles as a therapeutic approach towards tuberculosis, *Int J Antimicrob Ag*, **17**, 115–122.

84. ul-Ain, Q., Sharma, S., and Khuller, G. K. (2003). Chemotherapeutic potential of orally administered poly(lactide-*co*-glycolide) microparticles containing isoniazid, rifampin, and pyrazinamide against experimental tuberculosis, *Antimicrob Agents Chemother*, **47**, 3005–3007.

85. Pandey, R., Zahoor, A., Sharma, S., and Khuller, G. K. (2003). Nanoparticle encapsulated antitubercular drugs as a potential oral drug delivery system against murine tuberculosis, *Tuberculosis*, **83**, 373–378.

86. Semete, B., Booysen, L., Lemmer, Y., Kalombo, L., Katata, L., Verschoor, J., and Swai, H. S. (2010). *In vivo* evaluation of the biodistribution and safety of PLGA nanoparticles as drug delivery systems, *Nanomed Nanotechnol Biol Med*, **6**, 662–671.

87. Pandey, R., and Khuller, G. K. (2006). Oral nanoparticle-based antituberculosis drug delivery to the brain in an experimental model, *J Antimicrob Chemother*, **57**, 1146–1152.

88. Pandey, R., and Khuller, G. K. (2004). Subcutaneous nanoparticle-based antitubercular chemotherapy in an experimental model, *J Antimicrob Chemother*, **54**, 266–268.

89. Sharma, A., Sharma, S., Khuller, G. K. (2004) Lectin-functionalized poly (lactide-*co*-glycolide) nanoparticles as oral/aerosolized antitubercular drug carriers for treatment of tuberculosis, *J Antimicrob Chemother*, **54**, 761–766.

90. Kalluru, R., Fenaroli, F., Westmoreland, D., Ulanova, L., Maleki, A., Roos, N., Paulsen Madsen, M., Koster, G., Egge-Jacobsen, W., Wilson, S., Roberg-Larsen, H., Khuller, G. K., Singh, A., Nyström, B., and Griffiths, G. (2013). Poly(lactide-*co*-glycolide)-rifampicin nanoparticles efficiently clear *Mycobacterium bovis* BCG infection in macrophages and remain membrane-bound in phago-lysosomes, *J Cell Sci*, **126**, 3043–3054.

91. O'Hara, P., and Hickey, A. J. (2000). Respirable PLGA microspheres containing rifampicin for the treatment of tuberculosis: manufacture and characterization, *Pharm Res*, **17**, 955–961.

92. Pandey, R., Sharma, A., Zahoor, A., Sharma, S., Khuller, G. K., and Prasad, B. (2003). Poly(DL-lactide-*co*-glycolide) nanoparticle-based inhalable sustained drug delivery system for experimental tuberculosis, *J Antimicrob Chemother*, **52**, 981–986.

93. Yoshida, A., Matumoto, M., Hshizume, H., Oba, Y., Tomishige, T., Inagawa, H., Kohchi, C., Hino, M., Ito, F., Tomoda, K., Nakajima, T., Makino, K., Terada, H., Hori, H., and Soma, G.-I. (2006). Selective delivery of rifampicin incorporated into poly (DL-lactic-*co*-glycolic) acid microspheres after phagocytotic uptake by alveolar macrophages, and the killing effect against intracellular Mycobacterium bovis Calmettee-Guérin, *Microbes Infection*, **8**, 2484–2491.

94. Sona, Y.-J., and McConville, J. T. (2012). Preparation of sustained release rifampicin microparticles for inhalation, *J Pharm Pharmacol*, **64**, 1291–1302.

95. Pandey, R., and Khuller, G. K. (2007). Nanoparticle-based oral drug delivery system for an injectable antibiotic – streptomycin, *Chemotherapy*, **53**, 437–441.

96. Parikh, R., and Dalwadi, S. (2014). Preparation and characterization of controlled release poly-ε-caprolactone microparticles of isoniazid for drug delivery through pulmonary route, *Powder Technol*, **264**, 158–165.

97. Moretton, M. A., Glisoni, R. J., Chiappetta, D. A., and Sosnik, A. (2010). Molecular implications in the nanoencapsulation of the antituberculosis drug rifampicin within flower-like polymeric micelles, *Colloids Surf B: Biointerfaces*, **79**, 467–479.

98. Moretton, M. A., Hotch, C., Taira, C., and Sosnik, A.(2014). Encapsulation of rifampicin within flower-like polymeric micelles enhances the oral bioavailability in a pediatric fixed dose combination with isoniazid, *Nanomedicine (Lond)*, **9**, 1635–1650.

99. Moretton, M. A., Chiappetta, D. A., Andrade, F., das Neves, J., Ferreira, D., Sarmento, B., and Sosnik, A. (2013). Hydrolyzed galactomannan-modified nanoparticles and flower-like polymeric micelles for the active targeting of rifampicin to macrophages, *J Biomed Nanotechnol*, **9**, 1076–1087.

100. Kumar, G., Sharma, S., Shafiq, N., Pandhi, P., Khuller, G. P., and Malhotra, S. (2011). Pharmacokinetics and tissue distribution studies of orally administered nanoparticles encapsulated ethionamide used as potential drug delivery system in management of multi-drug resistant tuberculosis, *Drug Del*, **18**, 65–73.

101. Kumar, G., Malhotra, S., Shafiq, N., Pandhi, P., Khuller, G. P., and Sharma, S. (2011). *In vitro* physicochemical characterization and short term *in vivo* tolerability study of ethionamide loaded PLGA nanoparticles: potentially effective agent for multidrug resistant tuberculosis, *J Microencapsul*, **28**, 717–728.

102. Kumar, G., Sharma, S., Shafiq, N., Khuller, G. P., and Malhotra, S. (2012). Optimization, *in vitro–in vivo* evaluation, and short-term tolerability of novel levofloxacin-loaded PLGA nanoparticle formulation, *J Pharm Sci*, **101**, 2165–2176.

103. Palazzo, F., Giovagnoli, S., Schoubben, A., Blasi, P., Rossi, C., and Ricci, M. (2013). Development of a spray-drying method for the formulation of respirable microparticles containing ofloxacin–palladium complex, *Int J Pharm*, **440**, 273– 282.

104. Mosqueira, V. C., Legrand, P., and Barratt, G. (2006). Surface-modified and conventional nanocapsules as novel formulations for parenteral delivery of halofantrine, *J Nanosci Nanotechnol*, **6**, 3193–3202.

105. Mosqueira, V. C. F., Loiseau, P. M., Bories, C., Legrand, P. Devissaquet, J. P., and Barratt, G. (2004). Efficacy and pharmacokinetics of intravenous nanocapsule formulations of halofantrine in Plasmodium berghei-infected mice, *Antimicrob Agents Chemother*, **48**, 1222–1228.

106. Surolia, R., Pachauri, M., and Ghosh, P. C. (2012). Preparation and characterization of monensin loaded PLGA nanoparticles: *in vitro* anti-

malarial activity against *Plasmodium falciparum*, *J Biomed Nanotechnol*, **8**, 172–181.

107. Men, Y., Gander, B., Merklet, H. P., and Corradin, G. P. (1996). Induction of sustained and elevated immune responses to weakly immunogenic synthetic malarial peptides by encapsulation in biodegradable polymer microspheres, *Vaccine*, **14**, 1442–1450.

108. Rosas, J. E., Hernández, R. M., Gascon, A. R., Igartua, M., Guzman, F., Patarroyo, M. E., and Pedraz, J. L. (2001). Biodegradable PLGA microspheres as a delivery system for malaria synthetic peptide SPf66, *Vaccine*, **14**, 4445–4451.

109. Carcaboso, A. M., Hernández, R. M., Igartua, M., Gascón, A. R., Rosas, J. E., Patarroyo, M. E., and Pedraz, J. L. (2003). Immune response after oral administration of the encapsulated malaria synthetic peptide SPf66, *Int J Pharm*, **260**, 273–282.

110. Carcaboso, A. M., Hernández, R. M., Igartua, M., Rosas, J. E., Patarroyo, M. E., and Pedraz, J. L. (2004). Enhancing immunogenicity and reducing dose of microparticulated synthetic vaccines: single intradermal administration, *Pharm Res*, **21**, 121–126.

111. Carcaboso, A. M., Hernández, R. M., Igartua, M., Rosas, J. E., Patarroyo, M. E., and Pedraz, J. L. (2004). Potent, long lasting systemic antibody levels and mixed Th1/Th2 immune response after nasal immunization with malaria antigen loaded PLGA microparticles, *Vaccine*, **22**, 1423–1432.

112. Mata, E., Carcaboso, A. M., Hernández, R. M., Igartua, M., Corradin, G., and Pedraz, J. L. (2007). Adjuvant activity of polymer microparticles and Montanide ISA 720 on immune responses to Plasmodiumfalciparum MSP2 long synthetic peptides in mice, *Vaccine*, **25**, 877–885.

113. Mata, E., Igartua, M., Patarroyo, M. E., Pedraz, J. L., and Hernández, R. M. (2011). Enhancing immunogenicity to PLGA microparticulate systems by incorporation of alginate and RGD-modified alginate, *Eur J Pharm Sci*, **44**, 32–40.

114. Moon, J. J., Suh H., Polhemus, M. E., Ockenhouse, C. F., Yadava, A., and Irvine, D. J. (2012). Antigen-displaying lipid-enveloped PLGA nanoparticles as delivery agents for a Plasmodium vivax malaria vaccine, *PLoS One*, **7**, e31472.

Chapter 18

Polyester Nano- and Microtechnologies for Tissue Engineering

Namdev B. Shelke,[a,b,c] **Matthew Anderson,**[a,b,c] **Sana M. Idrees,**[a,b,c] **Jonathan Nip,**[a,b,c] **Sonia Donde,**[b,c] **Xiaojun Yu,**[f] **Gloria Gronowicz,**[g] **Sangamesh G. Kumbar,**[a,b,c,d,e]

[a]*Institute for Regenerative Engineering, UCONN HEALTH, Farmington, CT, USA.*
[b]*Raymond and Beverly Sackler Center for Biomedical, Biological, Physical and Engineering Sciences, UCONN HEALTH, Farmington, CT, USA.*
[c]*Department of Orthopaedic Surgery, UCONN HEALTH, Farmington, CT, USA.*
[d]*Department of Biomedical Engineering, University of Connecticut, Storrs, CT, USA.*
[e]*Department of Materials Science and Engineering, University of Connecticut, Storrs, CT, USA.*
[f]*Department of Chemistry, Chemical Biology and Biomedical Engineering Stevens Institute of Technology, Hoboken, NJ 07030, USA.*
[g]*Department of Surgery, UCONN HEALTH, Farmington, CT, USA.*
kumbar@uchc.edu

Polyesters such as poly(L-lactic acid) (PLLA), poly(L-lactic-*co*-glycolic acid) (PLGA), and poly(ε-caprolactone) (PCL) have been explored extensively to fabricate nano- and microdevices for tissue regeneration and drug delivery applications. They are FDA approved and their applications in tissue engineering span from bone

Handbook of Polyester Drug Delivery Systems
Edited by MNV Ravikumar

ISBN 978-981-4669-65-8 (Hardcover), 978-981-4669-66-5 (eBook)
www.panstanford.com

regeneration to muscle regeneration. The use of polyesters in tissue engineering (TE) voids the use of autografts and allografts, where these are considered as a gold standard for tissue regeneration purposes. Moreover, they are used widely for delivering growth factors. The benefits of using them for tissue engineering include suitable biocompatibility, biodegradability, mechanical properties, ease of processing for required size and duration of degradation, and cell response. This chapter focuses on nano- and microdevices of PLLA, PLGA, PCL, their copolymers, blends, and possible surface functionalization while discussing their applications for tissue engineering applications. Additionally, their applications with various cells and growth factors for bone, cartilage, ligament, tendon, nerve, and muscle are also discussed.

18.1 Introduction

Tissue engineering is a multidisciplinary field which combines biomaterial based scaffolds with cells and active agents such as growth factors. Tissue engineering can be used to repair the function of damaged tissue, maintain organ functionality, or to replace a fully damaged organ. There are nearly 200 million people that suffer from various musculoskeletal disorders [1] annually in the United States. It has become increasingly important to treat these disorders using tissue engineering (TE) applications. TE applications span from organ transplantation to drug delivery systems and over the past few decades have been used for the replacement of heart valves, ankle joints, and intervertebral discs [2–4]. The old practices of organ transplantation such as autografts and allografts have shown various limitations, including tissue availability, donor site morbidity, and possible disease transmission. Tissue engineering has evolved as an alternative discipline to replace damaged tissues and utilizes polymeric scaffolds, cells, and growth factors alone or in combination. As researchers learn more about the applicability of different polymer systems to various disease processes, the application of tissue engineering to medicine becomes increasingly boundless.

While tissue engineering is a viable alternative to current tissue replacement procedures, it is important to highlight the factors of tissue engineering that limit its effectiveness [5]. First, while transplanted stem cells have the potential to differentiate into many different cell types, it is important to have directed differentiation of these cells. In order to achieve this, scaffolds have been used to direct stem cell growth until their own extracellular membranes can guide self-growth [6]. Specifically, scaffold technology has been used in pediatric bladder repair, promotion of angiogenesis in wound healing [7], and regeneration of bone, articular cartilage, ligament, tendon, nerve, and muscle in musculoskeletal diseases [8–10]. Second, the cells need growth factors to help them differentiate into the appropriate cell type. Previously, boluses of growth factors have been used at the site of tissue regeneration, many of the growth factors diffuse out of the injection spot and their effect is not sufficient for the required biological activity or large doses of growth factors are required for efficacy, causing more side effects and greater cost [11–13].

The major challenge in constructing tissue scaffolds is their long-term stability, which is helpful for the improved vascularization and exchange of nutrients upon implantation. The applications and uses of biodegradable poly(L-lactic acid) (PLLA), poly(L-lactic-*co*-glycolic acid) (PLGA), and poly(ε-caprolactone) (PCL) devices such as sponges, nano-/microfibers, and porous scaffolds fabricated using various techniques for tissue engineering have far reaching implications for promoting human health in the future [14, 15]. Instead of replacing damaged tissue with nonnative materials such as grafts or engineered implants, the use of biological substitutes to regenerate tissues is being widely investigated. Targets for research include bone, cartilage, tendon, ligament, nerve, and muscle; polyesters are successful materials that have been shown to have excellent utility for these applications. Reasonably good biocompatibility, mechanical strength, ease of modulation, and biodegradability makes these polymers attractive for drug delivery and tissue regeneration applications. From applications ranging from nerve regeneration to muscle reconstruction, scaffolds can be made, and are used to create viable tissues that can help patients to recover from trauma, surgeries, and other injuries to the body.

To mimic the properties of biological substitutes using polyester scaffolds, cells including preosteoblasts, chondroblasts, and pluripotent stem cells are seeded onto them, where they mature and regenerate damaged tissues. During this time, scaffolds are expected to degrade at the same rate as tissue development to allow proper tissue regeneration and remodeling. Scaffolds are designed to be mechanically competent and biocompatible to support the cells as they mature. Once they fully develop, the tissues that are created are functional and viable for use by the body. To further enhance the success of the cells implanted with the scaffolds, growth factors are used to promote the enhanced tissue regeneration. Figure 18.1 represents therapeutic targets and sites of scaffold implantation for various tissue regeneration applications [16]. In many cases scaffolds alone may not be enough to support tissue regeneration but may need induction in the form of proteins and factors for the implanted

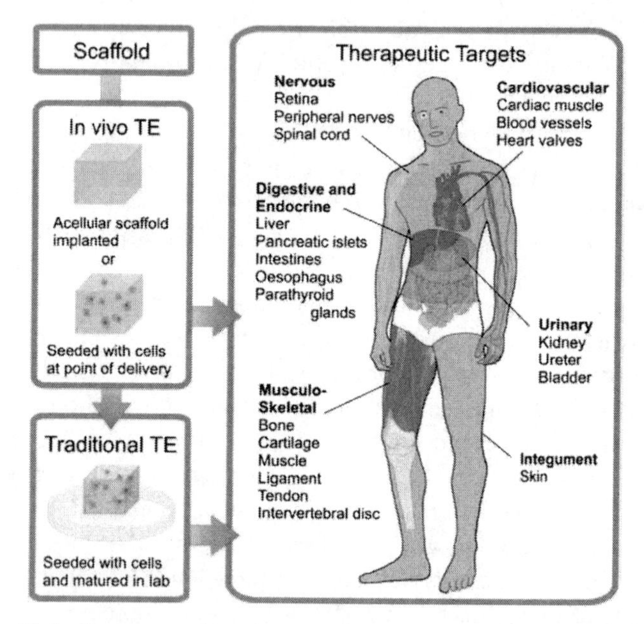

Figure 18.1 Representative image for various tissue engineering approaches for the implantation of scaffolds with or without cells. Traditional tissue engineering approach which has scaffolds seeded with cells are allowed to mature *in vitro* followed by implantation and in another approach scaffolds are implanted directly with or without seeding of the cells. The various sites for the therapeutic targets using the scaffolds for the implantations are also presented. Reproduced from Ref. 16, with permission of The Royal Society of Chemistry.

cells. The human body is an immensely complex system and multiple factors contribute to tissue remodeling and regeneration. Bioactive factors can also be incorporated into the scaffolds to help support the seeded cells as they grow. Growth factors are used to regulate cellular responses and guide the development and maturation of the cells seeded onto the scaffolds. Usually, growth factors are encapsulated within the bulk of the three-dimensional (3D) scaffold and are contained until the scaffold material degrades away. The encapsulation protects them from degradation for an extended period of time during tissue regeneration and are released gradually so that they do not to perfuse the tissue with too much factor at any given time. The timed release is calibrated to give the cells the optimal dosage for their growth as well as decreasing adverse side effects that may occur. Factors that are used include growth factors or hormones for promoting cell growth as well as antimicrobial agents for protecting the cellular environment against bacterial infection.

The aforementioned polyesters have a long-standing medical history as tissue substitutes and drug delivery devices because of their programmable mechanical and degradation properties. These polymers are adopted to produce devices to mimic the architecture of the natural human tissue at the nanometer scale, which allows for the development of new systems that mimic the complex, hierarchical structure of the native tissue [17, 18]. Polyesters can be processed into 3D structures using either solvent or melt processing to create scaffolds with high surface-area-to-volume ratios that facilitate cell adhesion, proliferation, migration, and differentiation [19]. Electrospinning has been acknowledged as a practicable and versatile technique for the fabrication of fibers in nanoscale dimension with large surface-to-volume ratio, high porosity, mechanical strength, and bio-mimicking of the physiologic microenvironment of extracellular matrix (ECM) [20]. Polymeric nano- and microstructured scaffolds produced using these polyesters by electrospinning have been popularly used for a variety of tissue regeneration applications and have produced the most promising results [21–23]. Polymeric nanofibers have been also used to modify the device surface to promote cellular activity [24]. Polyester nanocomposites with ceramics have been developed to improve mechanical and scaffold functional properties. In general, nanocomposite-based matrices at the optimized concentrations exhibit superior balance between strength and toughness, which often improves device characteristics as compared to neat polymeric matrices. Nanotopographies mimic the native ECM structurally and promote cell adhesion and spreading [16, 25]. Figure 18.2 illustrates the scaffold architecture and cellular events that can promote cell attachment and spreading. Figure 18.2 indicates that cells could attach and spread if the substrate surface is larger than the width of the cell, which presents two-dimensional (2D) surfaces to the cells, whereas in the case of fibrous scaffolds, where they present on the 3D surface, the cellular attachments and spreading are more influenced by the matrix interactions, which are similar to the native cell and ECM interactions at the site of implantation.

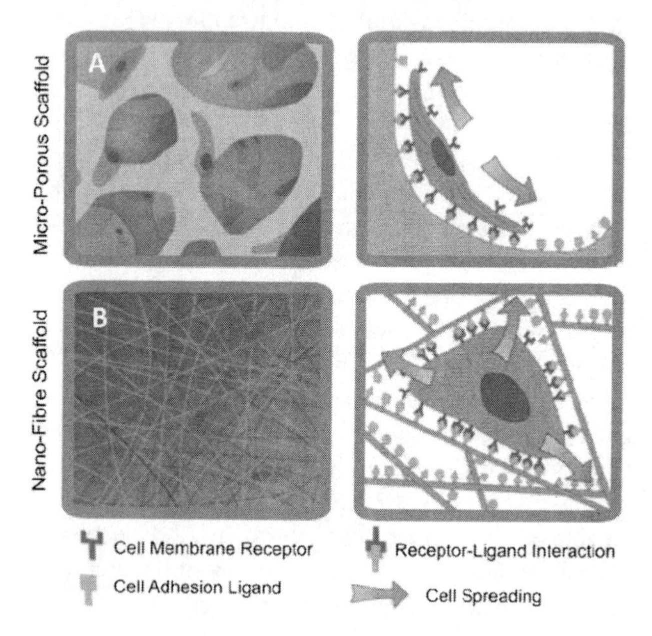

Figure 18.2 The architecture of the scaffold affects the cell binding and nature of the cell spreading. A. Scaffolds with much larger feature dimensions such as width than the cell acts as a 2D surface and helps for the attachment and spreading in flattened patterns. B. Scaffolds with fibrous architecture acts as a 3D surface to the cells and helps for cell attachment and spreading naturalistically. Reproduced from Ref. 16, with permission of The Royal Society of Chemistry.

Delivery of bioactive factors, including proteins, peptides, and genetic material, in scaffolds has provided multiple advantages over conventional pharmacotherapy. It is not appropriate to introduce growth factors as a bolus injection due to their relative short half-life (few minutes to hours), requiring frequent and high doses to achieve a therapeutic concentration at the target site. This approach also results in undesired side effects when introduced systematically and is economically not viable. Drug delivery systems have thus emerged as a novel solution to overcome the aforementioned limitations associated with conventional bolus injection. Polymeric scaffolds work as growth factor reservoirs and protect factors from degradation, release the desired quantity for an extended period locally at the desired site, and overcome the limitations associated with nonspecificity, dosing, and undesired side effects [11]. A variety of polyester-based systems have been developed and are constantly being modified to provide sustained release with increased availability of bioactive factors and drugs while decreasing the side-effect profile that plagues conventional drug therapies [26, 27]. These polyesters have a well-defined composition and molecular weight designed to deliver factors and to create scaffolds with mechanical properties and degradation parameters to suit the needs of tissue repair and regeneration. They have been popularly adopted for bone, cartilage, ligament, tendon, nerve, and muscle regeneration applications. The presence of hydrolytically labile ester linkages allow them to undergo hydrolytic degradations with the degradation products being metabolized. The thermoplastic nature possessed by these polyesters is the well-defined thermal feature that makes them attractive for the processing into different sizes and shapes using injection or compression molding in addition to the solution processing. Protocols have been also established to modify the surface of the polyester devices to immobilize bioactive factors and modify the surface hydrophilicity [28–31]. Along with the design of TE scaffolds and their interactions with the cells, other important factors such as physical forces also plays major role in the tissue regeneration. Numerous methodologies have been implemented using various polyesters for their use in TE applications. We have limited our discussions to the aforementioned polyester-based scaffolds with various cells and factors to repair and regenerate various musculoskeletal tissues and their interfaces. The micro- and nanofiber-based, micro- and nanoparticle-based, and micro- and nanoporous structure–based devices for the repair and regeneration of bone, cartilage, tendon, ligament, nerve, and muscle have been discussed in the further sections.

18.2 Applications of Nano-/Microfabricated Polyester Devices

18.2.1 Bone Tissue Engineering

The fundamental concepts of the molecular processes that govern bone regeneration and general requirements for delivery vehicles have revolutionized the field of regenerative medicine. Factor delivery strategies should mimic and aid molecular and cellular pathways mediating bone formation seen during embryogenesis, post-fracture healing, and surgical osseous fusion [32]. Bone healing following a trauma predominately occurs through intramembranous and/or endochondral ossification. In stable conditions, direct intramembranous ossification occurs when mesenchymal stem cells (MSCs) are transformed into osteoblasts, leading to the regeneration of lamellar bone. In most of the cases, however, post-fracture healing follows the endochondral ossification pathway [33]. It involves a complex cascade of spatiotemporally regulated interactions of multiple growth factors, cytokines, hormones, and recruitment of MSCs for the generation of a transitory cartilaginous callus. This callus is vascularized, calcified, and finally remodeled to form mature lamellar bone architecture [34]. Unfortunately, despite the considerable potential for regeneration, 5–10% of all fractures result in delayed healing or non-unions [35]. The gold standards of autografts, allografts, and vascularized grafts are limited to their application of bone tissue engineering due to requiring specific bone size, viability of the host site, i.e., donor site morbidity, loss of bone inductive factors, and major surgical procedures. Disease transmission and infections can also ensue. A bone tissue engineering approach relies on the use of polymeric bone graft substitutes termed scaffolds to fill the bone defect. The porous graft must have suitable mechanical properties, and the interconnected porous structure should facilitate cell migration and nutrient exchange. A representation of bone at various macroscopic and microscopic levels demonstrating strong calcified layers, osteons, coating of residential cells into the cell membrane receptors, and nanostructured ECM is illustrated [25]. In general, efforts are made to incorporate these features into the scaffold design which are fabricated using polyesters to mimic bone structure and composition.

It is imperative to understand the properties of polyesters for their applications to tissue engineering. Hydrolytically labile ester linkages in polyesters respond differently with respect to various buffering conditions. The degradation behavior of PLGA (75:25) *in vitro* was studied at various pH levels and found to mimic the degradation profile *in vivo* at pH 5.0, which simulates the acidic environment from activated macrophages, pH 7.4 of normal physiological fluid, and pH 6.4, an intermediate pH. For the first 4 months, the degradation profiles appear similar for all pH conditions except for scaffolds at pH 5, at which the degradation rate increased. Over 8 months, scaffolds maintained at a pH of 5 lost 90% of their weight, while those maintained in pH 7.4 and 6.4 lost about 30% of initial weight. The PLGA75/35 foams maintained their 3D structure up to 6 months at physiological pH, which is expedient for bone tissue engineering applications [36]. Numerous studies indicate that PLGA in various compositions and devices composed of PLGA with specific degradation rates could meet requirements of bone tissue engineering applications [37–39]. To obtain scaffolds for prolonged applications with slow degradation profiles and good mechanical strength, PLLA polymers are being explored. Compared to PLGA, PLLA polyesters degrade slowly and yield mechanically competent scaffolds. PLLA yields L-isomer of lactic acid upon degradation, which is a major biological metabolite and is already present in physiological fluid [40]. A three dimensional porous scaffold using the PLLA polymer has been tested for various tissue regeneration applications [41, 42]. The PCL polyesters possess highly elastic properties than other and are explored for the various applications in tissue engineering [43].

To develop more appropriate biological substitute for bone tissue engineering, stem cells and growth factors are used along with polymeric scaffolds. Stem cells seeded with the scaffolds differentiate into desired cell types upon implantation. Biomaterial scaffolds direct the stem cell growth and stem cells eventually supplement endogenous cell activity along with the secretion of bioactive factors, which increase bone regeneration at the repair site [44]. Addition of growth factors to scaffolds is an additional factor which can further enhance the tissue healing process. The controlled release of these growth factors over extended period of time are well known for the effective growth of healing tissues. The delivery of growth factors using microporous PGA scaffolds over 21 days enhances radiopacity at the implant site as compared to the control implant site in rat models [45]. PCL microporous scaffolds fabricated using selective laser sintering technique and loaded with morphogenetic protein-7 (BMP-7) with mechanical strengths close to lower range of human trabecular bone retains biological activity subcutaneously and promote ectopic bone formation [46]. Several common strategies have been adopted to incorporate osteoinductive factors with polyester scaffolds, including BMPs, fibroblast growth factor (FGF), insulin-like growth factors-1 (IGF), platelet-derived growth factors (PDGF), and vascular endothelial growth factors (VEGF) to promote bone regeneration [47–51]. Owing to their extraordinary pleiotropic effect and short systemic half-life, localized delivery vehicles with well-controlled kinetics are a basic criterion to guarantee safe and efficacious therapeutic application of growth factors [52]. BMP has been delivered for an extended period of time using a variety of polyesters, most commonly including PLLA, PLGA, and PCL [53–55]. The improved cell–biomaterial interaction at the nanoscale level and dramatic advances in biomaterial fabrication technologies are encouraging for developing nanoencapsulation methods for growth factor delivery. Encapsulation of rhBMP-2 in PLGA/HAp composite fibers could be achieved through electrospinning technique. Composite nanofibers with good morphology and mechanical strength of the nanofibers as well as sustained *in vivo* release and bioactivity can be achieved [56]. PLGA nanospheres loaded with rhBMP7 can also be used with controlled release kinetics and produce successful ectopic bone formation with nanofibers in an animal model [57]. Apart from the delivery of osteogenic factors, regeneration of large-size bone defects relies on a mature network of blood vessels for the transportation of essential nutrients, circulating biologic factors, stem cells, and oxygen. The role of FGF and VEGF is to stimulate endothelial cells to secrete proteases and plasminogen activators for the degradation of the basement membrane to facilitate entry of endothelial progenitor cells, additional angiogenic growth factors, and supporting cells for the formation of new vessels from preexisting vasculature [58]. For instance, the delivery of VEGF using PLGA scaffolds for an extended period of time enhances neovascularization and bone formation significantly compared to PLGA scaffolds used in the same manner [59].

Hydrophobic polymer devices with the enhanced hydrophilicity using various approaches are recognized for providing enhanced cell attachment, proliferation, and osteogenesis. Surface modifications which help enhance hydrophilic properties are created either by physical or by chemical means to generate or introduce functional groups on the surface [60]. For instance, introduction of hydroxylated glycolide, or polysaccharides such as cyclodextrin on PCL porous scaffolds results in enhanced MSC attachment, proliferation, and osteogenic differentiation as compared to control PCL [61, 62]. Apart from the use of functional hydrophilic polymers the presence of ionic groups such as phosphates also improves the surface hydrophilicity. The presence of phosphate groups results in increased amounts of mineral (hydroxyapatite (HA)) deposition and enhances the cell viability and metabolic activity of osteogenic and chondrogenic cell lines [63]. Similarly, modifications of these scaffolds with proteins also show enhanced cell adhesion and proliferation [64]. Addition of silica nanoparticles using layer-by-layer self-assembly techniques helps to improve the hydrophilicity of the PCL fiber matrix. This electrostatically assisted deposition improves the fiber wettability and surface roughness, resulting in enhanced cell attachment, proliferation, and alkaline phosphatase (ALP) expression [65]. Polyester blends along with other hydrophilic polymers have also been adopted in scaffold design to improve performance. They combine the benefits of the parent polymers and overcome the limitations associated with them. For instance, polymeric microfibers produced from the blends of PLGA and Pluronic® F-108 (PF) show increased fiber hydrophilicity with respect to increase in PF contents. MSCs stretch along the length of the fibers showing preferential cell adhesion. This also results in increased MSC adhesion, proliferation, infiltration, differentiation, and mineralization as compared to fibers of parent polymers [66]. Similarly, increased mineral depositions with respect to increase in surface hydrophilicities of the fiber meshes are noticed (Fig. 18.3).

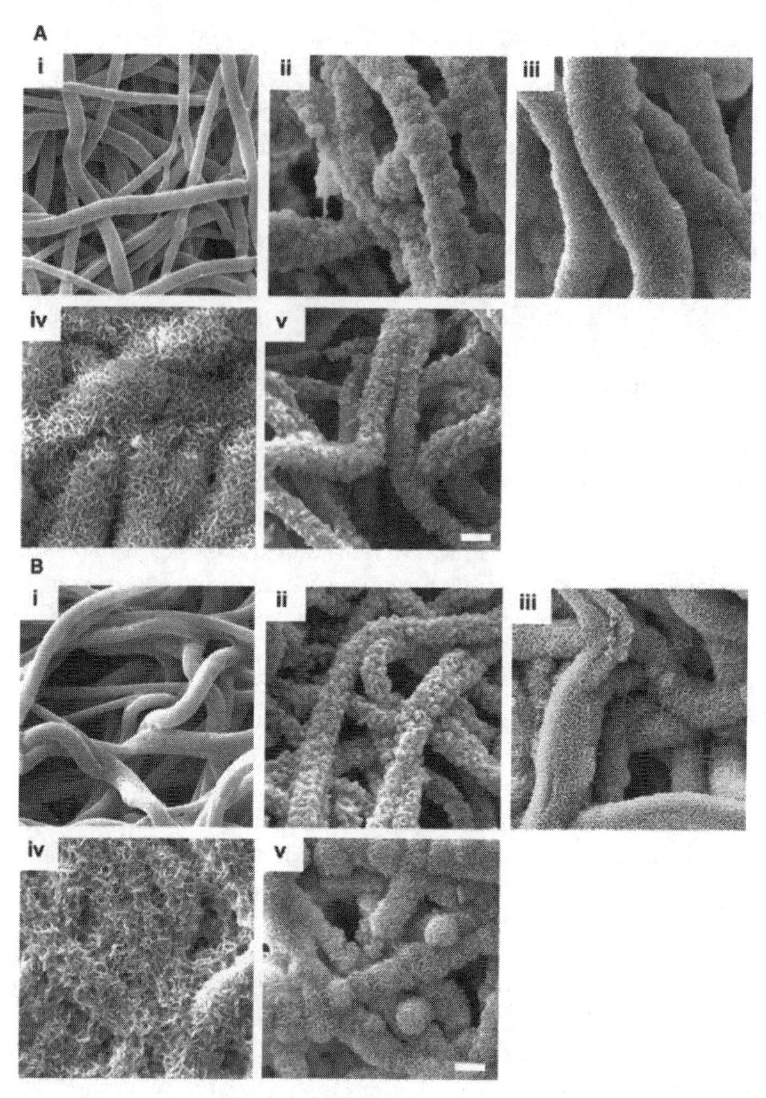

Figure 18.3 Scanning electron micrographs of electrospun PGA blended with Pluronic® F-108 (PF) fiber meshes with various compositions (i–v correspond to PLGA, 0.5 PF, 1 PF, 1.5 PF and 2 PF respectively). Blends of varying surface hydrophilicity incubated in 2.5 × simulated body fluid (SBF) for (A) 3 days, and (B) 6 days respectively increased mineral depositions with respect to increase in surface hydrophilicities (scale bar = 2 μm). Reprinted from Ref. 66, Copyright 2014, with permission from Elsevier.

Polysaccharide blends with polyesters have been reported to enhance *in vitro* mineralization as well as promote cell adhesion and proliferation [67, 68]. Electrospun nanofibers were created from a blend of PCL with chitosan in an effort to improve PCL hydrophilicity and provide free function groups to facilitate bioactive molecule tethering [8]. This PCL–chitosan blends nanofibers with tethered type I collagen via carbodiimide chemistry and shows significant increase in MSCs adhesion, spreading, and proliferation as compared to controls. Inclusion of an ECM protein results in elevated levels of osteogenic gene expression and mineralization as compared to control scaffolds. This approach would be helpful for cell growth and mineralization due to the compositional, structural, and biological similarities for bone tissue engineering applications [69, 70]. Ongoing modifications, such as addition of polyethylene glycol (PEG), development of plastic PEG, and insertion of *p*-dioxanone into a PLLA-PEG block improve its degradation characteristics and increase the osteogenic capabilities of recombinant rhBMP2 [53]. Thermosensitive polymeric hydrogel-based polyesters for biomedical applications have been used extensively because of their dual hydrophobic and hydrophilic nature [71]. Novel polymers including PLLA-DX-PEG exhibit unique temperature-dependent liquid–semisolid transitions that permit percutaneous injection of the delivery system to circumvent invasive surgical procedures [53]. The correct configuration of growth factor delivery carriers is as important as the selection of appropriate carrier materials and immobilization methods. Configuration of these osteogenic carrier materials ranges from complex 3D scaffolds, thermosensitive hydrogels, and micro-/nanoscale particles to nanofibers based on natural and/or synthetic polymers [10, 68, 72].

Polyester-based bioceramic composite materials also have been adopted for bone tissue engineering to obtain mechanical competent scaffolds. These polymeric composites are essentially an effort to mimic the native bone material. For instance, bone is a natural composite material comprised of collagen and HA mineral phase arranged in an intricate hierarchical fashion. Hydroxyapatite provides bioactivity and regulates the MSCs differentiation into the osteogenic lineage. It possesses high compressive strength and is also brittle in nature. Owing to its high crystallinity, HA undergoes remodeling very slowly in the bone healing environment.

Polyester composites with HA overcomes material brittleness and provides much desired bioactivity and hydrophilicity to the polyester scaffolds for bone healing applications [73–76]. PLGA-HA scaffolds implanted in a bone defect for 8 weeks in athymic mice demonstrated extensive bone formation [73]. A variety of nanofiber matrices with HA resulted in enhanced cell adhesion, proliferation, and osteoblast phenotypic expression *in vitro* [77, 78]. HA disks coated with PLGA polymers acquired compressive modulus in the range of 54–91 MPa and were found to be suitable for engineering cortical bone [79]. Zhang *et al.* [74] fabricated porous structured PCL-HA spiral scaffolds with various compositions of PCL:HA and tested effects of HA on bone regeneration properties. Figure 18.4 shows the scaffold design, which illustrates the micro-nano features. These studies reported enhanced cell proliferation and osteogenic differentiation that was dependent on HA content in the scaffold. For instance, spiral scaffolds with a PCL: HA ratio of 4:1 had optimal mineralization and gene expression as compared to other compositions. An example of osteoblast morphology and viability is on these spiral structures is presented in Fig. 18.5. To improve the mechanical strength of scaffolds various designs have

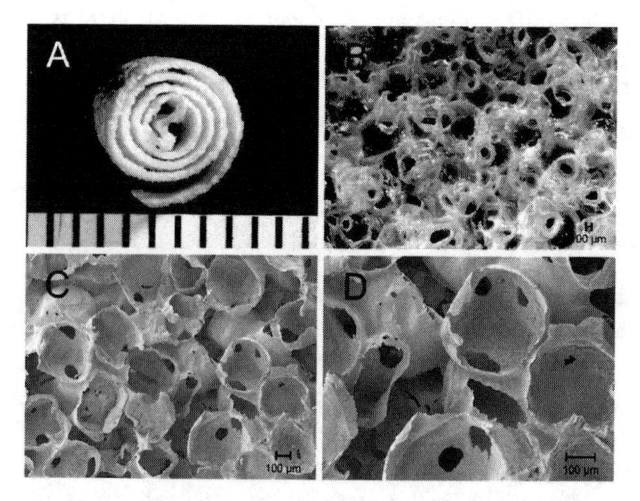

Figure 18.4 Representative photographs of nano-HA/PCL spiral scaffold. Gross view (A), stereomicroscope image (B), and SEM of nano-HA/PCL spiral scaffold indicating morphologies. Reprinted from Ref. 74, Copyright 2014 Zhang *et al.*

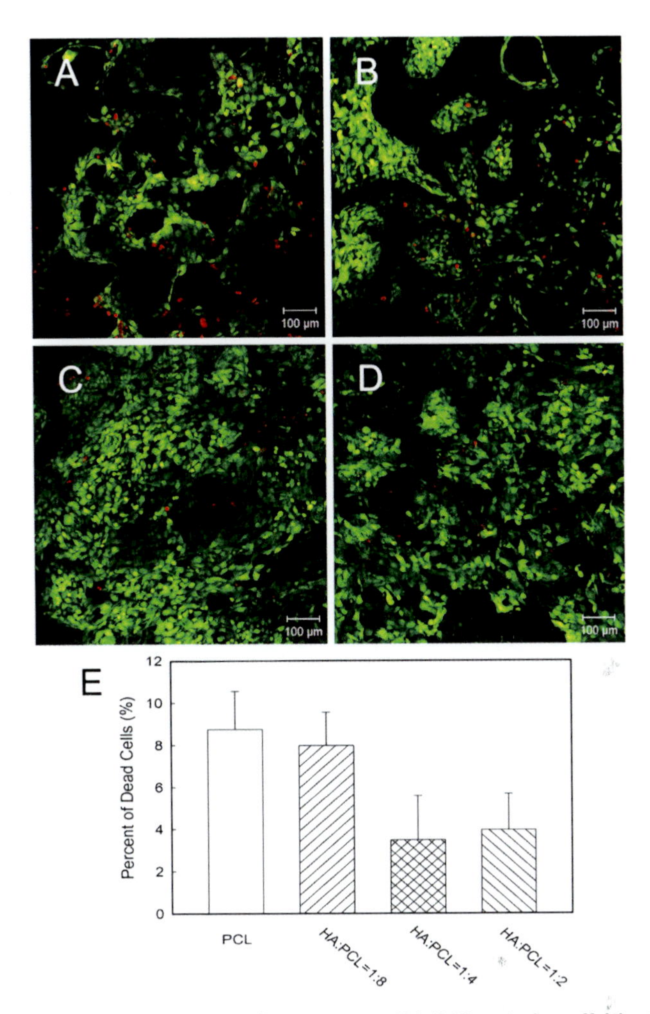

Figure 18.5 Cell viability studies on nano-HA/PCL spiral scaffolds. Spiral scaffolds fabricated using HA: PCL compositions of 0:1 (A), 1:8 (B), 1:4 (C), and 1:2 for 7 days and the quantitative analysis of percent live dead cells (E) within these spiral scaffolds. For live dead assay live cells were stained green, dead cells were stained red, and nuclei were stained blue. Reprinted from Ref. 74, Copyright 2014 Zhang *et al.*

been explored. Wang *et al.* [80] developed PCL micro-porous spiral scaffolds coated with nanofibers and tested mechanical strengths, cell attachments, proliferation, mineralization, and differentiation.

The porosities of these scaffolds were similar to human trabecular bone and could be structurally appropriate for cellular infiltration. The spiral structured scaffolds incorporated with nanofibers showed enhanced cellular proliferation, mineralization, and differentiation as compared to cylindrical scaffolds. Micro- and nanoparticle-based scaffolds have emerged as attractive delivery vehicles because of their minute dimension, high surface-area-to-volume ratio, drug-loading efficiency, and adaptability to environmental stimuli [81]. Borden *et al.* reported on thermally sintered PLGA microsphere scaffolds with a pore architecture bio-mimicking the porosity and mechanical strength of trabecular bone capable of effective osteoblast activities [82]. Tissue repair is not only limited to the design of scaffolds and topography but it is also stimulated because of the presence of external mechanical strains. The mechanical stimuli generated by ultrasound or cyclic strains *in vitro* are found useful in bone-healing processes [83]. These stimuli would mimic the *in vivo* conditions since the living tissues are exposed to various types of strains [84]. The mechanical stimuli have effects on both proliferation and differentiation. One of the designs that mimic the *in vivo* mechanical strains, bioreactors are used where cyclic strains and ultrasound are applied. The porous scaffolds composed of PCL and PLLA seeded with preosteoblasts and stimulated in presence of cyclic stain, ultrasound, and combined strain indicated a significant increase in the expression of osteogenic genes when compared to only mechanical strain [83].

18.2.2 Cartilage Tissue Engineering

The basic elements for cartilage tissue engineering are scaffolds providing a dynamic 3D support for the delivery of stem cells, and biochemical and biomechanical cues for the purpose of structural and functional restoration of the original tissue. Because they have similar mechanical properties to native articular cartilage, modifiable degradation rates, and established safety in other FDA-approved applications, synthetic PLGA microparticles have been widely utilized for the delivery of growth factors to promote chondrogenesis. For clinical applications, injectable delivery systems are often necessitated for focal chondral defects of irregular geometry. Star-shaped PLLA scaffolds self-assembled into nanofibrous hollow microspheres as an injectable cell carrier mimic the structural features of ECM [85, 86]. Inspired by favorable cell behavior at the nanoscale level, many processing techniques have been developed to fabricate fibrillar structures mimicking the native tissue architecture. Integration of biomimetic signals such as glycosaminoglycans (GAGs), proteins, or short peptides have also been reported to impact cell proliferation and differentiation. The porous PCL scaffolds loaded with type II collagen and chondroitin sulfate improves the surface wettability due to a larger amount of secreted GAGs and collagen, which results in promoted proliferation of chondroblasts [87].

Because of the poor intrinsic regenerative capacity of articular cartilage, much effort has been dedicated to identify optimal cell sources. Chondrocytes, fibroblasts, stem cells, and genetically modified cells have been examined for their potential as a feasible cell source. Use of progenitor cells, in particular MSCs, is a promising strategy due to their multi-lineage potential [88]. MSCs can be easily isolated from bone marrow, adipose tissue, synovial membrane, periosteum, trabecular bone, umbilical cord blood, and other sources [89]. Long-term clinical outcomes of articular lesions treated using mesenchymal progenitor cells with HA have been evaluated in randomized and controlled trials and demonstrate better results than with HA alone [90]. Addition of bone marrow-derived stem cells have been associated with the expression of multiple markers for hypertrophic chondrogenesis such as collagen type X, matrix metalloproteinase 13, and readily mineralizes when exposed to osteogenic stimuli [91]. The chondrogenic potential of stem cells derived from different tissues has been demonstrated to be highly dependent on the combination of different growth factors. For instance, the adipose-derived stem cells (ASCs) shows higher upregulation of aggrecan gene expression in response to BMP-6, while bone marrow-derived MSCs responds more favorably to TGF-β3 [92]. Various growth factor delivery scaffolds used for cartilage tissue engineering applications have improved effect on cell proliferation, differentiation, and production of factors. The current approaches for cartilage regeneration accentuate the application of "smart" scaffolds, providing a three-dimensional microenvironment similar to native cartilage tissue. Apart from growth factor and structure of the scaffolds, chondrocytes from different zones of cartilage responds differently to mechanical stimuli. Also growth factor are studied along with various mechanical strains for the chondrogenic gene expressions and have shown synergetic effect of these stimuli. Studies also have shown that combination of growth factor and mechanical stimuli have synergetic effect on collagen production [93].

PLLA, PLGA, and PCL nano-/microscaffolds seeded with chondrocytes are studied widely due to their suitable mechanical properties, ability to promote cell infiltration, biocompatibility, and biodegradability. Studies have also shown that the fiber diameter of these polymers plays an important role in cell attachment and proliferation. For instance, the biomechanical strength, cell attachment and neocartilage response were optimal for the polyglycolic acid (PGA) fibers in a range of 0.8-3 μm fiber constructs [94]. The cells distribute evenly throughout the nonwoven fibers in this range. Upon implantation of auricle-shaped scaffolds of these nanofibers in autogenous large animal models, these fibers demonstrate cartilage formation evenly all over the surface of nanofiber scaffolds. The cartilage formation is primarily present on outer surfaces of the nanofiber scaffolds of 0.8 and 20 μm fiber size. The gross morphology and safranin O staining of the auricle-type scaffold 20 weeks after implantation have been observed. These studies indicate that the PGA fibers of a specific size range could help the induction of cartilage regeneration.

Cartilage tissues are soft in nature and the use of flexible scaffolds mimic the native tissue's mechanical properties. Use of the PCL, which has elastic properties, makes scaffolds suitable for cartilage regeneration applications. PCL scaffolds implanted in rabbits for over three months to repair articular had similar mechanical properties compared to regular articular cartilage [95]. The nanostructured PCL scaffolds, which are flexible in nature and fabricated using a thermally induced phase separation and lyophilization technique, showed expression of chondrogenic genes such as sox9, collagen type I, II, and aggrecan after 6 days of seeding with rabbit chondrocytes. Also upon implantation of these scaffolds in New Zealand white rabbits, the femoral defective site indicated higher histological scores. When compared with commercially available collagen (chondro-Gide) scaffolds, the PCL nanostructured scaffolds showed significantly improved gene expression and histological scores respectively [96]. To improve the mechanical strength and flexibility for cartilage tissue regeneration PLCL, hyaluronan, and fibrin gel scaffolds with a 300–500 μm pore size range and with 80% porosity using a gel pressing method were fabricated [97–99]. These scaffolds showed a 94% extension without losing their original form. Upon subcutaneous implantation of scaffolds with chondrocytes in nude mice for 5–8 weeks, ECM accumulation, mature chondrocytes, and chondrocyte tissue are noticed. Similarly, highly flexible and porous PLCL sponge scaffolds with rabbit bone marrow-derived MSCs in rabbit knee joints show subchondral bone formation, cartilaginous matrix formation, and the presence of collagen and glycosaminoglycan [100]. These results indicate the usefulness of PLCL sponge scaffolds for osteochondral regeneration. Elastic PCL-based PLCL scaffolds provide a suitable environment to facilitate cartilage growth. Moreover, seeding of these scaffolds with scaffolds enhances the biological substitution ability. The perichondrocyte-PLLA composite porous scaffolds supports cartilaginous growth with firm cartilage formation in 96% of the animals when implanted in femoral condyles of adult white rabbits [101]. Additionally, the scaffolds' architecture is helpful for the alignment of the cells at the implantation site. The web-structured scaffolds of PLGA fabricated by a knitting process shows excellent mechanical support. Moreover, the presence of collagen within these scaffolds supports cell attachment and proliferation. When seeded with bovine chondrocytes, these scaffolds maintain cell morphology, and produce type II collagen and aggrecan as a part of cartilaginous ECM.

In addition, blends of polyesters with polysaccharides are explored for cartilage tissue engineering. Blend scaffolds of chitosan and PCL with various compositions produce glycosaminoglycan formation as evidenced by significant neo-cartilage formations [102]. Comparatively higher amounts of chitosan with respect to PCL in scaffolds produce higher amounts of glycosaminoglycan. Figure 18.6 describes the ECM formation over 21 days in PCL-blended chitosan fiber scaffolds. The surface modification of polyesters with RGD peptide and collagen seeded with chondrocytes enhances production of glycosaminoglycan, indicating neo-cartilage formation in the case of scaffolds modified with collagen type II [103]. Similarly, blends of PLLA-PLGA shows better cartilage formation than PLLA alone.

Moreover, PLLA-PLGA modified with collagen type II demonstrates cartilage repair after six months without any inflammation or any capsule formation. These studies demonstrate that the blending approach has potential to repair cartilage [104].

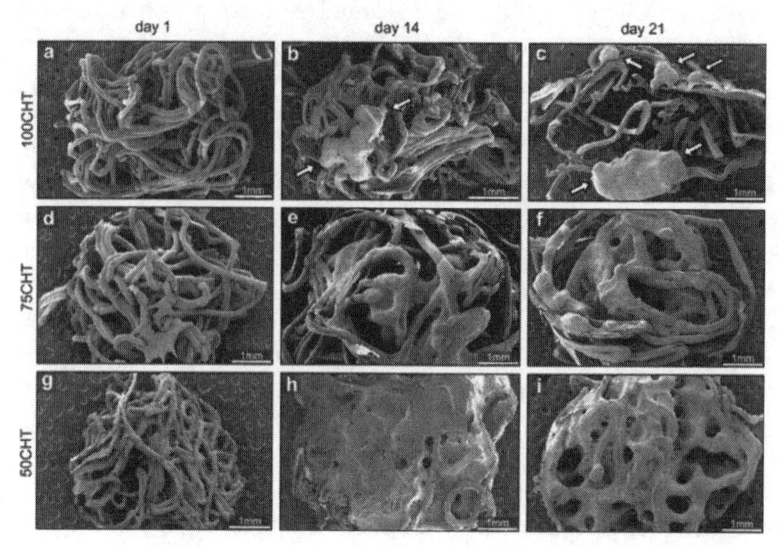

Figure 18.6 SEM micrographs of 100% chitosan (a), 25% PCL with chitosan (b), and 50% PCL with chitosan scaffolds showing ECM distribution over 1, 14, and 21 days in culture with differentiation medium. ECM aggregates are indicated using arrows. Reprinted from Ref. 102, Copyright 2011, with permission from Elsevier.

The repair and homeostasis of articular cartilage is mediated by a large number of growth factors such as TGF-β, FGF, IGF, and platelet-derived growth factor along with other soluble factors. Apart from TGF-β, the most studied growth factors for cartilage regeneration are BMP-2, BMP-7, and growth differentiation factor-5 (GDF-5). Multiple growth factors, either individually or in combination, have been explored in an attempt to enhance chondrogenesis. Connective tissue growth factor (CTGF), also called CCN2, has unique properties for promoting proliferation and differentiation of articular chondrocytes. Hence the damaged articular cartilage can be repaired using CTGF. It was also noticed that the level of CTGF increased in osteoarthritic rat models, which would indicate that CTGF may play an important role in repairing damaged cartilage.

Nishida *et al.* [105] studied therapeutic use of CTGF in rat models. The injection of CTGF within hydrogels effectively repairs the damaged cartilage. Studies using these growth factors, MSCs, or chondrocytes with polyesters show advances for cartilage regeneration. Zhu *et al.* transfected BMSCs with CTGF and showed the effects of the growth factor on cartilage repair. BMSCs transfected with CTGF, and BMSCs seeded onto PLGA, and sodium hydroxide (NaOH)-treated PLGA scaffolds significantly stimulated proliferation and chondrogenic differentiation *in vitro*. *In vivo* studies using this approach demonstrated that NaOH-treated PLGA scaffolds with CTGF-BMSCs could be the suitable for high-loading sites in large osteochondral defects and promote the formation of hyaline-like cartilage, similar to normal cartilage. Thus, stem cell proliferation and differentiation can be controlled to express specific cell lineages using a transfection and scaffold modification approach [106].

18.2.3 Ligament and Tendon Tissue Engineering

Polyesters have been very useful for biomedical applications in which retention of strength at the implantation site is required for an extended period of time. Such applications include ligaments, tendons, vascular stents, and urological stents [107]. The degradation period of polyesters varies from months to several years and similarly the mechanical strengths of these polyesters are variable with respect to degradation. For instance, 3D porous scaffolds of PLLA have been created for culturing different cell types used in various cell-based gene therapies for cardiovascular diseases, muscle tissues, bone and cartilage regeneration, and other treatments [42, 108, 109]. The PLLA may take between 10 months to 4 years to degrade, depending on microstructural factors, such as chemical composition, porosity, and crystallinity, that can also influence properties like tensile strength [110, 111]. These polymers have already shown favorable results in the fixation of fractures and osteotomies [112, 113].

Polyester scaffolds with a biodegradable, biocompatible structure and with suitable mechanical and structural properties that mimic the properties of patient's native tissues have been explored for ligament tissue engineering. The main advantage of using polyester-based scaffolds for ligament regeneration is the availability of scaffolds with needed mechanical properties, porosity, and biodegradation to promote minimal patient morbidity, minimal surgery and risk of infection, and rapid recovery. Materials like PLCL have been used to regenerate the interface between ligament and bone, providing a sturdy support against which a ligament can support itself [114]. Lee *et al.* fabricated PLCL porous scaffolds infused with a heparin-based hydrogel that contained the desired growth factor. Upon implantation, the mechanical properties of the PLCL scaffold matched the native tissue and the enhanced calcification of the fibrocartilage [114]. These studies indicate the suitability of PLCL polyesters for the tissues with high flexibility and strength.

Braiding of polyester fibers have emerged as a novel technique for various TE applications [115]. In particular, PLLA, PLGA, and PLCL scaffolds formed with braid, twist, and braid-twist fibers, which are highly flexible with good mechanical strength and can mimic the native tissue properties potentially for ACL reconstruction [116, 117]. The cellular response studies for ligament tissue regeneration using PLLA-collagen hybrid braid scaffolds indicated homogeneous cell distribution and proliferation. When tested for medial collateral ligament (MCL) regeneration in the ruptured MCL of the Wistar rat, cell migration as well as new blood vessel formation occurred in the scaffolds. PLLA braids with collagen showed improved cell migration and new blood vessel formation compared with PLLA braids without collagen. Chinks in the PLLA-collagen fibers were filled with collagen and comparatively less collagen was seen on PLLA-only fiber bundles (Fig. 18.7). Similarly, PLGA knitted mesh scaffolds with collagen and fibroblasts seeded on them promoted collagen formation (Fig. 18.8a–c) [117, 118]. Light and SEM images as shown in Fig. 18.9a,b respectively represent the adhesion and spreading of hASCs over time on to the PLLA and PCL blend nanofibers. In comparison, nanofiber scaffolds with 1:1 ratio of PLLA and PCL (1/1 NFs) supported cell attachments and proliferation very well [119]. Scaffolds composed of PLGA, which has the desired mechanical properties, including a high modulus of elasticity, high tensile strength, low ductility, and toughness, have shown advantages for various ligament regeneration applications [120]. The use of PLGA has been investigated for periodontal ligaments to facilitate guided tissue regeneration [121, 122]. To alter their physical and chemical properties to improve cell interactions or drug delivery profiles, PLLA, PLGA, and PCL have been modified using methoxy poly(ethylene glycol) (mPEG) and/or proteins such as fibronectin. Use of these hydrophilic polymers with polyesters lowers the glass transition temperature, thus increasing the elasticity and flexibility of the resulting copolymer while increasing hydrophilic nature of the polymers [123]. The hydrophilicity of PLGA for periodontal ligament tissue regeneration was altered by surface modifications using fibronectin [124]. The hydrophilic nature of fibronectin improved surface hydrophilicity and cell adhesion on the nanofibers without affecting the mechanical properties of nanofibers for periodontal ligament regeneration application.

Figure 18.7 The SEM images of PLLA and PLLA-collagen in the lower and upper micrographs respectively. As indicated in PLLA-collagen hybrid micrographs the chinks of the fiber bundle were filled with collagen (scale bar 100 μm, with 60 ×). Reprinted from Ref. 118, Copyright 2001, with permission from Elsevier.

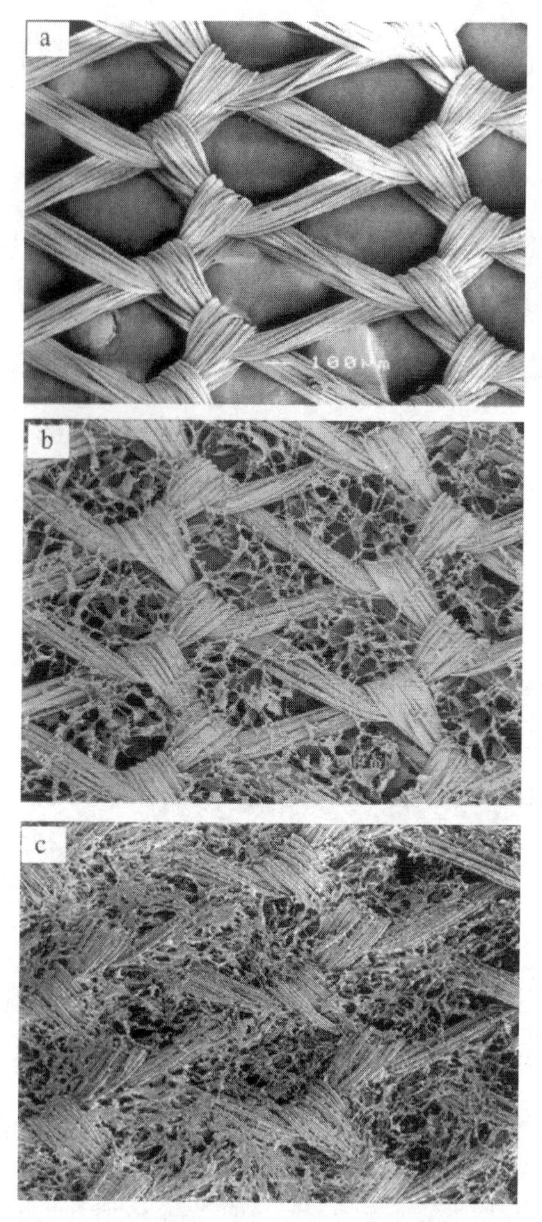

Figure 18.8 The comparative SEM images of knitted PLGA (a), PLGA ±collagen hybrid (b), and PLGA ±collagen mesh cultured with human fibroblasts for one day (c). Reprinted from Ref. 117, with permission from John Wiley and Sons.

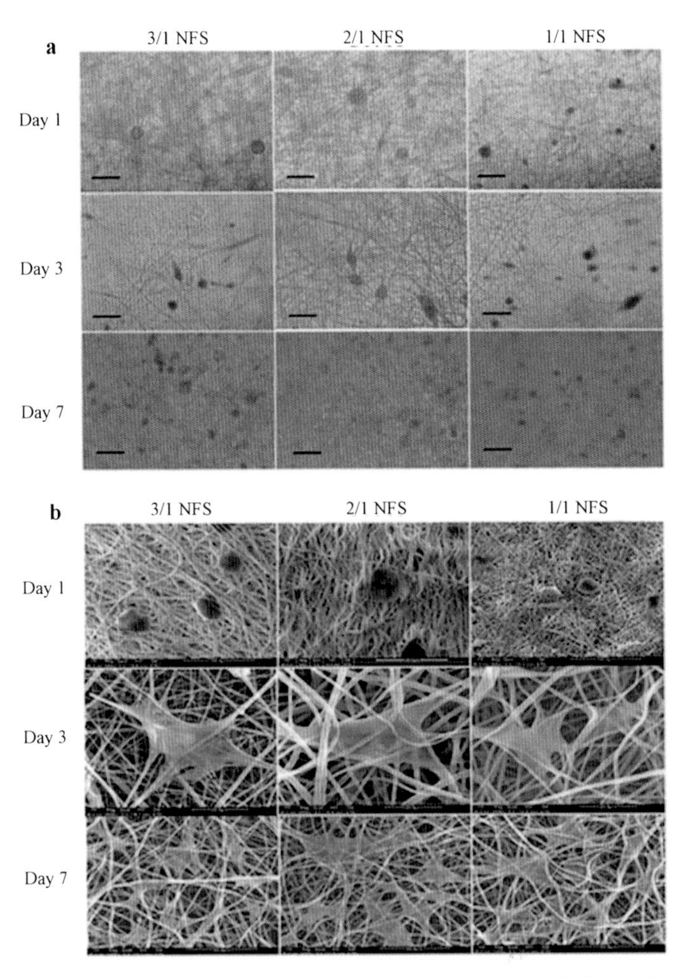

Figure 18.9 Light microscope and SEM images (a and b) indicating morphology of hASCs seeded on different PLLA: PCL blend nanofibers. Images with light microscopy (scale bar = 100 μm) and SEM (scale bar = 50 μm) indicates round shaped cells on day 1 after seeding on NFs. Cells extended on day 3 and day 7 after seeding on NFs for all blends. Also SEM for day 7 samples indicates that white film of hASCs with irregular shape attached on the fibers and grew in number from day 3 to day 7. Reprinted from Ref. 119, Copyright 2013 Chen *et al.*

The modulus of native crimp-like ACL is approximately 111 MPa [125]. Efforts have been made to determine the modulus of PLLA, and poly(L,D-lactic acid) (PLDLA) that can approach the crimp-like nature of native ACL during tissue regeneration. PLLA polymers with a slow degradation rate and suitable mechanical properties compared to PLGA are tested for ACL regeneration [126, 127]. These nanofiberous scaffolds turn into a crimp-like structure when treated in aqueous medium under heated conditions, which is similar to the native collagen structure at the implantation site. These scaffolds show a slight decrease in strength compared to the initial strength due to the polymer degradation after treatment of aqueous medium over 6 months. Also, the scaffolds fabricated using PLDLA show a slight decrease in elastic modulus as compared to the PLLA. These results demonstrate that scaffolds are stable upon implantation. Bovine fibroblast attachment, proliferation, and generation of tendon ECM fascicles are prevalent upon seeding with these nanofibers. The formation of ECM fascicles was also shown in native ligaments, indicating the potential of PLLA-based nanofibers for ACL regeneration. The response of these structures under strain is also an important factor when considering that the injury site undergoes strain during recovery. To understand these parameters, PLDLA scaffolds with a crimp-like structure are seeded with bovine fibroblasts and uniaxial tensions with 5%, 10%, and 20% of strain amplitudes was applied [128]. The strains show various effects on collagen synthesis and sulfated proteoglycan synthesis. Crimped scaffolds with a strain of 10% shows production of collagen with the down-regulation of proteoglycan whereas the scaffolds with 20% strain upregulates both proteins when compared to un-crimped fiber scaffolds and scaffolds without strain. Studies also indicate that the un-crimped nanofibers under strain enhances fibroblast proliferation compared to crimped nanofibers and nanofibers without any load and under dynamic mechanical stimulations enhances the ECM fascicle formation on crimped nanofiber scaffolds.

Scaffolds created with a 3D porous structure are a common strategy in tissue engineering where the optimal porous structure has optimum cell infiltration and nutrient exchange, increasing cell proliferation. The porosities of scaffolds have been optimized in many ways. Sahoo *et al.* [129] shows PLGA-knitted scaffolds and electrospun PLGA nanofibers coated onto them has large surface area to enhance cell infiltration. Seeding of porcine bone marrow stromal cells onto the nanofibers scaffolds shows promoted cell attachment, proliferation, and ECM synthesis as compared to PLGA knitted scaffolds coated with fibrin gel. Additionally, the expression of collagen I and genes such as decorin and biglycan were highly evident. These observations indicate the PLGA could be used optimally for ligament regeneration applications.

To support the growth of fibroblasts and fibro-chondrocytes to regenerate the ligament, growth factors such as BMP-2 are integrated into the scaffold. BMP-2 is a member of the transforming growth factor β (TGF-β) family and it is reported that use of BMP increases mechanical properties such as stiffness and tensile strengths, doubling them in a chicken model [130]. Studies also show that the coating of scaffolds with BMP-2 improves the osteo-integration which occurs between the ligament and bone [131]. BMP is osteogenic and induces osteoblastic differentiation of various cell types including embryonic stem cells, bone marrow stromal cells, and synovium-derived progenitor cells. It is also able to promote chondrogenic differentiation, maintain chondrocyte phenotypes in 3D systems, and stimulate cartilage proteoglycan synthesis. Basic fibroblast growth factor (bFGF) also has been used to improve the therapeutic efficacy of ACLs after surgery [132]. The plain woven braided scaffolds of PLLA loaded with bFGF in gelatin and then further wrapped in to a collagen membrane shows the enhanced mechanical strengths of regenerated ACL tissue as compared to the scaffolds without bFGF with the significant bone regeneration around the scaffold.

Protecting the tissues from infection is important after surgeries. Chemotherapeutic drugs are used to shield tissues from bacteria. Some of these drugs include penicilins, tetracyclines, nitro-imidazoles, and quinolone antibiotics. One example of a polyester ligament scaffold loaded with chemotherapeutic drugs is tetracycline, which is used to combat pathogens that hinder the wound healing process and to inhibit metalloproteases involved in tissue destruction. Tetracycline is the optimal example but any drug with antibacterial or anti-inflammatory characteristics can be used. This concept can be applied to scaffolds created for other purposes as well as for periodontal ligament scaffolds. Any set or combination of drugs that are most beneficial to the seeded cells can be loaded, as long as the scaffold is designed to disperse the drugs with an optimal release profile [122].

A tendon is a dense connective tissue between muscle and bone and plays an important role in musculoskeletal tissue engineering.

The use of polyesters including PLLA and its respective copolymers with PGA has been previously proposed for tendon tissue regeneration. The polyesters have adequate physical properties to sufficiently support growing tissue growth for these applications. As discussed previously, growth factors may be integrated into the scaffold to support and maintain tissue growth and development. Tendon tissue repair using biomaterials is widely preferred due to its reduced costs related to surgeries and reduced donor site damage. For instance, tendon surgical repair is required for many injuries including hand and wrist injuries, which account for almost one in five emergency room visits. Using a tissue engineering scaffold approach with or without growth factors could change the approach clinicians take towards healing injuries and thus possibly reduce costs in the future. The suitable mechanical strength of a scaffold upon implantation is important for tendon regeneration. Polyesters could be obtained in various mechanical strengths for these applications. The number of fiber strands in the tendon scaffold can also be changed. Lee *et al.* [133] tested mechanical strengths of PLCL scaffolds under strain in biological conditions. PLCL scaffolds seeded with tenocytes obtained from rabbit Achilles tendons via static and dynamic cell seeding methods shows improved mechanical strengths in dynamic cell culture due to enhanced ECM formation from enhanced cell proliferation and secretion of collagen type I as compared to the static cell culture method. From these studies it can be noted that the elastic and mechanical properties of PLCL polymers are maintained for a certain period of time upon implantation, which are useful for tendon regeneration. On other hand, the mechanical strengths of the scaffolds are obtained by varying the fiber design. For instance, the scaffolds of three-strand bound sutures of PLDLA yields suitable mechanical strength at repair site for active mobilization in an *ex vivo* model of porcine extensor tendons [134]. The effect of strands and strengths of these scaffolds for tendon repair *in vivo* are studied. Using the six-strand method for the flexor tendon, one can repair 36 lacerated fingers. In studies using six-strand scaffolds, 81% of all the lacerated fingers were treated to the acceptable level [135].

The major cause of shoulder pain after age of 65 is due to rotator cuff tears [136–138]. This rotator cuff injury is difficult to repair because it is susceptible later to retear and poor functionality.

Multilayered PCL electrospun nanofiber scaffolds are tested for rotator cuff tear repair. The scaffolds are coated with tendon-derived ECM and seeded with human adipose stem cells (hASCs). These scaffolds show significant collagen production with the stimulation of tenascin-c and decorin genes after 28 days in a culture without growth factors. Mechanical tests performed after implantation of scaffolds indicates the Young's modulus of these multilayered nanofiber PCL scaffolds remains unaffected. Tenogenic differentiation and suitable mechanical strength indicate the potential of multilayered PCL nanofiber scaffolds for tendon regeneration [136]. PDLGA-based nanofiber scaffolds which have suitable mechanical strength could similarly be useful for these applications [139]. Defect sites in infraspinatus tendons of Japanese white rabbits implanted with PDLGA scaffolds for 4 weeks and 8 weeks contained spindle-shaped cells and osteoblasts, respectively. And by 16 weeks the matured interface between scaffold and bone with the collagen expressions were prevalent. Mechanical assessment indicates normal stiffness as compared to control groups, which were reattached without making defects. These layered PLLA cell-free scaffolds used in the repair of infraspinatus tendon defects in a rabbit's rotator cuff showed cell migration and the production of type III collagen connective tissue that was connected to the bone [140]. Mechanical strength increased in a time-dependent manner and no statistical difference noticed between normal infraspinatus tendon and the scaffold group at 16 weeks. Braided scaffolds of PLLA are studied as a potential construct for tendon as well as for ligament regeneration [141]. Constructs fabricated by using aligned fiber bundles of PLLA nanofibers shows tri-phasic mechanical behaviors similar to the native tendons and ligaments. Seeding with hMSCs indicates cell attachment, actin cytoskeleton, and cell alignment parallel to the fibers, and increases expression of pluripotent genes required for stem cell pluripotency and self-renewal. Additionally, seeding of scaffolds with cells in the presence of tenogenic growth factors and mechanical strains increases the Scleraxis gene expression as evidence of tenogenic lineage differentiations of the hMSCs. These studies indicate that the PLLA braided nanofiberous scaffolds support tenocyte proliferation as well as differentiation. Poly(L-lactic) acid electrospun nanofiber matrix scaffolds, when directly integrated with bFGF-loaded dextran nanoparticles support tenocyte development and retain the bioactivity of the growth factor for a prolonged period of time. These scaffolds seeded with pluripotent C3H10T stem cells shows increased differentiation, proliferation, mechanical properties, cell adhesion, and drug release profiles. *In vivo*, the tissue that developed closely resembles the mature tenocytes and fibroblasts that developed were characteristically similar to fibroblasts found in tendons. The bioactivity of the encapsulated fibroblast growth factor was maintained, and *in vivo* studies found increased healing and cell proliferation capabilities.

PLGA interwoven nanofibers are explored for hand and wrist tendon repair. Scaffolds are developed using electrospun PLGA nanofibers interwoven with a heparin/fibrin-based delivery system for providing growth factors to the injury site. Alternating layers of PLGA fibers with layers of fibrin are constructed into a mat-shaped scaffold for the controlled growth factor release as well as to form stable platform on which the mechanical and structural properties of a tendon can be mimicked. These mats are biodegradable in an aqueous setting and are resistant to enzymatic degradation. These scaffolds seeded with adipose-derived MSCs and treated with BMP 14 display upregulated tenogenesis. The growth factor release occurs over the course of the scaffold degradation with no toxic effect on cells. The aim of delivering both cells and stimulating growth factors could be used in clinical settings in the future [142]. Viinikainen *et al.* [143] tested PLDLA sutures which were coated and braided for flexor tendon healing in rabbits. After 52 weeks, the PLDLA sutures disappeared and the flexor tendon appeared normal in histologic examination.

18.2.4 Nerve Tissue Engineering

Peripheral nerve injury is a phenomenon that has been steadily increasing in the global population, and it ranges from minor to long-term disability due to ineffective reparative techniques. Annually there are nearly 360,000 peripheral nerve injury cases in the United States and Europe [144]. Despite the various complications associated with use of autologous nerve grafts, it is still considered a gold standard for bridging nerve gaps. To repair peripheral nerve injuries using biomaterials, nerve conduits have been explored as an alternative to autologous nerve grafts. Numerous synthetic biomaterials have been investigated and polyesters such as PLLA, PLGA, polyhydroxybutyric acid, and PCL have also been studied for these applications. There are a variety of different physical and chemical properties for each of these polyesters, which can allow fabrication of suitable devices for nerve regeneration. Polyester-based neural conduits degrade through hydrolysis of ester bonds, releasing acidic degradation products. Fortunately, studies show that the acidic products produced through hydrolysis do not affect significantly the development of the neural conduits. This may be explained through the timing of the degradation, as only small amounts of acidic products are released in the early stages, are easily neutralized, and are removed by the body's circulation system [145].

To produce suitable nerve-like physical properties with polyester scaffolds, the scaffolds may be blended or copolymerized with other polymers. These variations can be used to create implants for different parts of the body that require either a more substantial framework for the implanted cells to proliferate or sites that require freedom of movement such as joints. Elasticity is an important consideration when designing a neural scaffold for joints to protect the integrity of the developing neural tube and providing an environment that allows both maturation of the neural tissue and freedom of movement of the joint [144, 146]. Specific pores in the grafts are important since they allow for entry of connective tissue but avoid scarring. Compared to nonporous nerve grafts, one with nano- or microporosities would be helpful for successful nerve regeneration. In addition the pores would facilitate the diffusion of growth factors through the circulation. The effect of porous and nonporous PCL nerve conduits for nerve growth regeneration studied by Vleggeert-Lankamp *et al.* [147] for rat sciatic nerve regeneration for 12 weeks of implantation indicates improved electrophysiological, and morphometrical properties in porous nerve conduits compared to the nonporous nerve conduit grafts. In addition, the porous PCL nerve conduits show bridged nerve gaps with myelinated nerve fibers, and high electrophysiological responses. Nerve grafts with a porosity of 1–10 μm pore size seems effective when compared with nerve grafts with porosity of 10–230 μm pore size. These studies indicate that in addition to the biocompatibility and mechanical properties, porosity is an important factor while developing nerve conduits.

Owing to the directional growth of neurons on aligned nanofibers, use of scaffolds fabricated using aligned nanofibers is an interesting strategy to develop devices for nerve regeneration. The seeding of neurite and Schwann cells on highly aligned PLLA nanofibers promote directional growth [148]. Increase in fiber density influences positively on the number of neurites without affecting neurite length as compared to low-density aligned nanofiber scaffolds. In addition, the fiber diameter also effects on the extension ability of neurites [149]. The migration of embryonic stage 9 (E9) dorsal root ganglia (DRG) and neurite extension are along the aligned nanofibers. The small fibers showed shorter neurite lengths compared to intermediate and large aligned fibers. Schwan cell migration was similar in the case of small- and large-diameter scaffolds. The slow migrations are noticed in case of intermediate diameter scaffolds. The cells were present around the surface of fibers and not between the fibers. The filopodial extensions were noted to grab and attach to the adjacent fibers. Figure 18.10 shows the growth of neurofilaments in the different directions on large-, intermediate-, and small-diameter fibers. These results indicate that fiber diameter plays an important role that needs to be considered for scaffold development for nerve regeneration.

PLGA-hMSC porous scaffolds implanted in a transected rat spinal cords indicated complete recovery of the spinal cord, which could not have been possible on its own [150]. The presence of hMSCs at implantation indicates an evidence of cell survival due to PLGA scaffolds compatibility. Nerve cells displaying axonal growth upon use of PLGA-hMSC scaffolds indicate the potential of PLGA scaffolds for spinal cord recovery.

Growth factors help assist cell growth and promote mature neural conduits when incorporated into the conduits. They can be incorporated into the conduits by conjugation to the walls or they can be absorbed and integrated into the material. Drug factors can also be integrated into gels that are attached to the lumen of the neural conduits, which would be useful for neuronal development. Growth factors that bind to collagen and laminin are used for growth enhancement as well and are integrated into gels. To achieve this dual purpose, growth factors are integrated into nanofibers placed in the lumen of the neural conduits for localized drug release as well as for mechanical guides for axonal growth with a downside of having rapid drug release profiles. Studies show that prolonged release of drug factors is viable through integrated neural polymer delivery devices [151–153].

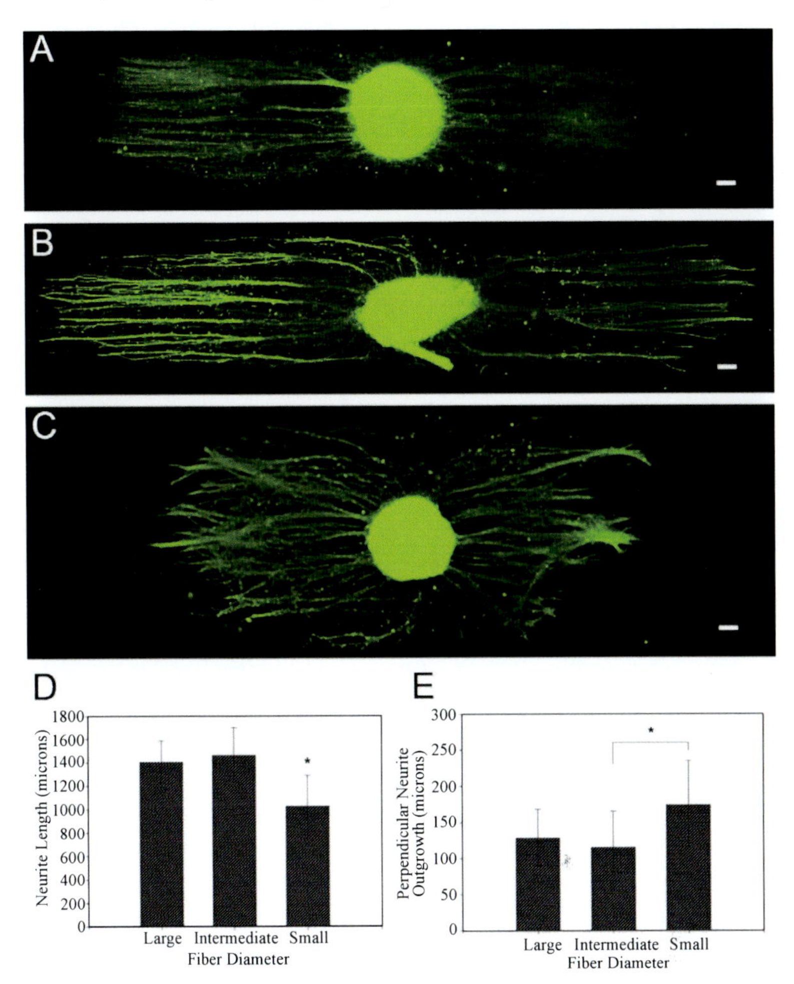

Figure 18.10 Effect of nanofiber size diameter on the neurite outgrowths. Neurofilament-stained image of DRG cultured for 5 days on large (A), intermediate (B), and small (C) diameter fibers. (D) The neurite outgrowth was in parallel to the direction of the nanofibers for A–C and the length of neurites was shorter on small fibers compared to large and intermediate. The neurite outgrowth which was perpendicular on each fibers reveals the extended growth in perpendicular direction for small diameter fibers (scale bar 100 μm). Reprinted from Ref. 149, Copyright 2010, with permission from Elsevier.

The presence of nerve growth factors at the site of nerve injury has advanced effects on nerve regeneration. When nerve injury occurs, neurotrophic growth factors are upregulated for 1 to 2 months by Schwann cells. The limited duration of nerve growth factor upregulation warrants the development of biomaterial-based devices that allow "smart" growth factor delivery as well as supporting the nerve at implantation site [151]. Polyesters have been used for the delivery of various growth factors safely with extended release profiles. Micro- and nanoscaffolds that have an initial high burst have been incorporated into hydrogels. The incorporation of nano-/microscaffolds into the hydrogels showed a dual advantage by minimizing initial release of active agents for effective nerve recovery. In these cases, the PLGA scaffolds loaded with glial cell-derived neurotrophic factor (GDNF) when embedded in fibrin gel extended drug release over months, greatly improving axonal regeneration and ultimately leading to restoration of muscle innervation and improved motor function outcomes [154]. This indicates that GDNF is an effective factor for axonal elongation when released for an extended period at the damaged nerve site. Double-walled microspheres loaded with GDNF using PLGA-PLLA and further incorporation into PCL fibers were highly effective for nerve repair [155]. Figure 18.11 is a representation of PCL disks, double-walled microspheres and microspheres incorporated into PCL nerve grafts. Implantation of these nerve guides into a rat sciatic nerve maintains mechanical integrity, improved tissue integration along the lumen of the implant, and increased Schwann cell numbers in the distal part of implant. Similarly, double-walled microspheres of PLGA-PCL loaded with GDNF further incorporated into the porous PCL nerve guides for the extended release of neurotrophic growth factors helped to improve tissue integration for bridging a peripheral nerve gap of 1.5 cm during 16 weeks in a Lewis rat model [156]. The nerve conduits fabricated with PLCL incorporated nerve growth factor and bovine serum albumin implanted in defective sciatic nerves in rats for 12 weeks displays myelination and nerve function similar to autografts. This could be due to the extended release of the bioactive agents using polyesters at the target site [157]. In spite of numerous efforts that are made, information on the therapeutic dose of NGF is not clear, indicating a need for more research trials. The success of the use of growth factors in neuronal regeneration depends on the method by which the growth factors are incorporated into the system, but so far they have not been as successful as using autologous nerve grafts. This may be attributed to reduced bioactivity, inadequate cellular support, insufficient release kinetics, and the use of single factors as opposed to using growth factor combinations [151].

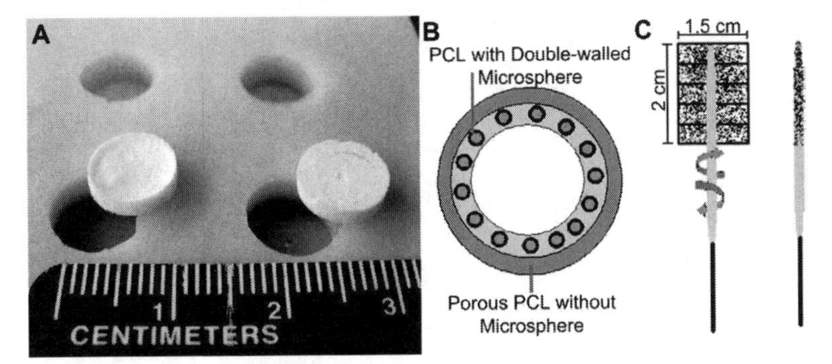

Figure 18.11 Photographs of (A) PCL disks fabricated using custom-made silicone mold, (B) schematic of polymer and microsphere orientation, and (C) schematic of technique for incorporation of double-walled microspheres into PCL nerve guide. Reprinted from Ref. 155, Copyright 2010, with permission from Elsevier.

18.2.5 Muscle Tissue Engineering

Muscle injuries result in severe long-term pain, causing discomfort during everyday tasks and commonly occur during sports [158]. The traditional methods including injections of nonsteroidal anti-inflammatory drugs, and corticosteroids do not optimally help muscle regeneration due to their immediate clearance from the site of injection and requirement of repeated injections. Other methods involving injections at the site of injury have been implemented to improve the recovery. Studies indicate that the vascularization of injured tissues significantly promotes recovery. Vascularization of the muscle is difficult because it is a thick and complex tissue. To overcome these difficulties, porous scaffolds have been tested successfully and are also being investigated for replacing skeletal muscle [159]. There are also many diseases that can cause loss of function to the muscle tissue or even destroy it. Proper muscle function can be impaired by traumatic injury both blunt and sharp, exposure to toxins, or myopathies that deteriorate muscle tissue and cause dystrophy. In the initial onset of injury, muscle tissue will try to repair itself but the proliferation capability of those cells is limited. Once exhausted, muscle regeneration ceases and the production of connective tissue forming scarring occurs. After implantation, scaffolds are infiltrated by mononuclear cells. As these cells settle into the scaffold, the scaffold degrades at the same rate as the cells' maturation. Tissues specific to the implantation site eventually develop. The overall goal is to develop a coordinated and organized temporal and spatial system that will develop a self-sustaining tissue. The scaffold must also act as a dual purpose material in not just providing a structural framework for cells, but also providing a system for the delivery of drugs and other factors that will promote cell growth and differentiation.

It is also possible to use the scaffold itself to deliver the cells needed for tissue regeneration of the affected site. Limitations involving cell-based approaches may include an inability to migrate away from the implantation site and a poor engraftment of cells onto the surrounding tissue. Thus the use of the implant scaffolds as a cell delivery system is being investigated. Materials such as hyaluronic acid, laminin, collagen, and fibrin have been tested as substrates. In particular, hyaluronic acid-based hydrogels and collagen scaffolds have shown to greatly facilitate cellular responses such as myogenesis. A major hurdle for the development of successful implants has been the development of sufficient vascularization to supply the new muscle tissue. Vascularization is important during tissue development because the new muscle fibers will not survive without an adequate blood supply. To avoid this phenomenon, porous scaffolds can be seeded with embryonic endothelial cells or with umbilical vein endothelial cells. Subsequent development of the tissue revealed formation of endothelium networks between the myoblasts and developing muscle tissue. Further cultivation showed improved integration with the tissue surrounding the implant and successful perfusion of the tissue from the surrounding vasculature.

Polyesters can serve as good candidates for muscle regeneration because they can be manipulated as per the need for degradation duration by varying monomer compositions, biocompatibility, and mechanical properties. The microporous polyester scaffolds are tested for these applications. PLGA, which undergoes fast degradation, could be blended with a slow degrading polymer such as PLLA. The use of PLGA would help to degrade scaffolds faster by leaving dominantly PLLA porous structures behind with suitable mechanical properties. In one of those approaches fabrication of PLLA and PLGA (50:50) with pore size range of 225–500 μm and 93% porous sponge shows degradation of PLGA in 3 weeks and PLLA takes about 6 months [159]. The co-seeding of myoblasts, embryonic fibroblasts, and endothelial cells on porous scaffolds indicates increased expression of vascular endothelial growth factors and endothelial vessel formations *in vitro*. As hypothesized, these porous scaffolds help vascularization, survival of tissue muscles, and blood perfusion *in vivo* when tested in three different animal models. The potential of 3D structured scaffolds when co-seeded using mouse myoblasts with human embryonic stem cell (hESC) or derived endothelial cells (EC) or human umbilical vein endothelial cells (HUVEC) with 3D porous PLLA and PLGA allows the formation of vessel networks in the scaffolds. The seeding of scaffolds with myoblasts and embryonic fibroblasts also produces enhanced vascularization. Upon implantation of these scaffolds into different mice models, increased vascularization and differentiation was noted. The seeding of scaffolds with hESC-cardiomyocytes (CM), hESC-EC, or HUVEC with or without embryonic fibroblasts allows the formation of human cardiac tissue with endothelial vessel networks. These studies demonstrate the potential of 3D scaffolds for vascularization and tissue regeneration [160]. As mentioned previously, differences in polymer compositions and fiber diameters have also had various effects on cellular responses [161]. For instance, PLCL fibers with 50:50 compositions and diameters of 0.3 μm or 1.2 μm showed HUVEC attachment and proliferation. On the other hand, fibers in range of 7 μm in diameter reduced cell attachment and proliferation. A representative image for the cell attachments with respect to fiber diameters is shown in Fig. 18.12.

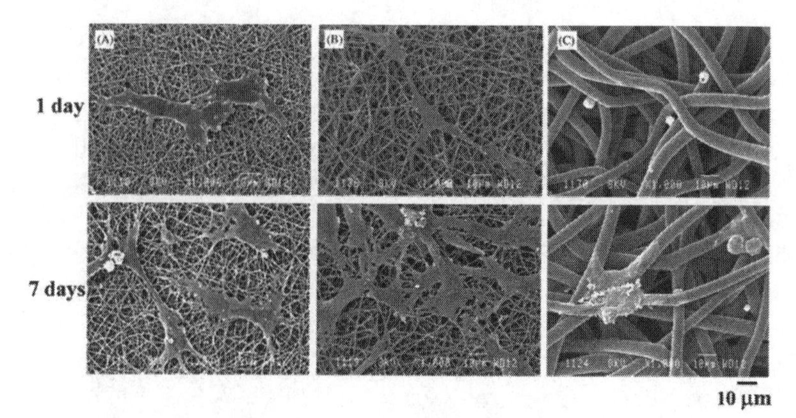

Figure 18.12 SEM images indicting HUVEC on electrospun PLCL (50/50) fibers of different diameter (A) 0.3 μm, (B) 1.2 μm, and (C) 7 μm for day 1 and 7. Reprinted from Ref. 161, Copyright 2005, with permission from Elsevier.

Apart from their uses in bone, cartilage, ligament, tendon, nerve, and muscle, they have been used for various other tissue engineering applications. The aligned blend nanofibers of PLLA-PCL (75:25 and 25:75) have been fabricated for cardiovascular applications [162]. The changes of PLLA and PCL compositions significantly affected the mechanical properties of the scaffolds. The scaffolds with higher content of PLLA have higher tensile strain and fiber density than those with less PLLA. Thromboresistivity, cell adhesion, and gene expression in both types of scaffolds were comparable. These studies suggest that PLLA-PCL scaffolds could be highly useful for vascular engineering applications. Owing to the flexibility of these blended scaffolds, they may be appropriate for the fabrication of periosteal sheets. Kouya *et al.* [163] fabricated blended microporous scaffold membranes of PLLA-PCL. The scaffolds containing 50% PCL had no fragility, and elongation properties were improved without affecting pore structure. Upon cultivation of periosteal bovine tissue segments, multilayered cell structures were noticed on asymmetric microporous membrane surfaces. In one of these studies, microporous foam scaffolds of PLLA-PCL (1:4) were fabricated using thermally induced phase separation techniques which allowed hepatic cells to enter the pores. When compared with only PCL and PLLA, the depth of cell penetration into the blend scaffolds was more. This study concludes that when scaffolds are fabricated with PCL or PLLA-PCL, the blended pores were larger and their sizes more controllable compared to only PCL scaffolds [164]. The method of microporous structure formation also impacts on the mechanical properties as well as on cellular responses. The porous scaffolds fabricated using thermally induced phase separation techniques had higher elastic and plastic modulus as compared to porous fibrous scaffolds using the same technique. These porous scaffolds resembled the native ECM structurally and could support human osteosarcoma cell growth (165).

The scaffold-to-tissue interface compatibilities are improved by surface functionalization using hydrophilic moieties such as mPEG, and or peptides. The polyester surface with significant penetration can be modified to express amine groups by immobilization with di-amines for conjugation with peptides using carbodiimide chemistry. When seeded with NIH3T3 cells, this modification produced enhanced cell adhesion [166]. Gloria *et al.* [103] developed PCL scaffolds using a 3D fiber deposition technique with surface-treated RGD moieties. Surface functionalization improved NIH3T3 fibroblast attachments without compromising the macromechanical properties in addition to reducing scaffold hardness due to the addition of hydrophilic ligands. These studies indicate that surface modification is an important approach to improving the biological responses of the scaffolds without affecting the mechanical properties. It is important to consider that properties including various porosities, fiber diameters, and surface functionality play an important role in the cellular responses. The PCL scaffolds created with sequential electrospinning yielded nano- and microfibers with various porosities [167]. The rat bone marrow stromal cell spreading was noticed uniformly due to the presence of nanofibers; however, cell attachment did not improve with increases in the amount of nanofibers. The increased thickness of nanofiber layers also restricted cell infiltration. These results indicate the possibility of structural modulations of scaffolds for better cellular responses. Nanofibers can be produced with linearly aligned patterns and circumferentially aligned patterns similar to those in knee menisci, which are crescent-shaped cartilaginous tissues made of collagen bundles that microscopically change directionality and help equalize the load distribution across the knee [168, 169]. Therefore, the circumferentially aligned PCL nanofibers could be helpful in mimicking these local tissue patterns. They changed their orientation over time and cellular growth occurred in the same direction as the nanofibers. Mechanical tests of these scaffolds indicated interactions with tissues, and tensile modulus decreased near the edges. These observations were supported by simulation studies. The circumferentially oriented aligned nanofibers of PCL may be useful for the regeneration of the meniscus.

The final and probably the most difficult challenge of these systems is in making the step from *in vitro* and *in vivo* models to clinical trials and translating these results into clinical applications. Despite ongoing research and development of tissue engineering devices and large number of growth factor carrier materials, only a handful have been approved for use in patients. These discrepancies in translational points necessitate critical integration of interdisciplinary efforts in order to make theory a clinical reality.

18.3 Conclusions

Polyesters have been explored for various tissue engineering and drug delivery applications over the past several decades. Owing to their inherent properties such biocompatibility, biodegradability, and suitable cellular responses, they have been widely accepted for human applications by the FDA. Interestingly, some byproducts of polyester degradation are part of regular human metabolites already present in physiological fluid. Moreover, properties of the scaffolds, including ease of processibility for suitable degradation rates and mechanical strengths, make their use highly useful for the fabrication of required devices. Numerous studies have been performed analyzing their use for tissue substrates in the form of porous scaffolds that voids the need of autografts and allografts. The advantage of using porous scaffolds is to obtain suitable cell infiltration, nutrient exchange, and increased surface area for suitable cellular responses. Creating porous scaffolds using polyesters produces mechanically competent scaffolds and, as mentioned, can be modulated as needed. These devices are further explored for tissue engineering applications by blending polyesters with hydrophilic polymers or nanocomposites to improve the cellular responses and may be surface-modified for similar purpose.

Use of these devices with various cells and growth factors unlocks their great potential for serving as a cell supplement and source of bioactive growth factors at the repair site upon implantation. Hence it is highly advantageous to use these polyesters with the aforementioned approaches for the development of scaffolds for tissue engineering and medical therapies.

Acknowledgments

The authors gratefully acknowledge funding from the Connecticut Regenerative Medicine Research Fund (Grant Number: 15-RMB-UCHC-08), National Science Foundation Award (Grant Numbers: IIP-1311907, IIP-1355327, EFRI-1332329), Department of Defense (Grant Number: OR120140), and the Raymond and Beverly Sackler Center for Biomedical, Biological, Physical and Engineering Sciences, Yale University, Connecticut, USA.

References

1. *Musculoskeletal Disorders and the Workplace: Low Back and Upper Extremities.* (National Academies Press, Washington, DC, 2001), p. 512.
2. Silva-Correia, J., Correia, S. I., Oliveira, J. M., Reis, R. L. (2013). Tissue engineering strategies applied in the regeneration of the human intervertebral disk, *Biotech Adv,* **31**, 1514.
3. Mack, M. (2014). Progress toward tissue-engineered heart valves, *J Am Coll Cardiol,* **63**, 1330.
4. Ohgushi, H., Kotobuki, N., Funaoka, H., Machida, H., Hirose, M., Tanaka, Y., Takakura, Y. (2005). Tissue engineered ceramic artificial joint—*ex vivo* osteogenic differentiation of patient mesenchymal cells on total ankle joints for treatment of osteoarthritis, *Biomaterials,* **26**, 4654.
5. Blackwood, K. A., Bock, N., Dargaville, T. R., Ann Woodruff, M. (2012). Scaffolds for growth factor delivery as applied to bone tissue engineering, *Int J Polym Sci,* **2012**, 25.
6. Chan, B. P., Leong, K. W. (2008). Scaffolding in tissue engineering: general approaches and tissue-specific considerations, *Eur Spine J,* **17**, 467.
7. Serbo, J. V., Gerecht, S. (2013). Vascular tissue engineering: biodegradable scaffold platforms to promote angiogenesis, *Stem Cell Res Ther,* **4**, 1.

8. Cheng, Y., Ramos, D., Lee, P., Liang, D., Yu, X., Kumbar, S. G. (2014). Collagen functionalized bioactive nanofiber matrices for osteogenic differentiation of mesenchymal stem cells: bone tissue engineering, *J Biomed Nanotechnol,* **10**, 287.
9. Kock, L., van Donkelaar, C. C., Ito, K. (2012). Tissue engineering of functional articular cartilage: the current status, *Cell Tissue Res,* **347**, 613.
10. Lee, P., Tran, K., Chang, W., Shelke, N. B., Kumbar, S. G., Yu, X. (2014). Influence of chondroitin sulfate and hyaluronic acid presence in nanofibers and its alignment on the bone marrow stromal cells: cartilage regeneration, *J Biomed Nanotechnol,* **10**, 1469.
11. Shi, J., Votruba, A. R., Farokhzad, O. C., Langer, R. (2010). Nanotechnology in drug delivery and tissue engineering: from discovery to applications, *Nano Lett,* **10**, 3223.
12. Lee, K., Silva, E. A., Mooney, D. J. (2011). Growth factor delivery-based tissue engineering: general approaches and a review of recent developments, *J R Soc Interface,* **8**, 153.
13. Aravamudhan, A., Ramos, D., Nip, J., Subramanian, A., James, R., D Harmon, M., Yu, X., G Kumbar, S. (2013). Osteoinductive small molecules: growth factor alternatives for bone tissue engineering, *Curr Pharm Des,* **19**, 3420.
14. Kumbar, S. G., Nukavarapu, S. P., James, R., Nair, L. S., Laurencin, C. T. (2008). Electrospun poly(lactic acid-*co*-glycolic acid) scaffolds for skin tissue engineering, *Biomaterials,* **29**, 4100.
15. Guadalupe, E., Ramos, D., Shelke, N. B., James, R., Gibney, C., Kumbar, S. G. (2015). Bioactive polymeric nanofiber matrices for skin regeneration, *J Appl Polym Sci,* **132**, 41879.
16. Place, E. S., George, J. H., Williams, C. K., Stevens, M. M. (2009). Synthetic polymer scaffolds for tissue engineering, *Chem Soc Rev,* **38**, 1139.
17. Armentano, I., Dottori, M., Fortunati, E., Mattioli, S., Kenny, J. M. (2010). Biodegradable polymer matrix nanocomposites for tissue engineering: a review, *Polym Degrad Stab,* **95**, 2126.
18. Dhandayuthapani, B., Yoshida, Y., Maekawa, T., Kumar, D. S. (2011). Polymeric scaffolds in tissue engineering application: a review, *Int J Polym Sci,* **2011**, 1.
19. Bolland, B. J., Kanczler, J. M., Ginty, P. J., Howdle, S. M., Shakesheff, K. M., Dunlop, D. G., Oreffo, R. O. (2008). The application of human bone marrow stromal cells and poly(D,L-lactic acid) as a biological bone graft extender in impaction bone grafting, *Biomaterials,* **29**, 3221.

20. Ghasemi-Mobarakeh, L., Prabhakaran, M. P., Balasubramanian, P., Jin, G., Valipouri, A., Ramakrishna, S. (2013). Advances in electrospun nanofibers for bone and cartilage regeneration, *J Nanosci Nanotechnol,* **13**, 4656.
21. Chen, H., Truckenmuller, R., Blitterswijk, C., Moroni, L. (2013). Fabrication of nanofibrous scaffolds for tissue engineering applications, in *Nanomaterials in Tissue Engineering: Fabrication and Applications,* Gaharwar, A. K., Sant, S., Hancock, M. J., and Hacking, S. A., eds. (Woodhead Publishing Limited), 158–183.
22. Kumbar, S. G., James, R., Nukavarapu, S. P., Laurencin, C. T. (2008). Electrospun nanofiber scaffolds: engineering soft tissues, *Biomed Mater,* **3**, 034002.
23. Jiang, T., Carbone, E. J., Lo, K. W. H., Laurencin, C. T. (2014). Electrospinning of polymer nanofibers for tissue regeneration, *Prog Polym Sci,* **46**, 1.
24. Bosco, R., Beucken, J. V. D., Leeuwenburgh, S., Jansen, J. (2012). Surface engineering for bone implants: a trend from passive to active surfaces, *Coatings,* **2**, 95.
25. Stevens, M. M., George, J. H. (2005). Exploring and engineering the cell surface interface, *Science,* **310**, 1135.
26. Langer, R. (1993). Polymer-controlled drug delivery systems, *Acc Chem Res,* **26**, 537.
27. Shelke, N. B., Kadam, R., Tyagi, P., Rao, V. R., Kompella, U. B. (2011). Intravitreal poly (L-lactide) microparticles sustain retinal and choroidal delivery of TG-0054, a hydrophilic drug intended for neovascular diseases, *Drug Deliv Transl Res,* **1**, 76.
28. Freed, L. E., Vunjak-Novakovic, G., Biron, R. J., Eagles, D. B., Lesnoy, D. C., Barlow, S. K., Langer, R. (1994). Biodegradable polymer scaffolds for tissue engineering, *Nat Biotech,* **12**, 689.
29. Mikos, A. G., Thorsen, A. J., Czerwonka, L. A., Bao, Y., Langer, R., Winslow, D. N., Vacanti, J. P. (1994). Preparation and characterization of poly(L-lactic acid) foams, *Polymer,* **35**, 1068.
30. Harris, L. D., Kim, B. S., Mooney, D. J. (1998). Open pore biodegradable matrices formed with gas foaming, *J Biomed Mater Res,* **42**, 396.
31. El-Fattah, A. A., Kenawy, E. R., Kandil, S. (2014). Biodegradable polyesters as biomaterials for biomedical applications, *Int J Chem Appl Biol Sci,* **1**, 2.
32. Vortkamp, A., Pathi, S., Peretti, G. M., Caruso, E. M., Zaleske, D. J., Tabin, C. J. (1998). Recapitulation of signals regulating embryonic bone

formation during postnatal growth and in fracture repair, *Mech Dev,* **71**, 65.

33. Scotti, C., Piccinini, E., Takizawa, H., Todorov, A., Bourgine, P., Papadimitropoulos, A., Barbero, A., Manz, M. G., Martin, I. (2013). Engineering of a functional bone organ through endochondral ossification, *Proc Natl Acad Sci U S A,* **110**, 3997.
34. Ai-Aql, Z. S., Alagl, A. S., Graves, D. T., Gerstenfeld, L. C., Einhorn, T. A. (2008). Molecular mechanisms controlling bone formation during fracture healing and distraction osteogenesis, *J Dent Res,* **87**, 107.
35. Marsell, R., Einhorn, T. A. (2010). Emerging bone healing therapies, *J Orthop Trauma,* **24**, S4.
36. Holy, C. E., Dang, S. M., Davies, J. E., Shoichet, M. S. (1999). *In vitro* degradation of a novel poly(lactide-*co*-glycolide) 75/25 foam, *Biomaterials,* **20**, 1177.
37. Lu, L., Peter, S. J., Lyman, M. D., Lai, H. L., Leite, S. M., Tamada, J. A., Uyama, S., Vacanti, J. P., Langer, R., Mikos, A. G. (2000). *In vitro* and *in vivo* degradation of porous poly(DL-lactic-*co*-glycolic acid) foams, *Biomaterials,* **21**, 1837.
38. Pamula, E., Menaszek, E. (2008). *In vitro* and *in vivo* degradation of poly(L-lactide-*co*-glycolide) films and scaffolds, *J Mater Sci Mater Med,* **19**, 2063.
39. Wu, L., Ding, J. (2005). Effects of porosity and pore size on *in vitro* degradation of three-dimensional porous poly(D,L-lactide-*co*-glycolide) scaffolds for tissue engineering, *J Biomed Mater Res A,* **75**, 767.
40. Kinoshita, Y., Maeda, H. (2013). Recent developments of functional scaffolds for craniomaxillofacial bone tissue engineering applications, *Sci World J,* **2013**, 21.
41. Papenburg, B. J., Liu, J., Higuera, G. A., Barradas, A. M., de Boer, J., van Blitterswijk, C. A., Wessling, M., Stamatialis, D. (2009). Development and analysis of multi-layer scaffolds for tissue engineering, *Biomaterials,* **30**, 6228.
42. Coutu, D. L., Yousefi, A. M., Galipeau, J. (2009). Three-dimensional porous scaffolds at the crossroads of tissue engineering and cell-based gene therapy, *J Cell Biochem,* **108**, 537.
43. Lowry, K. J., Hamson, K. R., Bear, L., Peng, Y. B., Calaluce, R., Evans, M. L., Anglen, J. O., Allen, W. C. (1997). Polycaprolactone/glass bioabsorbable implant in a rabbit humerus fracture model, *J Biomed Mater Res,* **36**, 536.

44. Caplan, A. I. (2009). New era of cell-based orthopedic therapies, *Tissue Eng Part B, Rev,* **15**, 195.

45. Whang, K., Tsai, D. C., Nam, E. K., Aitken, M., Sprague, S. M., Patel, P. K., Healy, K. E. (1998). Ectopic bone formation via rhBMP-2 delivery from porous bioabsorbable polymer scaffolds, *J Biomed Mater Res,* **42**, 491.

46. Williams, J. M., Adewunmi, A., Schek, R. M., Flanagan, C. L., Krebsbach, P. H., Feinberg, S. E., Hollister, S. J., Das, S. (2005). Bone tissue engineering using polycaprolactone scaffolds fabricated via selective laser sintering, *Biomaterials,* **26**, 4817.

47. Reguera-Nunez, E., Roca, C., Hardy, E., de la Fuente, M., Csaba, N., Garcia-Fuentes, M. (2014). Implantable controlled release devices for BMP-7 delivery and suppression of glioblastoma initiating cells, *Biomaterials,* **35**, 2859.

48. Yilgor, P., Hasirci, N., Hasirci, V. (2010). Sequential BMP-2/BMP-7 delivery from polyester nanocapsules, *J Biomed Mater Res A,* **93**, 528.

49. Devescovi, V., Leonardi, E., Ciapetti, G., Cenni, E. (2008). Growth factors in bone repair, *Chir Organi Mov,* **92**, 161.

50. Amstutz, H. C. (1995). A tribute to Mr. "BMP": Marshall Urist, *Clin Orthop Relat Res,* **313**, 286.

51. Axelrad, T. W., Einhorn, T. A. (2009). Bone morphogenetic proteins in orthopaedic surgery, *Cytokine Growth Factor Rev,* **20**, 481.

52. Luginbuehl, V., Meinel, L., Merkle, H. P., Gander, B. (2004). Localized delivery of growth factors for bone repair, *Eur J Pharm Biopharm,* **58**, 197.

53. Saito, N., Okada, T., Horiuchi, H., Murakami, N., Takahashi, J., Nawata, M., Ota, H., Nozaki, K., Takaoka, K. (2001). A biodegradable polymer as a cytokine delivery system for inducing bone formation, *Nat Biotech,* **19**, 332.

54. Lo, K. W. H., Ulery, B. D., Ashe, K. M., Laurencin, C. T. (2012). Studies of bone morphogenetic protein based surgical repair, *Adv Drug Deliv Rev,* **64**, 1277.

55. Miyamoto, S., Takaoka, K., Okada, T., Yoshikawa, H., Hashimoto, J., Suzuki, S., Ono, K. (1992). Evaluation of polylactic acid homopolymers as carriers for bone morphogenetic protein, *Clin Orthop Relat Res,* **278**, 274.

56. Fu, Y.-C., Nie, H., Ho, M.-L., Wang, C.-K., Wang, C.-H. (2008). Optimized bone regeneration based on sustained release from three-dimensional fibrous PLGA/HAp composite scaffolds loaded with BMP-2, *Biotechnol Bioeng,* **99**, 996.

57. Wei, G., Jin, Q., Giannobile, W. V., Ma, P. X. (2007). The enhancement of osteogenesis by nano-fibrous scaffolds incorporating rhBMP-7 nanospheres, *Biomaterials,* **28**, 2087.
58. Hsiong, S. X., Mooney, D. J. (2006). Regeneration of vascularized bone, *Periodontology 2000,* **41**, 109.
59. Kaigler, D., Silva, E. A., Mooney, D. J. (2013). Guided bone regeneration using injectable vascular endothelial growth factor delivery gel, *J Periodontol,* **84**, 230.
60. Yoo, H. S., Kim, T. G., Park, T. G. (2009). Surface-functionalized electrospun nanofibers for tissue engineering and drug delivery, *Adv Drug Deliv Rev,* **61**, 1033.
61. Seyednejad, H., Gawlitta, D., Dhert, W. J. A., van Nostrum, C. F., Vermonden, T., Hennink, W. E. (2011). Preparation and characterization of a three-dimensional printed scaffold based on a functionalized polyester for bone tissue engineering applications, *Acta Biomater,* **7**, 1999.
62. Zhan, J., Singh, A., Zhang, Z., Huang, L., Elisseeff, J. H. (2012). Multifunctional aliphatic polyester nanofibers for tissue engineering, *Biomatter,* **2**, 202.
63. Mahjoubi, H., Kinsella, J. M., Murshed, M., Cerruti, M. (2014). Surface modification of poly(D,L-lactic acid) scaffolds for orthopedic applications: a biocompatible, nondestructive route via diazonium chemistry, *ACS Appl Mater Interfaces,* **6**, 9975.
64. Xu, J., Li, S., Hu, F., Zhu, C., Zhang, Y., Zhao, W., Akaike, T., Yang, J. (2014). Artificial biomimicking matrix modifications of nanofibrous scaffolds by hE-cadherin-Fc fusion protein to promote human mesenchymal stem cells adhesion and proliferation, *J Nanosci Nanotechnol,* **14**, 4007.
65. Tang, Y., Zhao, Y., Wang, X., Lin, T. (2014). Layer-by-layer assembly of silica nanoparticles on 3D fibrous scaffolds: Enhancement of osteoblast cell adhesion, proliferation, and differentiation, *J Biomed Mater Res A,* **102**, 3803.
66. Thomas, M., Arora, A., Katti, D. S. (2014). Surface hydrophilicity of PLGA fibers governs *in vitro* mineralization and osteogenic differentiation, *Mater Sci Eng C,* **45**, 320.
67. Ciardelli, G., Chiono, V., Vozzi, G., Pracella, M., Ahluwalia, A., Barbani, N., Cristallini, C., Giusti, P. (2005). Blends of poly-(ε-caprolactone) and polysaccharides in tissue engineering applications, *Biomacromolecules,* **6**, 1961.
68. Shelke, N. B., James, R., Laurencin, C. T., Kumbar, S. G. (2014). Polysaccharide biomaterials for drug delivery and regenerative engineering, *Polym Adv Technol,* **25**, 448.

69. Lin, C.-C., Fu, S.-J., Lin, Y.-C., Yang, I. K., Gu, Y. (2014). Chitosan-coated electrospun PLA fibers for rapid mineralization of calcium phosphate, *Int J Biol Macromol,* **68**, 39.

70. Cui, W., Li, X., Xie, C., Chen, J., Zou, J., Zhou, S., Weng, J. (2010). Controllable growth of hydroxyapatite on electrospun poly(D,L-lactide) fibers grafted with chitosan as potential tissue engineering scaffolds, *Polymer,* **51**, 2320.

71. Gong, C., Qi, T., Wei, X., Qu, Y., Wu, Q., Luo, F., Qian, Z. (2013). Thermosensitive polymeric hydrogels as drug delivery systems, *Curr Med Chem,* **20**, 79.

72. Stile, R. A., Healy, K. E. (2002). Poly(N-isopropylacrylamide)-based semi-interpenetrating polymer networks for tissue engineering applications. 1. Effects of linear poly(acrylic acid) chains on phase behavior, *Biomacromolecules,* **3**, 591.

73. Kim, S.-S., Sun Park, M., Jeon, O., Yong Choi, C., Kim, B.-S. (2006). Poly(lactide-*co*-glycolide)/hydroxyapatite composite scaffolds for bone tissue engineering, *Biomaterials,* **27**, 1399.

74. Zhang, X., Chang, W., Lee, P., Wang, Y., Yang, M., Li, J., Kumbar, S. G., Yu, X. (2014). Polymer-ceramic spiral structured scaffolds for bone tissue engineering: effect of hydroxyapatite composition on human fetal osteoblasts, *PLoS One,* **9**, e85871.

75. Nie, L., Suo, J., Zou, P., Feng, S. (2012). Preparation and properties of biphasic calcium phosphate scaffolds multiply coated with HA/PLLA nanocomposites for bone tissue engineering applications, *J Nanomater,* **2012**, 11.

76. Wei, G., Ma, P. X. (2004). Structure and properties of nano-hydroxyapatite/polymer composite scaffolds for bone tissue engineering, *Biomaterials,* **25**, 4749.

77. Prabhakaran, M. P., Venugopal, J., Ramakrishna, S. (2009). Electrospun nanostructured scaffolds for bone tissue engineering, *Acta Biomater,* **5**, 2884.

78. Yu, H.-S., Jang, J.-H., Kim, T.-I., Lee, H.-H., Kim, H.-W. (2009). Apatite-mineralized polycaprolactone nanofibrous web as a bone tissue regeneration substrate, *J Biomed Mater Res A,* **88A**, 747.

79. Takeoka, Y., Hayashi, M., Sugiyama, N., Yoshizawa-Fujita, M., Aizawa, M., Rikukawa, M. (2015). *In situ* preparation of poly(L-lactic acid-*co*-glycolic acid)/hydroxyapatite composites as artificial bone materials, *Polym J,* **47** 164.

80. Wang, J., Valmikinathan, C. M., Liu, W., Laurencin, C. T., Yu, X. (2010). Spiral-structured, nanofibrous, 3D scaffolds for bone tissue engineering, *J Biomed Mater Res A,* **93A**, 753.
81. Kohane, D. S. (2007). Microparticles and nanoparticles for drug delivery, *Biotechnol Bioeng,* **96**, 203.
82. Borden, M., El-Amin, S. F., Attawia, M., Laurencin, C. T. (2003). Structural and human cellular assessment of a novel microsphere-based tissue engineered scaffold for bone repair, *Biomaterials,* **24**, 597.
83. Kang, K. S., Lee, S.-J., Lee, H., Moon, W., Cho, D.-W. (2011). Effects of combined mechanical stimulation on the proliferation and differentiation of pre-osteoblasts, *Exp Mol Med,* **43**, 367.
84. Mauney, J. R., Sjostorm, S., Blumberg, J., Horan, R., O'Leary, J. P., Vunjak-Novakovic, G., Volloch, V., Kaplan, D. L. (2004). Mechanical stimulation promotes osteogenic differentiation of human bone marrow stromal cells on 3-D partially demineralized bone scaffolds *in vitro, Calcif Tissue Int,* **74**, 458.
85. Liu, X., Jin, X., Ma, P. X. (2011). Nanofibrous hollow microspheres self-assembled from star-shaped polymers as injectable cell carriers for knee repair, *Nat Mater,* **10**, 398.
86. Barnes, C. P., Sell, S. A., Boland, E. D., Simpson, D. G., Bowlin, G. L. (2007). Nanofiber technology: designing the next generation of tissue engineering scaffolds, *Adv Drug Deliv Rev,* **59**, 1413.
87. Chang, K. Y., Hung, L. H., Chu, I., Ko, C. S., Lee, Y. D. (2010). The application of type II collagen and chondroitin sulfate grafted PCL porous scaffold in cartilage tissue engineering, *J Biomed Mater Res A,* **92**, 712.
88. Pittenger, M. F., Mackay, A. M., Beck, S. C., Jaiswal, R. K., Douglas, R., Mosca, J. D., Moorman, M. A., Simonetti, D. W., Craig, S., Marshak, D. R. (1999). Multilineage potential of adult human mesenchymal stem cells, *Science,* **284**, 143.
89. Orth, P., Rey-Rico, A., Venkatesan, J. K., Madry, H., Cucchiarini, M. (2014). Current perspectives in stem cell research for knee cartilage repair, *Stem Cells Cloning,* **7**, 1.
90. Saw, K.-Y., Anz, A., Siew-Yoke Jee, C., Merican, S., Ching-Soong Ng, R., Roohi, S. A., Ragavanaidu, K. (2013). Articular cartilage regeneration with autologous peripheral blood stem cells versus hyaluronic acid: a randomized controlled trial, *Arthroscopy,* **29**, 684.
91. Ye, K., Felimban, R., Moulton, S. E., Wallace, G. G., Bella, C. D., Traianedes, K., Choong, P. F., Myers, D. E. (2013). Bioengineering of articular cartilage: past, present and future, *Regen Med,* **8**, 333.

92. Diekman, B. O., Rowland, C. R., Lennon, D. P., Caplan, A. I., Guilak, F. (2009). Chondrogenesis of adult stem cells from adipose tissue and bone marrow: induction by growth factors and cartilage-derived matrix, *Tissue Eng Part A,* **16**, 523.

93. Zhang, L., Hu, J., Athanasiou, K. A. (2009). The role of tissue engineering in articular cartilage repair and regeneration, *Crit Rev Biomed Eng,* **37**, 1.

94. Itani, Y., Asamura, S., Matsui, M., Tabata, Y., Isogai, N. (2014). Evaluation of nanofiber-based polyglycolic acid scaffolds for improved chondrocyte retention and *in vivo* bioengineered cartilage regeneration, *Plast Reconstr Surg,* **133**, 805e.

95. Martinez-Diaz, S., Garcia-Giralt, N., Lebourg, M., Gomez-Tejedor, J. A., Vila, G., Caceres, E., Benito, P., Pradas, M. M., Nogues, X., Ribelles, J. L., Monllau, J. C. (2010). *In vivo* evaluation of 3-dimensional polycaprolactone scaffolds for cartilage repair in rabbits, *Am J Sports Med,* **38**, 509.

96. Christensen, B., Foldager, C., Hansen, O., Kristiansen, A., Le, D., Nielsen, A., Nygaard, J., Bünger, C., Lind, M. (2012). A novel nano-structured porous polycaprolactone scaffold improves hyaline cartilage repair in a rabbit model compared to a collagen type I/III scaffold: *in vitro* and *in vivo* studies, *Knee Surgery, Sport Traumatol Arthrosc,* **20**, 1192.

97. Jung, Y., Kim, S. H., You, H. J., Kim, Y. H., Min, B. G. (2008). Application of an elastic biodegradable poly(L-lactide-*co*-ε-caprolactone) scaffold for cartilage tissue regeneration, *J Biomater Sci Polym Ed,* **19**, 1073.

98. Jung, Y., Park, M. S., Lee, J. W., Kim, Y. H., Kim, S. H. (2008). Cartilage regeneration with highly-elastic three-dimensional scaffolds prepared from biodegradable poly(L-lactide-*co*-ε-caprolactone), *Biomaterials,* **29**, 4630.

99. Jung, Y., Kim, S. H., Kim, Y. H. (2010). The effect of hybridization of hydrogels and poly(L-lactide-*co*-ε-caprolactone) scaffolds on cartilage tissue engineering, *J Biomater Sci Polym Ed,* **21**, 581.

100. Xie, J., Han, Z., Naito, M., Maeyama, A., Kim, S. H., Kim, Y. H., Matsuda, T. (2010). Articular cartilage tissue engineering based on a mechano-active scaffold made of poly(L-lactide-*co*-ε-caprolactone): *in vivo* performance in adult rabbits, *J Biomed Mater Res B Appl Biomater,* **94**, 80.

101. Chu, C. R., Coutts, R. D., Yoshioka, M., Harwood, F. L., Monosov, A. Z., Amiel, D. (1995). Articular cartilage repair using allogeneic perichondrocyte seeded biodegradable porous polylactic acid (PLA): a tissue-engineering study, *J Biomed Mater Res,* **29**, 1147.

102. Neves, S. C., Moreira Teixeira, L. S., Moroni, L., Reis, R. L., Van Blitterswijk, C. A., Alves, N. M., Karperien, M., Mano, J. F. (2011). Chitosan/poly(ε-caprolactone) blend scaffolds for cartilage repair, *Biomaterials,* **32**, 1068.

103. Gloria, A., Causa, F., Russo, T., Battista, E., Della Moglie, R., Zeppetelli, S., De Santis, R., Netti, P. A., Ambrosio, L. (2012). Three-dimensional poly(ε-caprolactone) bioactive scaffolds with controlled structural and surface properties, *Biomacromolecules,* **13**, 3510.

104. Hsu, S. H., Chang, S. H., Yen, H. J., Whu, S. W., Tsai, C. L., Chen, D. C. (2006). Evaluation of biodegradable polyesters modified by type II collagen and Arg-Gly-Asp as tissue engineering scaffolding materials for cartilage regeneration, *Artif Organs,* **30**, 42.

105. Nishida, T., Kubota, S., Kojima, S., Kuboki, T., Nakao, K., Kushibiki, T., Tabata, Y., Takigawa, M. (2004). Regeneration of defects in articular cartilage in rat knee joints by CCN2 (connective tissue growth factor), *J Bone Miner Res,* **19**, 1308.

106. Zhu, S., Zhang, B., Man, C., Ma, Y., Liu, X., Hu, J. (2014). Combined effects of connective tissue growth factor-modified bone marrow-derived mesenchymal stem cells and NaOH-treated PLGA scaffolds on the repair of articular cartilage defect in rabbits, *Cell Transplant,* **23**, 715.

107. Dürselen, L., Dauner, M., Hierlemann, H., Planck, H., Claes, L. E., Ignatius, A. (2001). Resorbable polymer fibers for ligament augmentation, *J Biomed Mater Res,* **58**, 666.

108. Kellomäki, M., Niiranen, H., Puumanen, K., Ashammakhi, N., Waris, T., Törmälä, P. (2000). Bioabsorbable scaffolds for guided bone regeneration and generation, *Biomaterials,* **21**, 2495.

109. Papenburg, B. J., Liu, J., Higuera, G. A., Barradas, A., de Boer, J., van Blitterswijk, C. A., Wessling, M., Stamatialis, D. (2009). Development and analysis of multi-layer scaffolds for tissue engineering, *Biomaterials,* **30**, 6228.

110. Vainionpää, S., Rokkanen, P., Törmälä, P. (1989). Surgical applications of biodegradable polymers in human tissues, *Prog Polym Sci,* **14**, 679.

111. Lopes, M. S., Jardini, A. (2012). Poly (lactic acid) production for tissue engineering applications, *Procedia Eng,* **42**, 1402.

112. Bos, R. R., Boering, G., Rozema, F. R., Leenslag, J. W. (1987). Resorbable poly (L-lactide) plates and screws for the fixation of zygomatic fractures, *J Oral Maxillofac Surg,* **45**, 751.

113. Rokkanen, P. U. (1991). Absorbable materials in orthopaedic surgery, *Ann Med,* **23**, 109.

114. Lee, J., Choi, W. I., Tae, G., Kim, Y. H., Kang, S. S., Kim, S. E., Kim, S.-H., Jung, Y., Kim, S. H. (2011). Enhanced regeneration of the ligament–bone interface using a poly(L-lactide-*co*-ε-caprolactone) scaffold with local delivery of cells/BMP-2 using a heparin-based hydrogel, *Acta Biomater,* **7**, 244.

115. Martin, D. P., Williams, S. F. (2003). Medical applications of poly-4-hydroxybutyrate: a strong flexible absorbable biomaterial, *Biochem Eng J,* **16**, 97.

116. Boland, E. D., Wnek, G. E., Simpson, D. G., Pawlowski, K. J., Bowlin, G. L. (2001). Tailoring tissue engineering scaffolds using electrostatic processing techniques: a study of poly(glycolic acid) electrospinning, *J Macromol Sci A,* **38**, 1231.

117. Chen, G., Ushida, T., Tateishi, T. (2002). Scaffold design for tissue engineering, *Macromol Biosci,* **2**, 67.

118. Ide, A., Sakane, M., Chen, G., Shimojo, H., Ushida, T., Tateishi, T., Wadano, Y., Miyanaga, Y. (2001). Collagen hybridization with poly(L-lactic acid) braid promotes ligament cell migration, *Mater Sci Eng C,* **17**, 95.

119. Chen, L., Bai, Y., Liao, G., Peng, E., Wu, B., Wang, Y., Zeng, X., Xie, X. (2013). Electrospun poly(L-lactide)/poly(ε-caprolactone) blend nanofibrous scaffold: characterization and biocompatibility with human adipose-derived stem cells, *PLoS One,* **8**, e71265.

120. Inanç, B., Arslan, Y. E., Seker, S., Elçin, A. E., Elçin, Y. M. (2009). Periodontal ligament cellular structures engineered with electrospun poly(DL-lactide-*co*-glycolide) nanofibrous membrane scaffolds, *J Biomed Mater Res A,* **90A**, 186.

121. Inanc, B., Elcin, A. E., Elcin, Y. M. (2006). Osteogenic induction of human periodontal ligament fibroblasts under two- and three-dimensional culture conditions, *Tissue Eng,* **12**, 257.

122. Owen, G. R., Jackson, J. K., Chehroudi, B., Brunette, D. M., Burt, H. M. (2010). An *in vitro* study of plasticized poly(lactic-*co*-glycolic acid) films as possible guided tissue regeneration membranes: material properties and drug release kinetics, *J Biomed Mater Res A,* **95A**, 857.

123. Shelke, N. B., Rokhade, A. P., Aminabhavi, T. M. (2010). Preparation and evaluation of novel blend microspheres of poly (lactic-co-glycolic) acid and pluronic F68/127 for controlled release of repaglinide, *J Appl Polym Sci,* **116**, 366.

124. Campos, D. M., Gritsch, K., Salles, V., Attik, G. N., Grosgogeat, B. (2014). Surface entrapment of fibronectin on electrospun PLGA scaffolds for periodontal tissue engineering, *BioRes Open Access,* **3**, 117.

125. Vunjak-Novakovic, G., Altman, G., Horan, R., Kaplan, D. L. (2004). Tissue engineering of ligaments, *Annu Rev Biomed Eng,* **6**, 131.

126. Surrao, D. C., Waldman, S. D., Amsden, B. G. (2012). Biomimetic poly(lactide) based fibrous scaffolds for ligament tissue engineering, *Acta Biomater,* **8**, 3997.

127. Sambit, S., James Goh, C.-H., Toh, S.-L. (2007). Development of hybrid polymer scaffolds for potential applications in ligament and tendon tissue engineering, *Biomed Mater,* **2**, 169.

128. Surrao, D. C., Fan, J. C., Waldman, S. D., Amsden, B. G. (2012). A crimp-like microarchitecture improves tissue production in fibrous ligament scaffolds in response to mechanical stimuli, *Acta Biomater,* **8**, 3704.

129. Sahoo, S., Ouyang, H., Goh, J. C., Tay, T. E., Toh, S. L. (2006). Characterization of a novel polymeric scaffold for potential application in tendon/ligament tissue engineering, *Tissue Eng,* **12**, 91.

130. Lou, J., Tu, Y., Burns, M., Silva, M. J., Manske, P. (2001). BMP-12 gene transfer augmentation of lacerated tendon repair, *J Orthop Res,* **19**, 1199.

131. Chen, C. H., Liu, H. W., Tsai, C. L., Yu, C. M., Lin, I. H., Hsiue, G. H. (2008). Photoencapsulation of bone morphogenetic protein-2 and periosteal progenitor cells improve tendon graft healing in a bone tunnel, *Am J Sports Med,* **36**, 461.

132. Kimura, Y., Hokugo, A., Takamoto, T., Tabata, Y., Kurosawa, H. (2008). Regeneration of anterior cruciate ligament by biodegradable scaffold combined with local controlled release of basic fibroblast growth factor and collagen wrapping, *Tissue Eng Part C Methods,* **14**, 47.

133. Lee, J., Guarino, V., Gloria, A., Ambrosio, L., Tae, G., Kim, Y. H., Jung, Y., Kim, S. H. (2010). Regeneration of Achilles' tendon: the role of dynamic stimulation for enhanced cell proliferation and mechanical properties, *J Biomater Sci Polym Ed,* **21**, 1173.

134. Viinikainen, A. K., Goransson, H., Huovinen, K., Kellomaki, M., Tormala, P., Rokkanen, P. (2009). Bioabsorbable poly-L/D-lactide (PLDLA) 96/4 triple-stranded bound suture in the modified Kessler repair: an *ex vivo* static and cyclic tensile testing study in a porcine extensor tendon model, *J Mater Sci Mater Med,* **20**, 1963.

135. Savage, R., Risitano, G. (1989). Flexor tendon repair using a "six strand" method of repair and early active mobilisation, *J Hand Surg Br,* **14**, 396.

136. Chainani, A., Hippensteel, K. J., Kishan, A., Garrigues, N. W., Ruch, D. S., Guilak, F., Little, D. (2013). Multilayered electrospun scaffolds for tendon tissue engineering, *Tissue Eng Part A,* **19**, 2594.

137. Taylor, W. (2005). Musculoskeletal pain in the adult New Zealand population: prevalence and impact, *N Z Med J,* **118**, U1629.

138. Lewis, J. S. (2009). Rotator cuff tendinopathy, *Br J Sports Med,* **43**, 236.

139. Inui, A., Kokubu, T., Mifune, Y., Sakata, R., Nishimoto, H., Nishida, K., Akisue, T., Kuroda, R., Satake, M., Kaneko, H., Fujioka, H. (2012). Regeneration of rotator cuff tear using electrospun poly(D,L-lactide-*co*-glycolide) scaffolds in a rabbit model, *Arthroscopy,* **28**, 1790.

140. Inui, A., Kokubu, T., Fujioka, H., Nagura, I., Sakata, R., Nishimoto, H., Kotera, M., Nishino, T., Kurosaka, M. (2011). Application of layered poly (L-lactic acid) cell free scaffold in a rabbit rotator cuff defect model, *Sports Med Arthrosc Rehabil Ther Technol,* **3**, 1758.

141. Barber, J. G., Handorf, A. M., Allee, T. J., Li, W. J. (2013). Braided nanofibrous scaffold for tendon and ligament tissue engineering, *Tissue Eng Part A,* **19**, 1265.

142. Manning, C. N., Schwartz, A. G., Liu, W., Xie, J., Havlioglu, N., Sakiyama-Elbert, S. E., Silva, M. J., Xia, Y., Gelberman, R. H., Thomopoulos, S. (2013). Controlled delivery of mesenchymal stem cells and growth factors using a nanofiber scaffold for tendon repair, *Acta Biomater,* **9**, 6905.

143. Viinikainen, A., Goransson, H., Taskinen, H. S., Roytta, M., Kellomaki, M., Tormala, P., Rokkanen, P. (2014). Flexor tendon healing within the tendon sheath using bioabsorbable poly-L/D-lactide 96/4 suture. A histological *in vivo* study with rabbits, *J Mater Sci Mater Med,* **25**, 1319.

144. Wang, S., Cai, L. (2010). Polymers for fabricating nerve conduits, *Int J Polym Sci,* **2010**, 20

145. Yucel, D., Kose, G. T., Hasirci, V. (2010). Polyester based nerve guidance conduit design, *Biomaterials,* **31**, 1596.

146. Borkenhagen, M., Stoll, R. C., Neuenschwander, P., Suter, U. W., Aebischer, P. (1998). *In vivo* performance of a new biodegradable polyester urethane system used as a nerve guidance channel, *Biomaterials,* **19**, 2155.

147. Vleggeert-Lankamp, C. L. A. M., De Ruiter, G. C. W., Wolfs, J. F. C., Pêgo, A. P., Van Den Berg, R. J., Feirabend, H. K. P., Malessy, M. J. A., Lakke, E. A. J. F. (2007). Pores in synthetic nerve conduits are beneficial to regeneration, *J Biomed Mater Res A,* **80**, 965.

148. Wang, H. B., Mullins, M. E., Cregg, J. M., Hurtado, A., Oudega, M., Trombley, M. T., Gilbert, R. J. (2009). Creation of highly aligned electrospun poly-L-lactic acid fibers for nerve regeneration applications, *J Neural Eng,* **6**, 1741.

149. Wang, H. B., Mullins, M. E., Cregg, J. M., McCarthy, C. W., Gilbert, R. J. (2010). Varying the diameter of aligned electrospun fibers alters neurite outgrowth and Schwann cell migration, *Acta Biomater,* **6**, 2970.

150. Kang, K. N., Kim, D. Y., Yoon, S. M., Lee, J. Y., Lee, B. N., Kwon, J. S., Seo, H. W., Lee, I. W., Shin, H. C., Kim, Y. M., Kim, H. S., Kim, J. H., Min, B. H., Lee, H. B., Kim, M. S. (2012). Tissue engineered regeneration of completely transected spinal cord using human mesenchymal stem cells, *Biomaterials,* **33**, 4828.

151. Madduri, S., Gander, B. (2012). Growth factor delivery systems and repair strategies for damaged peripheral nerves, *J Control Release,* **161**, 274.

152. Sun, W., Sun, C., Lin, H., Zhao, H., Wang, J., Ma, H., Chen, B., Xiao, Z., Dai, J. (2009). The effect of collagen-binding NGF-beta on the promotion of sciatic nerve regeneration in a rat sciatic nerve crush injury model, *Biomaterials,* **30**, 4649.

153. Sun, W., Sun, C., Zhao, H., Lin, H., Han, Q., Wang, J., Ma, H., Chen, B., Xiao, Z., Dai, J. (2009). Improvement of sciatic nerve regeneration using laminin-binding human NGF-β, *PLoS One,* **4**, e6180.

154. Wood, M. D., Gordon, T., Kim, H., Szynkaruk, M., Phua, P., Lafontaine, C., Kemp, S. W., Shoichet, M. S., Borschel, G. H. (2013). Fibrin gels containing GDNF microspheres increase axonal regeneration after delayed peripheral nerve repair, *Regen Med,* **8**, 27.

155. Kokai, L. E., Ghaznavi, A. M., Marra, K. G. (2010). Incorporation of double-walled microspheres into polymer nerve guides for the sustained delivery of glial cell line-derived neurotrophic factor, *Biomaterials,* **31**, 2313.

156. Kokai, L. E., Bourbeau, D., Weber, D., McAtee, J., Marra, K. G. (2011). Sustained growth factor delivery promotes axonal regeneration in long gap peripheral nerve repair, *Tissue Eng Part A,* **17**, 1263.

157. Liu, J. J., Wang, C. Y., Wang, J. G., Ruan, H. J., Fan, C. Y. (2011). Peripheral nerve regeneration using composite poly(lactic acid-caprolactone)/nerve growth factor conduits prepared by coaxial electrospinning, *J Biomed Mater Res A,* **96**, 13.

158. Longo, U. G., Loppini, M., Berton, A., Spiezia, F., Maffulli, N., Denaro, V. (2012). Tissue engineered strategies for skeletal muscle injury, *Stem Cells Int,* **2012**, 175038.

159. Levenberg, S., Rouwkema, J., Macdonald, M., Garfein, E. S., Kohane, D. S., Darland, D. C., Marini, R., van Blitterswijk, C. A., Mulligan, R. C., D'Amore, P. A., Langer, R. (2005). Engineering vascularized skeletal muscle tissue, *Nat Biotech,* **23**, 879.

160. Kaully, T., Kaufman-Francis, K., Lesman, A., Levenberg, S. (2009). Vascularization--the conduit to viable engineered tissues, *Tissue Eng Part B,* **15**, 159.

161. Keun Kwon, I., Kidoaki, S., Matsuda, T. (2005). Electrospun nano- to microfiber fabrics made of biodegradable copolyesters: structural characteristics, mechanical properties and cell adhesion potential, *Biomaterials,* **26**, 3929.

162. Sankaran, K. K., Krishnan, U. M., Sethuraman, S. (2014). Axially aligned 3D nanofibrous grafts of PLA-PCL for small diameter cardiovascular applications, *J Biomater Sci Polym Ed,* **25**, 1791.

163. Kouya, T., Tada, S.-i., Minbu, H., Nakajima, Y., Horimizu, M., Kawase, T., Lloyd, D. R., Tanaka, T. (2013). Microporous membranes of PLLA/PCL blends for periosteal tissue scaffold, *Mater Lett,* **95**, 103.

164. Tanaka, T., Eguchi, S., Saitoh, H., Taniguchi, M., Lloyd, D. R. (2008). Microporous foams of polymer blends of poly(L-lactic acid) and poly(ε-caprolactone), *Desalination,* **234**, 175.

165. Molladavoodi, S., Gorbet, M., Medley, J., Ju Kwon, H. (2013). Investigation of microstructure, mechanical properties and cellular viability of poly(L-lactic acid) tissue engineering scaffolds prepared by different thermally induced phase separation protocols, *J Mech Behav Biomed Mater,* **17**, 186.

166. Causa, F., Battista, E., Della Moglie, R., Guarnieri, D., Iannone, M., Netti, P. A. (2010). Surface investigation on biomimetic materials to control cell adhesion: the case of RGD conjugation on PCL, *Langmuir,* **26**, 9875.

167. Pham, Q. P., Sharma, U., Mikos, A. G. (2006). Electrospun poly(ε-caprolactone) microfiber and multilayer nanofiber/microfiber scaffolds: characterization of scaffolds and measurement of cellular infiltration, *Biomacromolecules,* **7**, 2796.

168. Fisher, M. B., Henning, E. A., Söegaard, N., Esterhai, J. L., Mauck, R. L. (2013). Organized nanofibrous scaffolds that mimic the macroscopic and microscopic architecture of the knee meniscus, *Acta Biomater,* **9**, 4496.

169. Baker, B. M., Mauck, R. L. (2007). The effect of nanofiber alignment on the maturation of engineered meniscus constructs, *Biomaterials,* **28**, 1967.

Chapter 19

Polyester Particles for Curcumin Delivery

Murali M. Yallapu, Meena Jaggi, and Subhash C. Chauhan
Department of Pharmaceutical Sciences and the Center for Cancer Research, University of Tennessee Health Science Center, Memphis, TN 38163, USA
myallapu@uthsc.edu, schauha1@uthsc.edu

Curcumin is a phenolic molecule extracted from turmeric and is prevalently used in Asian cuisine. This molecule has also shown great pharmacological activities. The pleiotropic property of curcumin enables it to act on multiple targets and as a promising therapeutic agent for various diseases. However, the treatment paradigm of curcumin in the clinical setting is not well established because of its low solubility, poor adsorption, low bioavailability, nonlinear pharmacokinetics, extensive metabolism, and degradation/secretion. Particle-mediated delivery of curcumin can address such shortcomings and enhance the therapeutic activity of curcumin. Particle technology aimed at preferential accumulation of curcumin at the disease site, thereby reducing adverse side effects, makes curcumin ideal as an efficient delivery strategy. Therefore, this chapter discusses polyester particles for curcumin delivery based on structurally varied delivery systems, which are aimed at increasing

Handbook of Polyester Drug Delivery Systems
Edited by MNV Ravikumar

ISBN 978-981-4669-65-8 (Hardcover), 978-981-4669-66-5 (eBook)
www.panstanford.com

bioavailability and biological effects. This chapter also discloses possible translational recommendations for polyester based curcumin particles for developing alternative clinical formulation(s) for therapeutic applications.

19.1 Unmet Needs in Chemotherapy and Role of Curcumin

Chemotherapy is considered to be a major therapeutic strategy to treat a number of human diseases. Additionally, targeted delivery of drugs to the disease site is a multitask. Conventional chemotherapeutic modalities have been associated with significant side effects. Therefore, there is a high demand for developing safe chemotherapies that can offer minimal side effects and exponentially target the disease site. During the last two decades, the use of a naturally occurring molecule, such as curcumin, has increased due to its inherent anti-inflammatory, antimicrobial, antimalarial, anticarcinogen, and antioxidant properties [1–3]. Unlike many other chemotherapeutic agents, curcumin exhibits a wide range of pharmacological properties that is attributed to binding to a number of protein targets [4–11]. Such pleiotropic properties are capable of modulating a number of signaling pathways, including pro-inflammatory cytokines, apoptotic proteins, NF-κB, cyclooxygenase-2, 5-LOX, STAT3, C-reactive protein, prostaglandin E(2), prostate-specific antigen, adhesion molecules, phosphorylase kinase, transforming growth factor-β, triglyceride, ET-1, creatinine, HO-1, AST, and ALT [12–15]. Further, this modulation is possible due to efficient interaction with lipid bilayer membranes, perturbation of the plasma membrane and trans-membrane protein conformations [16–18]. In particular, curcumin shows superior anticancer properties through the altering of genetic and epigenetic mechanisms, including DNA methylation, histone modification, chromatin remodeling, and microRNA (miRNA) regulation [19].

19.1.1 Evidence of Curcumin's Medicinal Value within the Literature

Advances from academic/medical research have offered a more comprehensive understanding of the curcumin molecule and its medicinal value. Cumulative studies suggest that interplay of curcumin between various signaling pathways is involved in the regulation of a number of diseases or gene status/functions [2, 3, 14–16]. Because of its inherent parent properties, such as antioxidant, antibiotic, chemoprevention, growth inhibition, immune stimulation, mutagenesis, and insecticide, curcumin has been traditionally used since ancient times for beauty, chemoprevention, contraception, cough, wound healing, stomach, and liver problems [3, 13, 15] (Fig. 19.1). Recent medicinal use includes but is not limited to inflammation, cancer, diabetes, infection, neurological disorders, bowel disorders, and fibrosis (Fig. 19.1).

Limitation
Low solubility
Less physico-chemical stability
Poor pharmacokinetics
Poor bioavailability
Rapid metabolism
Less active targeting

Strength
PROPERTIES: Antioxidant, Antibiotic, Chemoprevention, Growth inhibitor, Immuno-stimulator, Mutagenesis, Insecticide/larvicide
TRADITIONAL USE: Food/curry, Beauty, Chemoprevention, Contraception, Cough, Wound healing, Stomach/liver problems
ANIMAL STUDIES: Chemo-/radio-protection, Alzheimer's disease, Parkinson's disease, Neurological disorders, Wound healing, Inflammation and cancer, Diabetes/aging/arthritis, Memory improvement, Thrombosis
MEDICINAL USE: Inflammation, Cancer, Diabetes, Infection, Neurological disorders, Bowel syndrome, Fibrosis

Opportunity
DEVELOPMENT OF CURCUMIN FORMULATIONS TO IMPROVE: Solubility/stability, Pharmacokinetics, Bioavailability, Degradation/metabolism, Active targeting
FORMULATIONS OF CURCUMIN FOR THERAPEUTIC/MEDICINAL APPLICATIONS: Conjugates/polymer conjugates, Polymer composites, Stabilizers/Micelles, Lipid/liposomes, Hydro/micronanogels, Nanoparticles

Figure 19.1 Schematic illustration/chart of curcumin, its weaknesses, strengths, and new opportunities for medicinal applications.

Additionally, a search of current literature (Google.com; PubMed, and Google Patents) and clinical trials (Clinicaltrials.gov) (Fig. 19.2) shows that curcumin has proven to be a promising therapeutic molecule that has been studied and implemented for various disease treatments. Advantages of curcumin for medicinal application include low risk of side effects, improved efficacy, low cost, and widespread availability [20, 21]. Significant progress in a number of clinical trials with curcumin and discoveries obtained through research has transformed basic research into the development of nutraceuticals or pharmaceuticals [2, 3, 14].

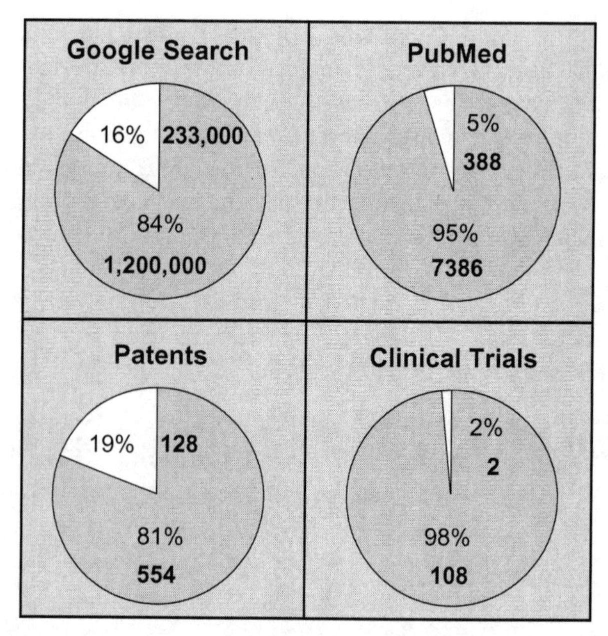

Figure 19.2 Pie chart demonstrating number or percent of literature belonging to curcumin (gray portions) and curcumin in the form of particle formulation (white portions). Data obtained from Google.com, PubMed, Google Patents, and Clinical trials (Clinicaltrials.gov) on March 1, 2015.

19.1.2 Safety and New Approaches

High-throughput pharmaceutical screening methods aimed at successful drug discovery are shifting the drug development paradigm from a single target to multiple-target approach [22]. Such multi-target approaches may offer an ideal master key that can be chosen to activate a number of locks in order to achieve major clinical benefits [19]. Selecting naturally occurring molecules that are designated as "Generally Recognized As Safe" (GRAS) material by the United States Food Drug Administration (USFDA) can prompt its use for medical applications [23, 24]. In such circumstances, the naturally occurring curcumin molecule is recognized as uncover a compound with novel mechanisms along with multi synergistic actions that reposition benefits in pharmaceutics. Based on the collective literature, it is clearly evident that among natural compounds, curcumin is one of the most widely used molecules and exhibits anticancer and chemopreventive properties and reverses chemo-/radiation resistance. However, in reality the translation of curcumin as a therapeutic molecule is hindered by a number of limitations (Fig. 19.1) [20, 21, 25]. These limitations extend to the lack of dosage instruction, regulation, large quantity and duration of treatment modalities. To overcome several of these drawbacks, research has focused on nanotechnology or particle based formulations for development of curcumin therapeutic particles (Fig. 19.2, white portions). Therapeutic particles of curcumin increase the availability and interaction with cellular targets for a better therapeutic benefit with enhanced bioequivalent activity [21]. A number of pioneering strategies are available to encapsulate or to load curcumin for future therapeutics.

19.1.3 Particles for Curcumin Delivery

The focus of this book chapter is to delineate the latest developments of curcumin particle formulations [26–28]. A number of curcumin based particle formulations have been under human use as oral pills/suspensions or tablets [14, 27]. Various forms of curcumin particle formulations can be considered as nutraceuticals. Although most of the particle based curcumin formulations may serve to improve normal health condition, no product/formulation at this time is approved by the USFDA for any specific medical condition. However, the key interest would be developing particle formulations for therapeutic applications, which is significantly lacking [29]. Generation of a particle formulation of curcumin with targeting ability at the disease site is highly important. To achieve this specifically, combination of a virtual library of curcumin composition, method of particle preparation with good stabilization, evaluation of physicochemical properties, testing *in vitro* and *in vivo* and clinical trials are highly recommended. In generating such a conceptual idea of particle formulation of curcumin, choosing a universal particle carrier is most important. Again, selection of biodegradable and USFDA approved polymer as particle carrier is of utmost importance.

19.2 Particle Technology

Particle- or nanoparticle-medicated drug delivery technology is currently vanguard in various disease treatments [30]. Particle technology plays a vital role in drug delivery as it results in delivering the drug at the right place and at the right concentration while reducing side effects, which minimizes the mortality rate [31–33]. Particle formulations of various therapeutic agents have been studied during that last two decades [34]. Various types of particle drug delivery systems, such as polymers, liposomes, lipid, gel particles, and other types of nano-/microparticles, have provided potential approaches. In fact, particle formulations with traditional anticancer drug (paclitaxel, docetaxel, etc.) delivery using Abraxane, Doxil, Genexol-PM, CALAA-01, MCC-465, MBP-426, and BIND-014 particle formulations are approved by the USFDA for clinical use or are under clinical trials. The basic role of particle technology in drug delivery is to increase solubility, stability, bioavailability, protect from systemic toxicity, physical/chemical degradation, enhanced pharmacological activity and tissue distribution [35]. In this chaper, we document various polyester particle carrier(s) that can be used to deliver curcumin in its active form for efficient therapeutic applications [30]. This includes generation of penetration peptides, modification of the nanoparticle's surface, and attachment of targeted motif [30]. Multiple drug(s) or therapeutic agent(s) loading along with curcumin [36, 37] will also be helpful for the wide range of therapies.

19.3 Polyester Particles for Curcumin Delivery

Extensive research and development in biodegradable polymers have led to curcumin's use for a wide variety of medical applications [38–40]. Biodegradable polymer nanoparticles based drug delivery is highly suitable for all therapeutic administration routes [41–43]. Over two decades, polyesters (polyester polymers) such as poly(lactic acid) (PLA), poly(glycolic acid) (PGA), their copolymers poly(lactic acid-*co*-glycolic acid) (PLGA), and poly(caprolactone) (PCL) (Fig. 19.3) have been approved for medical applications by the USFDA and European Medicines Agency (EMA) owing to their excellent biocompatibility and biodegradability [39]. Additionally, polyhydroxyvalerate (PHV) and polyhydroxybutyrate (PHB) are sought for polyester particles. So far, these polymers have been extensively used to formulate various curcumin nanoformulations following traditional to the most advanced techniques. Curcumin is dissolved, entrapped, encapsulated, or attached to a particle matrix during preparation of polyester particles. We have published extensive review articles on the general aspects of curcumin nanoparticle formulation [27, 28]. These articles have dealt with curcumin nanoparticle preparative methods, types of nanoparticles, *in vivo* fate and cancer therapeutic applications.

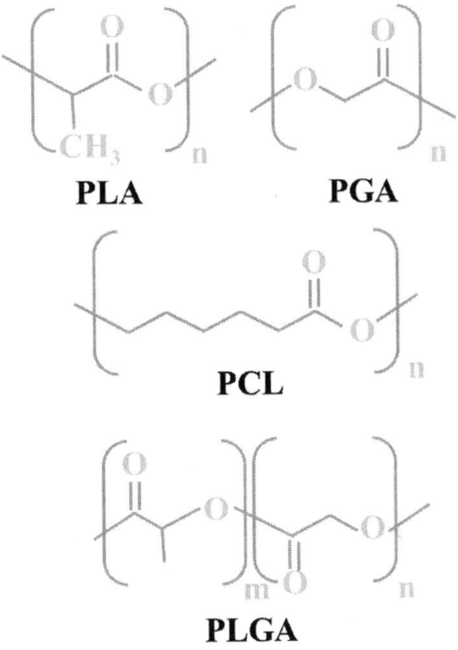

Figure 19.3 Chemical structures of commonly used polyester polymers, poly(lactic acid) (PLA), poly(glycolic acid) (PGA), poly(caprolactone) (PCL), and poly(lactic acid-*co*-glycolic acid) (PLGA) for the preparation of polyester particles for curcumin delivery.

The physicochemical and biological properties of particles widely vary and are highly influenced by polymer/copolymer molecular weight, method of preparation, type and amount of stabilizers/lyoprotectants, and curcumin loading ratio. Some proof-of-concept studies demonstrated efficient site-specific delivery of curcumin by nanoformulations [44–46]. Many *in vitro* and *in vivo* results support that polyester curcumin nanoformulations have demonstrated improved effects over free curcumin, through their cytotoxicity, improved cellular uptake, accumulation and retention, and targeting. All these characteristics were manipulated by altering the particle size, surface charge, surface modification, and overall hydrophobicity of polymer curcumin nanoformulations [47–50]. Oral, intraperitoneal, and intravenous administrations of these polyester curcumin nanoparticles exhibited improved pharmacokinetic, pharmacodynamic, and bioavailability profiles [51–56]. These particular aspects may improve significantly if controlled release of curcumin occurs in active from. The performance of a few polyester based curcumin nanoformulations have been tested *in vivo* in xenograft mouse models and facilitated curcumin drug accumulation for long-term release and achieved a greater degree of tumor regression and mice survival [27, 28].

Our laboratory interest includes generation of polyester NPs utilizing PLGA, which has shown promising results in both *in vitro* and preclinical studies. Our particle formulations have an outstanding capability to effectively encapsulate anticancer drugs. Drug loaded particle formulation(s) composed of a PLGA core is subsequently coated with poly(vinyl alcohol) (PVA), poly(L-lysine) (PLL) and/or pluronic polymer (F127) for efficient drug delivery and active targeting of tumors. This formulation has several unique properties: (i) the PLGA core is capable of loading drug molecules (including curcumin) and responsible for sustained release of drugs, (ii) an F127 polymer layer on PLGA particles provides stability overall to the particle formulation, (iii) the F127 polymer also reverses multi-drug resistance protein in cancer cells, (iv) the polyethylene glycol chains of pluronic F127 polymer act as a stealth polymer which diminishes the nonspecific uptake of particles, and (v) amine functional groups on particles are useful for antibody/aptamer conjugation through a PEG-linker, *N*-hydroxysuccinimide, NHS group for targeting tumor/cancer cells.

19.3.1 Design and Development

The goal of generation of polyester particle formulation(s) for curcumin delivery is to achieve the desired concentrations at the intended target site while minimizing toxicity to normal cells [57–60]. In this section, we document various types of polyester particles that have been designed and developed during the last two decades [27]. Usually, polyester particles are prepared predominantly by two methods: (i) dispersion of the preformed polymers (solvent evaporation, spontaneous emulsification-solvent diffusion, nanoprecipitation, salting-out/emulsion–diffusion method, supercritical fluid technologies, etc.) or (ii) polymerization of monomers [61, 62]. Each preparative method offers distinct properties to the formulation that includes but not limited to (i) improved stability, (ii) pharmacokinetics and bioavailability, (iii) enhanced efficacy and specificity, and (iv) targeted therapeutic efficacy. These unique properties can be achieved by controlling the formulation's physicochemical properties. Polyester-based curcumin particles and the common preparative methods and stabilizers used can regulate particle size, zeta potential, and stability of formulations. Additionally, most of these polyester based curcumin particle formulations are well studied in both *in vitro* and *in vivo* animal models.

It is widely accepted that PLGA, PCL, PLGA-PEG, PCL-PEG, and other combinational formulations have been used to deliver curcumin. From the available literature, it is apparent that most of these polyester curcumin particles can produce particle size of ~100–250 nm [48, 63–65]. It should be noted that particle size range achieved by all these formulations efficiently allows escape from macrophage uptake [66]. In general, these polyester curcumin nanoformulations are produced with a defined structure of a core of polyester polymer that is covered and stabilized with branched units of surfactants/polymers [67, 68]. These polyester particles offer an excellent depot for curcumin which varies based on the polyester molecular weight, composition of copolymers, stabilizer ratio to polyester, and so on [69–72].

The stabilizer used for preparation of polyester particles plays a major role. Introduction of PVA, PEG, Pluronics, Tween 20/80, and Vitamin E (surfactants/stabilizers/polymers) in polyester particle formulations lead to an improved stability. These stabilizer polymers offer low interfacial tension with biological fluids, which suggests longer circulation time in the body can be maintained and this may have great implications in the medical field. Different ratios of stabilizer with polyester polymer in the polyester particle composition determine the particle size of the formulation [73]. For example, curcumin polyester particles developed by Kumar's group show an average particle size of ~121 nm to 242 nm, while changing various surfactants in the composition: PVA (242 ± 2 nm), Pluronic F-68 (230 ± 6 nm), Vitamin E-TPGS (208 ± 2 nm), and CTPB (121 ± 3 nm) [56]. Pluronic polymers, di- or triblock copolymers, liposomes, phospholipids, and cholesterol are considered to be stabilizers for polyester curcumin particles because of their hydrophilic and hydrophobic units, which cooperatively bind with curcumin molecules through the self-assembly process [74–78]. The important roles of several other polymers including poly(styrene sulfonate), poly(allylamine hydrochloride), poly(glutamic acid), poly(L-lysine), dextran sulfate, protamine sulfate, carboxymethyl cellulose, and gelatin need to be investigated. Preparation of polyester particles involves a number of techniques, but the nanoprecipitation method has been popular [79, 80] because of the quick process to obtain a uniform size with high curcumin incorporation.

19.3.2 Curcumin Encapsulation/Loading and Release

Curcumin loading and release have direct implications on its activity that is associated with the type of polyester particle and its preparative method. Over the past five years, significant advances have been made in developing suitable polyester particle formulations for efficient curcumin delivery. A successful polyester particle formulation must have a high loading capacity, possess controlled release of curcumin and reduce the quantity of the carrier required for *in vivo* administration. Curcumin loading or encapsulation in a polyester particle can be achieved by either incorporating the curcumin during polyester particle preparation or later through the absorption/adsorption process by incubating curcumin in solution with polyester particles. It is known that a large number of reports have dealt with the entrapment of curcumin during preparation of polyester particles [76, 77]. Encapsulation by the absorption/adsorption process is an easy approach to allow tailor-made curcumin loading. However, in both processes, curcumin loading depends on the efficiency of binding with polyester polymer or with the stabilizer that was used in the particle formulations. Curcumin loading in polyester particles is usually determined by the estimation of un-bound curcumin fraction or the encapsulated in the formulation. The un-entrapped/unbound estimation gives an indirect while the encapsulated estimation quantifies a direct estimation of curcumin which exists in the polyester particles [72]. Our recent review article discusses more details about curcumin loading estimations in various particle formulations [27, 28].

Curcumin loading and release is also dictated by geometry and physicochemical characteristics of the particle formulation (Fig. 19.4). Most studies measure *in vitro* curcumin release using diffusion cells (with artificial or biological membranes), dialysis bags, reverse dialysis sac, ultrafiltration, centrifugal ultrafiltration, or ultracentrifugation. The dialysis method has been widely used because there are less technical difficulties yet separation is easy, it is less time consuming, and no major instrumentation is required. A detailed procedure can be referred to in our recent reports [88–90]. The release of curcumin from polyester particles occurs in two or three phases, depending on the polyester polymer composition. The initial release peak usually arises by the release of loosely bound curcumin on the surface of the particles, followed by low and sustained release due to diffusion, and the final release stage relates to the degradation of polyester polymer. This clearly suggests that surface morphology/surfactant chemical groups, particle networks, and degradability govern release of curcumin. More importantly, sustained release is particularly directed by the degradation characteristics of the overall particle formulation. This degradation behavior is depicted in Fig. 19.4b. The degradation characteristics mainly depend on the polymer ratio, composition of ingredients (surfactant/polymers/stabilizers), and pore size of particles (networks). Additionally, curcumin release strictly relies on the method of preparation and sterilization, molecular weight of the polyester, hydrophilicity/hydrophobicity, polyester matrice(s), morphology of particles and crystallinity, amount of curcumin loading, external additives and release conditions, etc.

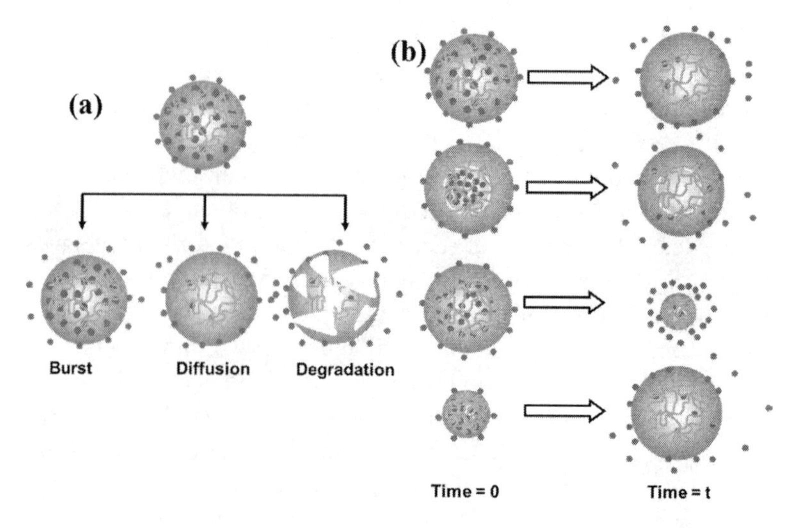

Figure 19.4 Curcumin release patterns from polyester particles. (a) Overall type and stages of curcumin release and (b) release of curcumin from polyester particles with distinct morphology, size, and shape under different stimuli conditions over time.

19.4 Therapeutic Implications

Targeting curcumin to the diseased organ or passing through the blood–brain barrier has been a major challenge. Achieving improved serum bioavailability and enhanced biodistribution are the key parameters in pursuit of the above measures. Additionally, possessing hemo-compatibility is another desired criteria for *in vivo* application of curcumin polyester particles. A summary of particle formulations that have both improved pharmacokinetics and bioavailability is documented in recent review articles [26–28]. Kumar's group [56] demonstrated for the first time that curcumin entrapped particles exhibited approximately ninefold increase in oral bioavailability compared to curcumin plus piperine (absorption enhancer) administration. Tsai *et al.* [53] proved that PLGA based curcumin improved curcumin bioavailability up to 22-fold over free curcumin, which is highly significant in terms of the clinical perspective. Improved bioavailability and distribution of biodegradable curcumin particle formulations potentiate a greater role in curing various diseases [54, 55, 65] (Table 19.1). We have not included here the use of curcumin nanoparticles in cancer treatment since there are many studies already presented in our earlier review articles [27, 28]. More significant research using curcumin nanoparticles dedicated to tackling the issue of chemo-/radiation damages is needed.

Table 19.1 Role of biodegradable polyester curcumin particles for various disease treatment

Biodegradable polymer	Disease	Outcome
PLGA [81]	Alzheimer's	Curcumin particles enhance neuronal differentiation and expression of neurogenic genes, which can activate Wnt/β-catenin signaling to reverse Aβ-mediated inhibitory effects to improve learning and memory in an AD condition.
PLGA [81]	Alzheimer's	Polyester curcumin particles encourage neurogenesis through activation of the canonical Wnt/β-catenin pathway, which could enhance a brain self-repair mechanism.
PLGA [82]	Brain injury	These curcumin particles induce neuroprotective effects by upward regulation of NF-κB (p65) and reduction of caspase-9a. Additionally, curcumin particles decrease CSF levels of TNF-α and IL-1β.
PLGA [83]	Diabetes	Nanocurcumin at a higher dose through encapsulated curcumin prevents or delays diabetic cataract.

(Continued)

Table 19.1 (*Continued*)

Biodegradable polymer	Disease	Outcome
PLGA [84]	Cystic fibrosis	Polyester curcumin nanoparticles partially correct subset of Cystic fibrosis symptoms in 129 background and C57/BL6 mouse strains and exhibits a more favorable outcome over free curcumin.
PLGA [85]	Cancer (Chemoprevention)	A single dose of curcumin microparticles resulted in hepatic GST and COX-2 activities in the liver and acts as a chemopreventive agent.
PLGA [86]	Wound healing	Curcumin particles significantly accelerate the wound closure by down regulating inflammatory responses.
PLGA [83]	Cataract	Ingestion of biodegradable curcumin particles not only improved oral bioavailability but also paved the way for a better efficacy in delaying diabetic cataract in rat study.
PLGA/PEG-PLA [87]	Morphine tolerance	Mice behavioral studies (tail-flick and hot-plate tests) were conducted to verify the effects of nanocurcumin on attenuating morphine tolerance. Significant analgesia was observed in mice during both tail-flick and hot-plate tests using orally administered nanocurcumin following subcutaneous injections of morphine. However, unformulated curcumin at the same dose showed no effect.

19.5 Steps toward Translation

Standard characterization measures are required to assess the pre-clinical curcumin particles formulation effectiveness for implementation as therapeutic purpose [88, 89]. In this regard, preclinical formulations of curcumin particles must be evaluated first for (i) purity, (ii) crystallinity, (iii) thermal properties, (iv) stability, (v) particle size, and (vi) zeta potential. Such tested curcumin particles formulations should be implemented for high throughput cytotoxicity assays in a number of panels of various cell line models drug discovery portfolio. Subsequently, examine their suitability/applicability in relevant *in vivo* (transgenic mouse models and primates) models. The role of such preclinical outcomes needs to be understood through their specific targeted mechanism, metabolism, and extent of potentials of curcumin particle formulations. Bioavailability, pharmacokinetics and tumor targeting are the main drivers that will determine curcumin particles' long-term therapeutic utility. Another major step is to examine systemic toxicity profiles of these polyester therapeutic curcumin nanoparticles. After all these, the next challenge is to select the particle formulation that suites an individual patient need or for a specific condition, i.e., paste/emulsions, implants, tablet (pill), gel capsules, or parental injectable formulations. Additionally, it is also necessarily to consider the frequency of treatment/combinational modality and age related concerns.

19.6 Conclusion

Curcumin is one of the most clinically relevant natural compounds and possesses extensive pharmacological activity that can be used to treat and prevent a wide variety of human diseases. This book chapter briefly summarizes the development of curcumin particle based therapeutic formulations from concept to clinical practice. Curcumin therapeutic particles will be important formulations in the effort to develop efficient therapeutic strategies that can significantly reduce the cost of treatments as well as improve patients' compliance and therapeutic outcomes. Further, multifunctional curcumin therapeutic particle formulations are important for the advancement of scientific understanding as well as mechanism driven clinical applications.

Acknowledgements

The authors thank Cathy Christopherson for editorial assistance. This work was partially supported by grants from the National Institutes of Health Research Project Grant Program (K22 CA174841 and U01 CA142736); and Department of Defense ARMY GRANT W81XWH-14-1-0154.

Declaration

Declared no conflicts of interest.

References

1. Aggarwal, B. B., Kumar, A., and Bharti, A. C. (2003). Anticancer potential of curcumin: preclinical and clinical studies. *Anticancer Res,* **23**, 363–398.
2. Goel, A., Kunnumakkara, A. B., and Aggarwal, B. B. (2008). Curcumin as "Curecumin": from kitchen to clinic. *Biochem Pharmacol,* **75**, 787–809.
3. Hatcher, H., Planalp, R., Cho, J., Torti, F. M., and Torti, S. V. (2008). Curcumin: from ancient medicine to current clinical trials. *Cell Mol Life Sci,* **65**, 1631–1652.
4. Beevers, C. S., Chen, L., Liu, L., Luo, Y., Webster, N. J., and Huang, S. (2009). Curcumin disrupts the Mammalian target of rapamycin-raptor complex. *Cancer Res,* **69**, 1000–1008.
5. Beevers, C. S., Li, F., Liu, L., and Huang, S. (2006). Curcumin inhibits the mammalian target of rapamycin-mediated signaling pathways in cancer cells. *Int J Cancer,* **119**, 757–764.
6. Dairaku, I., Han, Y., Yanaka, N., and Kato, N. (2010). Inhibitory effect of curcumin on IMP dehydrogenase, the target for anticancer and antiviral chemotherapy agents. *Biosci Biotechnol Biochem,* **74**, 185–187.
7. Sahebkar, A. (2014). Low-density lipoprotein is a potential target for curcumin: novel mechanistic insights. *Basic Clin Pharmacol Toxicol,* **114**, 437–438.

8. Singh, A. K., and Misra, K. (2013). Human papilloma virus 16 E6 protein as a target for curcuminoids, curcumin conjugates and congeners for chemoprevention of oral and cervical cancers. *Interdiscip Sci*, **5**, 112–118.

9. Trujillo, J., Granados-Castro, L. F., Zazueta, C., Anderica-Romero, A. C., Chirino, Y. I., and Pedraza-Chaverri, J. (2014). Mitochondria as a target in the therapeutic properties of curcumin. *Archiv der Pharmazie*, **347**, 873–884.

10. Xiao, Z., Zhang, A., Lin, J., Zheng, Z., Shi, X., Di, W., *et al.* (2014). Telomerase: a target for therapeutic effects of curcumin and a curcumin derivative in Abeta1-42 insult *in vitro*. *PLoS One*, **9**, e101251.

11. Zhao, F., Gong, Y., Hu, Y., Lu, M., Wang, J., Dong, J., *et al.* (2015). Curcumin and its major metabolites inhibit the inflammatory response induced by lipopolysaccharide: translocation of nuclear factor-kappaB as potential target. *Mol Med Rep*, **11**, 3087–3093.

12. Maheshwari, R. K., Singh, A. K., Gaddipati, J., and Srimal, R. C. (2006). Multiple biological activities of curcumin: a short review. *Life Sci*, **78**, 2081–2087.

13. Moghadamtousi, S. Z., Kadir, H. A., Hassandarvish, P., Tajik, H., Abubakar, S., and Zandi, K. (2014). A review on antibacterial, antiviral, and antifungal activity of curcumin. *BioMed Res Int*, **2014**, 186864.

14. Gupta, S. C., Patchva, S., and Aggarwal, B. B. (2013). Therapeutic roles of curcumin: lessons learned from clinical trials. *AAPS J*, **15**, 195–218.

15. Gupta, S. C., Kismali, G., and Aggarwal, B. B. (2013). Curcumin, a component of turmeric: from farm to pharmacy. *BioFactors*, **39**, 2–13.

16. Ingolfsson, H. I., Thakur, P., Herold, K. F., Hobart, E. A., Ramsey, N. B., Periole, X., *et al.* (2014). Phytochemicals perturb membranes and promiscuously alter protein function. *ACS Chem Biol*, **9**, 1788–1798.

17. Kopec, W., Telenius, J., and Khandelia, H. (2013). Molecular dynamics simulations of the interactions of medicinal plant extracts and drugs with lipid bilayer membranes. *FEBS J*, 280, 2785–805.

18. Chen, G., Chen, Y., Yang, N., Zhu, X., Sun, L., and Li, G. (2012). Interaction between curcumin and mimetic biomembrane. *Sci China Life Sci*, **55**, 527–532.

19. Goel, A., Jhurani, S., and Aggarwal, B. B. (2008). Multi-targeted therapy by curcumin: how spicy is it? *Mol Nutr Food Res*, **52**, 1010–1030.

20. Yang, C. S., Sang, S., Lambert, J. D., and Lee, M. J. (2008). Bioavailability issues in studying the health effects of plant polyphenolic compounds. *Mol Nutr Food Res*, **52** Suppl 1, S139–S151.

21. Anand, P., Kunnumakkara, A. B., Newman, R. A., and Aggarwal, B. B. (2007). Bioavailability of curcumin: problems and promises. *Mol Pharm*, **4**, 807–818.
22. Lu, J. J., Pan, W., Hu, Y. J., and Wang, Y. T. (2012). Multi-target drugs: the trend of drug research and development. *PLoS One*, **7**, e40262.
23. http://www.nutraingredients-usa.com/Suppliers2/Sabinsa-gets-FDA-no-objection-letter-for-GRAS-status-of-its-Curcumin-C3-Complex.
24. http://www.google.com/url?sa=t&rct=j&q=&esrc=s&source=web&cd=1&cad=rja&uact=8&ved=0CB8QFjAA&url=http%3A%2F%2Fwww.researchgate.net%2Fpublictopics.PublicPostFileLoader.html%3Fid%3D543f9169d5a3f25e128b45f3%26key%3D8e986fed-99cb-4b43-8d9b-be2c826c0e7c&ei=XAITVd27LsPOsQSlhoCwBQ&usg=AFQjCNGX7Z-SnJ2nnrcA0rLo3yO88LQ8fg&bvm=bv.89217033,d.cWc.
25. Burgos-Moron, E., Calderon-Montano, J. M., Salvador, J., Robles, A., and Lopez-Lazaro, M. (2010). The dark side of curcumin. *Int J Cancer*, **126**, 1771–1775.
26. Sun, M., Su, X., Ding, B., He, X., Liu, X., Yu, A., *et al.* (2012). Advances in nanotechnology-based delivery systems for curcumin. *Nanomedicine*, **7**, 1085–1100.
27. Yallapu, M. M., Jaggi, M., and Chauhan, S. C. (2012). Curcumin nanoformulations: a future nanomedicine for cancer. *Drug Discov Today*, **17**, 71–80.
28. Yallapu, M. M., Jaggi, M., and Chauhan, S. C. (2013). Curcumin nanomedicine: a road to cancer therapeutics. *Curr Pharm Des*, **19**, 1994–2010.
29. Ghalandarlaki, N., Alizadeh, A. M., and Ashkani-Esfahani, S. (2014). Nanotechnology-applied curcumin for different diseases therapy. *BioMed Res Int*, **2014**, 394264.
30. Petros, R. A., and DeSimone, J. M. (2010). Strategies in the design of nanoparticles for therapeutic applications. *Nat Rev Drug Discov*, **9**, 615–627.
31. De Jong, W. H., and Borm, P. J. (2008). Drug delivery and nanoparticles: applications and hazards. *Int J Nanomedicine*, **3**, 133–149.
32. Brigger, I., Dubernet, C., and Couvreur, P. (2002). Nanoparticles in cancer therapy and diagnosis. *Adv Drug Deliv Rev*, **54**, 631–651.
33. Portney, N. G., and Ozkan, M. (2006). Nano-oncology: drug delivery, imaging, and sensing. *Anal Bioanal Chem*, **384**, 620–630.

34. Torchilin, V. P. (2014). Multifunctional, stimuli-sensitive nanoparticulate systems for drug delivery. *Nat Rev Drug Discov,* **13**, 813–827.

35. Gunasekaran, T., Haile, T., Nigusse, T., and Dhanaraju, M. D. (2014). Nanotechnology: an effective tool for enhancing bioavailability and bioactivity of phytomedicine. *Asian Pac J Trop Biomed,* **4**, S1–S7.

36. Das, M., and Sahoo, S. K. (2012). Folate decorated dual drug loaded nanoparticle: role of curcumin in enhancing therapeutic potential of nutlin-3a by reversing multidrug resistance. *PLoS One,* **7**, e32920.

37. Scarano, W., Souza, P., and Stenzel, M. H. (2015). Dual-drug delivery of curcumin and platinum drugs in polymeric micelles enhances the synergistic effects: a double act for the treatment of multidrug-resistant cancer. *Biomater Sci,* **3**, 163–174.

38. Nair, L. S., and Laurencin, C. T. (2007). Biodegradable polymers as biomaterials. *Prog Polym Sci,* **32**, 762–798.

39. Ulery, B. D., Nair, L. S., and Laurencin, C. T. (2011). Biomedical applications of biodegradable polymers. *J Polym Sci B Polym Phys,* **49**, 832–864.

40. Kumar, N., Ravikumar, M. N., and Domb, A. J. (2001). Biodegradable block copolymers. *Adv Drug Deliv Rev,* **53**, 23–44.

41. Tiwari, G., Tiwari, R., Sriwastawa, B., Bhati, L., Pandey, S., Pandey, P., *et al.* (2012). Drug delivery systems: an updated review. *Int J Pharm Investig,* **2**, 2–11.

42. Maurya, S. K., Pathak, K., and Bali, V. (2010). Therapeutic potential of mucoadhesive drug delivery systems: an updated patent review. *Recent Pat Drug Deliv Formul,* **4**, 256–265.

43. Srivastava, R., and Pathak, K. (2011). An updated patent review on ocular drug delivery systems with potential for commercial viability. *Recent Pat Drug Deliv Formul,* **5**, 146–162.

44. Zhao, Y., Lin, D., Wu, F., Guo, L., He, G., Ouyang, L., *et al.* (2014). Discovery and *in vivo* evaluation of novel RGD-modified lipid-polymer hybrid nanoparticles for targeted drug delivery. *Int J Mol Sci,* **15**, 17565–17576.

45. Li, L., Xiang, D., Shigdar, S., Yang, W., Li, Q., Lin, J., *et al.* (2014). Epithelial cell adhesion molecule aptamer functionalized PLGA-lecithin-curcumin-PEG nanoparticles for targeted drug delivery to human colorectal adenocarcinoma cells. *Int J Nanomedicine,* **9**, 1083–1096.

46. Thamake, S. I., Raut, S. L., Gryczynski, Z., Ranjan, A. P., and Vishwanatha, J. K. (2012). Alendronate coated poly-lactic-*co*-glycolic acid (PLGA)

nanoparticles for active targeting of metastatic breast cancer. *Biomaterials*, **33**, 7164–7173.

47. Thamake, S. I., Raut, S. L., Ranjan, A. P., Gryczynski, Z., and Vishwanatha, J. K. (2011). Surface functionalization of PLGA nanoparticles by non-covalent insertion of a homo-bifunctional spacer for active targeting in cancer therapy. *Nanotechnology*, **22**, 035101.

48. Yallapu, M. M., Gupta, B. K., Jaggi, M., and Chauhan, S. C. (2010). Fabrication of curcumin encapsulated PLGA nanoparticles for improved therapeutic effects in metastatic cancer cells. *J Colloid Interface Sci*, **351**, 19–29.

49. Punfa, W., Yodkeeree, S., Pitchakarn, P., Ampasavate, C., and Limtrakul, P. (2012). Enhancement of cellular uptake and cytotoxicity of curcumin-loaded PLGA nanoparticles by conjugation with anti-P-glycoprotein in drug resistance cancer cells. *Acta Pharmacol Sin*, **33**, 823–831.

50. Yallapu, M. M., Khan, S., Maher, D. M., Ebeling, M. C., Sundram, V., Chauhan, N., *et al.* (2014). Anti-cancer activity of curcumin loaded nanoparticles in prostate cancer. *Biomaterials*, **35**, 8635–8648.

51. Tsai, Y. M., Chang-Liao, W. L., Chien, C. F., Lin, L. C., and Tsai, T. H. (2012). Effects of polymer molecular weight on relative oral bioavailability of curcumin. *Int J Nanomedicine*, **7**, 2957–2966.

52. Ghosh, D., Choudhury, S. T., Ghosh, S., Mandal, A. K., Sarkar, S., Ghosh, A., *et al.* (2012). Nanocapsulated curcumin: oral chemopreventive formulation against diethylnitrosamine induced hepatocellular carcinoma in rat. *Chem Biol Interact*, **195**, 206–214.

53. Tsai, Y. M., Jan, W. C., Chien, C. F., Lee, W. C., Lin, L. C., and Tsai, T. H. (2011). Optimised nano-formulation on the bioavailability of hydrophobic polyphenol, curcumin, in freely-moving rats. *Food Chem*, **127**, 918–925.

54. Xie, X., Tao, Q., Zou, Y., Zhang, F., Guo, M., Wang, Y., *et al.* (2011). PLGA nanoparticles improve the oral bioavailability of curcumin in rats: characterizations and mechanisms. *J Agric Food Chem*, **59**, 9280–9289.

55. Tsai, Y. M., Chien, C. F., Lin, L. C., and Tsai, T. H. (2011). Curcumin and its nano-formulation: the kinetics of tissue distribution and blood-brain barrier penetration. *Int J Pharm*, **416**, 331–338.

56. Shaikh, J., Ankola, D. D., Beniwal, V., Singh, D., and Kumar, M. N. (2009). Nanoparticle encapsulation improves oral bioavailability of curcumin by at least 9-fold when compared to curcumin administered with piperine as absorption enhancer. *Eur J Pharm Sci*, **37**, 223–230.

57. Monsuez, J. J., Charniot, J. C., Vignat, N., and Artigou, J. Y. (2010). Cardiac side-effects of cancer chemotherapy. *Int J Cardiol*, **144**, 3–15.

58. Carelle, N., Piotto, E., Bellanger, A., Germanaud, J., Thuillier, A., and Khayat, D. (2002). Changing patient perceptions of the side effects of cancer chemotherapy. *Cancer*, **95**, 155–163.

59. Conklin, K. A. (2000). Dietary antioxidants during cancer chemotherapy: impact on chemotherapeutic effectiveness and development of side effects. *Nutr Cancer*, **37**, 1–18.

60. Dodd, M. J. (1993). Side effects of cancer chemotherapy. *Annu Rev Nurs Res*, **11**, 77–103.

61. Csaba, N., and Alonso, M. J. (2014). 12. Biodegradable polymer nanoparticles as protein delivery systems: Original research articles: Design of biodegradable particles for protein delivery (2002), Chitosan nanoparticles as delivery systems for doxorubicin (2001); design of microencapsulated chitosan microspheres for colonic drug delivery (1998). *J Control Release*, **190**, 53–54.

62. Soppimath, K. S., Aminabhavi, T. M., Kulkarni, A. R., and Rudzinski, W. E. (2001). Biodegradable polymeric nanoparticles as drug delivery devices. *J Control Release*, **70**, 1–20.

63. Mukerjee, A., and Vishwanatha, J. K. (2009). Formulation, characterization and evaluation of curcumin-loaded PLGA nanospheres for cancer therapy. *Anticancer Res*, **29**, 3867–3875.

64. Braden, A. R. C., Vishwanatha, J. K., and Kafka, E. (2008). Formulation of active agent loaded activated PLGA nanoparticles for targeted cancer nano-therapeutics. In *Publication UPA*, editor. United States: University of North Texas Health Science Center at Fort Worth.

65. Anand, P., Nair, H. B., Sung, B., Kunnumakkara, A. B., Yadav, V. R., Tekmal, R. R., *et al.* (2010). Design of curcumin-loaded PLGA nanoparticles formulation with enhanced cellular uptake, and increased bioactivity *in vitro* and superior bioavailability *in vivo*. *Biochem Pharmacol*, **79**, 330–338.

66. Owens, D. E., 3rd, and Peppas, N. A. (2006). Opsonization, biodistribution, and pharmacokinetics of polymeric nanoparticles. *Int J Pharm*, **307**, 93–102.

67. Shi, W., Dolai, S., Rizk, S., Hussain, A., Tariq, H., Averick, S., *et al.* (2007). Synthesis of monofunctional curcumin derivatives, clicked curcumin dimer, and a PAMAM dendrimer curcumin conjugate for therapeutic applications. *Org Lett*, **9**, 5461–5464.

68. Babaei, E., Sadeghizadeh, M., Hassan, Z. M., Feizi, M. A., Najafi, F., and Hashemi, S. M. (2012). Dendrosomal curcumin significantly suppresses cancer cell proliferation *in vitro* and *in vivo*. *Int Immunopharmacol,* **12**, 226–234.

69. Yallapu, M. M., Vasir, J. K., Jain, T. K., Vijayaraghavalu, S., and Labhasetwar, V. (2008). Synthesis, characterization and antiproliferative activity of rapamycin-loaded poly(N-isopropylacrylamide)-based nanogels in vascular smooth muscle cells. *J Biomed Nanotechnol,* **4**, 16–24.

70. Yallapu, M. M., Ebeling, M. C., Chauhan, N., Jaggi, M., and Chauhan, S. C. (2011). Interaction of curcumin nanoformulations with human plasma proteins and erythrocytes. *Int J Nanomedicine,* **6**, 2779–2790.

71. Rejinold, N. S., Sreerekha, P. R., Chennazhi, K. P., Nair, S. V., and Jayakumar, R. (2011). Biocompatible, biodegradable and thermo-sensitive chitosan-g-poly (N-isopropylacrylamide) nanocarrier for curcumin drug delivery. *Int J Biol Macromol,* **49**, 161–172.

72. Bisht, S., Feldmann, G., Soni, S., Ravi, R., Karikar, C., and Maitra, A. (2007). Polymeric nanoparticle-encapsulated curcumin ("nanocurcumin"): a novel strategy for human cancer therapy. *J Nanobiotechnology,* **5**, 3.

73. Dandekar, P. P., Jain, R., Patil, S., Dhumal, R., Tiwari, D., Sharma, S., *et al.* (2010). Curcumin-loaded hydrogel nanoparticles: application in anti-malarial therapy and toxicological evaluation. *J Pharm Sci,* **99**, 4992–5010.

74. Sou, K., Inenaga, S., Takeoka, I., and Tsuchida, E. (2008). Loading of curcumin into macrophages using lipid-based nanoparticles. *Int J Pharm,* **352**, 287–293.

75. Mourtas, S., Canovi, M., Zona, C., Aurilia, D., Niarakis, A., La Ferla, B., *et al.* (2011). Curcumin-decorated nanoliposomes with very high affinity for amyloid-beta1-42 peptide. *Biomaterials,* **32**, 1635–1645.

76. Cui, J., Yu, B., Zhao, Y., Zhu, W., Li, H., Lou, H., *et al.* (2009). Enhancement of oral absorption of curcumin by self-microemulsifying drug delivery systems. *Int J Pharm,* **371**, 148–155.

77. Sahu, A., Kasoju, N., and Bora, U. (2008). Fluorescence study of the curcumin-case in micelle complexation and its application as a drug nanocarrier to cancer cells. *Biomacromolecules,* **9**, 2905–2912.

78. Gou, M., Men, K., Shi, H., Xiang, M., Zhang, J., Song, J., *et al.* (2011). Curcumin-loaded biodegradable polymeric micelles for colon cancer therapy *in vitro* and *in vivo*. *Nanoscale,* **3**, 1558–1567.

79. Liu, Z., Huang, Y., Jin, Y., and Cheng, Y. (2010). Mixing intensification by chaotic advection inside droplets for controlled nanoparticle preparation. *Microfluid Nanofluid,* **9**, 773–786.

80. He, Y., Huang, Y., and Cheng, Y. (2010). Structure evolution of curcumin nanoprecipitation from a micromixer. *Cryst Growth Des*, **10**, 1021–1024.

81. Tiwari, S. K., Agarwal, S., Seth, B., Yadav, A., Nair, S., Bhatnagar, P., *et al.* (2014). Curcumin-loaded nanoparticles potently induce adult neurogenesis and reverse cognitive deficits in Alzheimer's disease model via canonical Wnt/beta-catenin pathway. *ACS Nano*, **8**, 76–103.

82. Chang, C. Z., Wu, S. C., Lin, C. L., and Kwan, A. L. (2015). Curcumin, encapsulated in nano-sized PLGA, down-regulates nuclear factor kappaB (p65) and subarachnoid hemorrhage induced early brain injury in a rat model. *Brain Res*, **1608**, 215–224.

83. Grama, C. N., Suryanarayana, P., Patil, M. A., Raghu, G., Balakrishna, N., Kumar, M. N., *et al.* (2013). Efficacy of biodegradable curcumin nanoparticles in delaying cataract in diabetic rat model. *PLoS One*, **8**, e78217.

84. Cartiera, M. S., Ferreira, E. C., Caputo, C., Egan, M. E., Caplan, M. J., and Saltzman, W. M. (2010). Partial correction of cystic fibrosis defects with PLGA nanoparticles encapsulating curcumin. *Mol Pharm*, **7**, 86–93.

85. Shahani, K., and Panyam, J. (2011). Highly loaded, sustained-release microparticles of curcumin for chemoprevention. *J Pharm Sci*, **100**, 2599–2609.

86. Chereddy, K. K., Coco, R., Memvanga, P. B., Ucakar, B., des Rieux, A., Vandermeulen, G., *et al.* (2013). Combined effect of PLGA and curcumin on wound healing activity. *J Control Release*, 171, 208–215.

87. Shen, H., Hu, X., Szymusiak, M., Wang, Z. J., and Liu, Y. (2013). Orally administered nanocurcumin to attenuate morphine tolerance: comparison between negatively charged PLGA and partially and fully PEGylated nanoparticles. *Mol Pharm*, **10**, 4546–4551.

88. Grama, C. N., Venkatpurwar, V. P., Lamprou, D. A., and Ravikumar, M. N. (2013). Towards scale-up and regulatory shelf-stability testing of curcumin encapsulated polyester nanoparticles. *Drug Deliv Transl Res*, **3**, 286–293.

89. Ranjan, A. P., Mukerjee, A., Helson, L., and Vishwanatha, J. K. (2012). Scale up, optimization and stability analysis of Curcumin C3 complex-loaded nanoparticles for cancer therapy. *J Nanobiotechnology*, **10**, 38.

Chapter 20

Polyester Carriers for Enzyme Delivery

Raisa Kiseleva and Alexey Vertegel
Bioengineering Department, Clemson University, 301 Rhodes Research Center, Clemson, SC 29631, USA
vertege@clemson.edu, rkisele@clemson.edu

Owing to their high activity and specificity of binding to their targets, enzymes have important advantages as potential therapeutic agents [1]. Moreover, enzymes convert target substances by acting as catalytic agents. These features distinguish enzymes from all other types of drugs. Development of enzyme drugs for the treatment of various disorders is a growing field in pharmaceutical research [2]. Although there is a range of approved drugs for enzyme therapy, there are still a lot of challenges for broader application of enzyme-based drugs. Limitations include low storage stability and limited shelf life, poor penetration of physiological barriers due to large molecular weight, and potential immunogenicity [3]. Moreover, many, if not all, enzymatic drugs act at a certain location in the human body–organ, tissue, or cell compartment. By the time these drugs reach the targeted site, their therapeutic efficacy could be limited by either inhibitors or proteolytic degradation [4]. This usually

Handbook of Polyester Drug Delivery Systems
Edited by M. N. V. Ravi Kumar

ISBN 978-981-4669-65-8 (Hardcover), 978-981-4669-66-5 (eBook)
www.panstanford.com

leads to requirement of high doses of enzymatic drugs; in some cases such high doses can result in systemic toxicity [5]. Because of these reasons, a large number of newly developed enzymatic drugs are eventually rejected by researchers and will never benefit patients. Therefore, improving enzyme delivery systems remains an extremely important subject of scientific inquiry.

A number of approaches were developed to overcome the aforementioned limitations of enzymatic drugs [6–8]. These approaches include chemical modification [5], PEGylation [9], as well as encapsulation of therapeutic enzymes into a passive carrier [3]. In connection with the aim of minimizing side effects and maximizing bioavailability and stability levels, such drug delivery systems have been extensively studied in recent years [10–12]. Biodegradability and generally high biocompatibility are key properties of aliphatic polyesters that make them attractive for biomedical and drug delivery applications [10]. The purpose of this chapter is to give a brief overview on recent advances in targeted delivery of enzyme therapeutics by polyester carriers, yet challenging and not fully achieved area of research.

20.1 Therapeutic Enzymes

Research, development, and manufacturing of therapeutic enzymes are growing fields of pharmaceutical industry. There are several advantages that give enzymes a great potential to be used as drugs. Biological action of enzymes centers on their catalytic activity. It allows them to rapidly convert multiple target molecules into products, which significantly amplifies its effect compared to that of other small molecule drugs.

The history of using enzymes as therapeutic agents takes its routes from ancient times when proteolytic enzymes were parts of various mixtures of animal and plant materials [1]. For example, long before western discovery of North America, Native American healers applied fruits and leaves of the papaya plant to malignant tumors, thus empirically utilizing local enzyme therapy by a proteolytic enzyme papain. With scientific progress, advantageous properties of enzymes have encouraged investigators to discover specific enzymes to be used as therapeutic agents.

Since the beginning of the 20th century, scientists have attempted to use enzymes to modify and inhibit the activity of bacteria [13]. A groundbreaking research was done in the 1930s by Avery and Dubos [14], who succeeded in finding a specific microbial enzyme by using a soil enrichment technique. This enzyme not only degraded purified capsular polysaccharide of type III Pneumococcus *in vitro*, but also destroyed the capsules of the living organisms both *ex vivo* and in an animal body. Preparations of this enzyme were efficient in protecting mice against infection with virulent type III Pneumococcus [14].

Interestingly, this soil enrichment technique discovered by Dubos has later encouraged two British scientists, Howard Florey and Ernst Chain, to revive their stalled research on penicillin [1]. Enzyme technologies became the mainstream topic of many pharmaceutical findings with enzyme therapeutics covering a wide range of conditions outlined in Fig. 20.1.

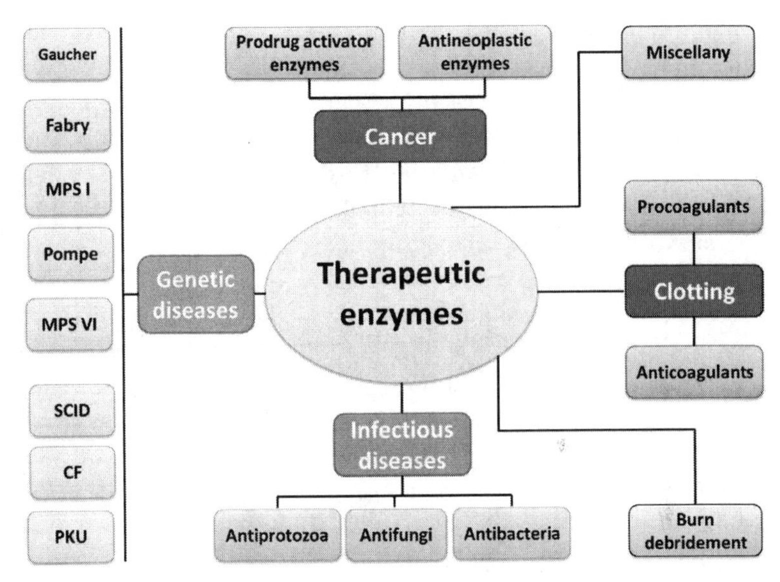

Figure 20.1 Therapeutic enzymes are used in the treatment of a variety of disorders and diseases. Reprinted from [15], Copyright 2003, with permission from Elsevier.

Enzymes have been found to be applicable as antineoplastic agents. First findings in this area were conducted around 1820, when Physick *et al.* used proteolytic enzymes for cancer treatment [16]. Physicians all over the world used preparations that mainly consisted of freshly prepared pancreatic extracts to achieve therapeutic effect. With the development of the techniques to manufacture purified and crystalline enzymes, a new era in enzyme applications has started. Since then, numerous discoveries in enzyme therapy were made. One of the recent examples of an FDA approved enzyme-based drug, already in clinical practice, is Oncaspar®. This drug has shown promising results in numerous studies on the treatment of children with newly diagnosed standard-risk acute lymphoblastic leukemia [9, 17–20].

A further application of enzymes as therapeutic agents can be illustrated by antibody-directed enzyme–prodrug therapy (ADEPT), related gene-directed enzyme–prodrug therapy (GDEPT), virus-directed enzyme–prodrug therapy (VDEPT), inhibitor-directed enzyme–prodrug therapy (IDEPT), and bacterial-directed enzyme prodrug therapy (BDEPT). These therapies are aimed to locally activate systemically administered "prodrugs" within the tumor in order to induce selective tumor destruction [21–24]. Biocompatible polymers can also passively localize at a tumor site; this observation led to the development of the relevant PDEPT approach [25].

Enactment of the Orphan Drug Act that was passed by the United States Congress in 1983 had great influence on the development of new enzyme drugs. This act is designed to encourage pharmaceutical companies to develop and commercialize treatments for diseases and conditions are that considered rare and affect relatively small numbers of individuals (less than 200,000). Drugs given that are eligible for an orphan drug status have many provisions and incentives and receive seven years of market exclusivity [15, 18].

A number of enzyme replacement therapies for certain rare genetic diseases developed as a direct outcome of the Orphan Drug Act have shown excellent performance. This topic is discussed in detail in review by Michael Vellard [15]. Examples of these enzyme therapies include such orphan drugs as Adagen® (pegadamase bovine), used for the treatment of severe combined immunodeficiency (SCID); Ceredase® (alglucerase injection), used for the treatment of Gaucher disease, a lysosomal storage disease (LSD); and Pulmozyme®, used to reduce mucous viscosity and enable the clearance of airway secretions in patients with cystic fibrosis.

Enzymes used as anticoagulant or coagulant agents, to promote fibrinolysis or proteolysis, etc., have been approved by the FDA. Most recently approved enzyme drugs include Lumizyme (alglucosidase alfa), used for treatment of patients with infantile-onset Pompe disease, and Vimizim (elosulfase alfa), the first FDA-approved treatment for the rare congenital enzyme disorder mucopolysaccharidosis type IVA (Morquio A syndrome). Even more enzyme drugs are currently under investigation and either have shown encouraging results in animal studies or have passed different levels of clinical trials. Human superoxide dismutase, an anti-inflammatory agent, is a great example of such a promising drug candidate still in research for various conditions [26].

Other type of enzymes with significant role in pharmaceutical industry is pancreatic enzyme products (PEPs). These drugs contain the active ingredient pancrelipase, a mixture of the digestive enzymes amylase, lipase, and protease. According to the FDA, by 2012 six PEPs were approved. Creon and Zenpep were approved for marketing in 2009, Pancreaze was approved in April 2010, Ultresa and Viokace were approved in March 2012, and Pertzye was approved in May 17, 2012. A wide variety of applications of enzymes as therapeutics can be also demonstrated by listing Activase® (alteplase; recombinant human tissue plasminogen activator), which is the first enzyme drug approved by the FDA in 1987. This enzyme is the only pharmacological treatment currently approved for patients with acute ischemic stroke [15].

In most cases targets of the approved enzymatic drugs are extracellular [3]. There are, however, many diseases for which treatment requires intracellular delivery. For instance, intracellular delivery to lysosomal compartments is required for effective enzyme replacement in treatment of genetic diseases such as lysosomal storage diseases. Intracellular enzyme-based treatments are expected to emerge, including inhibition or reversal of ubiquitination for cancer treatment or phosphorylation of Tau protein in treatment of Alzheimer's disease.

As we can see, enzymes have shown great potential in the treatment of various diseases [15]. Advancements in research have allowed scientists to produce safer, cheaper enzymes with enhanced potency and specificity. Along with this, changes in orphan drug regulations have been effective in facilitating efforts to develop new enzyme drugs [15, 18]. At the same time, modern research of enzyme biochemistry and therapy as well as their practical application depends on many factors [5]. In the following section we discuss the limitations of enzyme-based drugs and challenges in their development.

20.2 Challenges in Enzyme Drug Development and Methods to Overcome

Along with their unique properties outlined in Section 20.1, enzymes have some limiting factors that affect their performance as drugs. First, they must be exhaustively purified from potentially toxic impurities, such as endotoxins, resulting in high cost of purified enzymatic formulations. Second, enzymes are rapidly removed by the reticuloendothelial system or degraded by proteolytic enzymes *in vivo*. Third, given the large size of molecules and their complex tertiary or quaternary structure, enzymes sometimes have limited ability to reach their target organ in the human body [24]. Furthermore, the optimal therapeutic effect can often be achieved only at high local concentration of enzyme in particular tissue or organ, but soluble enzymes would entail dilution as a consequence of distribution throughout the body. Finally, enzymatic drugs obtained from nonhuman species could be immunogenic leading to even faster clearance rates [27]. Eventually immunogenicity may lead to decreased efficacy or severe hypersensitivity reaction.

With all these issues, a critical need exists to develop delivery systems that would improve performance of enzymatic drugs [27]. Different approaches have been designed in order to improve enzyme stability and limited distribution [9], reduce clearance rates [20], and eliminate or reduce immunogenicity [24]. The most common approaches include encapsulation [4] and introduction of drug delivery systems [28], such as liposomes [29], biodegradable particles [4] etc., or chemical modification of enzymes [30], replacement of labile amino acids, cyclization and increment of molecular mass by PEGylation or oligomerization or immobilization on biocompatible matrices [6–8].

A delivery system must possess certain characteristics. First, manufacturing conditions should be gentle enough to preserve maximized enzymatic activity. The size of delivery vehicle depends on the application, but should remain in the submicron range to penetrate tissues and internalization by cells. By these means, delivery vehicles would possess several benefits in comparison to free soluble enzyme formulations. These benefits include increased efficiency, higher stability, lower immune response and targeting capabilities. Examples of delivery vehicles that have been used with enzymatic drugs are shown in Table 20.1. Carefully designed and developed delivery systems allow researchers to extend drug's plasma half-life without chemical modification [31]. Several major types of delivery vehicles include microparticles [32, 33], nanoparticles [34], liposomes [12], and biological carriers (erythrocytes) [35].

Table 20.1 Examples of delivery vehicles for various enzymatic drugs

Type of carrier	Example of enzymatic drug	Suggested action against disease/ target	References
Liposomes	L-Asparaginase	Leukemia, hepatic tumors	[12, 36–40]
	Lysozyme	Antibacterial enzyme	[41, 42]
	Human tissue plasminogen activator (tPA)	Thrombolytic agent	[53]
	Superoxide dismutase (SOD)	Antioxidant drug	[29, 44]
Erythrocytes	Lysozyme	Antibacterial enzyme used as model compound	[45–47]
	L-Asparaginase	Leukemia, hepatic tumors	[48]
	Arginase	Hyperargininemia,	[49]
	β-Glucoserebrosidase, β-Glucosidase, β-Galactosidase	Lysosomal storage disease	[50]

(Continued)

Table 20.1 (*Continued*)

Type of carrier	Example of enzymatic drug	Suggested action against disease/ target	References
	Pegadamase bovine	Severe combined immunodeficiency disease	[51]
Micelles	Peroxidase	Surface-active enzyme	[32, 45]
Virus scaffolds	tPA, β-lactamase	Phage therapy	[52]
Tubular wall reactors	Phenylalanine ammonia-lyase	Extracorporeal therapy, substitution of enzyme deficiency	[53]
Polymeric microparticles	Anticancer therapies	ADEPT/MDEPT/ AMIRACS	[25]
Magnetic microcarriers	tPA	Thrombolytic agent	[34, 54, 55]

Erythrocyte carriers are schematically represented in Fig. 20.2 to show step-by-step formulation and drug loading. These are widely explored systems for oral, parenteral, and topical administrations both *in vitro* and *in vivo*.

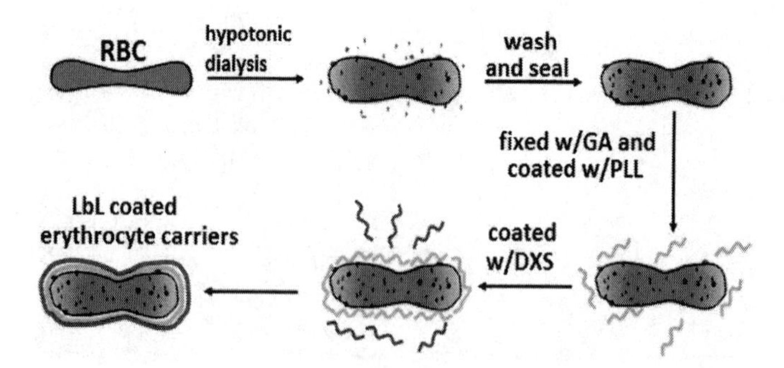

Figure 20.2 Schematic illustration of preparation of erythrocyte carriers coated with multilayer polyelectrolyte shell. Reprinted from [46], Copyright Springer Science+Business Media, LLC 2011, with permission of Springer.

Most delivery systems immobilize enzymes on the surface of another material or encapsulate them into a polymeric, lipid, or mesoporous material. Other approaches involve fusion of an enzyme to other proteins or peptides [56].

Surface modification is another widely spread technique to overcome enzyme's instability [20]. PEGylation, or covalent attachment of methoxypolyethyleneglycol (mPEG) to enzyme carrier or protein itself, is method used to increase circulating time, reduce immunogenicity and antigenicity of enzymes and at the same time retain their bioactivity. One example of successful PEGylation is modification of liposomes containing tPA [43].

20.3 Polyester Carriers for Enzymatic Drugs

Among all listed types of delivery vehicles polyesters play crucial role in the development of alternative routes of drug delivery. Aliphatic polyesters became one of the most important classes of synthetic polymers in biomedical industry largely due to their favorable features of biodegradability, biocompatibility, and cytocompatibility [25, 57]. Although some professionals insist on using the term "biodegradable" only for ecological polymers aiming at the protection of earth environments from plastic wastes [27], most often "biodegradable" is used for medical applications as well. In our review we discuss biodegradable polyesters of two types: those that require enzymes for hydrolytic or oxidative degradation and those that are soluble in aqueous media and are readily hydrolyzed in our body to the respective monomers and oligomers.

The history of aliphatic polyesters in biomaterial applications takes routes from pioneering works of DeBakey and coworkers [58] on Dacron™ (polyethylene terepthalate) for cardiovascular prostheses. Another example of groundbreaking work on aliphatic polyesters for tissue-engineering and drug-delivery was done in Robert Langer's lab [59]. A great variety of biodegradable polyesters that has been developed in the last two decades is nowadays available commercially [60]. They are being used in various applications such as degradable sutures, drug delivery systems [61], and tissue-engineering scaffolds [60, 62]. Biocompatibility of polyesters was shown in numerous studies of targeting drug delivery systems including cancer therapeutic applications [63, 64]. Some aliphatic polyesters with significant biomedical applications are listed in Table 20.2.

High hydrolysis of polyesters in human body makes them perfect drug carriers for controlled release devices. Polyester carriers have attracted much attention as delivery systems for small molecules, DNA, peptides and proteins. Most commonly studied ones are primarily represented by poly(ε-caprolactone) (PCL), poly(lactic acid) (PLA) and poly(lactic-*co*-glycolic acid) (PLGA) matrices.

20.3.1 Synthesis and Encapsulation of Enzymes in Polyester Nanoparticles

Methods of preparation of polyester nanoparticles for delivery of enzymatic drugs are the same as for any other type of drugs. Physicochemical and mechanical properties of polyester matrices can be modified via selection of polymer molecular weight, copolymerization, and functionalization. Consequently, structural organization of final formulation depends on the method of preparation. As it was noted earlier, two major approaches are generally used to formulate a delivery vehicle: enzyme is either entrapped inside the core of carrier or attached to the surface of the carrier [10].

The most common technique of drug entrapment is the emulsification-solvent evaporation method [11, 60]. First, the polymer and the drug are dissolved in an organic solvent (e.g., dichloromethane). Second, water and a surfactant (e.g., polysorbate-80, poloxamer-188) are added to the polymer solution. As a result, the emulsion of oil (O) in water (W), i.e., O/W, is formed. The next step consists of sonication or homogenization of emulsion leading to formation of nanosize droplets with the following solvent evaporation or extraction to produce pure nanoparticle suspension. A modification of this technique, the double emulsion W/O/W, is most commonly used to encapsulate hydrophilic drugs, such as peptides, proteins and nucleic acids. Schematic examples of enzyme-nanoparticle systems are presented on Fig. 20.3.

Table 20.2 Some examples of polyester materials for drug delivery applications

Type of polyester	Example of polyester	Methods of preparation	Example of application	References
Polylactides	Polyglycolide, Polylactides	Solvent evaporation, solvent displacement, salting out, solvent diffusion	Encapsulation of psychotic, restenosis drugs, hormones, oridonin and proteins.	[11, 63, 65–68]
Copolymers	Poly(lactic-*co*-glycolic acid) (PLGA)	Emulsification–diffusion, solvent emulsion–evaporation, interfacial deposition and nanoprecipitation method	Encapsulation of anticancer, diabetes, psychotic drugs, hormones	[11, 69–75]
Polylactones	Poly(ε-carpolactone)	Nanoprecipitation, solvent displacement and solvent evaporation	Encapsulation of anticancer, diabetes drugs, antifungal, clonezepam	[11, 64, 76–78]

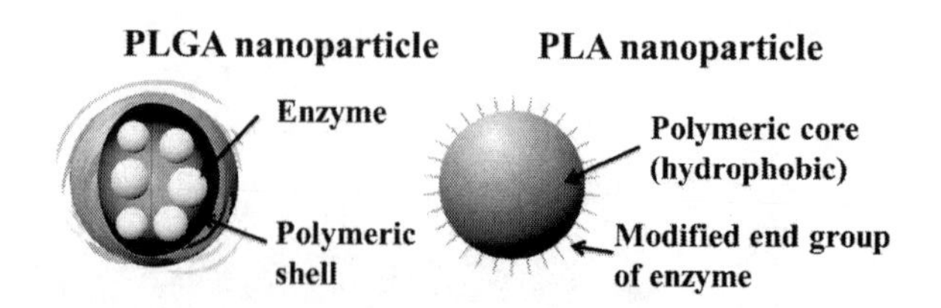

Figure 20.3 Schematic representation of polyester delivery systems.

Polyester nanocarriers can also be formed by nanoprecipitation method [10, 11, 60]. This method is also called the interfacial deposition method. During the synthesis both the polymer and the drug are first dissolved in an organic solvent (such as acetone) and then added by drops to water phase with or without emulsifier/stabilizer. The formulated emulsion is then left in the air so the organic solvent is evaporated. Following centrifugation is the final step when the pellets are collected.

Interfacial deposition method is most commonly used for the formation of both nanocapsule and nanospheres [10, 60]. Two basic steps are the following: particles are synthesized in the interfacial layer of water and organic solvent (water miscible) and then separated by centrifugation.

The salting-out technique is based on using salting-out agents (magnesium chloride, calcium chloride) to separate a water-miscible solvent from aqueous solution thus minimizing stress to protein encapsulants [60]. The summarized data on most extensively studied applications are listed in the Table 20.3.

Nanoparticles are colloidal systems; therefore determination of drug loading efficiency is complicated. Most reliable way to separate nanoparticles from the solution is to use ultracentrifugation or gel filtration. Overall, encapsulation efficiency of a particular method can be evaluated by using a simple formula [10]:

$$\text{Encapsulation efficiency (\%)} = \frac{\text{Amount of drug in nanoparticles}}{\text{Initial amount of drug}} \times 100\%$$

Table 20.3 Polyester carriers are used for various enzyme drugs

Enzyme drug	Type of polyester	Indication	References
Superoxide dismutase	PLGA in complex with alginate/chitosan	Anti-inflammatory	[79]
	PLGA	Neuroprotective efficacy	[80–82]
Tissue plasminogen activator	PLGA; Fe_3O_4-based PLGA; PLA-PEG	Fibrinolytic agent	[28, 34, 83–84]
L-asparaginase, catalase, glucose oxidase	poly(3-hydroxybutyrate-*co*-3-hydroxyvalerate)	Cancer treatment	[85]
Lysozyme Lysostaphin	poly(lactic-*co*-hydroxymethyl glycolic acid; PLA	Antibacterial	[31, 86]

As we can see from this short discussion, polyester nanoparticles have been synthesized using various methods. More detailed synthesis process is described elsewhere [60]. Each of the techniques listed above has its advantages and disadvantages. Moreover, encapsulation method has primary impact on drug's behavior, its bioavalability and release mechanisms. We will now discuss specific examples of polyester nanocarriers for particular enzymes in various applications.

20.3.2 Polyester Carriers for Delivery of Anti-inflammatory Enzymes

Generation of reactive oxygen species (ROS) is increasingly recognized as an important cellular process involved in numerous physiological and pathophysiological processes. Superoxide radicals are one of the most toxic ROS and their damaging effects lead to a variety of detrimental health conditions including cardiovascular diseases, neurodegenerative disorders, and extensive oxidative inflammations [3, 87–91]. The list of pathophysiological conditions that are associated with the overproduction of ROS expands every day.

The first line of defense to neutralize overproduction of ROS is natural antioxidant system. Superoxide dismutases (SOD) represent a group of key antioxidant enzymes responsible for conversion of superoxide radicals to much less reactive hydrogen peroxide [88], thus preventing the formation of highly aggressive compounds such as peroxynitrite ($ONOO^-$) and hydroxyl radical (HO) (Fig. 20.4). Role of SOD in mitigation of the oxidative stress was extensively studied by McCord and Fridovich [26, 92, 93].

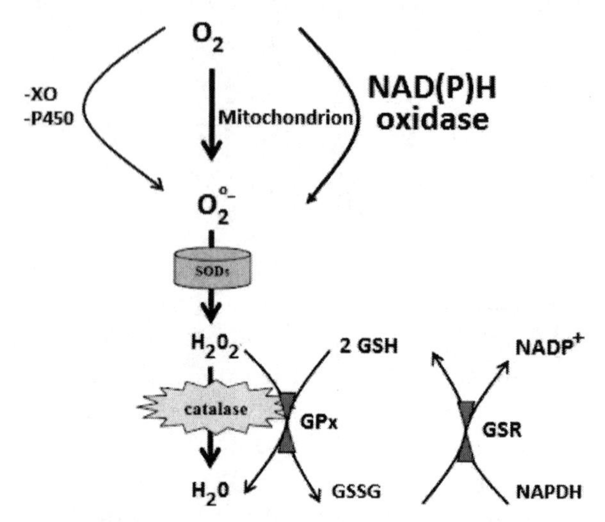

Figure 20.4 Pathways for the production and removal of reactive oxygen species. GSH, glutathione; GSSG, oxidized glutathione; GSR, glutathione reductase; XO, xanthine oxidase. Reproduced from [88], Copyright © 2007, Elsevier Masson SAS.

Mechanisms of action of SOD *in vivo* were established by extensive research over the past 45 years. It has been shown to play major role in balancing oxidant-antioxidant system in many disease conditions [94–97]. For instance, protective effect of SOD in inflammatory joint disease was shown in animal models of ischemia and inflammation [98, 99].

Despite these advantageous properties of SOD enzyme, it is still a big challenge to overcome rapid renal clearance and slow extravasation due to large molecular radius and charge density. These are the factors that affect the pharmacodynamics and pharmacokinetics of all enzymes used as drugs. Moreover, delicate balance between superoxide and SOD should be taken into account. When produced in proper amount, superoxide is a useful metabolite that serves as signaling molecule in important cell processes. However, if superoxide is overproduced, it can initiate lipid peroxidation, protein oxidation, and DNA damage, leading to cell dysfunction and death by apoptosis or necrosis [26].

To overcome these limitations various attempts on modification of SOD enzyme were made [42, 33, 26, 100]. Active targeting can be reached by the development of nanoparticulate systems and attaching antibodies for the utilization of their targeting properties [79, 81]. For instance, Ye Wang and coauthors reported development of PLGA–alginate–chitosan complex microspheres for sustained delivery of SOD (Fig. 20.5). They have shown that complex formulations not only prolong the release of drug but also decrease the burst release. They compared these findings with simple PLGA or alginate–chitosan microspheres and found that although alginate–chitosan microspheres showed much higher entrapment efficiency (91.08%+/−1.28%) than that of PLGA microspheres (36.42%+/−1.81%), the burst release of SOD from microspheres was better controlled in alginate-chitosan formulations. Authors studied encapsulation efficacy, SOD activity and drug release profile of constructed microspheres, but no *in vitro* or *in vivo* testing was not performed.

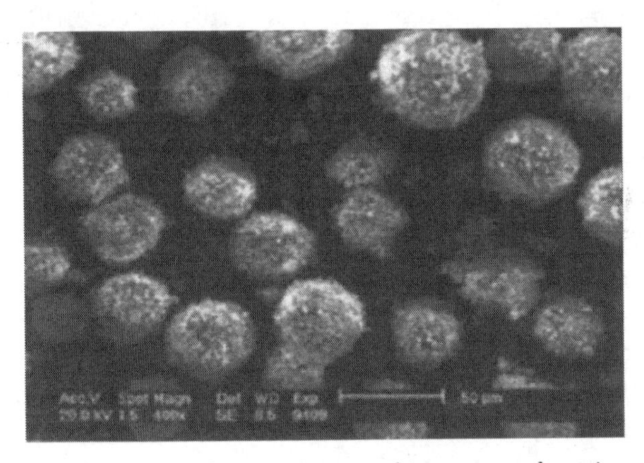

Figure 20.5 SEM photo of PLGA–alginate–chitosan complex microspheres. Reprinted from [79], Copyright 2006, with permission from John Wiley and Sons.

Naturally, highly selective permeability of the blood–brain barrier (BBB) restricts the delivery of drugs to the brain [101]. By using modified delivery systems it is possible to promote endocytosis and target the BBB receptors. This approach was reported by Reddy *et al.* [81]. In their study on using PLGA-based nanoparticles to target the BBB, the authors proposed the method to treat cerebral diseases by delivering SOD to the brain. Efficacy of free enzyme is limited by its short *in vivo* half-life (~6 min) and poor permeability across the BBB. *In vitro* model of primary cultures of human fetal neurons were challenged with hydrogen peroxide following by treatment with free SOD, PEGylated SOD, and SOD-loaded PLGA nanoparticles (~81±4 nm in diameter, 0.9% w/w SOD loading). Testing was performed in a dose- and time-dependent manner, and neuroprotective efficacy of SOD-loaded PLGA nanoparticles against oxidative stress was shown to be comparatively higher among all these treatment groups. Interestingly, the neuroprotective effect of SOD-NPs was seen up to 6 hours after H_2O_2-induced oxidative stress, but the effect lessened thereafter. Although authors were able to show the neuroprotective effect of encapsulated SOD in cultured human neurons is dose- and time-dependent, their study also demonstrated that the protective effect does not seem to last beyond 6 hours. Authors suggest that this could be due to the depletion of cellular endogenous catalase and GPx enzymes, which are crucial for the process of neutralization of H_2O_2 produced during dismutation of superoxide by SOD.

In a follow up *in vivo* study same authors reported their findings on using encapsulated enzyme to neutralize the deleterious effects of excessive production of ROS after cerebral ischemia and reperfusion [80]. Here, PLGA NPs loaded with SOD were compared to SOD in solution (SOD-Sol) and control NPs alone or mixed with SOD-Sol (Fig. 20.6). In ischemia-reperfusion model, animals receiving SOD-NPs (10,000 U of SOD/kg) have shown greater survival rate than animals treated with control samples (75% vs. 0% at 28 days). Moreover, SOD-NPs maintained BBB integrity, thus preventing edema, reduced the level of ROS formed following reperfusion, and protected neurons from undergoing apoptosis. The authors propose SOD-NPs to be an effective treatment option in conjunction with a thrombolytic agent for stroke patients.

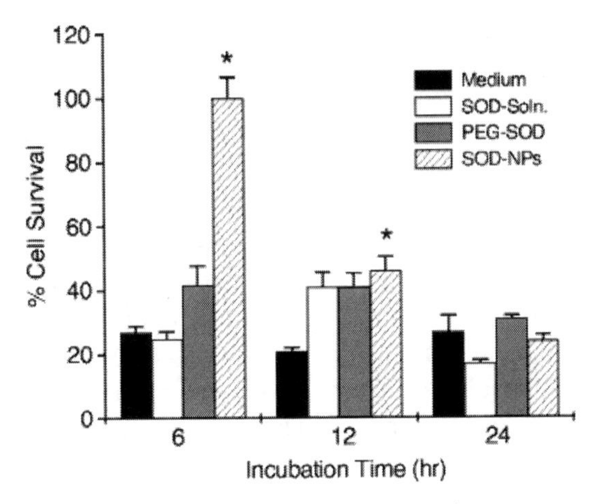

Figure 20.6 Neuroprotective efficacy of SOD in human neurons after H_2O_2-induced oxidative stress in the study of Reddy *et al.* [80]. Neuroprotective efficacy of SODat different time points in neurons after H_2O_2-induced oxidative stress and SOD treatment. Dose of SOD = 100 U. Data as mean ± SEM ($n = 3$), $^*P < 0.05$ against medium control.

One of the major causes of death and disability in the United States is stroke [82]. Ischemic injury is the result of decreased blood flow to the brain, and it often associates with the following extensive generation of ROS and reactive nitrogen species due to the reoxygenation during reperfusion that can further injure the tissue [94, 102]. This oxidative damage contributes to the long-term outcome of primary injury and eventually leads to ischemia, trauma, and degenerative disorders [103]. Xiang Yun *et al.* proposed to use various nanoparticulate carriers (liposomes, polybutylcyanoacrylate (PBCA), or PLGA) loaded with active SOD to determine the impact of these molecules in the mouse model of cerebral ischemia and reperfusion injury [82]. Nanoparticles were untagged or tagged with nonselective antibodies or antibodies directed against the *N*-methyl-D-aspartate (NMDA) receptor 1. Authors have shown that nanoparticles containing SOD protected primary neurons *in vitro* from oxygen-glucose deprivation. For *in vivo* studies, Yun *et al.* used fixed SOD dose of 25 U per injection, and therefore they had different concentrations of targeting antibody for different vehicles. Main data for animal studies were observed for PBCA NPs due to their highest anti-NR1 antibody concentration. Despite this fact, other types of carriers (liposomes, PLA and PLGA NPs) were examined as well to show similar protective efficacy *in vivo*. Therapeutic effects of these systems on infarct volumes in the mice after ischemia and reperfusion injury are represented on Fig. 20.7. Significant improvement in all treatment groups was observed in all samples treated with targeted nanoparticles coated by NR1 antibody. In comparison, plain nanoparticles were not effective in reducing infarct volume. All three delivery vehicles showed similar effect on the infarct volumes. Notably, nanoparticles did not significantly influence physiologic parameters such as mean arterial pressure, blood p02, pC02, CBF, and pH. These parameters were measured

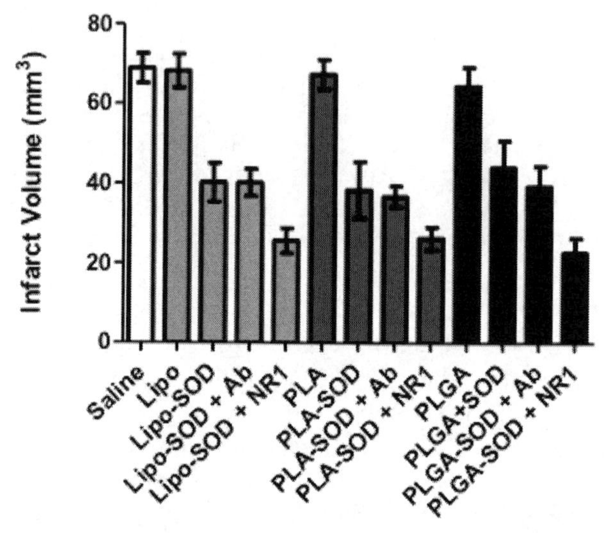

Figure 20.7 Effect of nanoparticles on infarct volume, neurologic score, and survival rate in mice subjected to ischemia and reperfusion injury (IRI). Mice were subjected to 1 hour ischemia and 24 hours of reperfusion. Immediately after ischemia, animals were administered nanoparticles and infarct volumes determined at 24 hours. $N = 10$ per group. $^{*}P < 0.01$; $^{**}P < 0.01$ compared with SOD and SOD-Ab for each group. Lipo, liposomes; PLA, polylactic acid; PLGA, poly (lactic-*co*-glycolic acid). Ab, non-specific antibody; NMDA receptor antibody; NR-1, NMDA, *N*-methyl-D-aspartate; PBCA, polybutylcyanoacrylate; SOD, superoxide dismutase. Reprinted with permission from [82], Copyright 2013 ISCBFM.

between the vehicle and treated mice at baseline, during ischemia, or after reperfusion. This, using animal model, authors observed protective efficacy of formulated nanoparticles with a 50% to 60% reduction in infarct volume when applied after injury. Studies reported by Yun *et al.* demonstrated the potential for using targeted nanoparticles loaded with antioxidant enzymes for treatment of stroke.

Antioxidant enzymes play essential role in neutralizing damages provoked by the overproduction of ROS. Extensive studies on developing polyester carriers to deliver antioxidants are now being conducted. Pioneering works on the investigation of their *in vivo* delivery were shortly discussed here [80, 82]. This research points out to the importance of nanoparticles with antioxidant enzymes for targeted delivery and their *in vivo* therapeutic applications. Further studies, especially those involving *in vivo* experiments, are needed to create robust and effective delivery systems.

20.3.3 Polyester Carriers for Thrombolytic Enzymes

Tissue plasminogen activator (tPA), or serine protease, is an enzyme that catalyzes the conversion of plasminogen to plasmin, a major enzyme responsible for blood clot breakdown. tPA is currently used in clinical practice to treat embolic or thrombotic strokes. tPA along with streptokinase (SK) represent great examples of enzymatic fibrinolytic agents used to lyse occlusive blood clots and restore blood perfusion [104]. Although these enzymes were successfully encapsulated in liposomes or PEG microparticles showing promising results in *in vitro* and *in vivo* animal studies [30], such delivery systems suffer from stability problems. Various attempts on designing SK- and tPA-encapsulated polyester nanoparticles were also performed and will be highlighted in the following discussion.

PLGA is a well-known polyester that can have various ratios of lactide and glycolide. It is a biocompatible, biodegradable, and only slightly toxic biomaterial [104]. It has been used as a carrier for drug delivery or as a scaffold in tissue engineering [104, 105]. Tze-Wen Chung *et al.* report on designing and delivering PLGA NPs with chitosan (CS) and CS-GRGD coatings [104]. They used a blood clot-occluded tube model to evaluate thrombolysis capabilities of designed nanosystems. FT-IR, a laser particle/zeta potential analyzer and HPLC were employed to determine characteristics and release profiles of tPA-encapsulated PLGA, PLGA/CS, and PLGA/CS-GRGD NPs. Clot lysis times were reported to be the shortest for PLGA/CS NPs (e.g., 20.7±0.7 min). PLGA/CS-GRGD NPs have the highest weight percentages of digested clots (e.g., 25.7±1.3 wt%). Future employment of these NPs in clinical studies authors associate with their significantly faster thrombolysis than tPA solution in a clot-occluded tube model.

Another approach was suggested by Yumei Xie *et al.* They synthesized drug carriers that contained tPA and magnetite within PLGA and poly(D,L-lactide)-*co*-poly(ethylene glycol) (PLA-PEG) copolymers [28] (Fig. 20.8A). Although nanospheres were larger than needed for systemic drug delivery, tPA encapsulation and release rates were higher than theoretical thrombolysis concentrations. With modified synthesis protocol that ensures reduced size of the spheres to less than the size of red blood cells authors hope to investigate new delivery method of magnetic microspheres containing thrombolytic concentration of tPA locally to the place of occlusion.

Later some modifications towards improved delivery of plasminogen activators were made by other authors [34, 83, 84]. For example, Jun Zhou *et al.* describe Fe_3O_4-based PLGA nanoparticles carrying recombinant tPA as a dual-function tool in the early detection of a thrombus and in the dynamic monitoring of the thrombolytic efficiency using MRI [34]. In a recent publication they were first to report the construction of PLGA nanoparticles loaded with an MR contrast agent for targeted detection of a thrombus *in vivo* as well as for the dynamic monitoring of the thrombolytic efficiency.

RGD grafting of PLGA nanoparticles mentioned earlier also finds its application in targeting of tumor endothelium [106]. In order to synthesize a chitosan (CS) film containing cyclic three-amino acid sequence, conjugated arginine–glycine–aspartate (cRGD), was grafted onto the CS surface using carbodiimide-mediated amide bond formation. To produce tPA encapsulated Fe_3O_4-based PLGA nanoparticles (Fe_3O_4–PLGA-rtPA/CS-cRGD), double emulsion solvent evaporation method (water in oil in water [W/O/W]) was used. Successfully constructed nanoparticles were proven to have a regular shape, a relatively uniform size, a high carrier rate of Fe_3O_4

and encapsulation efficiency of rtPA, and a relatively high activity of released rtPA (Fig. 20.8B). Transmission electron microscope (TEM), clinical MRI scanner, and *in vitro* and *in vivo* experiments on donor blood samples confirmed that the Fe_3O_4–PLGA-rtPA/CS-cRGD nanoparticles specifically accumulated on the edge of the thrombus and that they had a significant effect on the thrombolysis compared with the Fe_3O_4–PLGA, Fe_3O_4–PLGA-rtPA, and Fe_3O_4–PLGA-rtPA/CS nanoparticles and with free rtPA solution (Fig. 20.8C).

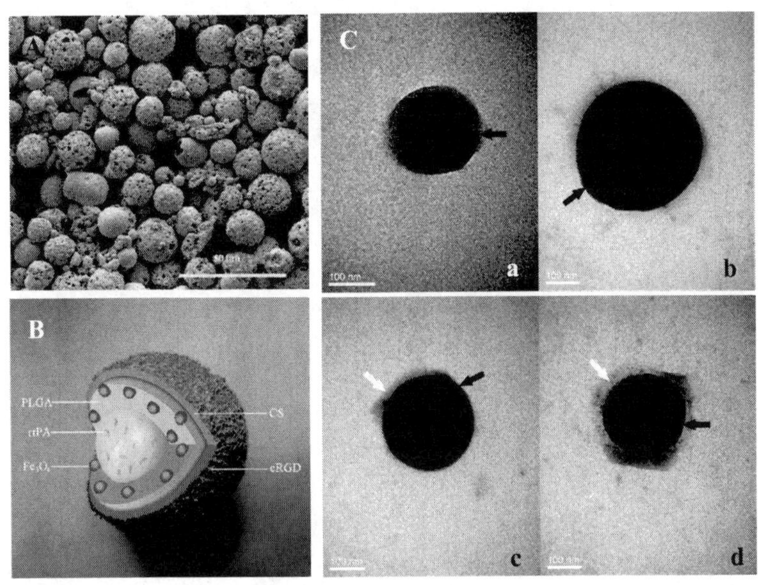

Figure 20.8 **A**. SEM micrographs of tPA magnetic carriers prepared with PLGA (5 kDa). Adapted from Xie [28]. **B**. Schematic representation of an Fe_3O_4–PLGA-rtPA/CScRGD nanoparticle. The structure of the nanoparticle is briefly described as follows: the core of the first emulsion was an rtPA solution encapsulated in a shell of PLGA and Fe_3O_4, and a CS-cRGD film was coated on the surface of the PLGA-Fe_3O_4 layer using the W/O/W method. **C.** TEM images of the (a) Fe_3O_4–PLGA, (b) Fe_3O_4–PLGA-rtPA, (c) Fe_3O_4–PLGA-rtPA/CS, and (d) Fe_3O_4–PLGA-rtPA/CS-cRGD nanoparticles. The iron oxide particles (black arrows) were relatively uniformly distributed in the nanospherical shell, and the CS film (white arrows) was observed around the nanoparticles that had been coated with a CS or CS-cRGD film. Reprinted with permission from [34], Copyright 2014 American Chemical Society.

Although all the aforementioned methods show promising future applications, they share common limitations such as insufficient encapsulation efficiency of the rtPA, the loss of enzyme activity, and the size of nanoparticles which is in most cases higher than it is required for systemic delivery. Acquired results must be confirmed on a larger group of samples as well as in *in vivo* studies on animal models.

20.3.4 Polyester Carriers and Cancer Treatment Enzymes

Cancer treatment enzymes have some good examples of successful application. For instance, Oncaspar®, PEGylated enzyme pegaspargase, is a clinically used drug for the treatment of children with newly diagnosed standard-risk acute lymphoblastic leukemia [15]; other examples include the earlier mentioned antibody-directed enzyme–prodrug therapy (ADEPT) and other prodrug therapies currently used for cancer treatment.

Increased drug encapsulation efficacy can be achieved by various modifications in the compositions of the phases of double emulsion [41, 107]. Baran *et al.* have developed poly(3-hydroxybutyrate-*co*-3-hydroxyvalerate) (PHBV) nanocapsules to be used as carriers in cancer therapy. Double emulsion-solvent evaporation procedure (W/O/W) was used for encapsulation of model enzymes (L-asparaginase, catalase, glucose oxidase) and bovine serum albumin [85]. By treating PHBV with sodium borohydride, low molecular weight polymer was synthesized. The adjustment of the second water phase to the isoelectric point of the proteins significantly increased the encapsulation yields of catalase, L-asparaginase and BSA. This study shows preliminary steps made towards improving the encapsulation efficiency of the enzyme drugs in PHBV nanocapsules for cancer therapy.

Despite the fact that enzymes themselves have great potential in cancer therapy, not many attempts on encapsulating enzyme drugs to the polymer carrier were made to increase drug's antitumor efficacy, while reducing systemic side-effects [108, 109]. Small number of studies in this area of research show promise in future development of nanoparticles with enzymatic drugs to be used as effective tools against cancer, but more intense studies are needed.

20.3.5 Polyester Carriers for Antibacterial Enzymes

Bacterial infections are one of the common issues related to post-operative complications. It is known that bacteria form multicellular communities called biofilms causes two thirds of all infections. Moreover, biofilm formation demonstrates a 10 to 1000 fold increase in adaptive resistance to conventional antibiotics [110]. For example, *Staphylococcus aureus* is a versatile pathogen, one of the major causes of mortality in patients with bacterial infections [111]. Unfortunately, there are no approved drugs that specifically target bacterial biofilms.

Different efforts were attempted to design smart antibacterial systems to be used as either independent drugs or coatings for biomedical devices. Lately, native enzymatic agents and peptides having antibiotic properties have been investigated in order to be used in the treatment and prevention of bacterial infections [112]. Great advantage of natural occurring enzymes lies in their low resistance rate.

Lysozyme is the enzyme that damages bacterial cell walls. Hydroxylated aliphatic polyester, poly(lactic-*co*-hydroxymethyl glycolic acid) (PLHMGA) was used to produce microspheres containing lysozyme, by means of double emulsion extraction-evaporation method. Importantly, the structure and functionality of the released lysozyme was preserved. This study of Ghassemu *et al.* [86] shows that PLHMGA microspheres are promising systems for the controlled release of pharmaceutical proteins.

Lysostaphin is another antibacterial enzyme. It is a zinc metalloenzyme which has a specific lytic action against *S. aureus* [112]. Lysostaphin has activities of three enzymes, namely, glycylglycine endopeptidase, endo-*β*-*N*-acetyl glucosamidase and *N*-acteyl muramyl-L-alanine amidase. Because of these properties, lysostaphin is extensively studied to be used as the treatment of antibiotic-resistant staphylococcal infections. Satishkumar *et al.* [31] reported design of delivery systems comprised of lysostaphin conjugated to biodegradable PLA nanoparticles with or without co-immobilized anti-*S. aureus* antibody (Fig. 20.9).

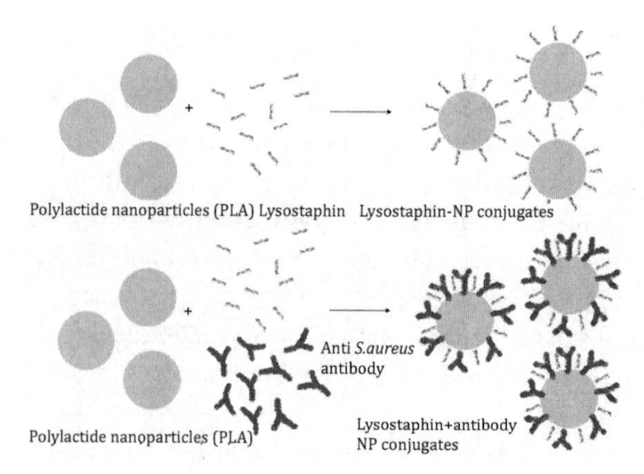

Figure 20.9 Schematic of the synthesis of protein–NP conjugates. Reprinted by permission from Macmillan Publishers Ltd: *Nanotechnology* [31], Copyright 2013.

Enhanced antimicrobial activity was observed for both enzyme-coated and enzyme–antibody-coated NPs for lysostaphin coatings corresponding to ~40% of the initial monolayer and higher in comparison to the free enzyme case ($p < 0.05$). Authors have demonstrated antimicrobial activity of designed systems in *in vitro* studies. Future *in vivo* studies are still needed to show stability and reliability of such nanoparticles loaded with novel therapeutic agents.

20.4 Future of Polyester Carriers for Delivery of Enzymatic Drugs

Nonfunctionalized aliphatic polyesters such as PLA or PLGA have been extensively studied for the controlled delivery of therapeutic enzymes in the past two decades. Polyester carriers have great potential advantages such as biodegradability, biocompatibility, and high encapsulation efficacy. Along with this, there are certain drawbacks associated with these systems, including incomplete protein release due to unfavorable protein interactions with the polymer [113], acidification of the matrix occurring within controlled-release depots [83], and difficulty in tailoring the release [84].

A number of approaches have been proposed to overcome these limitations. For example, protein stability issues during the formulation process have successfully been treated by using stabilizating additives. Another approach is to develop functionalized polyesters such as the mentioned earlier poly(L-lactide-*co*-hydroxymethyl glycolide) (poly(HMMG-L) [86]. Nevertheless, drug release/activity profiles as well as structural and functional integrity of therapeutic enzymes should be extensively studied both *in vitro* and *in vivo* in order to receive effective treatments.

There is currently certain level of skepticism regarding the safety of nanoparticulate systems. These issues can only be resolved after more *in vivo* and human studies are conducted. With the rising number of studies in this area, regulatory issues will become a greater concern and could eventually become a limiting factor in development of nanoscale drug carriers.

Another obstacle for widespread clinical studies of nanoparticle delivery systems is the difficulty to achieve high level of targeting. We see one of the promising future methods to solve this problem in using sustained drug release system by nontraditional delivery routes such as transdermal or intranasal administration.

References

1. Holcenberg, J. S. (1977). Enzymes as drugs. *Annu Rev Pharmacol Toxicol*, **17**(1), 97–116.
2. Godfrin, Y., and Bax, B. E. (2012). Enzyme bioreactors as drugs. *Drugs Fut*, **37**(4), 263–272.
3. Maximov, V., Reukov, V., and Vertegel, A. A. (2009). Targeted delivery of therapeutic enzymes. *J Drug Deliv Sci Technol*, **19**(5), 311–320.
4. Simone, E., *et al.* (2010). Synthesis and characterization of polymer nanocarriers for the targeted delivery of therapeutic enzymes. *Methods Mol Biol*, **610**, 145–164.
5. Vertegel, A. (2013). Targeted delivery of therapeutic antioxidant enzymes and enzyme mimetics. *2013 AIChE Annual Meeting*, San Francisco, CA.
6. Barig, S., *et al.* (2014). Dry entrapment of enzymes by epoxy or polyester resins hardened on different solid supports. *Enzyme Microb Technol*, **60**, 47–55.

7. Holcenberg, J. S., and Roberts, J. (1981). *Enzymes as Drugs*, New York: Wiley.
8. Werle, M., and Bernkop-Schnuerch, A. (2006). Strategies to improve plasma half life time of peptide and protein drugs. *Amino Acids*, **30**(4), 351–367.
9. Vieira Pinheiro, J. P., *et al.* (2002). Pharmacology of PEG-asparaginase in childhood acute lymphoblastic leukemia (ALL). *Blood*, **100**(5), 1923–1924.
10. Kumar, M. N. V. R. (2008). *Handbook of Particulate Drug Delivery*. Stevenson Ranch, Calif: American Scientific Publishers.
11. Kumari, A., Yadav, S. C., and Yadav, S. K. (2010). Biodegradable polymeric nanoparticles based drug delivery systems. *Colloids Surf B Biointerfaces*, **75**(1), 1–18.
12. Torchilin, V. P. (2005). Recent advances with liposomes as pharmaceutical carriers. *Nat Rev Drug Discov*, **4**(2), 145–60.
13. Löhnis, F. (1921). *Studies Upon the Life Cycles of the Bacteria*. United States Government Publishing Office.
14. Avery, O. T., and Dubos, R. (1931). The protective action of a specific enzyme against type III pneumococcus infection in mice. *J Exp Med*, **54**(1), 73–89.
15. Vellard, M. (2003). The enzyme as drug: application of enzymes as pharmaceuticals. *Curr Opin Biotechnol*, **14**(4), 444–450.
16. Wolf, M., and Ransberger, K. (1972). *Enzyme-Therapy*. New York: Vantage Press.
17. Avramis, V. I., *et al.* (2002). A randomized comparison of native Escherichia coli asparaginase and polyethylene glycol conjugated asparaginase for treatment of children with newly diagnosed standard-risk acute lymphoblastic leukemia: a Children's Cancer Group study. *Blood*, **99**(6), 1986–1994.
18. Dinndorf, P. A., *et al.* (2007). FDA drug approval summary: pegaspargase (oncaspar) for the first-line treatment of children with acute lymphoblastic leukemia (ALL). *Oncologist*, **12**(8), 991–998.
19. Douer, D., *et al.* (2014). Pharmacokinetics-based integration of multiple doses of intravenous pegaspargase in a pediatric regimen for adults with newly diagnosed acute lymphoblastic leukemia. *J Clin Oncol*, **32**(9), 905–911.
20. Rob, W., *et al.* (2007). PEGylated proteins: evaluation of their safety in the absence of definitive metabolism studies. *Drug Metab Dispos*, **35**(1), 9–16.

21. Lee, H. J., *et al.* (2014). Enzyme/prodrug gene therapy for pancreatic cancer using human neural stem cells encoding carboxylesterase. *Mol Ther*, **22**, S171–S171.

22. Lehouritis, P., Springer, C., and Tangney, M. (2013). Bacterial-directed enzyme prodrug therapy. *J Control Release*, **170**(1), 120–131.

23. Martin, S. E., *et al.* (2014). Development of inhibitor-directed enzyme prodrug therapy (IDEPT) for prostate cancer. *Bioconjug Chem*, **25**(10), 1752–1760.

24. Schellmann, N., *et al.* (2010). Targeted enzyme prodrug therapies. *Mini Rev Med Chem*, **10**(10), 887–904.

25. Dumitriu, S. (2001). *Polymeric Biomaterials, Revised and Expanded*, Hoboken: CRC Press.

26. McCord, J. M., and Edeas, M. A. (2005). SOD, oxidative stress and human pathologies: a brief history and a future vision. *Biomed Pharmacother*, **59**(4), 139–142.

27. Ikada, Y., and Tsuji, H. (2000). Biodegradable polyesters for medical and ecological applications. *Macromol Rapid Commun*, **21**(3), 117–132.

28. Xie, Y., *et al.* (2007). Physicochemical characteristics of magnetic microspheres containing tissue plasminogen activator. *J Magn Magn Mater*, **311**(1), 376–378.

29. Tong, S. R. (2003). *Studies on Preparation and Pharmacological Properties of Liposome-Entrapped SOD*. ProQuest, UMI Dissertations Publishing.

30. Vaidya, B., Agrawal, G. P., and Vyas, S. P. (2012). Functionalized carriers for the improved delivery of plasminogen activators. *Int J Pharm*, **424**(1–2), 1–11.

31. Satishkumar, R., and Vertegel, A. A. (2011). Antibody-directed targeting of lysostaphin adsorbed onto polylactide nanoparticles increases its antimicrobial activity against *S. aureus in vitro*. *Nanotechnology*, **22**(50), 505103.

32. Lee, M.-Y., *et al.* (2014). Enzyme attached on polymeric micelles as a nanoscale reactor. *Korean J Chem Eng*, **31**(2), 188–193.

33. Liu, L., Ge, Y., and Yuan, Q. S. (2004). Effect of excipients on stability and structure of rhCuZn-SOD encapsulated in PLGA microspheres. *Chem Res Chinese U*, **20**(3), 323–327.

34. Zhou, J., *et al.* (2014). Construction and evaluation of Fe_3O_4-based PLGA nanoparticles carrying rtPA used in the detection of thrombosis

and in targeted thrombolysis. *ACS Appl Mater Interfaces*, **6**(8), 5566–5576.

35. Muzykantov, V. R. (2001). Targeting of superoxide dismutase and catalase to vascular endothelium. *J Control Release*, **71**(1), 1–21.
36. Besic, S., and Minteer, S. D. (2010). Micellar polymer encapsulation of enzymes. *Methods Mol Biol*, **679**, 113–131.
37. Borman, S. (2014). Encapsulation for better enzymes. *Chem Eng News*, **92**(30), 35–35.
38. Caruso, F., *et al.* (2000). Enzyme encapsulation in layer-by-layer engineered polymer multilayer capsules. *Langmuir*, **16**(4), 1485–1488.
39. Matsuura, S., *et al.* (2012). Enzyme encapsulation using highly ordered mesoporous silica monoliths. *Mater Lett*, **89**, 184–187.
40. Yoshimoto, M. (2010). Stabilization of enzymes through encapsulation in liposomes. *Methods Mol Biol*, **679**, 9–18.
41. Douroumis, D., and Fahr, A. (2012). *Drug Delivery Strategies for Poorly Water-Soluble Drugs*, Chichester, West Sussex: John Wiley & Sons.
42. Giovagnoli, S., *et al.* (2004). Biodegradable microspheres as carriers for native superoxide dismutase and catalase delivery. *AAPS PharmSciTech*, **5**(4), 1–9.
43. Kim, J.-Y., *et al.* (2009). The use of PEGylated liposomes to prolong circulation lifetimes of tissue plasminogen activator. *Biomaterials*, **30**(29), 5751–5756.
44. Application of SOD in liposomes, PCT No. PCT/EP95/04352 Sec. 371 Jul. 1, 1997.
45. Li, X., and Jasti, B. R. (2006). *Design of Controlled Release Drug Delivery Systems*. New York: McGraw-Hill.
46. Luo, R., *et al.* (2012). Engineering of erythrocyte-based drug carriers: control of protein release and bioactivity. *J Mater Sci Mater Med*, **23**(1), 63–71.
47. Millán, C. G., *et al.* (2004). Drug, enzyme and peptide delivery using erythrocytes as carriers. *J Control Release*, **95**(1), 27–49.
48. Hamidi, M., *et al.* (2007). Applications of carrier erythrocytes in delivery of biopharmaceuticals. *J Control Release*, **118**(2), 145–160.
49. Aiswarya, S., Aneesh, T. P., and Viswanad, V. (2013). Erythrocytes as a potential carrier for chronic systemic diseases. *Int J Pharm Sci Res*, **4**(8), 2843–2848.

50. Ihler, G. M., Glew, R. H., and Schnure, F. W. (1973). Enzyme loading of erythrocytes. *Proc Natl Acad Sci U S A*, **70**(9), 2663–2666.

51. Bax, B. E., *et al.* (2000). *In vitro* and *in vivo* studies with human carrier erythrocytes loaded with polyethylene glycol-conjugated and native adenosine deaminase. *Br J Haematol*, **109**(3), 549–554.

52. Cardinale, D., Carette, N., and Michon, T. (2012). Virus scaffolds as enzyme nano-carriers. *Trends Biotechnol*, **30**(7), 369–376.

53. Pedersen, H. (1978). *Biomedical Applications of Immobilized Enzyme Tubes*. Yale University: Ann Arbor, p. 124.

54. Kaminski, M. D., *et al.* (2008). Encapsulation and release of plasminogen activator from biodegradable magnetic microcarriers. *Eur J Pharm Sci*, **35**(1), 96–103.

55. Mahmoodi, M., *et al.* (2010). Synthesis and release study of tissue plasminogen activators (tPA) loaded chitosan coated poly (lactide-*co*-glycolide acid) nanoparticles. *17th Iranian Conference of Biomedical Engineering (ICBME), 2010*, 1–4.

56. Andrady, C., Sharma, S. K., and Chester, K. A. (2011). Antibody-enzyme fusion proteins for cancer therapy. *Immunotherapy*, **3**(2), 193–211.

57. Bikiaris, D., Karavelidis, V., and Karavas, E. (2009). Novel biodegradable polyesters. Synthesis and application as drug carriers for the preparation of raloxifene HCl loaded nanoparticles. *Molecules (Basel, Switzerland)*, **14**(7), 2410–2430.

58. Debakey, M. E., *et al.* (1964). The fate of dacron vascular grafts. *Arch Surg*, **89**, 757–782.

59. Langer, R. (2000). Biomaterials in drug delivery and tissue engineering: one laboratory's experience. *Acc Chem Res*, **33**(2), 94–101.

60. Khemani, K. C., and Scholz, C. (2012). *Degradable Polymers and Materials: Principles and Practice*. Vol. 1114. Washington, DC: American Chemical Society.

61. Nasongkla, N., *et al.* (2004). cRGD-functionalized polymer micelles for targeted doxorubicin delivery. *Angew Chem Int Ed Engl*, **43**(46), 6323–6327.

62. Han, D. K., and Hubbell, J. A. (1996). Lactide-based poly(ethylene glycol) polymer networks for scaffolds in tissue engineering. *Macromolecules*, **29**(15), 5233–5235.

63. Leroux, J.-C., *et al.* (1996). Biodegradable nanoparticles—from sustained release formulations to improved site specific drug delivery. *J Control Release*, **39**(2), 339–350.

64. Zheng, D., *et al.* (2009). Study on docetaxel-loaded nanoparticles with high antitumor efficacy against malignant melanoma. *Acta Biochim Biophys Sin (Shanghai)*, **41**(7), 578–587.

65. Fishbein, I., *et al.* (2000). Nanoparticulate delivery system of a tyrphostin for the treatment of restenosis. *J Control Release*, **65**(1), 221–229.

66. Gao, H., *et al.* (2005). Synthesis of a biodegradable tadpole-shaped polymer via the coupling reaction of polylactide onto mono(6-(2-aminoethyl)amino-6-deoxy)-beta-cyclodextrin and its properties as the new carrier of protein delivery system. *J Control Release*, **107**(1), 158–173.

67. Matsumoto, J., *et al.* (1999). Preparation of nanoparticles consisted of poly (L-lactide)–poly(ethylene glycol)–poly (L-lactide) and their evaluation *in vitro*. *Int J Pharm*, **185**(1), 93–101.

68. Xing, J., Zhang, D., and Tan, T. (2007). Studies on the oridonin-loaded poly (D, L-lactic acid) nanoparticles *in vitro* and *in vivo*. *Int J Biol Macromol*, **40**(2), 153–158.

69. Danhier, F., *et al.* (2012). PLGA-based nanoparticles: an overview of biomedical applications. *J Control Release*, **161**(2), 505–522.

70. Fonseca, C., Simões, S., and Gaspar, R. (2002). Paclitaxel-loaded PLGA nanoparticles: preparation, physicochemical characterization and *in vitro* anti-tumoral activity. *J Control Release*, **83**(2), 273–286.

71. Hans, M. L., and Lowman, A. M. (2002). Biodegradable nanoparticles for drug delivery and targeting. *Curr Opin Solid State Mater Sci*, **6**(4), 319–327.

72. Kumar, P. S., *et al.* (2006). Influence of microencapsulation method and peptide loading on formulation of poly(lactide-*co*-glycolide) insulin nanoparticles. *Pharmazie*, **61**(7), 613–617.

73. Redhead, H. M., Davis, S. S., and Illum, L. (2001). Drug delivery in poly(lactide-*co*-glycolide) nanoparticles surface modified with poloxamer 407 and poloxamine 908: *in vitro* characterisation and *in vivo* evaluation. *J Control Release*, **70**(3), 353–363.

74. Rosenberg, B. (1985). Fundamental studies with cisplatin. *Cancer*, **55**(10), 2303–2316.

75. Teixeira, M., *et al.* (2005). Development and characterization of PLGA nanospheres and nanocapsules containing xanthone and 3-methoxyxanthone. *Eur J Pharm Biopharm*, **59**(3), 491–500.

76. Choi, C., Chae, S. Y., and Nah, J.-W. (2006). Thermosensitive poly(N-isopropylacrylamide)-b-poly(ε-caprolactone) nanoparticles for efficient drug delivery system. *Polymer*, **47**(13), 4571–4580.

77. Prabu, P., *et al.* (2009). Preparation, characterization, in-vitro drug release and cellular uptake of poly(caprolactone) grafted dextran copolymeric nanoparticles loaded with anticancer drug. *J Biomed Mater Res A*, **90**(4), 1128–1136.

78. Shah, L. K., and Amiji, M. M. (2006). Intracellular delivery of saquinavir in biodegradable polymeric nanoparticles for HIV/AIDS. *Pharm Res*, **23**(11), 2638–2645.

79. Wang, Y., *et al.* (2006). Biodegradable and complexed microspheres used for sustained delivery and activity protection of SOD. *J Biomed Mater Res B Appl Biomater*, **79**(1), 74–78.

80. Reddy, M. K., and Labhasetwar, V. (2009). Nanoparticle-mediated delivery of superoxide dismutase to the brain: an effective strategy to reduce ischemia-reperfusion injury. *FASEB J*, **23**(5), 1384–1395.

81. Reddy, M. K., *et al.* (2008). Superoxide dismutase-loaded PLGA nanoparticles protect cultured human neurons under oxidative stress. *Appl Biochem Biotechnol*, **151**(2–3), 565–577.

82. Yun, X., *et al.* (2013). Nanoparticles for targeted delivery of antioxidant enzymes to the brain after cerebral ischemia and reperfusion injury. *J Cereb Blood Flow Metab*, **33**(4), 583–592.

83. Sophocleous, A. M., Zhang, Y., and Schwendeman, S. P. (2009). A new class of inhibitors of peptide sorption and acylation in PLGA. *J Control Release*, **137**(3), 179–184.

84. Houchin, M. L., and Topp, E. M. (2008). Chemical degradation of peptides and proteins in PLGA: a review of reactions and mechanisms. *J Pharm Sci*, **97**(7), 2395–2404.

85. Baran, E. T. (2002). Poly(hydroxybutyrate-*co*-hydroxyvalerate) nanocapsules as enzyme carriers for cancer therapy: an *in vitro* study. *J Microencapsul*, **19**(3), 363–376.

86. Ghassemi, A. H., *et al.* (2009). Preparation and characterization of protein loaded microspheres based on a hydroxylated aliphatic polyester, poly(lactic-*co*-hydroxymethyl glycolic acid). *J Control Release*, **138**(1), 57–63.

87. Barnham, K. J., Masters, C. L., and Bush, A. I. (2004). Neurodegenerative diseases and oxidative stress. *Nat Rev Drug Discov*, **3**(3), 205–214.

88. Afonso, V., *et al.* (2007). Reactive oxygen species and superoxide dismutases: role in joint diseases. *Joint Bone Spine*, **74**(4), 324–329.

89. Zhang, N., Bradley, T. A., and Zhang, C. (2010). Inflammation and reactive oxygen species in cardiovascular disease. *World J Cardiol*, **2**(12), 408–410.

90. Carlo, B., *et al.* (2004). Oxygen, reactive oxygen species and tissue damage. *Curr Pharm Des*, **10**(14), 1611–1626.

91. Reuter, S., *et al.* (2010). Oxidative stress, inflammation, and cancer: how are they linked? *Free Radic Biol Med*, **49**(11), 1603–1616.

92. Joe, M. M., and Irwin, F. (1968). The reduction of cytochrome c by milk xanthine oxidase. *J Biol Chem*, **243**(21), 5753.

93. McCord, J. M., and Fridovich, I. (1969). Superoxide dismutase. An enzymic function for erythrocuprein (hemocuprein). *J Biol Chem*, **244**(22), 6049–6055.

94. Borgens, R. B., and Liu-Snyder, P. (2012). Understanding secondary injury. *Q Rev Biol*, **87**(2), 89–127.

95. Carillon, J., *et al.* (2013). Superoxide dismutase administration, a potential therapy against oxidative stress related diseases: several routes of supplementation and proposal of an original mechanism of action. *Pharm Res*, **30**(11), 2718–2728.

96. Kim, W., *et al.* (2012). Neuroprotective effects of PEP-1-Cu,Zn-SOD against ischemic neuronal damage in the rabbit spinal cord. *Neurochem Res*, **37**(2), 307–313.

97. Pan, J., *et al.* (2012). Protective effect of recombinant protein SOD-TAT on radiation-induced lung injury in mice. *Life Sci*, **91**(3–4), 89–93.

98. Fujimura, M., *et al.* (2000). The cytosolic antioxidant copper/zinc-superoxide dismutase prevents the early release of mitochondrial cytochrome c in ischemic brain after transient focal cerebral ischemia in mice. *J Neurosci*, **20**(8), 2817–2824.

99. Yu, D. H., *et al.* (2012). Over-expression of extracellular superoxide dismutase in mouse synovial tissue attenuates the inflammatory arthritis. *Exp Mol Med*, **44**(9), 529–535.

100. Muzykantov, V. R. (2001). Delivery of antioxidant enzyme proteins to the lung. *Antioxid Redox Signal*, **3**(1), 39–62.

101. Ian, G. T. (2012). Delivery of drugs to the brain via the blood–brain barrier using colloidal carriers. *J Microencapsul*, **29**(5), 475–486.

102. Yang, Y., and Rosenberg, G. A. (2011). Blood–brain barrier breakdown in acute and chronic cerebrovascular disease. *Stroke*, **42**(11), 3323–3328.

103. Antonino, T., *et al.* (2009). Neuron protection as a therapeutic target in acute ischemic stroke. *Curr Top Med Chem*, **9**(14), 1317–1334.

104. Chung, T.-W., Wang, S.-S., and Tsai, W.-J. (2008). Accelerating thrombolysis with chitosan-coated plasminogen activators

encapsulated in poly-(lactide-*co*-glycolide) (PLGA) nanoparticles. *Biomaterials,* **29**(2), 228–237.

105. Panyam, J., and Labhasetwar, V. (2003). Biodegradable nanoparticles for drug and gene delivery to cells and tissue. *Adv Drug Deliv Rev,* **55**(3), 329–347.

106. Danhier, F., *et al.* (2009). Targeting of tumor endothelium by RGD-grafted PLGA-nanoparticles loaded with Paclitaxel. *J Control Release,* **140**(2), 166–173.

107. Lanza, R., Langer, R., and Vacanti, J. (2014). *Tissue Engineering Intelligence Unit: Principles of Tissue Engineering* (4th Edition), Elsevier Science.

108. Chiu, Y.-R., *et al.* (2012). Enzyme-encapsulated silica nanoparticle for cancer chemotherapy. *J Nanopart Res,* **14**(4), 1–10.

109. Jain, P. K., El-Sayed, I. H., and El-Sayed, M. A. (2007). Au nanoparticles target cancer. *Nano Today,* **2**(1), 18–29.

110. Chaudhuri, A., Shekar, K., and Coulter, C. (2012). Post-operative deep sternal wound infections: making an early microbiological diagnosis. *Eur J Cardiothorac Surg,* **41**(6), 1304–1308.

111. Barnea, Y., *et al.* (2008). *Staphylococcus aureus* mediastinitis and sternal osteomyelitis following median sternotomy in a rat model. *J Antimicrob Chemother,* **62**(6), 1339–1343.

112. Kumar, J. K. (2008). Lysostaphin: an antistaphylococcal agent. *Appl Microbiol Biotechnol,* **80**(4), 555–561.

113. Paillard-Giteau, A., *et al.* (2010). Effect of various additives and polymers on lysozyme release from PLGA microspheres prepared by an s/o/w emulsion technique. *Eur J Pharm Biopharm,* **75**(2), 128–136.

Index